Nanotechnology in Food Packaging and Preservation

This book provides a comprehensive overview of the use of nanotechnology in food preservation, food packaging, and the development of novel preservation methods. The chapters focus on applications of nanotechnology in active food packaging, modified atmosphere packaging, and the development of nanocomposite films and coatings for food packaging. Additionally, the book further explores the potential of nanosensors and biosensors for real-time monitoring of food quality and safety and provides insightful case studies showcasing successful nanotechnology applications in the food industry. It also discusses the potential challenges and limitations of using nanotechnology in food preservation, as well as the regulatory and safety concerns that need to be addressed. Furthermore, the book examines consumer perceptions and acceptance of nanotechnology in food preservation and concludes by comparing nanotechnology-based methods with traditional approaches, paving the way for future directions and advancements in this field. This book is a valuable resource for researchers, practitioners, and policymakers in the food industry.

KEY FEATURES

- Provides an overview of the application of nanotechnology in food packaging and preservation
- Discusses the potential of nanotechnology in active food packaging, modified atmosphere packaging, and nanocomposite films for packaging
- Reviews the potential of nanosensors and biosensors for real-time monitoring of food quality and safety
- Explores potential challenges and limitations related to nanotechnology in food preservation
- Examines consumer perceptions and acceptance of nanotechnology in food preservation
- Presents real-world case studies on the use of successful nanotechnology implementations in the food industry

Nanotechnology in Food Packaging and Preservation

Principles and Practical Applications

Edited by
Shakeel Ahmed and Aman Ullah

CRC Press
Taylor & Francis Group
Boca Raton London New York

CRC Press is an imprint of the
Taylor & Francis Group, an **informa** business

Designed cover image: Shutterstock

First edition published 2026
by CRC Press
2385 NW Executive Center Drive, Suite 320, Boca Raton FL 33431

and by CRC Press
4 Park Square, Milton Park, Abingdon, Oxon, OX14 4RN

CRC Press is an imprint of Taylor & Francis Group, LLC

ISBN: 978-1-032-67740-8 (hbk)
ISBN: 978-1-032-67743-9 (pbk)
ISBN: 978-1-032-67742-2 (ebk)

DOI: 10.1201/9781032677422

Typeset in Times
by Apex CoVantage, LLC

Contents

Preface

One of the most promising applications of nanotechnology is in food packaging and preservation, emerging as a revolutionary force across various industries. As the global food production continues to grow, it has become increasingly important to develop innovative, sustainable, and efficient methods to ensure food safety, extend shelf life, and minimize waste in order to meet the demands of a growing population. While traditional food packaging systems are indispensable, they often fall short of effectively safeguarding the quality and freshness of food, preventing contamination, or monitoring food conditions. This is where nanotechnology offers groundbreaking solutions that have the potential to revolutionize the future of food preservation. The objective of this book, *Nanotechnology in Food Packaging and Preservation*, is to offer a thorough examination of the potential applications, challenges, and emerging trends of nanotechnology in the food industry. It explores the various methods by which nanomaterials and nanoscale technologies are being incorporated into packaging materials to increase antimicrobial activity, enhance their barrier properties, and even develop intelligent systems that can monitor food quality in real time. Additionally, it addresses the growing consumer and regulatory apprehensions regarding the safety, environmental impact, and ethical dilemmas of nanomaterials in food-related applications. The primary goal of this work is to offer academic researchers and industry professionals a practical and comprehensive comprehension of the current state of nanotechnology in food packaging, as well as its potential future trajectory. This book highlights the dynamic and evolving role of nanotechnology in food packaging by examining both scientific research and real-world case studies. *Nanotechnology in Food Packaging and Preservation* endeavours to enable its readers to explore innovative solutions that resolve the current challenges of food preservation, safety, and sustainability by bridging the gap between academic theory and practical application.

This book comprises 22 chapters that highlight the recent trends and research advancements in the application of nanotechnology to food packaging and preservation. Chapter 1 serves as a comprehensive introduction, providing an overview of the fundamental concepts of nanotechnology in the context of food preservation. Chapter 2 then delves into the transformative impact of nanotechnology, showcasing how these innovations are reshaping traditional practices in the food industry. Chapter 3 focuses on the significance of nanomaterials in coatings and nanocomposite films for food packaging. Subsequent chapters, Chapters 4 to 7, explore specialized applications of nanotechnology, examining its role in preserving various food categories including seafood, cereals and bakery products, paediatric foods, and meat and poultry. Chapters 8 and 9 highlight the applications of nanotechnology in the preservation of confectionery snacks and beverages, addressing both the challenges and solutions presented by these technologies. Chapters 10 and 11 examine advanced applications, such as the utilization of nanosensors and biosensors for real-time food quality monitoring, underscoring the importance of maintaining food safety and integrity. Chapters 12 through 14 explore the use of nanoparticles in packaging and preservation methods, including antimicrobial packaging. Chapter 15 examines the role of nanotechnology in supply chain management, emphasizing how these technologies can optimize logistics and traceability. Chapters 16 to 18 highlight the role of nanotechnology in the application of essential oils-based food preservation and development of delivery systems. Chapters 19 and 20 address the ethical, economic, and sustainability aspects of nanomaterials in food preservation, prompting critical reflection on the broader impact of these technologies. Finally, Chapters 21 and 22 discuss precautionary approaches in food packaging in the absence of regulation, as well as the challenges and limitations of nanotechnology in food preservation. This volume is the result of collaborative efforts from leading experts in the field, who have generously shared their valuable insights and research findings. We extend our heartfelt gratitude to the contributors, whose expertise has enriched this work, making it a comprehensive resource for researchers, educators, and students alike.

We envision this book as a valuable resource for those passionate about the advancement of the field of food packaging by utilizing cutting-edge nanotechnology, fostering interdisciplinary collaborations, and inspiring new ideas. The integration of nanotechnology into food packaging holds significant potential for creating a future where food is preserved more efficiently, distributed more safely, and consumed with greater confidence, especially as we face urgent global challenges related to environmental sustainability and food security.

Shakeel Ahmed
Aman Ullah
Editors

About the Editors

Dr. Shakeel Ahmed is working as an assistant professor of chemistry at the Higher Education Department, Government of Jammu and Kashmir and as an assistant professor at the Postgraduate Department of Chemistry, Government Postgraduate College Rajouri, Jammu and Kashmir, India. He obtained a first degree in general science from Government Postgraduate College Rajouri (University of Jammu) followed by master's degree and doctoral degree in chemistry from Jamia Millia Islamia, a central university, New Delhi. He gained post-doctoral experience in biocomposite materials at Indian Institute of Technology, Delhi. He has published several research publications in the areas of green nanomaterials and biopolymers for various applications including biomedical, packaging, and water treatment. He is regular member of American Chemical Society (ACS), USA; member of Royal Society of Chemistry (MRSC), UK; member of international association of advanced materials (IAAM), Sweden; and life member of Asian Polymer Association and Society of Materials Chemistry (India). He is an active reviewer and member of the editorial board of many reputed journals. He has published more than 40 books in the area of nanomaterials and green materials with publishers of international repute. He is a fellow of The Linnean Society of London and International Society for Development and Sustainability (ISDS), Japan.

Dr. Aman Ullah is currently working as a professor and associate dean of academics at the University of Alberta. Dr. Aman Ullah is a professor of Lipid Utilization: Renewable Resources for Industrial Processing in the Department of Agricultural, Food, and Nutritional Sciences, Faculty of Agricultural, Life, and Environmental Sciences at the University of Alberta, Canada. He obtained a PhD (with distinction) in chemical sciences and technologies from the University of Genoa, Italy, by working together as a visiting PhD scholar at Southern Methodist University, Dallas, TX, USA. Dr. Ullah's research is focused on the development of bioproducts and biomaterials from renewable resources for different applications, including packaging, drug delivery, wastewater and mine tailings remediation, and drinking water monitoring and remediation in developing countries and First Nations communities. Dr. Ullah has more than 150 publications and 9 patent applications (four granted). He has licensed two technologies to a new start-up biorefinery in Alberta. He delivered 63 invited talks in the United States/Canada, Europe, Asia, and Australia, including one plenary and ten keynote lectures. Dr. Ullah received several awards in his career, including the Canadian Rising Star Award by "Grand Challenges Canada", the Faculty of Agricultural, Life, and Environmental Sciences "Research and Innovation Award", and the ASTech Award winner for the "33rd ASTech Innovation Awards".

1 Introduction to Nanotechnology in Food Preservation

Amol D. Gholap[a,b], Pankaj R. Khuspe[c], Sadikali F. Sayyad[b], Navnath T. Hatvate[d], Machindra Chavan[e], and Md Faiyazuddin[f]

[a]Department of Pharmaceutics, St. John Institute of Pharmacy and Research, Palghar, Maharashtra, India

[b]Department of Pharmaceutics, Amrutvahini College of Pharmacy, Sangamner, Maharashtra, India

[c]Departement of Bioengineering integral University, College of Pharmacy, Paniv, Solapur, Maharashtra, India

[d]Institute of Chemical Technology Mumbai, Marathwada Campus, Jalna, Maharashtra India

[e]Department of Pharmacognosy, Amrutvahini College of Pharmacy, Sangamner, Maharashtra, India

[f]School of Pharmacy, Al–Karim University, Katihar, Bihar, India

1.1 INTRODUCTION

For millennia, food preservation has been an integral part of human civilization, motivated by the basic need to guarantee the availability of wholesome and safe food, no matter the time of year or place. The search to prolong the shelf life of perishable items has changed over time, from the ancient practices of sun drying and salting to the use of advanced processes in today's food business (A. D. Gholap, Sawant, et al., 2023). A new frontier in this age-old pursuit has surfaced in recent decades, and it is known as nanotechnology. The importance of food preservation is fundamental to maintaining public health, economic stability, and global food security (Amit et al., 2017; Díaz-Montes & Castro-Muñoz, 2021; Parfitt et al., 2010). Unprecedented urbanization, population growth, and rising consumer demand for a wide variety of food goods are occurring in our society. It is now extremely difficult to meet these needs, and there is an increasing need for creative and sustainable preservation techniques due to food waste, foodborne illnesses, and environmental changes as presented in Figure 1.1 (Martillanes et al., 2017; Schanes et al., 2018; Shankar Tumuluru, 2023). Even though they work well, traditional food preservation techniques frequently include adding chemicals as preservatives or using physical treatments. Concerns about these methods' effects on consumer preferences, the environment, and health have been highlighted (A. D. Gholap, Gupta, et al., 2024). The use of nanotechnology in food preservation has come to light as a potential remedy to these problems, providing a way to improve the caliber, security, and durability of our food supply (Kasza et al., 2022; Vågsholm et al., 2020).

Food preservation has a plethora of options because of nanotechnology, which functions at the nanoscale where materials display special qualities. In a number of ways, nanomaterials and nanoparticles have shown they have the power to completely transform industry. They make it possible to create cutting-edge food packaging materials with improved barrier qualities, which successfully lower the penetration of moisture, oxygen, and other impurities that cause spoiling (R. Biswas et al., 2022; Jagtiani, 2022). Incorporating nanoparticles with strong antibacterial capabilities into coatings and packaging can act as a barrier against foodborne microorganisms. Additionally, food waste can be decreased and consumer confidence increased by using nanosensors to track and report real-time data on the quality of food goods. This enables more accurate quality control and the prompt removal of tainted items. This book chapter investigates the science, uses, and ramifications of this cutting-edge topic as it relates to the intriguing confluence of nanotechnology and food preservation (T. Singh et al., 2017). It explores the application of antimicrobial nanoparticles, the creation of smart sensors, the integration of nanomaterials into packaging, and the moral and legal issues that support these developments. We will traverse the opportunities and difficulties that come with using nanotechnology to secure the survival of the world's expanding population as we set out on this voyage into the nanoworld of food preservation (Angelopoulou et al., 2022; Chadha et al., 2022).

Nanotechnology is at the vanguard of a world where scientific developments are happening at a apid pace, providing an exciting window into the future of science and innovation (Gholap & Musmade, 2011; Gholap & Rajurkar, n.d.; Mahajan & Gholap*, 2011; Mahajan et al., n.d.; Gholap & Mahajan, 2011; A. D. Gholap et al., 2011a, 2011b; Rajurkar, 2012; Magar & Gholap, 2011). This introduction provides a

DOI: 10.1201/9781032677422-1

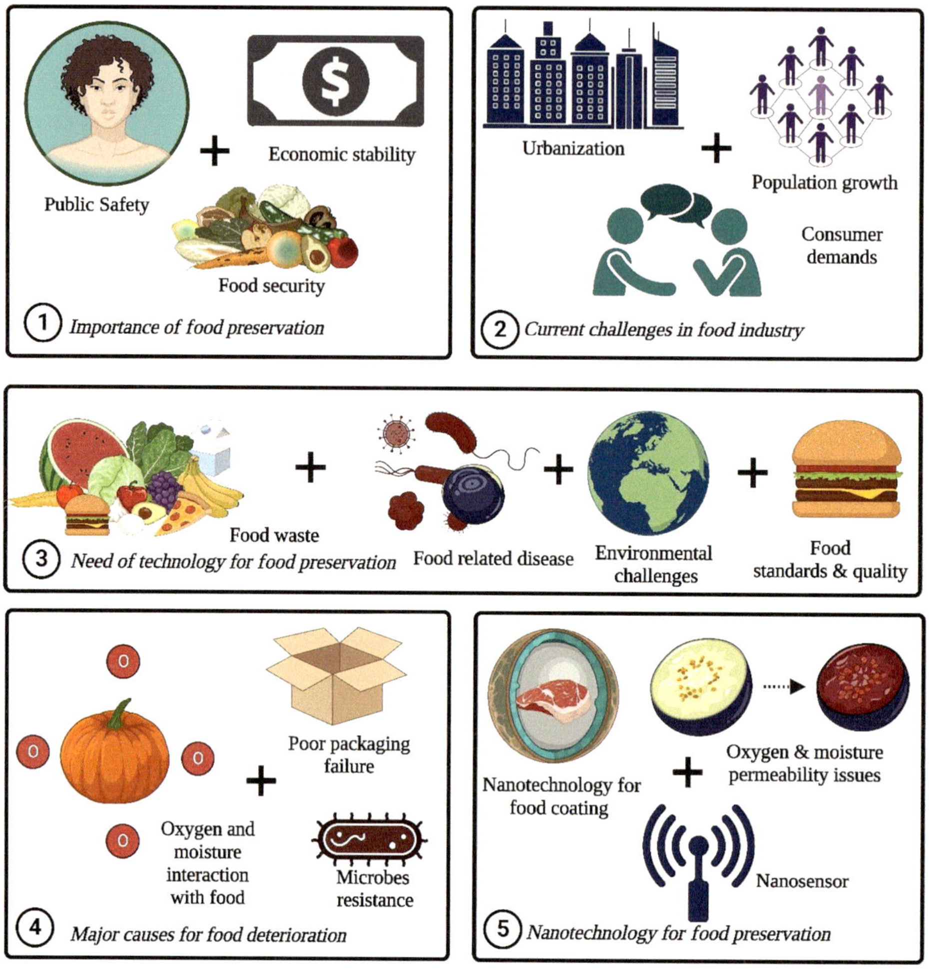

FIGURE 1.1 Summary of importance of nanotechnology into food preservation.

starting point for our investigation of the remarkable uses of nanotechnology in the food preservation industry, highlighting its potential to completely transform the way we protect our most valuable resource food (Baig et al., 2021; Pokrajac et al., 2021; Taha et al., 2022). By definition, nanotechnology works at the nanoscale, where unique characteristics and behaviors can be found in materials and structures. Because of this special quality, scientists, engineers, and innovators have been inspired to create new things, putting the sector in the forefront. Materials can develop new characteristics at the nanoscale, such as increased strength, reactivity, and conductivity (Gholap & Tupe, 2012a; A. D. Gholap, 2012; Gholap & Dighe, 2012b; Gholap & Rajurkar, 2012b, 2012c; Gholap & Tupe, 2012b; A. D. Gholap, 2012). This revolutionary potential has had far-reaching effects on a variety of industries, including electronics, medicine, and now—perhaps most intriguingly—food preservation (Bayda et al., 2019; Malik et al., 2023).

The food sector has a significant obstacle in maintaining food safety, freshness, and quality along the supply chain. The pursuit of novel approaches has resulted in a fascinating collaboration with nanotechnology, which holds the potential to transform the fundamental principles of food preservation. With the help of nanotechnology, food safety may be improved, spoiling can be minimized, and perishable food products can have their shelf lives significantly extended (Awuchi, 2023; T. Singh et al., 2017). The development of packaging based on nanomaterials, where nanoparticles and nanocomposites are prominent, is where this influence is most noticeable. These materials are designed to have outstanding barrier qualities, serving as watchful guardians against the entry of moisture, oxygen, and other impurities that usually hasten food deterioration. They thereby address a crucial issue in the food sector by helping to maintain the freshness and edibility of food for longer periods of time. Furthermore, the concept of antibacterial nanoparticles—such as zinc and silver oxide—is introduced by nanotechnology. These nanoparticles can be easily incorporated into coatings and materials used in food packaging, giving them strong antibacterial capabilities. This acts as a strong barrier to prevent the growth of bacteria and other pathogens that endanger the safety of food (Fajardo et al., 2022; Majid et al., 2018).

Especially, this novel method reduces the need for conventional chemical preservatives, providing an eco-friendly substitute and resolving the growing concerns about bacteria resistant to antibiotics. We shall delve further into the uses of nanotechnology in food preservation in the upcoming chapters (Mubeen et al., 2021). We will explore the creation of smart sensors for real-time quality monitoring, the scientific foundations of nanomaterials, and the ethical and legal issues that must be taken into account to properly steer these innovations. Both producers and consumers who depend on the preservation of their most important resource—food—stand to gain from the introduction of nanotechnology into the food preservation industry. This will create a more safe, sustainable, and effective food supply chain. Through exploration of this fascinating field, we will learn how nanotechnology can influence food preservation in the future (Gupta et al., 2019; B. P. Singh et al., 2023).

1.1.1 Nanomaterials in Food Packaging

Nanotechnology has revolutionized food packaging by enhancing safety, prolonging shelf life, and promoting sustainability. By manipulating materials at the nanoscale, packaging systems can offer superior barriers against gases, moisture, and UV light, thereby preserving food quality more effectively than conventional materials. Nanoparticles like silver, zinc oxide, and titanium dioxide impart antimicrobial properties, reducing spoilage and contamination. Moreover, active packaging can interact with the food environment to release or absorb substances that help maintain freshness, while smart packaging employs nanosensors to monitor factors such as spoilage and temperature in real-time. Biodegradable nanocomposites, which include materials like nanoclays or nanocellulose, contribute to eco-friendly solutions by enhancing the mechanical strength of biopolymers. Nonetheless, there are concerns about the potential migration of nanoparticles into food and their long-term health effects. Regulatory bodies stress the importance of thorough safety assessments to ensure consumer protection. Overall, nanotechnology offers a promising future for intelligent, safe, and sustainable innovations in food packaging (Khedkar et al., 2012a, 2012b; A. D. Gholap & Rajurkar, 2012; Gholap & Dighe, 2012a; Gholap & Rajurkar, 2012a; Bothiraja et al., 2014; Vilas Bhagat, 2012). By definition, nanotechnology works at the nanoscale, where unique characteristics and behaviors can be found in materials and structures (Joseph et al., 2023; Y. Khan et al., 2022). Because of this special quality, scientists, engineers, and innovators have been inspired to create new things, putting the sector in the forefront. Materials can develop new characteristics at the nanoscale, such as increased strength, reactivity, and conductivity. This revolutionary potential has had far-reaching effects on a variety of industries, including electronics, medicine, and now—perhaps most intriguingly—food preservation Tupe, D., & Gholap, A. D. 2012; Magar, T. V., Gholap, A. D., & D. S. M., 2011). The food sector has a significant obstacle in maintaining food safety, freshness, and quality along the supply chain. The pursuit of novel approaches has resulted in a fascinating collaboration with nanotechnology, which holds the potential to transform the fundamental principles of food preservation (Charitidis et al., 2014; Malik et al., 2023).

With the help of nanotechnology, food safety may be improved, spoiling can be minimized, and perishable food products can have their shelf lives significantly extended (Faiyazuddin et al., 2023; A. Gholap, Pardeshi, et al., 2023; A. D. Gholap, Hatvate, Khuspe, et al., 2023; A. D. Gholap, Hatvate, Sawant, et al., 2023; Pardeshi et al., 2023; Puri & Gholap, 2018). The development of packaging based on nanomaterials, where nanoparticles and nanocomposites are prominent, is where this influence is most noticeable (Nile et al., 2020; Sharma et al., 2017). These materials are designed to have outstanding barrier qualities, serving as watchful guardians against the entry of moisture, oxygen, and other impurities that usually hasten food deterioration. They thereby address a crucial issue in the food sector by helping to maintain the freshness and edibility of food for longer periods of time. Furthermore, the concept of antibacterial nanoparticles—such as zinc and silver oxide—is introduced by nanotechnology (A. D. Gholap, Bhowmik, et al., 2023; A. D. Gholap, Gupta, et al., 2023; A. D. Gholap, Hatvate, Khuspe, et al., 2023; A. D. Gholap, Rojekar, et al., 2024; A. D. Gholap, Said, et al., 2023; A. D. Gholap, Sayyad, et al., 2023; Vora et al., 2023). These nanoparticles can be easily incorporated into coatings and materials used in food packaging, giving them strong antibacterial capabilities. This acts as a strong barrier to prevent the growth of bacteria and other pathogens that endanger the safety of food. Especially, this novel method reduces the need for conventional chemical preservatives, providing an eco-friendly substitute and resolving the growing concerns about bacteria resistant to antibiotics (Barage et al., 2022; R. Biswas et al., 2022; Y. Huang et al., 2018).

The chapters will go more into the uses of nanotechnology in food preservation. We will explore the creation of smart sensors for real-time quality monitoring, the scientific foundations of nanomaterials, and the ethical and legal issues that must be taken into account to properly steer these innovations (R. Biswas et al., 2022; T. Singh et al., 2017). Both producers and consumers who depend on the preservation of their most important resource—food—stand to gain from the introduction of nanotechnology into the food preservation industry. This will create a more safe, sustainable, and effective food supply chain. Through exploration of this fascinating field, we will learn how nanotechnology can influence food preservation in the future (Saritha et al., 2022).

Nanotechnology has revealed a potent ally in the shape of nanomaterials in the constantly changing field of food preservation, especially with the creation of nanocomposites. These nanocomposites are changing the game because of their amazing ability to improve food packaging's barrier qualities, therefore increasing the shelf life of perishable items (R. Biswas et al., 2022; Hutapea et al., 2022). Although they can be somewhat efficient, traditional food packaging materials sometimes fall short when it comes to maintaining the safety and freshness of food goods over a prolonged period of time. The permeability of oxygen and moisture as well as the movement of pollutants present serious problems that have spurred research into creative fixes (Y. Huang et al., 2018). This is the exact situation in which nanocomposites are useful. Nanoparticles, usually at the nanoscale, are incorporated into the packaging material's matrix to create nanocomposites, which are designed materials. The material's qualities are dramatically changed by the addition of these nanoparticles, strengthening the packaging's barrier qualities by increasing its ability to stop the entry of dangerous substances (R. Biswas et al., 2022; T. Singh et al., 2017).

The main function of nanocomposites is to prevent moisture and oxygen from penetrating, as these are two important elements in food deterioration. Oxygen is well-known for its involvement in oxidative reactions, which cause lipids and other food ingredients to degrade, giving food an unpleasant taste, discolouration, and decreased nutritional value (R. Biswas et al., 2022; Hutapea et al., 2022; T. Singh et al., 2017). Contrarily, moisture encourages the growth of microorganisms, which can lead to spoiling and safety issues. In order to overcome these difficulties, nanocomposites build a strong barrier. Because of their large surface area, nanoparticles—like carbon nanotubes, graphene oxide, and clay-based nanomaterials—can adsorb substances and prevent moisture and oxygen from penetrating. This reduces the possibility of infection and spoiling while still maintaining the food's quality. Moreover, these nanocomposites can act as barriers against outside impurities like gases and volatile organic molecules, which can alter the flavor and aroma of food (Bharti et al., 2020; Domínguez et al., 2019; Ezati et al., 2023; Geng et al., 2023; Y. Huang et al., 2018; Sampels, 2013).

There are several benefits of using nanocomposites in food packaging. By reducing food waste and packaging materials, it can significantly increase product shelf life, lessen the need for chemical additives and preservatives, and advance sustainability (Bradley et al., 2011). These advantages are especially relevant in a world where more effective and environmentally responsible food preservation techniques are sought after. The potential uses of nanocomposites in food packaging are numerous and promise to transform the food preservation strategies used by the industry as nanotechnology develops. Because of their ability to improve barrier qualities, nanocomposites are opening the door to a more sustainable and safe food supply chain, allowing customers to enjoy safer and fresher food items for longer (de Sousa et al., 2023; El-Sayed & Youssef, 2023).

The incorporation of nanoparticles into packaging materials is a groundbreaking invention brought about by nanotechnology, which is helping to modernize food packaging and increase the shelf life of perishable foods. This tactic is proven to be essential for improving food preservation because it provides a strong barrier against the entry of hazardous substances, guaranteeing the safety and freshness of food items (El-Sayed & Youssef, 2023; Picazo-Vela & Ricardo Hernández, 2020). Though somewhat dependable, traditional food packaging materials are frequently compromised by pathogens, moisture, and oxygen seeping through. These elements greatly worsen the quality and safety of food, necessitating creative solutions to solve the problems. The use of nanoparticles into packaging materials presents an opportunity to address these issues directly. Silver and zinc oxide nanoparticles are well-known for their exceptional antibacterial qualities. They act as a strong barrier against spoiling microorganisms and foodborne pathogens when integrated into packing materials. This function is especially helpful in maintaining food safety since it prevents hazardous bacteria from growing and proliferating (Agriopoulou, Stamatelopoulou, Skiada, et al., 2020; T. Singh et al., 2017). Through this action, nanoparticles lessen the need for conventional chemical preservatives, which frequently give rise to questions over their long-term health implications and environmental impact. Additionally, the incorporation of nanoparticles offers consumers a safer and more environmentally friendly option for food preservation, addressing the growing worries about antibiotic-resistant microorganisms. In addition to being antibacterial, nanoparticles can improve the barrier qualities of materials used in packaging. They can produce an effective barrier that prevents moisture and oxygen from penetrating, two things that are known to cause food to deteriorate. This keeps food freshness and quality intact for extended periods of time, minimizing food waste and guaranteeing that customers receive goods that live up to their standards for safety and quality. Additionally, the presence of nanoparticles in packaging materials may help lessen the negative impacts of outside pollutants including gases and volatile organic compounds. These chemicals may alter the flavor and scent of food, which could negatively impact the whole dining experience (R. Biswas et al., 2022; A. Kumar et al., 2021).

The incorporation of nanoparticles into food packaging materials is a big step in the direction of a more efficient and sustainable food preservation method. In addition to meeting the demand for more effective and environmentally friendly preservation techniques, it increases food safety, prolongs the shelf life of food goods, and reduces the need for chemical additives (Bajpai et al., 2018; de Sousa et al., 2023; A. Kumar et al., 2021). The incorporation of nanoparticles into packaging materials shows promise for a more sustainable and safe food supply chain as nanotechnology develops. It guarantees that food goods are safe, fresh, and of excellent quality for longer periods of time, and it also lessens the environmental impact of packaging waste. In the field of food packaging, where shelf-life extension of perishable commodities is crucial, nanotechnology has shown to be a potent ally. Using cutting-edge nanomaterials in food packaging, we can greatly improve food product preservation, cut down on waste, and raise the general standard and safety of the goods we eat (Ahari & Soufiani, 2021; Iversen et al., 2022; Kim et al., 2022; Onyeaka et al., 2022).

While somewhat effective, traditional food packaging materials frequently fall short of offering a sufficient defense against the factors that endanger food's shelf life. Food quality can be negatively impacted by oxygen, moisture, and toxins, which calls for creative solutions. The application of nanotechnology in food packaging aims to address these issues and prolong the shelf life of goods (Czerwiński et al., 2021; Primožič et al., 2021; Wyser et al., 2016). The capacity of nanomaterials, including nanoparticles and nanocomposites, to improve the barrier qualities of food packaging materials is special. These materials can be designed to form a barrier that keeps out oxygen and moisture, two main causes of food deterioration. Nanotechnology efficiently extends the shelf life of the encapsulated food product by keeping these elements from penetrating the packaging. Additionally, by using nanoparticles into packaging, the product's quality and safety can be maintained by shielding it from contamination and spoiling. In the context of the global food supply chain, where goods may travel great distances and be exposed to a variety of climatic conditions, this is extremely crucial (Adeyemi & Fawole, 2023; Neme et al., 2021). By incorporating nanoparticles, nanotechnology can confer antibacterial qualities to packaging materials, in addition to augmenting their physical barriers. These nanoparticles, which include zinc oxide and silver, have strong antibacterial properties that prevent the growth of bacteria and other diseases. Food safety is further improved by incorporating them into packaging materials, which lessens the need for conventional chemical preservatives and additives (Adeyemi & Fawole, 2023; Sirelkhatim et al., 2015).

The way the food industry approaches food preservation is about to change dramatically with the application of nanotechnology in food packaging. Because items stay fresher for longer, it has the potential to reduce food waste. It also promotes sustainability by lowering the need for chemical preservatives and having a less negative environmental impact (R. Singh et al., 2023; Sridhar et al., 2021). As nanotechnology develops, there is a lot of potential for integrating nanoparticles into food packaging materials to guarantee that consumers receive high-quality, safe food products. We are making great progress toward a more sustainable and effective food supply chain that will benefit both producers and consumers by using nanotechnology to extend the shelf life of food (Angelopoulou et al., 2022; Sridhar et al., 2021).

1.2 ANTIMICROBIAL NANOPARTICLES

In the current period of increased emphasis on health and safety, the development of antimicrobial nanoparticles has shown to be a revolutionary breakthrough in the continuous fight against pathogenic bacteria as presented in Figure 1.2 (Lee et al., 2019). Because they offer a strong barrier against bacteria and infections, these small but mighty substances have the potential to completely transform a range of sectors, including healthcare and food preservation. Nanoparticles' amazing antibacterial characteristics, which function at the nanoscale, are the driving force behind this discovery. They differ from traditional antibacterial agents due to their special quality, which has created new opportunities for preserving public health and fostering a cleaner, safer environment. The ability of antimicrobial nanoparticles, such zinc and silver oxide nanoparticles, to impede the growth and spread of germs and diseases has drawn interest in them (Bruna et al., 2021; Vassallo et al., 2020). They accomplish this by taking advantage of their large surface area, which allows them to have molecular interactions with microbial cells. This contact damages the microbial cell membrane and ultimately causes it to die by upsetting essential cellular processes. In addition to being efficient, this focused strategy reduces the possibility of antibiotic resistance, which is becoming an increasing issue in healthcare and other fields. Antimicrobial nanoparticles have widespread applicability in a variety of sectors and contexts. To lower the risk of infections, they are being added to surfaces, medical equipment, and wound dressings in healthcare institutions. Their implementation may improve patient safety, lessen the demand for antibiotics, and decrease the rate of infections acquired in hospitals (Mamun et al., 2021; Ozdal & Gurkok, 2022).

The food industry has also adopted antimicrobial nanoparticles since foodborne illness and spoiling prevention are critical concerns. These nanoparticles serve as a strong barrier against dangerous microorganisms when incorporated into food packaging materials and coatings, guaranteeing the security and durability of food good (Fadiji et al., 2023; Pereda et al., 2018). This answers worries about antibiotic-resistant microorganisms in the food supply chain and lessens the demand for conventional chemical preservatives. Antimicrobial nanoparticles are used in industries such as water treatment, textile production, and environmental remediation in addition to healthcare and food preservation. They offer a comprehensive response to a broad range of problems that dangerous microbes present in our day-to-day existence (Anvar et al., 2021; Krishna et al., 2022). The ethical and environmental implications of using antimicrobial nanoparticles must be carefully considered as we continue to realize their promise. It's crucial to strike a balance between utilizing these potent molecules to our advantage and making sure their deployment is sustainable and responsible. Antimicrobial nanoparticles offer the possibility of a safer, healthier, and more sustainable future for all of us with continued study and creative uses (Hutapea et al., 2022; Mizielińska & Bartkowiak, 2023).

Silver nanoparticles and zinc oxide nanoparticles are two amazing agents that have been at the forefront of research and invention in the field of antibacterial nanoparticles. Operating at the nanoscale, these minuscule yet formidable fighters possess special qualities and attributes that render them invaluable in the struggle against bacteria and infections (Alhujaily et al., 2022; Kachoei et al., 2021). Silver nanoparticles' remarkable antibacterial properties have attracted a lot of attention. Because of their significantly larger surface area at the nanoscale, silver particles can interact with microbial cells in a manner that bulk silver cannot (Chadha et al., 2022; Nguyen et al., 2019). These interactions cause the microbial cell membrane to become compromised, which hinders essential cellular functions and ultimately results in the microorganism's death. Since they work well against a wide range of germs, silver nanoparticles have shown to be very useful in healthcare settings. To lower the risk of infections, they are included into textiles, wound dressings, and medical devices (Mohamad Sukri et al., 2023; Punjabi et al., 2018).

However, in terms of antibacterial defense, zinc oxide nanoparticles hold a comparable promise. Because these

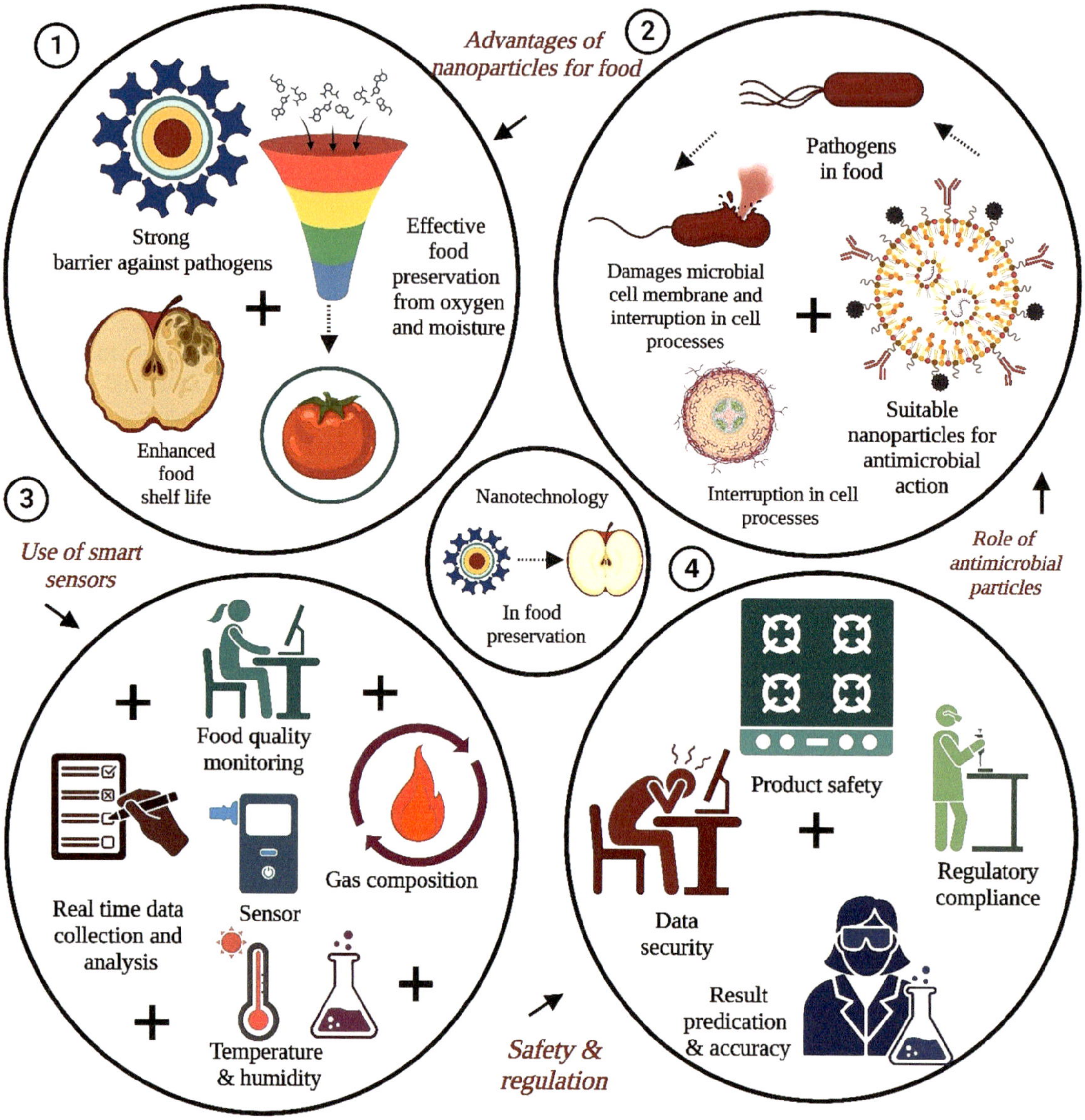

FIGURE 1.2 The impact of nanotechnology on that of the food industry.

nanoparticles produce reactive oxygen species when exposed to light or moisture, they have shown remarkable antibacterial capabilities. The bacterial cells become inactive as a result of this oxidative stress, which negatively impacts their structural and functional integrity (Sirelkhatim et al., 2015; Yang et al., 2023). In order to lower the risk of infections and increase safety, zinc oxide nanoparticles have found employment in a variety of industries, including health care. These products include sunscreens, wound care treatments, and antibacterial surface coatings. Due to their large surface area and capacity to damage microbial cells, zinc oxide and silver oxide nanoparticles have special properties that make them effective antibacterial agents (Jiang et al., 2020; Pati et al., 2014). Although both of these kinds of nanoparticles aim to fight bacteria and infections, they accomplish this through different methods, which makes them adaptable and appropriate for various uses (A. Biswas et al., 2022; Souza et al., 2019; Ye et al., 2022).

These antimicrobial nanoparticles are being investigated in a number of industries, including health care, food preservation, water treatment, and more, as research and innovation in nanotechnology expand. They lessen the need for conventional antibiotics and chemical preservatives, offering a possible route toward safer and more environmentally friendly procedures (Fernando et al., 2018; L. Wang et al., 2017). To guarantee their efficacy and reduce any hazards, these nanoparticles must be used and regulated responsibly. This will eventually make the environment safer, cleaner, and healthier for everyone. The use of antimicrobial nanoparticles in coatings and packaging materials has emerged as a major breakthrough in the field of food safety and preservation. This new method reduces the need for chemical additives while improving food products' freshness and safety by introducing a potent defense against dangerous bacteria (Nishu & Kumar, 2023; Sahoo et al., 2021). Despite their effectiveness, traditional food packing materials frequently fail to maintain food quality and prevent microbiological contamination. One major step toward overcoming these restrictions is the use of antimicrobial nanoparticles, like zinc and silver oxide, into coatings and packaging materials (Bajpai et al., 2018; R. Biswas et al., 2022).

These antimicrobial nanoparticles' main benefit is their ability to stop bacteria and infections from growing and spreading. They form an impenetrable barrier that actively fights microbiological risks when included in packing materials. This specific method damages the integrity of microbial cell membranes and throws off important cellular processes, which ultimately results in their death. In the context of food preservation, where contamination by pathogenic bacteria can lead to foodborne diseases and deterioration, this antimicrobial protection is invaluable (Baptista et al., 2018; Bruna et al., 2021). The need for conventional chemical preservatives and additives is lessened by antimicrobial nanoparticles, which provide a sustainable substitute. The use of nanoparticles in food packaging materials and coatings increases the shelf life of food items by forming a strong barrier that prevents pollutants, moisture, and oxygen from penetrating. Because oxygen and moisture can start oxidative processes and microbial development, these are well-known causes of food spoiling (Agarwal et al., 2023; Dakal et al., 2016; Su et al., 2020). Nanoparticles ensure that items stay fresher for longer lengths of time by blocking their entry, which lowers food waste and increases consumer confidence (Azeredo et al., 2019; Wu et al., 2021).

Additionally, food goods can be protected by antimicrobial nanoparticles against outside pollutants including gases and volatile organic compounds, which can affect the flavor and scent of the product. For products that are transported or stored for an extended period of time, this protection is extremely important (R. Biswas et al., 2022; Onyeaka et al., 2022). In the food business, the use of antimicrobial nanoparticles into coatings and packaging is a significant advancement. In addition to preserving and guaranteeing the quality and safety of food goods, it also satisfies the growing demand for sustainable and environmentally friendly preservation techniques. These nanoparticles provide a more ecologically friendly way to preserve food by reducing the need for chemical additions and preservatives. Antimicrobial nanoparticle integration into coatings and packaging materials is set to become a normal procedure in the food business as nanotechnology research and innovation continue to progress. It guarantees that consumers will obtain food items that meet their expectations for a more efficient and sustainable food supply chain in addition to being safe and fresh (Mahmud & Khan, 2018; Soung et al., 2021; Suvarna et al., 2022).

The development of antimicrobial nanoparticles has brought in a new age of food product protection in the ongoing fight against foodborne infections. Operating at the nanoscale, these tiny but potent molecules have shown themselves to be strong defenders against the threat posed by germs and diseases in the food business (Beyth et al., 2015; Bruna et al., 2021; L. Wang et al., 2017). Foodborne pathogen suppression is one of the main functions of antimicrobial nanoparticles, such as zinc and silver oxide nanoparticles. These nanoparticles are very good at stopping the growth and spread of dangerous germs that can contaminate food products because they have strong antimicrobial qualities. This inhibitor has a very interesting and well-defined mechanism. Because of their minuscule size, antimicrobial nanoparticles have a far larger surface area, which enables them to engage in complex molecular interactions with microbial cells (Dakal et al., 2016; Slavin et al., 2017). The pathogens suffer damage to their cell membranes because of this interaction, which ultimately causes them to perish by upsetting essential cellular processes. By using a targeted strategy, these nanoparticles are guaranteed to be effective against a wide range of foodborne pathogens, providing a promising means of improving food safety (Zorraquín-Peña et al., 2020).

Using antimicrobial nanoparticles in the food business is especially important since foodborne bacteria can seriously harm food quality and public health. Foodborne infections brought on by contaminated food can result in a variety of health issues (Agriopoulou, Stamatelopoulou, Sachadyn-Król, et al., 2020; Awuchi, 2023). The food industry has discovered a practical way to lower this risk and improve food safety: adding antimicrobial nanoparticles to coatings and materials used in food packaging. Moreover, the application of antimicrobial nanoparticles offers a sustainable substitute for conventional chemical preservatives and additives, satisfying the growing need for more environmentally friendly food preservation techniques. This lowers the possible health concerns linked to chemical substances while simultaneously improving food safety. Adding antimicrobial nanoparticles to food preservation procedures is a big step in solving the problems caused by foodborne bacteria. The application of these nanoparticles holds the potential to create a food supply chain that is safer, healthier, and more sustainable as long as research and innovation in nanotechnology continue. Antimicrobial nanoparticles have the potential to significantly improve food quality and public health by blocking foodborne bacteria (Anvar et al., 2021; Bajpai et al., 2018; Suvarna et al., 2022).

1.3 SMART SENSORS FOR FOOD QUALITY MONITORING

The food industry is always changing, and the use of smart sensors to check food quality is bringing the industry into a new era of accuracy, security, and customer pleasure. With their state-of-the-art technology, these creative sensors allow for real-time data collection and analysis, greatly enhancing food quality assurance and monitoring (Hassoun et al., 2023; I. Kumar et al., 2021). Food quality is a broad term that includes things like flavor, safety, freshness, and nutritional value. Manual inspection and sporadic sampling are common components of the old techniques of evaluating these characteristics, which can lead to inaccuracies and the omission of important data points. This is where the process can be revolutionized by smart sensors (Abideen et al.,

2021; P. W. Khan et al., 2020). A wide range of factors that affect food quality can be detected and measured by smart sensors used for food quality monitoring. Temperature, humidity, gas composition, pH, and even the existence of particular pollutants or diseases are examples of these factors. There are various benefits associated with their ability to gather data continually and send it in real time to monitoring systems (Sonwani et al., 2022; X. Wang et al., 2022).

The primary benefit of these sensors is their unmatched precision. The freshness and safety of food goods can be considerably impacted by even the smallest variations in variables like temperature and humidity, which they are able to detect. This accuracy lowers the chance of food waste and spoiling by enabling prompt remedial action. Furthermore, proactive quality management is made easier by smart sensors' ability to collect data in real-time. Rapid detection of unfavorable alterations in food quality enables the prompt extraction of tainted products from the supply chain (da Costa et al., 2022; Skawińska & Zalewski, 2022). This reduces the possibility that customers may be exposed to inferior or potentially hazardous items while simultaneously improving food safety. The food supply chain's manufacturing and distribution phases are not the only places where smart sensors are used. This technology also has benefits for consumers. For example, real-time information regarding a product's safety and freshness can be given to customers through smart packaging that has sensors integrated into it. In addition to increasing consumer confidence, this openness gives people the ability to choose the food they eat with greater knowledge (S. Chen et al., 2020; Zuo et al., 2022).

Furthermore, it is possible to evaluate the data gathered by smart sensors to find patterns and possible areas for process optimization. Better supply chain management and production techniques may result from this realization, which will ultimately help both manufacturers and customers. A major development in the food sector is the use of smart sensors to monitor food quality (L. He et al., 2020; Javaid et al., 2021). They provide previously unheard-of precision, improve safety, and encourage supply chain openness. The incorporation of smart sensors offers an increasingly safe and effective food supply chain as technology develops, benefiting both producers and consumers. Nanosensors are emerging as the game-changers in the rapidly evolving field of food quality monitoring, setting new benchmarks for accuracy and real-time data collection. These tiny, yet mighty gadgets are about to revolutionize food quality monitoring by providing previously unthinkable levels of precision and reactivity. Food quality is a complicated, multidimensional notion that depends on a number of factors, including humidity, temperature, gas composition, and contaminant levels. Historically, manual inspections and recurrent sample were needed to monitor these characteristics, which could be error-prone and labor-intensive. However, by enabling continuous, real-time data collection, nanosensors are turning the game around (Chauhan et al., 2022; Cui, 2018; Yang & Duncan, 2021).

Because they function at the nanoscale, as their name implies, nanosensors are extremely sensitive and responsive. They are made to pick up on even the smallest alterations in the surrounding conditions, such as a tiny change in temperature, a modification in the composition of the gas, or the presence of particular pollutants (X. Huang et al., 2021). The capacity of nanosensors to identify and react to changes in the gas composition within food packaging is one of many amazing uses in the monitoring of food quality. For instance, they can identify variations in the concentrations of carbon dioxide and oxygen, two essential markers of food deterioration and freshness. Nanosensors could identify deviations and send out real-time alerts, enabling prompt remedial steps to ensure the safety and quality of the product (Mohamed & Awad, 2022).

Nanosensors can also be used to track variables such as humidity, which is important for food preservation. They are very good at spotting changes in humidity, which can result in spoiling from moisture or the growth of microorganisms. Nanosensors allow for exact control of environmental conditions, ensuring food products remain at the highest possible quality by giving real-time data on humidity. Nanosensors are important in ways that go beyond the food supply chain's distribution and manufacturing phases. This technology immediately benefits consumers, particularly when it is incorporated into intelligent packaging. Nanosensors not only guarantee food safety and quality, but they also provide suppliers and producers with insightful information. The ability to continuously gather and analyze data makes it possible to see trends and possible areas for process optimization. In the end, both producers and consumers stand to gain from this data-driven approach's potential to improve supply chain management and production techniques. The precision and responsiveness of nanosensors for real-time food quality monitoring have advanced significantly. They improve food safety, encourage openness, and provide an abundance of information for process improvement. The application of nanosensors in the food sector promises to improve the security and efficiency of the food supply chain, benefiting both producers and consumers as nanotechnology develops (Chelliah et al., 2021; Mathivanan, 2021).

The use of smart sensors is revolutionizing the accuracy and efficiency of tracking critical parameters in the field of food quality monitoring. The core function of these sensors is to monitor temperature, humidity, and gas composition. This provides a comprehensive method for guaranteeing the quality, safety, and freshness of food goods. An essential component of evaluating food quality is temperature monitoring (Kuswandi et al., 2011; Ma et al., 2023). Variations in temperature can hasten the deterioration of perishable food items and impact their texture and flavor. Smart sensors are excellent at tracking temperature over time, giving real-time information that makes sure goods are transported and stored within the ideal temperature range. These sensors detect deviations and send out notifications so that quick corrective action may be taken to preserve product quality. Food preservation depends heavily on humidity, especially when it comes to regulating moisture content to stop microbiological growth and preserve product texture. Humidity variations can be detected by smart sensors, which enable accurate environmental control and stop spoiling due to moisture. In sectors where texture and freshness are crucial, like baking and confectionery, this degree of oversight is particularly crucial (W. Huang et al., 2023; Osmólska et al., 2022).

Monitoring gas composition is yet another essential component of intelligent food quality sensors. Food packaging's oxygen and carbon dioxide contents have a direct bearing on shelf life and freshness. For instance, microbial activity and product deterioration are indicated by a drop in oxygen content and an increase in carbon dioxide (Rodrigues et al., 2021). Smart sensors offer real-time insights into the status of packed food by continuously measuring its gas composition. This allows for early actions to maintain quality. These sensors have advantages at every stage of the food supply chain, including manufacturing, shipping, and retail. Smart sensors in the production process can guarantee that food is cooked, prepared, and kept in the best possible ways, lowering the chance of contamination and spoiling. They protect food items in transit by continuously monitoring environmental parameters, which enables prompt correction of situations that can affect quality (Meng et al., 2014). By assisting in the identification and removal of tainted products from the shelves at the retail level, these sensors can improve food safety. Smart sensors' accuracy and reactivity greatly improve food safety and quality, cut down on food waste, and eventually increase consumer trust. These sensors provide a comprehensive approach to food quality assurance by monitoring temperature, humidity, and gas composition. This allows the food sector to meet consumer expectations for high standards while simultaneously improving efficiency and sustainability across the supply chain (Ma et al., 2022).

1.4 SAFETY AND REGULATORY CONSIDERATIONS

It is critical to address the safety and regulatory concerns related to this revolutionary technology as the food sector welcomes the integration of smart sensors for quality monitoring. While there are many advantages to these sensors, such as improved quality control and decreased food waste, there are also significant concerns about data security, product safety, and regulatory compliance. Data security is one of the main issues with safety. Sensitive information regarding food products, such as temperature, humidity, and gas composition, is gathered by smart sensors (Siddiqui et al., 2022; Watson et al., 2021). Due to its frequent electronic transmission and storage, this data is vulnerable to security lapses. It is crucial to make sure that this data is protected because any breach could result in improper quality control, the loss of confidential data, or even product tampering. Clear criteria for data security in smart food quality sensors must be established by industry stakeholders and regulatory organizations in order to allay these worries. To protect data from unwanted access, strong encryption techniques, access controls, and authentication procedures should be put in place. It takes regular security audits and assessments to find weaknesses and keep data protection at an optimal level (Alam et al., 2021; Ali et al., 2023).

The possibility of inaccurate or malfunctioning sensors is another safety concern. A broken sensor could result in inaccurate quality evaluations, endangering consumers and affecting the safety of food. Creating redundancy systems and fail-safe mechanisms is essential to making sure that the accuracy and functionality of the sensors are regularly maintained and monitored. Regulatory considerations are also extremely important (Kajmakovic et al., 2022; Schukat et al., 2016). The implementation of intelligent food quality sensors demands conformity to current laws and guidelines controlling food safety and quality. To guarantee accuracy and dependability, these sensors need to go through extensive testing and validation. To ensure continued compliance, regulatory agencies must set standards for the calibration and validation of these sensors as well as the frequency of testing (Damaševičius et al., 2023).

Furthermore, the information produced by smart sensors is frequently essential for adhering to regulations. Food goods may be shown to meet safety and quality criteria with the help of this data. As a result, it is crucial that the information gathered by these sensors is precisely documented, safely kept, and made available for regulatory authorities to review as needed. Regulatory agencies should update their policies to reflect these technologies as smart sensor integration in the food business progresses. Regulatory agencies and industry partners must work together to create comprehensive frameworks that strike a balance between safety, innovation, and compliance (Javaid et al., 2022; Pech et al., 2021). Regulatory compliance and safety must come first, even though smart food quality sensors provide many advantages. To guarantee that these sensors have a favorable impact on food safety and quality, data security, sensor dependability, and compliance with current laws are essential. Through proactive attention to these factors, the food industry may fully utilize smart sensor technology to protect customers and maintain industry standards. The use of nanomaterials in food has opened a number of creative possibilities, especially when it comes to food quality monitoring. It also highlights the urgent need to address any possible health risks connected to these nanoparticles (Dhanaraju et al., 2022). It is critical to guarantee the safety and legal compliance of nanotechnology in food to safeguard consumers and encourage responsible adoption. The possibility of harmful health effects from nanoparticles in food is one of the main worries. Because of their small size and high reactivity, nanomaterials have unique qualities that make us wonder how they interact with the human body.

Specifically, thorough research is needed to determine the possibility of nanoparticles accumulating in tissues, overcoming biological barriers, and having unanticipated health effects (Mohammad et al., 2022).

1.5 CASE STUDIES AND APPLICATIONS

Innovative solutions resulting from the application of nanotechnology in food preservation are having a noticeable effect on the food business. This section examines case studies and applications from the real world that demonstrate how nanotechnology can be used practically to improve food quality, safety, and sustainability. The creation of food packaging based on nanomaterials is one of the most common uses of nanotechnology in food preservation. For example, by mixing nanoparticles such as clay and silver into packaging materials, researchers have created nanocomposite films (Barage et al., 2022; H. Chen et al., 2014; Duncan, 2011; Mustafa & Andreescu, 2020; Naseer et al., 2018). These coatings have excellent barrier qualities that keep moisture and oxygen out. One example is the meat packing industry's usage of films coated with silver nanoparticles, which has proven to be remarkably effective in increasing the shelf life of fresh meat products. When it comes to real-time quality monitoring of food products, nanosensors are essential. Nanosensors are used in the dairy industry to track pH levels and identify signs of spoiling. Producers are able to remove contaminated items from the supply chain when producers receive signals from nanosensors when the pH of dairy products changes, suggesting possible spoiling. This guarantees that consumers receive safer and fresher dairy products while also reducing food waste (Kiss, 2020; McClements, 2020; Mei & Wang, 2020).

The use of nanocoatings to keep fruits and vegetables fresh has grown in popularity. To increase the shelf life of fresh food, for example, edible nanocoatings containing nanoparticles like chitosan and zinc oxide have been used. By reducing the rate of moisture loss and preventing the growth of spoilage bacteria, these coatings serve as a protective barrier (S. Kumar et al., 2020; Xing et al., 2019). Applications in the real world include the preservation of leafy greens' texture and apples' crispness using nanocoatings. The practical applications of silver and zinc oxide nanoparticles' antibacterial capabilities include their ability to suppress foodborne microorganisms. Silver nanoparticles have been added to ice during the processing of seafood to lessen bacterial infection during storage and transit. By reducing the possibility of foodborne infections, the release of silver ions from the nanoparticles contributes to the preservation of seafood products' quality and safety (Pham et al., 2023; Suriati, 2022).

In the meat sector, smart packaging that incorporates nanosensors has become popular. Within the meat packaging, these packets track the temperature, humidity, and gas composition. Nanosensors identify deviations and send out signals so that quick corrective action can be taken to maintain the safety and quality of the meat. This device increases consumer confidence in the freshness of meat products while also reducing food waste. Case studies such as these demonstrate the practical and wide-ranging uses of nanotechnology in food preservation (Ahmed et al., 2018; Chelliah et al., 2021). Nanotechnologies are enabling more safe, sustainable, and effective food supply chains, which will benefit both farmers and consumers. These benefits range from packaging to quality control. These actual cases demonstrate how nanotechnology is still altering food preservation techniques. The application of nanotechnology to food preservation has produced a number of success stories in addition to difficulties that the sector is still facing. This section explores several facets, illuminating the advancements achieved and the challenges that require further consideration (Mohebi & Marquez, 2015; Ramachandraiah et al., 2014).

The lengthening of shelf life for different products is one of the notable success stories in the application of nanotechnology to food preservation. Packaging made of nanomaterials has greatly decreased food deterioration, especially in fresh produce, meat, and dairy products. These packaging materials offer better barrier qualities against oxygen, moisture, and pollutants, ensuring that items stay fresh over extended periods of time (Lamri et al., 2021; Sagar et al., 2022). They are frequently supplemented with nanoparticles like silver and clay. The use of nanotechnology has been essential in reducing food waste. Smart packaging with embedded nanosensors can quickly respond to even the smallest changes in temperature, humidity, or gas composition. This has decreased the amount of contaminated product disposal while simultaneously improving food safety (Ashfaq et al., 2022; Perera et al., 2023). Such waste reduction is consistent with both economic efficiency and sustainability objectives. Incorporating antimicrobial nanoparticles such as zinc oxide and silver has been effective in preventing the growth of foodborne bacteria. For instance, applying silver nanoparticles to lower bacterial contamination during storage and transportation has shown impressive benefits for the seafood industry. By using this application, seafood goods are guaranteed to reach customers with a lower risk of contracting foodborne illnesses (Angelopoulou et al., 2022; Barage et al., 2022).

1.6 CHALLENGES IN INDUSTRY

Concerns about the possible health implications of foodborne nanoparticles persist. Even if a number of studies support the safety of these materials, more thorough investigation is still required to fully understand the long-term health effects of nanotechnology in food. In order to guarantee consumer safety, regulatory agencies and researchers need to work together (Ferdous & Nemmar, 2020; Martirosyan & Schneider, 2014). Although it is changing, the regulatory environment around nanotechnology in food is still complicated. The industry needs a standardized set of rules that are both clear and uniform. Since regional laws differ, it is difficult to get an international consensus on these requirements (Onyeaka et al., 2022).

Notwithstanding the advantages of nanotechnology for food preservation, issues with transparency and customer knowledge still exist. Clear labeling is necessary to notify consumers about the presence of nanoparticles in food products. Increasing awareness and trust need constant work. Environmental issues are raised by the manufacture and disposal of nanomaterials (Adeyeye, 2019; Jagtiani, 2022). Minimizing the ecological impact of nanotechnology in food preservation is an issue facing the business. It is necessary to improve sustainable methods further, such as recycling and ethical disposal of nanoparticles. In conclusion, there has been a noticeable improvement in food safety, food waste reduction, and shelf life due to the use of nanotechnology in food preservation. Nonetheless, issues with environmental effect, safety, transparency, and regulations still exist. The ethical and sustainable development of nanotechnology in the food sector will depend critically on overcoming these obstacles and building on the achievements of the past (R. Biswas et al., 2022; X. He et al., 2019).

1.7 FUTURE TRENDS AND EMERGING TECHNOLOGIES

The field of food preservation nanotechnology is characterized by ongoing innovation and research initiatives meant to meet the changing demands of the food sector. Current research is examining a number of exciting avenues that could fundamentally alter how we maintain and guarantee the quality of food items (Pandey, 2018; Ravichandran, 2010). The creation of nanocarriers to encapsulate and transport bioactive substances including probiotics, vitamins, and antioxidants is the subject of ongoing study. These sensitive chemicals are better protected from degradation and have increased bioavailability thanks to these nanocarriers. This can prolong the shelf life of food products and improve their nutritional content. Researchers are hard at work developing smart nanomaterials that adapt to their surroundings. One such example is packaging made of nanomaterials that alerts consumers to the freshness of the product by changing color when food deterioration indications are present. Consumers and the food sector could benefit from real-time information provided by such responsive nanoparticles (Dikshit et al., 2021; Monirul Hasan Tipu et al., 2022).

One important topic of continuing research is the creation of increasingly sophisticated nanosensors. By quickly identifying pollutants and dangerous pathogens, these nanosensors lower the possibility of foodborne illnesses and guarantee food safety. It is anticipated that the use of these sensors in the food supply chain would increase, providing speedier and more precise evaluations of quality. Researchers are looking into food packaging materials made of environmentally friendly nanoparticles (Rahmati et al., 2020; Thakur et al., 2022). For example, nanocellulose is being researched as a biodegradable substitute for conventional packaging materials. These advancements correspond with the increasing need for packaging options that are both environmentally friendly and biodegradable. One interesting area of study is the targeted delivery of preservatives and antibacterial agents using nanoparticles. When particular parameters, like pH or temperature, are satisfied, these nanoparticles may release preservatives selectively. This strategy may lessen the use of artificial preservatives in food items. The application of nanotechnology to food authenticity and traceability is another area of ongoing research. Labels and packaging can be made uniquely identifiable and tamper-evident by using nanomaterials. In order to prevent food fraud and guarantee the authenticity of high-end food products, these markers are being investigated (Das & Ganguly, 2023; H. Kumar et al., 2020; Reddy et al., 2022).

Regulatory science is changing to meet the special problems brought about by food nanotechnology. The goal of ongoing research is to create reliable safety evaluations and legal requirements that guarantee the responsible usage of nanomaterials. In this regard, cooperation between the scientific community and regulatory agencies is essential. Food quality, safety, and sustainability could be further enhanced as nanotechnology in food preservation advances through continuous research and new technologies (Chadha et al., 2022; Mishra et al., 2018). These advancements are anticipated to have a significant impact on the food industry's future by offering creative solutions that satisfy the changing needs of the contemporary food supply chain and benefit producers as well as consumers. Future advancements in the rapidly evolving field of nanotechnology in food preservation hold the potential to revolutionize the methods now used to preserve food and guarantee its quality. There are a number of intriguing opportunities and cutting-edge research directions that could provide us with some idea of this technology's possible future (Amenta et al., 2015; Nile et al., 2020; Onyeaka et al., 2022). The development of highly precise delivery systems for preservatives and bioactive substances is one of the anticipated advances. The capacity to design nanocarriers that can target particular regions of food products is made possible by nanotechnology. Preservatives, antioxidants, and other useful substances may become more effective as a result of this precise distribution (Foulkes et al., 2020; Tinkle et al., 2014).

It is anticipated that advancements in nanotechnology will result in the production of biodegradable and sustainable nanomaterials for food packaging. Particularly nanocellulose exhibits potential as a flexible and eco-friendly substitute for traditional packaging materials. These developments are in line with the growing need for environmentally friendly and sustainable packaging options. More sensitive and particular sensing systems are anticipated as a result of the advancement of nanosensor technology (Fahmy et al., 2020; T. Singh et al., 2017). With the use of these nanosensors, food quality may be continuously monitored, even at trace levels, for signs of contamination and spoiling.

The quality control procedures could be completely changed by integrating these sensors into the food supply chain. Personalized nutrition could be made possible by nanotechnology, which could allow food products to be customized to meet dietary requirements and personal tastes. Subsequent advancements could entail the utilization of nanocarriers to introduce customized combinations of flavors and nutrients into meals, guaranteeing that consumers acquire goods that are catered to their needs (Barage et al., 2022; Silva et al., 2020). Research on using nano-enzymes in food processing is still ongoing. These nanoscale enzymes have the ability to quicken biological processes like fermentation and ripening, which can improve process efficiency and shorten production timelines. The food processing sector could undergo a revolution because of such advancements. There is potential to improve food safety through the identification and synthesis of new nanomaterials possessing remarkable antibacterial capabilities. In order to lessen the need for chemical preservatives and lower the danger of foodborne illnesses, researchers are looking at the potential of these materials to impede the growth of a variety of pathogens (Chadha et al., 2022; Mishra et al., 2018).

The application of nanotechnology to personalized medicine—wherein therapeutically effective bioactive substances are delivered via nanoparticles—is an emerging concept. This strategy might be used to create customized food items and individualized dietary recommendations for those with certain medical requirements (Han et al., 2023; Kashyap et al., 2023). An intriguing new frontier in the field of food preservation nanotechnology is represented by these possible discoveries and upcoming advances. Future breakthroughs that improve food quality, safety, and sustainability are anticipated as scientists and inventors continue to push the boundaries of nanotechnology, which will benefit the food sector as well as consumers (T. Singh et al., 2017).

1.8 CONCLUSION

The incorporation of nanoparticles into food packaging has greatly increased the perishable foods' shelf lives, decreasing food waste and improving product freshness. Silver and zinc oxide, two antimicrobial nanoparticles, have shown promise in suppressing foodborne bacteria and enhancing food safety. Nanotechnology-enabled smart sensors offer real-time monitoring of vital indicators like humidity and temperature, enhancing quality control and reducing food waste. But applying nanotechnology to food preservation also presents a number of difficulties, including safety issues, intricate legal requirements, and the need for openness and consumer education. Furthermore, when applying nanotechnology to food, ethical issues highlight the significance of openness, fair access, environmental sustainability, and long-term health impacts. Further advancements in food preservation techniques could come from continued study and developing technologies, which could lead to the development of precise delivery systems, environmentally friendly packing materials, sophisticated nanosensors, tailored diets, and many other breakthroughs. The application of nanotechnology in the food business is a dynamic and developing field that has the potential to improve food sustainability, safety, and quality, which will help producers and consumers alike in their pursuit of a more reliable and effective food supply chain. Future implications of nanotechnology for the food preservation sector seem endless. Future developments in food preservation techniques could include targeted delivery methods, environmentally friendly packaging, sophisticated nanosensors, customized diets, and more. These advancements could come from ongoing research as well as from developing technologies. The food industry's use of nanotechnology is a dynamic and changing field that has the potential to drastically change how we think about food preservation.

REFERENCES

Abideen, A. Z., Sundram, V. P. K., Pyeman, J., Othman, A. K., & Sorooshian, S. (2021). Food Supply Chain Transformation Through Technology and Future Research Directions—a Systematic Review. *Logistics*, *5*(4), 83. https://doi.org/10.3390/logistics5040083

Adeyemi, J. O., & Fawole, O. A. (2023). Metal-Based Nanoparticles in Food Packaging and Coating Technologies: A Review. *Biomolecules*, *13*(7), 1092. https://doi.org/10.3390/biom13071092

Adeyeye, S. A. O. (2019). Food Packaging and Nanotechnology: Safeguarding Consumer Health and Safety. *Nutrition & Food Science*, *49*(6), 1164–1179. https://doi.org/10.1108/NFS-01-2019-0020

Agarwal, A., Shaida, B., Rastogi, M., & Singh, N. B. (2023). Food Packaging Materials with Special Reference to Biopolymers-Properties and Applications. *Chemistry Africa*, *6*(1), 117–144. https://doi.org/10.1007/s42250-022-00446-w

Agriopoulou, S., Stamatelopoulou, E., Sachadyn-Król, M., & Varzakas, T. (2020). Lactic Acid Bacteria as Antibacterial Agents to Extend the Shelf Life of Fresh and Minimally Processed Fruits and Vegetables: Quality and Safety Aspects. *Microorganisms*, *8*(6), 952. https://doi.org/10.3390/microorganisms8060952

Agriopoulou, S., Stamatelopoulou, E., Skiada, V., Tsarouhas, P., & Varzakas, T. (2020). Emerging Nanomaterial Applications for Food Packaging and Preservation: Safety Issues and Risk Assessment. *The 1st International Electronic Conference on Food Science and Functional Foods*, 7. https://doi.org/10.3390/foods_2020-07747

Ahari, H., & Soufiani, S. P. (2021). Smart and Active Food Packaging: Insights in Novel Food Packaging. *Frontiers in Microbiology*, *12*. https://doi.org/10.3389/fmicb.2021.657233

Ahmed, I., Lin, H., Zou, L., Li, Z., Brody, A. L., Qazi, I. M., Lv, L., Pavase, T. R., Khan, M. U., Khan, S., & Sun, L. (2018). An Overview of Smart Packaging Technologies for Monitoring Safety and Quality of Meat and Meat Products. *Packaging Technology and Science*, *31*(7), 449–471. https://doi.org/10.1002/pts.2380

Alam, A. U., Rathi, P., Beshai, H., Sarabha, G. K., & Deen, M. J. (2021). Fruit Quality Monitoring with Smart Packaging. *Sensors*, *21*(4), 1509. https://doi.org/10.3390/s21041509

Alhujaily, M., Albukhaty, S., Yusuf, M., Mohammed, M. K. A., Sulaiman, G. M., Al-Karagoly, H., Alyamani, A. A.,

Albaqami, J., & AlMalki, F. A. (2022). Recent Advances in Plant-Mediated Zinc Oxide Nanoparticles with Their Significant Biomedical Properties. *Bioengineering*, *9*(10), 541. https://doi.org/10.3390/bioengineering9100541

Ali, O., Abdelbaki, W., Shrestha, A., Elbasi, E., Alryalat, M. A. A., & Dwivedi, Y. K. (2023). A Systematic Literature Review of Artificial Intelligence in the Healthcare Sector: Benefits, Challenges, Methodologies, and Functionalities. *Journal of Innovation & Knowledge*, *8*(1), 100333. https://doi.org/10.1016/j.jik.2023.100333

Amenta, V., Aschberger, K., Arena, M., Bouwmeester, H., Botelho Moniz, F., Brandhoff, P., Gottardo, S., Marvin, H. J. P., Mech, A., Quiros Pesudo, L., Rauscher, H., Schoonjans, R., Vettori, M. V., Weigel, S., & Peters, R. J. (2015). Regulatory Aspects of Nanotechnology in the Agri/Feed/Food Sector in EU and Non-EU Countries. *Regulatory Toxicology and Pharmacology*, *73*(1), 463–476. https://doi.org/10.1016/j.yrtph.2015.06.016

Amit, S. K., Uddin, M. M., Rahman, R., Islam, S. M. R., & Khan, M. S. (2017). A Review on Mechanisms and Commercial Aspects of Food Preservation and Processing. *Agriculture & Food Security*, *6*(1), 51. https://doi.org/10.1186/s40066-017-0130-8

Angelopoulou, P., Giaouris, E., & Gardikis, K. (2022). Applications and Prospects of Nanotechnology in Food and Cosmetics Preservation. *Nanomaterials*, *12*(7), 1196. https://doi.org/10.3390/nano12071196

Anvar, A. A., Ahari, H., & Ataee, M. (2021). Antimicrobial Properties of Food Nanopackaging: A New Focus on Foodborne Pathogens. *Frontiers in Microbiology*, *12*. https://doi.org/10.3389/fmicb.2021.690706

Ashfaq, A., Khursheed, N., Fatima, S., Anjum, Z., & Younis, K. (2022). Application of Nanotechnology in Food Packaging: Pros and Cons. *Journal of Agriculture and Food Research*, *7*, 100270. https://doi.org/10.1016/j.jafr.2022.100270

Awuchi, C. G. (2023). HACCP, Quality, and Food Safety Management in Food and Agricultural Systems. *Cogent Food & Agriculture*, *9*(1). https://doi.org/10.1080/23311932.2023.2176280

Azeredo, H. M. C., Otoni, C. G., Corrêa, D. S., Assis, O. B. G., de Moura, M. R., & Mattoso, L. H. C. (2019). Nanostructured Antimicrobials in Food Packaging—Recent Advances. *Biotechnology Journal*, *14*(12). https://doi.org/10.1002/biot.201900068

Baig, N., Kammakakam, I., & Falath, W. (2021). Nanomaterials: A Review of Synthesis Methods, Properties, Recent Progress, and Challenges. *Materials Advances*, *2*(6), 1821–1871. https://doi.org/10.1039/D0MA00807A

Bajpai, V. K., Kamle, M., Shukla, S., Mahato, D. K., Chandra, P., Hwang, S. K., Kumar, P., Huh, Y. S., & Han, Y.-K. (2018). Prospects of Using Nanotechnology for Food Preservation, Safety, and Security. *Journal of Food and Drug Analysis*, *26*(4), 1201–1214. https://doi.org/10.1016/j.jfda.2018.06.011

Baptista, P. V., McCusker, M. P., Carvalho, A., Ferreira, D. A., Mohan, N. M., Martins, M., & Fernandes, A. R. (2018). Nano-Strategies to Fight Multidrug Resistant Bacteria—"A Battle of the Titans." *Frontiers in Microbiology*, *9*. https://doi.org/10.3389/fmicb.2018.01441

Barage, S., Lakkakula, J., Sharma, A., Roy, A., Alghamdi, S., Almehmadi, M., Hossain, M. J., Allahyani, M., & Abdulaziz, O. (2022). Nanomaterial in Food Packaging: A Comprehensive Review. *Journal of Nanomaterials*, *2022*, 1–12. https://doi.org/10.1155/2022/6053922

Bayda, S., Adeel, M., Tuccinardi, T., Cordani, M., & Rizzolio, F. (2019). The History of Nanoscience and Nanotechnology: From Chemical–Physical Applications to Nanomedicine. *Molecules*, *25*(1), 112. https://doi.org/10.3390/molecules25010112

Beyth, N., Houri-Haddad, Y., Domb, A., Khan, W., & Hazan, R. (2015). Alternative Antimicrobial Approach: Nano-Antimicrobial Materials. *Evidence-Based Complementary and Alternative Medicine*, *2015*, 1–16. https://doi.org/10.1155/2015/246012

Bharti, S. K., Pathak, V., Alam, T., Arya, A., Basak, G., & Awasthi, M. G. (2020). Materiality of Edible Film Packaging in Muscle Foods: A Worthwhile Conception. *Journal of Packaging Technology and Research*, *4*(1), 117–132. https://doi.org/10.1007/s41783-020-00087-9

Biswas, A., Kar, U., & Jana, N. R. (2022). Cytotoxicity of ZnO Nanoparticles Under Dark Conditions via Oxygen Vacancy Dependent Reactive Oxygen Species Generation. *Physical Chemistry Chemical Physics*, *24*(22), 13965–13975. https://doi.org/10.1039/D2CP00301E

Biswas, R., Alam, M., Sarkar, A., Haque, M. I., Hasan, M. M., & Hoque, M. (2022). Application of Nanotechnology in Food: Processing, Preservation, Packaging and Safety Assessment. *Heliyon*, *8*(11), e11795. https://doi.org/10.1016/j.heliyon.2022.e11795

Bothiraja, C., Gholap, A. D., Shaikh, K. S., & Pawar, A. P. (2014). Investigation of Ethyl Cellulose Microsponge Gel for Topical Delivery of Eberconazole Nitrate for Fungal Therapy. *Therapeutic Delivery*, *5*(7), 781–794. https://doi.org/10.4155/tde.14.43

Bradley, E. L., Castle, L., & Chaudhry, Q. (2011). Applications of Nanomaterials in Food Packaging with a Consideration of Opportunities for Developing Countries. *Trends in Food Science & Technology*, *22*(11), 604–610. https://doi.org/10.1016/j.tifs.2011.01.002

Bruna, T., Maldonado-Bravo, F., Jara, P., & Caro, N. (2021). Silver Nanoparticles and Their Antibacterial Applications. *International Journal of Molecular Sciences*, *22*(13), 7202. https://doi.org/10.3390/ijms22137202

Chadha, U., Bhardwaj, P., Selvaraj, S. K., Arasu, K., Praveena, S., Pavan, A., Khanna, M., Singh, P., Singh, S., Chakravorty, A., Badoni, B., Banavoth, M., Sonar, P., & Paramasivam, V. (2022). Current Trends and Future Perspectives of Nanomaterials in Food Packaging Application. *Journal of Nanomaterials*, *2022*, 1–32. https://doi.org/10.1155/2022/2745416

Charitidis, C. A., Georgiou, P., Koklioti, M. A., Trompeta, A.-F., & Markakis, V. (2014). Manufacturing Nanomaterials: From Research to Industry. *Manufacturing Review*, *1*, 11. https://doi.org/10.1051/mfreview/2014009

Chauhan, S., Singh, R., Gehlot, A., Akram, S. V., Twala, B., & Priyadarshi, N. (2022). Digitalization of Supply Chain Management with Industry 4.0 Enabling Technologies: A Sustainable Perspective. *Processes*, *11*(1), 96. https://doi.org/10.3390/pr11010096

Chelliah, R., Wei, S., Daliri, E. B.-M., Rubab, M., Elahi, F., Yeon, S.-J., Jo, K. hee, Yan, P., Liu, S., & Oh, D. H. (2021). Development of Nanosensors Based Intelligent Packaging Systems: Food Quality and Medicine. *Nanomaterials*, *11*(6), 1515. https://doi.org/10.3390/nano11061515

Chen, H., Seiber, J. N., & Hotze, M. (2014). ACS Select on Nanotechnology in Food and Agriculture: A Perspective on Implications and Applications. *Journal of Agricultural and Food Chemistry*, *62*(6), 1209–1212. https://doi.org/10.1021/jf5002588

Chen, S., Brahma, S., Mackay, J., Cao, C., & Aliakbarian, B. (2020). The Role of Smart Packaging System in Food

Supply Chain. *Journal of Food Science*, *85*(3), 517–525. https://doi.org/10.1111/1750-3841.15046

Cui, Y. (2018). Supply Chain Innovation with IoT. In *Multi-Criteria Methods and Techniques Applied to Supply Chain Management*. InTech. https://doi.org/10.5772/intechopen.74155

Czerwiński, K., Rydzkowski, T., Wróblewska-Krepsztul, J., & Thakur, V. K. (2021). Towards Impact of Modified Atmosphere Packaging (MAP) on Shelf-Life of Polymer-Film-Packed Food Products: Challenges and Sustainable Developments. *Coatings*, *11*(12), 1504. https://doi.org/10.3390/coatings11121504

da Costa, T. P., Gillespie, J., Cama-Moncunill, X., Ward, S., Condell, J., Ramanathan, R., & Murphy, F. (2022). A Systematic Review of Real-Time Monitoring Technologies and its Potential Application to Reduce Food Loss and Waste: Key Elements of Food Supply Chains and IoT Technologies. *Sustainability*, *15*(1), 614. https://doi.org/10.3390/su15010614

Dakal, T. C., Kumar, A., Majumdar, R. S., & Yadav, V. (2016). Mechanistic Basis of Antimicrobial Actions of Silver Nanoparticles. *Frontiers in Microbiology*, *7*. https://doi.org/10.3389/fmicb.2016.01831

Damaševičius, R., Bacanin, N., & Misra, S. (2023). From Sensors to Safety: Internet of Emergency Services (IoES) for Emergency Response and Disaster Management. *Journal of Sensor and Actuator Networks*, *12*(3), 41. https://doi.org/10.3390/jsan12030041

Das, T. K., & Ganguly, S. (2023). Revolutionizing Food Safety with Quantum Dot–Polymer Nanocomposites: From Monitoring to Sensing Applications. *Foods*, *12*(11), 2195. https://doi.org/10.3390/foods12112195

de Sousa, M. S., Schlogl, A. E., Estanislau, F. R., Souza, V. G. L., dos Reis Coimbra, J. S., & Santos, I. J. B. (2023). Nanotechnology in Packaging for Food Industry: Past, Present, and Future. *Coatings*, *13*(8), 1411. https://doi.org/10.3390/coatings13081411

Dhanaraju, M., Chenniappan, P., Ramalingam, K., Pazhanivelan, S., & Kaliaperumal, R. (2022). Smart Farming: Internet of Things (IoT)-Based Sustainable Agriculture. *Agriculture*, *12*(10), 1745. https://doi.org/10.3390/agriculture12101745

Díaz-Montes, E., & Castro-Muñoz, R. (2021). Edible Films and Coatings as Food-Quality Preservers: An Overview. *Foods*, *10*(2), 249. https://doi.org/10.3390/foods10020249

Dikshit, P., Kumar, J., Das, A., Sadhu, S., Sharma, S., Singh, S., Gupta, P., & Kim, B. (2021). Green Synthesis of Metallic Nanoparticles: Applications and Limitations. *Catalysts*, *11*(8), 902. https://doi.org/10.3390/catal11080902

Domínguez, R., Pateiro, M., Gagaoua, M., Barba, F. J., Zhang, W., & Lorenzo, J. M. (2019). A Comprehensive Review on Lipid Oxidation in Meat and Meat Products. *Antioxidants*, *8*(10), 429. https://doi.org/10.3390/antiox8100429

Duncan, T. V. (2011). Applications of Nanotechnology in Food Packaging and Food Safety: Barrier Materials, Antimicrobials and Sensors. *Journal of Colloid and Interface Science*, *363*(1), 1–24. https://doi.org/10.1016/j.jcis.2011.07.017

El-Sayed, S. M., & Youssef, A. M. (2023). Eco-Friendly Biodegradable Nanocomposite Materials and Their Recent Use in Food Packaging Applications: A Review. *Sustainable Food Technology*, *1*(2), 215–227. https://doi.org/10.1039/D2FB00021K

Ezati, P., Khan, A., Priyadarshi, R., Bhattacharya, T., Tammina, S. K., & Rhim, J.-W. (2023). Biopolymer-Based UV Protection Functional Films for Food Packaging. *Food Hydrocolloids*, *142*, 108771. https://doi.org/10.1016/j.foodhyd.2023.108771

Fadiji, T., Rashvand, M., Daramola, M. O., & Iwarere, S. A. (2023). A Review on Antimicrobial Packaging for Extending the Shelf Life of Food. *Processes*, *11*(2), 590. https://doi.org/10.3390/pr11020590

Fahmy, H. M., Salah Eldin, R. E., Abu Serea, E. S., Gomaa, N. M., AboElmagd, G. M., Salem, S. A., Elsayed, Z. A., Edrees, A., Shams-Eldin, E., & Shalan, A. E. (2020). Advances in Nanotechnology and Antibacterial Properties of Biodegradable Food Packaging Materials. *RSC Advances*, *10*(35), 20467–20484. https://doi.org/10.1039/D0RA02922J

Faiyazuddin, M., Sophia, A., Ashique, S., Gholap, A. D., Gowri, S., Mohanto, S., Karthikeyan, C., Nag, S., Hussain, A., Akhtar, M. S., Bakht, M. A., Ahmed, M. G., Rustagi, S., Rodriguez-Morales, A. J., Salas-Matta, L. A., Mohanty, A., Bonilla-Aldana, D. K., & Sah, R. (2023). Virulence Traits and Novel Drug Delivery Strategies for Mucormycosis Post-COVID-19: A Comprehensive Review. *Frontiers in Immunology*, *14*. https://doi.org/10.3389/fimmu.2023.1264502

Fajardo, C., Martinez-Rodriguez, G., Blasco, J., Mancera, J. M., Thomas, B., & De Donato, M. (2022). Nanotechnology in Aquaculture: Applications, Perspectives and Regulatory Challenges. *Aquaculture and Fisheries*, *7*(2), 185–200. https://doi.org/10.1016/j.aaf.2021.12.006

Ferdous, Z., & Nemmar, A. (2020). Health Impact of Silver Nanoparticles: A Review of the Biodistribution and Toxicity Following Various Routes of Exposure. *International Journal of Molecular Sciences*, *21*(7), 2375. https://doi.org/10.3390/ijms21072375

Fernando, S., Gunasekara, T., & Holton, J. (2018). Antimicrobial Nanoparticles: Applications and Mechanisms of Action. *Sri Lankan Journal of Infectious Diseases*, *8*(1), 2. https://doi.org/10.4038/sljid.v8i1.8167

Foulkes, R., Man, E., Thind, J., Yeung, S., Joy, A., & Hoskins, C. (2020). The Regulation of Nanomaterials and Nanomedicines for Clinical Application: Current and Future Perspectives. *Biomaterials Science*, *8*(17), 4653–4664. https://doi.org/10.1039/D0BM00558D

Fuertes, G., Soto, I., Vargas, M., Valencia, A., Sabattin, J., & Carrasco, R. (2016). Nanosensors for a Monitoring System in Intelligent and Active Packaging. *Journal of Sensors*, *2016*, 1–8. https://doi.org/10.1155/2016/7980476

Geng, L., Liu, K., & Zhang, H. (2023). Lipid Oxidation in Foods and its Implications on Proteins. *Frontiers in Nutrition*, *10*. https://doi.org/10.3389/fnut.2023.1192199

Gholap, A. D. (2012). Translational Nanomedicine: Status Assessment and Opportunities. *International Journal of Research in Pharmaceutical and Biomedical Sciences*, *3*(4), 1454–1463.

Gholap, A. D., Bhowmik, D. D., Deshmukh, A. Y., & Hatvate, N. T. (2023). Quintessential Impact of Dendrimer Bioconjugates in Targeted Drug Delivery. https://doi.org/10.1016/bs.coac.2023.02.009

Gholap, A. D., Borude, S. S., & Mahajan, A. M. (2011a). Smart Crystals Technology: A Review. *Pharmacologyonline*, *3*, 238–243.

Gholap, A. D., Borude, S. S., Mahajan, A. M., & Gholap, M. A. D. (2011b). Aquasomes: A Potential Drug Delivery Carrier. *Pharmacologyonline*, *3*, 230–237.

Gholap, A. D., & Dighe, S. B. (2012a). Dandruff: Pathogenesis And Treatment. *Inventi Rapid: Cosmeceuticals [ISSN 0976-3864]*, *4*, 1–4.

Gholap, A. D., & Dighe, S. B. (2012b). Role of Sebum in Human Health & Acne. *Inventi Rapid: Cosmeceuticals [ISSN 0976-3864]*, *4*, 1–3.

Gholap, A. D., Gupta, J., Kamandar, P., Bhowmik, D. D., Rojekar, S., Faiyazuddin, M., Hatvate, N. T., Mohanto, S., Ahmed,

M. G., Subramaniyan, V., & Kumarasamy, V. (2024). Harnessing Nanovaccines for Effective Immunization—a Special Concern on COVID-19: Facts, Fidelity, and Future Prospective. *ACS Biomaterials Science & Engineering*, *10*(1), 271–297. https://doi.org/10.1021/acsbiomaterials.3c01247

Gholap, A. D., Gupta, J. S., Kamandar, P. A., Banchhod, G. V., & Hatvate, N. T. (2023). Antibody-Drug Conjugates for Cancer Therapy: An Up-to-Date Review on the Chemistry and Pharmacology. https://doi.org/10.1016/bs.coac.2023.02.007

Gholap, A. D., Hatvate, N. T., Khuspe, P. R., Mandhare, T. A., Kashid, P., & Gaikwad, V. D. (2023). Proteomics in Oncology: Retrospect and Prospects. In *Novel Technologies in Biosystems, Biomedical & Drug Delivery* (pp. 243–269). Springer Nature Singapore. https://doi.org/10.1007/978-981-99-5281-6_10

Gholap, A. D., Hatvate, N. T., Sawant, S. D., Khemka, P. N., & Vora, L. K. (2023). Unlocking the Potential of Nanobiohybrids to Combat Environmental Pollution. In *Nanobiohybrids for Advanced Wastewater Treatment and Energy Recovery* (pp. 169–190). IWA Publishing. https://doi.org/10.2166/9781789063592_0169

Gholap, A. D., & Musmade, D. S. (2011). Advances in the Field of Nanooncology: An Overview. *Inventi Impact: NDDS [ISSN 0976-3791]*, *4*, 1–3.

Gholap, A., Pardeshi, S., & Giram, P. (2023). Preparation, Antibacterial and Antiviral Activity Measurements and Detection Methods (pp. 33–64). https://doi.org/10.1021/bk-2023-1458.ch002

Gholap, A. D., & Rajurkar, V. G. (2012a). Lipid Nanocapsules: A Novel Promising Tool as Drug Delivery System. *Inventi Rapid: NDDS [ISSN 0976-3791]*, *4*, 1–4.

Gholap, A. D., & Rajurkar, V. G. (2012b). New Threat of Clinical Toxicities Evoked from Nanocarrier Systems. *Inventi Rapid: NDDS [ISSN 0976-3791]*, *4*, 1–4.

Gholap, A. D., & Rajurkar, V. G. (2012c). Potential Applications of Superparmagnetic Nanoparticles Technology. *Inventi Rapid: NDDS [ISSN 0976-3791]*, *4*, 1–3.

Gholap, A. D., & Rajurkar, V. G. (n.d.). Potential Applications of Capillary Electromigration Methods for the Improvement of Separation Efficiency. *Inventi Rapid: Pharm Analysis & Quality Assurance [ISSN 0976-3813]*, *2012*(4), 1–3.

Gholap, A. D., Rojekar, S., Kapare, H. S., Vishwakarma, N., Raikwar, S., Garkal, A., Mehta, T. A., Jadhav, H., Prajapati, M. K., & Annapure, U. (2024). Chitosan Scaffolds: Expanding Horizons in Biomedical Applications. *Carbohydrate Polymers*, *323*, 121394. https://doi.org/10.1016/j.carbpol.2023.121394

Gholap, A. D., Said, R. P., Pawar, R. D., Ambore, G. S., & Hatvate, N. T. (2023). Importance of Carbohydrate-Drug Conjugates in Vaccine Development: A Detailed Review. https://doi.org/10.1016/bs.coac.2023.03.002

Gholap, A. D., Sawant, S. D., Sayyad, S. F., Hatvate, N. T., Chavan, M., Rojekar, S., & Faiyazuddin, M. (2023). Food Biopackaging for Human Benefits: Status and Perspectives. In *Biobased Packaging Materials* (pp. 1–35). Springer Nature Singapore. https://doi.org/10.1007/978-981-99-6050-7_1

Gholap, A. D., Sayyad, S. F., Hatvate, N. T., Dhumal, V. V., Pardeshi, S. R., Chavda, V. P., & Vora, L. K. (2023). Drug Delivery Strategies for Avobenzone: A Case Study of Photostabilization. *Pharmaceutics*, *15*(3), 1008. https://doi.org/10.3390/pharmaceutics15031008

Gholap, A. D., & Tupe, D. (2012a). Potential Applications of Botanical Extract. *Inventi Rapid: Ethnopharmacology [ISSN 0976-3805]*, *4*, 1–4.

Gholap, A. D., & Tupe, D. (2012b). Artificial Viral Envelopes: Emerging Vehicle for Gene Delivery. *Pharmatutor-Art-1439*, 1–3.

Gupta, A., Mumtaz, S., Li, C.-H., Hussain, I., & Rotello, V. M. (2019). Combatting Antibiotic-Resistant Bacteria Using Nanomaterials. *Chemical Society Reviews*, *48*(2), 415–427. https://doi.org/10.1039/C7CS00748E

Han, M., Liu, K., Liu, X., Rashid, M. T., Zhang, H., & Wang, M. (2023). Research Progress of Protein-Based Bioactive Substance Nanoparticles. *Foods*, *12*(16), 2999. https://doi.org/10.3390/foods12162999

Hassoun, A., Marvin, H. J. P., Bouzembrak, Y., Barba, F. J., Castagnini, J. M., Pallarés, N., Rabail, R., Aadil, R. M., Bangar, S. P., Bhat, R., Cropotova, J., Maqsood, S., & Regenstein, J. M. (2023). Digital Transformation in the Agri-Food Industry: Recent Applications and the Role of the COVID-19 Pandemic. *Frontiers in Sustainable Food Systems*, *7*. https://doi.org/10.3389/fsufs.2023.1217813

He, L., Xue, M., & Gu, B. (2020). Internet-of-Things Enabled Supply Chain Planning and Coordination with Big Data Services: Certain Theoretic Implications. *Journal of Management Science and Engineering*, *5*(1), 1–22. https://doi.org/10.1016/j.jmse.2020.03.002

He, X., Deng, H., & Hwang, H. (2019). The Current Application of Nanotechnology in Food and Agriculture. *Journal of Food and Drug Analysis*, *27*(1), 1–21. https://doi.org/10.1016/j.jfda.2018.12.002

Huang, W., Wang, X., Xia, J., Li, Y., Zhang, L., Feng, H., & Zhang, X. (2023). Flexible Sensing Enabled Agri-Food Cold Chain Quality Control: A Review of Mechanism Analysis, Emerging Applications, and System Integration. *Trends in Food Science & Technology*, *133*, 189–204. https://doi.org/10.1016/j.tifs.2023.02.010

Huang, X., Zhu, Y., & Kianfar, E. (2021). Nano Biosensors: Properties, Applications and Electrochemical Techniques. *Journal of Materials Research and Technology*, *12*, 1649–1672. https://doi.org/10.1016/j.jmrt.2021.03.048

Huang, Y., Mei, L., Chen, X., & Wang, Q. (2018). Recent Developments in Food Packaging Based on Nanomaterials. *Nanomaterials*, *8*(10), 830. https://doi.org/10.3390/nano8100830

Hutapea, S., Ghazi al-Shawi, S., Chen, T.-C., You, X., Bokov, D., Abdelbasset, W. K., & Suksatan, W. (2022). Study on Food Preservation Materials Based on Nano-Particle Reagents. *Food Science and Technology*, *42*. https://doi.org/10.1590/fst.39721

Iversen, L. J. L., Rovina, K., Vonnie, J. M., Matanjun, P., Erna, K. H., 'Aqilah, N. M. N., Felicia, W. X. L., & Funk, A. A. (2022). The Emergence of Edible and Food-Application Coatings for Food Packaging: A Review. *Molecules*, *27*(17), 5604. https://doi.org/10.3390/molecules27175604

Jagtiani, E. (2022). Advancements in Nanotechnology for Food Science and Industry. *Food Frontiers*, *3*(1), 56–82. https://doi.org/10.1002/fft2.104

Javaid, M., Abid Haleem, Pratap Singh, R., Rab, S., & Suman, R. (2021). Upgrading the Manufacturing Sector via Applications of Industrial Internet of Things (IIoT). *Sensors International*, *2*, 100129. https://doi.org/10.1016/j.sintl.2021.100129

Javaid, M., Haleem, A., Singh, R. P., & Suman, R. (2022). Enhancing Smart Farming Through the Applications of Agriculture 4.0 Technologies. *International Journal of Intelligent Networks*, *3*, 150–164. https://doi.org/10.1016/j.ijin.2022.09.004

Jiang, S., Lin, K., & Cai, M. (2020). ZnO Nanomaterials: Current Advancements in Antibacterial Mechanisms and Applications. *Frontiers in Chemistry*, *8*. https://doi.org/10.3389/fchem.2020.00580

Joseph, T., Kar Mahapatra, D., Esmaeili, A., Piszczyk, Ł., Hasanin, M., Kattali, M., Haponiuk, J., & Thomas, S. (2023). Nanoparticles: Taking a Unique Position in Medicine. *Nanomaterials*, *13*(3), 574. https://doi.org/10.3390/nano13030574

Kachoei, M., Divband, B., Rahbar, M., Esmaeilzadeh, M., Ghanizadeh, M., & Alam, M. (2021). A Novel Developed Bioactive Composite Resin Containing Silver/Zinc Oxide (Ag/ZnO) Nanoparticles as an Antimicrobial Material Against *Streptococcus mutans*, *Lactobacillus*, and *Candida albicans*. *Evidence-Based Complementary and Alternative Medicine*, *2021*, 1–8. https://doi.org/10.1155/2021/4743411

Kajmakovic, A., Diwold, K., Römer, K., Pestana, J., & Kajtazovic, N. (2022). Degradation Detection in a Redundant Sensor Architecture. *Sensors*, *22*(12), 4649. https://doi.org/10.3390/s22124649

Kashyap, B. K., Singh, V. V., Solanki, M. K., Kumar, A., Ruokolainen, J., & Kesari, K. K. (2023). Smart Nanomaterials in Cancer Theranostics: Challenges and Opportunities. *ACS Omega*, *8*(16), 14290–14320. https://doi.org/10.1021/acsomega.2c07840

Kasza, G., Veflen, N., Scholderer, J., Münter, L., Fekete, L., Csenki, E. Z., Dorkó, A., Szakos, D., & Izsó, T. (2022). Conflicting Issues of Sustainable Consumption and Food Safety: Risky Consumer Behaviors in Reducing Food Waste and Plastic Packaging. *Foods*, *11*(21), 3520. https://doi.org/10.3390/foods11213520

Khan, P. W., Byun, Y.-C., & Park, N. (2020). IoT-Blockchain Enabled Optimized Provenance System for Food Industry 4.0 Using Advanced Deep Learning. *Sensors*, *20*(10), 2990. https://doi.org/10.3390/s20102990

Khan, Y., Sadia, H., Ali Shah, S. Z., Khan, M. N., Shah, A. A., Ullah, N., Ullah, M. F., Bibi, H., Bafakeeh, O. T., Khedher, N. Ben, Eldin, S. M., Fadhl, B. M., & Khan, M. I. (2022). Classification, Synthetic, and Characterization Approaches to Nanoparticles, and Their Applications in Various Fields of Nanotechnology: A Review. *Catalysts*, *12*(11), 1386. https://doi.org/10.3390/catal12111386

Khedkar, A., Tupe, D., & Gholap, A. (2012a). Cubosomes: Emerging Drug Delivery System. *Inventi Rapid: NDDS [ISSN 0976-3791]*, *4*, 1–3.

Khedkar, A., Tupe, D., & Gholap, A. (2012b). Pharmacosomes: An Emerging Vesicular System. *Inventi Rapid: NDDS [ISSN 0976-3791]*, *4*, 1–3.

Kim, I., Viswanathan, K., Kasi, G., Thanakkasaranee, S., Sadeghi, K., & Seo, J. (2022). ZnO Nanostructures in Active Antibacterial Food Packaging: Preparation Methods, Antimicrobial Mechanisms, Safety Issues, Future Prospects, and Challenges. *Food Reviews International*, *38*(4), 537–565. https://doi.org/10.1080/87559129.2020.1737709

Kiss, É. (2020). Nanotechnology in Food Systems: A Review. *Acta Alimentaria*, *49*(4), 460–474. https://doi.org/10.1556/066.2020.49.4.12

Krishna, A. R., Gurumoorthy, S., Elayappan, P., Sakthivadivel, P., Kumaran, S., & Pushparaj, P. (2022). A Review on the Application of Nanotechnology in Food Industries. *Current Research in Nutrition and Food Science Journal*, *10*(3), 871–883. https://doi.org/10.12944/CRNFSJ.10.3.5

Kumar, A., Choudhary, A., Kaur, H., Mehta, S., & Husen, A. (2021). Metal-Based Nanoparticles, Sensors, and Their Multifaceted Application in Food Packaging. *Journal of Nanobiotechnology*, *19*(1), 256. https://doi.org/10.1186/s12951-021-00996-0

Kumar, H., Kuča, K., Bhatia, S. K., Saini, K., Kaushal, A., Verma, R., Bhalla, T. C., & Kumar, D. (2020). Applications of Nanotechnology in Sensor-Based Detection of Foodborne Pathogens. *Sensors*, *20*(7), 1966. https://doi.org/10.3390/s20071966

Kumar, I., Rawat, J., Mohd, N., & Husain, S. (2021). Opportunities of Artificial Intelligence and Machine Learning in the Food Industry. *Journal of Food Quality*, *2021*, 1–10. https://doi.org/10.1155/2021/4535567

Kumar, S., Mukherjee, A., & Dutta, J. (2020). Chitosan Based Nanocomposite Films and Coatings: Emerging Antimicrobial Food Packaging Alternatives. *Trends in Food Science & Technology*, *97*, 196–209. https://doi.org/10.1016/j.tifs.2020.01.002

Kuswandi, B., Wicaksono, Y., Jayus, Abdullah, A., Heng, L. Y., & Ahmad, M. (2011). Smart Packaging: Sensors for Monitoring of Food Quality and Safety. *Sensing and Instrumentation for Food Quality and Safety*, *5*(3–4), 137–146. https://doi.org/10.1007/s11694-011-9120-x

Lamri, M., Bhattacharya, T., Boukid, F., Chentir, I., Dib, A. L., Das, D., Djenane, D., & Gagaoua, M. (2021). Nanotechnology as a Processing and Packaging Tool to Improve Meat Quality and Safety. *Foods*, *10*(11), 2633. https://doi.org/10.3390/foods10112633

Lee, N.-Y., Ko, W.-C., & Hsueh, P.-R. (2019). Nanoparticles in the Treatment of Infections Caused by Multidrug-Resistant Organisms. *Frontiers in Pharmacology*, *10*. https://doi.org/10.3389/fphar.2019.01153

Ma, M., Yang, X., Ying, X., Shi, C., Jia, Z., & Jia, B. (2023). Applications of Gas Sensing in Food Quality Detection: A Review. *Foods*, *12*(21), 3966. https://doi.org/10.3390/foods12213966

Ma, Y., Yang, W., Xia, Y., Xue, W., Wu, H., Li, Z., Zhang, F., Qiu, B., & Fu, C. (2022). Properties and Applications of Intelligent Packaging Indicators for Food Spoilage. *Membranes*, *12*(5), 477. https://doi.org/10.3390/membranes12050477

Magar, T. V., Gholap, A. D., & Deepak Musmade (2011). Elastic Vesicles as Drugs Carriers Through the Skin. *Inventi Rapid: NDDS [ISSN 0976-3791]*, *4*, 1–2.

Mahajan, A. M., Gholap, A. D., & Borude, S. S. (2011). POEMS Syndrome: A Review. *Pharmacologyonline*, *3*, 265–274.

Mahajan, A. M., Gholap*, A. D., & Borude, S. S. (n.d.). Recent Updates on Q Fever: Diagnosis and Treatment. *Pharmacologyonline*, *3*, 254–264.

Mahmud, J., & Khan, R. A. (2018). Characterization of Natural Antimicrobials in Food System. *Advances in Microbiology*, *08*(11), 894–916. https://doi.org/10.4236/aim.2018.811060

Majid, I., Ahmad Nayik, G., Mohammad Dar, S., & Nanda, V. (2018). Novel Food Packaging Technologies: Innovations and Future Prospective. *Journal of the Saudi Society of Agricultural Sciences*, *17*(4), 454–462. https://doi.org/10.1016/j.jssas.2016.11.003

Malik, S., Muhammad, K., & Waheed, Y. (2023). Nanotechnology: A Revolution in Modern Industry. *Molecules*, *28*(2), 661. https://doi.org/10.3390/molecules28020661

Mamun, M. M., Sorinolu, A. J., Munir, M., & Vejerano, E. P. (2021). Nanoantibiotics: Functions and Properties at the Nanoscale to Combat Antibiotic Resistance. *Frontiers in Chemistry*, *9*. https://doi.org/10.3389/fchem.2021.687660

Martillanes, S., Rocha-Pimienta, J., Cabrera-Bañegil, M., Martín-Vertedor, D., & Delgado-Adámez, J. (2017). Application of Phenolic Compounds for Food Preservation: Food Additive and Active Packaging. In *Phenolic Compounds—Biological Activity*. InTech. https://doi.org/10.5772/66885

Martirosyan, A., & Schneider, Y.-J. (2014). Engineered Nanomaterials in Food: Implications for Food Safety and Consumer Health. *International Journal of Environmental Research and Public Health*, *11*(6), 5720–5750. https://doi.org/10.3390/ijerph110605720

Mathivanan, S. (2021). Perspectives of Nano-Materials and Nanobiosensors in Food Safety and Agriculture. In *Novel Nanomaterials*. IntechOpen. https://doi.org/10.5772/intechopen.95345

McClements, D. J. (2020). Nanotechnology Approaches for Improving the Healthiness and Sustainability of the Modern Food Supply. *ACS Omega*, *5*(46), 29623–29630. https://doi.org/10.1021/acsomega.0c04050

Mei, L., & Wang, Q. (2020). Advances in Using Nanotechnology Structuring Approaches for Improving Food Packaging. *Annual Review of Food Science and Technology*, *11*(1), 339–364. https://doi.org/10.1146/annurev-food-032519-051804

Meng, X., Kim, S., Puligundla, P., & Ko, S. (2014). Carbon Dioxide and Oxygen Gas Sensors-Possible Application for Monitoring Quality, Freshness, and Safety of Agricultural and Food Products with Emphasis on Importance of Analytical Signals and Their Transformation. *Journal of the Korean Society for Applied Biological Chemistry*, *57*(6), 723–733. https://doi.org/10.1007/s13765-014-4180-3

Mishra, R. K., Ha, S. K., Verma, K., & Tiwari, S. K. (2018). Recent Progress in Selected Bio-Nanomaterials and Their Engineering Applications: An Overview. *Journal of Science: Advanced Materials and Devices*, *3*(3), 263–288. https://doi.org/10.1016/j.jsamd.2018.05.003

Mizielińska, M., & Bartkowiak, A. (2023). Overview of Food Antimicrobial Packaging. In *Food Processing and Packaging Technologies—Recent Advances*. IntechOpen. https://doi.org/10.5772/intechopen.108666

Mohamad Sukri, S. N. A., Shameli, K., Teow, S.-Y., Chew, J., Ooi, L.-T., Lee-Kiun Soon, M., Ismail, N. A., & Moeini, H. (2023). Enhanced Antibacterial and Anticancer Activities of Plant Extract Mediated Green Synthesized Zinc Oxide-Silver Nanoparticles. *Frontiers in Microbiology*, *14*. https://doi.org/10.3389/fmicb.2023.1194292

Mohamed, E. F., & Awad, G. (2022). Development of Nano-Sensor and Biosensor as an Air Pollution Detection Technique for the Foreseeable Future (pp. 163–188). https://doi.org/10.1016/bs.coac.2021.11.003

Mohammad, Z. H., Ahmad, F., Ibrahim, S. A., & Zaidi, S. (2022). Application of Nanotechnology in Different Aspects of the Food Industry. *Discover Food*, *2*(1), 12. https://doi.org/10.1007/s44187-022-00013-9

Mohebi, E., & Marquez, L. (2015). Intelligent Packaging in Meat Industry: An Overview of Existing Solutions. *Journal of Food Science and Technology*, *52*(7), 3947–3964. https://doi.org/10.1007/s13197-014-1588-z

Monirul Hasan Tipu, M., Baroi, A., Rana, J., Islam, S., Jahan, R., Shipon Miah, M., & Asaduzzaman, M. (2022). Potential Applications of Nanotechnology in Agriculture: A Smart Tool for Sustainable Agriculture. In *Agricultural Development in Asia—Potential Use of Nano-Materials and Nano-Technology*. IntechOpen. https://doi.org/10.5772/intechopen.101142

Mubeen, B., Ansar, A. N., Rasool, R., Ullah, I., Imam, S. S., Alshehri, S., Ghoneim, M. M., Alzarea, S. I., Nadeem, M. S., & Kazmi, I. (2021). Nanotechnology as a Novel Approach in Combating Microbes Providing an Alternative to Antibiotics. *Antibiotics*, *10*(12), 1473. https://doi.org/10.3390/antibiotics10121473

Mustafa, F., & Andreescu, S. (2020). Nanotechnology-Based Approaches for Food Sensing and Packaging Applications. *RSC Advances*, *10*(33), 19309–19336. https://doi.org/10.1039/D0RA01084G

Naseer, B., Srivastava, G., Qadri, O. S., Faridi, S. A., Islam, R. U., & Younis, K. (2018). Importance and Health Hazards of Nanoparticles Used in the Food Industry. *Nanotechnology Reviews*, *7*(6), 623–641. https://doi.org/10.1515/ntrev-2018-0076

Neme, K., Nafady, A., Uddin, S., & Tola, Y. B. (2021). Application of Nanotechnology in Agriculture, Postharvest Loss Reduction and Food Processing: Food Security Implication and Challenges. *Heliyon*, *7*(12), e08539. https://doi.org/10.1016/j.heliyon.2021.e08539

Nguyen, V., Vu, V., Nguyen, T., Nguyen, T., Tran, V., & Nguyen-Tri, P. (2019). Antibacterial Activity of TiO2-and ZnO-Decorated with Silver Nanoparticles. *Journal of Composites Science*, *3*(2), 61. https://doi.org/10.3390/jcs3020061

Nile, S. H., Baskar, V., Selvaraj, D., Nile, A., Xiao, J., & Kai, G. (2020). Nanotechnologies in Food Science: Applications, Recent Trends, and Future Perspectives. *Nano-Micro Letters*, *12*(1), 45. https://doi.org/10.1007/s40820-020-0383-9

Nishu, & Kumar, S. (2023). Smart and Innovative Nanotechnology Applications for Water Purification. *Hybrid Advances*, *3*, 100044. https://doi.org/10.1016/j.hybadv.2023.100044

Onyeaka, H., Passaretti, P., Miri, T., & Al-Sharify, Z. T. (2022). The Safety of Nanomaterials in Food Production and Packaging. *Current Research in Food Science*, *5*, 763–774. https://doi.org/10.1016/j.crfs.2022.04.005

Osmólska, E., Stoma, M., & Starek-Wójcicka, A. (2022). Application of Biosensors, Sensors, and Tags in Intelligent Packaging Used for Food Products—a Review. *Sensors*, *22*(24), 9956. https://doi.org/10.3390/s22249956

Ozcakir, G. (2023). Applications of Nanomaterials in Food Industry: A Review. *IOCN 2023*, 22. https://doi.org/10.3390/IOCN2023-14470

Ozdal, M., & Gurkok, S. (2022). A Recent Advances in Nanoparticles as Antibacterial Agent. *ADMET and DMPK*. https://doi.org/10.5599/admet.1172

Pandey, G. (2018). Challenges and Future Prospects of Agri-Nanotechnology for Sustainable Agriculture in India. *Environmental Technology & Innovation*, *11*, 299–307. https://doi.org/10.1016/j.eti.2018.06.012

Pardeshi, S., Gholap, A., More, M., Togre, N., Rebello, N., & Giram, P. (2023). *Dendrimers Based Antibacterial and Antiviral Materials* (pp. 139–169). https://doi.org/10.1021/bk-2023-1458.ch005

Parfitt, J., Barthel, M., & Macnaughton, S. (2010). Food Waste Within Food Supply Chains: Quantification and Potential for Change to 2050. *Philosophical Transactions of the Royal Society B: Biological Sciences*, *365*(1554), 3065–3081. https://doi.org/10.1098/rstb.2010.0126

Pati, R., Mehta, R. K., Mohanty, S., Padhi, A., Sengupta, M., Vaseeharan, B., Goswami, C., & Sonawane, A. (2014). Topical Application of Zinc Oxide Nanoparticles Reduces Bacterial Skin Infection in Mice and Exhibits Antibacterial Activity by Inducing Oxidative Stress Response and Cell Membrane Disintegration in Macrophages. *Nanomedicine: Nanotechnology, Biology and Medicine*, *10*(6), 1195–1208. https://doi.org/10.1016/j.nano.2014.02.012

Pech, M., Vrchota, J., & Bednář, J. (2021). Predictive Maintenance and Intelligent Sensors in Smart Factory: Review. *Sensors*, *21*(4), 1470. https://doi.org/10.3390/s21041470

Pereda, M., Marcovich, N. E., & Ansorena, M. R. (2018). Nanotechnology in Food Packaging Applications: Barrier Materials, Antimicrobial Agents, Sensors, and Safety Assessment. In *Handbook of Ecomaterials* (pp.

1–22). Springer International Publishing. https://doi.org/10.1007/978-3-319-48281-1_30-1

Perera, K. Y., Hopkins, M., Jaiswal, A. K., & Jaiswal, S. (2023). Nanoclays-Containing Bio-Based Packaging Materials: Properties, Applications, Safety, and Regulatory Issues. *Journal of Nanostructure in Chemistry*. https://doi.org/10.1007/s40097-023-00525-5

Pham, T. T., Nguyen, L. L. P., Dam, M. S., & Baranyai, L. (2023). Application of Edible Coating in Extension of Fruit Shelf Life: Review. *AgriEngineering*, *5*(1), 520–536. https://doi.org/10.3390/agriengineering5010034

Picazo-Vela, S., & Ricardo Hernández, L. (2020). Technology, Science and Culture—a Global Vision, Volume II. In *Technology, Science and Culture—a Global Vision, Volume II*. IntechOpen. https://doi.org/10.5772/intechopen.90099

Pokrajac, L., Abbas, A., Chrzanowski, W., Dias, G. M., Eggleton, B. J., Maguire, S., Maine, E., Malloy, T., Nathwani, J., Nazar, L., Sips, A., Sone, J., van den Berg, A., Weiss, P. S., & Mitra, S. (2021). Nanotechnology for a Sustainable Future: Addressing Global Challenges with the International Network4Sustainable Nanotechnology. *ACS Nano*, *15*(12), 18608–18623. https://doi.org/10.1021/acsnano.1c10919

Primožič, M., Knez, Ž., & Leitgeb, M. (2021). (Bio)Nanotechnology in Food Science—Food Packaging. *Nanomaterials*, *11*(2), 292. https://doi.org/10.3390/nano11020292

Punjabi, K., Mehta, S., Chavan, R., Chitalia, V., Deogharkar, D., & Deshpande, S. (2018). Efficiency of Biosynthesized Silver and Zinc Nanoparticles Against Multi-Drug Resistant Pathogens. *Frontiers in Microbiology*, *9*. https://doi.org/10.3389/fmicb.2018.02207

Puri, A. V., & Gholap, A. D. (2018). Comparative Evaluation of Papaya and Banana Starch as Disintegrant in Diclofenac Sodium Tablet Formulation. *International Journal of Pharmacy and Biological Sciences*, *8*(3), 512–522.

Rahmati, F., Hosseini, S. S., Mahuti Safai, S., Asgari Lajayer, B., & Hatami, M. (2020). New Insights into the Role of Nanotechnology in Microbial Food Safety. *3 Biotech*, *10*(10), 425. https://doi.org/10.1007/s13205-020-02409-9

Rajurkar, V. G., & Gholap, A. D. (2012). Recent Technologies in Nanomedicine for Treatment of Cancer. *Inventi Rapid: Pharm Tech [ISSN 0976–3783]*, *4*, 1–3.

Ramachandraiah, K., Han, S. G., & Chin, K. B. (2014). Nanotechnology in Meat Processing and Packaging: Potential Applications—a Review. *Asian-Australasian Journal of Animal Sciences*, *28*(2), 290–302. https://doi.org/10.5713/ajas.14.0607

Ravichandran, R. (2010). Nanotechnology Applications in Food and Food Processing: Innovative Green Approaches, Opportunities and Uncertainties for Global Market. *International Journal of Green Nanotechnology: Physics and Chemistry*, *1*(2), P72–P96. https://doi.org/10.1080/19430871003684440

Reddy, B. L., Jatav, H. S., Rajput, V. D., Minkina, T., Ranjan, A., Harikrishnan, A., Veena, V. K., Chauhan, A., Kumar, S., Prakash, A., & Prasad, R. (2022). Nanomaterials Based Monitoring of Food-and Water-Borne Pathogens. *Journal of Nanomaterials*, *2022*, 1–13. https://doi.org/10.1155/2022/9543532

Rodrigues, C., Souza, V. G. L., Coelhoso, I., & Fernando, A. L. (2021). Bio-Based Sensors for Smart Food Packaging—Current Applications and Future Trends. *Sensors*, *21*(6), 2148. https://doi.org/10.3390/s21062148

Sagar, N. A., Kumar, N., Choudhary, R., Bajpai, V. K., Cao, H., Shukla, S., & Pareek, S. (2022). Prospecting the Role of Nanotechnology in Extending the Shelf-Life of Fresh Produce and in Developing Advanced Packaging. *Food Packaging and Shelf Life*, *34*, 100955. https://doi.org/10.1016/j.fpsl.2022.100955

Sahoo, M., Vishwakarma, S., Panigrahi, C., & Kumar, J. (2021). Nanotechnology: Current Applications and Future Scope in Food. *Food Frontiers*, *2*(1), 3–22. https://doi.org/10.1002/fft2.58

Sampels, S. (2013). Oxidation and Antioxidants in Fish and Meat from Farm to Fork. In *Food Industry*. InTech. https://doi.org/10.5772/53169

Saritha, G. N. G., Anju, T., & Kumar, A. (2022). Nanotechnology—Big impact: How Nanotechnology Is Changing the Future of Agriculture? *Journal of Agriculture and Food Research*, *10*, 100457. https://doi.org/10.1016/j.jafr.2022.100457

Schanes, K., Dobernig, K., & Gözet, B. (2018). Food Waste Matters—a Systematic Review of Household Food Waste Practices and Their Policy Implications. *Journal of Cleaner Production*, *182*, 978–991. https://doi.org/10.1016/j.jclepro.2018.02.030

Schukat, M., McCaldin, D., Wang, K., Schreier, G., Lovell, N. H., Marschollek, M., & Redmond, S. J. (2016). Unintended Consequences of Wearable Sensor Use in Healthcare. *Yearbook of Medical Informatics*, *25*(01), 73–86. https://doi.org/10.15265/IY-2016–025

Shankar Tumuluru, J. (2023). Introductory Chapter: Food Processing, Preservation, and Packaging—a Brief Overview. In *Food Processing and Packaging Technologies—Recent Advances*. IntechOpen. https://doi.org/10.5772/intechopen.110229

Sharma, C., Dhiman, R., Rokana, N., & Panwar, H. (2017). Nanotechnology: An Untapped Resource for Food Packaging. *Frontiers in Microbiology*, *8*. https://doi.org/10.3389/fmicb.2017.01735

Siddiqui, S. A., Zannou, O., Bahmid, N. A., Fidan, H., Alamou, A.-F., Nagdalian, A. A., Hassoun, A., Fernando, I., Ibrahim, S. A., & Arsyad, M. (2022). Consumer Behavior Towards Nanopackaging—a New Trend in the Food Industry. *Future Foods*, *6*, 100191. https://doi.org/10.1016/j.fufo.2022.100191

Silva, F. A. G. S., Dourado, F., Gama, M., & Poças, F. (2020). Nanocellulose Bio-Based Composites for Food Packaging. *Nanomaterials*, *10*(10), 2041. https://doi.org/10.3390/nano10102041

Singh, B. P., Rohit, Manju, K. M., Sharma, R., Bhushan, B., Ghosh, S., & Goel, G. (2023). Nano-Conjugated Food-Derived Antimicrobial Peptides As Natural Biopreservatives: A Review of Technology and Applications. *Antibiotics*, *12*(2), 244. https://doi.org/10.3390/antibiotics12020244

Singh, R., Dutt, S., Sharma, P., Sundramoorthy, A. K., Dubey, A., Singh, A., & Arya, S. (2023). Future of Nanotechnology in Food Industry: Challenges in Processing, Packaging, and Food Safety. *Global Challenges*, *7*(4). https://doi.org/10.1002/gch2.202200209

Singh, T., Shukla, S., Kumar, P., Wahla, V., Bajpai, V. K., & Rather, I. A. (2017). Application of Nanotechnology in Food Science: Perception and Overview. *Frontiers in Microbiology*, *8*. https://doi.org/10.3389/fmicb.2017.01501

Sirelkhatim, A., Mahmud, S., Seeni, A., Kaus, N. H. M., Ann, L. C., Bakhori, S. K. M., Hasan, H., & Mohamad, D. (2015). Review on Zinc Oxide Nanoparticles: Antibacterial Activity and Toxicity Mechanism. *Nano-Micro Letters*, *7*(3), 219–242. https://doi.org/10.1007/s40820-015-0040-x

Skawińska, E., & Zalewski, R. I. (2022). Economic Impact of Temperature Control During Food Transportation—A COVID-19 Perspective. *Foods*, *11*(3), 467. https://doi.org/10.3390/foods11030467

Slavin, Y. N., Asnis, J., Häfeli, U. O., & Bach, H. (2017). Metal Nanoparticles: Understanding the Mechanisms Behind Antibacterial Activity. *Journal of Nanobiotechnology*, *15*(1), 65. https://doi.org/10.1186/s12951-017-0308-z

Sonwani, E., Bansal, U., Alroobaea, R., Baqasah, A. M., & Hedabou, M. (2022). An Artificial Intelligence Approach Toward Food Spoilage Detection and Analysis. *Frontiers in Public Health*, *9*. https://doi.org/10.3389/fpubh.2021.816226

Soung, S.-H., Lee, S., Lee, S.-H., Kim, H.-J., Lee, N.-R., & Lee, C.-H. (2021). Metabolomic-Based Comparison of Traditional and Industrial Doenjang Samples with Antioxidative Activities. *Foods*, *10*(6), 1377. https://doi.org/10.3390/foods10061377

Souza, R. C. de, Haberbeck, L. U., Riella, H. G., Ribeiro, D. H. B., & Carciofi, B. A. M. (2019). Antibacterial Activity of Zinc Oxide Nanoparticles Synthesized by Solochemical Process. *Brazilian Journal of Chemical Engineering*, *36*(2), 885–893. https://doi.org/10.1590/0104-6632.20190362s20180027

Sridhar, A., Ponnuchamy, M., Kumar, P. S., & Kapoor, A. (2021). Food Preservation Techniques and Nanotechnology for Increased Shelf Life of Fruits, Vegetables, Beverages and Spices: A Review. *Environmental Chemistry Letters*, *19*(2), 1715–1735. https://doi.org/10.1007/s10311-020-01126-2

Su, C., Huang, K., Li, H.-H., Lu, Y.-G., & Zheng, D.-L. (2020). Antibacterial Properties of Functionalized Gold Nanoparticles and Their Application in Oral Biology. *Journal of Nanomaterials*, *2020*, 1–13. https://doi.org/10.1155/2020/5616379

Suriati, L. (2022). Nano Coating of Aloe-Gel Incorporation Additives to Maintain the Quality of Freshly Cut Fruits. *Frontiers in Sustainable Food Systems*, *6*. https://doi.org/10.3389/fsufs.2022.914254

Suvarna, V., Nair, A., Mallya, R., Khan, T., & Omri, A. (2022). Antimicrobial Nanomaterials for Food Packaging. *Antibiotics*, *11*(6), 729. https://doi.org/10.3390/antibiotics11060729

Taha, T. B., Barzinjy, A. A., Hussain, F. H. S., & Nurtayeva, T. (2022). Nanotechnology and Computer Science: Trends and advances. *Memories—Materials, Devices, Circuits and Systems*, *2*, 100011. https://doi.org/10.1016/j.memori.2022.100011

Thakur, M., Wang, B., & Verma, M. L. (2022). Development and Applications of Nanobiosensors for Sustainable Agricultural and Food Industries: Recent Developments, Challenges and Perspectives. *Environmental Technology & Innovation*, *26*, 102371. https://doi.org/10.1016/j.eti.2022.102371

Tinkle, S., McNeil, S. E., Mühlebach, S., Bawa, R., Borchard, G., Barenholz, Y. (Chezy), Tamarkin, L., & Desai, N. (2014). Nanomedicines: Addressing the Scientific and Regulatory Gap. *Annals of the New York Academy of Sciences*, *1313*(1), 35–56. https://doi.org/10.1111/nyas.12403

Tupe, D., & Gholap*, A. D. (2012). The Threat of Photoaging: Clinical Conditions, Evaluation and Treatment. *Inventi Rapid: Cosmeceuticals [ISSN 0976-3864]*, *4*, 1–3.

Vågsholm, I., Arzoomand, N. S., & Boqvist, S. (2020). Food Security, Safety, and Sustainability—Getting the Trade-Offs Right. *Frontiers in Sustainable Food Systems*, *4*. https://doi.org/10.3389/fsufs.2020.00016

Vassallo, A., Silletti, M. F., Faraone, I., & Milella, L. (2020). Nanoparticulate Antibiotic Systems as Antibacterial Agents and Antibiotic Delivery Platforms to Fight Infections. *Journal of Nanomaterials*, *2020*, 1–31. https://doi.org/10.1155/2020/6905631

Vilas Bhagat, A. G. (2012). Gold Nanorods: A Promising Drug Delivery Vehicle. *Inventi Rapid: NDDS [ISSN 0976-3864]*, *4*, 1–4.

Vora, L. K., Gholap, A. D., Jetha, K., Thakur, R. R. S., Solanki, H. K., & Chavda, V. P. (2023). Artificial Intelligence in Pharmaceutical Technology and Drug Delivery Design. *Pharmaceutics*, *15*(7), 1916. https://doi.org/10.3390/pharmaceutics15071916

Wang, L., Hu, C., & Shao, L. (2017). The Antimicrobial Activity of Nanoparticles: Present Situation and Prospects for the Future. *International Journal of Nanomedicine*, *Volume 12*, 1227–1249. https://doi.org/10.2147/IJN.S121956

Wang, X., Luo, Y., Huang, K., & Cheng, N. (2022). Biosensor for Agriculture and Food Safety: Recent Advances and Future Perspectives. *Advanced Agrochem*, *1*(1), 3–6. https://doi.org/10.1016/j.aac.2022.08.002

Watson, N. J., Bowler, A. L., Rady, A., Fisher, O. J., Simeone, A., Escrig, J., Woolley, E., & Adedeji, A. A. (2021). Intelligent Sensors for Sustainable Food and Drink Manufacturing. *Frontiers in Sustainable Food Systems*, *5*. https://doi.org/10.3389/fsufs.2021.642786

Wu, F., Misra, M., & Mohanty, A. K. (2021). Challenges and New Opportunities on Barrier Performance of Biodegradable Polymers for Sustainable Packaging. *Progress in Polymer Science*, *117*, 101395. https://doi.org/10.1016/j.progpolymsci.2021.101395

Wyser, Y., Adams, M., Avella, M., Carlander, D., Garcia, L., Pieper, G., Rennen, M., Schuermans, J., & Weiss, J. (2016). Outlook and Challenges of Nanotechnologies for Food Packaging. *Packaging Technology and Science*, *29*(12), 615–648. https://doi.org/10.1002/pts.2221

Xing, Y., Li, W., Wang, Q., Li, X., Xu, Q., Guo, X., Bi, X., Liu, X., Shui, Y., Lin, H., & Yang, H. (2019). Antimicrobial Nanoparticles Incorporated in Edible Coatings and Films for the Preservation of Fruits and Vegetables. *Molecules*, *24*(9), 1695. https://doi.org/10.3390/molecules24091695

Yang, H., Zhang, J., Li, Z., Huang, J., Wu, J., Zhang, Y., Ge, H., & Zhao, Y. (2023). Antibacterial Effect of Low-Concentration ZnO Nanoparticles on Sulfate-Reducing Bacteria Under Visible Light. *Nanomaterials*, *13*(14), 2033. https://doi.org/10.3390/nano13142033

Yang, T., & Duncan, T. V. (2021). Challenges and Potential Solutions for Nanosensors Intended for Use with Foods. *Nature Nanotechnology*, *16*(3), 251–265. https://doi.org/10.1038/s41565-021-00867-7

Ye, K., Huang, M., He, X., An, Z., & Qin, H. (2022). Synergistic Antibacterial Effect of Zinc Oxide Nanoparticles and Polymorphonuclear Neutrophils. *Journal of Functional Biomaterials*, *13*(2), 35. https://doi.org/10.3390/jfb13020035

Zorraquín-Peña, I., Cueva, C., Bartolomé, B., & Moreno-Arribas, M. V. (2020). Silver Nanoparticles Against Foodborne Bacteria. Effects at Intestinal Level and Health Limitations. *Microorganisms*, *8*(1), 132. https://doi.org/10.3390/microorganisms8010132

Zuo, J., Feng, J., Gameiro, M. G., Tian, Y., Liang, J., Wang, Y., Ding, J., & He, Q. (2022). RFID-Based Sensing in Smart Packaging for Food Applications: A Review. *Future Foods*, *6*, 100198. https://doi.org/10.1016/j.fufo.2022.100198

2 Nanotechnology Revolutionizing Food Preservation

A Wide-Ranging Exploration

Hajira Bibi[a], *Abdul Wahab*[b,c], *Farwa Batool*[d], *Murad Muhammad*[c,e], *Gholamreza Abdi*[f], *Muhammad Awais Ashraf*[c,g], *Muhammad Nabeel Amjad*[c,g], *and Wajid Zaman*[h]

[a]Department of Biochemistry, Quaid-i-Azam University Islamabad, Islamabad, Pakistan

[b]Shanghai Center for Plant Stress Biology, CAS Center for Excellence in Molecular Plant Sciences, Chinese Academy of Sciences, Shanghai

[c]University of Chinese Academy of Sciences, Beijing, China

[d]Department of Botany, Lahore College for Women University, Lahore, Pakistan

[e]State Key Laboratory of Desert and Oasis Ecology, Key Laboratory of Ecological Safety and Sustainable Development in Arid Lands, Xinjiang Institute of Ecology and Geography, Chinese Academy of Sciences, Urumqi, China

[f]Department of Biotechnology, Persian Gulf Research Institute, Persian Gulf University, Bushehr, Iran

[g]CAS Key Laboratory of Molecular Virology & Immunology, Institutional Center for Shared Technologies and Facilities, Pathogen Discovery and Big Data Platform, Shanghai Institute of Immunity and Infection, Chinese Academy of Sciences, Shanghai, China

[h]Department of Life Sciences, Yeungnam University, Gyeongbuk, Republic of Korea

2.1 INTRODUCTION

2.1.1 Background

Traditional food preservation methods such as heat treatment, chemical additives, and a controlled atmosphere have been indispensable in preserving food and prolonging the shelf life of fruits and vegetables [1]. However, these methods have limitations, such as changes in taste and texture, nutrient degradation, and environmental consequences. Nanotechnology is an emerging and promising alternative in this terrain, offering solutions to overcome these limitations and enhance food preservation efficiency [2]. The challenge of food preservation, particularly perishable food products such as vegetables and fruits, has prevailed throughout human history. Ancient civilizations had developed several methods, such as salting, drying, smoking, and fermenting, to enhance the shelf life of food supplies. Those methods have been adequate to some extent but often come with limitations affecting preserved food quality, safety, and sustainability [3]. Recently, chemical additives, such as salts, vinegar, and sugar, have been playing a significant role in food preservation, employed to inhibit microbial growth in food and enhance the shelf life of foods such as vegetables and fruits [4]. Although these methods achieved their food preservation goals, they also raised serious concerns about taste alteration, high sodium intake, and texture changes. Heat treatments revolutionized food preservation by using high temperatures to kill pathogens in pasteurization and canning [5]. However, this method of heat treatment had drawbacks such as degradation of essential nutrients, alteration in taste and texture of food, loss of vibrant colors, and overcooking or undercooking by uniform heating, further compromising food quality [6]. They recently developed a controlled storage atmosphere that altered the atmospheric conditions around vegetables and fruits to slow their ripening and reduce their decay. While this process preserved the nutritional content and flavor of the food better than other methods, it demanded a controlled environment and specialized equipment, making it less accessible and costly for many producers [7].

As the global population continues to grow, the demand for fresh fruits and vegetables increases, thus intensifying the food preservation and sustainability challenges. Food wastage has also gained prominence with the world's population increase [8]. According to the United Nations Food and Agriculture Organization (UNFAO), approximately 1.3 billion metric tons of food are wasted annually, one-third of all the food produced by humans for consumption [9]. This food wastage has many economic implications

DOI: 10.1201/9781032677422-2

and exacerbates environmental sustainability concerns, including resource depletion and greenhouse emissions. Nanotechnology has emerged as an alternative revolutionary technology in response to these challenges to ensure effective food preservation. Nanotechnology manipulates matter at the nanoscale, where the case exhibits unique behaviors and properties that can be utilized for several applications, including food safety and preservation [10]. Nanotechnology in food preservation is based on developing and employing nanomaterials engineered at the nanoscale with specific properties that address the challenges and limitations of traditional food preservation methods [11]. For example, silver nanoparticles exhibit exceptional antimicrobial properties due to their tiny size and large surface area, effectively inhibiting the growth of fungi, bacteria, and many other microorganisms [12]. Microbial growth and contamination in food can be significantly reduced by employing silver nanoparticles in food items directly or in food packaging, extending the shelf life of food without chemical additives [13]. Another facet of nanotechnology is presented by nanoemulsions that enable the encapsulation and regulated, controlled release of essential nutrients and preservative agents. Encapsulating food materials in nanoscale droplets precisely control their freedom, preventing their nutrient loss and premature spoilage. The precisely controlled kinetic release guarantees fruits' and vegetables' freshness and nutrient value over an extended period [14]. Biodegradable polymers or natural materials like cellulose can create nanofibers as a physical barrier to prevent microbial infiltration. Fruits and vegetables can be protected from microbial contamination by being wrapped in coverings made of nanofibers. The shelf life of perishable food, such as fruits and vegetables, is extended by thin nanofiber coverings [15]. In conclusion, the historical background of food preservation, the difficulties and the drawbacks of conventional techniques, and the revelation of nanotechnology as a critical means of overcoming these obstacles highlight the importance of investigating the possibilities of nanotechnology in improving the effectiveness of food preservation and reducing the issues of sustainability, food waste, and environmental degradation [16].

2.1.2 Significance of Nanotechnology

There is immense significance in employing nanotechnology to preserve and enhance the shelf life of vegetables and fruits that decay quickly. One-third of the food produced worldwide spoils and is wasted due to decay and decomposition, adversely affecting the environment, sustainability, greenhouse gas emissions, and resource depletion. An additional cost of the yearly worldwide food waste is the depletion of expensive transportation, agriculture, and distribution resources [17]. Furthermore, it nettles the economic disparity as consumers have to pay more for food owing to food shortages caused by food wastage. Environmental sustainability concerns are equally critical as food wastage, which leads to ecological degradation due to the disproportionate use of water, energy, and land resources [18]. By preserving fruits, vegetables, and other food items for more extended periods with optimal quality and edibility, minimizing resource depletion, economic loss, and environmental degradation, and promoting environmental sustainability and stewardship in global food systems, nanotechnology offers practical approaches to these problems [19].

2.1.3 Objectives and Scope

The objectives of the chapter are centered on understanding the role of nanotechnology in preserving vegetables and fruits. Firstly, the chapter categorizes nanomaterials such as silver nanoparticles, gold nanoparticles, nanoemulsions, and nanofibers, providing a broad perspective on available tools. Secondly, it explains the scientific mechanisms of how these nanomaterials work, clarifying how the nanoparticles disrupt bacteria or how nanoemulsions gradually release preservatives. Thirdly, the chapter elaborates on the practical applications with real-time examples, presenting how nanotechnology expands the freshness and taste of fruits and vegetables. Lastly, the chapter assesses the safety and regulatory aspects, guaranteeing that these nano techniques meet rigid and elevated standards. This chapter aims to comprehensively grasp the contributions of nanotechnology to preserving fruits and vegetables and emphasize safety and practicality.

2.2 TYPES OF NANOMATERIALS USED IN PRESERVATION

2.2.1 Silver Nanoparticles

The use of silver nanoparticles has gained ground and popularity in several fields, prominently in antimicrobial applications. Due to their nanoscale dimensions and size, this growing interest in silver nanoparticles can be linked to their high antibacterial activity and unique features [20]. Compared to their volume, nanoparticles have a large surface area due to their small size, which is a crucial factor in their high antimicrobial efficacy. Extensive interactions of silver nanoparticles are facilitated with bacteria, fungi, and other microorganisms by their larger surface area, qualifying them as highly effective antimicrobial agents [21]. The significant antimicrobial efficacy of silver nanoparticles is ascribed to certain vital factors: Silver nanoparticles attach to the membranes of microorganisms easily and penetrate them, leading to leakage of cellular contents and resulting in microbial cell death [22]. Silver nanoparticles can derange the respiratory chain proteins of microorganisms, thus hampering their ability to generate cellular energy vital for their survival and growth [23]. Silver nanoparticles also interact with microorganisms' genetic material by binding to their DNA, inhibiting the DNA transcription and translation process in microbial cells, which is essential for the biochemical process, survival, growth, and reproduction of microorganisms. Nanoparticles efficiently prevent microbial infections by inhibiting the development of central dogma in microbial cells [24].

The synthesis method of silver nanoparticles is the crucial perspective of their application, and various methods can be executed to synthesize them. These methods involve the formation of silver nanoparticles (Ag^0) from silver ions (Ag^+) [25]. The standard method used for synthesizing silver nanoparticles is the reduction method, which involves the reduction of Ag^+ ions into silver nanoparticles Ag^0. The reducing agents used typically are sodium borohydride or citrate. In this process, Ag+ ions are reduced to silver atoms Ag^0 that aggregate and form silver nanoparticles [26]. In the green synthesis method, biological agents, such as microorganisms or plant extracts, are employed as reducing agents for synthesizing silver nanoparticles. The green synthesis is more sustainable and environmentally friendly than chemical synthesis methods [27]. The physical techniques for synthesizing silver nanoparticles are based on specific biological processes, such as evaporation and condensation [28]. Silver nanoparticles have many applications, and one of the most significant applications is in food preservation, where the potent antifungal and antibacterial properties make them ideal for this use [29]. The essential aspects of their application in food preservation are as follows: Silver nanoparticles can be embedded effectively in food packaging materials, whereby incorporating AgNPs helps control the growth of food-spoiling pathogens and extend the shelf life of food products [30]. Silver nanoparticles effectively inhibit microbial growth when applied to the surfaces of food items. This method is beneficial, particularly for food items that are easily perishable, such as fruits, meats, vegetables, and dairy products [31]. Many pathogens, including Gram-negative and Gram-positive bacteria, are strongly inhibited by silver nanoparticles. This broad range of antibacterial action is significant in the context of antibiotic-resistant and multidrug-resistant microorganisms [32]. Silver nanoparticles are vital in decreasing food spoilage and food wastage by inhibiting the growth of microorganisms that cause food spoilage. This application of AgNPs is critical in sustainability and global food security [33]. Silver nanoparticles in food preservation also maintain food safety and quality during transportation and storage, extending the shelf life of food items [34].

Silver nanoparticles are valuable tools in several applications owing to their antimicrobial efficacy and unique properties, particularly in food preservation. Silver nanoparticles interfere with cellular processes, disrupt the cellular membranes, and inhibit DNA expression and replication in microorganisms. These abilities make them highly effective agents against a wide range of pathogens. The synthesis methods of silver nanoparticles, such as chemical, green, or physical methods, further enhance their applications in diverse fields and sectors [35]. In food preservation, silver nanoparticles present a promising solution to improve the shelf life of food, minimize food wastage, and maintain the optimal quality and safety of food items [36]. Silver nanoparticles' broad-spectrum antimicrobial activity, particularly against multiple drug-resistant bacterial strains, highlights their significance in guaranteeing food safety and public health in the ongoing efforts [37].

2.2.2 Gold Nanoparticles

Gold nanoparticles have specific catalytic effects, making them distinct from other nanoparticles and thus recognized globally despite their higher cost of production [38]. Gold nanoparticles are known for their particular property of surface plasmon resonance, in which the nanoparticle surface electrons oscillate when exposed to light [39]. Owing to this characteristic, gold nanoparticles offer extraordinary catalytic activities, qualifying them as best-fitted for several chemical reactions [40]. The catalytic efficacy of gold nanoparticles is attributed to their surface plasmon resonance phenomenon; thus, they act as catalysts in different chemical reactions, which include the chemical reactions needed for food preservation [41]. When gold nanoparticles interact with light, there is an increased energy transfer, thus catalyzing chemical reactions that would otherwise be impossible or less efficient [42]. In the food industry, the catalytic activity of gold nanoparticles is significantly advantageous in preserving food by inhibiting enzymatic reactions causing food spoilage. When food items such as vegetables and fruits are exposed to oxygen, enzymatic food browning is caused by polyphenol oxidase. Gold nanoparticles inhibit the activity of such enzymes through certain chemical reactions, thus halting the process of food browning [43]. This is a significant application of gold nanoparticles in preserving food items, such as vegetables and fruits, with optimal aesthetic and nutritional quality. As gold nanoparticles have optical characteristics owing to their unique surface plasmon resonance property, they can effectively respond to changes in the packaged food environment [44]. Gold nanoparticles are significantly helpful in detecting contaminants and specific compounds in food due to their plasmon resonance property. They bestow them with another property of sensing potential and quality control, guaranteeing food safety and integrity [45].

Despite the efficacy of AuNPs, the widespread use of these nanoparticles in food preservation is constrained by their very high cost of production. However, their unique properties present notable advantages that advocate their use in premium food products where cost is considered secondary to the shelf life extension and quality [46]. Moreover, the concurrent research and developments in the synthesis of nanoparticles might lead to cost-effective production strategies for gold nanoparticles, thus broadening the application scope for gold nanoparticles in preserving food and quality enhancement [47]. In summary, gold nanoparticles present a promising solution in food preservation and safety niches with their unique catalytic activities and optical properties. Their ability to halt fruit and vegetable browning due to enzymatic activity and serve as food quality sensors underscores their potential for the safety of perishable food items and prolong their shelf life. Though the cost considerations of gold nanoparticles currently limit their widespread use, concurrent advancements in nanotechnology might soon discover broader applications of gold nanoparticles in food safety and food industries [48].

2.2.3 Nanoemulsions Composition and Controlled Release

Nanoemulsions present yet another advancement in the nanotechnology field, particularly in the realm of food safety and food preservation. Nanoemulsions are acceptable oiled-in-water emulsions where the droplet size is typically in the nanometer range, mostly less than 100 nanometers. During the nanoemulsion formation process, preservative agents or bioactive compounds are encapsulated within minuscule oiled droplets, fostering their stability and size for increased efficacy in food preservation [49]. Nanoemulsions consist of tiny oil droplets suspended in a water-based solution. The preparation process of nanoemulsions involves blending water with oil and surfactant. Surfactants are required to stabilize the emulsion, preventing oil and water separation. The nanoscale droplets are a critical feature distinguishing nanoemulsions from conventional ones [50]. Nanoemulsions are generally prepared by high-pressure homogenization that involves subjecting the mixture of water, oil, and surfactant to high pressure and passing it through a narrow valve. The shear forces and intense pressure generated during this process break the oil droplets into nano-sizes, thus forming a uniform and stable emulsion [51].

The significant advantage of nanoemulsions in food preservation is based on their regulated and controlled release capabilities: Nanoemulsions can gradually release encapsulated preservatives and bioactive compounds over time [51]. This gradual and sustained release mechanism guarantees a prolonged and consistent effect, thus maintaining the quality and freshness of food products throughout their shelf life [52]. The controlled and regulated release of food preservatives and bioactive compounds is beneficial for liquid products more prone to rapid spoilage, such as dairy products and fruit juices [53]. By gradually releasing the antioxidants, preservatives, and antimicrobial agents, nanoemulsions can effectively enhance the shelf life of food products and preserve their nutritional value and taste [54]. Nanoemulsions present the flexibility to encapsulate a range of different compounds. This nanoemulsion versatility allows customization based on the peculiar preservation needs of various food products, such as antioxidant effects, enhancing antimicrobial properties, and other functional attributes [55]. Employing nanoemulsions in food preservation offers a promising step forward in food technology. Their ability to sustain and control the release of various bioactive compounds and food preservatives guarantees long-term food preservation with optimal safety and quality.

Furthermore, the nanoscale droplets' size increases the interaction and activity of encapsulated compounds with food molecules, significantly improving their preservation effectiveness [56]. In summary, nanoemulsions present a promising food safety and preservation solution with their characteristic preparation method, unique composition, and ability to control release mechanisms [57]. Their ability to encapsulate versatile bioactive compounds and preservatives and gradually release them makes them significantly effective in enhancing the shelf life of food products and maintaining their quality and taste. With the advancement of research and development in this area, nanoemulsions are more likely to play a vital role in addressing the challenges of food wastage and spoilage [58].

2.2.4 Nanofibers

Nanofibers mark a significant advancement in nanotechnology employed in food safety and preservation. Nanofibers are extremely thin, having diameters from a few nanometers to 100 nanometers [59]. Despite their ultra-thin and lightweight structure, these nanofibers' minuscule size bestows their unique physical strength properties. The primary role of nanofibers in food preservation is to provide a physical barrier that offers an innovative approach to enhancing the shelf life of food [60]. Various materials such as polypropylene, polyethylene, and polylactic acid can synthesize nanofibers. The type of synthetic material used is based on the desired role and properties of nanofibers [61] Usually, natural substances, such as cellulose, which is nontoxic, eco-friendly, and biodegradable, are used to synthesize nanofibers.

Due to this property, cellulose is considered a promising and attractive choice in the food industry [62]. The material and method chosen for nanofiber synthesis drastically affect nanofibers physical and chemical properties. Nanofibers offer various edges and different applications in the food industry for preserving food: Nanofibers are applied as coatings on the surfaces of vegetables, fruits, and various perishable foods, acting as efficient physical barriers [63]. These physical barriers play a crucial role in preventing the entry of pathogens into food, preventing its spoilage and wastage [64]. Nanofibers may be tailored to specific preservation requirements. They can gradually release antibacterial and food preservation chemicals or offer a physical barrier against moisture and oxygen, a unique and important advancement [65]. Such a customizable ability of nanofibers qualifies them to meet the preservation requirements of different food products. Nanofibers are considered safe for their applications and consumption in food items. Their vital aspect is their nontoxic nature, making them suitable for food preservation applications in the food industry and ensuring that they do not compromise the safety or quality of food items [66]. Using nanofibers in food preservation presents a notable step forward in prolonging food items' shelf life and safety. The versatile nature of nanofibers and their ability to serve as practical physical barriers against pathogens and contaminants make them valuable tools for minimizing food spoilage and food wastage [67]. With the progress in research and innovation in this field, nanofibers' benefits and potential applications in food preservation and safety will likely expand further, promising significant solutions to some of humanity's most severe challenges in food preservation and protection [66].

In conclusion, the various nanomaterials, such as nanofibers, gold nanoparticles, silver nanoparticles, and nanoemulsions, collectively offer significant contributions to innovative approaches in the realm of food safety and food preservation [68]. Unique properties and mechanisms are brought to the table by these nanomaterials, increasing the safety, quality, and shelf life of a broad range of perishable food items. Their continuous application and developments in the food industry are vital in addressing and treating global concerns of food spoilage, wastage, and safety [69]. The evaluation of silver nanoparticles in the field of food preservation presents a convincing illustration of nanotechnology mechanisms that can prolong the shelf life of perishable food products and ensure their safety. Silver nanoparticles are known to be significant for their multifaceted antimicrobial properties in multiple academic studies and scientific literature [70].

2.3 ANTIBACTERIAL ACTION

2.3.1 Silver Nanoparticles: A Multifaceted Approach

Through diverse interaction mechanisms, silver nanoparticles demonstrate considerable antibacterial capabilities. These many contact mechanisms contribute to their effectiveness in preventing microbial development. By interacting with the lipid bilayer of the microbial cell membrane, silver nanoparticles cause structural damage to the microbial cell membrane, resulting in enhanced permeability. Because the integrity of the microbial cell membrane has been disrupted, the essential components of the microbial cell seep out, resulting in cell death [71]. Silver nanoparticles can also be used to target proteins in bacterial respiratory chains. They impair the energy production mechanisms in bacterial cells by binding to respiratory chain proteins and enzymes. This protein interaction significantly reduces bacterial cells' capacity to produce energy required for life activities, survival, and replication [72]. Silver nanoparticles also interact with bacterial DNA, which is another efficient action of silver nanoparticles. Silver nanoparticles halt the process of bacterial DNA expression and replication by binding to it, thus arresting the bacterial life cycle effectively. This genomic inhibition of bacterial cells is vital in preventing bacterial growth [73]. These mechanisms collectively result in the strong antibacterial action of silver nanoparticles, making them highly effective in hampering bacterial growth and annihilating the existing bacterial colonies. Thus, the application of silver nanoparticles in food preservation significantly enhances the shelf life of perishable food items via their multifaceted mechanisms of action in halting microbial growth in such food items [74].

2.3.2 Validation from Academic Research Implications for Food Preservation

The effectiveness of silver nanoparticles is well-documented by academic research in food preservation and food safety. The antimicrobial mechanisms of action of silver nanoparticles have been explored by numerous studies and validated in peer-reviewed journals [75]. These studies present empirical evidence supporting the role of silver nanoparticles in preserving and enhancing the shelf life of food items via their antimicrobial action [76]. The employment of silver nanoparticles in food preservations is a significant advancement to tackle the challenge of food wastage. The ability of silver nanoparticles to kill bacterial cells by several mechanisms ensures food preservation with optimal food safety and quality [77]. The potential of silver nanoparticles is specifically focused on global efforts to achieve food security and minimize food wastage and its consequent environmental hazards [78]. In summary, the multifaceted antibacterial properties of silver nanoparticles advocate their use in food preservation. Their potential to disrupt cellular respiration and the cellular membrane of bacteria and inhibit DNA expression and replication provides an effective tool to control food spoilage caused by bacterial contamination [79]. The significant role of silver nanoparticles in preserving food, enhancing its shelf life, and guaranteeing its safety is reinforced by rigorous and extensive academic studies and peer-reviewed literature [80].

2.3.3 Gold Nanoparticles: Targeted Enzyme Deactivation and Selective Catalysis

The use of gold nanoparticles in food preservation is a notable example of how nanotechnology can be utilized to prolong the shelf life of perishable food items and enhance their quality [81]. The peculiar catalytic properties of gold nanoparticles, specifically their role in deactivating and targeting specific food spoilage enzymes, highlight their ability to preserve food. Gold nanoparticles act as catalysts primarily, referring to the enzymatic reactions that result in food spoilage [82]. Gold nanoparticles mainly target the enzyme polyphenol oxidase. This enzyme is a key factor contributing to the browning process and a key sign of spoilage of perishable food items such as fruits and vegetables [83]. Gold nanoparticles effectively prevent or slow down the browning process of fruits and vegetables by inhibiting the enzyme polyphenol oxidase, thus enhancing food items' shelf life and freshness [84]. Due to the nanoscale size of gold nanoparticles, they can interact with active sites of enzymes such as polyphenol oxidase [85].

Additionally, their surface plasmon resonance property enhances gold nanoparticles' catalytic activity. In surface plasmon resonance, the surface electrons of nanoparticles resonate in response to the incident light energy, contributing to the ability of gold nanoparticles to interact with enzymes and deactivate them [86]. Gold nanoparticles have very selective catalytic activity, for example, inhibiting enzymes implicated in food deterioration while not interfering with other critical biochemical processes in the food [87]. This specificity is essential to preserving the food's flavours, nutrients, and texture. This results in the maintenance of qualitative attributes of the food products, making gold nanoparticles a promising and effective option for preserving the quality of perishable food products [88].

2.3.3.1 Implications for Food Preservation

The employment of gold nanoparticles in food safety and preservation presents several advantages: Gold nanoparticles significantly enhance the shelf life of perishable food items such as fruits and vegetables by halting the enzymes that cause food spoilage [89]. The use of gold nanoparticles in food preservation guarantees the food items' sensory qualities and nutritional value due to their specific characteristic of inactivating food spoilage enzymes [90]. Gold nanoparticles might be particularly well-suited for premium or high-value food products owing to their high cost, where quality preservation is paramount [91]. In conclusion, gold nanoparticles provide a sophisticated method of food preservation. Their unique catalytic qualities and capacity to selectively deactivate odd enzymes that cause food deterioration make them valuable instruments for maintaining the freshness and quality of fruits, vegetables, and other perishable food products [92]. The application of gold nanoparticles in this realm is evidence of their potential in nanotechnology to elevate food preservation techniques, offering a pathway to minimize food wastage while optimizing food quality standards [93].

2.3.4 Nanoemulsions: Engineered, Controlled Release Systems for Sustained Preservation

The employment of nanoemulsions in preserving food products demonstrates the integration of nanotechnology in addressing the global challenges of food wastage caused by the rapid spoilage of perishable products [94]. The ability of nanoscale emulsions to encapsulate food preservative agents and act as controlled releasing systems makes them particularly noteworthy, presenting an innovative strategy for enhancing the shelf life of food items while maintaining their quality [95]. Nanoemulsions are recognized and appraised by their peculiar, controlled release mechanism: Nanoemulsions can encapsulate a variety of preservative agents in their nanoscale droplets. The encapsulation of various preservative agents is a vital aspect of the potential and effectiveness of nanoemulsions in preserving food items [96]. Preservative agents are released via nanoemulsions in a well-controlled method. This controlled release feature ensures that preservative chemicals are released gradually and sustainably, and it is best matched with the kinetics of food deterioration mechanisms in many food items. To maximize the shelf life of food products, these preservative agents must have this particular attribute of nanoemulsions [97].

Nanoemulsions are commonly synthesized from materials that are safe for food contact. Food-grade material is essential to guarantee nanoemulsions safety so that they do not compromise the quality and safety of food items [98]. The designing and synthesis of nanoemulsions allows the customization of their release profile, which means that the duration and rate of the release of preservatives can be tuned to the desired effects and peculiar needs of a variety of food products, depending on their required shelf life and susceptibility to spoiling [99]. Nanoemulsions present several advantages in preserving food: Nanoemulsions maintain the high quality, freshness, and aesthetic qualities of food products over a prolonged period because of their controlled and measured release of preservatives [100]. This property of nanoemulsions to release preservative agents in a regulated and controlled manner also preserves the food's nutritional value, which is compromised in perishable food items during storage and transportation [101]. Nanoemulsions can be customized according to the requirements of various food items, qualifying them as highly adaptable tools with multifaceted significance in food preservation [102]. They can be customized according to the specific preservation needs of different food items, such as fresh fruits and beverages [103]. The researcher concludes that nanoemulsions offer significant innovation in food preservation, offering a promising solution to the challenges of rapid food spoilage. Nanoemulsions are efficient solutions for enhancing the shelf life of perishable food items with optimum quality due to their capability to encase the food-grade preservatives and release them in a controlled manner [104]. The practical, flexible customization and versatile nature of nanoemulsions qualify them as potential agents in reforming techniques for food preservation, ensuring minimum food wastage and maximum food safety [105].

2.3.5 Nanofibers: Multilayered Physical Barriers for Enhanced Protection

Using nanofibers in food preservation underscores the multidimensional potential of nanotechnology in developing better protection strategies for fruits and vegetables [106]. Nanofibers, having customizable and physical properties, act as practical physical barriers that can be designed according to specific preservation needs. Nanofibers are novel and advanced solutions in food preservation, owing to their role as practical physical barriers [107]. As nanofibers are commonly synthesized from various materials such as chitosan, alcohol, and other polymers, they offer flexibility regarding their physical properties and chemical composition [108]. Due to their capacity to be tailored for varied thicknesses, pore sizes, and chemical functions, nanofibers may create barriers customized to effectively battle various microbiological contaminants [109]. Nanofibers significantly enhance the shelf life of fruits and vegetables, maintaining their quality and freshness for an extended period by hampering microbial and other forms of contamination [110].

A detailed and elaborated schematic presentation in Figure 2.1 can effectively illustrate the modes and mechanisms of action of different nanomaterials employed in food preservation. The diagram would serve as a valuable academic tool, assisting the understanding of multifaceted, complex biochemical interactions and nanoscale phenomena based on nanotechnology efficacy in food preservation.

2.3.5.1 Implications for Academic Dialogue and Future Research

The elaborate examination of nanotechnology's mechanisms in food preservation aims to enrich the research and academic dialogue in the burgeoning field of nanotechnology [111]. With its thorough explanation of the functions and uses of various nanomaterials, this presentation offers a broad reference for upcoming research endeavors in nanotechnology and its diverse applications [112]. The study of nanofibers and various nanomaterials highlights nanotechnology's potential to revolutionize food preservation and security techniques [109]. With the continuously evolving research in this area, new advancements are expected, further elevating the scope and effectiveness of nanotechnology applications in food preservation, guaranteeing food security, reducing food wastage, and contributing to global food security efforts [112]. In summary, nanofibers offer notable advantages and contributions to food preservation and food security as a part of the broader category of nanomaterials. This role of nanofibers is customizable, acting as a physical barrier against microbial invasion and contamination, highlighting the novel approaches being developed in nanotechnology to tackle the challenges of food preservation, particularly the protection of fruits and vegetables [113]. An inclusive understanding of these innovative technologies is essential for conducting advanced academic research and real-time applications in the food preservation and food industry [114].

2.4 ANALYSIS OF NANOTECHNOLOGY MECHANISMS IN FRUIT PRESERVATION

The practical use of nanotechnology in food preservation—especially in apple preservation—highlights the expanding relationship between nanomaterials and sustainable agricultural methods. Silver nanoparticles and nanoemulsions have demonstrated the multidisciplinary effectiveness of nanomaterials in prolonging the shelf life and maintaining the quality of apples [115].

2.4.1 Role of Silver Nanoparticles and Nanoemulsions for Sustainable Agriculture

The employment of silver nanoparticles and nanoemulsions in preserving apples embraces several key advantages: Both silver nanoparticles and nanoemulsions are widely known for their highly significant antimicrobial properties [116]. They effectively inhibit bacterial and fungal growth, which is usually a reason for the spoilage of apples. This antimicrobial effect of silver nanoparticles and nanoemulsions is vital in enhancing the shelf life of apples [117]. These nanomaterials maintain essential physiochemical properties of apples, such as moisture content, ascorbic acid concentrations, and firmness [118]. Retaining these physiochemical

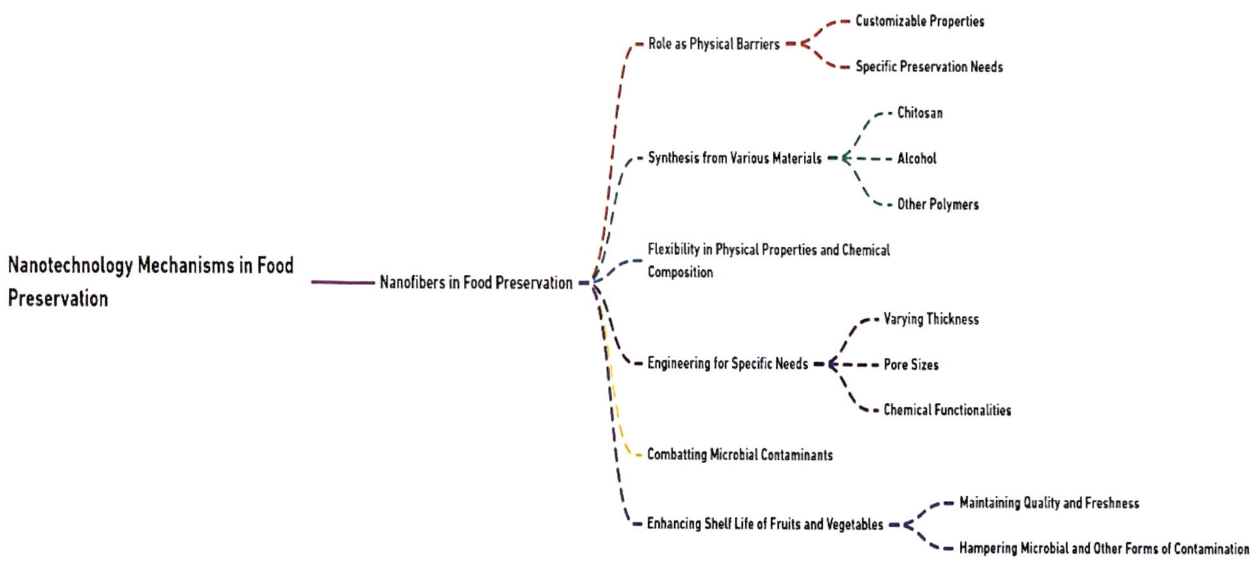

FIGURE 2.1 Comprehensive schematic diagram of nanotechnology mechanisms. Created with Whimsical.com (2024).

properties guarantees that the apples remain nutritious and fresh over a prolonged shelf life [119]. Sensory assessments have illustrated that employing silver nanoparticles and nanoemulsions does not adversely affect the apple's aroma, flavor, and overall quality [120]. The efficiency of these nanomaterials uncovers the possibility of depending on synthetic chemical preservatives to preserve apples [121]. The potential swing toward nanomaterial preservation methods can lead to more sustainable and environmentally friendly practices in agriculture and the food industry [122]. The use of nanotechnology in the preservation of apples promises agricultural sustainability by enhancing their shelf life and effectively reducing global food wastage, a vital concern in global food systems [123]. Nanomaterials such as silver nanoparticles and nanoemulsions can potentially replace synthetic chemical preservatives; thus, they offer natural and safer methods of food preservation [124]. Preserving apples' nutritional and sensory quality ensures consumers' access to safe, healthy, and high-quality food [125].

In conclusion, apples may be preserved with heterogeneous effectiveness by using silver nanoemulsions and nanoparticles. This is a promising first step toward the potential applications of nanotechnology in enhancing food security and preservation with little food waste in the future. Nanomaterials provide a substantial alternative to traditional food preservation techniques by preventing food decomposition and preserving food quality. The application of nanotechnology in apple preservation demonstrates its promise for sustainable agriculture practices with reduced food waste and ensured customer health. A viable avenue for the development of sustainable food preservation methods is highlighted by this paradigm of nanotechnology research and applications [126].

2.4.2 Gold Nanoparticles as Multifunctional Agents

The employment of gold nanoparticles in the preservation of grapes demonstrates the multifaceted potential of nanotechnology in amplifying the shelf life and quality of perishable food items. Gold nanoparticles serve as potent agents in grapes, controlling both microbial spoilage and enzymatic browning while maintaining the vital quality attributes of grapes [127].

The employment of AuNPs in the preservation of grapes offers various significant advantages: AuNPs efficiently disrupt the biochemical enzymatic reactions that cause polyphenol oxidation in grapes. This action is vital in hampering the food browning process in grapes, a common challenge compromising the sensory and nutritional quality and aesthetics [128]. Besides addressing the challenge of enzymatic browning, AuNPs also present significant antimicrobial properties. AuNPs can reduce fungal and bacterial colonies on grapes, preventing and significantly delaying their spoilage. The dual job of halting both the biochemical enzymatic and microbial spoilage is specifically irreplaceable in preserving the freshness of grapes [129]. Concurrent studies reveal that AuNPs can also play a role in maintaining the texture of grapes. This is a promising development as the texture of fruits is a crucial characteristic that impacts the satisfaction and acceptance of consumers [130]. The ability of AuNPs to preserve other fruit qualities, such as sweetness and acidity, opens up new possibilities for grape preservation. Preserving the tastes is essential to ensure the grapes' overall quality and attractiveness.

2.4.2.1 Implications for Grape Preservation and Beyond

The diverse range of uses for AuNPs in grape preservation is considerably greater. By tackling several issues at once, AuNPs can significantly extend the shelf life of grapes. The grapes will stay appealing to customers for an extended time since AuNPs can preserve essential quality traits, including sweetness, texture, and acidity. Since gold nanoparticles effectively preserve grapes, there may be more opportunities to use them to protect other fruits, particularly those more vulnerable to enzymatic browning and microbial deterioration. In conclusion, AuNPs offer a multifunctional, heterogeneous, and effective solution to the challenge of grape spoilage. Their ability to halt microbial growth and enzymatic browning and maintain the quality features of grapes simultaneously highlights the potential nanomaterials and nanotechnology in revolutionizing the methods of food preservation. The concurrent exploration and continued application of AuNPs in the food industry promises to prolong the shelf life and enhance the quality of numerous perishable food products, significantly contributing to advanced sustainable food preservation and minimizing food wastage globally [131].

2.4.3 Nanofibers and Their Comprehensive Protective Effects

Employing nanofibers to preserve mango presents a significant development in using nanotechnology to enhance shelf life and maintain the quality of fruits. Nanofibers offer a diversified protective effect on mango preservation, manipulating the properties of physical barriers and the ability to incorporate effective antimicrobial agents [132].

2.4.3.1 Nanofibers Present Various Key Advantages in Mango Preservation

Nanofibers, synthesized from ceramic or polymeric materials, form a physical barrier around the mango by forming a microencapsulation. This physical barrier of microencapsulation effectively hampers the exposure of the mango to bacterial and fungal contaminants. Nanofibers help minimize the risk of fruit spoilage by preventing microbial invasion [133]. A vital feature of nanofiber technology in preserving perishable food items is the flexibility to synthesize these fibers to enclose natural antimicrobial agents, such as essential oils. Incorporating biological antimicrobial agents offer an additional protective layer against microorganisms that cause food spoilage [134]. The amalgamation of the properties of physical barriers and the containing of antimicrobial agents in nanofibers results in a synergistic effect. This perspective not only hampers the biological contamination of microorganisms but also effectively combats their growth, thus significantly enhancing the shelf life of mangoes [135]. One of the key advantages of employing nanofibers for preserving mangoes is their potential to maintain fruits' sensory and nutritional properties. By offering protection against food spoilage without adversely affecting the inherent properties of fruits, nanofibers guarantee that the mangoes are nutritious, fresh, and appealing to consumers [136].

2.4.3.2 Implications for Mango Preservation

The use of nanofibers in the preservation of mangoes has various implications: Nanofibers can primarily enhance the shelf life of mangoes, minimizing food waste and improving storage and distribution efficiency. By integrating natural antimicrobial agents into their structure, nanofibers present an effective and safe method for preserving mangoes, reducing the use of synthetic chemical preservatives. The ability to engineer nanofibers for specific preservation needs makes them versatile for application in various fruit preservation scenarios, including other tropical fruits susceptible to rapid spoilage. Nanofibers offer promising technology for preserving mangoes, offering a multifaceted protective effect through their physical barrier properties and incorporating natural antimicrobial agents. This nanotechnology not only enhances the shelf life of perishable food items but also guarantees the preservation of the sensory and nutritional attributes of the fruit. The concurrent development and employment of nanofibers in food preservation carry notable benefits and potential for amplifying the efficiency and sustainability of food supply chains globally, specifically in the realm of perishable food items such as mangoes [137].

2.4.4 Application of Zinc Oxide Nanoparticles

Utilizing zinc oxide nanoparticles (ZnO NPs) to preserve berries symbolizes an exciting and promising advancement in nanotechnology and food preservation. This area of research is specifically aimed at utilizing the antifungal properties of ZnO NPs to inhibit fungal growth, which is a common cause of berries spoilage [138]. The application of zinc oxide nanoparticles in preserving berries encompasses several promising aspects: The utilization of ZnO NPs in berries preservation encompasses various promising features. The potent antifungal capabilities of ZnO NPs qualify them as significant agents that effectively inhibit the growth of fungi that cause spoilage in berries. This feature of ZnO NPs is vital in enhancing the shelf life of berries, which are highly susceptible to mycotic (fungal) contamination [139]. Precursory studies reveal that ZnO NPs may also stabilize the nutritional profile and color of berries, which includes the retention of vital antioxidants, such as anthocyanin, that are crucial for the sensory qualities of berries and their health benefits [140]. By preserving the nutritional value, color, and texture of berries, ZnO NPs ensure the maintenance of appeal and overall quality of the fruit, which is essential for consumer acceptance.

Proposed as an investigative tool, Figure 2.2 provides a comprehensive and versatile quantitative comparative analysis of different nanomaterials' efficiency in preserving

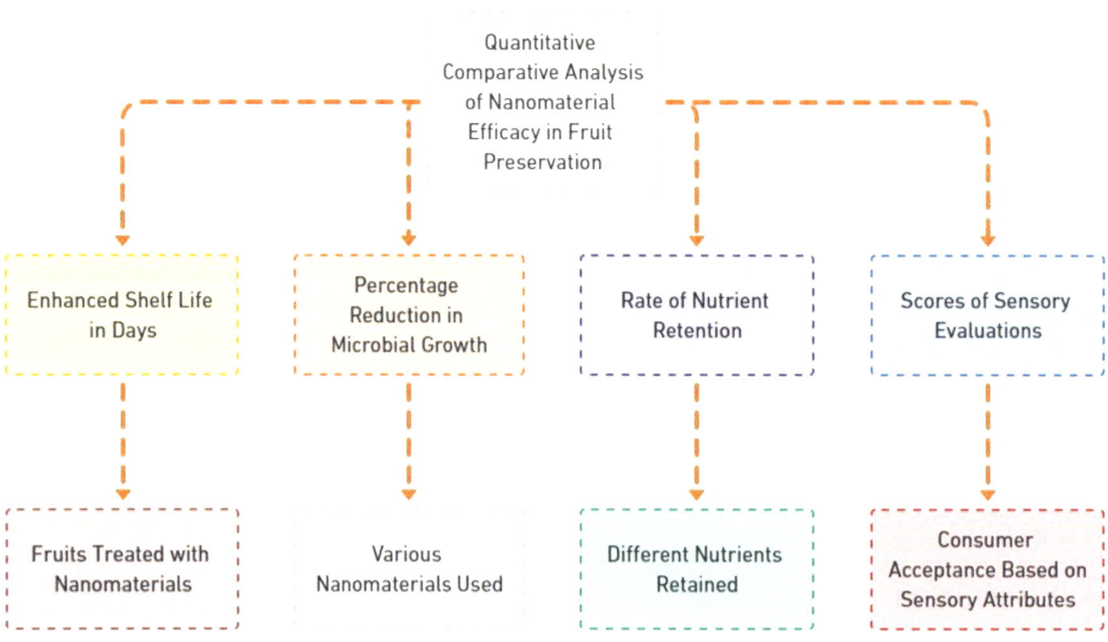

FIGURE 2.2 Quantitative comparative analysis of nanomaterial efficacy in fruit preservation across diverse case studies. Created with Whimsical.com (2024).

fruits. Information regarding the rate of nutrient retention for different nutrients underscores the effectiveness of multiple nanomaterials in maintaining the nutritional value of fruits. Scores of sensory evaluations based on consumer acceptance evaluate the effect of nanomaterial employment on the sensory attributes of fruits.

2.4.4.1 Implications for Future Research and Industrial Applications

The integration of this multifaceted data in Figure 2.2 can aid in various significant purposes: It lays a foundation for mapping out strategies for future research to discover the most potent nanomaterials for preserving specific fruit types. It can aid the decision-making process for industrial employment of nanotechnology by presenting an understanding of the most influential and consumer-friendly nanotechnology methods in fruit preservation. It also might offer guidance on the most effective and sustainable preservation techniques, contributing to the global efforts to reduce food wastage and develop advanced food safety measures. In a nutshell, the employment of ZnO NPs in the preservation method for berries and the presented quantitative comparison analysis of the efficacy of different nanomaterials present a valuable understanding of the capacity of nanotechnology in fruit preservation [141]. These discoveries enhance the shelf life of fruits, maintain their quality, and present invaluable data for directing future research and real-time applications in the food industry. The development in the application of nanotechnology for fruit preservation highlights nanomaterials' potential for sustainable and novel solutions in meeting the global challenges of food wastage and spoilage [142].

2.5 MECHANISMS OF NANOTECHNOLOGY IN PRESERVATION APPLICATIONS FOR VEGETABLES

2.5.1 Multifaceted Contributions of Gold Nanoparticles

The application of gold nanoparticles in lettuce preservation illustrates an astonishing innovation in food preservation technology. These nanoparticles effectively disrupt and inactivate the activity-specific enzymes, such as polyphenol oxidases and peroxidases. These enzymes catalyze oxidation reactions, resulting in browning and spoilage of lettuce [143]. By halting these enzymatic activities, gold nanoparticles maintain lettuce's nutritional value and textural integrity and preserve its aesthetic attributes. Furthermore, the employment of gold nanoparticles might result in reduced dependency and use of synthetic chemical preservatives, contributing to healthier and more organic food availability for consumers [144]. Preliminary studies also reveal that nanoparticles might increase the resistance of lettuce leaves to microbial spoilage, although further academic investigation is needed to realize these claims [145].

2.5.2 Advanced Applications of Nanoemulsions in Nutrient and Freshness Retention

The role of nanoemulsions in spinach preservation extends beyond mere nutrient preservation. These nanoparticles also act as carriers for antioxidants that can be invigorated with antimicrobial agents, hence resulting in a multifaceted food preservation approach. This versatile approach maintains and enhances the bioavailability of vitamins and iron, making spinach a nutritionally superior, longer-lasting food product [146]. Furthermore, nanoemulsions can be designed and synthesized for the controlled release of essential nutrients, realizing sustained nutrient availability during storage and after consumption, having far-reaching consequences for health and well-being [147].

2.5.3 Comprehensive Antimicrobial and Preservative Attributes of Silver Nanoparticles

The effective antimicrobial combating properties of silver nanoparticles in carrots must be considered. These nanoparticles present broad-spectrum antimicrobial activity, qualifying them significantly effective against several microbial pathogens, including bacteria and fungi, and strains resistant to concurrent antimicrobial agents. This enhances the carrots' shelf life and minimizes the risk of foodborne diseases [147]. Furthermore, silver nanoparticles can complement existing food preservation technologies. For example, synergizing silver nanoparticles with altered atmosphere packaging could result in a more resilient food preservation method, effectively reducing microbial spoilage and oxidative degradation of carrots [148].

2.6 EXPLORING THE UPCOMING POSSIBILITY OF NANOTECHNOLOGY IN FOOD PROTECTION AND CARE

With the advancement of nanotechnology in food preservation, the transition of experiments has become increasingly crucial, from small laboratory scales to larger-scale industrial applications. There is an increasing need to develop cost-effective methods for the large-scale production of nanomaterials and their incorporation into the existing food supply chains, ensuring that the benefits of nanotechnology-based food preservation reach a broad consumer base. The concurrent acceptance of nanotechnology in food preservation is deeply linked to public perception [149]. Conduct research with social scientists to understand consumer concerns and attitudes regarding this technology. Develop effective communication strategies to make the public aware of nanotechnology's safety measures and benefits in food preservation.

Furthermore, several key areas must be focused on in future research, including comprehensive studies to analyze the long-term effect of nanomaterials on human health when consumed with food, ascertain the environmental impact of nanotechnology used in food preservation, and develop economical methods to produce and employ nanomaterials in food. Collaboration with regulatory bodies is crucial to establish standards and clear guidelines for using nanotechnology in food preservation, promising efficacy, safety, and environmental responsibility. These efforts are necessary to realize the full potential of nanotechnology in food preservation, utilization, sustainability, and safety [150].

2.7 CONCLUSION

Nanotechnology has an immense potential to amplify food security, minimize food wastage, and optimize the quality of perishable food items, as highlighted by the study of nanotechnology in preserving vegetables and fruits. Throughout this chapter, nanotechnology and nanomaterials such as gold and silver nanoparticles, nanofibers, and nanoemulsions present promising solutions, elaborating the mechanisms of action and applications in food preservation. The discussion has also encompassed nanotechnology's crucial regulatory and safety aspects, highlighting the comprehensive evaluation of these nanomaterials to ensure consumer health and well-being. Though nanotechnology has promising potential in the food industry, its application must be imperatively accompanied with caution. More understanding and knowledge about the long-term effects of nanotechnology and nanomaterials on human health and the environment needs to be gained. Further research must tackle these concerns collaboratively, involving scientists, industry, and regulatory bodies. This collaborative approach, with an ethical and vigorous framework, is crucial for the accountable application of nanotechnology for food preservation. In short, an exciting glance is offered into the future of reduced food wastage and sustainable food production globally by employing nanotechnology in preserving vegetables and fruits. With the continuous evolution of nanotechnology, it is crucial to balance innovations and safety standards and adhere to ethical regulations. This balanced approach will fulfil the needs of future generations and cultivate a sustainable and secure food system globally by ensuring resilient and efficient food supply chains.

Author contribution: All authors have contributed equally.

Conflict of interest

The authors declare that they have no conflict of interest.

REFERENCES

1. Bajpai, V.K., et al., Prospects of using nanotechnology for food preservation, safety, and security. *Journal of Food and Drug Analysis*, 2018. **26**(4): p. 1201–1214.
2. Malik, S., K. Muhammad, and Y. Waheed, Nanotechnology: A revolution in modern industry. *Molecules*, 2023. **28**(2): p. 661.
3. Kim, D.Y., et al., Recent developments in nanotechnology transforming the agricultural sector: A transition replete with opportunities. *Journal of the Science of Food and Agriculture*, 2018. **98**(3): p. 849–864.
4. Sharma, S., et al., Nanotechnology for food: Regulatory issues and challenges. *Advances in Agri-Food Biotechnology*, 2020: p. 367–389.
5. Wang, T., et al., Nanozyme: An emerging tool for food packaging. *Food Control*, 2023: p. 110104.
6. Nura, A., Advances in food packaging technology-a review. *Journal of Postharvest Technology*, 2018. **6**(4): p. 55–64.
7. Bashir, O., et al., Nano-engineered materials for sensing food pollutants: Technological advancements and safety issues. *Chemosphere*, 2022. **292**: p. 133320.
8. BS, Mitchelle Nissintha Daisy, Lavudiya Ramesh Babu, Ambigalla Ebenezer Paul, Sevanan Murugan, and Rajiv Periakaruppan. "Nanozymes as catalytic marvels for biomedical and environmental concerns: a chemical engineering approach." *Journal of Cluster Science*, 2024. **35**(3): 715–740.
9. Assadpour, E., C. Dima, and S.M. Jafari, Fundamentals of food nanotechnology, in *Handbook of food nanotechnology*. 2020, Elsevier. p. 1–35.
10. Banerjee, A.N., Green syntheses of graphene and its applications in internet of things (IoT)—a status review. *Nanotechnology*, 2022. **33**(32): p. 322003.
11. Arshad, R., et al., Nanotechnology for therapy of zoonotic diseases: A comprehensive overview. *ChemistrySelect*, 2022. **7**(21): p. e202201271.
12. Ray, M.K., et al., Nanotechnology as a promising tool against phytopathogens: A futuristic approach to agriculture. *Agriculture*, 2023. **13**(9): p. 1856.
13. Ukkund, S.J., et al., Role of microbial nanotechnology in energy devices, in *Handbook of Microbial Nanotechnology*. 2022, Elsevier. p. 517–547.
14. Dulta, K., et al., Nanotechnology and applications, in *Applications of computational intelligence in multi-disciplinary research*. 2022, Elsevier. p. 129–141.
15. Huang, B., et al., Advances in targeted pesticides with environmentally responsive controlled release by nanotechnology. *Nanomaterials*, 2018. **8**(2): p. 102.
16. Gupta, V., et al., Nanotechnology in cosmetics and cosmeceuticals—a review of latest advancements. *Gels*, 2022. **8**(3): p. 173.
17. Sampathkumar, K., K.X. Tan, and S.C.J. Loo, Developing nano-delivery systems for agriculture and food applications with nature-derived polymers. *Iscience*, 2020. **23**(5).
18. Gayathiri, E., et al., Investigating the ecological implications of nanomaterials: Unveiling plants' notable responses to nanopollution. *Plant Physiology and Biochemistry*, 2023: p. 108261.
19. Dixit, S., Shukla, A., Singh, V. and Upadhyay, S.K. (2021). Bioprospecting of Natural Compounds for Industrial and Medical Applications. In *Bioprospecting of Plant Biodiversity for Industrial Molecules* (eds S.K. Upadhyay and S.P. Singh). https://doi.org/10.1002/9781119718017.ch3
20. Nurazzi, N., et al., *Nanofillers for sustainable applications*. 2023: CRC Press.
21. Bag, T., Socio-economic impacts of scientific-technological advancements. *International Journal of Multidisciplinary Educational Research*, 2023: 12.
22. Bhandi, S., et al., Antimicrobial efficacy of silver nanoparticles as root canal irrigant's: A systematic review. *Journal of Clinical Medicine*, 2021. **10**(6): p. 1152.
23. Adam, R.Z. and S.B. Khan, Antimicrobial efficacy of silver nanoparticles against Candida albicans: A systematic review protocol. *PLoS One*, 2021. **16**(1): p. e0245811.
24. Devanesan, S., et al., Cytotoxic and antimicrobial efficacy of silver nanoparticles synthesized using a traditional

phytoproduct, asafoetida gum. *International Journal of Nanomedicine*, 2020: p. 4351–4362.
25. Qamer, S., et al., Systematic review on biosynthesis of silver nanoparticles and antibacterial activities: Application and theoretical perspectives. *Molecules*, 2021. **26**(16): p. 5057.
26. Quintero-Quiroz, C., et al., Optimization of silver nanoparticle synthesis by chemical reduction and evaluation of its antimicrobial and toxic activity. *Biomaterials Research*, 2019. **23**: p. 1–15.
27. Abdelghany, T., et al., Recent advances in green synthesis of silver nanoparticles and their applications: About future directions. A review. *BioNanoScience*, 2018. **8**: p. 5–16.
28. Some, S., et al., Biosynthesis of silver nanoparticles and their versatile antimicrobial properties. *Materials Research Express*, 2018. **6**(1): p. 012001.
29. Chen, J., et al., Antimicrobial activity of pomegranate peel and its applications on food preservation. *Journal of Food Quality*, 2020. **2020**: p. 1–8.
30. Zhang, W., et al., Molybdenum disulfide nanosheets loaded with chitosan and silver nanoparticles effective antifungal activities: In vitro and in vivo. *Materials Science and Engineering: C*, 2019. **97**: p. 486–497.
31. Fontecha-Umaña, F., et al., Antimicrobial activity and prevention of bacterial biofilm formation of silver and zinc oxide nanoparticle-containing polyester surfaces at various concentrations for use. *Foods*, 2020. **9**(4): p. 442.
32. Sheng, Y., et al., In vitro and in vivo efficacy of green synthesized AgNPs against Gram negative and Gram positive bacterial pathogens. *Process Biochemistry*, 2022. **112**: p. 241–247.
33. Kumar, A., et al., Metal-based nanoparticles, sensors, and their multifaceted application in food packaging. *Journal of Nanobiotechnology*, 2021. **19**(1): p. 256.
34. Sridhar, A., et al., Food preservation techniques and nanotechnology for increased shelf life of fruits, vegetables, beverages and spices: A review. *Environmental Chemistry Letters*, 2021. **19**: p. 1715–1735.
35. Verma, P. and S.K. Maheshwari, Applications of silver nanoparticles in diverse sectors. *International Journal of Nano Dimension*, 2019. **10**(1): p. 18–36.
36. Kraśniewska, K., S. Galus, and M. Gniewosz, Biopolymers-based materials containing silver nanoparticles as active packaging for food applications–a review. *International Journal of Molecular Sciences*, 2020. **21**(3): p. 698.
37. Ahmad, S., et al., Green nanotechnology: A review on green synthesis of silver nanoparticles—an ecofriendly approach. *International Journal of Nanomedicine*, 2019: p. 5087–5107.
38. Sarfraz, N. and I. Khan, Plasmonic gold nanoparticles (AuNPs): Properties, synthesis and their advanced energy, environmental and biomedical applications. *Chemistry–An Asian Journal*, 2021. **16**(7): p. 720–742.
39. Hammami, I. and N.M. Alabdallah, Gold nanoparticles: Synthesis properties and applications. *Journal of King Saud University-Science*, 2021. **33**(7): p. 101560.
40. Qin, L., et al., "Gold rush" in modern science: Fabrication strategies and typical advanced applications of gold nanoparticles in sensing. *Coordination Chemistry Reviews*, 2018. **359**: p. 1–31.
41. Kang, H., et al., Stabilization of silver and gold nanoparticles: Preservation and improvement of plasmonic functionalities. *Chemical Reviews*, 2018. **119**(1): p. 664–699.
42. Lou-Franco, J., et al., Gold nanozymes: From concept to biomedical applications. *Nano-Micro Letters*, 2021. **13**: p. 1–36.
43. Zhang, M., et al., Recent advances in polymers and polymer composites for food packaging. *Materials Today*, 2022. **53**: p. 134–161.
44. Chen, H., K. Zhou, and G. Zhao, Gold nanoparticles: From synthesis, properties to their potential application as colorimetric sensors in food safety screening. *Trends in Food Science & Technology*, 2018. **78**: p. 83–94.
45. Khurana, K. and N. Jaggi, Localized surface plasmonic properties of Au and Ag nanoparticles for sensors: A review. *Plasmonics*, 2021. **16**(4): p. 981–999.
46. Nile, S.H., et al., Nanotechnologies in food science: Applications, recent trends, and future perspectives. *Nano-Micro Letters*, 2020. **12**: p. 1–34.
47. Ali, S.S., et al., Nanobiotechnological advancements in agriculture and food industry: Applications, nanotoxicity, and future perspectives. *Science of the Total Environment*, 2021. **792**: p. 148359.
48. Singh, R., et al., Future of nanoTechnology in food industry: Challenges in processing, packaging, and food safety. *Global Challenges*, 2023. **7**(4): p. 2200209.
49. Okeke, E.S., et al., Antibiotic resistance in aquaculture and aquatic organisms: A review of current nanotechnology applications for sustainable management. *Environmental Science and Pollution Research*, 2022. **29**(46): p. 69241–69274.
50. Anand, U., et al., Safer plant-based nanoparticles for combating antibiotic resistance in bacteria: A comprehensive review on its potential applications, recent advances, and future perspective. *Science of the Total Environment*, 2022. **821**: p. 153472.
51. Håkansson, A., Fabrication of nanoemulsions by high-pressure valve homogenization, in *Nanoemulsions*. 2018, Elsevier. p. 175–206.
52. Leong, T.S.H., et al., *Ultrasonic production of nanoemulsions for bioactive delivery in drug and food applications*. 2018: Springer.
53. Amiri, S., et al., Natural protective agents and their applications as bio-preservatives in the food industry: An overview of current and future applications. *Italian Journal of Food Science*, 2021. **33**(SP1): p. 55–68.
54. Batiha, G.E.-S., et al., Application of natural antimicrobials in food Preservation: Recent views. *Food Control*, 2021. **126**: p. 108066.
55. Manzoor, A., et al., Recent insights into polysaccharide-based hydrogels and their potential applications in food sector: A review. *International Journal of Biological Macromolecules*, 2022. **213**: p. 987–1006.
56. Saifullah, M., et al., Micro and nano encapsulation, retention and controlled release of flavor and aroma compounds: A critical review. *Trends in Food Science & Technology*, 2019. **86**: p. 230–251.
57. Akhavan, S., et al., Lipid nano scale cargos for the protection and delivery of food bioactive ingredients and nutraceuticals. *Trends in Food Science & Technology*, 2018. **74**: p. 132–146.
58. Liu, Q., et al., Food-grade nanoemulsions: Preparation, stability and application in encapsulation of bioactive compounds. *Molecules*, 2019. **24**(23): p. 4242.
59. Kumar, P., et al., Nanotechnology and its challenges in the food sector: A review. *Materials Today Chemistry*, 2020. **17**: p. 100332.
60. Rizvi, S.S., et al., Nanotechnology and food safety, in *Ensuring global food safety*. 2022, Elsevier. p. 325–340.
61. Abioye, A., et al. Review of the development of biodegradable plastic from synthetic polymers and selected

synthesized nanoparticle starches, in *Journal of Physics: Conference Series*. 2019, IOP Publishing.
62. Moohan, J., et al., Cellulose nanofibers and other biopolymers for biomedical applications. A review. *Applied Sciences*, 2019. **10**(1): p. 65.
63. Jafarzadeh, S., et al., Application of bio-nanocomposite films and edible coatings for extending the shelf life of fresh fruits and vegetables. *Advances in Colloid and Interface Science*, 2021. **291**: p. 102405.
64. Kumar, L., et al., Edible films and coatings for food packaging applications: A review. *Environmental Chemistry Letters*, 2022: p. 1–26.
65. Topuz, F. and T. Uyar, Antioxidant, antibacterial and antifungal electrospun nanofibers for food packaging applications. *Food Research International*, 2020. **130**: p. 108927.
66. Leidy, R. and Q.-C.M. Ximena, Use of electrospinning technique to produce nanofibres for food industries: A perspective from regulations to characterisations. *Trends in Food Science & Technology*, 2019. **85**: p. 92–106.
67. Rostami, M., et al., Recent advances in electrospun protein fibers/nanofibers for the food and biomedical applications. *Advances in Colloid and Interface Science*, 2023. **311**: p. 102827.
68. Cheng, H., et al., Starch-based biodegradable packaging materials: A review of their preparation, characterization and diverse applications in the food industry. *Trends in Food Science & Technology*, 2021. **114**: p. 70–82.
69. Goyal, R., et al., Nanoparticles and nanofibers for topical drug delivery. *Journal of Controlled Release*, 2016. **240**: p. 77–92.
70. Biswas, R., et al., Application of nanotechnology in food: Processing, preservation, packaging and safety assessment. *Heliyon*, 2022. **8**(11).
71. Durán, N., et al., Silver nanoparticles: A new view on mechanistic aspects on antimicrobial activity. *Nanomedicine: Nanotechnology, Biology and Medicine*, 2016. **12**(3): p. 789–799.
72. Wang, L., C. Hu, and L. Shao, The antimicrobial activity of nanoparticles: Present situation and prospects for the future. *International Journal of Nanomedicine*, 2017: p. 1227–1249.
73. Gupta, K., et al., Green silver nanoparticles: Enhanced antimicrobial and antibiofilm activity with effects on DNA replication and cell cytotoxicity. *RSC Advances*, 2014. **4**(95): p. 52845–52855.
74. Pandey, R.P., et al., Potential of nanoparticles encapsulated drugs for possible inhibition of the antimicrobial resistance development. *Biomedicine & Pharmacotherapy*, 2021. **141**: p. 111943.
75. Tran, Q.H. and A.-T. Le, Silver nanoparticles: Synthesis, properties, toxicology, applications and perspectives. *Advances in Natural Sciences: Nanoscience and Nanotechnology*, 2013. **4**(3): p. 033001.
76. Luksiene, Z., Nanoparticles and their potential application as antimicrobials in the food industry, in *Food preservation*. 2017, Elsevier. p. 567–601.
77. Nunes, C., et al., Edible coatings and future trends in active food packaging–fruits' and traditional sausages' shelf life increasing. *Foods*, 2023. **12**(17): p. 3308.
78. Omerović, N., et al., Antimicrobial nanoparticles and biodegradable polymer composites for active food packaging applications. *Comprehensive Reviews in Food Science and Food Safety*, 2021. **20**(3): p. 2428–2454.
79. Avila-Quezada, G.D. and G.P. Espino-Solis, Silver nanoparticles offer effective control of pathogenic bacteria in a wide range of food products. *Pathogenic Bacteria*, 2020: p. 203.
80. Matthews, K.R., K.E. Kniel, and T.J. Montville, *Food microbiology: An introduction*. 2017: John Wiley & Sons.
81. Lira, A.L., et al., Allosteric inhibition of α-thrombin enzymatic activity with ultrasmall gold nanoparticles. *Nanoscale Advances*, 2019. **1**(1): p. 378–388.
82. Xi, Y., et al., Growth of Au nanoparticles on phosphorylated zein protein particles for use as biomimetic catalysts for cascade reactions at the oil–water interface. *Chemical Science*, 2021. **12**(11): p. 3885–3889.
83. Beshai, H., et al., Freshness monitoring of packaged vegetables. *Applied Sciences*, 2020. **10**(21): p. 7937.
84. Mattila, P., et al., Underutilized Northern plant sources and technological aspects for recovering their polyphenols. *Advances in Food and Nutrition Research*, 2021. **98**: p. 125–169.
85. Amina, S.J. and B. Guo, A review on the synthesis and functionalization of gold nanoparticles as a drug delivery vehicle. *International Journal of Nanomedicine*, 2020: p. 9823–9857.
86. Nejati, K., et al., Biomedical applications of functionalized gold nanoparticles: A review. *Journal of Cluster Science*, 2021: p. 1–16.
87. Lin, Y., et al., Multicolor hydrogen sulfide sensor for meat freshness assessment based on Cu2+-modified boron nitride nanosheets-supported subnanometer gold nanoparticles. *Food Chemistry*, 2022. **381**: p. 132278.
88. Huang, Y., J. Ren, and X. Qu, Nanozymes: Classification, catalytic mechanisms, activity regulation, and applications. *Chemical Reviews*, 2019. **119**(6): p. 4357–4412.
89. Rai, M., et al., Smart nanopackaging for the enhancement of food shelf life. *Environmental Chemistry Letters*, 2019. **17**: p. 277–290.
90. Prakash, J., et al., Application of nanoparticles in food preservation and food processing. *Journal of Food Hygiene and Safety*, 2019. **34**(4): p. 317–324.
91. Chellaram, C., et al., Significance of Nanotechnology in food industry. *APCBEE Procedia*, 2014. **8**: p. 109–113.
92. Akgönüllü, S., H. Yavuz, and A. Denizli, SPR nanosensor based on molecularly imprinted polymer film with gold nanoparticles for sensitive detection of aflatoxin B1. *Talanta*, 2020. **219**: p. 121219.
93. Panda, M.K., Singh, Y.D., Behera, R.K., Dhal, N.K. (2020). Biosynthesis of Nanoparticles and Their Potential Application in Food and Agricultural Sector. In: Patra, J., Fraceto, L., Das, G., Campos, E. (eds) Green Nanoparticles. *Nanotechnology in the Life Sciences*. Springer, Cham. https://doi.org/10.1007/978-3-030-39246-8_10
94. Katiyar, V. and T. Ghosh, *Nanotechnology in edible food packaging*. 2021: Springer.
95. Pushparaj, K., et al., Nano-from nature to nurture: A comprehensive review on facets, trends, perspectives and sustainability of nanotechnology in the food sector. *Energy*, 2022. **240**: p. 122732.
96. Suyanto, A., et al. Nano-emulsion and nano-encapsulation of fruit flavor, in *IOP conference series: Earth and environmental science*. 2019. IOP Publishing.
97. Jamali, S.N., et al., Natural antimicrobial-loaded nanoemulsions for the control of food spoilage/pathogenic microorganisms. *Advances in Colloid and Interface Science*, 2021. **295**: p. 102504.
98. Zambrano-Zaragoza, M., et al., Nanoemulsions and nanosized ingredients for food formulations, in *Handbook of food nanotechnology*. 2020, Elsevier. p. 207–256.
99. Salvia-Trujillo, L., et al., Physicochemical characterization and antimicrobial activity of food-grade emulsions and nanoemulsions incorporating essential oils. *Food Hydrocolloids*, 2015. **43**: p. 547–556.

100. Patil, V., R. Shams, and K.K. Dash, Techno-functional characteristics, and potential applications of edible coatings: A comprehensive review. *Journal of Agriculture and Food Research*, 2023: p. 100886.
101. Aguilar-Pérez, K.M., et al., Insight of nanotechnological processing for nano-fortified functional foods and nutraceutical—opportunities, challenges, and future scope in food for better health. *Critical Reviews in Food Science and Nutrition*, 2023. **63**(20): p. 4618–4635.
102. Trukhanov, A., P. Thakur, and A. Thakur, Nanotechnological applications in food and agriculture, in *Integrated nanomaterials and their applications*. 2023, Springer. p. 393–417.
103. Saari, N.H.M. and L.S. Chua, Nano-based products in beverage industry, in *Nanoengineering in the beverage industry*. 2020, Elsevier. p. 405–436.
104. Ahmed, I., et al., A comprehensive review on the application of active packaging technologies to muscle foods. *Food Control*, 2017. **82**: p. 163–178.
105. Sohail, M.I., et al., Environmental application of nanomaterials: A promise to sustainable future, in *Comprehensive analytical chemistry*. 2019, Elsevier. p. 1–54.
106. Gupta, A.K., et al., Novel food materials: Fundamentals and applications in sustainable food systems for food processing and safety. *Food Bioscience*, 2023: p. 103013.
107. Suhag, D., A. Thakur, and P. Thakur, *Integrated nanomaterials and their applications*. 2023: Springer. p. 1–17.
108. Elsabee, M.Z., H.F. Naguib, and R.E. Morsi, Chitosan based nanofibers, review. *Materials Science and Engineering: C*, 2012. **32**(7): p. 1711–1726.
109. Lin, A.C., Size matters: Regulating nanotechnology. *Harvard Environmental Law Review*, 2007. **31**: p. 349.
110. Huang, Y., et al., Scalable manufacturing and applications of nanofibers. *Materials Today*, 2019. **28**: p. 98–113.
111. Hullmann, A., *European activities in the field of ethical, legal and social aspects (ELSA) and governance of nanotechnology*. DG Research, Brussels: European Commission, 2008.
112. Lead, J.R., et al., Nanomaterials in the environment: Behavior, fate, bioavailability, and effects—an updated review. *Environmental Toxicology and Chemistry*, 2018. **37**(8): p. 2029–2063.
113. Tekade, R.K., et al., Nanotechnology for the development of Nanomedicine, in *Nanotechnology-based approaches for targeting and delivery of drugs and genes*. 2017, Elsevier. p. 3–61.
114. Michelson, E.S. and D. Rejeski, Transnational nanotechnology governance: A comparison of the US and China, in *Nanotechnology & society: Current and emerging ethical issues*. 2009, Springer. p. 281–299.
115. Liu, W., M. Zhang, and B. Bhandari, Nanotechnology–a shelf life extension strategy for fruits and vegetables. *Critical Reviews in Food Science and Nutrition*, 2020. **60**(10): p. 1706–1721.
116. Prasad, M., et al., The importance of nanomedicine in prophylactic and theranostic intervention of bacterial zoonoses and reverse zoonoses in the Era of microbial resistance. *Journal of NanoScience and Nanotechnology*, 2021. **21**(6): p. 3404–3452.
117. Alsaif, M.A., et al., Pouteria caimito fruit derived nanoparticles inhibited the Apple ring rot disease as well as extended the shelf-life of sliced apples. *Saudi Journal of Biological Sciences*, 2023. **30**(9): p. 103744.
118. Pilon, L., et al., Chitosan nanoparticle coatings reduce microbial growth on fresh-cut apples while not affecting quality attributes. *International Journal of Food Science & Technology*, 2015. **50**(2): p. 440–448.
119. Liu, Z., et al., Chitosan films incorporating litchi peel extract and titanium dioxide nanoparticles and their application as coatings on watercored apples. *Progress in Organic Coatings*, 2021. **151**: p. 106103.
120. Lakshmayya, N., et al., Essential oils-based nano-emulsion system for food safety and preservation: Current status and future prospects. *Biocatalysis and Agricultural Biotechnology*, 2023: p. 102897.
121. Faheem, F., et al., Uncovering the industrial potentials of lemongrass essential oil as a food preservative: A review. *Antioxidants*, 2022. **11**(4): p. 720.
122. Arserim-Uçar, D.K. and B. Çabuk, Emerging antibacterial and antifungal applications of nanomaterials on food products, in *Nanotoxicity*. 2020, Elsevier. p. 415–453.
123. Zhang, B.Y., et al., Assessment of carbon footprint of nano-packaging considering potential food waste reduction due to shelf life extension. *Resources, Conservation and Recycling*, 2019. **149**: p. 322–331.
124. Gulati, S., Kumar, S., Goyal, K., Singh, A. (2021). Plant-Based Nanomaterials: Novel and Highly Effectual Preservatives for Food. In: *Handbook of Consumer Nanoproducts*. Springer, Singapore. https://doi.org/10.1007/978-981-15-6453-6_95-1
125. Shikha, P., Arvind, K. and Gupta, A. (2021). Technological Advancement in Food Additives and Preservatives. In *Food Chemistry*, M. Sen (Ed.). https://doi.org/10.1002/9781119792130.ch12
126. Gulati, S., et al., Plant-based nanomaterials: Novel and highly effectual preservatives for food, in *Handbook of consumer nanoproducts*. 2022, Springer. p. 797–823.
127. Foujdar, R., et al., Role of nanomaterials for food processing and preservation, in *Nanotechnology horizons in food process engineering*. 2023, Apple Academic Press. p. 47–83.
128. Afzal, M., A. Redha, and R. AlHasan, Anthocyanins potentially contribute to defense against Alzheimer's disease. *Molecules*, 2019. **24**(23): p. 4255.
129. Raymundo-Pereira, P.A., et al., Polyphenol oxidase-based electrochemical biosensors: A review. *Analytica Chimica Acta*, 2020. **1139**: p. 198–221.
130. Xu, L.Q., K.-G. Neoh, and E.-T. Kang, Natural polyphenols as versatile platforms for material engineering and surface functionalization. *Progress in Polymer Science*, 2018. **87**: p. 165–196.
131. Ali, S., et al., The avenue of fruit wastes to worth for synthesis of silver and gold nanoparticles and their antimicrobial application against foodborne pathogens: A review. *Food Chemistry*, 2021. **359**: p. 129912.
132. Jan, N.A., et al., Nano-fiber enabled regulated release of hexanal vapor and its impact on shelf life of mango fruits. *Polymer Bulletin*, 2023. **80**(1): p. 865–881.
133. Huang, L., et al., Nanocellulose-based polymeric nanozyme as bioinspired spray coating for fruit preservation. *Food Hydrocolloids*, 2023. **135**: p. 108138.
134. Kumar, T.S.M., et al., A comprehensive review of electrospun nanofibers: Food and packaging perspective. *Composites Part B: Engineering*, 2019. **175**: p. 107074.
135. Xiao, J., et al., Development and characterization of an edible chitosan/zein-cinnamaldehyde nano-cellulose composite film and its effects on mango quality during storage. *Lwt*, 2021. **140**: p. 110809.
136. Cao, Z., et al., The role of essential oils in maintaining the postharvest quality and preservation of peach and other fruits. *Journal of Food Biochemistry*, 2022. **46**(12): p. e14513.

137. Martins, R., et al., Food wastes and microalgae as sources of bioactive compounds and pigments in a modern biorefinery: A review. *Antioxidants*, 2023. **12**(2): p. 328.
138. Abou El-Nasr, M.K., et al., Using zinc oxide nanoparticles to improve the color and berry quality of table grapes cv. Crimson seedless. *Plants*, 2021. **10**(7): p. 1285.
139. Kumar, U.J., et al., Effect of different concentrations of iron oxide and zinc oxide nanoparticles on growth and yield of strawberry (Fragaria x ananassa Duch) cv. Chandler. *International Journal of Current Microbiology and Applied Sciences*, 2017. **6**(8): p. 2440–2445.
140. Gao, Y., et al., Green synthesis of zinc oxide nanoparticles using Citrus sinensis peel extract and application to strawberry preservation: A comparison study. *Lwt*, 2020. **126**: p. 109297.
141. La, D.D., et al., Effects of antibacterial ZnO nanoparticles on the performance of a chitosan/gum arabic edible coating for post-harvest banana preservation. *Progress in Organic Coatings*, 2021. **151**: p. 106057.
142. Cisneros, L., F. Gao, and A. Corma, Silver nanocluster in zeolites. Adsorption of ethylene traces for fruit preservation. *Microporous and Mesoporous Materials*, 2019. **283**: p. 25–30.
143. Rodríguez-González, V., C. Terashima, and A. Fujishima, Applications of photocatalytic titanium dioxide-based nanomaterials in sustainable agriculture. *Journal of Photochemistry and Photobiology C: Photochemistry Reviews*, 2019. **40**: p. 49–67.
144. Mahajan, S., et al., Application of silver nanoparticles in in-vitro plant growth and metabolite production: Revisiting its scope and feasibility. *Plant Cell, Tissue and Organ Culture (PCTOC)*, 2022. **150**(1): p. 15–39.
145. Srivastava, S., et al., Dynamic bioactive properties of nutritional superfood *Moringa oleifera*: A comprehensive review. *Journal of Agriculture and Food Research*, 2023: p. 100860.
146. Tiwari, P., Nanotechnologies and sustainable agriculture for food and nutraceutical production: An update, in *Plant and nanoparticles*. 2022, Springer. p. 315–337.
147. Quiñonero Muñoz, F.J., et al., Vegetal-derived bioactive compounds as multidrug resistance modulators in colorectal cancer. *Applied Sciences*, 2023. **13**(4): p. 2667.
148. Ribeiro, A., I. Zheng, and M. Alves, Synthesis of metal-oxide nanoparticles from vegetables and their waste materials for diverse applications, in *Nanomaterials from agricultural and horticultural products*. 2023, Springer. p. 31–48.
149. Thiruvengadam, M., G. Rajakumar, and I.-M. Chung, Nanotechnology: Current uses and future applications in the food industry. *3 Biotech*, 2018. **8**: p. 1–13.
150. Chaudhary, P., F. Fatima, and A. Kumar, Relevance of nanomaterials in food packaging and its advanced future prospects. *Journal of Inorganic and Organometallic Polymers and Materials*, 2020. **30**: p. 5180–5192.

3 Nanocomposite Films and Coatings for Food Packaging

Alisha Kakkar and Sudesh Kumar[a]
[a]Department of Chemistry, Banasthali University, Banasthali Vidyapith, Rajasthan, India

3.1 INTRODUCTION

Customer needs for different kinds of food over the years, and a priority for easiness, have stimulated remarkable advancements in food preservation, ensuring the accessibility of non-harmful and nutritious food. The main purpose of food packaging is to segregate food products from the surroundings, reducing or protecting contact with deterioration elements like the impacts of microbes, air, heat, and water to prevent the degradation of nutrients and quality, hence increasing the life cycle. However, food preserving and wrapping serve several other vital purposes, like giving handiness, engaging with customers, and promoting the wrapped item. Artificial plastic materials derived from fossils have historically controlled the food wrapping sector, making up 37% of the overall marketplace for wrapping products. This is primarily because of their affordable price, convenient transportability, reduced mass, high tensile strength, stiffness, effective shielding capabilities, capability for thermal sealing, and flexibility in shaping. Nevertheless, these artificial polymers impose detrimental impacts on the ecosystem (Cox, Davies, Juanes, & Dudas, 2019). Ensuring nutritious and secure food devoid of artificial substances has emerged as a pivotal obstacle for food producers and food enterprises. In recent years, rising public apprehension regarding ecological balance and human well-being has heightened researchers' attention toward the utilization of bio-derived wrapping materials as a substitute for artificial polymers in wrapping food items (Rodríguez-Rojas, Rodríguez-Vélez, & Arana-Florez, 2019). Decomposable substitutes encompass carbohydrates like cellulose, chitosan, starch, and agar including casein, offering viable alternatives to synthetic polymers. Eco-friendly plastics break down under suitable humidity and heat levels and the presence of air without generating any harmful by-products. Certain eco-friendly plastics decompose within mere weeks, while the decomposition of artificial plastics spans many years, contingent upon the kind and source of the plastic. Bacteriostatic wrapping has garnered significant interest from the food sector due to the rise in customer preference for natural, secure food that is devoid of artificial agents. Bacteriostatic wrapping encompasses methods like gas-flushed packaging or wrapping supplies incorporating antibacterial substances like organic antibacterials, nanostructures, etc., which are employed to manage the proliferation and diminish the activity of microorganisms. The fundamental objectives of an antibacterial wrapping system encompass ensuring nontoxic food, preserving the quality of food, and extending the life cycle. Recently, diverse endeavors have been undertaken to create natural polymer-based antibacterial wrapping items to enhance the durability and excellence of food. Within natural plastic materials, chitosan has been gaining traction as a potential substance for antibacterial purposes in the food sector due to its decomposable, less harmful, and bio-friendly properties, along with its affordability and plentiful natural presence. Mixed membranes comprising chitosan and other organic polymers like proteins or sugar polymers have been created through methods like extrusion, coating of a solution, sequential layering, and various other processes. These films have been analyzed for their physical, chemical, operational, and antibacterial characteristics, evaluating their suitability as materials for food wrapping purposes. Chitosan compounds incorporating various organic free radical scavengers, antibacterial elements, and nanostructures have garnered considerable attention recently as shown in Figure 3.1.

Recently, nanocomposite materials based on organic polymers have garnered significant attention due to their exceptional structural, chemical, and decomposable characteristics (Cesur, Köroğlu, & Yalçın, 2018). Usually, carbohydrate-clay nanomaterials consist of minuscule mineral clay particles, as additives distributed within a polysaccharide framework. When employed as wrapping components or combined with antibacterial agents, clay-polysaccharide nanocomposite materials can be transformed into dynamic wrapping solutions, demonstrating antibacterial traits for various purposes in the security of the food (Sharma, Jafari, & Sharma, 2020). A majority of the uses of chitosan rely on its ionically charged behavior and complexation capability, which stem from the plentiful NH_2 groups present in its structure. Broadly, chitosan does not dissolve in water-based or organic solutions; it can be exclusively dissolved in a weak acidic solution. Coverings and membranes crafted solely from chitosan encounter drawbacks such as fragility, substantial humidity, and air diffusion, as well as limited physiologically active capabilities (Mujtaba et al., 2019). Integrating nanoscale clays represents a potential technique to enhance the physical properties of chitosan, endowing it with enhanced attributes for real-world purposes. Nanoscale clays are part of a

DOI: 10.1201/9781032677422-3

category of mineral compounds, either organic or artificial, that can disperse within the chitosan framework, forming three distinct shapes (infusion, tactoid, and layer removal) as shown in Figure 3.2 (El Assimi et al., 2020). In the tactoid state, nanoscale clays are merely deemed traditional additives, as their sheets remain unseparated, and the entire nanoclay particles are distributed within the chitosan framework. Optimal clay–chitosan composite materials exhibit exfoliation or infused configurations. Infused configurations take place when 1.0 to 3.0% of layers of chitosan are introduced within the clay sheets. Even as the gaps between layers widen, the plates of clays maintain their parallel arrangement within a meticulously organized multilayer framework. In the configuration of exfoliation, clay and chitosan undergo blending through diverse techniques (such as stirring, sonication, blending, etc.), causing the separate palates to disperse within the chitosan framework, thus losing their parallel alignment (Wang et al., 2021).

3.2 CHITOSAN AND ITS CHARACTERISTICS

3.2.1 Origin and Framework

Chitin ranks as the second most prevalent carbohydrate in the world, following cellulose. It comprises three types of chemically responsive groups: an NH_2 group at position C-2, and both 1° and 2° alcohol groups at positions C-3 and C-6, respectively. Chitin is primarily found in sea animals and insects, where it serves as a significant component of the outer shell. It is also present in certain mycota, forming a part of their plasma membranes. Sea animals like crabs, lobsters, shrimps, and oysters are annually eaten as they are rich in protein. However, the outer shells and other non-consumable portions of these marine animals make up approximately 50% of their total weight and are commonly disposed of as debris. These thrown remnants serve as a significant origin of chitin. Chitosan is a complex carbohydrate composed of acetylglucosamine and glucosamine molecules, primarily derived from the partial deacetylation of chitin. Chitosan is commercially produced by incomplete deacetyl modification of chitin, resulting in the generation of an acetylglucosamine-glucosamine polymer. Chitosan is a dissolvable derivative of chitin and has been employed in diverse applications, especially in food conservation and wrapping.

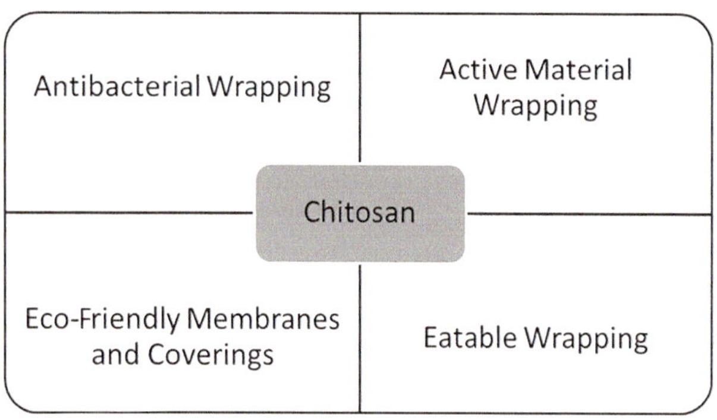

FIGURE 3.1 Different uses of chitosan in packaging of food items.

3.2.2 Deacetylation Level and Dispersibility

The extent of removal of an acetyl group stands as a crucial chemical trait of chitosan. Throughout the deacetylation process, the acetyl moiety is substituted by an amino moiety derived from the polymer, leading to the creation of a polyblend comprising glucosamine and acetyl glucosamine. Polymers with over 50% glucosamine molecules are commonly referred to as chitosan. The extent of removal of an acetyl group in a chitosan molecule is determined by the proportion of no. of glucosamine units to the total of acetyl glucosamine and glucosamine molecules within the structure. Therefore, if chitosan has a degree of deacetylation of 70%, it signifies that 70% of its composition consists of D-glucosamine molecules and 30% comprises N-acetyl-glucosamine molecules within the chitosan framework.

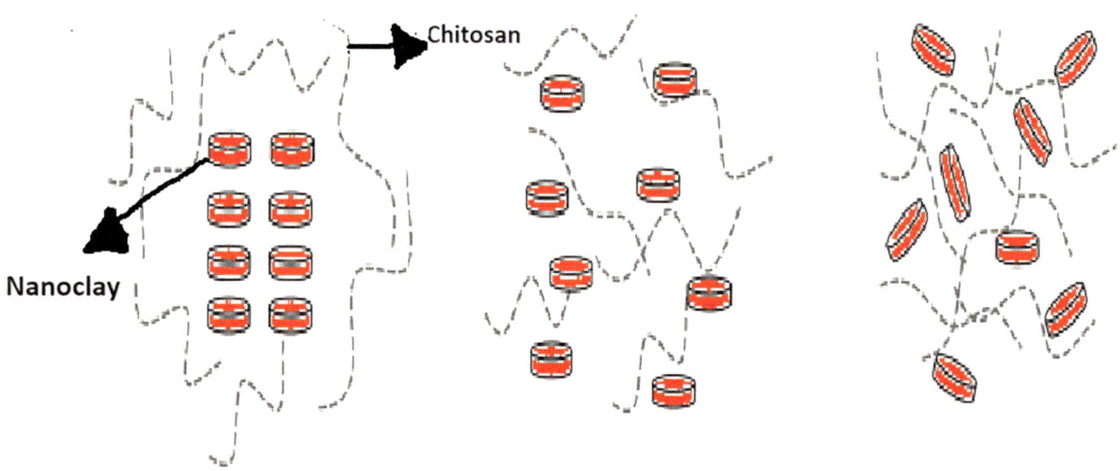

FIGURE 3.2 Illustrative overview of chitosan–clay nanocomposite configurations: (A) tactoid, (B) infusion, and (C) exfoliation.

3.2.3 Antibacterial Performance

Chitosan represents a naturally occurring polymer material renowned for its effective antibacterial properties against a diverse range of microbes, including positive and negative bacterial organisms, mycelium, and fungus. Numerous scientists have validated chitosan's antibacterial characteristics. However, the precise working of microbes remains unclear. Many hypotheses explaining antibacterial performances have been suggested, but the most widely used are as follows: (i) the engagement of bacterial cell wall, carrying a negative charge, with the cationic NH_2 groups in chitosan, resulting in a modification of the cellular barrier characteristics. This alteration causes the release of cellular components, eventually culminating in cellular demise (Hosseinnejad & Jafari, 2016). (ii) The alternative suggested process mechanism relies on the binding attributes of the chitosan polymer. Chitosan specifically interacts with metallic elements, impeding different biological catalysts within bacterial membranes by obstructing their reactive sites and diminishing bacterial proliferation. (iii) The molecular mass of chitosan can additionally impact its antibacterial effectiveness. Chitosan with a large mass can create a non-porous polymer coating on the exterior of the bacterial membrane, altering the cell porosity and eventually hindering the intake of minerals intracellularly, while less mass of polymer might infiltrate the cytosol and attach to the cellular genetic code. This results in disruption of messenger ribonucleic acid and protein synthesis, culminating in cell demise. In circumstances with low pH levels, the polyvalent cationic characteristic of chitosan stems from the NH_2 moiety, which carries a positive charge. These NH_2 groups subsequently engage with anionic elements on the membranes of bacterial cells, causing cell fatality. This is the widely acknowledged mechanism.

3.2.4 Radical-Quenching Performance

Food items are regularly subjected to various forms of air decay. The antioxidant activity includes their capacity to scavenge reactive oxygen species and binding of ionic metals providing radical-scavenging properties to organic polymers (Anraku et al., 2018). The radical-quenching properties of the polymer can be enhanced by incorporating antioxidants obtained from phytochemicals. The plant chemical constituents in green tea (for instance, theaflavins compounds, flavonols, and condensed tannins) exhibit commendable radical-quenching effects. Siripatrawan and co-authors documented that membranes made of chitosan (2% chitosan solution with 95% deacetylation in 1% ethanoic acid) infused with the essence of green tea demonstrated enhanced operational characteristics (such as stretching resistance and extension before breaking) in the hybrid film in contrast to mixtures involving fats or proteins.

3.3 Chitosan and Its Combination with Other Natural Polymers

Many organic polymeric compounds have been thoroughly investigated for crafting eco-friendly, ingestible, and antibacterial membranes in the realm of food wrapping. These natural polymers encompass cellulose, starch, their modified forms, proteins, and fats. To utilize chitosan in the configuration of membranes and coverings, and to enhance its effectiveness as a material for wrapping food items (Zhu, Wu, & Sun, 2019), it needs to be amalgamated with diverse natural polymers, including carbohydrates, fats, and proteins. Broadly, amalgamations of complex carbohydrates offer numerous benefits (K. Li, Guan, & Wu, 2019).

3.3.1 Chitosan Combinations with Polysaccharides

Chitosan is integrated with diverse carbohydrates including starch, alginic acid salts, Irish moss, fruit fiber, and bacterial nanocellulose. Both alginic acid salts and chitosan are carbohydrates sourced from marine animals, harmonious with one another, and their amalgamation is more fitting for food wrapping purposes compared to the amalgamations of chitosan polymer with fats or protein. Carrageenan is a negatively charged straight-chain carbohydrate derived from red algae. This sugar molecule has previously been utilized as coverings on vegetables and fruits to reduce air decay and prevent water content. Similar to alginic acid salts, carrageenan also binds with chitosan polymer to create an electrostatic polymer blend. Most investigations concerning this chitosan combination have been centered on enclosure purposes in the food and medical sectors. A handful of examinations regarding films created from the carrageenan-chitosan amalgamation indicated inadequate flexibility and substandard water control characteristics of the composite membranes (Shahbazi, Ettelaie, & Rafe, 2016).

3.3.2 Chitosan Complexes with Proteins

Films created from a mixture of chitosan and protein compounds like fibrous protein, sodium caseinate, and gel have been thoroughly researched. Both chitosan and gel are water-loving polymers that exhibit strong attraction and harmonious integration. Their combinations, namely both chitosan-gelatin mixtures and chitosan-gelatin double-layer films, have been scrutinized. Both methods of mixing yield membranes exhibit enhanced prevention characteristics against water penetration and result in a substantial increase in extension before breaking. Regarding the combination with polymer chitosan, gel polymer sourced from marine animals has been predominantly employed. Coating of a solution using chitosan solutions prepared in water and gelatin at an acidity of pH 4 and temperature 60 °C represents a straightforward approach for fabricating the amalgamated film. Similar to the gel-chitosan amalgamation, a blend of the chitosan-collagen membrane can be generated through the coating of a solution technique, employing a

1:1 combination of the two organic polymers. The amalgamation transpires at a microscopic scale transforming the framework of collagen through hydrogen interactions to seamlessly integrate with the polymer chitosan.

3.4 POLYMER ANTIBACTERIAL COMPOUNDS MADE FROM CHITOSAN AND ITS AMALGAMATION

Chitosan inherently exhibits potent antibacterial properties against numerous microbes. Nonetheless, particularly in the context of food protection and wrapping purposes, subsequent enhancement in chitosan's antibacterial attributes is necessary. In the pursuit of amplifying its antibacterial features, organic antibacterial agents and nano-sized antimicrobial agents (including metallic nanostructures) have been applied (Matharu, Ciric, & Edirisinghe, 2018).

3.4.1 Naturally Occurring Antibacterial Agents

These antibacterial agents are primarily derived from plants, animals, and microorganisms. Naturally occurring antibacterial agents are biomaterial-friendly and decomposable, with low detrimental effects. Their antibacterial efficacy is equivalent to their artificial alternatives, showcasing antibacterial properties against a diverse spectrum of microbes, encompassing food-spoiling microbes. Naturally occurring antibacterial agents are utilized for three primary objectives: to regulate inherent decay mechanism (i.e., food protection), to sustain or enhance nutrient content (i.e., food standards), and to suppress/halt the proliferation of microbes. The frequently utilized organic antibacterial agents available in the commercial market comprise natural oils sourced from plants, antibacterials sourced from microbial origins (such as bacterial toxins), and those obtained from animals (like egg lysozyme). Indeed, organic antibacterial agents can aid in preserving looks, flavor, excellence, and prolonged life cycles by conserving the nourishment composition. Extracts from plants like natural oils exhibit outstanding radical-quenching properties, coupled with antibacterial characteristics, employed in the production of antibacterial compounds for food protection and wrapping purposes.

3.4.2 Antibacterial Agents Utilizing Metal and Its Oxides

Metal-based nanostructures, including copper and silver, have been firmly recognized as wide-ranging antimicrobial agents, owing to their exceptional efficacy against positive and negative bacterial organisms, yeast, pathogens, and microbes (Hoseinnejad, Jafari, & Katouzian, 2018). Nano-sized antibacterial agents possess the ability to electrically adhere to microorganism exteriors, altering the microbial cell membrane capability. This alteration results in harm to the cell membrane, subsequently leading to compromised breathing, disrupted transport equilibrium, disrupted energy transference, and breakdown of the cell. Ultimately, this sequence of events culminates in cell demise. An alternative approach for antibacterial agents involves generating oxidative radicals through nanostructures. Depending on the physical and chemical attributes of these nanostructures and the specific bacterial strains, certain nanoparticles may trigger swelling. This swelling can potentially transform less harmful oxidizing agents, hydrogen peroxide and oxygen free radicals, into reactive species like hydroxy radicals. Moreover, when exposed to light as an ecological element, numerous semiconductors can produce reactive species through the interaction between activated electron-hole duos and aqueous molecules. This process is harmful to microbes. Reactive species can induce significant oxidative damage, resulting in genetic material impairment, protein modification, fatty acid oxidation, and biological enzyme suppression, ultimately causing the cellular demise of microbes. Additional impacts of nano-sized antibacterial agents encompass the immediate hindrance of vital bacterial catalysts, the creation of nitrogen radicals, and the initiation of apoptosis have been observed. The dimensions, configuration, surface properties, and surface compositions of nanomaterials primarily impact the production of oxygen radicals and antibacterial efficacy. Nanoparticles of smaller dimensions can readily infiltrate cellular walls and additional microbial barricades. The quantity of absorption diminishes with the escalation in particle dimension. Moreover, the liberation of metallic ions from nanostructures like copper, silver, and zinc may interact with oxygen, engendering oxygen radicals, which subsequently impose antibacterial impacts.

3.4.3 Synthesis Methods for Nanostructured Coverings/Films

3.4.3.1 Solvent Casting

The solvent casting technique stands as one of the most extensively employed approaches for fabricating chitosan-derived coverings and membranes due to its ease of use. Solvent mixing constitutes a procedure wherein chitosan and its polymer blend are solvated in a solvent, subsequently succeeded by vaporization. It is a comparatively cost-effective and straightforward approach, wherein intermolecular ionic and hydrogen interactions lead to alterations in the polymeric arrangement (Muxika, Guerrero, & de la Caba, 2017). Nevertheless, these molecular associations also render the membrane fragile. Various softeners, including polyhydric alcohols such as glycerin, dextrose, saccharose, glucitol, polyethylene alcohol, and fats, are incorporated to enhance the tensile strength and stiffness of the membranes. While coating a solvent method remains straightforward and relatively inexpensive, upscaling the coating procedure using this approach could present hurdles and require additional refinement and advancement.

3.4.3.2 Sequential Layering

This sequential layering technique is a flexible method that has been thoroughly investigated in the creation of nanostructured membranes, enabling precise control over physical characteristics and capabilities. This approach enables the amalgamation of operational attributes from various polymeric materials. It is a technique employed for the creation of multifaceted membranes that don't require any advanced apparatus. In this method, surface alteration predominantly relies on the settlement and reciprocal allure of interchanging ionic polymers with contrasting overall charges on a solid surface (Tu et al., 2019). The ionic polymers might include natural polymer materials like amino acids, carbohydrates, artificial polymers, and others, which carry an overall charge. Depositing can be accomplished either by immersing the base in interchanging ionic polymer solutions or by applying solutions onto the support through spraying.

3.4.3.3 Extrusion

This pressing method is primarily employed for fabricating traditional industrial polymer wrapping membranes. This method is frequently favored over the solvent coating technique because of its quicker manufacturing duration and reduced energy usage (Estevez-Areco, Famá, Candal, & Goyanes, 2019). Several stages are engaged in the pressing method. Forming compositions involves employing diverse components and combining them. This blend is mixed in a molder to granulate all essential components necessary for membrane processing. The extruded material is then later sliced into granules using a granulator, following which the pellets are dehydrated in an oven. Afterward, the granules are pressed out into layers with the other processor, and ultimately, the blended polymers are blown into a membrane using a film-blowing machine. This pressing method frequently imparts membranes with satisfactory tensile strength and robust thermal resilience. While this method shows potential for membrane production, there exists a restricted amount of research concerning the utilization of this technique for chitosan membrane production.

3.4.3.4 Coverings/Atomizing

Frequently, a covering is administered to the exterior of food items like vegetables and fruits, seafood, etc. to prolong their storage duration. Coverings can be executed by disseminating, misting, submerging, or soaking food substances in chitosan-infused blends. Dissemination coating is conducted with the aid of specific devices, such as a brush. The covering is an efficient method to control the proliferation of microbes, which contributes to prolonging the life cycle, and thus aids in preserving the food standards. The coating procedure encompasses various stages: composing raw components by employing suitable ratios of additives and chitosan polymer including radical-quenching agents, antibacterial agents, fortifying agents, etc.; preparing samples for covering with processes like blending, warming, radiation, and steam flash processing; sanitizing food items with a bleach solution; disseminating chitosan-infused blends over food items to create a consistent covering by an antiseptic applicator; dehydrating in specific environments like uniform ventilation flow; and wrapping and preserving in normal or other suitable environments (vacuum-sealed, frozen, moisture-regulated, or other environment settings).

3.5 FOOD WRAPPING USES OF COVERINGS AND MEMBRANES

3.5.1 Prolonging the Freshness of Vegetables and Fruits

Vegetables and fruits demonstrate ongoing breathing and evaporation even after being cropped, leading to a brief life cycle and deterioration in quality. As a result, numerous scientists have explored organic, non-harmful compounds that can delay the development process or decrease the spoilage of vegetables and fruits post-gathering. One particular research, as detailed by researchers, explored the implementation of a chitosan-infused edible covering comprising wild thyme essential oil. This application notably prolonged the life cycle of strawberries for a span of 15 days under refrigeration. Covering of biodegradable edible membranes presents an economical, uncomplicated, and efficient technique for preventing water content, thereby decreasing decay and breathing speed (Harish Prashanth & Tharanathan, 2007). Specifically, chitosan is becoming increasingly significant, particularly in the context of food wrapping for the formation of membranes and coverings because of its natural antibacterial properties, coating capability, and eco-friendliness.

3.6 APPLICATIONS OF CHITOSAN–CLAY COMPOSITES IN FOOD WRAPPING

Utilizing chitosan–clay composites in food packaging offers an economical and environmentally friendly solution. Moreover, decomposability can be preserved as both clay and chitosan are organic substances, polymers, and naturally occurring mineral compounds. Particularly when these composites spoil, no harmful or artificial substances are released into the surroundings (Laaraibi et al., 2018). Hence, these eco-friendly and degradable substances demonstrate practical physical and operational characteristics, offering a viable alternative to conventional fossil fuel-derived polymers. Prior studies have explored its uses in medical, biomaterial engineering, electrochemistry, and water purification. Instances of these characteristic improvements primarily encompass enhanced physical properties, superior prevention characteristics, and antibacterial and antioxidant features. Table 3.1 illustrates diverse chitosan–clay frameworks and their viable physical and chemical attributes.

TABLE 3.1
Chitosan–Clay Composites for Application in the Food Sector

Framework	Variable/Blended Component	Efficiency	Reference
Chitosan-montmorillonite	Enzyme fixation Incorporation of flexibilizers Incorporation of radical-quenchers/antibacterial nanostructures	Improved antibacterial effectiveness Increased water repellency, air-blocking capabilities, mechanical strength, thermal resistance, and chemical characteristics Decreased moisture penetrability and enhanced antioxidant properties of membranes Preserved food surface texture while inhibiting loss of weight	(Benucci et al., 2018; Cacciotti, Lombardelli, Benucci, & Esti, 2019)
Chitosan-graphene oxide nanoparticles	Zinc ion engraved polymer/ introduction of antibacterials	Improved antibacterial effectiveness Increased water repellency, air-blocking capabilities, mechanical strength, thermal resistance, and chemical characteristics Decreased moisture penetrability Reduced UV transmission Extended shelf life of food Selective and precise detection of zinc in food items	(Ahmed, Mulla, & Arfat, 2017; Barra et al., 2019)
Chitosan-laponite	Introduction of antibacterials	Increased water repellency, air-blocking capabilities, mechanical strength, and thermal resistance Extended shelf life of food Restricted discharge of silver particles	(Wu, Huang, Li, Xiao, & Wang, 2018)
Chitosan-vermiculite	Nanocomposite-treated polyethylene terephthalate	Augmented oxygen and helium blocking	(Essabti, Guinault, Roland, Régnier, Ettaqi, & Gervais, 2018)

3.7 CONCLUSION

Various organic polymeric compounds have been recently investigated for their appropriateness in food wrapping uses. Chitosan is among the extensively researched polymers known to enhance the life cycle of food items such as vegetables, fruits, and seafood. Introducing nanostructures and plant compounds in chitosan enhances its antibacterial, physical, and prevention characteristics in the formed membranes and coverings, resulting in enhancements in food standards and extended life cycles. Various nanostructures like zinc oxide, silver, and copper have been widely employed as primary options for producing chitosan blends. Additional polymers like gel, alginic acid salts, carrageenan, and so on are utilized alongside chitosan to form blended coverings and membranes. Chitosan frameworks function as conveyors of reactive substances like nanostructures, natural oils, fruit essence, and other plant-based compounds. Implementing chitosan as a natural polymer in food wrapping extends the life cycle of diverse food items and farm produce, while concurrently diminishing the dependence on artificial plastics and supplements, thus advancing nutritious food and fostering wholesome surroundings. Integrating nanoscale clays represents a potential technique to enhance the physical properties of chitosan, endowing it with enhanced attributes for real-world purposes. Subsequent studies could focus on extensive production and industrialization of chitosan-infused food wrapping solutions, aiming to alleviate the impact of artificial plastics on our communities and ecosystems.

REFERENCES

Ahmed, J., Mulla, M., & Arfat, Y. A. (2017). Mechanical, thermal, structural, and barrier properties of crab shell chitosan/graphene oxide composite films. *Food Hydrocolloids*, 71, 141–148.

Anraku, M., Gebicki, J. M., Iohara, D., Tomida, H., Uekama, K., Maruyama, T., Hirayama, F., & Otagiri, M. (2018). Antioxidant activities of chitosans and its derivatives in vitro and in vivo studies. *Carbohydrate Polymers*, 199, 141–149.

Barra, A., Ferreira, N. M., Martins, M. A., Lazar, O., Pantazi, A., Jderu, A. A., & Ferreira, P. (2019). Eco-friendly preparation of electrically conductive chitosan-reduced graphene oxide flexible bionanocomposites for food packaging and biological applications. *Composites Science and Technology*, 173, 53–60.

Benucci, I., Liburdi, K., Cacciotti, I., Lombardelli, C., Zappino, M., Nanni, F., & Esti, M. (2018). Chitosan/clay nanocomposite films as supports for enzyme immobilization: An innovative green approach for winemaking applications. *Food Hydrocolloids*, 74, 124–131.

Cacciotti, I., Lombardelli, C., Benucci, I., & Esti, M. (2019). Clay/chitosan biocomposite systems as novel green carriers for covalent immobilization of food enzymes. *Journal of Materials Research and Technology*, 8(4), 3644–3652.

Cesur, S., Köroğlu, C., & Yalçın, H. T. (2018). Antimicrobial and biodegradable food packaging applications of polycaprolactone/organic nano clay/chitosan polymeric composite films. *Journal of Vinyl and Additive Technology*, 24(4), 376–387.

Cox, K. D., Covernton, G. A., Davies, H. L., Dower, J. F., Juanes, F., & Dudas, S. E. (2019). Human consumption of microplastics. *Environmental Science & Technology*, 53(12), 7068–7074.

El Assimi, T., Lakbita, O., El Meziane, A., Khouloud, M., Dahchour, A., Beniazza, R., & Lahcini, M. (2020). Sustainable coating material based on chitosan-clay composite and paraffin wax for slow-release DAP fertilizer. *International Journal of Biological Macromolecules*, 161, 492–502.

Essabti, F., Guinault, A., Roland, S., Régnier, G., Ettaqi, S., & Gervais, M. (2018). Preparation and characterization of poly (ethylene terephthalate) films coated by chitosan and vermiculite nanoclay. *Carbohydrate Polymers*, 201, 392–401.

Estevez-Areco, S., Guz, L., Famá, L., Candal, R., & Goyanes, S. (2019). Bioactive starch nanocomposite films with antioxidant activity and enhanced mechanical properties obtained by extrusion followed by thermo-compression. *Food Hydrocolloids*, 96, 518–528.

Harish Prashanth, K. V., & Tharanathan, R. N. (2007). Chitin/chitosan: Modifications and their unlimited application potential—an overview. *Trends in Food Science & Technology*, 18(3), 117–131.

Hosseinnejad, M., & Jafari, S. M. (2016). Evaluation of different factors affecting antimicrobial properties of chitosan. *International Journal of Biological Macromolecules*, 85, 467–475.

Hoseinnejad, M., Jafari, S. M., & Katouzian, I. (2018). Inorganic and metal nanoparticles and their antimicrobial activity in food packaging applications. *Critical reviews in microbiology* 44, 161–181. Taylor & Francis.

Laaraibi, A., Moughaoui, F., Damiri, F., Ouakit, A., Charhouf, I., Hamdouch, S., & Bennamara, A. (2018). Chitosan-clay based (CS-NaBNT) biodegradable nanocomposite films for potential utility in food and environment. In *Chitin-ChitosanMyriad Functionalities in Science and Technology* (pp. 45–69). London: IntechOpen.

Li, K., Zhu, J., Guan, G., & Wu, H. (2019). Preparation of chitosan-sodium alginate films through layer-by-layer assembly and ferulic acid crosslinking: Film properties, characterization, and formation mechanism. *International Journal of Biological Macromolecules*, 122, 485–492.

Matharu, R. K., Ciric, L., & Edirisinghe, M. (2018). Nanocomposites: Suitable alternatives as antimicrobial agents. *Nanotechnology* 29, 282001.

Mujtaba, M., Morsi, R. E., Kerch, G., Elsabee, M. Z., Kaya, M., Labidi, J., & Khawar, K. M. (2019). Current advancements in chitosan-based film production for food technology; A review. *International Journal of Biological Macromolecules*, 121, 889–904.

Muxika, A., Etxabide, A., Uranga, J., Guerrero, P., & de la Caba, K. (2017). Chitosan as a bioactive polymer: Processing, properties and applications. *International Journal of Biological Macromolecules*, 105, 1358–1368.

Rodríguez-Rojas, A., Arango Ospina, A., Rodríguez-Vélez, P., & Arana-Florez, R. (2019). What is the new about food packaging material? A bibliometric review during 1996–2016. *Trends in Food Science & Technology*, 85, 252–261.

Shahbazi, M., Rajabzadeh, G., Ettelaie, R., & Rafe, A. (2016). Kinetic study of κ-carrageenan degradation and its impact on mechanical and structural properties of chitosan/κ carrageenan film. *Carbohydrate Polymers*, 142, 167–176.

Sharma, R., Jafari, S. M., & Sharma, S. (2020). Antimicrobial bio-nanocomposites and their potential applications in food packaging. *Food Control*, 112, Article 107086.

Tu, H., Wu, G., Yi, Y., Huang, M., Liu, R., Shi, X., & Deng, H. (2019). Layer-by-layer immobilization of amphoteric carboxymethyl chitosan onto biocompatible silk fibroin nanofibrous mats. *Carbohydrate Polymers*, 210, 9–16.

Wang, Y., Yi, S., Lu, R., Sameen, D. E., Ahmed, S., Dai, J., & Liu, Y. (2021). Preparation, characterization, and 3D printing verification of chitosan/halloysite nanotubes/tea polyphenol nanocomposite films. *International Journal of Biological Macromolecules*, 166, 32–44.

Wu, Z., Huang, X., Li, Y.-C., Xiao, H., & Wang, X. (2018). Novel chitosan films with laponite immobilized Ag nanoparticles for active food packaging. *Carbohydrate Polymers*, 199, 210–218.

Zhu, J., Wu, H., & Sun, Q. (2019). Preparation of crosslinked active bilayer film based on chitosan and alginate for regulating the ascorbate-glutathione cycle of postharvest cherry tomato (*Lycopersicon esculentum*). *International Journal of Biological Macromolecules*, 130, 584–594.

4 Nanotechnology in Seafood Preservation during Cold Storage

Xiaoyu Luo[a]
[a]Food Science and Technology Program, Beijing Normal–Hong Kong Baptist University, Zhuhai, Guangdong, China

4.1 INTRODUCTION

With the continuous population growth in the world, the global consumption of seafood products has increased dramatically in the past few decades, climbing from 28 million tonnes in 1961 to 158 million tonnes in 2019, as shown in Figure 4.1 (FAO 2023). Such an increasing trend may be attributed to economic growth, change of lifestyle, elevated nutrition demands, etc. As shown in Figure 4.1(B), Asian countries are among the largest markets for aquatic food consumption, due to the large population density, the expansion of fishery industry, and the plentiful seawater resources. In comparison, the landlocked countries showed less seafood consumption, potentially because of a lack of marine water resources and cold chains for seafood preservation.

Seafood products mainly include the meat flesh from fish, shellfish, and aquatic mammal species, as well as their derivatives such as caviar and other animal by-products. In general, these products contribute to approximately 17% of edible meat in the global market, and it is expected to rise by 12–25% by the year of 2050 (Costello et al. 2020). Marine fish species vary with compositions, while they can be classified briefly as white and oily fish. Mainly found close to the bottom of the sea, the white fish contain more lean meats and exhibit colorless fish flesh, such as sole, cod, seabass, sea bream, red snapper, and so on. In comparison, the oily fish possess higher fat content that can be up to 20%, with representative examples of tuna, salmon, eel, and mackerel (Prabhakar et al. 2020). Shellfish can be divided into three main subcategories, namely crustaceans, molluscs, and echinoderms. Crustaceans are invertebrates with hard shells and jointed appendages, with daily life food examples of shrimp, crab, and lobster. Univalve and bivalve shellfish are common molluscs that contain a soft body but no internal muscle skeleton in the shell, e.g., clam, oyster, scallop, mussel, and abalone. Cephalopods are shellfish that do not possess hard shells, where the representative examples are octopus and squid. The other shellfish species, such as sea cucumber, sea urchin, and sea star, belong to echinoderms.

The nutritional benefits of seafood products are amongst one of the most important factors contributing to their tremendous production and consumption worldwide. The major nutritional components and health-related values of seafoods are shown in the conceptual chart (Figure 4.2), where the nutritional features are categorized by their presence of high-quality proteins, ample essential amino acids, high levels of unsaturated fatty acids, and the existence of multiple minerals and fat-soluble vitamins that can meet daily dietary requirements. Seafood products can be regarded as a high-protein and low-fat food source, where the proximate compositions vary greatly, but they approximately include 60–80% moisture, 15–20% proteins, 0.2–20% fats, and 0.2–5% carbohydrates (Pateiro et al. 2022; Rittenschober et al. 2013). It is well-known that fish oil is rich in unsaturated fatty acids, where long-term consumption contributes to risk reduction of several chronic diseases, especially cardiovascular diseases (Hosomi et al. 2012). The most prominent example is omega-3 fatty acids, primarily docosahexaenoic acid and eicosapentaenoic acid (Rittenschober et al. 2013; Shalders et al. 2022), which makes seafood a promising animal-based dietary source for unsaturated fats. Due to the relatively low carbohydrate and fat contents in seafood products, these products are beneficial for weight control and reducing the risk of developing noncommunicable diseases related to obesity and being overweight (Hosomi et al. 2012). Fish and shellfish proteins are considered high quality, due to their high degree of digestibility and abundance of essential amino acids that human beings must obtain from food sources. These amino acids can be then used for synthesizing enzymes, hormones, and cellular components to support muscle contraction, tissue repair, and other physiological functions of human body (Shalders et al. 2022). Moreover, seafood products are rich in some fat-soluble vitamins, such as vitamins E (i.e., tocopherols) and D, where the former possesses substantial antioxidant activities that can help quench free radicals, and the latter contributes to the development and maintenance of bones and cartilages, thereby reducing the risk of atherosclerosis in the elderly population. Fish are also rich in vitamin A and carotenoids, which are known to contribute to eye health and immune functions. The macrominerals involving sodium, potassium, calcium, magnesium, and phosphorus obtained from seafood products can help regulate body electrolyte concentration and cellular growth and development (Fraga 2005). On the other hand, some micro-minerals can be found in appreciable quantities involving zinc, selenium, and iodine. It is noteworthy that seafoods are natural sources of iodine that can contribute

DOI: 10.1201/9781032677422-4

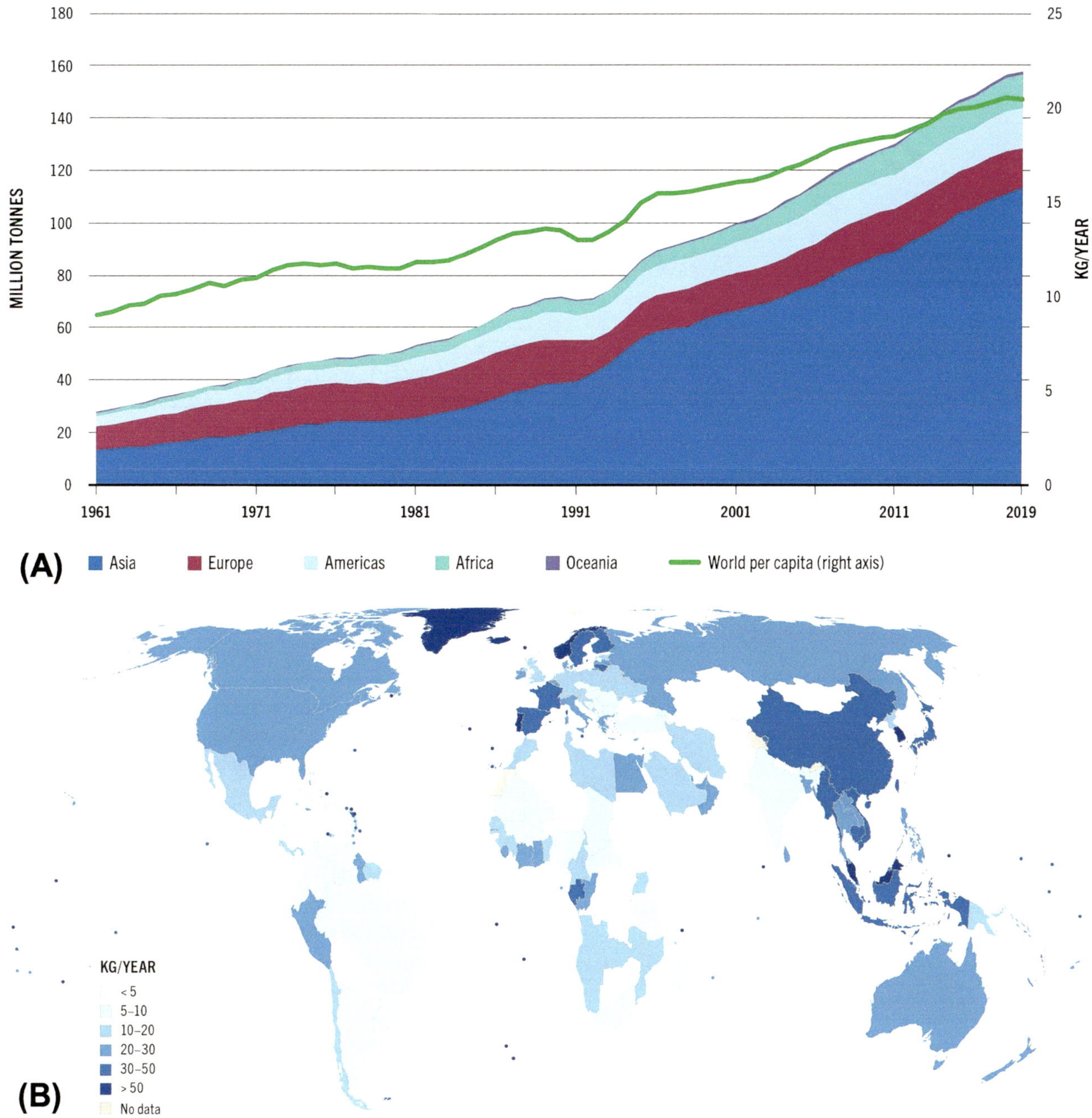

FIGURE 4.1 Global consumption of aquatic foods: (A) annual consumption by continent in the years of 1961–2019; and (B) annual consumption map by nations in the years of 2017–2019 (adopted from FAO 2023).

to thyroid functionality, whereas this trace element cannot be obtained from other terrestrial animal food sources. Overall, the nutritional composition and sensory properties of seafoods are dependent on the habitat and seawater quality, which are highly affected by the changing climate (Shalders et al. 2022).

Due to the plentiful macro-nutrients, high-moisture content, and low acidity, seafood moieties are ideal environments for microbial growth and biochemical degradations. Such deteriorative reactions result in the development of undesirable sensory attributes that can be either assessed by analytical instruments or perceived directly by sensory evaluation. For instance, the prolonged fat oxidation of oily fish fillets exhibits apparent soppy odor and off-flavor, and the microbial activities can form an polysaccharide-rich slimy film on the surface of fish muscles, which can be readily detected by the seafood retailors or consumers (Luo et al. 2022). The spoilage-indicating gas of many marine fish species is trimethylamine (TMA), which imparts the typical spoiled fish odor and showcases the unacceptability of stored fish fillets (Luo et al. 2022; Walayat et al. 2023). The discoloration of fish eye and shrimp skin caused by pigment changes, along with the softening of meat muscles as a result of protein degradation, are also common practices

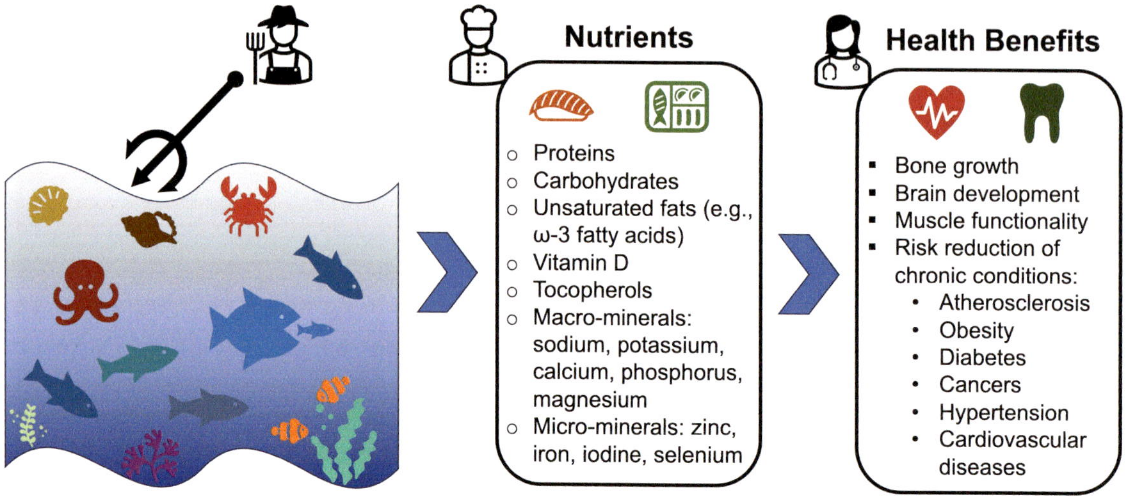

FIGURE 4.2 Conceptual chart of the main nutritional components and health benefits of seafood products.

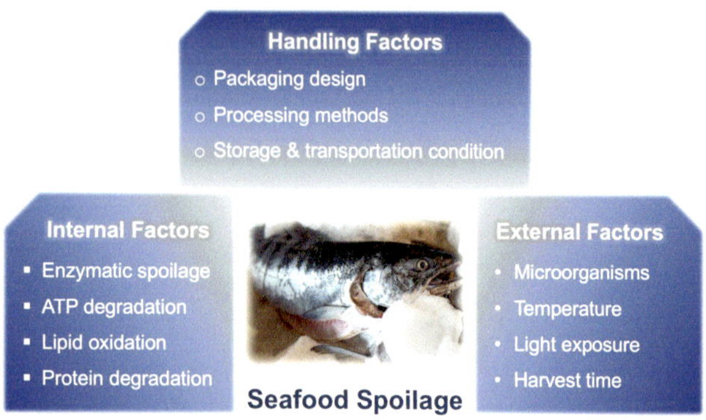

FIGURE 4.3 Significant spoilage factors (internal, external, and handling) of seafood products.

suggesting the deterioration of stored seafoods (Peng et al. 2022; Walayat et al. 2023).

As shown in Figure 4.3, the common spoilage factors include the internal, external, and manual handling factors that may accelerate or decline seafood deterioration. The intrinsic factors are directly associated with the physiological and biochemical changes within seafoods since harvesting, involving enzymatic breakdown of macromolecules, decomposition of ATP, oxidative rancidity of lipids, and deterioration of proteins (Luo et al. 2022; Nie et al. 2022). In practice, these internal spoilage factors can be controlled rather than prevented, through the change of external environmental conditions or the implementation of various handling protocols. For example, the rate of enzymatic spoilage can be slowed down by changing the environmental pH, temperature, ionic strength, and/or through the application of enzyme inhibitors. The protection towards the growth of pathogenic microorganisms can be achieved by determining the initial microbial load followed by discrete food hygiene procedures during seafood processing and packaging. Storing and transporting seafoods with strong sunlight exposure should be avoided due to its accelerating effect on certain chemical reactions, especially on lipid oxidation. Moreover, the storage time and temperature, gas composition of the in-package headspace, as well as the incorporation of exogenous agents all significantly influence the kinetics of deteriorative reactions. For instance, the improper cooling rate and prolonged frozen storage may cause ice crystal formation and subsequent protein denaturation, finally resulting in an increased drip loss upon thawing (Walayat et al. 2023). The available oxygen gas within the package can favor the growth of aerobic microorganisms and oxidation of proteins and lipids, hence leading to accelerated spoilage rate. To this end, the gas permeability of selected packaging material is essential in controlling the gas exchange with the ambient environment (Luo et al. 2022). In addition, the processing of seafoods, such as salting, marinating, smoking, drying, and heat treatments can substantially improve the storage quality.

The freshness decay of seafoods can be assessed by multiple approaches. First, sensory analysis is a prevalent quality evaluation protocol for determining the real-time quality change via the determination of seafood color, appearance, texture, and odor. The Quality Index Method is the most frequently used, where such sensory attribute changes are quantified and then correlated to the shelf life of the corresponding seafood species (Freitas et al. 2021). Besides, the physical and chemical analyses of seafoods can indicate the generation of metabolites during the storage period. For example, the release of TMA and other volatile amines from seafood tissues can be either directly quantified through analytical instruments, or indirectly reflected via the determination of total volatile basic nitrogen (TVBN), pH, and titratable acidity changes (Luo et al. 2021, 2022). The degree of lipid oxidation can be obtained from the assessment of peroxide value or the quantification of short-chain aldehydes that are the main secondary metabolites of fat autoxidation in seafoods (Walayat et al. 2023). The microbiological spoilage on seafoods can be unraveled by investigating the alteration of fish and shellfish spoilage microbiota, such as the determination of total viable counts and the evaluation on the dominant bacterial strains. Even the most representative spoilage bacteria in seafoods are psychrotrophic, the specific dominant species vary depending on different species and their living habitats.

The appropriate means of preservation is essential in conserving the desirable sensory properties and nutritional quality of seafoods, considering their perishability is higher than many other food categories. The conventional methods for seafood preservation highly relies on low temperature storage that can slow down the deteriorative chemical reactions, inhibit microbial growth, and inactivate endogenous enzymes. Such cold temperature treatments include chilling with refrigeration temperatures (0–4°C), super-chilling that attains the cold temperature ranging from –2.5°C to –1°C, and freezing that reaches a low temperature condition of –18°C or below (Walayat et al. 2023). Another commonly applied preservation technique is via the deliberate modification of the in-package atmosphere, namely vacuum and modified atmosphere packaging, where the gas mixtures of N_2, O_2, and CO_2 can be prepared in the latter case (Nie et al. 2022; Peng et al. 2022). Specifically, the reduction or elimination of oxygen gas is favorable for controlling the proliferation of aerobic microorganisms. Other supplementary methods, such as the pressure modification and development of active packaging materials, are evolving, but most of the research studies still remain at the laboratory stage.

Nanotechnology in food packaging refers to the development of any packaging materials or compartments that are assessed and described at the nanoscale. Active packaging in seafood targets to incorporate active components inside the package in order for shelf-life extension and quality conservation (Luo et al. 2022). The invention or design of nanomaterials is one of the leading research fields in food preservation since it can help deliver the active components into the food matrix, enhance their interactions, and provide protection to the active compounds during the prolonged storage period (Qiu et al. 2022). The developed nanomaterials, usually containing one carrier or protective matrix and one or more active components, can be an alternative means to preserve seafood, through the conveyed antibacterial, antifungal, or antioxidant activities (Çiçek and Özoğul 2023; Qiu et al. 2022).

This chapter aims to provide a comprehensive elaboration on the spoilage mechanisms of seafood products collectively, and recent studies in the preservation of fresh seafood quality via the application of various types of nanomaterials, such as nanoemulsions, nanofibers, nanocomposite films, and so on. The insights into the future application of nanomaterials and the associated challenges are also discussed accordingly, in order to develop more compatible, effective, and economical quality-enhancing materials for seafood products.

4.2 SEAFOOD SPOILAGE MECHANISMS

The intrinsic seafood spoilage mechanisms can be classified into three processes, namely enzymatic autolysis, oxidative degradation, and microbial enumeration. The autolytic reaction happens as soon as the fish or shellfish is harvested, where the other two takes place concurrently in the latter phase of the deteriorative process during storage.

4.2.1 Enzymatic Autolysis

Autolysis can happen to plants, animals, or microorganism, and it refers to the cell's self-degradation through the activities of endogenous enzymes (Huang et al. 2023). Resulting from the uncomfortable environmental conditions, animal tissues release such endogenous enzymes, e.g., protease, lipase, phosphatase, amylase, cellulase, and gastrointestinal tract enzymes, from inside to the outside of the cell (Huang et al. 2023). Indeed the onset of enzymatic autolysis of fish and shellfish occurs immediately after harvest, so the macromolecules from muscle meat are decomposed gradually, leading to textural changes (Qiu et al. 2022). This suggests the rate of autolytic process can impact on the shelf life of seafoods, even without generation of apparent off-flavors and off-odors (Ghaly et al. 2010).

The compositions of endogenous enzymes vary depending on different seafood products, but it is generally believed that autolysis occurs starting from the liver and the digestive organs (Huang et al. 2023). The common endogenous enzymes with their respective targeted substrates and corresponding results in seafoods are shown in Figure 4.4. At the beginning of the postmortem period, glycogen phosphorylase and debranching enzyme cleave off the sugar unit from glycogen molecules, forming glucose-1-phosphate, which can be readily isomerized into glucose-6-phosphate and enter glycolytic pathway. The presence of glycolytic enzymes in seafood tissues demonstrates the capability of degrading glucose and glucose-6-phosphate into pyruvates, which can be further transformed into lactic acid through anaerobic respiration. The aforementioned process results in the postmortem pH decline in fish or shellfish meat, and the rate or extent of pH reduction is a significant factor that can alter the eating quality and chemical stability (Ghaly et al. 2010). In practice, the control of animal's pre-slaughter stress will have a substantial impact on the ultimate pH, which is approximately 6.0 but varies with fish species (Chauhan and England 2018). The ATP produced from the glycolytic pathway and the well-known phosphagen system is decomposed by the action of the nucleotide enzymes such as ATPase and AMP deaminase (Chauhan and England 2018). The final products of ATP breakdown are inosine and hypoxanthine (Hong et al. 2017). In the fishery industry, one of the common fish-freshness assessment protocol is to determine the K-value, which indicates the percentage of the final products formation (Prabhakar et al. 2020). The lower the K-value, the less amount of degradation products are formed from ATP, suggesting the fresher seafood

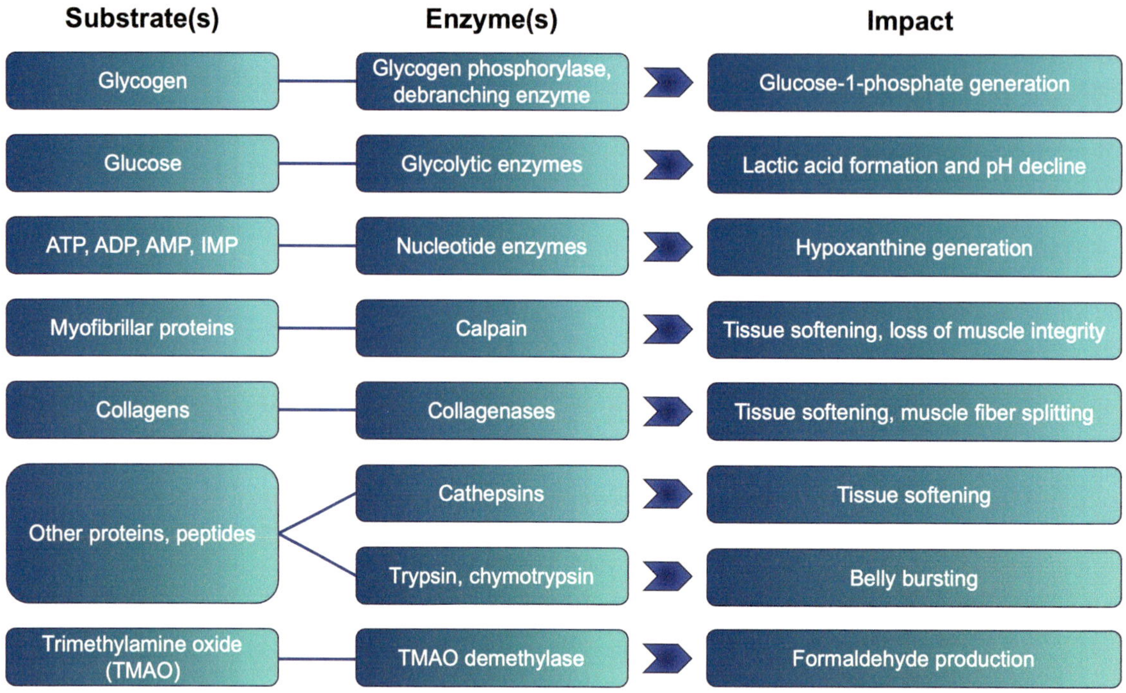

FIGURE 4.4 Main endogenous enzymes involved in the autolytic process of seafoods and the effects of the corresponding catalyzed reactions.

products (Ehira and Uchiyama 1987). Calpain and collagenases are common endogenous enzymes in animal muscles due to their specificity towards the main proteins in muscle fibers and connective tissues. The decomposition of myofibrillar proteins and collagens into peptides or amino acids reduces the intermolecular interactions within the protein matrices and increases water solubility, thereby resulting in softer meats that tend to tear apart. Cathepsins are a class of lysosomal proteases that can result in similar muscle softening effect by degrading troponin, tropomyosin, and myosin, and it was found that the proteolytic activities of cathepsins B and L were more prominent in fish flesh than ruminant meat (Chéret et al. 2007; Ghaly et al. 2010). Trypsin and chymotrypsin are alkaline proteases in fish intestine, where the former is more specific to cleave the lysine and arginine residues from the carboxyl side, and the latter is targeting more on the aromatic amino acid residues (Solovyev et al. 2023). The release of trypsin and chymotrypsin from the intestinal tract to the meat muscles result in belly bursting, which can be prevented with appropriate processing and storage temperature conditions (Veliyulin et al. 2007). In fish and shellfish muscle and viscera, trimethylamine oxide (TMAO) is a naturally occurring compound to tolerate osmotic pressure under deep seawater (Ghaly et al. 2010). During the early stages of autolytic process, the demethylase breaks down TMAO into dimethylamine (DMA) and formaldehyde as the primary metabolites (Sikorski and Kostuch 1982). During enzymatic autolysis, the production of such volatile compounds is not reaching high concentrations for inducing apparent spoilage-indicating odors.

4.2.2 Oxidative Degradation

Since the lipids in seafoods are rich in unsaturated fatty acids (e.g., oleic acid, linoleic acid, linolenic acid, arachidonic acid, and docosahexaenoic acid), the oxidation of lipids occurs readily during the postmortem period, representing significant quality loss due to the creation of off-flavors and the loss of lipophilic vitamins and other phytochemicals. The premortem stress and the postmortem factors such as the ultimate pH, temperature, and processing methods play essential roles in affecting the rate and extent of lipid oxidation in seafoods (Secci and Parisi 2015). In general, lipid oxidation takes place with the participation of oxygen and is catalyzed by the presence of transition metals, light exposure, or other reactive species, with the unsaturated double bonds in fatty acid molecules being oxidized. The primary lipid oxidation products are hydroperoxides, and the secondary products are mostly consisting of short-chain aldehydes, alcohols and ketones, particularly malonaldehyde, hexanal, and pentanal (Mariutti and Bragagnolo 2017).

Lipid oxidation in seafoods can proceed enzymatically or non-enzymatically. The enzymatic lipid oxidation in seafoods is mainly driven by the hydrolytic activity of lipases, for example, phospholipases and triacyl lipases, which breaks down phospholipids and triglycerides, respectively (Ghaly et al. 2010). These lipases may include the endogenous enzymes that naturally exist in muscle tissues and visceral organs, as well as the ones produced through the actions of psychrotrophic bacterial species. The release and

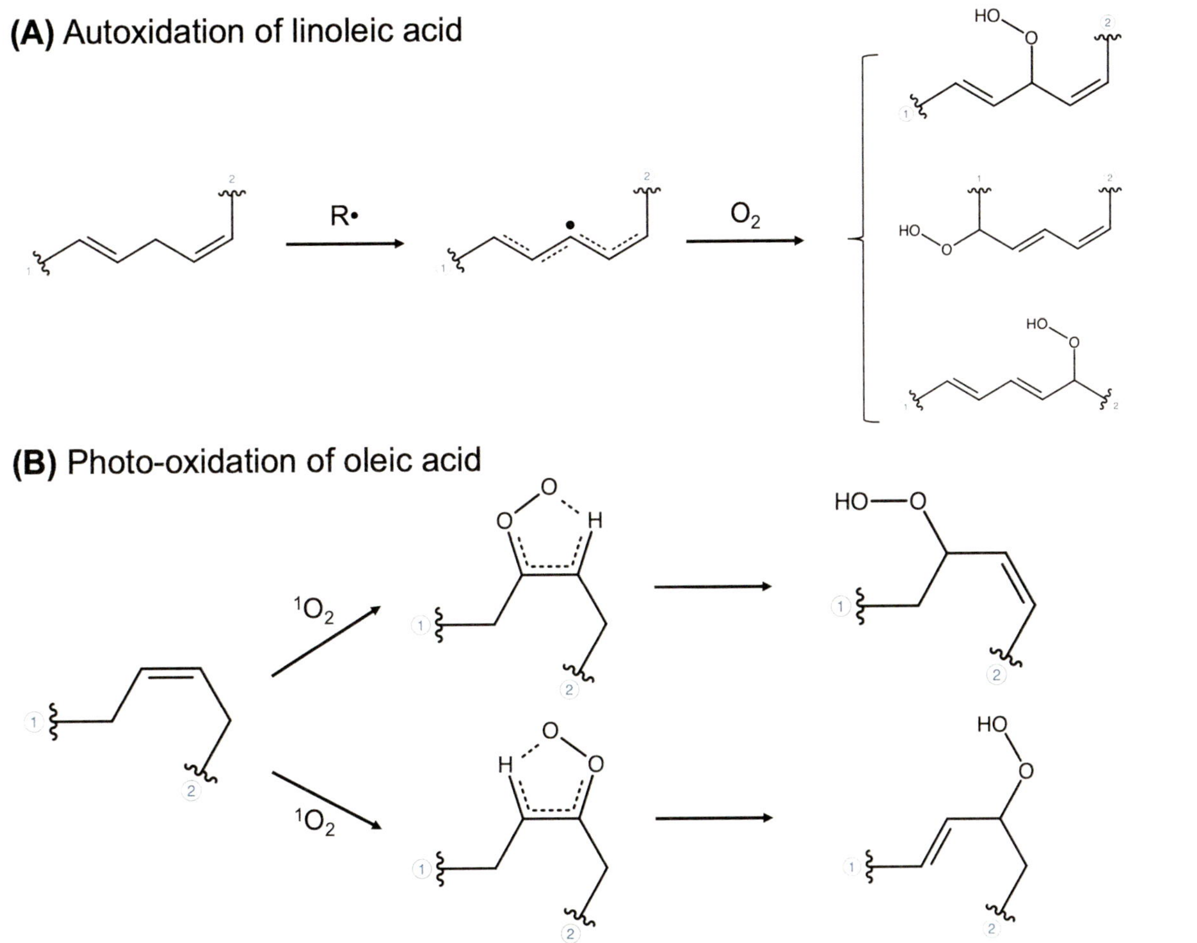

FIGURE 4.5 Mechanism of hydroperoxide generation from (A) the autoxidation of linoleic acid and (B) the photo-oxidation of oleic acid.

accumulation of free fatty acids impart the rancid off-flavor of spoiled fish and shellfish.

In comparison, the non-enzymatic oxidation of seafoods accounts for more of the postmortem quality loss than that of the enzymatic process. There are two mechanisms of non-enzymatic oxidation, namely autoxidation and photo-oxidation, which differ in the oxygen status in generating different hydroxide products (Figure 4.5). Autoxidation is one of the most noticeable seafood quality indicators, where molecular oxygen or triplet oxygen is reacting with fatty acids, with three phases: initiation, propagation, and termination (Ghaly 2010; Mariutti and Bragagnolo 2017). Initiation causes the conversion of fatty acid molecules to free radicals, induced by the transition metals, heat, light, and so on. Propagation is the stage where oxygen reacts with the lipid free radical, with the production of peroxyl radicals. In the termination phase, all the pre-formed radicals combine with each other and become non-radical hydroperoxide products (Figure 4.5[A]). The oxygen reactant under the photo-oxidation mechanism is singlet oxygen, which is more hydrophilic and reactive than triplet oxygen, thereby resulting in a faster oxidation process (Mariutti and Bragagnolo 2017). The singlet oxygen reacts with the unsaturated fatty acids by altering the configuration of the double bond, followed by oxygen addition and formation of a 5-membered ring structure (Mariutti and Bragagnolo 2017). The hydroperoxides can be generated accordingly, as shown in Figure 4.5(B). The aforementioned primary lipid oxidation products from both mechanisms, hydroperoxides, can be assessed by determining the peroxide value (Secci and Parisi 2015). However, hydroperoxides can be further decomposed into short-chain hydrocarbons through β-cleavage. The resulting volatile aldehyde molecules are the chemicals mainly imparting the undesirable fat oxidation odors in fish and shellfish (Mariutti and Bragagnolo 2017). The secondary lipid oxidation products can be determined by the use of thiobarbituric acid reactive substances (TBARS), which react with one of the short-chain aldehydes, malonaldehyde (Secci and Parisi 2015).

According to literature studies, various premortem and postmortem factors can impact on the degree of lipid oxidation in seafoods. For instance, the pre-slaughter stress of fish can be alleviated by feeding them with antioxidant-fortified diets, thereby reducing the TBARS levels in fish flesh

(Hernández et al. 2014). During the postmortem period, the reactive species that initiates lipid oxidation can be decreased by inhibited by the addition of antioxidants and chelating agents, which suggests one alternative way to control seafood lipid oxidation to supplement cold temperature storage.

4.2.3 Microbial Activities

The microbial spoilage happens at the latter stage in seafood deterioration since the initial microbial count is usually minimal or low. The major microorganisms responsible for microbial spoilage of seafoods are bacteria. The initial microbial loads on fresh seafoods are dependent on the habitat, such as the water, soil, sediment, or sand that the fish and shellfish live in. Fish and shellfish species, harvest season, and climate changes also affect the microbiota. Overall, the main composition of these bacteria include the *Pseudomonas*, *Photobacterium*, *Shewanella*, *Psychrobacter*, *Aeromonas*, *Acinetobacter*, and *Vibrio* species (Gram and Dalgaard 2002; Parlapani 2021). Through advanced techniques such as 16S gene RNA sequencing and polymerase chain reaction (PCR)-based methods, the bacterial species in spoiled seafoods can be identified and quantified, as shown in Table 4.1, where recent studies on the dominant spoilage bacteria from different cold stored seafoods are summarized. Depending on the living environment, biological origin, and processing methods, the dominant species may vary greatly. Generally the fish living in temperate zones are rich in Gram-negative species, *Pseudomonas* and *Shewanella*, while similar microbial species are commonly found in tropical fish but with higher levels (Gram 2009). The dominant bacteria in spoiled shrimp farmed in temperate regions are the *Psychrobacter* and *Acinetobacter* species (Heinsz et al. 1988; Parlapani et al. 2020), but the ones in tropical shrimp are mainly *Shewanella* (Benner et al. 2004). The microbiota profile in cephalopods and molluscs are similar to fish and shrimp, even with considerable body structure and composition differences (Anagnostopoulos et al. 2022).

Besides, the microbiota of seafoods can be also modified by the postharvest handling protocols, essentially during processing, packaging, and transportation where

TABLE 4.1
Dominant Spoilage Bacteria Found in Various Seafood Products During Cold Storage

Seafood	Storage Condition	Dominant Bacteria	Reference
Anchovy	Stored at 0, 4, and 10ºC, under air, vacuum, and modified atmosphere	*Pseudoalteromonas*, *Vibrio*, *Pseudomonas*, *Psychrobacter*, *Shewanella*, *Photobacterium*	Sequino et al. (2024)
Brackish river prawn	Stored at 0, 4, 10 and 25ºC	*Citrobacter*, *Pseudomonas*, *Shewanella*	Dabadé et al. (2023)
Bighead carp	Stored at –2°C with polyethylene bags	*Shewanella*, *Lactococcus*, *Flavobacterium*	Li et al. (2023)
Pacific white shrimp	Stored at 2°C and with –2°C ice slurry	*Pseudoalteromonas*, *Psychromonas*, *Psychrobacter*	Liu et al. (2023)
American lobster	Stored at 4°C	*Photobacterium phosphoreum*	Rolin et al. (2023)
Sturgeon fillets	Stored at 4°C	*Aeromonas sobria*	Tan et al. (2023)
Monkfish	Stored at 4°C under vacuum and ambient air conditions	*Carnobacterium*, *Myroides*, *Shewanella*, *Psychrobacter*, and *Vagococcus*	Wang et al. (2023b)
Yellow catfish	Stored at 4°C	*Shewanella putrefaciens* and *Pseudomonas*	Wang et al. (2023a)
Large yellow croaker fillets	Stored at high CO_2, refrigerated condition	*Shewanella*	Yan et al. (2023)
Grouper fillets	Vacuumed packaged and stored at 4°C	*Hafnia paralvei*, *Aeromonas allosaccharophila*	Yang et al. (2023)
Greater amberjack	Stored at 20, 4, –1, and –3°C	*Pseudomonas*	Zhong et al. (2023)
Grass carp	Stored at 4°C and packaged in HDPE bags	*Pseudomonas putida*, *Aeromonas rivipollensis*, *Shewanella putrefaciens*	Zhuang et al. (2022)
Rose shrimp	Stored at 0 and 4°C	*Psychrobacter*, *Carnobacterium*	Parlapani et al. (2020)
Mussels	Stored at 4°C	*Psychrobacter*	Parlapani et al. (2020)
Finfish fillets	Stored at –20°C on ice packs	*Pseudomonas*, *Psychrobacter*, *Photobacterium*	Shehata et al. (2020)
Atlantic cod	Stored at 0 and 3°C, under air or modified atmosphere	*Pseudomonas*, *Psychrobacter*, *Carnobacterium*, *Rahnella*	Sørensen et al. (2020)
European sea bass	Stored in 0°C ice with insulated boxes	*Psychrobacter*	Syropoulou et al. (2020)
Largemouth bass fillets	Stored at 4°C	*Aeromonas*, *Pseudomonas*	Zhuang et al. (2020)
Live mussels	Stored at 4°C under modified atmosphere	*Synechococcus*, *Shewanella*, *Acidaminococcus*, *Psychromonas*, *Acinetobacter*	Odeyemi et al. (2019)
Blue crab	Stored at 4 and 10°C	*Candidatus*, *Vibrio*, Rhodobacteraceae family	Parlapani et al. (2019)
Salted grass carp	Stored at 4°C, under air, vacuum and modified atmosphere	*Lactococcus*, *Acinetobacter*, *Pseudomonas*	Zhang et al. (2019)

microbiological hazards are the main concern, such as in contamination by *Escherichia coli*, *Salmonella*, *Shigella*, *Yersinia*, *Staphylococcus*, and so on (Parlapani 2021). Even numerous bacteria are present in cold stored fish and shellfish; it is noticeable that the spoilage-causing bacterial strains are mainly attributed to the Gram-negative, psychrotrophic species (Gram and Huss 2000). Also, seafoods contaminated by pathogenic bacteria, e.g., *Salmonella enterica*, *Listeria monocytogene*, *Clostridium botulinum*, and Shiga toxin-producing *E. coli*, pose higher risks of causing food poisoning upon human consumption (Parlapani 2021).

The growth of bacterial species is relying on the breakdown or fermentation of fish and shellfish nutrients, eventually leading to undesirable sensory attributes that can be readily detected. The production of volatile compounds from microbial activities accounts for most of the spoilage aroma compounds, involving trimethylamine (TMA), DMA, ammonia, hydrogen sulfide, and numerous volatile organic compounds (VOCs) (Luo et al. 2021; Odeyemi et al. 2018). Imparting the main spoilage fishy odor, TMA is produced from the microbial degradation of TMAO, involving the activities by the *Shewanella*, *Pseudomonas*, and *Vibrio* species (Ghaly et al. 2010), hence the quantification of TMA contents in stored fish or shellfish becomes a universal method to indicate the degree of seafood freshness (Luo and Lim 2019). The rate and extent of VOC production depends on the seafood species, storage condition, and packaging design. For instance, seafoods under low-oxygen packaging environments tend to suppress the aerobic bacteria, thereby influence the corresponding spoilage gas generation. The growth of bacteria can also result in color defects in spoiled fish, e.g., the greenish fluorescent color produced from *Pseudomonas fluorescens* (Wang et al. 2020b). To control all microbe-induced seafood spoilage, conventional cold storage is the mainstream preservation. Advanced methods can consider the inclusion of antimicrobial compounds in seafood preservation, where nanotechnology could be potentially applied to encapsulate such active components in food packages. For instance, Özogul et al. (2022) developed a nanoemulsion containing laurel essential oil, which displayed significant antimicrobial property towards fish spoilage bacteria, with the strongest inhibitory effect against *Pseudomonas luteola*. Likewise, the nanoemulsion formulated by Yazgan et al. (2019) utilized lemon essential oil as the active component, which considerably suppressed the growth of pathogenic bacteria and fish spoilage bacteria, such as *Enterococcus faecalis* and *Photobacterium damselae*, respectively.

4.3 TYPE OF NANOMATERIALS IN SEAFOOD PRESERVATION

The application of nanotechnology is an emerging innovative approach to prolonging the shelf life of perishable foods. Depending on different purposes and applications, the materials prepared by nanotechnology can be diverse, such as the development of nanoemulsions, electrospun nanofibers, nanoparticles, and multilayer nanofilms, for the preservation of seafoods stored at refrigerated temperature. Table 4.2 summarizes the recent studies in such innovations in delaying seafood spoilage with respective laboratory findings.

TABLE 4.2
Applications of Nanomaterials for Cold Storage Preservation of Seafood Products

Type of Nanomaterial	Targeted Seafood Product	Active Component(s)	Storage Condition	Main Finding	Reference
Nanoemulsion					
	Flathead mullet fillet	Grape and cinnamon essential oil	2°C for 14 days	The nanoemulsion containing both essential oils showed prolonged shelf life of fish fillet with less lipid oxidation and microbial growth.	Ameur et al. (2022)
	Shrimp (*Crocus sativus* L.)	Saffron essential oil	4°C or 8°C for 14 days	The nanoemulsion containing 5% saffron essential oil slowed down the lipid oxidation and microbial growth of *E. coli* and *S. aureus*.	Aboutorab et al. (2021)
	Caviar (*Acipenser persicus*)	Silver nanofilm	70 days	The nanoemulsion containing silver nanofilms showed significant antimicrobial activities by assessing total viable counts.	Asl et al. (2021)
	Hairtail (*Trichiurus haumela*)	Eugenol	4°C for 18 days	The nanoemulsion coated fish exhibited increasing shelf stability and conserved sensory properties.	Liu et al. (2021)
	Rainbow trout fillet	Curcumin and rosemary oil	4°C for 9 days	The growth of psychrophilic bacteria in stored fish was greatly controlled by the presence of either rosemary oil or curcumin.	Ceylan et al. (2020)

(Continued)

TABLE 4.2 (*Continued*)
Applications of Nanomaterials for Cold Storage Preservation of Seafood Products

Type of Nanomaterial	Targeted Seafood Product	Active Component(s)	Storage Condition	Main Finding	Reference
	Asian seabass fillet	Cinnamon oil	4°C for 8 days	The nanoemulsion encompassing cinnamon oil prolonged the fish shelf life to 6–8 days.	Chuesiang et al. (2020)
	Rainbow trout fillet	Essential oils of lemon, orange, mandarin, and grapefruit	4°C for 16 days	The nanoemulsion with all citrus essential oils maintained the TVBN level below the maximum allowable level within 12 days.	Durmuş (2020)
	Rainbow trout fillet	*Zataria multiflora* Boiss essential oil	4°C for 16 days	Microbial growth was largely inhibited by the nanoemulsions with active components.	Khanzadi et al. (2020)
	Rainbow trout fillet	Citrus essential oil	4°C for 16 days	The production of histamine and other biogenic amines in fish were suppressed by the prepared nanoemulsion.	Kosker (2020)
	Rainbow trout	Curcumin	4°C for 15 days	The use of 5% curcumin-laden nanoemulsion greatly controlled the growth of spoilage bacteria.	Khoshbouy Lahidjani et al. (2020)
	Sturgeon (*Acipenser stellatus*) fillet	Rosemary, cumin, and *Zataria multiflora Boiss* essential oil	4°C and 10°C for 14 days	The essential oil-loaded nanoemulsions showed significant antioxidant and antibacterial properties on stored fish fillets.	Nasiri et al. (2020)
	Hake burger	Oregano essential oil	4°C for 15 days	The nanoemulsion containing oregano essential oil preserved the fatty acids and slowed down the TVBN generation.	Asensio et al. (2019)
	Rainbow trout fillet	Black cumin essential oil	4°C for 12 days	The chitosan-based nanoemulsion showed considerable inhibition on the growth of *E. coli* O157:H7.	Kazemeini et al. (2019)
	Rainbow trout fillet	Thyme oil	4°C for 9 days	The nanoemulsion containing thyme oil showed significant inhibition on the growth of fish spoilage bacteria.	Marel et al. (2019)
Electrospun nanofiber					
	Salmon	Sesame oil	4°C for 8 days	The coating with sesame oil-based nanofiber can significantly delay the oxidation process and suppress the growth of bacteria, yeasts, and molds.	Ceylan et al. (2023)
	Rainbow trout	Grape seed oil	4°C for 9 days	The coated grape seed oil was effective due to the significant inhibition on the growth of microorganisms and autoxidation.	Ceylan et al. (2021)
	Seabass	Gold	4°C for 8 days	The gold-loaded nanofiber can effectively inhibit the growth of mesophilic bacteria in fish.	Çetinkaya et al. (2021)
	Rainbow trout	Nisin	4°C for 10 days	The nisin-embedded nanofiber showed protective effect on stored fish fillets, with preserved sensory attributes and limited microbial enumeration.	Oner et al. (2021)
	Shrimp	Cinnamon essential oil	4°C for 7 days	The essential oil phytosome was embedded in PVA-based nanofibers, which demonstrated strong antimicrobial effects.	Nazari et al. (2019)
	Rainbow trout fillet	Nisin and curcumin	4°C for 10 days	The nanofiber containing nisin and curcumin combined with the use of aluminum foil were effective in limiting microbial growth in fish fleshes.	Ceylan (2019)

(*Continued*)

TABLE 4.2 (Continued)
Applications of Nanomaterials for Cold Storage Preservation of Seafood Products

Type of Nanomaterial	Targeted Seafood Product	Active Component(s)	Storage Condition	Main Finding	Reference
	Rainbow trout and mackerel	*Lactobacillus reuteri*	4°C for 7 days	The bacteria-encapsulated nanofibers improved the oxidative stability using DPPH and FRAP assays.	Ceylan et al. (2019)
Nanoparticle					
	Grass carp	Curcumin and nisin	4°C for 10 days	The pectin-based nanoparticles showed significant inhibitory effect on *Pseudomonas* growth in fish fillets.	Zhang et al. (2023)
	Rainbow trout fillet	Peppermint essential oil	4°C for 12 days	The prepared lipid nanoparticles preserved the chemical stability and sensory attributes of stored trout fillets.	Safaeian Laein et al. (2023)
	Shrimp (*Penaeus vannamei*)	Metal oxides (zinc, silicon, and copper)	4°C for 12 days	The combination of ZnO and SiO showed the highest antimicrobial effect on the spoilage bacteria.	Shao et al. (2021)
	Red seabream fillet	Ferric oxide	4°C for 24 h	The application of nanoparticles along with microwave thawing and infrared thawing methods showed decreased bacterial growth of *Pseudomonas* species.	Cai et al. (2020)
	Grey mullet	Chitosan	4°C for 9 days	The chitosan-based nanoparticles showed strengthened antibacterial activity and declined myoglobin oxidation in stored fish surimi.	Hajji et al. (2019)
	Blue-spined trevally (*Carangoides coeruleopinnatus*)	Chitosan	4°C for 12 days	The nanoparticles were found to substantially suppress bacterial growth and lipid oxidation during the testing period.	Alboghbeish and Khodanazary (2019)
Nanocomposite film					
	Salmon fillet	Curcumin, capsaicin, silver nanoparticles	4°C for 12 days	The composite film exhibited inhibitory effects on bacterial growth, but not on lipid oxidation.	Grebieniarz et al. (2023)
	Sea bream	Silver montmorillonite	4°C for 15 days	The controlled release of silver from the nanocomposite film can provide continuous antibacterial property towards fish storage.	Seray et al. (2022)
	Rainbow trout	Copper nanoparticles	4°C for 14 days	The composite film was found to suppress the growth of both Gram-positive and negative bacteria, as well as lipid oxidation products.	Azari et al. (2022)
	Red sea bream fillet	Cinnamon and perilla essential oil	4°C for 14 days	The shelf life of film-wrapped fish fillets was prolonged to up to 8 days during cold storage.	Zhao et al. (2022)
	Rainbow trout fillet	Sodium montmorillonite	4°C for 25 days with different packaging conditions	The film containing nanoclay along with vacuum and modified atmosphere packaging showed substantial preservative effect on fish quality.	Bahmani et al. (2021)
	Tilapia (*Oreochromis niloticus*)	Metal oxides (silver, copper, and titanium)	4°C and −20°C for 10 days	The composite film containing all three metal oxide nanoparticles showed the highest antimicrobial effect during cold storage of fish.	Efatian et al. (2021)

4.3.1 Nanoemulsions

Nanoemulsions are the most widely used nanomaterials in seafood preservation to date. After the blending of a mixture of chemical components, an emulsion containing a continuous phase and a dispersed phase is formed, with water-in-oil (W/O) and oil-in-water (O/W) being the two main types. The latter is more commonly formulated for food preservation purposes, with the use of an emulsifier whose nonpolar tails are embedded in oil droplets and the polar heads are interacting with water phase (Zhao et al. 2023). In O/W nanoemulsion, the water phase is mostly contributing to the ionic strength, polarity, and overall rheological properties of the emulsion system (Aswathanarayan and Vittal 2019). For seafood preservation purposes, hydrophilic biopolymers can be added into the water phase, such as maltodextrin (Ceylan et al. 2020) and chitosan (Liu et al. 2021), in order to facilitate the encapsulation of active components. Due to the capability to form porous structures, these biopolymers are versatile for protective effects as well as the controlled release of any active compounds embedded inside. The oil phase in the O/W nanoemulsion is altering the density, viscosity, and interfacial surface tension of the emulsion (Aswathanarayan and Vittal 2019). To stabilize the O/W nanoemulsion, a surfactant (i.e., emulsifier) can be added to reduce the surface tension of the interface. The type of surfactants is selected based on the compatibility with other components in the emulsion. Based on literature studies in fish and shellfish preservation, some examples of surfactants for nanoemulsion preparation include glycerin, tween 80, tween 20, and span 80 (Çorapci 2022; Nasiri et al. 2020).

The active components targeted to delay seafood spoilage can be diverse, where the substances possessing significant antioxidant and antimicrobial characteristics are the most popular. Amongst these compounds, common food additives, plant extracts, and essential oils are representative examples due to their accessibility, high cost-effectiveness, and ease of operation. Multiple studies have reported the significant seafood preserving effect with the incorporation of essential oils in nanoemulsions. For instance, the addition of cinnamon, saffron, and oregano essential oils prepared in different nanoemulsion mixture were found to reduce the postmortem lipid oxidation and inhibiting bacterial growth in cold stored flathead mullet (Ameur et al. 2022), shrimp (Aboutorab et al. 2021), and hake burger (Asensio et al. 2019), respectively. Regardless of the effectiveness, it should be noticeable that most essential oils contain volatile hydrocarbons such as terpenes and terpenoids, which may interfere with the fresh seafood flavor if they are present in high concentrations in the packaging headspace.

As elaborated by Zhao et al. (2023), the mechanisms of bacterial attack by the prepared nanoemulsions are shown in Figure 4.6. When coming in contact with a bacterial cell, the active components can be delivered and released into the bacterial cytoplasm via various types of interactions between the nanoemulsion and the bacterial membrane. Once reaching the intracellular space, the active components generally induce three mechanisms for controlling bacterial growth, namely through (i) physical damage on cell wall and plasma membrane, (ii) chemical interference with bacterial metabolism, and (iii) inhibition of biosynthesis of macromolecules in bacterial cell (Çiçek and Özoğul 2023; Zhao et al. 2023). It was conveyed that the physical impairment on bacterial cell membrane was because of the declined phospholipid bilayer synthesis with the presence of active components (Hai et al. 2022). The encapsulated active compounds can also hinder the process of nutrient metabolism, by inhibiting the enzyme activities involved in the glycolytic reactions and citric acid cycle (de Oliveira et al. 2023), as well as in amino acid catabolism (Hai et al. 2022).

4.3.2 Electrospun Nanofibers

The production of fibrous materials in the nanoscale is usually achieved by the implementation of electrospinning, solution blow spinning, sonochemical, polymerization processes, and so on (Çiçek and Özoğul 2023), where electrospinning is the most employed technique in literature studies. Due to the stretching of the polymer molecules through high voltage, the generated nanofibers allow for the great potential in encapsulating active components to fight against food spoilage. The substrates to be electrospun can be diverse, involving food polysaccharides and proteins such as chitosan, methylcellulose, and whey protein, and non-food polymers such as poly(vinyl alcohol) (PVA) and poly(ethylene oxide) (PEO) (Luo and Lim 2020; Luo et al. 2021). One advantage of electrospun nonwoven is its enlarged surface area, indicating the ease of chemical release and interaction with the food matrix (Luo et al. 2021). Thus, any bioactive compounds that can control seafood spoilage factors are of interest to be encapsulated inside nanofibers. For example, the sesame oil-loaded electrospun fiber produced by using PVA as the carrier, with the diameter range of 286–656 nm, exhibited excellent antimicrobial properties, with limited growth of psychrophilic bacteria, molds, and yeasts in cold stored salmon (Ceylan et al. 2023). In another study conducted by Oner et al. (2021), nisin was encapsulated into PVA-based nanofibers for preserving rainbow trout, and significant antimicrobial effects were found towards lactic acid bacteria and mesophilic bacteria, with growth inhibition of 38% and 31%, respectively, in fish fillets stored under refrigerated temperature.

4.3.3 Nanoparticles

Nanoparticles are solid particles in the nanoscale (1–100 nm diameter), with exceptional large surface area to volume ratio (Çiçek and Özoğul 2023). Depending on the raw material and synthesis methods, nanoparticles can be categorized as metal-based, ceramic-based, carbon-based, lipid-based, and semiconductor nanoparticles (Çiçek and Özoğul 2023). Nanoparticles' versatility of use in food preservation is

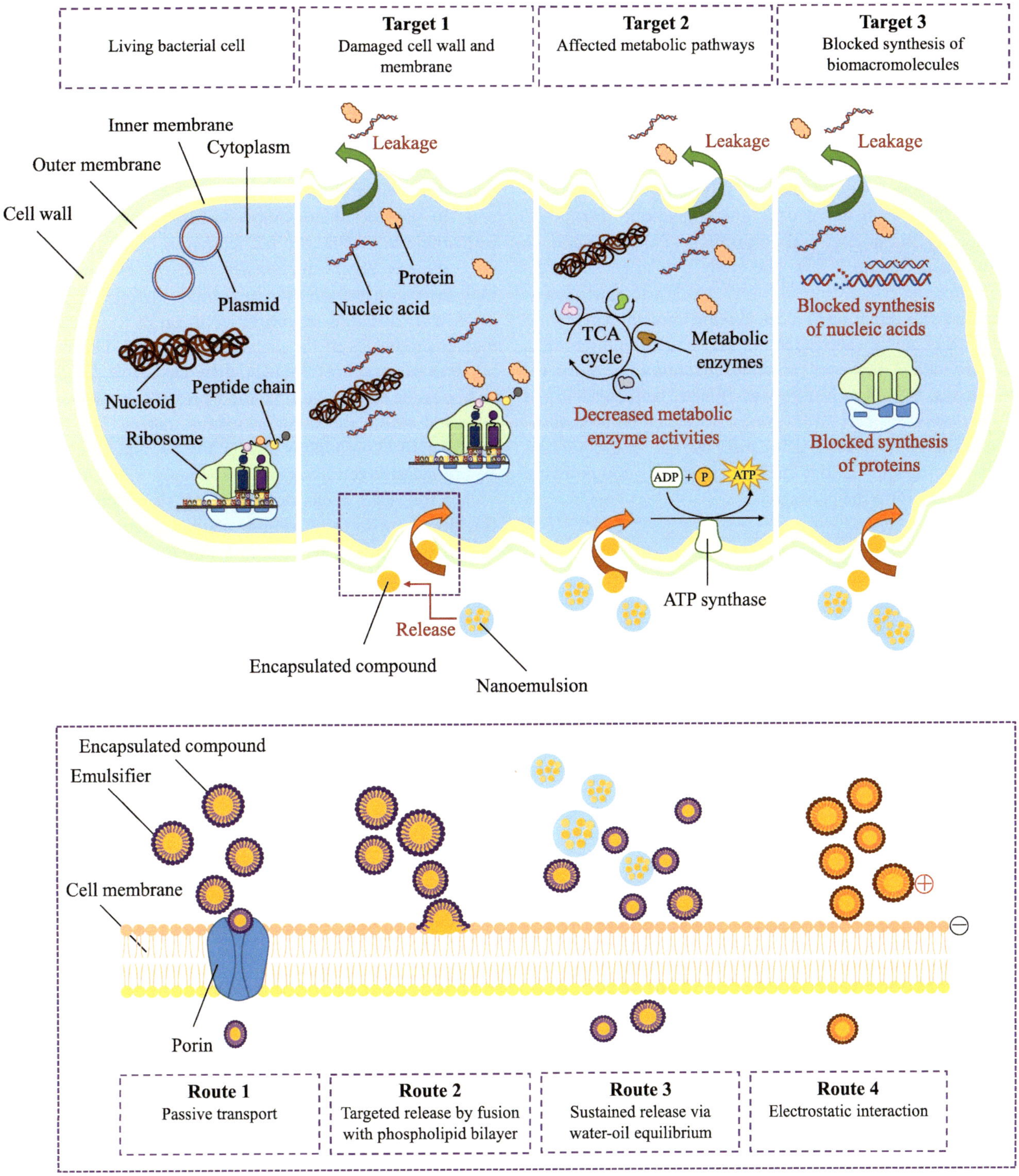

FIGURE 4.6 Antibacterial mechanism of nanoemulsions encapsulated with active components (adapted from Zhao et al. 2023).

attributed to their well-controlled size and permeability, oxygen-absorbing capability, and antimicrobial properties in general (Omerović et al. 2021). The organic nanoparticles, such as those made from lipids, can be digested by the human body, revealing their compatibility with direct food contact applications (Sagar et al. 2022). However, currently most of the developed nanoparticles for seafood preservation in literature is based on the essential antimicrobial and antioxidant characteristics of metal-based inorganic materials, involving titanium oxide (TiO_2), copper oxide (CuO), zinc oxide (ZnO), silver oxide (Ag_2O), ferric oxide (Fe_3O_4), and gold (Au), since they can also enhance the barrier properties and mechanical strength (Çiçek and Özoğul 2023; Sagar et al. 2022). The effectiveness of metal-based nanoparticles can be attributed to their capability in producing reactive oxygen species (ROS) in the form of hydroxide radicals and hydrogen peroxide (Kessler et al. 2022). The generation of significant amounts of hydrogen peroxide can scavenge the oxygen gas inside the packaging atmosphere, thereby the oxidative rancidity of stored seafoods. In addition, the free radicals are highly reactive in damaging the bacterial cellular components, as well as inhibiting the intracellular enzymes such as catalase and superoxide dismutase, suggesting the potential in controlling microbial spoilage (Çiçek and Özoğul 2023). The study by Cai et al. (2020) evaluated the effectiveness of ferric oxide nanoparticles on the freshness retention of red seabream upon thawing by far infrared or microwave radiation. The results on microbiota determination showed significant reduction of total microbial loads, as well as less growth of lactic acid bacteria and *Pseudomonas* species. However, it should be noted that the use of inorganic nanoparticles still poses food safety issues in packaging applications due to their toxicity and potential migration within the package.

4.3.4 Nanocomposite Films

Edible films and coatings are conventional materials for incorporating unstable small molecules in inert carriers for active packaging applications in seafoods (Koirala et al. 2023). Nanocomposite films are generated from polymer core molecules with additives and other active components. Numerous biopolymers possess film-forming properties upon the dissolution in the aqueous phase followed by the subsequent drying. The film formation can be conducted in either dry or wet formats, while the latter is the most prevalent, with solvent casting on a flat surface is the most studied from literature (Luo et al. 2022). To this end, the prepared nanocomposite film may not be present as a single layer, and the fabrication of multilayered films may enhance their functionality. On the one hand, such multilayered films are able to compile individual polymer films with different permeability in order to improve the barrier properties of the packaging material. On the other hand, in each layer of the film, different active components can be laden so that multiple food-preserving effects can be achieved within the same packaging matrix. The types of loaded active compounds can be diverse, and they can also be pre-synthesized nanoparticles. Figure 4.7 demonstrates such a four-layer nanocomposite film developed based on furcellaran, with the addition of silver nanoparticles, capsaicin, curcumin, and montmorillonite. This film exhibited substantial antioxidant capacity as determined by the DPPH and FRAP assays, with considerable antimicrobial activities contributed by curcumin, capsaicin, and silver nanoparticles (Grzebieniarz et al. 2023).

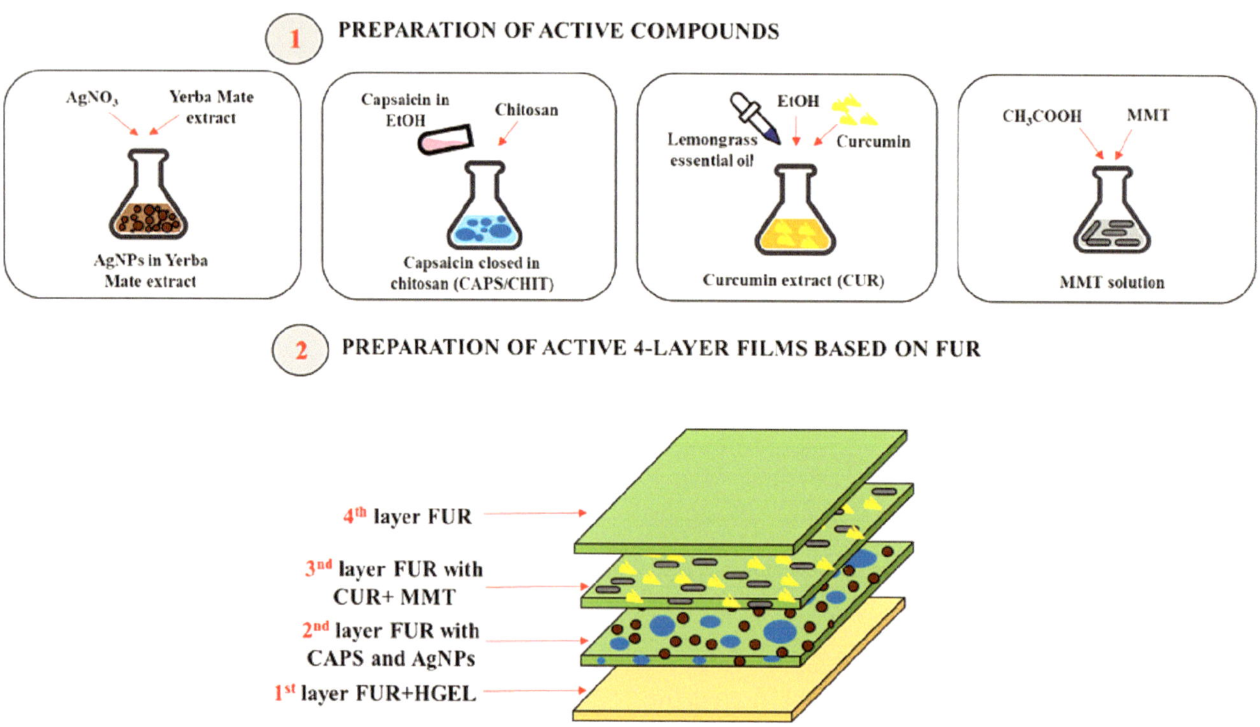

FIGURE 4.7 Design of a four-layer nanocomposite film intended for preserving fresh seafood products (adapted from Grzebieniarz et al. 2023).

4.4 CURRENT CHALLENGES AND FUTURE PROSPECTS

Overall, the application of nanomaterials in the preservation of perishable foods is still in the development stage due to technical, economical, and regulative challenges before moving forward to large-scale production. Risks of utilization of nanomaterials in food preservation is always a safety concern, regardless of their potential as preservatives. For instance, due to the toxic effects, the leakage and transmission of the metal-based nanoparticles into seafoods may raise the risk of human illnesses upon consumption (Adeyeye 2019). In particular, the ROS generated by such nanoparticles can attack human cells by damaging plasma membrane and the enclosed organelles, thereby influencing various metabolic activities such as biosynthesis of macromolecules and ATP production (Mahmud et al. 2022). From another standing point, the safety issues with the use of nanomaterials may be attributed to the potential migration inside the seafood packaging via mass transfer, although a few studies showed limited quantity of nanoparticles from food packaging materials were detected in foods (Wang et al. 2020a). Nevertheless, to be aligned with food laws and regulations in each country, the quantification measures of nanoparticles must be performed, and allowable levels should not be exceeded.

Another limitation is attributed by the multiple external dynamic factors that change the rate and extent of seafood spoilage, such as environmental pH, temperature, light exposure, and relative humidity. These factors alter the degree of all spoilage mechanisms as well as the efficacy of a number of active components being employed for seafood preservation. Currently there are no universal means to incorporate such nanomaterials into seafood packaging materials, with conceived stability and effectiveness at the market level. Thus, more validation studies are needed to consolidate the application of specific types of active components with their compatible nanomaterials, so that the gaps between the laboratory innovations and commercialization can be shortened.

In the future, more types of nanomaterials can be investigated for seafood preservation purposes to develop more versatile, effective, and economic packaging materials. In particular, the cost-effectiveness of newly developed nanomaterials should be evaluated in order to improve their affordability in underdeveloped regions. The ways for nanomaterials to be incorporated inside the package can be upgraded, preferably with no direct contact with food matrix by developing controlled release systems that can gradually release preservatives into the packaging headspace within the food's durable life. Another consideration is the creation of dual-function nanomaterials by incorporating the intelligent packaging concepts so that the real-time food quality can be disclosed to the stakeholders of the fishery supply chain. Last, but not least, the development of more biodegradable nanomaterials is important in further studies since the tremendous plastic packaging materials are raising the Earth's burden with prolonged decomposition rates.

4.5 CONCLUSION

In conclusion, this review summarizes the recent advances in preserving seafood products via the implementation of developed nanomaterials in different formats: nanoemulsions, electrospun nanofibers, nanoparticles, and nanocomposite films. As highly perishable foods, the spoilage of fish and shellfish occur immediately after harvest, and the deteriorative rate is affected by the intrinsic factors, such as the rate of enzymatic autolysis, lipid oxidation, and the initial microbial loads, but is also influenced by postmortem storage and preservation methods. The spoiled seafoods exhibit significant sensory profile changes, which are closely related to the postmortem pH decline, degradation of muscle tissues, generation of secondary lipid oxidation products, and off-flavors and color defects resulted from the proliferation of bacteria, particularly Gram-negative psychrotrophic species. The aforementioned seafood spoilage processes can be suppressed by the active component-embedded nanomaterials, which commonly possess antioxidant, antimicrobial, and enzyme-inhibiting characteristics. While a number of literature studies discussed their potential in seafood preservation under cold storage conditions, the potential interactions with fish and shellfish must be investigated in order to ensure the consumption safety and minimal impact on the alteration of seafood flavors. Meanwhile a more organized, well-defined regulative framework is needed to settle the standards on the official utilization of nanomaterials in the global fishery industry.

REFERENCES

Aboutorab M, Ahari H, Allahyaribeik S, et al (2021) Nano-emulsion of saffron essential oil by spontaneous emulsification and ultrasonic homogenization extend the shelf life of shrimp (*Crocus sativus* L.). *J Food Process Preserv* 45. https://doi.org/10.1111/jfpp.15224

Adeyeye SAO (2019) Food packaging and nanotechnology: Safeguarding consumer health and safety. *Nutr Food Sci* 49:1164–1179. https://doi.org/10.1108/NFS-01-2019-0020

Alboghbeish H, Khodanazary A (2019) The comparison of quality characteristics of refrigerated *Carangoides coeruleopinnatus* fillets with chitosan and nanochitosan coating. *Turkish J Fish Aquat Sci* :19957–967. https://doi.org/10.4194/1303-2712-v19_11_07

Ameur A, Bensid A, Özoğul F, et al (2022) Application of oil-in-water nanoemulsions based on grape and cinnamon essential oils for shelf-life extension of chilled flathead mullet fillets. *J Sci Food Agric* :102105–112. https://doi.org/10.1002/jsfa.11336

Anagnostopoulos DA, Parlapani FF, Boziaris IS (2022) The evolution of knowledge on seafood spoilage microbiota from the 20th to the 21st century: Have we finished or just begun? *Trends Food Sci Technol* :120236–247. https://doi.org/10.1016/j.tifs.2022.01.004

Asensio CM, Quiroga PR, Huang Q, et al (2019) Fatty acids, volatile compounds and microbial quality preservation with an oregano nanoemulsion to extend the shelf life of hake (*Merluccius hubbsi*) burgers. *Int J Food Sci Technol* :54149–160. https://doi.org/10.1111/ijfs.13919

Asl NM, Ahari H, Moghanjoghi AAM, Paidari S (2021) Assessment of nanochitosan packaging containing silver NPs on improving the shelf life of caviar (*Acipenser persicus*) and evaluation of nanoparticles migration. *J Food Meas Charact* :155078–5086. https://doi.org/10.1007/s11694-021-01082-7

Aswathanarayan JB, Vittal RR (2019) Nanoemulsions and their potential applications in food industry. *Front Sustain Food Syst* 3:95. https://doi.org/10.3389/fsufs.2019.00095

Azari A, Ahari H, Anvar AA (2022) Increased shelf life of *Oncorhynchus mykiss* (Rainbow trout) through Cu-Clay nanocomposites. *Food Sci Biotechnol* :31295–309. https://doi.org/10.1007/s10068-022-01031-0

Bahmani Z, Hosseini SV, Amanpour A (2021) Effect of nanocomposite clay/low-density polyethylene film on the quality of rainbow trout (*Oncorhynchus mykiss*) fillets stored with four different packaging conditions. *J Aquat Food Prod Technol* 30:1315–1329. https://doi.org/10.1080/10498850.2021.1988790

Benner RA, Staruszkiewicz WF, Otwell WS (2004) Putrescine, cadaverine, and indole production by bacteria isolated from wild and aquacultured penaeid shrimp stored at 0, 12, 24, and 36°C. *J Food Prot* :67124–133. https://doi.org/10.4315/0362-028X-67.1.124

Cai L, DAi Y, Cao A, Cao M (2020) The effects of CS@Fe_3O_4 nanoparticles combined with microwave or far infrared thawing on microbial diversity of red seabream (*Pagrus major*) fillets based on high-throughput sequencing. *Food Microbiol* 91:103511. https://doi.org/10.1016/j.fm.2020.103511

Çetinkaya T, Ceylan Z, MeRal R, et al (2021) A novel strategy for Au in food science: Nanoformulation in dielectric, sensory properties, and microbiological quality of fish meat. *Food Biosci* 41:101024. https://doi.org/10.1016/j.fbio.2021.101024

Ceylan Z (2019) A new cost-effective process for limitation of microbial growth in fish fleshes: Wrapping by aluminum foil coated with electrospun nanofibers. *J Food Saf* 39:e12697. https://doi.org/10.1111/jfs.12697

Ceylan Z, Atıcı C, Unal K, et al (2023) A novel material for the microbiological, oxidative, and color stability of salmon and chicken meat samples: Nanofibers obtained from sesame oil. *Food Res Int* 170:112952. https://doi.org/10.1016/j.foodres.2023.112952

Ceylan Z, Kutlu N, MeRal R, et al (2021) Protective effect of grape seed oil-loaded nanofibers: Limitation of microbial growth and lipid oxidation in kashar cheese and fish meat samples. *Food Biosci* 42:101076. https://doi.org/10.1016/j.fbio.2021.101076

Ceylan Z, Meral R, KoSe S, et al (2020) Characterized nano-size curcumin and rosemary oil for the limitation microbial spoilage of rainbow trout fillets. *LWT* 134:109965. https://doi.org/10.1016/j.lwt.2020.109965

Ceylan Z, Uslu E, İspirli H, et al (2019) A novel perspective for *Lactobacillus reuteri*: Nanoencapsulation to obtain functional fish fillets. *LWT* 115:108427. https://doi.org/10.1016/j.lwt.2019.108427

Chauhan SS, England EM (2018) Postmortem glycolysis and glycogenolysis: Insights from species comparisons. *Meat Sci* :144118–126. https://doi.org/10.1016/j.meatsci.2018.06.021

Chéret R, Delbarre-Ladrat C, LaMballerie-Anton M, Verrez-Bagnis V (2007) Calpain and cathepsin activities in post mortem fish and meat muscles. *Food Chem* 101:1474–1479. https://doi.org/10.1016/j.foodchem.2006.04.023

Chuesiang P, Sanguandeekul R, Siripatrawan U (2020) Phase inversion temperature-fabricated cinnamon oil nanoemulsion as a natural preservative for prolonging shelf-life of chilled Asian seabass (*Lates calcarifer*) fillets. *LWT* 125:109122. https://doi.org/10.1016/j.lwt.2020.109122

Çiçek S, Özoğul F (2023) Nanotechnology-based preservation approaches for aquatic food products: A review with the current knowledge. *Crit Rev Food Sci Nutr* :633255–3278. https://doi.org/10.1080/10408398.2022.2096563

Çorapci B (2022) The effect of rosehip seed oil nanoemulsion on some physical, chemical, and microbiological properties of sea bass fillets stored at 4±1°C. *J Aquat Food Prod Technol* :31672–685. https://doi.org/10.1080/10498850.2022.2094207

Costello C, Cao L, Gelcich S, et al (2020) The future of food from the sea. *Nature* :58895–100. https://doi.org/10.1038/s41586-020-2616-y

Dabadé DS, Yessoufou N, Adido L, et al (2023) Quality changes, potential spoilage organisms, and shelf-life prediction of brackish river prawn (*Macrobrachium macrobrachion*) at different storage temperatures. *Int J Food Microbiol* 405:110344. https://doi.org/10.1016/j.ijfoodmicro.2023.110344

de Oliveira TS, Costa AMM, Cabral LMC, et al (2023) Anthracnose controlled by essential oils: Are nanoemulsion-based films and coatings a viable and efficient technology for tropical fruit preservation? *Foods* 12:279. https://doi.org/10.3390/foods12020279

Durmuş M (2020) The effects of nanoemulsions based on citrus essential oils (orange, mandarin, grapefruit, and lemon) on the shelf life of rainbow trout (*Oncorhynchus mykiss*) fillets at 4 ± 2°C. *J Food Saf* 40. https://doi.org/10.1111/jfs.12718

Efatian H, Ahari H, ShahbazzaDeh D, et al (2021) Fabrication and characterization of LDPE/silver-copper/titanium dioxide nanocomposite films for application in Nile Tilapia (*Oreochromis niloticus*) packaging. *J Food Meas Charact* 15:2430–2439. https://doi.org/10.1007/s11694-021-00836-7

Ehira S, Uchiyama H (1987) Determination of fish freshness using the K value and comments on some other biochemical changes in relation to freshness. In: *Seafood Quality Determination: Proceedings of the International Symposium on Seafood Quality Determination*, edited by DE Kramer, J Liston. Elsevier Science Publishers B. V., Amsterdam.

Food and Agriculture Organization of the United Nations [FAO] (2023) *The state of world fisheries and aquaculture 2022*. Available at https://www.fao.org/3/cc0461en/online/sofia/2022/consumption-of-aquatic-foods.html. Accessed 20 November 2023

Fraga CG (2005) Relevance, essentiality and toxicity of trace elements in human health. *Mol Aspects Med* :26235–244. https://doi.org/10.1016/j.mam.2005.07.013

Freitas J, Vaz-Pires P, Câmara JS (2021) Quality Index Method for fish quality control: Understanding the applications, the appointed limits and the upcoming trends. *Trends Food Sci Technol* :111333–345. https://doi.org/10.1016/j.tifs.2021.03.011

Ghaly AE, Budge S, Brooks M (2010) Fish spoilage mechanisms and preservation techniques: Review. *Am J Appl Sci* :7859–877. https://doi.org/10.3844/ajassp.2010.859.877

Gram L (2009) Microbiological spoilage of fish and seafood products. In: *Compendium of the Microbiological Spoilage of Foods and Beverages*. Springer New York, New York, NY, pp. 87–119.
Gram L, Dalgaard P (2002) Fish spoilage bacteria—problems and solutions. *Curr Opin Biotechnol* :13262–266. https://doi.org/10.1016/S0958-1669(02)00309-9
Gram L, Huss HH (2000) Fresh and processed fish and shellfish. In: *The Microbiological Safety and Quality of Food*. Aspen Publishers, Gaithersburg, MD, pp. 472–506.
Grzebieniarz W, Tkaczewska J, Juszczak L, et al (2023) Improving the quality of multi-layer films based on furcellaran by immobilising active ingredients and impact assessment of the use of a new packaging material. *Food Chem* 428:136759. https://doi.org/10.1016/j.foodchem.2023.136759
Hai Y, Zhou D, Lam YLN, et al (2022) Nanoemulsified clove essential oils-based edible coating controls Pseudomonas spp.-causing spoilage of tilapia (*Oreochromis niloticus*) fillets: Working mechanism and bacteria metabolic responses. *Food Res Int* 159:111594. https://doi.org/10.1016/j.foodres.2022.111594
Hajji S, Hamdi M, Boufi S, et al (2019) Suitability of chitosan nanoparticles as cryoprotectant on shelf life of restructured fish *surimi* during chilled storage. *Cellulose* :266825–6847. https://doi.org/10.1007/s10570-019-02555-1
Heinsz LJ, Harrison MA, Leiting VA (1988) Microflora of brown shrimp (*Penaeus aztecus*) from Georgia coastal waters. *Food Microbiol* :5141–145. https://doi.org/10.1016/0740-0020(88)90012-3
Hernández A, García García B, Jordán MJ, Hernández MD (2014) Improved conservation of gilthead seabream (*Sparus aurata*) in ice storage. The influence of doses of rosemary extract added to feed. *Aquaculture* 426–427:31–40. https://doi.org/10.1016/j.aquaculture.2014.01.018
Hong H, Regenstein JM, Luo Y (2017) The importance of ATP-related compounds for the freshness and flavor of post-mortem fish and shellfish muscle: A review. *Crit Rev Food Sci Nutr* 00–00. https://doi.org/10.1080/10408398.2014.1001489
Hosomi R, Yoshida M, Fukunaga K (2012) Seafood consumption and components for health. *Glob J Health Sci* 4. https://doi.org/10.5539/gjhs.v4n3p72
Huang X, Wang H, Tu Z (2023) A comprehensive review of the control and utilization of aquatic animal products by autolysis-based processes: Mechanism, process, factors, and application. *Food Res Int* 164:112325. https://doi.org/10.1016/j.foodres.2022.112325
Kazemeini H, Azizian A, Shahavi MH (2019) Effect of chitosan nano-gel/emulsion containing *Bunium Persicum* essential oil and nisin as an edible biodegradable coating on *Escherichia coli* O157:H7 in rainbow trout fillet. *J Water Environ Nanotechnol* :4343–349. https://doi.org/10.22090/JWENT.2019.04.008
Kessler A, Hedberg J, Blomberg E, Odnevall I (2022) Reactive oxygen species formed by metal and metal oxide nanoparticles in physiological media-a review of reactions of importance to nanotoxicity and proposal for categorization. *Nanomaterials* 12(11):1922. https://doi.org/10.3390/nano12111922
Khanzadi S, Keykhosravy K, HasheMi M, Azizzadeh M (2020) Alginate coarse/nanoemulsions containing *Zataria multiflora* Boiss essential oil as edible coatings and the impact on microbial quality of trout fillet. *Aquac Res* :51873–881. https://doi.org/10.1111/are.14418
Khoshbouy Lahidjani L, Ahari H, Sharifan A (2020) Influence of curcumin-loaded nanoemulsion fabricated through emulsion phase inversion on the shelf life of *Oncorhynchus mykiss* stored at 4°C. *J Food Process Preserv* 44. https://doi.org/10.1111/jfpp.14592
Koirala P, Nirmal NP, Worapravote W, et al (2023) Nano-engineered edible films and coatings for seafood products. *Food Packag Shelf Life* 38:101135. https://doi.org/10.1016/j.fpsl.2023.101135
Kosker AR (2020) The effects of nanoemulsions based on citrus essential oils on the formation of biogenic amines in trout fillets stored at 4 ± 2°C. *J Food Saf* 40. https://doi.org/10.1111/jfs.12762
Li L, Liu D, Li X, et al (2023) The dynamic changes of microbial diversity and biogenic amines in different parts of bighead carp (*Aristichthys nobilis*) head during storage at −2 °C. *J Food Compos Anal* 118:105228. https://doi.org/10.1016/j.jfca.2023.105228
Liu J, Shao Y, Yuan C, et al (2021) Eugenol-chitosan nanoemulsion as an edible coating: Its impact on physicochemical, microbiological and sensorial properties of hairtail (*Trichiurus haumela*) during storage at 4 °C. *Int J Biol Macromol* 183:2199–2204. https://doi.org/10.1016/j.ijbiomac.2021.05.183
Liu T, Liu J, Wang P, et al (2023) Effect of slurry ice on quality characteristics and microbiota composition of Pacific white shrimp during refrigerated storage. *J Agric Food Res* 14:100792. https://doi.org/10.1016/j.jafr.2023.100792
Luo X, Ho I, Brankovan S, Lim L-T (2021) Inkjet-printed gradient colorimetric indicators for monitoring fish freshness. *Food Packag Shelf Life* 29:100719. https://doi.org/10.1016/j.fpsl.2021.100719
Luo X, Lim L-T (2019) Cinnamil-and quinoxaline-derivative indicator dyes for detecting volatile amines in fish spoilage. *Molecules* 24:3673. https://doi.org/10.3390/molecules24203673
Luo X, Lim L-T (2020) Curcumin-loaded electrospun nonwoven as a colorimetric indicator for volatile amines. *LWT—Food Sci Technol* 128. https://doi.org/10.1016/j.lwt.2020.109493
Luo X, Zaitoon A, Lim L-T (2022) A review on colorimetric indicators for monitoring product freshness in intelligent food packaging: Indicator dyes, preparation methods, and applications. *Compr Rev Food Sci Food Saf*. https://doi.org/10.1111/1541–4337.12942
Mahmud G, Zhang H, Douglas JF (2022) The dynamics of metal nanoparticles on a supporting interacting substrate. *J Chem Phys* 157(11):114505. https://doi.org/10.1063/5.0105208
Mariutti LRB, Bragagnolo N (2017) Influence of salt on lipid oxidation in meat and seafood products: A review. *Food Res Int* :9490–100. https://doi.org/10.1016/j.foodres.2017.02.003
Meral R, Ceylan Z, Kose S (2019) Limitation of microbial spoilage of rainbow trout fillets using characterized thyme oil antibacterial nanoemulsions. *J Food Saf* 39. https://doi.org/10.1111/jfs.12644
Nasiri M, Ahari H, Sharifan A, et al (2020) Nanoemulsion production techniques upgrade bioactivity potential of nanoemulsified essential oils on *Acipenser stellatus* filet preserving. *Int J Food Prop* 23:2174–2188. https://doi.org/10.1080/10942912.2020.1844749
Nazari M, Majdi H, Milani M, et al (2019) Cinnamon nanophytosomes embedded electrospun nanofiber: Its effects on microbial quality and shelf-life of shrimp as a novel packaging. *Food Packag Shelf Life* 21:100349. https://doi.org/10.1016/j.fpsl.2019.100349

Nie X, Zhang R, Cheng L, et al (2022) Mechanisms underlying the deterioration of fish quality after harvest and methods of preservation. *Food Control* 135:108805. https://doi.org/10.1016/j.foodcont.2021.108805

Odeyemi OA, Burke CM, Bolch CCJ, Stanley R (2018) Seafood spoilage microbiota and associated volatile organic compounds at different storage temperatures and packaging conditions. *Int J Food Microbiol* :28087–99. https://doi.org/10.1016/j.ijfoodmicro.2017.12.029

Odeyemi OA, Burke CM, Bolch CCJ, Stanley R (2019) Spoilage microbial community profiling by 16S rRNA amplicon sequencing of modified atmosphere packaged live mussels stored at 4°C. *Food Res Int* :121568–576. https://doi.org/10.1016/j.foodres.2018.12.017

Omerović N, Djisalov M, Živojević K, et al (2021) Antimicrobial nanoparticles and biodegradable polymer composites for active food packaging applications. *Compr Rev Food Sci Food Saf* 20:2428–2454. https://doi.org/10.1111/1541–4337.12727

Oner B, Meral R, Ceylan Z (2021) Determination of some quality indices of rainbow trout fillets treated with nisin-loaded polyvinylalcohol-based nanofiber and packed with polYethylene package. *LWT* 149:111854. https://doi.org/10.1016/j.lwt.2021.111854

Özogul Y, El Abed N, Özogul F (2022) Antimicrobial effect of laurel essential oil nanoemulsion on food-borne pathogens and fish spoilage bacteria. *Food Chem* 368:130831. https://doi.org/10.1016/j.foodchem.2021.130831

Parlapani FF (2021) Microbial diversity of seafood. *Curr Opin Food Sci* :3745–51. https://doi.org/10.1016/j.cofs.2020.09.005

Parlapani FF, FerrocIno I, Michailidou S, et al (2020) Microbiota and volatilome profile of fresh and chill-stored deepwater rose shrimp (*Parapenaeus longirostris*). *Food Res Int* 132:109057. https://doi.org/10.1016/j.foodres.2020.109057

Parlapani FF, Michailidou S, Anagnostopoulos DA, et al (2019) Bacterial communities and potential spoilage markers of whole blue crab (*Callinectes sapidus*) stored under commercial simulated conditions. *Food Microbiol* :82325–333. https://doi.org/10.1016/j.fm.2019.03.011

Parlapani FF, Syropoulou F, TsiArtsafis A, et al (2020) HRM analysis as a tool to facilitate identification of bacteria from mussels during storage at 4 °C. *Food Microbiol* 85:103304. https://doi.org/10.1016/j.fM.2019.103304

Pateiro M, Domínguez R, Munekata PES, et al (2022) Marine sources: Fish, shellfish, and algae. In: *Food Lipids*. Elsevier, pp. 51–68.

Peng S, Wei H, Zhan S, et al (2022) Spoilage mechanism and preservation technologies on the quality of shrimp: An overview. *Trends Food Sci Technol* :129233–243. https://doi.org/10.1016/j.tifs.2022.09.024

Prabhakar PK, VatSa S, Srivastav PP, Pathak SS (2020) A comprehensive review on freshness of fish and assessment: Analytical methods and recent innovations. *Food Res Int* 133:109157. https://doi.org/10.1016/j.foodres.2020.109157

Qiu L, Zhang M, Bhandari B, Yang C (2022) Shelf life extension of aquatic products by applying nanotechnology: A review. *Crit Rev FooD Sci Nutr* 62:1521–1535. https://Doi.org/10.1080/10408398.2020.1844139

Rittenschober D, Nowak V, Charrondiere UR (2013) Review of availability of food composition data for fish and shellfish. *Food Chem* :1414303–4310. https://doi.org/10.1016/J.foodchem.2013.07.007

Rolin J, Horricks RA, Stephen J, et al (2023) Changes in bacterial communities in chilled American lobster (Homarus *americanus*) tiSsues following mortality. *LWT* 185:115136. httpS://doi.org/10.1016/j.lwt.2023.115136

Safaeian Laein S, Khanzadi S, HasheMi M, et al (2023) Improving quality of trout fillet using gelatin coating-contain peppermint essential oil loaded solid lipid NAnoparticles (PEO-SLN). *J Food Meas Charact*. https://doi.org/10.1007/s11694-023-02157-3

Sagar NA, KumaR N, ChoudhaRy R, et al (2022) Prospecting the role of nanotechnology in extending the shelf-life of fresh produce and in developing advanced packaGing. *Food PackaG Shelf Life* 34:100955. https://doi.orG/10.1016/j.fpsl.2022.100955

Secci G, Parisi G (2016) From farm to fork: Lipid oxidation in fish products. A review. *Ital J Anim Sci* :15124–136. https://doi.orG/10.1080/1828051X.2015.1128687

Sequino G, VAlentino V, Esposito A, et al (2024) Microbiome dynamics, antibiotic resistance gene patterns and spoilage-associated genomic potential in fresh anchovies stored in different conditions. *Food Res Int* 175:113788. https://doi.org/10.1016/j.foodres.2023.113788

Seray M, Hadj-Hamou AS, Uzunlu S, Benhacine F (2022) Development of active packaging films based on poly (butylene adipate-co-terephthalate) and silver–montmorillonite for shelf life extension of sea bream. *Polym Bull* :793573–3594. https://doi.org/10.1007/s00289-021-03671-4

Shalders TC, Champion C, ColeMAn MA, Benkendorff K (2022) The nutritional and sensory quality of seafood in a changing climate. *Mar Environ Res* 176:105590. https://doi.org/10.1016/J.marenvres.2022.105590

Shao J, Wang L, Wang X, Ma J (2021) Enhancing microbial management and shelf life of shrimp *Penaeus vannamei* by using nanoparticles of metallic oxides as an alternate active packaging tool to synthetic chemicals. *Food Packag Shelf Life* 28:100652. https://doi.org/10.1016/j.fpsl.2021.100652

Shehata HR, Mitterboeck TF, Hanner R (2020) Characterization of the microbiota of commercially traded finfish fillets. *Food Res Int* 137:109373. https://doi.org/10.1016/j.foodres.2020.109373

Sikorski Z, Kostuch S (1982) Trimethylamine N-oxide demethylase: Its occurrence, properties, and rôle in technological changes in frozen fish. *Food CheM* :9213–222. https://doi.org/10.1016/0308-8146(82)90099-1

Solovyev M, Kashinskaya E, Gisbert E (2023) A meta-analysis for assessing the contributions of trypsin and chymotrypsin as the two major endoproteases in protein hydrolysis in fish intestine. *Comp Biochem Physiol Part A Mol Integr Physiol* 278:111372. https://doi.org/10.1016/j.cbpa.2023.111372

Sørensen JS, ØrNfeld-Jensen O, BøkNæs N, et al (2020) Thawed and chilled Atlantic cod (*Gadus morhua* L.) From Greenland—Options For improved distribution. *LWT* 131:109473. https://doi.org/10.1016/j.lwt.2020.109473

Syropoulou F, ParlapanI FF, BosmalI I, et al (2020) HRM and 16S rRNA gene sequenCing reveal the Cultivable miCrobiota of the European sea bass during iCe storage. *Int J Food MiCrobiol* 327:108658. https://doi.org/10.1016/j.ijfoodmiCro.2020.108658

Tan C, Li P, Shang N (2023) Novel perspective on the spoilage metabolism of refrigerated sturgeon fillets: Nonspecific spoilage dominant organisms play an important rolE. *LWT* 173:114292. https://doi.org/10.1016/j.lwt.2022.114292

Veliyulin E, Felberg H, Digre H, MartiNez I (2007) Non-destructive Nuclear magNetic resoNance image study of belly burstiNg iN herriNg (*Clupea hareNgus*). *Food Chem* 101:1545–1551. https://doi.org/10.1016/j.foodchem.2006.04.007

Walayat N, Tang W, Wang X, et al (2023) Quality evaluation of frozen and chilled fish: A review. *eFood* 4:. https://doi.org/10.1002/efd2.67

Wang J, Li J, Zhang Y, et al (2023a) Changes in physicochemical properties, moisture distribution and volatile flavor compounds of yellow catfish: The roles of dominant spoilage bacteria. *Food Biosci* 56:103331. https://doi.org/10.1016/j.fbio.2023.103331

Wang M, Kong Q, Yu H, et al (2023b) Identification of dominant spoilage bacteria, bacterial diversity analysis, and antibacterial exploration of chilled monkfish. *Food Biosci* 56:103121. https://doi.org/10.1016/j.fbio.2023.103121

Wang W, Yu Z, Alsammarraie FK, et al (2020a) Properties and antimicrobial activity of polyvinyl alcohol-modified bacterial nanocellulose packaging films incorporated with silver nanoparticles. *Food Hydrocoll* 100:105411. https://doi.org/10.1016/j.foodhYd.2019.105411

Wang Y, Hong X, Liu J, et al (2020b) Interactions between fish isolates *Pseudomonas fluorescens* and *Staphylococcus aureus* in dual-species biofilms and sensitivity to carvacrol. *Food Microbiol* 91:103506. https://doi.org/10.1016/j.fM.2020.103506

Yan Q, Guo M, Chen B, et al (2023) Molecular characterization of spoilage microbiota in high CO_2 refrigerated large yellow croaker (*Larimichthys crocea*) fillets using metagenomic and metabolomic approaches. *Food Biosci* 56:103227. https://doi.org/10.1016/j.fbio.2023.103227

Yang Z, Chu Y, Zhang C, et al (2023) Changes in the physicochemical properties of grouper (*Epinephelus coioides*) fillets stored under vacuum packaging at chilly temperature contributing witH tHe spoilage bacteria. *Food Biosci* 55:103017. Https://doi.org/10.1016/j.fbio.2023.103017

Yazgan H, Ozogul Y, Kuley E (2019) Antimicrobial influence of nanoemulsified lemon essential oil and pure lemon essential oil on food-borne pathogens and fish spoilage bacteria. *Int J Food Microbiol* 306:108266. https://doi.org/10.1016/J.ijfoodmicro.2019.108266

Zhang J, Li Y, Liu X, et al (2019) Characterization of the microbial composition and quality of lightly salted grass carp (*Ctenopharyngodon idellus*) fillets with vacuum or modified atmosphere packaging. *Int J Food MicrobioL* :29387–93. https://doi.org/10.1016/j.ijfoodmicro.2018.12.022

Zhang L, Yu D, Xu Y, et al (2023) Changes in quality and microbial diversity of refrigerated carp fillets treated by chitosan/zein bilayer film with curcumin/nisin-loaded pectin nanoparticles. *Food Biosci* 54:102941. https://doi.org/10.1016/J.fbio.2023.102941

Zhao J, Lan W, Xie J (2023) Recent developments in nanoemulsions against spoilage in cold-stored fish: A Review. *Food Chem* 429:136876. https://doi.oRg/10.1016/j.foodchem.2023.136876

Zhao R, Guan W, Zheng P, et al (2022) Development of edible composite film based on chitosan nanoparticles and their application in packaging of fresh red sea bream fillets. *Food Control* 132:108545. Https://doi.org/10.1016/j.foodcont.2021.108545

Zhong H, Wei S, Kang M, et al (2023) Effects of different storage conditions on microbial community and quality changes of greater amberjack (*Seriola dumerili*) fillets. *LWT* 179:114640. httpS://doi.org/10.1016/j.lwt.2023.114640

Zhuang S, Li Y, Hong H, et al (2020) Effects of ethyl lauroyl arginate hydrochloride on microbiota, quality and biochemical changes of container-cultured largemouth bass (*Micropterus salmonides*) fillets during storage at 4 °C. *Food Chem* 324:126886. httpS://doi.org/10.1016/j.foodchem.2020.126886

Zhuang S, Tan Y, Hong H, et al (2022) Exploration of the roles of spoilage bacteria in degrading grass carp proteins during chilled storage: A combined metagenomic and metabolomic approach. *Food Res Int* 152:110926. https://doi.org/10.1016/j.foodres.2021.110926

5 Nanotechnology in Cereal and Bakery Products Preservation

Sameer Ahmad[a], *Sarushi Rastogi*[b], *Mohsin Vahid Khan*[a], *Dawrul Islam*[c], *and Arun Kumar Gupta*[d]

[a]Departement of Bioengineering integral University, Integral University Lucknow, India

[b]Department of Nutrition and Dietetics, School of Allied Health Sciences, Manav Rachna International Institute of Research and Studies, Faridabad, India

[c]National Institute of Food Technology Entrepreneurship and Management, Kundli, Haryana, India

[d]Department of Food Science & Technology, Graphic Era (Deemed to be University), Dehradun, Uttarakhand, India

5.1 INTRODUCTION

The utilization of active packaging in bakery items has significant promise due to the swift advancement of nanotechnology. This might potentially avoid product deterioration and prolong food shelf life. The important fungi responsible for the baked products' spoilage are *penicillium* and *rhizopous*; *Wallemia* specifically in bread packaging can be controlled by nanopackaging. In general, bread is the most widely used baked product, however it's the fungi that lead to its shortened shelf life, Moreover, using artificial food additives and chemical preservatives in bakery products is constantly under challenges by the food regulatory organizations; additionally, it leads the risk of chronic and non-chronic diseases, as reported by Smith et al. (2004); therefore it is needed to look for alternatives such as natural biopreservatives nanocomposites, which are getting popular nowadays with the public health concern. The nanocomposites made of natural preservatives such as essential oils derived from oregano and cloves up to (4% w/v) when incorporated on bread packaging revealed a significant drop in mold and the yeast count. The wheat bread packaging comprised of Ag/TiO_2 nanocomposite (Ag/TiO_2-P), developed from the sol-gel method, showed better shelf life comparison to HDPE packaged bread, even when the bread was stored without a controlled storage system (Mihaly Cozmuta et al., 2015). Thus nanotechnology has the potential to help the bakery industry in prolonging the shelf life of food items made from grain and bakeries by enhancing food safety; moreover, the preservation mechanism co-related to the antioxidant activity, better control of water vapor, and oxygen retention activities of nanopackaging (Mihaly Cozmuta et al., 2015).

Cereals are a convenient and affordable supply of plant-based proteins, which are essential for food's functionality and nutritional value. If not properly stored, cereals and cereal products are perishable and have a short shelf life. Food products have been preserved using methods like radiofrequency heating and irradiation to extend their shelf life. Nanotechnology is becoming a significant innovation with the potential to significantly improve sustainability. The growing demand for nanotechnological applications has given rise to new uses for nanotechnology in different sectors of food science such as food development, food processing, food packaging, and food safety.

Some of the nanotechnological innovations applied in the preservation of cereals and cereal-based products are discussed next. Nanofibers derived from starch showed properties of nanocarriers, as in baked goods like bread and cookies, there is a problem of mold growth after 5th day. However, synthetic food additives have limited effects and sometimes may lead to adverse health effects, therefore the need for natural antifungal agents like carvacrol incorporated in nanofibers showed antifungal activity against *Penicillium* sp. and *Aspergillus flavus* (Fonseca et al., 2021).

5.2 NANOENCAPSULATION FOR THE PRESERVATION OF CEREALS-BASED FOOD

Nanotechnology is an emerging technique in almost every sector from food to agriculture and medicine to mechanics. In the food industries nanotechnology provides a wide array of opportunities for developing new applications and products. The main targets of nanotechnology in food industries can be to increase the delivery and absorption of intended bioactive compounds and nutrients and enhance product shelf life, organoleptic characteristics, and sensory properties.

5.2.1 Traditional Preservation and Quality Enhancement Techniques in Bakery Food

In breads the preservative should show superior antifungal properties to increase shelf life. Lactic acid bacteria strains and yeast used in microbial fermentation have

DOI: 10.1201/9781032677422-5

garnered the most attention among these methods. For example, bread slices treated for up to 21 days with strains of *Lactobacillus* bacteria suppressed fungal growth. Furthermore, plant extracts and essential oils like garlic and lemongrass oil have demonstrated the potential to lower the growth of fungi on bread and other bakery goods. In conclusion, newer biopreservation techniques that have attracted attention recently are also included, including the application of modified atmospheric packaging, active packaging, nanofibers, and whey. These methods show promise for addressing problems with bread spoiling and enhancing the durability and safety of bread products.

5.2.2 Enzymatic Assistance (Bread Manufacturing)

Growing consumer health consciousness has made traditional preservatives less popular, which has forced the bakery sector to look for safer substitutes. Because of their antibacterial qualities, essential oils have become recognized as viable alternatives, either as ingredients or in packaging systems.

5.2.2.1 Breakdown of Starches by Amylases

Among the starch hydrolytic enzymes, amylases are the enzymatic method that reduces bread staling the fastest (Grewal et al., 2015). Amylases degrade molecules of starch into dextrin and other unbranched short-chain molecules. The α-amylases are extensively used enzymes for bread making and play a vital role because of their diverse function related to quality improvement in baking industries. The α-amylases are responsible for the catalysis of endohydrolysis of α–1–4–D-glycosidic bond in the amylopectin and amylose molecules present in the flour starch. It can also behave as α-exo-amylase causing the generation of α-maltose from the non-reducing end. Hence, the varying length of maltose, oligosaccharides, and α-configured glucose released in increasing amounts causing changes in the structure of starch (Rebholz et al., 2021). Adding α-amylase to bread dough improves fermentation by providing additional substrates for yeast cells, including maltose and malto-oligosaccharides. This results in a longer and more advantageous fermentation process because yeast cells can ferment these sugars, shifting the metabolism from glucose to maltose.

There are three reasons that G4 amylase is attributed to the anti-staling effect. First, they significantly decrease the starch polymer molecular size. The shorter external amylopectin chains would contribute to a smaller amount of retrogradation. Second, the greatly hygroscopic nature of maltotetraose can retain molecules of water that surround starch molecules. This impedes the interaction between starch–starch molecules. Last, the dimension of maltotetraose might be preferable to inhibit the interaction between polymers of starch. The steric interference by this molecule would interfere with the formation of intermolecular hydrogen bonds.

5.2.2.2 Protein Modification by Proteases

Proteases have been widely investigated and used for improving bread quality. They enhance the dough volume and decrease the hardness of the bread. The bacterial proteases can hydrolyze the protein matrix that is water soluble in which granules of starch are embedded. This leads to the discharge of starch granules to form a tougher starch gel during baking (Renzetti & Arendt, 2009). Gluten, also known as an insoluble protein, controls the mass expansion capacity and viscoelastic properties throughout the process of baking. The enzyme proteases are utilized for the modification of gluten. These enzymes hydrolyze or proteolysis the protein by cleaving their peptide bonds to smaller peptides resulting increase in spread ratio and a decrease in gluten elasticity (Fallahi et al., 2018). Proteases are utilized in bakery industries to reduce the texture and mixing time, assure dough consistency, enhance flavor, and retain the gluten networking in the bread. The incorporation of proteases aids in breaking down gluten protein for more extensible and softer dough. Renzetti and Arendt (2009) explained that treating brown rice with protease resulted in enhanced baking and textural characteristics of bread made from brown rice. The specific volume of bread increased significantly, whereas chewiness and crumb hardness decreased. Furthermore, using *Bacillus stearothermophilus* protease to make gluten-free rice bread resulted in higher bread quality, such as a nice look of crumbs, large volume, low rate of stalling, and soft texture, as compared to the control sample of bread. Sarabhai et al. (2021) revealed that the use of protease enzymes in the manufacturing of gluten-free foxtail millet bread significantly reduces crumb hardness, cohesiveness, and chewiness. Nanomaterials have been incorporated into food packaging materials to help combat foodborne pathogens by causing their death through interaction with nanoparticles. The use of nanostructured antimicrobial materials is advantageous due to their death through interaction with nanoparticles. In contrast to microencapsulation, which preserves antimicrobial chemicals against degradation or evaporation, the huge surface area to volume ratio of nanoencapsulation systems permits the concentration of antimicrobials in food areas where bacteria are preferentially located. Baked products such as bread and cakes are highly susceptible to molds, yeasts, and bacteria; preservatives are required to increase their shelf life. In a study, zien nanocapsules were formed and loaded with thyme and oregano essential oils and were tested for their antibacterial efficacy on bread. It was found that with no colonies of mold or yeast, the product's shelf life was increased by 21 days upon the inclusion of the loaded nanocapsules containing the essential oils. Zein nanocapsules were also capable of shielding the encapsulated oils from baking. Even after the bread was cooked at 200°C, the oils from the oregano and thyme did not deteriorate (Sganzerla et al., 2023). Nowadays nanotechnology is being used to enhance the shelf life of bakery items by nanoencapsulation of antibacterial compounds. Encapsulation, in general, regulates the distribution of bioactive agents and shields them from unfavorable chemical reactions and interactions with dietary ingredients.

5.3 NANOPARTICLE COATED ANTIMICROBIALS

The utilization of small particles known as nanoparticles in place of antibiotics for fighting bacteria is advancing. Bacterial infections can be effectively treated with the use of this nanotechnology. For instance, coatings that prevent bacteria from developing on items like medical gadgets are created by scientists using nanoparticles. They also aid in the quicker healing of wounds. Antibiotics can also be directly delivered to the microorganism cell where they are required to treat illnesses using nanoparticles. They're even utilized in devices that enable rapid bacterial detection, which aids medical professionals in determining the cause of a patient's illness. Additionally, researchers are creating active packaging material inhibition of bacteria with nanoparticles. Entirely new approaches to combat bacterial infections are provided by nanotechnology.

This method of producing nanoparticles extracellularly might be beneficial for the fermentation sector. Bacterial strains that have contaminated materials could endanger fermentation processes. Utilized in fermentation procedures, yeast strains can generate nanoparticles as a self-sufficient defense against harmful microorganisms that stop the fermentation process.

Economic, environmental, and safety concerns have prompted industry and research groups to explore alternatives to petrochemical-based polymers in the form of active antimicrobial and biodegradable food packaging designs. The primary object of such packaging is to protect food and its products from environmental factors, prevent the growth of microbes on food surfaces, and prolong the shelf life of food while maintaining its quality and safety following international standards.

Due to such concerns about artificial chemical additions, foods preserved using natural ingredients have grown in popularity. Antimicrobials can be introduced directly into product formulations coated on product surfaces or integrated into packaging materials to prevent the growth of unwanted microbes in food. Bacterial populations are immediately but temporarily reduced when active agents are directly added to food; however, antimicrobial coatings remain active for extended periods. As antimicrobials, chitosan, citrus peel essential oil, mustard, lecithin, thymol, lemon extract, and ascorbic acid have all been employed to preserve cereal-based goods (Deseta et al., 2021).

Many of these natural substances are more costly, less effective than synthetic additions, or may interact adversely with dietary ingredients (Lucera et al., 2012). From this perspective, encapsulation techniques provide antimicrobial compounds with effective protection against chemical reactions and unfavorable interactions with other food ingredients, as well as increased solubility, reduced migration, and bioactive compound stability during processing and storage. Moreover, the encapsulation of bioactive substances enhances absorption and bioavailability while controlling administration.

However, nanoencapsulation systems, or systems with a nanometre size smaller than 100 nm, can improve passive cellular absorption mechanisms, resulting in higher antimicrobial activity and increasing the concentration of antimicrobials in specific food areas where microorganisms are preferably located. Furthermore, the use of nanoencapsulation techniques facilitates the addition of antimicrobial compounds to materials used in food packaging (Blanco-Padilla et al., 2014). Lipid nanoparticles, nanoemulsions, nanocapsules, and the shelf life of sourdough bread was enhanced using a food-grade coating composed of quinoa protein, chitosan, cinnamon, and lemongrass essential oils encased in 6% w/v mesoporous silica nanoparticles. Under accelerated shelf-life studies, the coated bread demonstrated low hardness, which inhibited bread staling while delaying color loss and microbiological deterioration. The coated bread was found to be palatable and demonstrated antifungal properties against *Rhizophus stolonifer* (Bautista-Espinoza et al., 2023). As shown in Figure 5.1 the use of nanostructured antimicrobial materials is advantageous due to their higher surface area to volume ratio when compared to their large-scale counterparts. Recent studies have shown that the incorporation of ZnO nanoparticles with polymeric matrices and composite materials can provide improved packaging properties (Sivaprakash et al., 2022).

A recent study looks at the effects of ferric oxide nanoparticles (NPs) on APO activity in seedlings of durum and bread wheat. The responses of the durum wheat seedlings vary, as evidenced by the results: Some showed enhanced APO activity, while others showed a decrease or no change. Mixed results were also observed in bread wheat seedlings. These results imply that different wheat cultivars react to NPs in different ways, which may aid in the selection of cultivars that are more resistant to environmental stresses (Soltani et al., 2023). To improve the shelf life of rye bread, researchers created a film-based solution containing titanium dioxide nanoparticles (TiO_2 NPs), chitosan (CS), and polylactic acid (PLA). The water vapor permeability (WVP) (0.322 × 10–7 g m–1 s–1 Pa–1), water solubility (2.83%), and oxygen permeability (OP) (1.85 × 10–7 g–1 s–1 Pa–1) metrics were all reduced for the film-coated rye bread with PLA-CS 3%-TiO_2 5%.

5.4 NANOCOMPOSITES

Nowadays consumers have a keen interest in food products that don't include artificial preservatives; moreover, without adding chemical preservatives directly, bread's microbial shelf life can be effectively extended by employing active packaging as a treatment. Superior barrier performance against volatiles, moisture, UV light, carbon dioxide, and

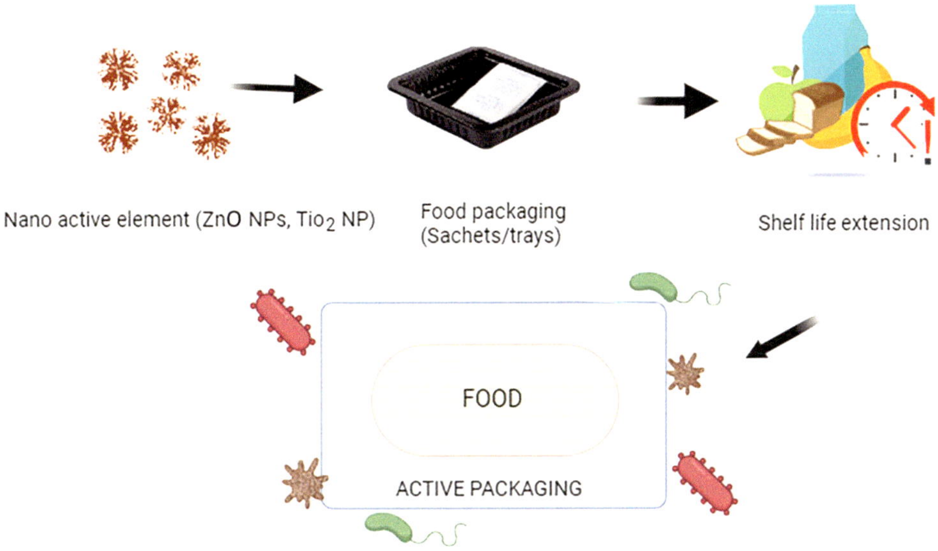

FIGURE 5.1 Nanoparticle ZnO TiO_2 mechanism of action as an active packaging to extend the shelf life of bread.

oxygen is delivered by nanocomposites. Moreover, they may release oxygen and other enzymes while preventing their entry, degrade ripening gasses like ethylene, and have antibacterial properties. Therefore, food shelf life can be increased by using nanocomposites, which would signify using less artificial preservatives in food.

Nanocomposite materials are made of two or more different materials, each having unique properties. The mixture of these components results in a material with the same set of characteristics as the component materials. To create nanocomposites, which are multiphase materials, the continuous phase of a polymer is combined with a discontinuous phase of nano-dimensional material such as nanoparticles, nanoclays, or nanofibers. These materials can take the form of organic or inorganic fibers, flakes, spheres, or particulates and are commonly referred to as "fillers". Thus, nanoparticles and conventional packaging polymers combine to form nanocomposites. Nanocomposites possess favorable optical qualities such as particle size transparency, a high surface-to-volume ratio, and flexibility without compromising strength or abrasion resistance (Hutapea et al., 2021). The reason behind this was their superior moisture, oxygen, and microbial barrier qualities; these materials help cereal grains keep their freshness and prolong their shelf life. In a research conducted by Noshirvani et al. (2017) a novel nanocomposite film and coating based on chitosan-carboxymethyl cellulose-oleic acid (CMC-CH-OL) was proposed as an active packaging material to extend the shelf life of sliced wheat bread. The nanoparticles, of zinc oxide (ZnO NPs) were incorporated at different concentrations (0.5%, 1%, and 2%) in the film to inhibit fungal growth. Moreover, incorporating ZnO NPs along with oleic acid decreased the water vapor permeability of the films.

By combining 5.0% (w/w) tamarind kernel powder with varying amounts of ZnO nanoparticles (0.5%, 1.0%, and 2.0% w/w), they created the nanocomposite films and evaluated their characteristics. The film's mechanical characteristics were enhanced by the addition of 0.5% (w/w) ZnO, however, the water vapor transfer rate and water solubility were decreased by the addition of 1.0% and 2.0% ZnO concentrations. Enhanced surface roughness was also seen at higher concentrations of nanoparticles. FTIR spectra analysis detected weak hydrogen bonds among the components of the films, while X-ray diffractogram confirmed the reduction of crystallinity in the tamarind kernel powder/ZnO nanocomposite films. The tamarind kernel powder-based films showed strong antimicrobial efficacy against *Bacillus cereus* and *Escherichia coli* due to the innate antimicrobial ability of ZnO nanoparticles. Food packaging applications might take advantage of the nanocomposite film containing 1.0% (w/w) ZnO, as it demonstrates the desired features of active packaging materials. Safe to use, zinc oxide (ZnO) nanoparticles (NPs) can be added to biopolymer packaging materials to increase their antibacterial and antioxidant properties; these films are widely used in bakery goods (Hoque et al., 2022).

Mihaly Cozmuta et al. (2015) increased the microbiological safety of bread by composing a packaging material composed of Ag/TiO_2 (silver/titanium dioxide). Experimental findings revealed that Ag/TiO_2-packaging greatly enhanced the shelf life and microbiological safety of bread compared with common plastic packaging. As a result of the water and oxygen molecules in the Ag/TiO_2-P's headspace being scavenged by the photo-activated Ag/TiO_2 nanocomposite, the rate of lipid breakdown decreased hence further increasing the shelf life of bread. The coated bread samples containing 3% AgNPs were found to maintain their moisture content and water activity shown in Figure 5.2, minimize bread firmness and staleness, and preserve the lightness index of the crust/crumb. All the actively coated treatments, particularly those with AgNPs at concentrations of 3% or higher, exhibited strong antimicrobial

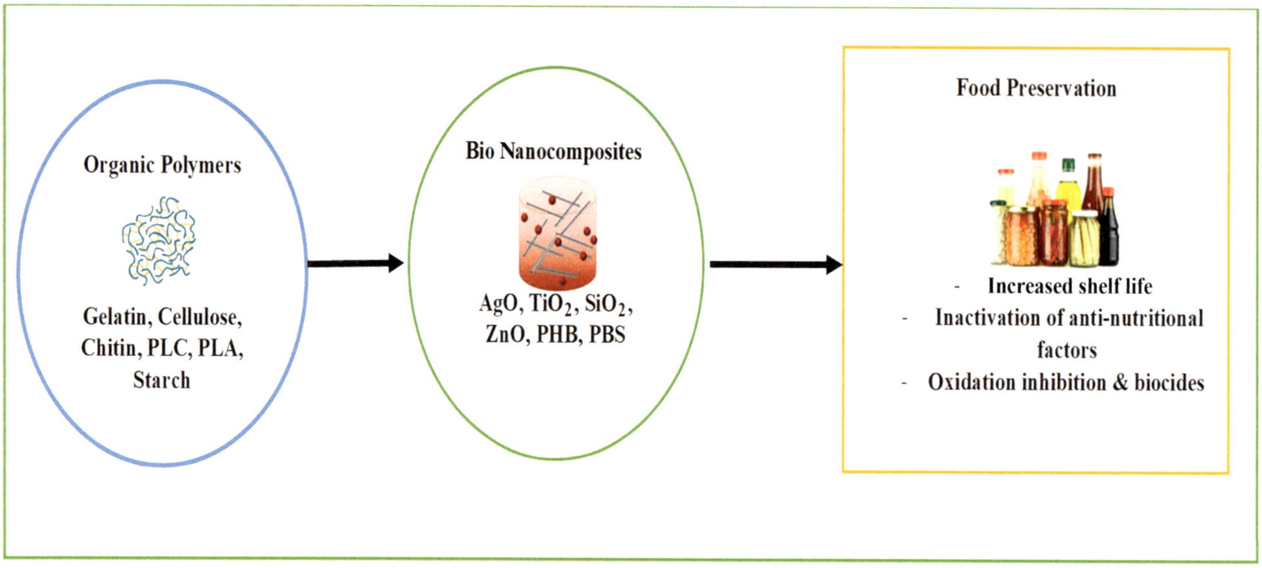

FIGURE 5.2 Applications of cellulose in development of bio-nanocomposite.

TABLE 5.1
Nanocomposites Application in Bakery Products

Nanocomposites	Characteristics	Applications in Bakery Products	References
Gelatin nanocomposite and gelatin emulsion (GNE)	-Antifungal	Extend the shelf life of sponge cake	(Sahraee et al., 2020)
PLA packaging films containing CSNPs-g-pSMA	-Antibacterial	Extend the shelf life of sliced bread	(Kongkaoroptham et al., 2021)
Nano-hybrid (LDH-sorbate) Polypropylene (PP) coated	-Antibacterial	Extend the shelf life of bread	(Viscusi et al., 2021)
ε-PL biofilms	-Antifungal -Inhibits toxin production	Extend the shelf life of bread by inhibiting mycotoxigenic fungi	(Luz et al., 2018)
Sulphur quantum dots-Alginate film	-Antibacterial activity -Antifungal activity	Extend the shelf life of bread by active packaging	(Riahi et al., 2022)
Ag/TiO_2 Nanocomposite	-Antibacterial	Extend the shelf life and preserve of bread	(Mihaly Cozmuta et al., 2015)
Gliadin films	-Antifungal activity	Extend the shelf life of sliced bread and cheese spread incorporated in bakery products	(Balaguer et al., 2013b)
Sachet containing allyl isothiocyanate (AITC)	-Antibacterial -Antiaflatoxin	Extend the shelf life of pizza crust	(Quiles et al., 2015)
Polypropylene/sodium propionate (PP/SP) composite films	-Antioxidant activity -Antifungal activity	Enhanced shelf life by applying the composite film on bread	(Thanakkasaranee et al., 2018)
Zinc oxide nanoparticles (ZnO NPs)	-Produces reactive oxygen activity -Damage bacterial cell wall	Enhanced the shelf life of bakery products	(Emamifar et al., 2010)
Poly(lactic acid) (PLA) and poly(butylene adipate-co-terephthalate) (PBAT)	-Antibacterial activity -Antifungal activity	Extend the shelf life of bread by inhibiting microbial growth	(Srisa & Harnkarnsujarit, 2020)
Ternary blend of carbohydrates and grapefruit seed extract (GSE) paper	-Antibacterial activity	Minced cake	(Shankar & Rhim, 2018)
CMC-PVA-CEO active films	-Antioxidant activity -Antifungal activity	Extend the shelf life of bread	(Fasihi et al., 2019)
Poly(lactic acid)/poly(butylene-succinate-co-adipate) (PLA/PBSA) blend films	-Antifungal -Microbial growth delayer	Extend the shelf life and composite alternative to natural preservative of bread	(Suwanamornlert et al., 2020)

(*Continued*)

TABLE 5.1 (*Continued*)
Nanocomposites Application in Bakery Products

Nanocomposites	Characteristics	Applications in Bakery Products	References
Chitosan-natamycin	-Antifungal	Extend the shelf life of fresh phyllo pastry	(Tsiraki et al., 2018)
Cashew gum (CG) and gelatin (G) films	-Antibacterial	Extend the shelf life of bread	(Oliveira et al., 2020)
Agro-industrial residues cassava inner bark (CIB) and grape stalks (GS)	-Antibacterial	Extend the shelf life and maintain freshness carrot cake and cherry tomatoes	(Engel et al., 2022)
Silver ions incorporated into polyvinyl chloride (PVC)	-Antibacterial	Extend the shelf life of bread	(Braga et al., 2018)

activity by preventing bacterial and mold growth during the storage period (Balaguer et al., 2013).

In a study by Deseta et al. (2021) egg white nanoparticles were used to encapsulate carvacrol (CAR), thymol (THY), and trans-cinnamaldehyde and demonstrated its antifungal efficacy in extending the shelf life of bread.

Egg white protein nanoparticles were found to be encapsulated bioactive compounds that showed improved antifungal properties against *A. niger* compared to pure BC ethanol solutions. Coat-forming solution based on egg white protein bioactive compound nano complexes was applied on preservative-free bread to preserve its physicochemical properties. Coatings based on egg white protein thymol and egg white protein carvacrol nano complexes showed the highest antifungal efficacy, extending the shelf life of bread for 7 days. The technology developed here is an advance in nanotechnological strategies for the biocontrol and conservation of baked goods (Deseta et al., 2021). The characteristics of composite films for use in bread packaging are made of polypropylene (PP) and sodium propionate (SP). The fabrication and characterization of these composite films are the primary focus of the study, which additionally looks at aspects including mechanical strength, barrier qualities, and antibacterial activity. The study aims to determine whether PP/SP composite films are appropriate for prolonging the shelf life of bread products, as shown in Table 5.1 (Thanakkasaranee et al., 2018).

5.5 APPLICATION OF NANOCOMPOSITE IN ACTIVE PACKAGING IN BREAD

Active packaging is a promising technology, deliberately incorporating active ingredients that can release or absorb substances into or out of the food packaging or its surroundings. Active packaging is more important for increasing the shelf life of food items and enhancing and preserving slices of bread and semi-perishable products. Nanocomposite coatings on edible films loaded with P-TiO_2 and P-Ag-TiO_2 demonstrated increased water vapor permeability, hydrophobicity, roughness, photocatalytic activity, and opacity on bread packaging films when compared to plain bread.

For the majority of people around the world, bread is one of the earliest and most common parts of breakfast and diet in different forms. However, the shelf life and retrogradation of bread is a major issue. In terms of chemical and microbiological stability of bread, the composition of bread leads to instability comprising high amounts of sugar, yeasts, total fat lipid hydroperoxides, protein, and molds. Moreover, consumers now prefer food items without any artificial preservatives. The use of active packaging is an effective means to address this issue and enhance the bread's microbiological shelf life without using food additives directly.

Similarly, Ag/TiO_2-based packaging preserves the microbiological safety of bread; inhibits the growth of yeast, molds, *B. cereus*, and *B. subtilis*; and increases bread's shelf life by slowing down the velocity at which nutritious components deteriorate. The shelf life of bakery goods may also be significantly extended by using similar packaging. To further reduce degradation rates and microorganism multiplication, more molecules of oxygen and water in the package's headspace could be scavenged using increased concentration of Ag/TiO_2 nanocomposite amount and optimized Ag concentrations, in line with food safety laws (Istiqola et al., 2020).

Apart from mold spoilage, another problem that reduces bread's shelf life is staling. The white bread packed with HDPE compared with the nanocomposite packaging material composed of polyethylene layers manufactured by the sol-gel method of Ag/TiO_2 showed extended shelf life (Mihaly Cozmuta et al., 2015).

Similarly, nanocomposite active coatings packaging maintained their moisture content more effectively than the control group (bread without coating) and exhibited controlled moisture content and water activity. Noshirvani et al. (2017) found that the addition of oleic acid with the ZnO NPs reduced the water vapor permeability of edible films was also able to control the moisture loss rate from the bread.

5.6 NANO MICROENCAPSULATION IN CEREAL FLOURS AND BAKED GOODS

Recent developments in nanotechnology have led to the "nanoencapsulation" of bioactive within nanocarriers. Nano- or microencapsulation is one of the more promising methods for tackling some persistent and emerging issues of healthy and functional bakery food products. A lot of researchers' attention is on nanoencapsulation and microencapsulation in the emerging sector of food like bakery products such as bread and cookies, specifically to develop new functional food products from bioactive and functional ingredients incorporation.

5.7 FOOD PACKAGING USING STARCH-BASED NANOSTRUCTURED MATERIALS

Starch nanoparticles have a significant stabilizing effect, and they are a suitable nanomaterial for both edible and non-edible packaging materials. Starch bionanomaterials are promising materials that have wider applications because they are nontoxic, have thickener properties and rheology enhancers, and have film-forming properties and that can bind water, stabilize emulsions, generate foam, and posess other characteristics.

5.8 NANOFIBERS AS BIOPRESERVATIVES LOWER THE AMOUNT OF FUNGAL DETERIORATION IN BREAD

Carvacrol nanofibers exhibit enormous promise as natural food biopreservatives (Fonseca et al., 2021). To reduce bread spoiling, the nanofibers were included in bread dough or utilized to create active packages. Moreover, it was found that free carvacrol showed significant antifungal behavior against *Aspergillus flavus* and *Penicillium sp.*

5.9 APPLICATION OF NANOFIBERS IN PREVENTING BREAD SPOILAGE

Bread shelf life is frequently extended by the application of food additives and preventing spoiling. Synthetic food additives have the possibility that people could acquire chronic, noncommunicable diseases as they accept daily intake and RDI not yet set by food safety agencies; however they are continuously inspected (Fonseca et al., 2021). Carvacrol was added to starch nanofibers at concentrations of around 22%, 31%, and 42% (dry basis, v/w). Results from in vitro studies showed the antifungal activity against the *Penicillium* sp. *Aspergillus flavus* on the third day showed inhibition; moreover, the seventh day the inhibition was reduced.

5.10 ENZYMES AND NANOMATERIALS

Nowadays, there is a growing trend toward the use of specifically designed enzyme mixtures to enhance the quality of baked goods. Studies have shown that bread can be made with a combination of xylanase, lipase, and amylase to extend its shelf life and volume. You can also enhance the qualities of bread by adding α-amylase and lipase to a commercial enzyme mixture. It has been observed that using an industrial preparation containing hemicellulose, glucokinase, and hexose oxidase in frozen prebaked French bread enhances the oven spring, cut opening, height, and proofing time.

Enzyme-incorporated nanomaterials exhibit significant promise across diverse domains, serving as a substitute for both synthetically and biologically produced hybrid structures. Better bread can be produced by combining adjuvants like xylanases and amylases, crosslink promoter enzymes like glucose oxidase and transglutaminases, and gluten degraders like proteases. When manufacturing bread, the most common enzyme combination is xylanases with glucose oxidases.

5.11 MISCELLANEOUS APPLICATIONS

In addition, the use of nanotechnology presents several other opportunities for the food industry, including the emergence of novel goods and the growth of food systems applications. It can be applied to enhance food safety and quality by identifying food pathogens. By encapsulating food ingredients, nutritional supplements, and additives into nanocapsules, functional foods can be created with improved taste masking, controlled release, and enhanced bioavailability. Nanotechnology can also address food-related illnesses, develop targeted nutrition therapies, and create new products through food fortification. Electrospun nanofibers are also being explored for use in food packaging (Sahoo et al., 2020).

5.12 REGULATION OF NANOPARTICLES

The silver nanoparticles can enhance food packaging technology. Several regulations have been established by the USFDA and the EFSA for their use. Silver nanoparticles combined with various polymers can increase food quality, prolong its shelf life, and stop it from spoiling. However, their potential toxicity and risk must be carefully considered. More research is needed to evaluate their safety and economic feasibility before they can be widely implemented (Istiqola & Syafiuddin, 2020).

5.13 CONCLUSION

The integration of nanotechnology in cereal preservation offers innovative solutions to extend shelf life while addressing food safety and sustainability concerns. Nanofibers, nanoencapsulation, and nanocomposites demonstrate efficacy in preserving cereal-based products by inhibiting microbial growth and enhancing barrier properties. These advancements pave the way for reduced reliance on synthetic additives and prolonged freshness of cereals, aligning with consumer preferences for natural and sustainable food options. Subsequently, it has the potential to improve the quality and safety of cereal products by continued research and the application of nanotechnological breakthroughs, which will strengthen and sustain the food supply chain.

REFERENCES

Balaguer, M. P., Lopez-Carballo, G., Catala, R., Gavara, R., & Hernández-Muñoz, P. (2013). Antifungal properties of gliadin films incorporating cinnamaldehyde and application in active food packaging of bread and cheese spread foodstuffs. *International Journal of Food Microbiology*, *166*(3), 369–377. https://doi.org/10.1016/J.IJFOODMICRO.2013.08.012

Bautista-Espinoza, P. I., Di Pierro, P., Amaya-Llano, S. L., García-Almendarez, B. E., Mares-Mares, E., Escamilla-García, M., Granados-Arvizu, J. A., Velázquez de la Cruz, G., & Regalado-González, C. (2023). Impact of edible coatings reinforced with mesoporous silica nanoparticles added with cinnamon and lemongrass essential oils applied on sourdough bread. *Food Bioscience*, *55*, 102992. https://doi.org/10.1016/J.FBIO.2023.102992

Blanco-Padilla, A., Soto, K. M., Hernández Iturriaga, M., & Mendoza, S. (2014). Food antimicrobials nanocarriers. *Scientific World Journal*, *2014*. https://doi.org/10.1155/2014/837215

Braga, L. R., Rangel, E. T., Suarez, P. A. Z., & Machado, F. (2018). Simple synthesis of active films based on PVC incorporated with silver nanoparticles: Evaluation of the thermal, structural and antimicrobial properties. *Food Packaging and Shelf Life*, *15*, 122–129. https://doi.org/10.1016/J.FPSL.2017.12.005

Deseta, M. L., Sponton, O. E., Erben, M., Osella, C. A., Frisón, L. N., Fenoglio, C., Piagentini, A. M., Santiago, L. G., & Perez, A. A. (2021). Nanocomplexes based on egg white protein nanoparticles and bioactive compounds as antifungal edible coatings to extend bread shelf life. *Food Research International (Ottawa, Ont.)*, *148*. https://doi.org/10.1016/J.FOODRES.2021.110597

Emamifar, A., Kadivar, M., Shahedi, M., & Soleimanian-Zad, S. (2010). Evaluation of nanocomposite packaging containing Ag and ZnO on shelf life of fresh orange juice. *Innovative Food Science & Emerging Technologies*, *11*(4), 742–748. https://doi.org/10.1016/J.IFSET.2010.06.003

Engel, J. B., Luchese, C. L., & Tessaro, I. C. (2022). Making the reuse of agro-industrial wastes a reality for starch-based packaging sector: A storage case study of carrot cake and cherry tomatoes. *International Journal of Biological Macromolecules*, *206*, 740–749. https://doi.org/10.1016/J.IJBIOMAC.2022.03.074

Fallahi, P., Habte-Tsion, H. M., & Rossi, W. (2018). Depolymerizating enzymes in human food: Bakery, dairy products, and drinks. Bakery, dairy products, and drinks. In *Enzymes in Human and Animal Nutrition: Principles and Perspectives* (pp. 211–237). https://doi.org/10.1016/B978-0-12-805419-2.00010-1

Fasihi, H., Noshirvani, N., Hashemi, M., Fazilati, M., Salavati, H., & Coma, V. (2019). Antioxidant and antimicrobial properties of carbohydrate-based films enriched with cinnamon essential oil by Pickering emulsion method. *Food Packaging and Shelf Life*, *19*, 147–154. https://doi.org/10.1016/J.FPSL.2018.12.007

Fonseca, L. M., Souza, E. J. D., Radünz, M., Gandra, E. A., Zavareze, E. da R., & Dias, A. R. G. (2021). Suitability of starch/carvacrol nanofibers as biopreservatives for minimizing the fungal spoilage of bread. *Carbohydrate Polymers*, *252*. https://doi.org/10.1016/J.CARBPOL.2020.117166

Grewal, N., Faubion, J., Feng, G., Kaufman, R. C., Wilson, J. D., & Shi, Y. C. (2015). Structure of waxy maize starch hydrolyzed by maltogenic α-amylase in relation to its retrogradation. *Journal of Agricultural and Food Chemistry*, *63*(16), 4196–4201. https://doi.org/10.1021/JF506215S

Hoque, M., Sarkar, P., & Ahmed, J. (2022). Preparation and characterization of tamarind kernel powder/ZnO nanoparticle-based food packaging films. *Industrial Crops and Products*, *178*, 114670. https://doi.org/10.1016/J.INDCROP.2022.114670

Hutapea, S., Ghazi Al-Shawi, S., Chen, T. C., You, X., Bokov, D., Abdelbasset, W. K., & Suksatan, W. (2021). Study on food preservation materials based on nano-particle reagents. *Food Science and Technology*, *42*, e39721. https://doi.org/10.1590/FST.39721

Istiqola, A., Chemical, A. S.-J. of the C., & 2020, undefined. (2020). A review of silver nanoparticles in food packaging technologies: Regulation, methods, properties, migration, and future challenges. *Wiley Online Library A Istiqola, A Syafiuddin Journal of the Chinese Chemical Society, 2020•Wiley Online Library*, *67*(11), 1942–1956. https://doi.org/10.1002/jccs.202000179

Istiqola, A., & Syafiuddin, A. (2020). A review of silver nanoparticles in food packaging technologies: Regulation, methods, properties, migration, and future challenges. *Journal of the Chinese Chemical Society*, *67*(11), 1942–1956. https://doi.org/10.1002/JCCS.202000179

Kongkaoroptham, P., Piroonpan, T., & Pasanphan, W. (2021). Chitosan nanoparticles based on their derivatives as antioxidant and antibacterial additives for active bioplastic packaging. *Carbohydrate Polymers*, *257*, 117610. https://doi.org/10.1016/J.CARBPOL.2020.117610

Lucera, A., Costa, C., Conte, A., & Del Nobile, M. A. (2012). Food applications of natural antimicrobial compounds. *Frontiers in Microbiology*, *3*(AUG), 28087. https://doi.org/10.3389/FMICB.2012.00287/BIBTEX

Luz, C., Calpe, J., Saladino, F., Luciano, F. B., Fernandez-Franzón, M., Mañes, J., & Meca, G. (2018). Antimicrobial packaging based on ε-polylysine bioactive film for the control of mycotoxigenic fungi in vitro and in bread. *Journal of Food Processing and Preservation*, *42*(1). https://doi.org/10.1111/JFPP.13370

Mihaly Cozmuta, A., Peter, A., Mihaly Cozmuta, L., Nicula, C., Crisan, L., Baia, L., & Turila, A. (2015). Active packaging system based on Ag/TiO2 nanocomposite used for extending the shelf life of bread. Chemical and microbiological investigations. *Packaging Technology and Science*, *28*(4), 271–284. https://doi.org/10.1002/PTS.2103

Noshirvani, N., Ghanbarzadeh, B., Rezaei Mokarram, R., & Hashemi, M. (2017). Novel active packaging based on carboxymethyl cellulose-chitosan-ZnO NPs nanocomposite for increasing the shelf life of bread. *Food Packaging and Shelf Life*, *11*, 106–114. https://doi.org/10.1016/J.FPSL.2017.01.010

Oliveira, M. A., Gonzaga, M. L. C., Bastos, M. S. R., Magalhães, H. C. R., Benevides, S. D., Furtado, R. F., Zambelli, R. A., & Garruti, D. S. (2020). Packaging with cashew gum/gelatin/essential oil for bread: Release potential of the citral. *Food Packaging and Shelf Life*, *23*, 100431. https://doi.org/10.1016/J.FPSL.2019.100431

Quiles, J. M., Manyes, L., Luciano, F., Mañes, J., & Meca, G. (2015). Influence of the antimicrobial compound allyl isothiocyanate against the Aspergillus parasiticus growth and its aflatoxins production in pizza crust. *Food and Chemical Toxicology*, *83*, 222–228. https://doi.org/10.1016/J.FCT.2015.06.017

Rebholz, G. F., Sebald, K., Dirndorfer, S., Dawid, C., Hofmann, T., & Scherf, K. A. (2021). Impact of exogenous α-amylases on sugar formation in straight dough wheat bread. *European Food Research and Technology*, *247*(3), 695–706. https://doi.org/10.1007/S00217-020-03657-Y

Renzetti, S., & Arendt, E. K. (2009). Effect of protease treatment on the baking quality of brown rice bread: From textural and rheological properties to biochemistry and microstructure. *Journal of Cereal Science*, *50*(1), 22–28. https://doi.org/10.1016/J.JCS.2009.02.002

Riahi, Z., Priyadarshi, R., Rhim, J. W., Lotfali, E., Bagheri, R., & Pircheraghi, G. (2022). Alginate-based multifunctional films incorporated with sulfur quantum dots for active packaging applications. *Colloids and Surfaces B: Biointerfaces*, *215*, 112519. https://doi.org/10.1016/J.COLSURFB.2022.112519

Sahoo, M., Vishwakarma, S., Panigrahi, C., & Kumar, J. (2020). Nanotechnology: Current applications and future scope in food. *Wiley Online Library M Sahoo, S Vishwakarma, C Panigrahi, J Kumar Food Frontiers, 2021•Wiley Online Library*, *2*(1), 3–22. https://doi.org/10.1002/fft2.58

Sahraee, S., Milani, J. M., Ghanbarzadeh, B., & Hamishehkar, H. (2020). Development of emulsion films based on bovine gelatin-nano chitin-nano ZnO for cake packaging. *Food Science & Nutrition*, *8*(2), 1303–1312. https://doi.org/10.1002/FSN3.1424

Sarabhai, S., Tamilselvan, T., & Prabhasankar, P. (2021). Role of enzymes for improvement in gluten-free foxtail millet bread: It's effect on quality, textural, rheological and pasting properties. *LWT*, *137*, 110365. https://doi.org/10.1016/J.LWT.2020.110365

Sganzerla, W. G., Castro, L. E. N., da Rosa, C. G., Almeida, A. da R., Maciel-Silva, F. W., Kempe, P. R. G., de Oliveira, A. L. R., Forster-Carneiro, T., Bertoldi, F. C., Barreto, P. L. M., Veeck, A. P. de L., & Nunes, M. R. (2023). Production of nanocomposite films functionalized with silver nanoparticles bioreduced with rosemary (*Rosmarinus officinalis* L.) essential oil. *Journal of Agriculture and Food Research*, *11*, 100479. https://doi.org/10.1016/J.JAFR.2022.100479

Shankar, S., & Rhim, J. W. (2018). Antimicrobial wrapping paper coated with a ternary blend of carbohydrates (alginate, carboxymethyl cellulose, carrageenan) and grapefruit seed extract. *Carbohydrate Polymers*, *196*, 92–101. https://doi.org/10.1016/J.CARBPOL.2018.04.128

Sivaprakash, G., Raja, R., Mohanrasu, K., Dinesh, G. H., & Arun, A. (2022). *Microbial nanotechnology in food industry: Antimicrobial packaging*. Elsevier. Retrieved April 26, 2024, from https://www.sciencedirect.com/science/article/pii/B9780128234266000024

Smith, J. P., Daifas, D. P., El-Khoury, W., Koukoutsis, J., & El-Khoury, A. (2004). Shelf life and safety concerns of bakery products—a review. *Critical Reviews in Food Science and Nutrition*, *44*(1), 19–55. https://doi.org/10.1080/10408690490263774

Soltani, Z., Tavakolipour, H., & Tabari, M. (2023). The influence of chitosan and titanium dioxide nanoparticles incorporated with polylactic acid on prolonging rye bread shelf life. *Journal of Food Measurement and Characterization*, *17*(2), 1806–1816. https://doi.org/10.1007/S11694-022-01728-0

Srisa, A., & Harnkarnsujarit, N. (2020). Antifungal films from trans-cinnamaldehyde incorporated poly(lactic acid) and poly(butylene adipate-co-terephthalate) for bread packaging. *Food Chemistry*, *333*, 127537. https://doi.org/10.1016/J.FOODCHEM.2020.127537

Suwanamornlert, P., Kerddonfag, N., Sane, A., Chinsirikul, W., Zhou, W., & Chonhenchob, V. (2020). Poly(lactic acid)/poly(butylene-succinate-co-adipate) (PLA/PBSA) blend films containing thymol as alternative to synthetic preservatives for active packaging of bread. *Food Packaging and Shelf Life*, *25*, 100515. https://doi.org/10.1016/J.FPSL.2020.100515

Thanakkasaranee, S., Kim, D., & Seo, J. (2018). Preparation and characterization of polypropylene/sodium propionate (PP/SP) composite films for bread packaging application. *Packaging Technology and Science*, *31*(4), 221–231. https://doi.org/10.1002/PTS.2369

Tsiraki, M. I., El-Obeid, T., Yehia, H. M., Karam, L., & Savvaidis, I. N. (2018). Effects of chitosan and natamycin on vacuum-packaged phyllo: A pastry product. *Journal of Food Protection*, *81*(12), 1982–1987. https://doi.org/10.4315/0362-028X.JFP-18-236

Viscusi, G., Bugatti, V., Vittoria, V., & Gorrasi, G. (2021). Antimicrobial sorbate anchored to layered double hydroxide (LDH) nano-carrier employed as active coating on Polypropylene (PP) packaging: Application to bread stored at ambient temperature. *Future Foods*, *4*, 100063. https://doi.org/10.1016/J.FUFO.2021.100063

6 Role of Nanotechnology in Pediatric Foods

Nibir Ghosh[a], *Srabani Podder*[a], *Sumee Ireen*[a], *and Laiba Parveen*[a]
[a]Eminent College of Pharmaceutical Technology, Department of Pharmacology, Kolkata, West Bengal

6.1 INTRODUCTION

Nanotechnology has emerged as a transformative force, not only in the food industry but also in the pharmaceutical and medical realms, where it is rapidly evolving and garnering significant attention. Particularly noteworthy is its role in drug delivery and nutrient delivery systems, where nanoparticles have emerged as highly impactful tools in contemporary health care. A myriad of nanosystems, including carbon dots, paramagnetic dendrimers, and nanoemulsions, have been introduced, each offering unique capabilities and applications. The shape and properties of nanoparticles are intricately tied to the physicochemical parameters of the starting materials and the chosen preparation techniques. Key methods utilized for manufacturing nanocarriers include the dispersion of premade polymers, coacervation, polymerization, nano-spray drying, and supercritical fluid technology. These methodologies play a pivotal role in shaping the physical stability and biological efficacy of drug-loaded nanoparticles, with particle size, surface charge, surface hydrophobicity, and drug release kinetics emerging as critical variables [1]. Nanotechnology, with its ability to engineer functional systems at the molecular level, holds immense promise across diverse fields, spanning from materials science to biology, owing to its unique physical, electrical, and optical properties [2]. Among the most prominent areas of nanotechnology research is nanomedicine, which focuses on the development of highly targeted pharmaceutical interventions for disease detection, prevention, and treatment. In recent years, the field of nanomedicine has witnessed a surge in research activity globally, underscoring its growing significance in revolutionizing healthcare practices and addressing pressing medical challenges. This research activity is currently turned into a commercialization activity around the world. Various kinds of drug delivery systems and nutrition delivery systems are now dominated by nanotechnology. There are several techniques that have been used to incorporate nanoparticles in food, drug delivery, and delivery of nutrition. Nanoparticles can also be made by fabrication methods. Nanoparticles have emerged as powerful tools in both nutrition and drug delivery systems, offering a range of applications in transporting essential nutrients and pharmaceutical compounds [3]. The design of nanoparticles as delivery systems entails meticulous control over various factors, including particle size, surface properties, drug transport mechanisms, and the release of active pharmaceutical ingredients (APIs). These parameters are essential for ensuring targeted pharmacological action at specific sites within the body, while also maintaining an appropriate therapeutic rate and dosage schedule [4, 5].

6.2 ADVANTAGE OF UTILIZING NANOTECHNOLOGY IN PHARMACEUTICAL TECHNOLOGY

In a drug delivery system nanoparticles played an important role. The two primary characteristics of nanoparticles as drug carriers are their small size and the fact that they are mostly made of biodegradable materials, which provide advantages. Particle size plays a crucial role in determining the effectiveness of drug delivery methods, with smaller particles offering significant advantages. Nanoparticles, owing to their large surface area and minute size, exhibit enhanced bioavailability and improved solubility of drugs [6]. Furthermore, their ability to traverse various physiological barriers, including the pulmonary system, the blood-brain barrier, endothelial linings of tumors, and the tight junctions in skin endothelial cells, adds to their utility [7]. The nanoscale dimensions of these particles facilitate targeted accumulation of medications at specific sites and efficient uptake by different cell types. In the realm of drug delivery, nanoparticles present distinct advantages compared to traditional microparticles, particularly in intravenous administration. The size of nanoparticles is critical, especially in the context of circulation through the body's smallest capillaries, which typically have diameters ranging from 5.6 to 7 micrometers. To prevent the risk of embolism, it is essential for circulating particles to be significantly smaller than 5 micrometers. Nanoparticles meet this requirement, ensuring safer intravenous delivery routes by circumventing the potential for vascular blockages. To leverage the benefits of nanoparticles further, both natural and synthetic biodegradable polymers are utilized in their formulation. These polymers play a pivotal role in achieving targeted drug delivery, enhancing bioavailability, and sustaining release kinetics [8]. By carefully selecting and modifying these polymers, drug-loaded nanoparticles can evade enzymatic degradation, thereby safeguarding the integrity of the drug payload and prolonging its therapeutic effects. This

DOI: 10.1201/9781032677422-6

tailored approach to nanoparticle design and optimization holds tremendous potential for revolutionizing pharmaceutical applications. Through meticulous engineering and refinement, nanoparticle-based drug delivery systems represent a groundbreaking advancement in the pharmaceutical industry. By capitalizing on their unique properties and customizable characteristics, these nanocarriers offer a promising avenue for enhancing drug efficacy, improving patient outcomes, and addressing various therapeutic challenges.

6.3 VARIOUS TYPES OF PHARMACEUTICAL NANOTECHNOLOGY

6.3.1 Pharmaceutical Nanosystems

a. **Polymeric Nanoparticles**—The size range of polymeric nanoparticles is between 10–1000 nm. It has biodegradable and biocompatible properties. Due to those properties, it provides complete protection to drugs [9].
b. **Dendrimers**—Dendrimers are generated by carefully regulated polymerization and have a size of less than 10 nm. These are monodispersing polymer systems with multiple branches. Dendrimers are utilized to deliver drugs to the liver and macrophages selectively and under control [10].
c. **Metallic Nanoparticles**—Metallic nanoparticles, such as gold and silver colloids, possess dimensions smaller than 100 nm. Their diminutive size results in a significantly increased surface area, leading to exceptional bioavailability and stability. These characteristics render them highly suitable for pharmaceutical applications. They have applications in increased radiation, sensitive diagnostic testing, thermal ablation, and medication and gene delivery [11].
d. **Polymeric Micelles**—These nanoparticles typically range in size from 10 to 100 nm, exhibiting high biocompatibility and a propensity for drug entrapment. They are employed for both active and passive targeted medication administration and have a high diagnostic value [12].
e. **Lipsome**—These are 50–100 nm in size phospholipid vesicles with good biocompatibility and entrapment effectiveness. Genes, proteins, and peptides can be delivered passively or actively using them [13].

6.3.2 Pharmaceutical Nanomaterials

Various nanomaterials have been used in the advancement of pharmaceutical nanotechnology. Those nanomaterials are as follows:

a. **Nanocrystaline Materials**—These are made to take the place of materials that have poor solubility, bioavailability, and other characteristics [14].
b. **Nanostructured Materials**—These are machined structures with particular forms and purposes, such as microarrays, nano and micro electromechanical systems, and microfluidicsm [15].

6.4 VARIOUS APPLICATION OF PHARMACEUTICAL NANOTECHNOLOGY

a. **Drug Delivery System**—Pharmaceutical nanotechnology enhances drug delivery in pediatrics, addressing issues of poor adherence, cytotoxicity, and high doses. Nano-sized carriers deliver drugs more efficiently to specific sites, reducing side effects and offering controlled release for sustained therapeutic effects, benefiting long-term treatment regimens in pediatric medicine. The delivery system has been developed in accordance with pharmaceutical nanotechnology concepts [16].
b. **Diagnostic Approaches**—Pharmaceutical nanotechnology aids in diagnostic approaches for pediatric medicine by utilizing molecular imaging to describe and quantify biological processes in animals, including gene expression, protein-protein interactions, signal transduction, cellular metabolism, and intra-and intercellular transport [17].
c. **Drug Discovery**—Nanotechnology significantly contributes to pediatric drug discovery by employing nano-sized carriers or nanoparticles to enhance the solubility and bioavailability of active pharmaceutical ingredients and excipients [18]. By reducing particle size and modifying surface properties, nanotechnology enables better absorption of drugs in the body, leading to improved therapeutic outcomes in pediatric patients. This approach addresses challenges related to drug administration in children, such as dosing accuracy, palatability, and safety concerns, ultimately facilitating the development of more effective pediatric medications.

6.5 THE IMPACT OF NANOTECHNOLOGY ON THE FOOD INDUSTRY

Nanotechnology has revolutionized the food industry over recent decades, offering innovative solutions and materials with unique properties. Working with atoms, molecules, or macromolecules ranging from 1 to 100 nm in size, nanotechnology has enabled the creation of nanomaterials with distinct characteristics [19]. These materials exhibit novel physiochemical features such as color, solubility, strength, diffusivity, toxicity, and more, owing to their high surface-to-volume ratio and nanoscale dimensions. This technological advancement has sparked a global industrial revolution, prompting significant investments from both developed and emerging nations. Nanotechnology presents numerous opportunities for the development and utilization of novel structures and materials across various sectors, including food, medicine, and agriculture. Researchers are focusing on enhancing food quality while preserving nutritional value, addressing consumer concerns about health and food quality [20]. Nanoparticle-based materials, many of which are nontoxic and nutrient-rich, have seen a surge in demand within the food industry. Their stability under high temperatures and pressures makes them valuable for food production, processing, and packaging. Nanotechnology offers comprehensive solutions for food safety, quality enhancement, and health benefits. In the food industry, nanotechnology finds applications primarily in two categories: food nanostructured ingredients and food nanosensing. Nanostructured substances are used in food processing and packaging, serving as additives, carriers for nutrient delivery, antimicrobial agents, and more, to improve mechanical strength and durability [21]. Food nanosensing, on the other hand, enhances food quality and safety assessment. Numerous institutions, scholars, and businesses are actively developing advanced approaches and products utilizing nanotechnology in food science [22, 23].

6.6 NANOTECHNOLOGY ROLE IN FOOD PROCESSING

Food ingredients with nanostructures are being created with the promise of better consistency, taste, and texture. Nanotechnology extends the shelf life of certain food ingredients and reduces the amount of food wasted as a result of microbial contamination. These days, food additives are delivered to food products using nanocarriers without affecting their fundamental morphology [24]. The distribution of any bioactive substance to different locations within the body may be directly impacted by particle size, as evidenced by the observation that in certain cell lines only submicron nanoparticles can be absorbed effectively while larger size microparticles cannot. Nanomaterials have an ideal delivery property. They are able to deliver an active compound to a specific target site precisely and also ensure the availability along with storage conditions for a long period of time. When used for the creation of association colloids, encapsulation, emulsions, biopolymer matrices, and simple solutions, nanotechnology provides effective delivery methods with all the aforementioned characteristics. In food packaging, nano polymers are attempting to take the place of traditional materials [4]. Foodborne pollutants, mycotoxins, and microbes can all be detected with nanosensors when it comes to encapsulation and release efficiency, nanoparticles outperform conventional encapsulation technologies. Furthermore, because of their tiny size, these delivery methods can easily penetrate deeper into tissues, facilitating the effective administration of active substances to specific locations inside the body [25, 26].

6.7 PEDIATRIC FOOD

A well-balanced and nourishing diet plays a crucial role in the growth and development of a child. It not only establishes a strong foundation for their overall health but also instills healthy eating habits from an early age. Early childhood nutrition offers three key benefits: boosting the immune system to combat various diseases, supporting the development of a healthy brain and other organs, and improving a child's cognitive abilities and level of physical activity. Therefore, ensuring that children receive a nutritious diet is essential as it sets them on the path to a long and balanced life. The Global Nutrition Report 2018 reveals that more than half of all child deaths in India under the age of five are primarily caused by under nutrition. Inadequate nutrition during the first 1000 days of life can impede a child's growth, leading to stunted growth, which is associated with cognitive decline and subpar performance in both educational and professional settings [27]. Malnutrition in children can be attributed to a variety of factors including the family environment, dietary habits, hygiene practices like hand washing, infections such as diarrhea, poverty, and maternal health literacy. Out of 119 eligible countries, India is ranked 102nd in the 2019 Global Hunger Index. India receives a score of 30.3, indicating a severe case of hunger. Four out of ten children in India, or nearly 47 million, are chronically undernourished or stunted, preventing them from reaching their full potential. The kids come from homes where the parents work as small-scale business owners, manufacturing workers, or daily wage laborers. Children enrolled in Smile's nutrition program can have freshly prepared, hot meals at school at noon with their friends, negating the need for them to bring a lunch. The kids look forward to these delectable meals a lot because they don't always have access to three full square meals a day. It's interesting to note that there is a rise in consistent attendance, a decline in dropout rates, and an improvement in the general academic and performance of the kids. The availability of a wholesome meal for their children at least once a day makes parents happy and appreciative. Therefore, without a balanced diet, a child's development cannot be considered complete. After a child has been taught the value of a healthy diet, it is up to her or him to mentor them in leading happy, healthy, and sustainable lives [28].

6.7.1 Importance of Nutrition in Pediatric Food

The nutritional requirements for children mirror those of adults, encompassing essential nutrients such as vitamins, minerals, carbohydrates, proteins, and fats. These vital components are collectively referred to as nutrients, with specific amounts varying depending on the age of the child [26]. The ideal feeding schedule takes the child's age, activity level, and other factors into account for their growth and development. Examine the basic principles of child nutrition outlined in the latest Dietary Guidelines for Americans.

Nutrient-dense foods are characterized by their minimal to nonexistent content of added sugar, saturated fat, or salt and are rich in essential nutrients. By focusing on nutrient-dense foods, children can meet their nutritional needs while consuming fewer calories in total [29, 30].

6.7.2 Types of Food Needed for Children

a. **Protein—**The meal should contain a sufficient amount of protein like seafood, meat and eggs, beans, peas, milk, nuts, and seeds.
b. **Fruit—**It is important to motivate your child to consume a diverse range of fresh, canned, frozen, or dried fruits. Opt for canned fruits labeled as light or packed in their own juice to ensure they are low in added sugar. Remember that 1/4 cup of dried fruit is equivalent to one serving of fruit.
c. **Vegetables—**Provide an assortment of fresh, canned, frozen, or dried vegetables. Alternate between peas or beans, as well as colorful vegetables every week. Select canned or frozen vegetables that are low in sodium.
d. **Grains—**Meals for every child should contain whole grains, such as bread, pasta, oatmeal, quinoa, popcorn, or brown rice.
e. **Dairy Products**—A child must consume good fat contained in dairy products, such as milk, yogurt, and cheese.

6.8 APPLICATION OF NANOTECHNOLOGY IN PEDIATRIC FOOD

Nanotechnology plays a crucial role in the advancement and enhancement of pediatric foods, providing numerous benefits in terms of nutrition, safety, and the overall quality of the products. The following are key aspects that emphasize the significance of nanotechnology in the realm of pediatric foods.

6.8.1 Texture, Appearance, and Taste of Food

A variety of methods for raising food quality and taste are made possible by nanotechnology. The delivery of culinary balance and enhanced taste release and retention have been achieved through the widespread usage of nanoencapsulation technology for the unstable and extremely reactive plant pigments known as anthocyanins, which exhibit a range of biological functions. The thermal and light stability of apo recombinant soybean seed H-2 subunit ferritin (rH-2) was improved by encapsulating cyanidin-3-O-glucoside (C3G) molecules within the inner cavity. This led to the development of versatile nanocarriers for the transportation and safeguarding of bioactive compounds. Despite rutin being a widely known dietary flavonoid with notable medicinal benefits, its low solubility hinders its application in the food industry. The solubility, thermal stability, and UV radiation stability of rutin were enhanced when it was encapsulated within ferritin nanocages, as compared to free rutin [31]. Nanoemulsions are commonly utilized for distributing lipid-soluble bioactive chemicals due to their natural food ingredients and simple production techniques. These nanoemulsions can be tailored to enhance water dispersion and bioavailability. Nanoparticles, with their subcellular size, provide an effective means of improving the bioavailability of nutraceutical components compared to larger particles that release encapsulated compounds at a slower rate over extended periods. Various metallic oxides like silicon dioxide (SiO_2) and titanium dioxide have been traditionally employed in food items as colorants or flow agents [32, 33]. One of the most popular food nanomaterials for encapsulating scents or flavors in food products is SiO_2.

6.8.2 Preservation of Shelf Life

While bioactive ingredients in functional foods are exposed to harsh environments, they frequently degrade and ultimately become inactive. However, when these bioactive ingredients are nanoencapsulated, the degradation processes are slowed down or even stopped until the product reaches its intended destination, extending its shelf life. Edible nanocoatings applied to various food components can also prolong the freshness of prepared meals, even after the packaging has been opened. These coatings act as a protective barrier against moisture and gas exchange, while also adding flavors, colors, antioxidants, enzymes, and anti-browning agents [34]. It is often feasible to decelerate chemical degradation processes by enclosing functional components within droplets through the modification of the interfacial layer characteristics. Turmeric (*Curcuma longa*) contains curcumin, the most active and least stable bioactive component. Curcumin was found to have lower antioxidant activity, be stable during pasteurization, and change in ionic strength when it was encapsulated [35].

6.8.3 Extended Shelf Life by Nanocomposite Packaging

One of the cutting-edge, emerging technologies of this century is nanotechnology. Applications of nanotechnology could overcome the drawbacks of conventional food preservation techniques. Materials generated using either "bottom-up" or "top-down" methods are known as nanomaterial because they have at least one nanoscale dimension. Fresh produce's ability to remain fresh longer on the shelf life is possible by using methods that postpone ripening, regulate transpiration, and inhibit microbial growth [36]. The antimicrobial films are prepared by cellulose and lignin, which improve the shelf life of product without hampering the physiochemical, biological, and phytochemical constituents. Nanomaterials have the ability to improve the shelf life of pediatric food in various ways. Strangeness, biodegradability, and the ability to let in gas and moisture are all necessary qualities in a packaging material. In contrast to conventional packaging methods, nanotechnology-based "smart" and "active" food packaging provides numerous advantages. These encompass improved mechanical durability, superior barrier properties, antimicrobial films, and the capability to identify pathogens through nanosensing, thereby informing consumers about the food's safety. This section basically covers the nanomaterial's use in food packaging and their role in maintaining the shelf life. There are several nanomaterials used in food packaging, and through different mechanism they increased the shelf life. The various nanomaterials are titanium dioxide (TiO_2), zinc oxide (ZnO), gold nanoparticles (AuNP), silver nanoparticles (AgNP), cooper oxide (CuO), and silicon dioxide (SiO_2) [37].

6.8.3.1 Titanium Dioxide (TiO_2)

Titanium dioxide nanoparticles have been studied for the beneficial activity in pediatric food materials' shelf-life enhancements due to their antioxidant and antibacterial property. TiO_2 has the ability to reduce food hazards in the food industry. TiO_2 nanoparticles have the ability to maintain the physiochemical property of foods. TiO_2 inhibits the growth of bacteria and fungi in the food materials. This property helps to reduce spoilage of food. Its antioxidant property also helps to prevent oxidative degradation, which helps to maintain the quality of food. Different concentration of TiO_2 have the ability to increase the hydrophobicity and thermal properties. Studies suggest that TiO_2 incorporated films have the ability to increase or maintain the shelf life of banana-flavored containing foods [38].

6.8.3.2 Zinc Oxide (ZnO)

ZnO's strong optical, electrical, piezoelectrical, semiconducting, and chemical sensing properties make it useful in a variety of applications. Development of antibacterial properties against food pathogens is a current research topic. ZnO acts as a crucial antibacterial agent in terms of food preservation. The size of ZnO nanoparticles play an important role in bacterial death; the size reduction of ZnO nanoparticles increase the bactericidal property. The nanoparticles interact with both the surface and the core of the bacteria. ZnO also has an antioxidative property. The antibacterial property also depends on reactive oxygen species (ROS) formation. Antioxidant property of zinc oxide helps to prevent the oxidative detoriation, which has further prevented the color degradation. ZnO nanoparticles maintain the nutritional value by preventing the damage of vitamins and sensitive compounds present in pediatric food [39].

6.8.3.3 Gold Nanoparticles (AuNP)

Gold nanoparticles (AuNP) are known as gold colloid. AuNP has several medicinal activities. It has also been used for diagnostic purposes. In the food industry it plays a vital role both in shelf life and food packaging. The antibacterial activity of AuNP depends on their size and shape. Encapsulating gold nanoparticles in various polymers plays a significant role in food shelf life preservation as well as packaging. Studies suggest that incorporation of AuNP with other polymers can improve the shelf life of flavored food (banana). AuNP possesses the capability to hinder the infiltration of gas and moisture, thereby aiding in the preservation of food quality and freshness.

A unique feature of AuNP is that it can be used in smart packaging by preparing sensors, which will help to detect the pathogens or the shelf life degradation. AuNP is also used to maintain the packaging strength, which is helpful to resist damage during transport [40].

6.8.3.4 Silver Nanoparticles (AgNP)

A silver nanoparticle has the great phenomenon to protect against antimicrobial activity of food. Several studies have been performed to check the ability of AgNP in terms of food shelf-life enhancement. Studies show that AgNP can extend the shelf life of mushroom-containing pediatric food for 16 days. There are several foods like cheese where the presence of AgNP also ensures food safety by preventing the growth of microorganisms. It helps with storage stability by preventing the food from spoilage [41].

6.8.3.5 Copper Oxide (CuO)

CuONPs are the metal oxides commonly employed in food packaging due to their effective antimicrobial properties. These nanoparticles have shown significant potential in reducing bacterial growth and fungus in food. The antimicrobial property of cooper oxide nanoparticles depends on the surface area, morphology, oxidative state, and size. CuONPs gradually release antimicrobial agents, which give further protection against microbial contamination. By adding CuONPs there is no need to add chemical additives to increase the shelf life. This property can be beneficial for pediatric food. CuONPs can enhance the shelf life of food products by preserving their freshness and quality, thus preventing spoilage [42].

6.8.3.6 Silicon Dioxide (SiO_2)

Silicon is an element available in silica and silicates. It has been observed that SiO_2 has a good antimicrobial property. In addition with other polymers it has ability to increase the shelf life of food. It is also added in pediatric foods to prevent the clumping of powder. SiO_2 nanoparticles protect the food from moisture due to the presence of silica. It also maintains shelf life by maintaining the texture of the food particles [43].

6.8.4 Nutrient Delivery and Bioavailability Enhancement of Pediatric Food

6.8.4.1 Various Approaches to Nutrient Delivery by Using Nanomaterials

Nanotechnology is a cutting-edge technology whose primary goals are to produce materials with unique structures and improve food quality at the molecular level. Although not entirely new, nanomaterials have been applied to a wide range of industries, including biotechnology, medicine delivery, nutrition, diagnostics, the chemical industry, and biomedicine. The basic concept of nanomaterials is that they have to provide satisfactory nutrients necessary for growth of children [44]. Nano nutrients can show good results compared to other macronutrients. Nanoparticles are classified by various types according to their chemical properties. There are several types including organic, inorganic, and emulsions. Organic nanoparticles have been used to deliver nutrition in pediatric food through feeding; additionally, organic nanoparticles do not hamper the taste and appearance of the product. They play a significant role in nutrition delivery due to their unique properties. Through encapsulation of organic nanoparticle vitamins, antioxidants, and nutrients can be delivered to children. Organic nanoparticles also enhance bioavailability, increasing the solubility and stability of hydrophobic products. Through organic nanoparticles nutrients can be develop for specific sites of the body. Organic nanoparticles play as a shield in the protection of sensitive compounds. They protect the product from degradation of sensitive nutrients.

On the other hand, inorganic nanoparticles such as titanium dioxide and silver nanoparticles play an important role in nutrient delivery. In terms of nutrition, inorganic nanomaterials have significant roles, like bioavailability enhancement, minerals supplementation, targeted delivery, coloring agent, and stability enhancement. In terms of bioavailability, these nanoparticles increase the solubility. Through inorganic nanoparticles the release of nutrients can control rate. They are designed to deliver nutrients in specific tissues or cells to maintain the therapeutic effectiveness [45]. Titanium dioxide nanoparticles are basically used as food colorant; they are also used to protect the nutrient from UV damage. In terms of silver nanoparticles, they are used to protect the shelf life of nutrients. Digestion of nano micronutrient occurs via the gastrointestinal tract [46]. Due to the smaller size, they can easily diffuse in the gastrointestinal tract. There are various kind of nanomaterials used as pediatric nutrition delivery systems, like nano suspension and nano crystals, liposomes, and solid lipid nanomaterial.

a. **Nano Suspension and Nano Crystals**—The extended loading, minimal steric, and electrostatic surface area for stabilization and high drug concentration that can be combined with a slow dissolution to prevent possible toxicity are the advantages over conventional nanoparticles. Nano crystals are used in various ways to deliver nutrients. They are beneficial for the delivery of fat-soluble vitamins and poorly water-soluble nutrients. Nano crystals allow low dosages of nutrients due to their bioavailability enhancing phenomenon [47].
b. **Liposomes**—Liposomes are basically part of the lipid-based delivery. The nutritional industry has been using the liposome delivery system for many years. There is no doubt that liposome based nutrient delivery will grow more in the next 5 to 10 years. The advantages of using liposome in nutrient delivery are their bioavailability enhancement property, targeted delivery, and extended release. Lipsome-based nutrient delivery is already being used in a number of dietary supplements according to clinical investigation. Liposomal Vitamin C and Liposomal Curcumin are the most promising nutrient supplements available [48].
c. **Solid Lipid Nanomaterials**—Solid lipid nanoparticles (SLNs) are another promising agent in the nutritional delivery for children. SLNs are made up of triglycerides and fatty acids. The biocompatibility of SLNs is very good because they are made up of natural lipids so that they can be easily absorbed in the body. Due to their easy breakdown property, SLNs are less toxic for children's food. SLNs have shown effective bioavailability enhancement. Studies have shown that SLNs loaded with vitamin C and curcumin are potent nutritional supplements. Overall SLNs are a beneficial nanomaterial for nutraceutical industry [49].

6.8.4.2 Role of Nanomaterials in Bioavailability Enhancement of Pediatric Foods

In the field of nutrition research, nanotechnology is crucial because it can help with precise information gathering regarding the location of nutrient absorption and utilization. Bioavailability, as defined by the FDA (Food and Drug Administration), is the rate and degree of absorption of active ingredients or therapeutic moieties and their availability at the site of action. Nowadays bioavability of nutrients is a big concern [50]. Bioavailability may change as a result of interactions between bioactive components or changes in the medium in which they are found. Bioavability is the important point to ensure the bio-efficacy of nutrients. Involvement of nanomaterials in pediatric food preparation can increase the bioavability in several ways. It increases the surface area due to the increased absorption. It also increases the solubility. Nanoparticles have protected the nutrient from degradation and allow more absorption in blood stream. There are several studies that have shown that modified nutrients as nanoparticles have shown better results than free nutrient molecule [51]. Examples are mentioned next.

a. **Vitamin A**—Knowledge of the bioavailability and physiological mechanisms governing the uptake of vitamin A is crucial in the development of vitamin A nanomaterials. Research has demonstrated that nano vitamin A exhibits higher bioavailability compared to its free form [52].
b. **Vitamin B_{12}**—In recent times, a novel hydrophilic bioactive element transport system has emerged in the form of protein lipid composite nanoparticles.

These nanoparticles consist of an inner hydrophilic portion and a three-layer assembly comprising of phospholipid, protein, and tocopherol layers. This innovative development holds great potential for various applications. The vitamin B_{12} uptake efficacy was increased by converting it into nanoparticles [53].

c. **Folic Acid**—The encapsulated folic acid nanoparticle has shown better bioavailability than free folic acid [30].

d. **Iron**—Iron with solid lipid nanoparticles has shown better bioavability compared to normal iron particles. Iron nanoparticles also show promising results in target-specific nutrient delivery [54].

6.8.5 Improved Texture and Taste by Encapsulation of Flavors and Colors in Pediatric Food

The food manufacturing industry encounters a wide array of challenges, including handling specific chemical pollutants, addressing microbiological concerns, and ensuring the desired physicochemical and organoleptic characteristics of the final product, all in the context of food safety and product presentation. Nanotechnology emerges to improve the end product's taste and safety in order to address these problems. The creation and application of materials, processes, or systems at the nanoscale scale is known as nanotechnology [55]. The term "nanomaterials" refers to substances having particles smaller than 100 nm that are unique in comparison to their macroscale equivalents, mostly because of their high surface-to-volume ratios. Recently, there has been a study on the design of nanostructured assemblies for delivering food additives. These assemblies, referred to as nanocapsules, are nano-sized carrier systems that can be used as dietary supplements, liposomes, micelles, or protein-based carriers. They offer several benefits such as enhancing bioavailability, masking unpleasant tastes, and improving the dispersion of insoluble additives without the need for surfactants or emulsifiers. The process of nanoencapsulation involves enclosing food additive molecules in a polymer nanocomposite. The controlled release of octenyl succinic acid is achieved through polylysine. Additionally, the anticancer activity of curcumin has been found to increase when it is encapsulated in hydrophobic modified starch. Furthermore, lipid-based nanoencapsulation methods like nanooliposomes, nanochocolateates, and archaeosomes have been identified as effective nano-delivery mechanisms for food additives, pharmaceuticals, enzymes, and microorganisms. Inorganic nano-sized additives and composites such as silver, iron, calcium, magnesium, selenium, and silicon have also been utilized as preservatives and flavor enhancers. Furthermore, for long-lasting antibacterial activity, silicagel microspheres combined with silicathiosulfate are utilized [56]. An illustration of nanoparticles is the food-grade polypeptide polylysine, which functions as an antioxidant to shield oil from oxidation. Because these polylysin nanoparticles are so much smaller than phytoglycogenoctenyl succinate nanoparticles, they can fill in the spaces between them. However, there is rising interest in the application of encapsulated nanomaterials agriculture for the manufacture of nanoformulated agricultural chemicals used in veterinary medication, fertilizer, insecticides, and fertilizers. These nanocapsules have shown to be more reliable, strong, and helpful in releasing active ingredients gradually. Nanocapsules can be used in animal feeds that are intended to include antimicrobials, detoxifying agents, and nanoadditives. Using nanosensors to identify animal diseases is one application of nanoparticles in agriculture. Recently, the manufacturing of nanopesticides—i.e., products comprising perfectly tailored nanoparticles, such as metals, metal oxides, and nanolays—have been carried out using nanocapsules and nanoemulsions. By creating novel delivery systems for the synthesis of nanoformulated agrochemicals, enhancing nutritional values, and creating new products through bioactive encapsulation, nanotechnology has completely transformed the food industry. The nanoencapsulation technique improves the nutritional value as well as the texture of the food. Food ingredients can be processed at the nanoscale to improve flavors, textures, and aromas while also improving nutritional content. Compared to traditional encapsulation techniques, nanoparticles offer better product encapsulation properties and higher release efficiency. Through nanoencapsulation techniques food industries are maintaining the texture of food in various ways [57]. In terms of pediatric food they are used to enclose bioactive compounds in tiny particles to enhance bioavability and stability. They protect the sensitive ingredients and help to improve the taste and texture of the food. This technology has potential benefits for the manufacturing of nutrition-enhanced food and child-friendly food.

6.8.6 Functional Foods

6.8.6.1 Role of Nanotechnology in Functional Food Development

Since the late 20th century, there has been an increasing recognition of the significant connection between diet and human health. Consequently, a novel classification of food called "functional foods" has emerged. Functional foods encompass a wide range of substances, including nutrients and non-nutrients, that have the potential to provide health benefits beyond what traditional diets offer. In simpler terms, these foods have the ability to impact various bodily processes associated with well-being and good health, while also reducing the risk of diseases [58]. In practical terms, a functional food can be defined as a natural and wholesome food that contains beneficial components in sufficient quantities. Numerous naturally existing substances, like lycopene in tomatoes, omega-3 fatty acids in fish, and soluble fiber in oats, are present in a wide range of fruits, vegetables, grains, and fish, providing health advantages that go beyond basic nourishment. Special conditions for growth or breeding techniques can be utilized to improve the functional components, as demonstrated by the development of rice rich in β-carotene, broccoli enriched with vitamins, and soybeans. The diet of the animal can change the composition of meat, poultry, fish, and eggs. For example, meat and milk products may contain higher levels of omega-3 or conjugated linoleic acid. Genetic engineering is probably going to be necessary to deliver truly unique health benefits, and consumers are not yet comfortable with that. Certain foods might be created with particular ingredients in order to offer health advantages [59]. Yogurts with probiotic bacteria added, margarines enhanced with plant sterols, flour enhanced with folic acid, and bread enhanced with omega-3 fatty acids are a few examples. A food can also be categorized as functional if a component has been eliminated to lessen the food's negative health effects (such as a decrease in saturated fat acids). Furthermore, a food can be considered functional if one or more of its constituents have undergone processing to alter their nature in order to promote health (e.g., the release of bioactive compounds by heating or shearing the food matrix, or the hydrolysis of protein by enzymes). The food industry's functional foods sector is projected to be valued at roughly 168 billion dollars and is expanding at a rate of roughly 9% annually [60]. The rise in metabolic disorders (such as heart disease, obesity, diabetes, and arthritis) is becoming increasingly prevalent among individuals, particularly in Western nations. This can be attributed to various factors, including an aging population with a higher incidence of age-related ailments and the availability of scientific knowledge regarding the correlation between diet and overall well-being. The food industry faces a number of significant challenges in order to meet the growing demand for functional foods. These challenges include determining the potential bioactivity of beneficial compounds, determining optimal intake levels, and creating suitable food delivery matrices and product formulations. The creation of functional foods is likewise beset with numerous regulatory obstacles. "Bioactive compounds" are food ingredients that have positive health effects; these ingredients are usually found in trace amounts in specific foods. Numerous bioactive substances have been thoroughly investigated to assess their impact on human health, specifically their ability to prevent cancer, cardiovascular disease, and hypertension. Lycopene in fruits and vegetables, for instance, is believed to offer protection against cancers of the prostate and other organs. Omega-3 fatty acids sourced from fish have been proven to possess cardioprotective qualities and could potentially safeguard against mental and immune response disorders like rheumatoid arthritis, diabetes, and inflammatory bowel disease. Obtaining sufficient quantities of many of these compounds by consuming large quantities of a diverse range of fruits, vegetables, whole grains, legumes, oils, and nuts rich in bioactive compounds may not always be feasible. Consequently, the food industry displays a keen interest in supplying consumers with adequate quantities of bioactive compounds through the process of "fortifying" everyday food items (such as milk, bread, and beverages) with bioactive compounds derived from natural food sources. However, adding bioactive compounds to food products' fortification causes a lot of technological issues because these compounds react with other food ingredients in various ways during processing, storage, and transportation [61]. The bioactive substance may occasionally have poor solubility in an aqueous solution. It might be susceptible to unfavorable changes in color, flavor, and odor due to its sensitivity to oxygen, light, temperature, and shear. In other situations, the bioactive ingredient might be firmly ensnared in the food matrix, preventing it from being easily liberated during the brief time it spends inside the digestive system. Delivering these substances in an aqueous medium is frequently beneficial since it improves their bioactivity, acceptability, and palatability. When these compounds are added to a beverage or high-moisture food, a delivery system is necessary because they are not easily dissolved in water and are typically prone to oxidation. A significant research challenge is how to add bioactive compounds to food to address stability and sensory issues without sacrificing health benefits. The development of functional foods may benefit from new approaches offered by nanoscience and nanotechnology, particularly when it comes to adding bioactive ingredients without impairing the consumer's sensory experience and enhancing the absorption of specific ingredients [62]. Nanoscience is the scientific study of phenomena and the manipulation of materials at the atomic, molecular, and macromolecular scales, where properties differ from those at a larger scale. The development of functional foods could greatly benefit from the innovative approaches provided by nanoscience and nanotechnology, especially in terms of incorporating bioactive ingredients without compromising the sensory experience of consumers and improving the absorption of specific nutrients. Nanotechnology refers to the manipulation of materials at the nanoscale to design, manufacture, and utilize systems, devices, and structures. This field specifically focuses on structures that are smaller than 100 nm in one dimension. Nanotechnology methods have sparked significant interest in developing advanced delivery systems for protecting and preserving biologically active components in food. Biocompatible and biodegradable nanoparticles can be utilized to carry various nutrients, bioactive compounds, and phytochemicals, improving their solubility, stability, availability, and circulation time in the body. Nano-sized delivery systems provide a larger surface area compared to micro-sized systems produced through traditional microencapsulation methods. They may also improve controlled release, increase bioavailability, improve solubility, and allow for more precise targeting of compounds that are entrapped. It appears to be feasible to create nanoparticles with distinct surface characteristics and to deliver active ingredients to the right locations by carefully selecting the molecular building blocks. But a lot of this work is done in laboratories, and some of it needs basic research to properly define the parameters that control the system. This section discusses some emerging nanoencapsulation technologies that are beneficial for functional food development [63].

6.8.6.2 Nanoemulsion

A nanoemulsion is a dispersion of at least one immiscible liquid and/or liquid crystal in another, characterized by discrete droplets with an average diameter typically ranging from 20 to 200 nm. These droplets give nanoemulsions their optical transparency or translucency, owing to their nanometer-scale size. However, nanoemulsions are thermodynamically unstable systems. There are two main types of nanoemulsions: oil-in-water (O/W) and water-in-oil (W/O). Nanoemulsions are considered metastable, meaning they are non-equilibrium systems that have a tendency to separate into their individual phases over time. Compared to macroemulsions, nanoemulsions generally exhibit higher kinetic stability against creaming, flocculation, and coalescence. This is primarily due to their low viscosity, narrow droplet size distribution, and the dominant Brownian motion between the droplets, which counteracts the weak gravitational separation force. Despite their inherent instability, nanoemulsions may find applications in personal and medical care products. However, the use of nanoemulsions in the food industry remains largely unexplored. One of the major challenges in developing food nanoemulsions is finding suitable food-grade surfactants, as most surfactants used in other industrial applications are not suitable for food use. Additionally, the high production cost poses another obstacle to the widespread adoption of food nanoemulsions [64].

6.8.6.3 Microemulsion

Thermodynamically stable combinations of water, oil, and amphiphilic molecules that spontaneously come together to form nanoscale droplets are referred to as microemulsions. Suitable food-grade surfactants for microemulsions include phospholipids and ethoxylated mono- and diacylglycerides. In some cases, ethanol may be required as a co-surfactant to dissolve long-chain triglycerides. O/W microemulsions offer a practical approach for incorporating lipophilic substances like lutein and lycopene into aqueous solutions. Due to their versatility, microemulsions have found applications in various fields such as medicine, cosmetics, and the modeling of biological membranes. However, the use of certain surfactants in microemulsion formation is limited in food applications. Food products cannot contain a high number of surfactants, and even the ones that can be added must be present in very small quantities. Additionally, solubilizing long-chain triglycerides, which are essential for food applications, is more challenging compared to short- or medium-chain triglycerides. While short-chain alcohols are commonly used as co-surfactants, they should not be added to food. Moreover, many microemulsion systems tend to break down at higher water concentrations and cannot be diluted, which restricts their use in food systems [65].

6.8.6.4 Solid Lipid Nanoparticle

SLNs are composed of lipids that solidify at room temperature, resulting in either a crystalline or amorphous undercooled matrix. These nanoparticles have a diameter ranging from 50 to 1000 nm and incorporate bioactive compounds within their matrix. To stabilize the particles, a layer of surfactants, typically a mixture, is applied. The process of creating SLNs involves dissolving the bioactive compound in melted lipid to form a "melt," which can then be transformed into nanoparticles using various methods. The most commonly used method involves dispersing the melt in a heated surfactant aqueous solution and homogenizing the mixture at temperatures higher than the melting point of the lipid phase. Upon cooling, some or all of the lipids crystallize out of the mixture. Alternatively, the melt can be cooled and ground into small particles. These particles are then homogenized and dispersed in a cold surfactant solution, where they generate sufficient cavitation forces to break down into nanoparticles. SLNs can be produced from microemulsions or nanoemulsions, resulting in particles with different properties and sizes ranging from 100 to 200 nm. By controlling the lipid crystallization process, various internal matrix structures can be achieved, including a homogeneous matrix, a core-shell structure with an outer shell enriched in bioactive compounds, and a core-shell structure with an inner core enriched in bioactive compounds. While SLNs have been extensively used for drug delivery, their application in the food industry has been limited. However, by encapsulating labile hydrophobic compounds within a structured solid matrix, SLNs may enhance their stability. Furthermore, by designing systems with different melting points, the release of these compounds can be controlled [66].

The influence of nanotechnology is set to bring about notable progress in the development of innovative delivery systems for bioactive compounds and nutraceuticals within the functional food industry. This presents a considerable opportunity to enhance the effectiveness and efficiency of bioactive compounds in promoting the overall health and wellness of children.

6.8.7 Nanotechnology-Based Drug Delivery for Medical Nutrition

The food and nutraceutical industries are becoming more open to a rapidly developing field of technology known as "nano science." By granting modified properties to nanomaterials and applying them to the creation of nanoformulations, dietary supplements, and the food industry, nanotechnology has defined a new frontier. In case where children have specific medical condition or nutritional deficiencies, nanotechnology can be employed for delivery of nutrition. It can increase the ways of delivering nutrients to specific areas [67]. There are several nanomaterials in nanotechnology that can be used to deliver nutrients. Those techniques are nanoemulsion, polymeric nanoparticles, polymeric micelles, lipid nanoparticles, and supercritical fluid technology. Like nanomaterials there are various nanomaterial-based technologies used in nutraceuticals. Those are composite based nanomaterials, metal oxide nanoparticles, inorganic-based nanomaterials, carbon-based nanomaterials, organic-based nanomaterials, and beta lactoglobulin based nanomaterials [68].

6.8.7.1 Emerging Role of Nutraceutical and Nanotechnology

Although they are still relatively new, requests for nanotechnology in the food or nutraceutical industries are predicted to increase over the next few years. The applications of antioxidants, antimicrobials, biosensors, and other nanomaterials to food ingredients are a focus of research in food nanobiotechnology [69]. Numerous opportunities for the advancement of novel products and requests in the food system are jeopardized by nanobiotechnology. It can be applied to the site-specific delivery of active ingredients or nutraceuticals using a variety of delivery methods, including nanodispersions and nanoemulsions. It has multiple applications such as a self-cleaning layer, antibacterial coating and wrapping (smart and useful), food pathogen detection, and nutrition decomposition monitoring using nanosensors [70].

6.8.7.2 Benefits of Nanomaterials in Nutritional Delivery System

a. Herbal nutraceuticals aid in the maintenance of health, aid in the treatment of acute or chronic illnesses triggered by nutrition, and enhance life expectancy, quality of life, and health [71].
b. Because of their potential nutritional value, safety, and therapeutic effects, nutraceuticals have generated a lot of interest [71].
c. The growing trends in nutraceuticals include environmental friendliness, local availability, cultivation and processing from renewable sources, and public education [72].
d. Effective solubilization, stabilization, encapsulation, and delivery of bioactives and nutraceuticals using nanotechnology result in improved therapeutic index, decreased frequency of dose administration, and increased absorption at low concentrations [72].
e. Enzyme, analyte, and related dye interactions can be regulated by nanosensors. Packaging incorporates nanosensors designed to detect and attract changes in microbiology and biochemistry [73].

6.8.8 Nanosensor and Food Safety

Food products are susceptible to contamination by microorganisms due to their rich nutritional content. These microorganisms can generate toxins as a defense mechanism, resulting in food contamination that may lead to various illnesses in humans. The adulteration of food can have detrimental health consequences, causing symptoms such as paralysis, vomiting, diarrhea, and, in severe cases, death. Food safety is a global concern as it is a fundamental requirement for human survival. When toxins or chemicals are utilized in warfare, the health risks are exacerbated. Hence, there is a critical need to create efficient and precise techniques for identifying hazardous chemicals, bacteria, and associated toxins [74]. Nanoparticles are increasingly utilized in the creation of detection methods for identifying food material contamination, adulteration, and freshness due to their unique properties. Properly packaged food items are protected from moisture, contamination, and spoilage. Passive packaging systems act as a barrier against moisture, dust, and air. On the other hand, intelligent packaging systems can detect the quality of food products, ensuring the longevity of shelf life and brand reputation. Unlike passive packaging systems, intelligent packaging systems are capable of addressing the rising issues related to food safety and bioterrorism [75]. Different types of intelligent packaging utilize smart devices like biosensors, gas indicators, time-temperature indicators, and barcodes. Moving forward, advanced tags using nanoparticles may be employed to authenticate food products, as traditional barcode protection tags are vulnerable to tampering [76].

6.8.8.1 Nanosensor for Detection of Pathogens in Food

6.8.8.1.1 Detection of Bacteria

The identification of the genetic composition of pathogenic bacteria or the complete bacterial cell is the primary approach used to detect such bacteria in food materials [77]. Compared to other conventional methods, the deoxyribonucleic acid (DNA) isolation and bacteria detection method aided by nanoparticles was more sensitive and required less time. *Listeria monocytogenes*, a milk pathogenic bacterium, has been isolated from its DNA using magnetic iron oxide nanoparticles. For quantifying DNA plolymerase chain reaction was used, which was extracted from the milk sample contaminated with *L. monocytogenes*. For PCR-based microbial detection, the 16s ribosomal ribonucleic acid is frequently employed as a selective marker. The microarray method based on 16s ribosomal ribonucleic acid is not sensitive, and the assay based on PCR is expensive. On other hand, the 16s ribosomal ribonucleic acid detection method based on nanoparticles is simpler and more accurate [78].

6.8.8.1.2 Detection of Toxin

A group of toxic and carcinogenic substances called aflatoxins can be found in food that has been contaminated with *Aspergillus flavus* and *Aspergillus parasiticus*. To detect the presence of aflatoxin M1 in milk, a method utilizing gold nanoparticles and immunochromatographic strips has been employed. When aflatoxin M1 is absent, a red color band is visible on the strip, while a colorless zone indicates the presence of aflatoxin M1 in the milk sample. Palytoxin, a marine toxin, is commonly found in contaminated seafood.

For the highly sensitive detection of palytoxin in mussel meat, electrochemiluminescent sensors based on carbon nanotubes have been developed [79].

6.8.8.1.3 *Detection of Chemicals and Pesticides*

Organophosphates are the most widely used type of pesticide among others. Organophosphorus and carbamate pesticides have been detected using colorimetric and fluorometric methods using gold nanoparticles as sensors. For the pesticide paraoxon sensing, cadmium selenide and cadmium selenide–zinc sulfide core-shell quantum dots have been investigated. To create voltammetric biosensors, phosphate groups containing the pesticide parathion have been selectively bound to a zirconium dioxide/gold nanocomposite film electrode. It has been documented that potentiometer sensors based on MWCNT–silica nanocomposite can detect harmful cadmium ions. The overuse of fertilizers raises similar concerns as pesticides [80]. Fertilizers like melamine are used to adulterate products high in protein, like eggs, biscuits, candies, and coffee beverages. Fluorescence sensors based on gold nanoparticles were able to identify melamine concentrations in infant formulas and cow milk down to the picomolar level.

In terms of pediatric food, safety is very important, so nanosensors played a key role in various ways to maintain the safety of food.

6.8.9 Personalized Nutrition

Nanotechnology can be integrated into diagnostics for assessing children's nutritional status allowing for the customization of pediatric food products to meet individual dietary requirements. In the United States, over two-thirds of adults are overweight or obese. Over 2.8 million people die each year as a result of obesity, which is only becoming more prevalent. On the other hand, eating a balanced diet lowers the chance of developing chronic illnesses, improves mental health, and encourages healthy pregnancies. In actuality, fast food is indeed fast. Nowadays, convenience is the ultimate factor, and unhealthy foods are indisputable when it comes to affordability and accessibility. Healthy foods are perceived as tasteless and costly, such as fruits, vegetables, and non-fried chicken. Fast food is not only bad for you, but it also plays a big role in addictive behaviors. Recent research has demonstrated that fast food's lack of variety and nutritional value inhibits gut microbiomes [81]. This affects a person's susceptibility to addiction because the gut, through the so-called "gut–brain axis," communicates directly with the brain; a person who has a poor variety of gut microbes is more likely to develop a food addiction. This explains why popular diets don't work in the long run because the gut microbiota adjusts to the food being consumed. Foodlysense uses data straight from the gut to create a personalized nutrition plan using AI and nanosensors. There are several applications like this made by using nanotechnology, which helps to personalize the nutrition [82].

6.8.9.1 Role of 3D Nanofoods in Personalized Nutrition

It's necessary to have good eating habits as you age. Nonetheless, chronic illnesses, mental health issues, dental issues, and decreased appetite are the main reasons why older adults are susceptible to malnutrition. It is critical to create innovative, customized eating plans that provide a more enjoyable and nutrient-dense eating experience. The 3D-NANOFOOD project offers a range of innovative methods, such as food enrichment, nano-delivery systems, and 3D printing, for producing customized foods. A flexible and new technology called 3D printing can be used to quickly produce food while also creating goods with customized textures and/or specific nutritional requirements. Food quality can be raised by enriching food with functional ingredients like flavorings and bioactive substances. Bioactive ingredients can improve the food's ability to promote health, and flavorings can enhance the taste or smell of printed foods, making them more palatable. Nevertheless, there are technological difficulties in the food fortification process, which have an impact on the food's qualities. The ultimate objective of 3D-NANOFOOD is to offer practical and inventive methods for producing customized foods for senior citizens that improve their sensory qualities and nutritional value [83].

One of the most exciting areas of medicine right now is personalized nutrition. Many scientists have argued that individually tailored interventions represent a paradigm shift in nutritional research. Available information indicates that a customized nutritional intervention may be useful in preventing a number of illnesses, including obesity and its effects, which are similar to the idea of precision medicine, demonstrating the practical challenges in implementing this method of dealing with young children. In fact, although gene–diet interactions have been extensively documented, there is still a dearth of information regarding direct interventions, especially for the pediatric age. This is despite the fact that the use of nanotechnology may be relevant to improve precision and accuracy of predictions in clinical settings [84]. Furthermore, the effectiveness of personalized nutrition has only been demonstrated in adult populations, and it is still unclear how children would react to these kinds of interventions.

6.8.10 Fortification of Food by Nanotechnology

Adding supplemental elements or health-promoting and nutrition-boosting additives to everyday foods has long been a common practice to improve food functionality through fortification. When it comes to improving functionality and creating functional foods, nanotechnology is extremely important. The functionality of foods is changed by the addition of competent nanoparticles, which also enhance the delivery and bioavailability of the fortification additives within the host body. Amazingly diverse nanomaterials have been created and used to carry various bioactive fortification agents in a wide range of foods, effectively increasing the bioavailability and enhancing immunity. Several nanoparticle treatments' effects on a variety of additives and other bioactive substances, as well as the addition of the NP-treated additives to a range of staple foods, have been discussed. The use of nanoparticles for food fortification has been extensively studied in relation to their capacity to carry nutrients, their warranted biocompatibility based on toxicological studies, and their safety evaluation to allow human consumption. To encourage continued advancement in this exciting field of nanofortification, the remaining obstacles as well as the wide range of potential applications have been clearly outlined [85].

6.8.11 Taste Masking by Nanotechnology

When it comes to medications that are not very palatable, taste is the most important organoleptic parameter that influences patient compliance. One of the biggest obstacles facing the pharmaceutical industry in developing orally ingested active pharmaceutical constituents is taste masking. Many traditional taste-masking methods have been thoroughly investigated. Parallel to this, one of the main concerns of traditional taste-masking techniques is that they may alter the drug's solubility or release. Numerous nanocarrier systems have surfaced recently, touting the benefit of efficient taste masking without interfering with the drug's release or solubility [86, 87]. We will introduce novel approaches to taste masking in this review, including methods that make use of nanocarrier systems like liposomes, solid lipid nanoparticles and polymeric micelles, submicron lipid emulsions, and nanogels. We will primarily focus on the makeup of these systems and how they are used to successfully mask the taste of bitter molecules in oral therapeutic delivery systems. In pediatric food it is very important to mask the taste of food for consumption; better taste will increase the food intake as well as nutritional value.

6.8.11.1 Various Nanocarriers Used for Taste Masking

Liposomes, nanoparticles, nanosuspensions, nanomicelles, and nanoemulsions are examples of nanocarrier systems. In addition, nanocarriers can be classified as hard or soft depending on their physical state; an intermediate form exists between the two. Lipid or polymeric nanoparticles are examples of the hard form; they are defined by their lack of elasticity and flexibility. Soft forms with the capacity to deform and reform include liposomes, polymeric micelles, and nanoemulsions.

6.8.11.1.1 Liposomes

The biodegradability, biocompatibility, and ability to ensnare hydrophilic or hydrophobic molecules have made liposomes a subject of interest in the field of drug delivery. Phospholipids and other extra lipids such as cholesterol make up the majority of the lipids in liposomes. Liposomes have the capacity to ensnare hydrophobic materials within the lipid bilayer and to enclose or entrap an aqueous payload inside the formed vesicle. Because liposomes interact with the lipids in cell membranes, they can deliver drugs to targets without requiring either active or passive uptake of the nano-vehicle by the target cells. By choosing the appropriate phospholipids and/or preparation technique, liposomes can be tailored based on the payload that needs to be entrapped and the release pattern. The drug's entrapment efficiency during preparation is highly influenced by the preparation technique. The primary factors influencing the preparation technique selection are the ease of scaling up the process, the required particle size, polydispersity, and the physicochemical properties of the entrapped drugs [88, 89].

6.8.11.1.2 Polymeric Nanoparticles

The polymeric nanoparticles, which range in size from 10 to 1000 nm and are made of biocompatible and biodegradable polymers, enable the drug to be encapsulated, dissolved, entrapped, or attached to the nanoparticle matrix. The preparation technique for the nanoparticle system dictates whether the resulting particles are nanospheres or nanocapsules. The device is referred to as a nanocapsule when the medication is trapped inside the cavity and encircled by a special polymer membrane. However, the term "nanosphere" refers to the uniform and physical dispersion of a drug within a polymer matrix [90].

6.8.11.1.3 Solid Lipid Nanoparticles

At room temperature or body temperature, SLNs are physically solid in the natural world. They are a colloidal system made up of lipids that have been surfactant-stabilized. While avoiding certain drawbacks from these systems, SLNs combine the advantages of liposomes, polymeric nanoparticles, and emulsions. The solid, lipophilic core matrix of SLNs is encased in a phospholipid coat. Because SLNs have a lipophilic core matrix encircled by phospholipid that adheres to the lipophilic core via its lipophilic portion, they are able to entrap lipophilic molecules in greater quantities than liposomes. SLNs have the capacity to protect the entrapped molecule as well as release drugs over an extended period of time. SLNs are more stable because they are less prone to aggregation and coalescence. They are better than other nanocarrier systems to prepare and use in taste-masking bitter molecules because of their nontoxic composition, biodegradability, affordability of production, and ease of scaling up [91].

6.8.11.1.4 Nanostructured Lipid Carriers

In order to create NLCs, a liquid lipid was used to partially replace the solid lipids in SLNs, resulting in the formation of a drug that was integrated into the liquid lipid matrix [92]. NLCs have a higher loading capacity than SLNs, making them potential carriers due to their superior composition characteristics.

6.8.11.1.5 Submicron Lipid Emulsion

Because of their many advantages, emulsions are widely used as delivery systems in the cosmetics and pharmaceutical industries. Emulsifiers are typically used to help disperse immiscible liquids into another liquid in the form of tiny droplets to create emulsions. The other liquid is the external, or continuous phase, and the scattered droplets are the internal phase. Emulsion stability is achieved by the co-surfactant and surfactant, which keep the dispersed droplets from coalescing or separating [93]. There are two types of emulsions: oil-in-water (O/W) and water-in-oil (W/O). Water is the internal or dispersed phase in (W/O), whereas oil is the external or continuous phase, and the opposite is true for (O/W). Pharmaceutical and cosmetic companies are particularly interested in microemulsions and nanoemulsions with higher stability because they have better bioavailability and can be tailored for passive drug targeting. The primary distinction between macro- and microemulsions is the size of the droplets; in the case of macro type, the droplet size is greater than 100 nm, whereas in the case of micro type, the droplet size varies between 5 and 100 nm. The emulsion is more stable the smaller the size of the particles. The macro type is opaque and has a higher viscosity, whereas the micro type is translucent and has a lower viscosity.

6.8.11.1.6 Nanogels

When cross-linked polymer networks are placed in an appropriate solvent, they expand into particles that are nanoscale in size, creating "nanogels." The delivery of polynucleotides via cross-linked non-ionic and poly-ionic polymer networks is referred to as "Nanogel" (NanoGelTM). The goal of advances in polymer science is to create smart nanosystems that ensure successful treatment with nanogels. The drug could be delivered to a particular site by the nanogels in a controlled and sustainable way [94].

6.8.11.1.7 Nanosponges

Nanosponges (NSs) can be thought of as colloidal structures that hold various molecules within them. NSs are lipophilic by nature, but because of their capacity to disperse in water, they can aid in disguising the taste that are not acceptable by solidifying liquid compounds. This lessens the medication's interaction with the tongue's sensory receptors. The cross-linking agent and cyclodextrins could interact to prepare NSs. In their review, Caldera et al. 2017, divided NSs into four generations. Urethane, carbonate, ether, and ester are in the first one. There are certain types of polymers in the second generation, like charged side chains or fluorescence. Neurons that respond to stimuli are found in the third generation. This kind of NSs can exhibit varying behaviors in response to environmental changes, such as variations in temperature, oxidative or reducing factors, pH, or other variables [95]. The fourth one is highly selective to the particular drug that is included.

6.8.11.1.8 Inclusion Complex Formation

In aqueous solutions, various cyclodextrin complexes may come together to form nanoscale aggregates. Previous research determined that the best type of cyclodextrin for inclusion complexation is β-cyclodextrin. Two different methods are the physical mixing technique, which involves mixing the drug with the cyclodextrin, and the wet kneading method, which creates an inclusion complex. The outside of cyclodextrin contains alcoholic hydroxyl moieties, which give the material a hydrophilic exterior. The central cavity of the material is filled with carbon and oxygen atoms derived from glucose residues. Because of its nature, cyclodextrin can take up molecules and host them in its cavities, a process called the inclusion complex. By concealing the bitter active molecule inside the cyclodextrin cavity, the inclusion complex can improve the taste of the rejected bitter active constituent. It can also improve the stability of the included molecule by halting its degradation and serving as a solubilizing agent [96].

6.8.11.1.9 pH-Responsive Coordination Polymer Coated Mesoporous Silica Nanoparticles

The medication delivery method that was created was easy to use and effective at masking oral tastes. The mechanism of the system involves controlling the release of the bitter medication by altering the pH in the gastrointestinal tract. Fe-4, 4′-bipyridine (Fe-bipy) complex, a pH-sensitive metal-organic coordination polymer (CP), functions as a taste masker in this system.

This emerging technique of nanotechnology helps to mask the taste of drugs along with nutraceuticals. This will help administration of drugs to children along with food [97].

6.9 POSSIBLE DETRIMENTAL EFFECTS OF NANOPARTICLES UTILIZED IN THE PEDIATRIC FOOD SECTOR

6.9.1 Nanotoxicity

The number of publications published in the field of nanotoxicology—the study of the harmful effects of nanomaterials on human health and the environment—has increased by roughly 600% during the past 10 years. The majority of these published investigations are in vitro investigations that are mainly carried out by delivering a certain concentration of nanomaterial to cells that are developing at the plate's bottom. Additionally, the cells' reactions are tracked to determine the impact of that specific nanomaterial on the cells. Studies cannot be compared to one another since they are frequently conducted using arbitrary settings and poorly described nanomaterials. One of the special characteristics of nanoparticles is their larger surface area per unit volume. As a result, they behave entirely differently from their counterparts in bulk. These special qualities allow gold nanoparticles to be used in a wide range of processes, including medication delivery and diagnostic imaging. On the other hand, nanoparticles have a higher propensity to interact with different biological entities like proteins, lipids, and entire cells. Nanoparticles have the ability to penetrate cell membranes, access different organs, and trigger immunological responses, including inflammatory ones [98].

6.9.2 Other Possible Harmful Effects of Nanoparticle

Despite the many benefits, using nanotechnology in food may have negative effects on the environment, society, and human health. This is because it is thought that these particles enter the ecosystem through pesticide delivery in agriculture or through application in the processed food industry, such as packaging, which raises questions about their toxicity. Because smaller particles are more bioavailable to our bodies and have a higher reactivity, nanoengineered particles carry a larger danger of long-term pathological repercussions. Nanomaterials can enter into the food chain via direct incorporation of nanomaterials in various forms like nanoemulsion, nanocapsulation, and the use of nanomaterials in food manufacturing and processing related to this [99]. Human exposure to nanoparticles varies substantially depending on the food industry application area and usage concentration. The danger of exposure is highest in areas where nanomaterials are applied directly to food products as carriers of innovative food additives. Nanoparticles play a role in oxidative stress due to that chances of inflammation that can be high. Also, there are several adverse effects of nanoparticles, i.e., renal problems, Parkinson's syndrome, and Alzheimer's disease [100, 101].

6.10 CONCLUSION

The role of nanotechnology in pediatric foods presents a transformative landscape for enhancing nutritional quality, safety, and sensory appeal. Through the precise manipulation of materials at the nanoscale, nanotechnology offers a range of opportunities to improve children's dietary intake and overall health. Nanotechnology facilitates the encapsulation of essential nutrients, vitamins, and functional ingredients, ensuring enhanced bioavailability and targeted delivery, addressing specific nutritional needs in children. It also contributes to the development of innovative food products with improved taste, texture, and visual appeal, making them more attractive and palatable for young consumers. Furthermore, the use of nanosensors and quality control methods enhances the safety and monitoring of pediatric foods, assuring compliance with stringent standards and reducing risks associated with contaminants or spoilage. Nanocomposite packaging extends shelf life, minimizing food waste and ensuring product safety. While these advancements hold immense promise, careful consideration must be given to the ethical, regulatory, and safety aspects of nanotechnology in pediatric foods. Addressing concerns regarding the potential risks and ensuring transparent communication about the use of nanomaterials is crucial for gaining public trust. Moving forward, continued research, collaboration, and regulatory oversight are essential to harness the full potential of nanotechnology in pediatric foods, ensuring that these innovations benefit children's health while maintaining the highest safety standards. The responsible and informed application of nanotechnology will play a pivotal role in shaping the future of pediatric nutrition, offering improved, safe, and appealing food choices for the betterment of children's well-being.

REFERENCES

1. Chellaram, C., Murugaboopathi, G., John, A. A., Krithika, S., & Jayabarathi, R. (2014). Significance of nanotechnology in food industry. *APCBEE Procedia*, *8*, 109–113.
2. Grumezescu AM, Holban AM, eds. *Impact of Nanoscience in the Food Industry*, vol. 12. London: Academic Press; 2018.
3. Chaudhry Q, Watkins R, Castle L. Nanotechnologies in food: What, why and how? In: *Nanotechnologies in Food*. Cambridge: Royal Society of Chemistry; 2017.
4. Nile SH, et al. Nanotechnologies in food science: Applications, recent trends, and future perspectives. *Nano-Micro Lett*. 2020;12(1):1–34.
5. Singh T, et al. Application of nanotechnology in food science: Perception and overview. *Front Microbiol*. 2017;8:1501.
6. Lombardo D, Kiselev MA, Caccamo MT. Smart nanoparticles for drug delivery application: Development of versatile nanocarrier platforms in biotechnology and nanomedicine. *J Nanomater*. 2019;12:1–26.
7. Farokhzad OC, Langer R. Nanomedicine: Developing smarter therapeutic and diagnostic modalities. *Adv Drug Deliv Rev*. 2006;58:1456–1459.
8. Buzea C, Pacheco II, Robbie K. Nanomaterials and nanoparticles: Sources and toxicity. *Biointerphases*. 2007;2:MR17–MR71.
9. Torchilin V. Targeted polymeric micelles for delivery of poorly soluble drugs. *Cell Mol Life Sci CMLS*. 2004;61(19–20):2549–2559. https://doi.org/10.1007/s00018-004-4153-5.
10. Huang D, Wu D. Biodegradable dendrimers for drug delivery. *Mater Sci Eng*. 2018;90:713–727. https://doi.org/10.1016/j.msec.2018.03.002.
11. Cuenca AG, Jiang H, Hochwald SN, Delano M, Cance WG, Grobmyer SR. Emerging implications of nanotechnology on cancer diagnostics and therapeutics. *Cancer*. 2006;107(3):459–466. https://doi.org/10.1002/cncr.22035.
12. Sahoo S, Misra SR, Parveen S. Nanoparticles: A boon to drug delivery, therapeutics, diagnostics and imaging. *Nanomedicine*. 2017;8(2):147–166. https://doi.org/10.1016/j.nano.2011.05.016.
13. Nekkanti V, Kalepu S. Recent advances in liposomal drug delivery: A review. *Pharm Nanotechnol*. 2015;3(1):35–55. https://doi.org/10.2174/2211738503666150709173905.
14. Suryanarayana, C. Nanocrystalline materials. *Int Mater Rev*. 1995;40:41–64.
15. Langer, R., & Tirrell, D. A. (2004). Designing materials for biology and medicine. *Nature*, *428*(6982), 487–492.
16. Kumar R, Lal S. Synthesis of organic nanoparticles and their applications in drug delivery and food nanotechnology: A review. *J Nanomater Mol Nanotechnol*. 2014;3:4.
17. Cheng F, Vijaykumar S. Applications of artificial neural network modeling in drug discovery. *Clin Exp Pharmacol*. 2012;2:e113.
18. Medina-Franco JL. Drug discovery with novel chemical libraries. *Drug Des*. 2012;1:e105.
19. Athinarayanan J, Periasamy VS, Alsaif MA, Al-Warthan AA, Alshatwi AA. Presence of nanosilica (E551) in commercial food products: TNF-mediated oxidative stress and altered cell cycle progression in human lung fibroblast cells. *Cell Biol Toxicol*. 2014;30:89–100. https://doi.org/10.1007/s10565-014-9271-8.

20. Bouwmeester H, Dekkers S, Noordam MY, Hagens WI, Bulder AS, Heer C, et al. Review of health safety aspects of nanotechnologies in food production. *Reg Toxicol Pharmacol*. 2009;53:52–62. https://doi.org/10.1016/j.yrtph.2008.10.008.
21. *Cientifica Report: Nanotechnologies in the Food Industry*. Cientifica Ltd., Aug. 2006.
22. Bradley EL, Castle L, Chaudhry Q. Applications of nanomaterials in food packaging with a consideration of opportunities for developing countries. *Trends Food Sci Technol*. 2011;22:603–610. https://doi.org/10.1016/j.tifs.2011.01.002.
23. Pathakoti K, Manubolu M, Hwang H-M. Nanostructures: Current uses and future applications in food science. *J Food Drug Anal*. 2017;25(2):245–253.
24. Gallocchio F, Belluco S, Ricci A. Nanotechnology and food: Brief overview of the current scenario. *Procedia Food Sci*. 2015;5:85–88.
25. Singh T, et al. Application of nanotechnology in food science: Perception and overview. *Front Microbiol*. 2017;8:1501.
26. Frisancho AR. *Anthropometric Standards: An Interactive Nutritional Reference of Body Size and Body Composition for Children and Adults*. 2nd ed. Ann Arbor: University of Michigan Press; 2008.
27. Entongo, T. Growth assessment and monitoring. In: M Corkins, ed. *The A.S.P.E.N. Pediatric Nutrition Core Curriculum*. Silver Spring, MD: American Society for Parenteral and Enteral Nutrition; 2010.
28. Leonberg BL, ed. *Academy of Nutrition and Dietetics Pocket Guide to Pediatric Nutrition Assessment*. 2nd ed. Chicago, IL: Academy of Nutrition and Dietetics; 2013.
29. Fenton TR, Kim JH. A systematic review and meta-analysis to revise the Fenton growth chart for preterm infants. *BMC Pediatr*. 2013;13:59.
30. Augustin MA, Hemar Y. Nano- and micro-structured assemblies for encapsulation of food ingredients. *Chem Soc Rev*. 2009;38:902912.
31. Gupta A, Silver S. Molecular genetics: Silver as a biocide: Will resistance become a problem? *Nat Biotechnol*. 1998;16:888.
32. Rhim J-W, Hong S-I, Park H-W, Ng PKW. Preparation and characterization of chitosan-based nanocomposite films with antimicrobial activity. *J Agric Food Chem*. 2006;54:58145822.
33. Renton A. *Welcome to the world of nano foods*. 2006. http://observer.guardian.co.uk/foodmonthly/futureoffood/story/0,1971266,00.html.
34. Sari P, Mann B, KumaR R, Singh RRB, ShaRma R, Bhardwaj M, et al. Preparation and characterization of nanoemulsion encapsulating curcumin. *Food Hydrocol*. 2015;43:540–546. https://doi.org/10.1016/j.foodhyd.2014.07.011; Singh Sekhon B. Nanotechnology in agri-food production: An overview. *Nanotechnol Sci Appl* 2014;7:31–53. https://doi.org/10.2147/NSA.
35. Lagarón JM, Cabedo L, Cava D, Feijoo JL, GavaRa R, Gimenez E. Improving packaged food quality and safety. Part 2: Nanocomposites. *Food Addit Contam*. 2005;22:994–998. https://doi.org/10.1080/02652030500239656.
36. Kumar S, Boro JC, Ray D, Mukherjee A, Dutta J. Bionanocomposite films of agar incorporated with ZnO nanoparticles as an active packaging material for shelf life extension of green grape. *Heliyon*. 2019;5:e01867. https://doi.org/10.1016/j.heliyon.2019.e01867.
37. Jeng HA, Swanson J. Toxicity of metal oxide nanoparticles in mammalian cells. *J Environ Sci Heal—Part A Toxic/Hazardous Subst Environ Eng*. 2006;41:2699–2711. https://doi.org/10.1080/10934520600966177.
38. Zawrah MF, Abd El-Moez SI. Antimicrobial activities of gold nanoparticles against major foodborne pathogens. *Life Sci J*. 2011;8:37–44.
39. Kraśniewska K, Pobiega K, Gniewosz M. Pullulan-biopolymer with potential for use as food packaging. *Int J Food Eng*. 2019;15:15. https://doi.org/10.1515/ijfe-2019-0030
40. Grigore ME, Biscu ER, Holban AM, Gestal MC, Grumezescu AM. Methods of synthesis, properties and biomedical applications of CuO nanoparticles. *Pharmaceuticals*. 2016;9:75. https://doi.org/10.3390/ph9040075.
41. Toğrul H, Arslan N. Moisture sorption isotherms and thermodynamic properties of walnut kernels. *J Stored Prod Res*. 2007;43:252–264.
42. Vance ME, Kuiken T, Vejerano EP, McGinnis SP, Hochella MF, Jr., Rejeski D, Hull MS. Nanotechnology in the real world: Redeveloping the nanomaterial consumer products inventory. *Beilstein J Nanotechnol*. 2015;6:1769–1780. https://doi.org/10.3762/bjnano.6.181.
43. Roco MC, Harthorn B, Guston D, Shapira P. Innovative and responsible governance of nanotechnology for societal development. In: Roco MC, Mirkin CA, Hersam MC, eds. *Nanotechnology Research Directions for Societal Needs in 2020*. Dordrecht, The Netherlands: Springer; 2011. pp. 561–617.
44. López-Lorente AI, Mizaikoff B. Recent advances on the characterization of nanoparticles using infrared spectroscopy. *TrAC Trends Anal Chem*. 2016;84:97–106. https://doi.org/10.1016/j.trac.2016.01.012.
45. Kumar K, Sarkar P. Encapsulation of bioactive compounds using nanoemulsions. *Environ Chem Lett*. 2018;16:59–70. https://doi.org/10.1007/s10311-017-0663-x.
46. Reineccius GA. Liposomes for controlled release in the food industry. In: Risch SJ, Reineccius GA, eds. *Encapsulation and Controlled Release of Food Ingredients*. Washington, DC: ACS Symposium Series; 1995. pp. 113–131. https://doi.org/10.1021/bk-1995–0590.ch011.
47. Mukherjee S, Ray S, Thakur RS. Solid lipid nanoparticles: A modern formulation approach in drug delivery system. *Indian J Pharm Sci*. 2009;71:349–358. https://doi.org/10.4103/0250-474X.57282.
48. He X, Deng H, Hwang H. The current application of nanotechnology in food and agriculture. *J Food Drug Anal*. 2019;27:1–21. https://doi.org/10.1016/J.JFDA.2018.12.002.
49. Jafari SM, MCClements DJ. Nanotechnology approaches for increasing nutrient bioavailability. *Adv Food Nutr Res*. 2017;81:1–30. https://doi.org/10.1016/bs.afnr.2016.12.008.
50. Faulks RM, Southon S. Challenges to understanding and measuring carotenoid bioavailability. *Biochim Biophys Acta*. 2005;1740(2):95–100. https://doi.org/10.1016/j.bbadis.2004.11.012.
51. Liu G, Yang J, Wang Y, Liu X, Chen L. Protein-lipid composite nanoparticles for the oral delivery of vitamin B12: Impact of protein succinylation on nanoparticle physicochemical and biological properties. *Food Hydrocoll*. 2019;92:189–197. https://doi.org/10.1016/j.foodhyd.2018.12.020.
52. Penalva R, Esparza I, González-Navarro CJ, Quincoces G, Peñuelas I, Irache JM. Zein nanoparticles for oral folic acid delivery. *J Drug Deliv Sci Tech*. 2015;30:450–457. https://doi.org/10.1016/j.jddst.2015.06.012.
53. Katuwavila NP, Perera AD, Dahanayake D, Karunaratne V, Amaratunga GA, Karunaratne DN. Alginate nanoparticles protect ferrous from oxidation: Potential iron delivery system. *Int J Pharm*. 2016;513(1–2):404–419.
54. Morris VJ, Woodward NC, Gunning AP. Atomic force microscopy as a nano science tool in rational food design. *J Sci Food Agric*. 2011;91:2117–2125.

55. Vidhyalakshmi R, Bhakyaraj R, Subhasree RS. Encapsulation the future of probiotics are view. *Adv Biol Res*. 2009;34:6103.
56. IFT. *Functional Foods: Opportunities and Challenges*. Washington, DC: IFT Expert Report Institute of Food Technology; 2005. pp. 7–10.
57. Frewer L, Scholderer J, Lambert N. Consumer acceptance of functional foods: Issues for the future. *Br Food J*. 2003;105:714–731. https://doi.org/10.1108/00070700310506263.
58. Heasman M, Mellentin J. *The Functional Foods Revolution: Healthy People, Healthy Profits?* London, UK: Earthscan Publications Ltd; 2001.
59. Contor L. Functional food science in Europe. *Nutr Metab Cardiovasc Dis*. 2001;11:20–23.
60. Hasler CM. Functional foods: Benefits, concerns and challenges–a position paper from the American Council on Science and Health. *J Nutr*. 2002;132:3772–3781.
61. Bigliardi B, Galati F. Innovation trends in the food industry: The case of functional foods. *Trends Food Sci Technol*. 2013;31:118–129. https://doi.org/10.1016/j.tifs.2013.03.006.
62. Tadros T, Izquierdo P, Esquena J, Solans C. Formation and stability of nano-emulsions. *Adv Colloid Interface Sci*. 2004;108–109:303–318. https://doi.org/10.1016/j.cis.2003.10.023.
63. Garti N, Yaghmur A, Aserin A, Spernath A, Elfakess R, Ezrahi S. Solubilization of active molecules in microemulsions for improved environmental protection. *Colloids Surf A*. 2003;230:183–190. https://doi.org/10.1016/j.colsurfa.2003.09.020.
64. Müller RH, Radtke M, Wissing SA. Solid lipid nanoparticles (SLN) and nanostructured lipid carriers (NLC) in cosmetic and dermatological preparations. *Adv Drug Deliv Rev*. 2002;54:S131–S155. https://doi.org/10.1016/S0169-409X(02)00118-7.
65. Aschberger K, Gottardo S, Amenta V, Arena M, Botelho Moniz F, Bouwmeester H, et al. Nanomaterials in food-current and future applications and regulatory aspects. *J Phy Conf Ser*. 2015;617:012032. https://doi.org/10.1088/1742-6596/617/1/012032.
66. Kumar AP, Depan D, SiNgh Tomer N, Singh RP. Nanoscale particles for polymer degradation and stabilization—trends and future perspectives. *Prog Polym Sci*. 2009;34:479–515. https://doi.org/10.1016/j.progpolymsci.2009.01.002.
67. Bouwmeester H, Brandhoff P, Marvin HJP, Weigel S, Peters RJB. State of the safety assessment and current use of nanomaterials in food and food production. *Trends Food Sci Technol*. 2014;40:200–210. https://doi.org/10.1016/j.tifs.2014.08.009.
68. Ameta SK, Rai AK, Hiran D, Ameta R, Ameta SC. Use of nanomaterials in food science. *Biog Nano-Particles Use Agro Ecosyst*. 2020;457:24. https://doi.org/10.1007/978-981-15-2985-6_24.
69. Magnuson BA, Jonaitis TS, Card JW. A brief review of the occurrence, use, and safety of food-related nanomaterials. *J Food Sci*. 2011;76:R126–R133. https://doi.org/10.1111/j.1750-3841.2011.02170.x.
70. Kumar LY. Role and adverse effects of nanomaterials in food technology. *J Toxicol Heal*. 2015;2:2. https://doi.org/10.7243/2056-3779-2-2.
71. Luo Y, Wang Q, Zhang Y. Biopolymer-based nanotechnology approaches to deliver bioactive compounds for food applications: A perspective on the past, present, and future. *J Agric Food Chem*. 2020;68:12993–3000. https://doi.org/10.1021/acs.jafc.0c00277.
72. Chen H, Zhou K, Zhao G. Gold nanoparticles: From synthesis, properties to their potential application as colorimetric sensors in food safety screening. *Trends Food Sci Technol*. 2018;78:83–94.
73. Alamer S, Eissa S, Chinnappan R, Herron P, Zourob M. Rapid colorimetric lactoferrin-based sandwich immunoassay on cotton swabs for the detection of foodborne pathogenic bacteria. *Talanta*. 2018;185:275–280.
74. Priyadarshini E, Pradhan N. Gold nanoparticles as efficient sensors in colorimetric detection of toxic metal ions: A review. *Sens Actuators B*. 2017;238:888–902.
75. Hadush A, Pal M. Detection of Campylobacter jejuni from food and its epidemiology. *J Public Health Epidemiol*. 2013;5:357–361.
76. Hu J, Jiang Y-Z, Tang M, Wu L-L, Xie H-Y, Zhang Z-L, et al. Colorimetric-fluorescent-magnetic nanosphere-based multimodal assay platform for Salmonella detection. *Anal Chem*. 2018;91:1178–1184.
77. Xie H, Dong J, Duan J, Hou J, Ai S, Li X. Magnetic nanoparticles-based immunoassay for aflatoxin B1 using porous g-C3N4 nanosheets as fluorescence probes. *Sens Actuators B*. 2019;278:147–152.
78. Yu X, He Y, Jiang J, Cui H. A competitive immunoassay for sensitive detection of small molecules chloramphenicol based on luminol functionalized silver nanoprobe. *Anal Chim Acta*. 2014;812:236–242.
79. Vu TQ, Maddipati R, Blute TA, Nehilla BJ, NusbLat L, Desai TA. Peptide-conjugated quantum dots activate neuronal receptors and initiate downstream signaling of neurite growth. *Nano Lett*. 2005;5:603–607.
80. Vu TQ, Liu HY. Quantum dot hybrid gel blotting: A technique for identifying quantum dot-protein/protein-protein interactions. *Methods Mol Biol*. 2009;544:381–391.
81. Ferrari CKB. Functional foods, herbs and nutraceuticals: Towards biochemical mechanisms of healthy aging. *Biogerontology*. 2004;5:275–289.
82. Godoi F, Prakash S, Bhandari BR. 3D printing technologies applied for food design: Status and prospects. *J Food Eng*. 2016;179:44–54.
83. Dary O, Hurrell R. *Guidelines on Food Fortification with Micronutrients*. Geneva, Switzerland: World Health Organization, Food and Agricultural Organization of the United Nations; 2006. pp. 1–376.
84. Beck TK, Jensen S, Bjoern GK, Kidmose U. The masking effect of sucrose on perception of bitter compounds in Brassica vegetables. *J Sens Stud*. 2014;29:190–200.
85. Lankpord BL, Becker CH. The use of some imitation flavors for masking distasteful drugs. II. Quinine hydrochloride. *J Am Pharm Assoc*. 1951;40:83–86.
86. Kulkarni SB, Betageri GV, Singh M. Factors affecting microencapsulation of drugs in liposomes. *J Microencapsule*. 1995;12:229–246.
87. Lasic DD. Novel applications of liposomes. *Trends Biotechnol*. 1998;16:307–321.
88. Rao JP, Geckeler KE. Polymer nanoparticles: Preparation techniques and size-control parameters. *Prog Polym Sci*. 2011;36:887–913.
89. Mehnert W, Mäder K. Solid lipid nanoparticles: Production, characterization and applications. *Adv Drug Deliv Rev*. 2001;47:165–196.
90. Akhoond Zardini A, Mohebbi M, FaRhoosh R, Bolurian S. Production and characterization of nanostructured lipid carriers and solid lipid nanoparticles containing lycopene for food fortification. *J Food Sci Technol*. 2018;55:287–298.

91. Simovic S, Hui H, Song Y, Davey AK, Rades T, Prestidge CA. An oral delivery system for indomethicin engineered from cationic lipid emulsions and silica nanoparticles. *J Control Release*. 2010;143:367–373.
92. Yin Y, Hu B, Yuan X, Cai L, Gao H, Yang Q. Nanogel: A versatile nano–delivery system for biomedical applications. *Pharmaceutics*. 2020;12:290.
93. Caldera F, Tannous M, Cavalli R, Zanetti M, Trotta F. Evolution of cyclodextrin nanosponges. *Int J Pharm*. 2017;531:470–479.
94. Stojanov M, Wimmer R, Larsen KL. Study of the inclusion complexes formed between cetirizine and α-, β-, and γ-cyclodextrin and evaluation on their taste-masking properties. *J Pharm Sci*. 2011;100:3177–3185.
95. Bao GM, Wang L, Yuan HQ, Wang XY, Mei TX, Qu MR. Taste masking of a drug by pH–responsive coordination polymer-coated mesoporous silica nanoparticles. *RSC Adv*. 2016;6:109453–109459.
96. Donaldson K, Stone V, Tran CL, Kreyling W, Borm PJ. Nanotoxicology. *Occup Environ Med*. 2004;61:727–728.
97. Oberdörster G, Oberdörster E, Oberdörster J. Nanotoxicology: An emerging discipline evolving from studies of ultrafine particles. *Environ Health Perspect*. 2005; 113:823–839.
98. Li X, Liu W, Sun L, Aifantis KE, Yu B, Fan Y, Feng Q, Cui F, Watari F. Effects of physicochemical properties of nanomaterials on their toxicity. *J Biomed Mater Res A*. 2015;103:2499–2507.
99. Morimoto Y, Horie M, Kobayashi N, Shinohara N, Shimada M. Inhalation toxicity assessment of carbon-based nanoparticles. *Acc Chem Res*. 2013;46:770–781.
100. Lam CW, James JT, McCluskey R, Hunter RL. Pulmonary toxicity of single-wall carbon nanotubes in mice 7 and 90 days after intratracheal instillation. *Toxicol Sci*. 2004;77:126–134.
101. Liu Y, Xu Z, Li X. Cytotoxicity of titanium dioxide nanoparticles in rat neuroglia cells. *Brain Inj*. 2013;27:934–939.

7 Nanotechnology in Meat and Poultry Preservation

Deepak Sharma[a], *Shyamli Thakur*[b], *Samir Bhargava*[c], *and Deepa Gupta*[d]
[a]Institute of Pharmacy, Assam Don Bosco University, Tapesia, Assam, India
[b]School of Pharmaceutical & Populations Health Informatics, DIT University, Dehradun, India
[c]School of Pharmaceutical & Populations Health Informatics, DIT University, Dehradun, India
[d]Department of Pharmaceutical Chemistry, LBS College of Pharmacy, Jaipur, Rajasthan, India

7.1 INTRODUCTION

Demands from consumers for high-quality food and their focus on well-being and health have resulted in notable developments and breakthroughs in the processing and packaging of food, which includes the meat production sector (Taherimehr et al., 2021). Many additives and functional elements with both nutritious and functional properties have been researched to recover the sensory, visual, and aromatic quality features of food products while also preventing spoilage (Fang et al., 2017; McMillin, 2017).

In the process of processing food, including meat products, various additives are added to enhance the final product's quality or highlight its functional properties. These additives include antioxidants, binders, thickeners, humectants, flavor enhancers, and tenderizing enzymes (Khan et al., 2016; Das et al., 2021; Gagaoua et al., 2021; Nanda et al., 2021). Certain ingredients that get employed by food and meat producers to extend the shelf life and sustain the quality of their products may cause health problems when consumers are exposed to increased levels of additives (Dainelli et al., 2008). Due to these factors, some of the traditional packaging that the industry uses may actually pose a health risk to consumers when artificial antioxidants seep into the food that is packaged (Salgado et al., 2021).

Finding safer and less expensive alternatives for the processing of meat products was therefore urgently needed. For the improvement of the overall quality of fresh meat products, the meat industry must adopt innovative strategies to increase the availability of these additives and bioactive components by altering their size and shape (Khan et al., 2016). In the meat production chain, nanotechnology has emerged as an innovative substitute that is being used more and more to ensure an extensive shelf life alongwith better quality of food (Ramachandraiah et al., 2015; Bayda et al., 2020). It can be characterized as a field of technology focused on developing nanomaterials, or materials smaller than 100 nm, with special and innovative qualities (Taherimehr et al., 2021; Khan et al., 2016). The increased surface area-to-volume proportion of nanoscale ingredients permits them to spread, thereby turning out to be more effective on their target in minute small concentrations (Khan et al., 2019).

7.2 DIFFERENT TECHNIQUES OF NANOPARTICLES

The wide range of materials referred to as nanoparticles (NPs) includes particulate materials with a dimension of at least 100 nm. As shown in Figure 7.1, NPs are distinguished into three groups based on their proportions, shape, and chemical features (Khan et al., 2019). NPs can be characterised into two primary clusters: inorganic and organic nanoparticles, which can be formed using organic materials like biopolymers and carbon, etc. into nanotubes, nanofibers, nanoemulsions, etc. (Ramachandraiah et al., 2015). Nanoparticles can also be produced by biological sources like microorganisms like fungi, bacteria, etc. (Singh et al., 2015). The kind of nanoparticles that are intended to be synthesized determines which synthesis method should be used. On the basis of their physicochemical properties, the following sections give a detailed description of the most widely used categories of nanoparticles.

7.2.1 Inorganic Nanoparticles

An inorganic particle-filled core sets apart inorganic nanoparticles. Magnetic, metallic, ceramic, and quantum dot NPs are a few types of these NPs (Istiqola et al., 2020). To ensure that the particle shape and size when synthesized are properly controlled, a particular method must be followed when creating inorganic nanoparticles that combine electronic and magnetic particles. Primary synthesis techniques are hydrothermal

DOI: 10.1201/9781032677422-7

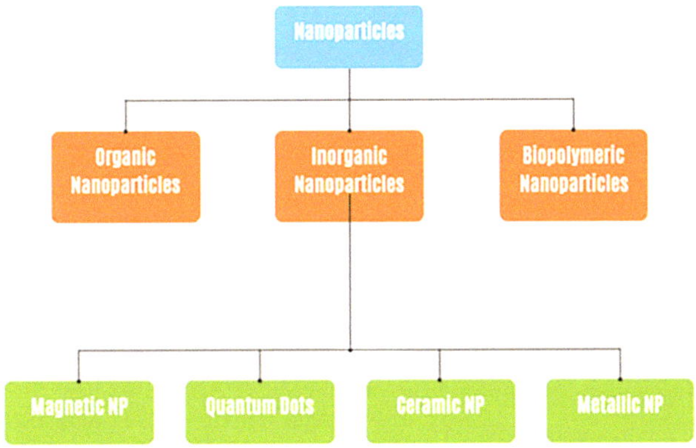

FIGURE 7.1 Classification of nanoparticles.

synthesis, microemulsions, salt precipitation in an aqueous medium, decomposition in an organic medium, polyol processes, and aerosol pyrolysis. Intriguing magnetic, optical, electrical, or fluorescent characteristics can be found in inorganic NPs (Rane et al., 2018; Salas et al., 2014; Abbasi et al., 2016). Magnetic nanoparticles are primarily produced by precipitating salt in a liquid medium. Other popular methods for creating magnetic nanoparticles include the polyol technique, disintegration in organic solution, etc. Compared to bulk methods, inorganic nanomaterials with a narrower size have recently been synthesized using microfluidic technology (Shen et al., 2020). Nebulizing on an agitated liquid surface, on the other hand, produced a solid phase of nanoparticles (Reverberi et al., 2016). It is significant to say that flame spray pyrolysis takes longer and is less effective than aerosol-assisted wet chemical methods (Lähde et al., 2014).

7.2.2 Organic Nanoparticles

Lipid or polymeric substances are used to make organic nanoparticles that are environmentally friendly, cost-effective, and suitable for biological applications (Kumar and Lal, 2014). The process of emulsification, followed by nanoprecipitation and drying, can result in the formation of organic nanoparticles. Following the emulsification process, which involves dissolving organic materials to produce nanodroplets of a particular size, nanoparticles are formed by a variety of techniques, including polymerization and precipitation. The synthesis of organic NPs also made use of spray drying, piezoelectric technology, and supercritical fluid (Lee et al., 2011; Wright et al., 2010).

Polymeric nanoparticles, micelles, liposomes, and capsules are just a few examples of systems that can be formed with the help of synthetic chemistry and self-organization. Organic nanoparticles may find use in the delivery of specific molecules due to their ability to physically encapsulate molecules or through surface or core conjugation. In this framework, a research by Pabast et al. used nanoliposomes loaded with Satureja essential oil (SKEO) to create a novel coating structure for lamb meat (Pabast et al., 2018). The researchers discovered that coating lamb meat with a chitosan mixture that included nanoencapsulated SKEO maintained the meat's premium characteristics, improved oxidative stability, and boosted microbiological safety. Similar findings from earlier studies demonstrated that essential oils used in liposome encapsulation had higher levels of antioxidant and antimicrobial activity than the oils did in their unreconstituted forms (Rai et al., 2017; Ghaderi-Ghahfarokhi et al., 2017). Thus, this would account for the decreased evaporation characteristics and the ease with which it penetrates the cellular wall.

7.2.3 Biopolymeric Nanoparticles

Initially, non-biodegradable polymers such as polyacrylamide and polymethyl acrylate were combined with biopolymers such as albumin to form biopolymeric nanoparticles (Sundar et al., 2010). They can be loaded with a variation of biopolymers, with particle sizes ranging from 1 to 1000 nanometers (Zielińska et al., 2020). The primary goal of formulating biopolymer nanoparticles is to evade the non-biodegradable polymers, posing harm to both beings and the ecology. These nanoparticles are primarily derived from lipids, proteins, and carbohydrates. As nanomaterials, proteins like whey and soybean are used in a variety of food packaging applications, including the packing of muscle meats (Tang et al., 2019; Oymaci et al., 2016; Hosseini et al., 2021). Researchers are also interested in the complex carbohydrate chitosan because of its possible application as an efficient and active packaging material (Kumar et al., 2020b). A number of lipids that are nanoencapsulated have made headlines due to their enhanced antioxidant properties in nanocomposite films, which are intended to better the stability of the products (Zhang et al., 2017; Boskovic et al., 2019). Combining two or more biopolymer nanoparticles into their matrices can also boost the thermal and physical attribures of different polymer packages that are commercially available (Tang et al., 2019). A variety of synthesis methods, including coacervation, emulsification, precipitation, and cross-linking, can be used to create these molecules (Zielińska et al., 2020). Additionally, the use of chitosan-based bio nanoparticles is growing (Sanchez-Alvarado et al., 2018). Chitosan fibers were created by electrospinning anionic biodegradable polyvinyl alcohol (PVOH) with an ethanolic NaOH solution used to stabilize the fibers. Furthermore, by thermally treating biopolymers, Okoroh et al. generated nanoparticles of zinc ferrite that were capped with what is an easy, affordable, and environmentally friendly process (Okoroh et al., 2018).

7.3 APPLICATION OF NANOTECHNOLOGY IN MEAT PRODUCTS

Similarly, a variety of foods such as meat can benefit from the use of nanoparticles and nanomaterials as shown in Figure 7.2. Adding nanoparticles in the form of ingredients and additives can improve flavor, taste, texture, and bioavailability. Additionally, they can be supplemented to packaging materials to increase their

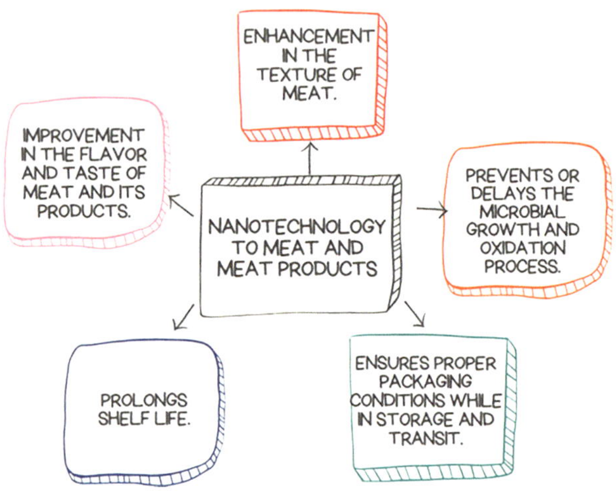

FIGURE 7.2 Advantages of applications of nanotechnology to meat and meat products.

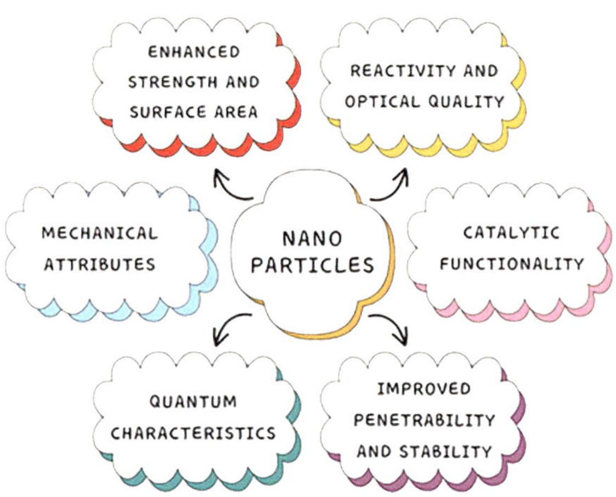

FIGURE 7.3 Functional properties and characteristics of nanoparticles.

life by slowing down the rate of multiplication of microorganisms and oxidation processes and guarantee that the packaging is in good condition during storage and transportation (Chaudhry and Castle, 2011; Duncan, 2011; Cushen et al., 2012; Baltic et al., 2013; Benetti and Venturini, 2015).

By means of food additives and complements of ingredients encapsulated in nanoparticles is an cutting-edge technique to lessen the amount of lipid, salt, sugar, or preservers used in food products while also enhancing the palatability of fat-soluble additives, according to Chaudhry and Castle (2011) and Baltic et al. (2013). Application of nanomaterials containing antioxidants and antimicrobials, as well as nanoencapsulated antioxidants, colors, savors, essence and stabilisers, may be advantageous for treated meats (Ozimek et al., 2010). Research is required to determine the effectiveness of oil-in-water emulsions of nanomaterials in intricated food mediums, like meat products, according to Salminen et al. (2013) and Ramachandraiah et al. (2015). As represented in Figure 7.3, nanoparticles are a whole new realm of application because of their enhanced strength, surface area, reactivity, optical, mechanical, catalytic, and quantum characteristics, as well as their improved penetrability and stability. The beneficial properties of nanotechnology will lead to its application in food packaging as well.

Baltic et al. (2013) state that to improve product quality of the product, reactive particles like nanoparticles and nanomaterials can be incorporated into packaging as nanosensors. Physical stimuli can be detected and responded to by nanosensors, which are nanoscale devices (Joyner and Kumar, 2015). Nanosensors that detect pollutants, chemicals, heavy metals, particulates, nutrients, allergens, gases, and pathogens and spoilage microorganisms should be utilised for improvement and ensuring safety of food. Additionally, they respond to fluctuations in the atmosphere surrounding the processed food in diverse ways (Duncan, 2011; Baltic et al., 2013; Joyner and Kumar, 2015; Ramachandraiah et al., 2015).

A variety of food production stages, such as raw material inspection, online process control, and storage condition monitoring, can benefit from the use of nanosensors (Joyner and Kumar, 2015). In terms of sensitivity, specificity, and testing speed, nanosensors are superior to conventional sensors, according to Jin et al. (2003). Table 7.1 depicts the three major categories of nanosensors (Joyner and Kumar, 2015).

7.4 ROLE OF NANOTECHNOLOGY IN MEAT-BORNE PATHOGEN CONTROL

Products of meat get easily contaminated with dangerous bacteria, which is why there are a lot of outbreaks of foodborne illnesses every year. In addition to other forms of meat and meat derivatives, significant sources of this pathogen include pork, eggs, and poultry (Boskovic et al., 2016). Many raw foods, including raw meats, can contain

TABLE 7.1
Major Categories of Nanosensors

Type of Nanosensor	Principle of Their Usage
Nanoparticles-Based Sensors	• Utilised in packaging to find microbes, gases, organic compounds, and moisture.
	• These sensors comprise various types of nanoparticles with intricate nanoscale structures.
Electrochemical Nanosensors	• It uses selective antibodies to bind to conductive nanomaterials such as nanowire, carbon nanotube, or nanocantilever.
	• When the target analyte attaches to the antibodies, the material's conductivity changes and is observed. This kind of detection mechanism is used in these devices.
Optical Nanosensors	• It primarily uses surface plasmon resonance or fluorescence-based mechanisms for pathogen detection

L. monocytogenes, as can meat that has been processed but tainted during the process. Furthermore, ready-to-eat meat products are frequently contaminated by this bacterium (Long et al., 2014).

Bacteria, such as *S. aureus* and *Bacillus* species, pose a serious health and economical hazards due to their increased rates of morbidity and death in developed and developing nations. Therefore, in order to detect, manage, and prevent foodborne agents in meat products, advanced techniques are needed.

Much research has confirmed the effectiveness of nanoparticles against a range of pathogens that are important for food safety, particularly in relation to the safety of meat. One of the potential applications in the fight against foodborne illness is antimicrobial packaging, as per Neethirajan and Jayas (2011), Silvestre et al. (2011), Baltic et al. (2013), and Wesley et al. (2014), is made to progressively release antimicrobial particles that are nano-sized, which reduces or prevents the growth of microorganisms and maintains food safety. Because it stops bacteria from growing on or near the food product's surface, antimicrobial nanopackaging can be especially important for unsterilized food products (Coma, 2008; Ramachandraiah et al., 2015; Tornuk et al., 2015).

Nano Ag has shown antibacterial activity counter to a variety of foodborne bacteria, like *E. coli*, in a number of studies when used either on its own or in amalgamation with other antibacterials (Shahverdi et al., 2007; Damm et al., 2008; Ahmad et al., 2015). Fedotova et al. (2010) stated that an Ag nanoparticle-containing sausage-shaped cellulose and collagen covering exhibited robust antibacterial and antifungal activities. The broad range of antimicrobial action of silver nanoparticles at low concentrations makes them potentially useful against resistant microbes (Duncan, 2011; Ramachandraiah et al., 2015). Silvestre et al. (2011), Azeredo et al. (2013), Baltic et al. (2013), and Azeredo (2009) all highlight the significance of temperature steadiness and little volatility in Ag, which render it appropriate for processing and real-world use. Apart from silver nanoparticles, research is being conducted on the possible antimicrobial properties of gold and zinc nanoparticles (Silvestre et al., 2011).

7.5 ANTIOXIDANT EFFECT OF THE NANOSYSTEMS APPLIED ON MEAT PRODUCTS

Proteolysis and oxidation of lipids and degradation are the main reasons for meat products' storage-related deterioration. In addition to producing rancidity, the changes in producing an off-flavor, the creation of lethal compounds, and loss of essential nutrients, the oxidation of polyunsaturated fatty acids also distresses color, nutrient profile, quality, consumer acceptance, texture, and shelf life (Domínguez et al., 2019). Natural antioxidants like plant extracts and essential oils can be added to meat products to slow down the oxidation of lipids. These ingredients enhancing flavor are termed as Generally Recognized as Safe (GRAS) (Jayasena, 2014). Natural antioxidants do, however, have a drawback in that they are extremely sensitive to several environmental factors, including temperature, pH, light, and gas concentrations.

When processing and storing fresh meat and meat products, nanosystems can shield, stabilize, and lessen these issues. A few of the benefits of nanometric-size coatings are their potential to have a natural additive at a particular place and their increased contact area, which enhances functionality. They can be given right away to meat products or utilized as an animal's dietary supplement (Bao et al., 2019; Estévez, 2021). The section providing specifics of each is followed by Figure 7.4, which illustrates the use of nanosystems on meat products for antioxidant effects.

7.5.1 Chitosan–Vitamin C Nanoparticle Complexes (CS–VC NP)

Meat was cured for 21 days at 4 °C after being dipped in CS–VC NP. The meat samples treated with CS–VC NP due to their antioxidant properties showed a good outcome on retarding peroxidation by prolonging the period of induction (Elbarbary et al., 2015).

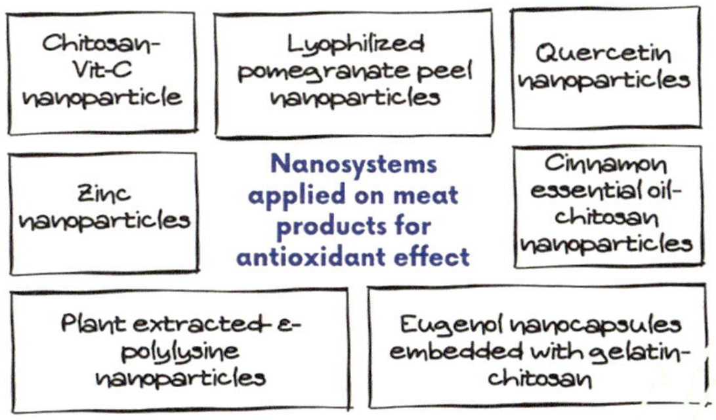

FIGURE 7.4 Nanosystems applied on meat products for antioxidant effect.

7.5.2 ε-Polylysine Nanoparticles (ε-PLN)

Due to its strong antioxidant properties after 45 days at 4 °C in refrigeration, ε-PLN combined with an extract of stinging nettle, green tea, and olive leaf could be utilized as an alternative to nitrite in sausages made in the frankfurter style. The TBARS values of the sausages containing ε-PLN and the extracted mixture were reduced after 45 days. When ε-PLN sausage samples were put into storage, their final TBARS value (2.39) fell between allowable limits which is 2–2.5 mg MDA/kg. No sign of rancidity was observed (Alirezalu et al., 2021).

7.5.3 Cinnamon Essential Oil–Chitosan Nanoparticles (CEO–CS NP)

The efficiency of CEO encapsulation as a natural antioxidant that stops primary and secondary oxidation of lipids in beef patties was evaluated using PV and TBARS values. The lipid oxidation of beef patties can be effectively delayed by CEO encapsulation, even at reduced concentrations, which were 0.05 and 0.1%, for eight days at 4 °C. The 0.1 CEO–CS NP solution technique showed the highest resistance to oxidation among the treatments in terms of PV and TBARS values at the termination of storage tenure (Ghaderi-Ghahfarokhi et al., 2016).

7.5.4 Zinc Nanoparticles

Zinc nanoparticles (Zn NPs) were found to reduce the MDA content of breast meat in chicken broilers, suggesting that the broilers' enhanced antioxidant status was preventing oxidative damage and lipid peroxidation (Eskandani et al., 2021).

7.5.5 Lyophilized Pomegranate Peel Nanoparticles (LPP NP)

While samples treated with LPP NP were kept at 4 °C for 15 days, they showed reduced PV and TBARS ranges. The samples were exposed during the meatball-forming procedure (Morsy et al., 2018).

7.5.6 Eugenol Nanocapsules Embedded with Gelatin-Chitosan (Eug-Gel-CS NC)

Eug-Gel-CS NC was used in submerging cured meat of pork. Comparing this technique to control samples, there was a lower rise in TBARS. In addition to mediating the protection against muscle oxidation caused by hydroperoxide, Eug also functions as an efficient natural antioxidant. For 15 days, the storage temperature was 4 °C (Wang et al., 2020).

7.5.7 Quercetin Nanoparticles (QT NP)

There were five different dietary treatments that included varying amounts of QT NPs in the diet (0, 100, 200, 300, and 400 mg/kg). The reactive oxygen species (ROS) and malondialdehyde (MDA) in muscle produced were significantly reduced by high levels of nutritional QT NP supplement, indicating reduced free radical production and thereby less of the damage of layer of lipid. QT NP-supplemented diets reduced the ROS by

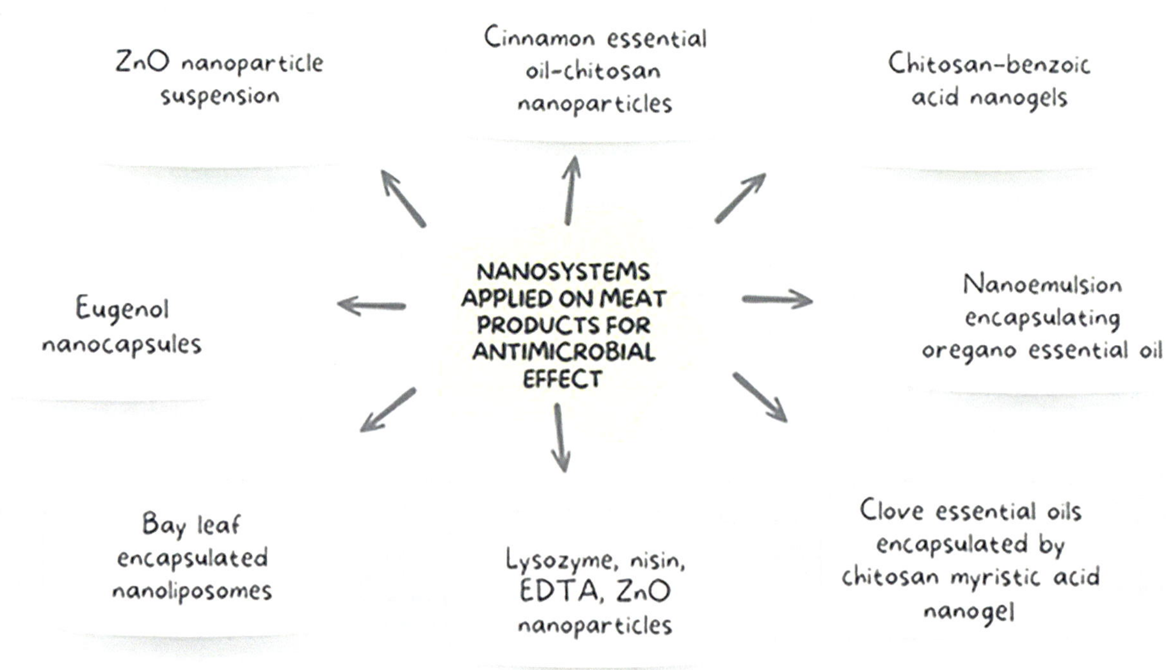

FIGURE 7.5 Nanosystems applied on meat products for antimicrobial effect.

fortifying cells' resistance to oxidation. Lipid peroxidation decreased as a result, and the number of healthy cells increased (Ibrahim et al., 2021).

7.6 ANTIMICROBIAL EFFECT OF THE NANOSYSTEMS APPLIED ON MEAT PRODUCTS

As listed in Figure 7.5, nanosystems have benefited the meat products in terms of the antimicrobial effect it imposes.

7.6.1 ZnO Nanoparticle Suspension

L. monocytogenes and *S. aureus* (7 log CFU/g) were used to inoculate the meat samples. The meat samples were then sprayed with 1% acetic acid and ZnO NP solutions at different concentrations (6 mM and 8 mM) and kept refrigerated. As compared to controls, high concentrations of ZnO–AA NP significantly reduced the number of microbes (Mirhosseini et al., 2014).

7.6.2 Cinnamon Essential Oil–Chitosan Nanoparticles (CEO–CS NP)

As compared to the control (6.85 log CFU/g), the meat treated with nanoparticles got 3.99 and 4.59 CFU/g (0.05 and 0.1% of encapsulated CEO). The authors report that there were no appreciable differences between the treatments at the start of storage. However, in the days that followed and at the termination of the storage period, encapsulated CEO demonstrated a unique pattern of *S. aureus* inhibition. The usage of CS NP as a CEO encapsulation seems to have good potential because there was less microbial growth. The meat was mixed with nanoparticles as part of the formulation, and samples were kept for eight days at 4 °C (Ghaderi-Ghahfarokhi et al., 2017).

7.6.3 Chitosan–Benzoic Acid Nanogels (CS–BA NG)

After 12 days of refrigeration for beef cutlet at 4 °C, the antimicrobic efficacy of REO-CS–BA NG in contrast to *S. typhimurium* was assessed in vivo. The surface was sprayed with 1 ml NC REOs at different concentrations of entrapped REOs. From day one, the population of *S. typhimurium* was immediately reduced by REO-CS–BA NG. Furthermore, following a 12-day period of storage, the layered samples exhibited a suggestively reduced population of bacteria (Hadian et al., 2017).

7.6.4 Nanoemulsion Encapsulating Oregano Essential Oil (OEO NE)

In relation to *E. Coli* contamination of chicken pâté, the OEO NE antibacterial effect was significantly increased ($p < 0.05$) than the other treatments on the final day of analysis. The growth inhibition effects of pure and nanoencapsulated oregano oil on *S. aureus* were found to be similar, while the most effective method of reducing bacterial growth was to treat the area with nitrite, a synthetic food preservative (Moraes-Lovison et al., 2017).

7.6.5 Clove Essential Oils Encapsulated by Chitosan Myristic Acid Nanogel (CEO–CS–MA NG)

When 1 and 2 mg of CEO–CS–MA NG were applied topically to beef, after three days of storage period, the Salmonella population were significantly reduced by 0.58 and 0.68 log CFU/g, respectively. These findings suggest that the encapsulation of bioactive composites can improve the antimicrobic activity of essential oils by encouraging their dispersal in an aqueous medium, triggering inactive adsorption mechanisms (Moraes-Lovison et al., 2017).

7.6.6 Lysozyme, Nisin, EDTA, ZnO Nanoparticles

An antimicrobial medium was evenly layered to the surface of meat post attachment, and the meat was then packaged and stored at 4 °C for 15 days. The treated and controls were supplemented with *E. Coli*, *L. monocytogenes*, and *B. cereus* at 7 log CFU/cm^2. When LNEZ was applied to *E. coli* O157:H7 after 15 days of storage, the bacteria's population decreased (to 2.7 log CFU/cm^2), against the control samples' population (8.1 log CFU/cm^2). The control samples reached concentrations exceeding 8 log CFU/cm^2, whereas *B. cereus* and *L. monocytogenes* showed final concentrations of 2 and 4 log CFU/cm^2, respectively, after 15 days in a refrigerator (Morsy et al., 2018).

7.6.7 Bay Leaf Encapsulated Nanoliposomes

After spraying the meat with 1000 and 1500 ppm of nano-extracted solution, it was marinated for 12 days at 4 °C. On most days, control exhibited the highest levels of *Staphylococcus aureus*, while on the 4th day of storage, no bacteria were detected in the solutions containing the nano-encapsulated extracted solution at 1500 ppm non-extract. *E. coli* responded better to higher concentrations of the extract and to solutions comprising the NC extract; and by 8th day, presence of *E. Coli* was not detected in the NC 1500 ppm extract (Tometri et al., 2020).

7.6.8 Eugenol Nanocapsules

The results demonstrated that the total bacteria count (TBC) of the meat in each treatment group increased with an increase in storage time. The Eug-Gel-CS NP set had the reduced TBC growth rate. It was established that the sustained release effect of the Eug nanoparticles on Eug may increase its antimicrobial qualities. Meat was soaked in the Eug-Gel-CS NCs solution and stored for 15 days 4 °C (Wang et al., 2020).

7.6.9 ε-Polylysine Nanoparticles

When frankfurter-style sausages were formulated with ε-PLN and stored for 45 days at 4 °C, their shelf life increased significantly. This resulted from the total viable count microorganisms only reaching 4.03 Log CFU/g after 30 days, indicating no discernible increase during this period. When it came to combating *E. coli*, ε-PLN showed better effectiveness than the other groups. The control samples exhibited 0.21 Log CFU/g, while the ε-PLN sausages showed no detectable values after 45 days in storage (Alirezalu et al., 2021).

7.7 ANTIMICROBIAL MECHANISM OF NANOTECHNOLOGY

Pateiro et al. (2021) have recommended that the efficiency of antimicrobial activity of nanoemulsions is related with their small particle size and high stability. They function by releasing the cytoplasmic contents when the bacterial membrane is damaged by the active compounds in the small droplets. The inhibitory nature of metal oxide nanoparticles (MONP) is dependent on their dimension, form, and concentration (Liu et al., 2009). According to Ali et al. (2021), the main antimicrobial and antibacterial effect of MONPs is the production of ROS (O_2–, OH•, H_2O_2). Most notably, this inhibits respiratory enzymes. The carboxylic acid groups on proteins in bacteria give their membranes anionic charge, whereas MONPs have a cationic charge. Due to electrostatic attraction brought on by the variance in charges between the bacterial outer membrane and the MONPs, the MONPs can gather on the cell surface until they are able to pass through the bacteria (Raghunath et al., 2017). Furthermore, by altering the permeability and structure of the outer plasma membrane, MONPs draw additional MOPNs into the cell cytoplasm and simultaneously cause cellular components leakage, toxicity, and damage to nucleic acid composition (Ali et al., 2021). The peptidoglycan layer's single-bonded SH groups interact with positively charged metal ions to change the helical nature of strands and bind to DNA, ultimately rupturing the cell wall. In addition to impairing DNA replication, cell division, and respiration, MONPs can also cause problems when they attach to mesosomes (Raghunath et al., 2017; Ali et al., 2021).

7.8 NANOTECHNOLOGY APPLICATION METHODS FOR MEAT PRESERVATION

During the industrial manufacturing and processing of meat products, additives that hasten the processes, color stabilization, and improvement of flavor are required. Products can be made healthier by adding prebiotics, probiotics, and other functional nutrients. Additionally, meat components can interact with nanosystems to reduce fat content without sacrificing flavor (Singh et al., 2016). Additionally, they can reduce salt concentrations of phosphates, sodium, and nitrates (Khan et al., 2016). Consumers often associate the color of meat products with freshness and flavor, making it the most important factors in their eyes (Purslow et al., 2020). Consumers cannot fully appreciate the flavor, juiciness, or tenderness of meat until it has been cooked, despite the fact that meat can be judged by its color when it is purchased. The pigment found in meat, lamb, pork, and poultry is called myoglobin, a sarcoplasmic protein (Dong et al., 2019). Color changes are caused by redox reactions of myoglobin due to variables like pH, the muscle part, antioxidants, lipid oxidation, and attributes of mitochondria (Purslow et al., 2020). Handling, nutrition, age at slaughter, pH, and heredity are additional variables that can impact color. Exogenous factors include variables such as oxygen, light, and the meat's original microbiological load (Suman and Joseph, 2013).

7.8.1 Marinating

Marinating is the utmost widely used technique for preserving meat and involves infusing the food in a liquid infusion to enhance their flavor, texture, and moistness. To enhance the appearance of the final product, marinades can also contain ingredients like salt, organic acids, water, sugar, flavours, spices and condiments, food enhancers, antioxidant or antimicrobial agents, and colorants. Process control, ingredient selection for marinade preparations, and continual scientific approach refinement are all necessary to advance the quality attributes of the product (Yusop et al., 2012).

7.8.2 Edible Coatings

To guarantee that they degrade readily once they serve their purpose, edible film materials are composed of proteins, polysaccharides, or lipids as depicted in Figure 7.6. Edible antimicrobial films usually have better mechanical properties, a low water barrier function, and a high moisture sensitivity; therefore, plasticizers are often combined with biopolymers for increasing the efficacy of processing, flexibility, and lower the glass transition temperature. Polyethylene glycol, sorbitol, mannitol, fructose, and mannose are the plasticizers that are most frequently used. The low gas permeability of polysaccharide-based films allows them to prolong food products' shelf lives without fostering anaerobic, i.e. absence of oxygen conditions. Yet because of the hydrophilic groups, they are moisture-sensitive. Pectin, chitosan, starch, alginate, cellulose, and carrageenan are examples of commonly used polysaccharides. Even though proteins derived from whey, soy, etc. can generally form coatings with improved mechanical and barrier parameters than polysaccharides, protein-based edible coatings are not as strong as synthetic versions (Umaraw et al., 2020). These edible films' properties are determined by the functional attributes of the nanomaterials that were used to arrange the layers that comprise them. Although they have little mechanical resistance and cannot stop gases from escaping,

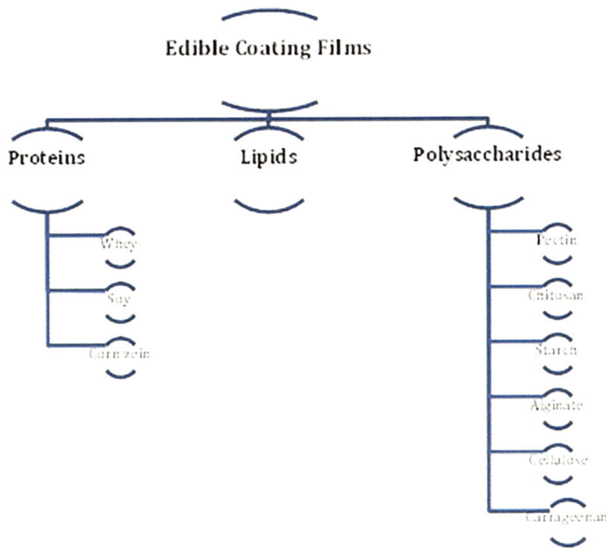

FIGURE 7.6 Composition of edible coating films.

lipid layers serve as barriers against moisture. Protein and polysaccharide-based coatings, on the other hand, function as naturally occurring edible barriers that provide nutritional advantages and longer shelf lives by effectively blocking gases but not moisture (Ramachandraiah et al., 2015).

7.8.3 Processing Ingredients

Meat products undergo biochemical alterations like proteolysis, lipolysis, and oxidation during processing, which lead to the acquisition of volatile aromatic composites throughout the curing process and fermentation. Although oxidation is considered to be the chief source of meat quality degradation during processing and storage, it is also a necessary reaction that contributes to the unique flavour of meat. It has been possible to lower product costs and increase product functionality by utilizing probiotics, dietary fibers, herbs, spices, and vegetable proteins (Ramachandraiah and Hong, 2021; Khan et al., 2016). This suggests that active components that function well during processing might likewise be encapsulated with nanosystems. For example, adding essential oil of oregano (*Origanum vulgare*) to nanoemulsions inhibited the growth of microoragnaisms thereby extending the shelf life of chicken. Cinnamon essential oil have also been demonstrated to function in concert with one another in fresh sausage made of pork and beef (Pateiro et al., 2021).

7.9 NANOTECHNOLOGY IN MEAT HAZARD CONTROL AND DETECTION

The high-water content (aw > 0.99), and nutritional composition of meat make it an perfect medium for bacterial growth (Buncic et al., 2014). Bacteria from the animal's skin or digestive tract can contaminate meat during the slaughter and processing procedures. Meat deterioration is a complicated process that involves numerous bacterial species and is dependent on a number of variables (Buncic et al., 2014). To increase the shelf life of meat and meat products, novel approaches to eliminate or reduce spoilage microorganisms should be used alongside current techniques (Burt, 2004; Baltic et al., 2013).

Essential oils with antimicrobic properties may be an improved alternative to chemical additives for organic and "natural" foods (Burt, 2004; Bošković et al., 2013). The primary hindrance with the usage of EOs is that some of their constituents may impair the organoleptic qualities of meat. Rather than directly adding EOs to the meat, this problem can be solved by incorporating them into the packaging.

The growth of bacteria can get suppressed by metal oxide nanoparticles. TiO_2 nanoparticles have demonstrated antibacterial properties against *Pseudomonas fluorescens* and *Macrococcus caseolyticus*, two bacteria that are commonly associated with the decaying of chicken, according to Wang et al. (2014). Additionally, ZnO/LDPE nanocomposite coverings also demonstrated antibacterial properties against bacteria on chicken breast fillets (Panea et al., 2014). According to the findings of these studies, using LDPE nanocomposite films containing different antibacterial nanoparticles may improve the safety and shelf life of meat.

Absorbent pads are extensively used in modern packaging techniques to shield the packaging systems from unhygienic meat extracts, preserve the freshness of meat or meat products, and create a visually pleasing package. By absorbing meat extracts, the pads get contaminated, which facilitates the growth of bacteria and serves as a reservoir for them. Utilizing cellulose absorbent pads that had been formerly dipped in 1% Ag nanoparticle solution, *Pseudomonas* spp., *Enterobacteriaceae*, and total aerobic bacteria were among the spoilage microorganisms that were successfully reduced in modified atmosphere-packaged (MAP) red meat (Fernandez et al., 2010). The microbiological quality of meat, the likelihood of foodborne bacterial growth, and the amount of time that spoiling can all be affected by the saturation of absorber pads with diverse nanoparticles.

Oxygen nanosensors are found to be helpful in this regard because oxygen is necessary for bacterial metabolism. Variations in the oxygen content of vacuum packaging and MAP may indicate bacterial activity and the start of spoiling, according to Valdés et al. (2009). The reduction of meat's shelf life is caused by oxidation, in addition to the result of bacterial growth. Lipid oxidation, according to Falowo et al. (2014), degraded the quality of meat and resulted in the buildup of dangerous amalgams that might be harmful to consumers' health. Nanosensors can be used to visually alert customers to changes in a package's oxygen content and to monitor a package's oxygen content without destroying it (Baltic et al., 2013). Photo-activated indicator ink for detecting oxygen has been established based on a reaction between nano-sized TiO_2 or SnO_2 units and methylene blue. Azeredo (2009) and Duncan (2011) claim that when oxygen levels drop, the indicator gradually changes color. It is blue when fully exposed and bleached when there is no exposure; the degree of change is dependent on the amount of O_2 exposure.

Li et al. (2004) stated that nanocrystalline TiO_2 was successfully placed on glass and acetate films to create deoxygenated closed atmospheres. Thus, nanocomponents can be utilized as oxygen scavengers to eradicate or diminish the amount of oxygen in the packaging to a desired or acceptable level. SiO_2 gallic acid nanoparticles in a similar manner demonstrated the ability to scavenge the free radical diphenyl picrylhydrazyl (DPPH), suggesting that they may find application as nanoantioxidants, which can be put into novel use (Deligiannakis et al., 2012; He and Hwang, 2016). Nanoparticles can be used as carriers of antioxidants to stop oxidation. For example, He and Hwang (2016) suggest that encasing bioactive substances such as vitamins and flavonoids in polymeric nanoparticles is a viable option. Silicate or other nanoparticle-enhanced packaging film also condensed the ingress of oxygen and different gases and increased the film's permeability, preventing or delaying deterioration (Neethirajan and Jayas, 2011).

7.10 GREEN NANOTECHNOLOGY

Due to the world's rapid population growth, the global food system—which includes safe and secure food supplies, sustainable resource use, and nutrient-dense diets—is severely strained. Developing new green nanotechnology-based technologies is becoming crucial to solving food security and sustainability problems. Due to their extensive use in the food packaging industry, synthetic nanomaterials have the likelihood to end up in landfills, water sources, soil, and the environment (Guimarães et al., 2017). Most nanomaterials, both organic and inorganic, have remained unidentified and may be potential pollutants because analytical techniques have limited our ability to identify them. Environmental sustainability is currently being promoted by green techniques in nanotechnology, which use nontoxic, mildly reacting precursors (Pateiro et al., 2021). Green synthesis of nanoparticles seemed to be an easy, affordable, eco-friendly, and readily replicable substitute (Domínguez et al., 2019). Biomaterial-based methods do not require harsh or toxic chemicals, making them a more environmentally friendly chemical process for producing exceptional quality materials.

It is encouraging that plant-based ingredients can be used to produce nanoparticles made up of metal because harmful chemicals are not needed for the decrease of metal ions. These techniques have been found out to be affordable, eco-friendly, and scalable (Devatha et al., 2018). Nanoparticles of gold and silver and other elements could be created by combining extracts from plant sources like black tea, green tea, eucalyptus leaf, and neem leaf (Iravani, 2011; Jadoun et al., 2021; Parveen et al., 2016). The reduction of substances by algae, fungi, bacteria, and viruses aided in the synthesis of safe and environmentally friendly metallic nanoparticles such as cadmium, gold, and metal oxides such as titanium oxide (TiO) and zinc oxide (ZnO) (Jadoun et al., 2021). Recent studies on green-synthesized AgNPs have demonstrated that the addition of plant extract(s) or phytochemicals during the synthesis process may enhance the antimicrobial activities when compared to AgNPs synthesized through non-biological routes (Oliver et al., 2018). It is common practice to prepare nanoparticles in a less dangerous manner while retaining their biogenicity by using extracts from plants, vitamins, and amino acids (Hussain et al., 2016; Singh et al., 2018). The amalgamation of metal nanoparticles has been observed to be commonly aided by the presence of glucose or fructose in plant extracts. Furthermore, it has been found that proteins extracted from plants like the soy protein and even gluten work well for making nanoparticles (Salas et al., 2012; Teng et al., 2013).

Value-adding of agro-industrial wastes is now required, and synthesizing nanoparticles with the least amount of hazardous chemicals and solvents is essential (Sinha et al., 2021; Ebrahimzadeh et al., 2020). Potential sustainable nanotechnology demands the application of green methods under benign circumstances without the use of hazardous precursors (Zhu et al., 2019). There abundance of space for revolution to grow in addition to expanding up new horizons, even in the face of notable advancements in several industries.

7.10.1 Applications of Green Nanoparticles in Meat Preservation

Since nanoparticles have better qualities than native counterparts with the same molecular makeup, their application in the food industry has grown recently. These properties include surface energy, electrical, mechanical, and optical performance. Both organic and inorganic nanoparticles (such as nanofiber and nanoemulsion) are used in food processing as food carriers for nutraceuticals or as color and flavor enhancers or preservatives. In food packaging, the main applications for both inorganic and organic nanoparticles are as antimicrobial agents and nanosensors found in films or coating solutions. An increasingly popular technique for improving the shelf life of meat and meat products is nanotechnology, which has an impact on every step of the production process and preservation till packaging (He et al., 2019).

7.10.1.1 Processing

In order to improve the nutritional value of meat products while lowering the amount of saturated fatty acids and potentially salt, Weiss et al. (2010) suggested substituting plant-based ingredients such as fiber of oats, citrus fruits, and soy seed varieties like linseed, flaxseed, and for fat. Here, the impact of such substitutes can be increased by employing nanoparticles, which can be a helpful tactic to improve the antioxidant mechanism (Lech and Pospiech, 2010).

According to Singh et al. (2011), producing meat products with natural qualities at a reasonable cost may be facilitated by using non-chemical ingredients in the synthesis of green technology-generated nanoparticles. The meat products' marinating ability and sensory appeal were enhanced by the addition of paprika nanoparticles during processing (Yusop et al., 2012).

While artificial meat preservatives, such as nitrites, can stop undesired deviations in meat products, they are harmful to the user. Natural preservatives, those with nanoscale plant origins like nanoemulsions, can help with this problem. Here, the combinations of essential oils and their nanoemulsions served as antioxidants without changing the technological characteristics of the meat product (Pinelli et al., 2021).

The quality and shelf life of frozen lamb loins were getting affected by safflower oil nanoemulsion and cumin EO in 2021 in combination with O_2 absorber packaging. The most effective systems for effectively encapsulating hydrophobic compounds and incorporating them into meat-active packaging have been found to be nanoemulsions. The creation of thymol chitosan nanoemulsions and stable thyme EO are examples of recent advancements (Liu and Liu, 2020). These nanoemulsions served as novel, all-natural antibacterial packaging materials, which made meat preservation effective.

7.10.1.2 Preservation

From the addition of salt to drying, refrigeration, and nanotechnology, meat preservation methods have changed over time (Djenane et al., 2020). Alirezalu et al. (2021) have utilized ε-polylysine and ε-polylysine nanoparticles derived from extracts of plants, specifically green tea and olive, to create sausages that do not contain nitrile. The shelf life of sausages was extended, and their microbiological safety was greatly improved by the addition of nanoparticles.

Indeed, microbes including bacteria, viruses, along with protozoa—both Gram-positive and negative—are known to be susceptible to the antimicrobial effects of silver (Ag) ions (Kumar et al., 2005). More specifically, it has been shown that cellulose pads containing AgNPs, which release silver ions gradually and derived from silver ions in situ, reduce the microbial levels of exudates from beef meat stored in packaging under a modified atmosphere (Fernandez et al., 2010). Ag nanoparticles made from tea leaf have been shown to possess antibacterial qualities (Loo et al., 2018). This can be used to shield meat products from microbiological harm brought on by Gram-negative microbes.

Ghaderi-Ghahfarokhi et al. (2016) stated improvements in the antioxidant potential, acceptability, and sensory features of beef burgers enhanced with thyme phenolic-loaded chitosan nanoparticles during refrigeration. When sunflower oil-based nanoemulsion was applied to Indo-Pacific king mackerel steaks, microbiological growth was inhibited and the steaks' shelf life was increased (Joe et al., 2012). Moreover, the attractive antimicrobial activity of nanoparticles is due to their improved specific surface area and correspondingly higher reactivity of surface (Kumar et al., 2020a).

When paired with chitosan-gelatin-based films, tarragon EO-charged nanocapsules can effectively prevent the quality of pork slices from declining. This is because the process of nanoencapsulation allows tarragon EOs to be released gradually, enhancing their antibacterial, antioxidant, and sensory qualities (Zhang et al., 2020). In an innovative move, Esmaeili et al. have effectively extended the shelf life of vacuum-packed sausages by adding nanoencapsulated garlic essential oil into edible films (Esmaeili et al., 2020).

7.11 CHALLENGES OF NANOTECHNOLOGY

The rapid advancement of nanotechnology in the international food industry has underscored the significance of creating dependable safety evaluation tools and enforcing stringent protocols to guarantee optimal performance and safeguards for consumers. While there is much hope that nanotechnology can enhance the microbiological safety and quality of meats, there are worries about the unknown health risks, environmental impact, and toxicological effects (Alfei et al., 2020). A growing number of publications and patents demonstrate how quickly this topic is expanding in the agro-food industry, and various companies are using nanotechnology in the development of their products. To ensure that nanomaterials have no negative effects and can be safely used on foods and surfaces that come into contact with food, more research is necessary before this innovative technology can be widely adopted in a number of food preservation and safety-related areas.

Due to direct application in food processing, animal husbandry, and product storage, nanomaterials may find their way into the final product. Sophisticated analytical techniques are needed for the identification, characterization, and risk assessment of nanomaterials, including those in meat products and used in nanopackages, due to the current dearth of literature. Despite the product's great ability to increase the safety and shelf life of food products, researchers are especially concerned about the issue of packaging material caused by nanoparticles. Reports state that after being stored for 15 days, chicken sausages layered with Ag nanoparticles exhibited symptoms of rapid oxidation (Marchiore et al., 2017).

It is possible that the interaction between meat proteins and Ag nanoparticles caused significant changes in parameters like chewiness and cohesiveness after 15 days of storage, but processing and cooking steps resulted in a lower concentration in sausages. This result validates a prior investigation (Ramos et al., 2017).

It is interesting to note that after being exposed to ionic and nanoparticulated silver suspensions, Ag was discovered in tissues including the glomeruli, epidermis, and intestines (Hadrup et al., 2014). To determine the amount of silver that migrated into the packaged food under normal home storage conditions, an AgNP-enhanced commercial food packaging intended to package chicken meat was evaluated (Gallocchio et al., 2016). Reassuringly slow migration was detected. A recent report by Liu et al. found no evidence of a measurable toxicity relationship between nanomaterials and their use in the food industry (Liu et al., 2019). These examples are supported by Kittler et al.'s study, which discovered that AgNPs dissolve slowly into ions over a period of several days (a trial of 125 days at 5, 25, and 37 °C) (Kittler et al., 2010).

According to this study, the biological effects of aged and freshly prepared nanoparticles appear to differ significantly because of the variations in the quantity of ions released. Thus, the degree of toxicity would be linearly related to the degree of exposure and the association with other nanoparticles (Kittler et al., 2010). It is well-known that nanosensors are useful instruments for identifying the existence of gases, chemicals, or pathogens. For example, food packaging with electronic tongues equipped with nanosensors could identify the degree of freshness and even detect the release of gases from spoilt food (Bowles et al., 2014). Similarly, the combination of carbon nanotubes and titanium dioxide (TiO_2) enhanced the antimicrobial activity of silver nanoparticles against *Escherichia coli* and *Bacillus cereus* (Berekaa, 2015). Thus, for both processed and fresh meat products, packaging with nanoparticles can guard against toxicity and spoiling (Berekaa, 2015; Kuuliala et al., 2015).

The absence of established techniques for evaluating nanotoxicity led to inconsistent data across published research, which hindered the creation of strong plans for assessing the risk associated with nanoparticles. More specifically, because of their higher surface-to-mass ratio and surface reactivity, nanoparticles may be able to pass through biological barriers and acquire new, potentially toxicological properties (Lamri et al., 2021).

7.12 NANOTECHNOLOGY RISK ASSESSMENT AND GOVERNANCE

The United States, China, Germany, and Japan are the topmost nations using their knowledge of nanotechnology to develop food products, including meat products (Helmut Kaiser Consultancy, 2009). Nonetheless, there are a lot of regulatory obstacles in developed countries like the United States due to the complexity of nanotechnology. The absence of a single, comprehensive framework that ensures consumer safety exacerbates these issues (Corley et al., 2009). As of right now, neither the FDA nor Europe have any explicit regulations controlling the use of nanotechnology in food (EFSA, 2009). The FDA says that "particle sizing is not the issue" and emphasizes that all materials will still go through the standard battery of safety evaluations, regardless of how sophisticated the technology used to create new materials is.

However, dimension is an issue, according to the Institute of Food Sentients and Technologist (IFST), an organization based in the United Kingdom made up of food scientists and engineers. Until safety is established, they advise handling nanoparticles as dangerous resources. Insufficient data, inadequate models, and concerns voiced by governmental bodies were among the flaws Kuzma et al. (2008) found in the risk analysis. Regulations concerning nanomaterials must also be put into effect immediately because risk assessment programs are being implemented so quickly.

Analyzing the safety of novel food packaging materials has become crucial for ensuring food safety. Furthermore, the pace of advancement in food products incorporating nanoparticles is not meeting the consumers' expectations concerning the products' safety (Cushen et al., 2012). Regulating nanoparticles with new and altered characteristics, defining them broadly, holding products accountable, and determining acceptable concentrations of the nanomaterials are all necessary for the regulations to have any value in food systems (Smolander and Chaudhry, 2010).

7.13 FUTURE PARADIGM OF NANOTECHNOLOGIES IN THE MEAT INDUSTRY

Meat has an active system with a highly reduced shelf life because its sensory qualities can alter during storage due to changes in the chemical, physical, or microbiological environment (Cenci-Goga et al., 2020). In order to improve sensory attributes, add barrier and antimicrobial qualities, and encapsulate bioactive compounds in nanoparticles, meat processing and packaging are utilizing nanotechnology (Morsy et al., 2018).

Because of its enormous impact, the food industry is finding new uses for nanotechnology, particularly in the processing of meat products. The objective is to formulate new, chemical-free, environmentally friendly ingredients for low-cost, high-performing, easily adjustable processes (Ulloa-Saavedra et al., 2022).

Considering this, a few factors can affect the freshness, safety, or quality of packaged goods. These factors include pH, temperature, and variations in gas levels. With the aid of nanoscale fillers like that made up of cellulose, and silicates, nanocomposites can significantly contribute to improved packaging by strengthening the polymer matrix and preventing spoiling while maintaining product quality (Silvestre et al., 2011). Concerns concerning possible environmental effects have been raised by the usage of nanocomposites in the food industry because they do not naturally degrade. More exhaustive research is required to ensure safety because the ecotoxicity of the nanoparticles has not received much attention up to this point (Rai et al., 2019; Störmer et al., 2017).

Consumers are driving the development of new techniques for creating nanosystems, along with producers, industrialists, and researchers. This is because consumers want safe, high-quality meat products and, most importantly, foods that contain specific nutrients that support human health. However, customers will have to shell out more money for these advantages in return for the higher value of meat products. Their readiness and preparedness to do so will lead to the advancement in the supply and demand system that will develop this technology (Ulloa-Saavedra et al., 2022). The application of nanotechnology concerning its toxicity has not been fully investigated to date. Surface methodology can be used to intelligently quantify the toxicity of the nanoparticles utilized in food production process.

7.14 CONCLUSIONS

Nanotechnology has the potential to be used in the food chain for a number of purposes, from packaging to processing and preserving fresh or processed meat products, especially in the meat production industry. To confirm the ultimate quality and protection of the products, it is imperative choosing the nanomaterials with great care. Certain theories predict that advances in nanotechnology will bring about a whole new digital research era, referred to as big data information. Additionally, using carbon nanotubes in meat product packaging materials will make it easier for scientists to identify harmful proteins and spoilage organisms in the future.

Nanosensors are useful tools for identifying all pathogens, even those present in trace amounts. Although consumers are less likely to accept nanomaterials as ingredients in meat products, they are more likely to accept them as packaging materials. This results from a lack of understanding regarding the makeup and function of nanoparticles in food as well as general unfamiliarity with them. The establishment of the protection of nanomaterials used in food processing and packaging may be a prerequisite for the public's acceptance of this technology. Public education regarding the benefits, health risks, safety concerns, and environmental effects of using nanotechnology in the food trade is imperative.

The food trade is expected to enhance various commercial applications of nanomaterials due to their unique and inventive qualities. To maintain security and safety, thorough research on the toxicity of the nanomaterials is essential. In order to give guidelines for evaluating the risks, standard regulation is in fact necessary. Making safer, more sustainable, and healthier nanoparticles will be a challenge that future developments in nanotechnology will tackle (Lamri et al., 2021).

REFERENCES

Abbasi, E.; Milani, M.; Fekri Aval, S.; Kouhi, M.; Akbarzadeh, A.; Tayefi Nasrabadi, H.; Nikasa, P.; Joo, S.W.; Hanifehpour, Y.; Nejati-Koshki, K.; et al. Silver nanoparticles: Synthesis methods, bio-applications and properties. *Crit. Rev. Microbiol.* **2016**, *42*, 173–180. [**Google Scholar**]

Ahmad, R.; Mohsin, M.; Ahmad, T.; Sardar, M. Alpha amylase assisted synthesis of TiO_2 nanoparticles: Structural characterization and application as antibacterial agents. *J. Hazard. Mater.* **2015**, *283*, 171–177.

Alfei, S.; Marengo, B.; Zuccari, G. Nanotechnology application in food packaging: A plethora of opportunities versus pending risks assessment and public concerns. *Food Res. Int.* **2020**, *137*, 109664. [**Google Scholar**] [**CrossRef**] [**PubMed**]

Ali, S.S.; Moawad, M.S.; Hussein, M.A.; Azab, M.; Abdelkarim, E.A.; Badr, A.; Sun, J.; Khalil, M. Efficacy of metal oxide nanoparticles as novel antimicrobial agents against multi-drug and multi-virulent staphylococcus aureus isolates from retail raw chicken meat and giblets. *Int. J. Food Microbiol.* **2021**, *344*, 109116. [**Google Scholar**] [**CrossRef**]

Alirezalu, K.; Pirouzi, S.; Yaghoubi, M.; Karimi-Dehkordi, M.; Jafarzadeh, S.; Khaneghah, A.M. Packaging of beef fillet with active chitosan film incorporated with ε-polylysine: An assessment of quality indices and shelf life. *Meat Sci.* **2021**, *176*, 108475. [**Google Scholar**] [**CrossRef**]

Arnout, F.; Frewer, L.; Bennett, D.; Morris, V.; Smolander, M.; Kampers, F.; Bouwmeester, H.; Gergely, A.; Bowman, D.; Tran, L. *Nanotechnologies in Food*; Chaudhry, Q.; Castle, L.; Watkins, R.; O'Brien, P.; Craighead, H.; Kroto, H., Eds.; The Royal Society of Chemistry, 2010. https://doi.org/10.1039/9781847559883.

Azeredo, B.P.; Sadhu, J.; Ma, J.; Jacobs, K.; Kim, J.; Lee, K.; Ferreira, P. Silicon nanowires with controlled sidewall profile and roughness fabricated by thin-film dewetting and metal-assisted chemical etching. *Nanotechnology* **2013**, *24*, 225–305.

Azeredo, H.M. Nanocomposites for food packaging applications. *Food Res Int* **2009**, *42*, 1240–1253.

Baltic, M.Z.; Bošković, M.; Ivanovic, J.; Dokmanovic, M.; Janjic, J.; Loncina, J.; Baltic, T. Nanotechnology and its potential applications in meat industry. *Tehnologija Mesa* **2013**, *54*, 168–175.

Bao, W.; Deng, Z.; Zhang, S.; Ji, Z.; Zhang, H. Next-generation composite coating system: Nanocoating. *Front. Mater* **2019**, *6*, 456324. https://doi.org/10.3389/fmats.2019.00072

Bayda, S.; Adeel, M.; Tuccinardi, T.; Cordani, M.; Rizzolio, F. The history of nanoscience and nanotechnology: From chemical–physical applications to nanomedicine. *Molecules* **2020**, *25*, 112. [**Google Scholar**] [**CrossRef**]

Benetti, F.; Venturini, M. Nanotechnology in the food sector. *Nutrafoods* **2015**, *14*, 107–108

Berekaa, M.M. Nanotechnology in food industry; advances in food processing, packaging and food safety. *Int. J. Curr. Microbiol. App. Sci.* **2015**, *4*, 345–357. [**Google Scholar**]

Boskovic, M.; Glisic, M.; Djordjevic, J.; Vranesevic, J.; Djordjevic, V.; Baltic, M.Z. Preservation of meat and meat products using nanoencapsulated thyme and oregano essential oils.

IOP Conf. Ser. Earth Environ. Sci. **2019**, *333*, 012038. [**Google Scholar**] [**CrossRef**]
Boskovic, M.; Zdravkovic, N.; Ivanovic, J.; Djordjevic, J.; Janjic, J.; Pavlicevic, N.; Baltic, M.Z. Inhibitory effect of thyme and oregano essential oils and some essential oil components on Salmonella Senftenberg and Salmonella Give. *Meat Technol./Tehnologija Mesa* **2016**, *57*, 67–72.
Bowles, M.; Lu, J. Removing the blinders: A literature review on the potential of nanoscale technologies for the management of supply chains. *Technol. Forecast. Soc. Chang.* **2014**, *82*, 190–198. [**Google Scholar**] [**CrossRef**]
Buncic, S.; Nychas, G.J.; Lee, M.R.; Koutsoumanis, K.; Hébraud, M.; Desvaux, M.; Chorianopoulos, N.; Bolton, D.; Blagojevic, B.; Antic, D. Microbial pathogen control in the beef chain: Recent research advances. *Meat Sci.* **2014**, *97*, 288–297.
Burt, S. Essential oils: Their antibacterial properties and potential applications in foods—a review. *Int. J. Food Microbiol.* **2004**, *94*, 223–253.
Cenci-Goga, B.T.; Iulietto, M.F.; Sechi, P.; Borgogni, E.; Karama, M.; Grispoldi, L. New trends in meat packaging. *Microbiol. Res.* **2020**, *11*, 56–67. [**Google Scholar**] [**CrossRef**]
Chaudhry, Q.; Castle, L. Food applications of nanotechnologies: An overview of opportunities and challenges for developing countries. *Trends Food Sci. Technol.* **2011**, *225*, 95–603.
Coma, V. Bioactive packaging technologies for extended shelf life of meat-based products. *Meat Sci.* **2008**, *78*, 90–103.
Corley, E.; Scheufele, D.A.; Hu, Q. Of risks and regulations: How leading US nanoscientists form policy stances about nanotechnology. *Nanopart. Res.* **2009**, *11*(7), 1573–1585. https://doi.org/10.1007/s11051-009-9671-5
Cushen, M.; Kerry, J.; Morris, M.; Cruz-Romero, M.; Cummins, E. Nanotechnologies in the food industry—recent developments, risks and regulation. *Trends Food Sci. Technol.* **2012**, *24*, 30–46.
Dainelli, D.; Gontard, N.; Spyropoulos, D.; Zondervan-van den Beuken, E.; Tobback, P. Active and intelligent food packaging: Legal aspects and safety concerns. *Trends Food Sci. Technol.* **2008**, *19*, S103–S112. [**Google Scholar**]
Damm, C.; Münstedt, H.; Rösch, A. The antimicrobial efficacy of polyamide 6/silver-nano- and microcomposites. *Mater. Chem. Phys.* 2008, *108*, 61–66.
Das, A.K.; Nanda, P.K.; Chowdhury, N.R.; Dandapat, P.; Gagaoua, M.; Chauhan, P.; Pateiro, M.; Lorenzo, J.M. Application of pomegranate by-products in muscle foods: Oxidative indices, colour stability, shelf life and health benefits. *Molecules* **2021**, *26*, 467. [**Google Scholar**] [**CrossRef**] [**PubMed**]
Deligiannakis, Y.; Sotiriou, G.A.; Pratsinis, S.E. Antioxidant and antiradical SiO2 nanoparticles covalently functionalized with gallic acid. *ACS Appl. Mater. Interfaces.* **2012**, *12*, 6609–6617.
Devatha, C.P.; Thalla, A.K. Chapter 7—Green synthesis of nanomaterials. In *Synthesis of Inorganic Nanomaterials*; Mohan Bhagyaraj, S.; Oluwafemi, O.S.; Kalarikkal, N.; Thomas, S., Eds.; Woodhead Publishing: Duxford, UK, 2018; pp. 169–184. [**Google Scholar**] [**CrossRef**]
Djenane, D.; Aboudaou, M.; Djenane, F.; García-Gonzalo, D.; Pagán, R. Improvement of the shelf-life status of modified atmosphere packaged camel meat using nisin and *Olea europaea* Subsp. laperrinei leaf extract. *Foods* **2020**, *9*, 1336. [**Google Scholar**] [**CrossRef**]
Domínguez, R.; Pateiro, M.; Gagaoua, M.; Barba, F.J.; Zhang, W.; Lorenzo, J.M. A comprehensive review on lipid oxidation in meat and meat products. *Antioxidants* **2019**, *8*, 429. [**Google Scholar**] [**CrossRef**][**Green Version**]
Dong, J.; Zhou, Y.; Lu, Y.; Lv, Y.; Chi, Y.; He, Q. Effect of tea polyphenols on the oxidation and color stability of porcine hemoglobin. *J. Food Sci.* **2019**, *84*, 2086–2090. [**Google Scholar**] [**CrossRef**]
Duncan, T.V. Applications of nanotechnology in food packaging and food safety: Barrier materials, antimicrobials and sensors. *J. Colloid Interface Sci.* **2011**, *363*, 1–24.
Ebrahimzadeh, M.A.; Naghizadeh, A.; Amiri, O.; Shirzadi-Ahodashti, M.; Mortazavi-Derazkola, S. Green and facile synthesis of Ag nanoparticles using *Crataegus pentagyna* fruit extract (CP-AgNPs) for organic pollution dyes degradation and antibacterial application. *Bioorg. Chem.* **2020**, *94*, 103425. [**Google Scholar**] [**CrossRef**] [**PubMed**]
EFSA. The potential risks arising from nanoscience and nanotechnologies on food and feed safety. *EFSA J.* **2009**, *7*(3). Wiley-Blackwell Publishing Ltd. https://doi.org/10.2903/j.efsa.2009.958.
Elbarbary, A.M.; El-Sawy, N.M.; Hegazy, E.S.A. Antioxidative properties of irradiated chitosan/vitamin C complex and their use as food additive for lipid storage. *J. Appl. Polym. Sci.* **2015**, *132*, 42105. [**Google Scholar**] [**CrossRef**]
Eskandani, M.; Janmohammadi, H.; Mirghelenj, S.A.; Ebrahimi, M.; Kalanaky, S. Effects of zinc nanoparticles on growth performance, carcass characteristics, immunity, and meat quality of broiler chickens. *Iran. J. Appl. Anim. Sci.* **2021**, *11*, 135–146. [**Google Scholar**]
Esmaeili, H.; Cheraghi, N.; Khanjari, A.; Rezaeigolestani, M.; Basti, A.A.; Kamkar, A.; Aghaee, E.M. Incorporation of nanoencapsulated garlic essential oil into edible films: A novel approach for extending shelf life of vacuum-packed sausages. *Meat Sci.* **2020**, *166*, 108135. [**Google Scholar**] [**CrossRef**]
Estévez, M. Critical overview of the use of plant antioxidants in the meat industry: Opportunities, innovative applications and future perspectives. *Meat Sci.* **2021**, *181*, 108610. [**Google Scholar**] [**CrossRef**]
Falowo, A.B.; Fayemi, P.O.; Muchenje, V. Natural antioxidants against lipid–protein oxidative deterioration in meat and meat products: A review. *Food Res. Int.* **2014**, *64*, 171–181.
Fang, Z.; Zhao, Y.; Warner, R.D.; Johnson, S.K. Active and intelligent packaging in meat industry. *Trends Food Sci. Technol.* **2017**, *61*, 60–71. [**Google Scholar**]
Fedotova, A.V.; Snezhko, A.G.; Sdobnikova, O.A.; Samoilova, L.G.; Smurova, T.A.; Revina, A.A.; Khailova, E.B. Packaging materials manufactured from natural polymers modified with silver nanoparticles. *Int. Polym. Sci. Technol.* **2010**, *37*, 59.
Fernandez, A.; Picouet, P.; Lloret, E. Reduction of the spoilage-related microflora in absorbent pads by silver nanotechnology during modified atmosphere packaging of beef meat. *J. Food Prot.* **2010**, *73*, 2263–2269. [**Google Scholar**] [**CrossRef**]
Gagaoua, M.; Dib, A.L.; Lakhdara, N.; Lamri, M.; Botineştean, C.; Lorenzo, J.M. Artificial meat tenderization using plant cysteine proteases. *Curr. Opin. Food Sci.* **2021**, *38*, 177–188. [**Google Scholar**] [**CrossRef**]
Gallocchio, F.; Cibin, V.; Biancotto, G.; Roccato, A.; Muzzolon, O.; Carmen, L.; Simone, B.; Manodori, L.; Fabrizi, A.; Patuzzi, I.; et al. Testing nano-silver food packaging to evaluate silver migration and food spoilage bacteria on chicken meat. *Food Addit. Contam. Part A.* **2016**, *33*, 1063–1071. [**Google Scholar**] [**CrossRef**]
Ghaderi-Ghahfarokhi, M.; Barzegar, M.; Sahari, M.A.; Ahmadi Gavlighi, H.; Gardini, F. Chitosan-cinnamon essential oil nano-formulation: Application as a novel additive for

controlled release and shelf life extension of beef patties. *Int. J. Biol. Macromol.* **2017**, *102*, 19–28. [**Google Scholar**] [**CrossRef**] [**PubMed**]

Ghaderi-Ghahfarokhi, M.; Barzegar, M.; Sahari, M.A.; Azizi, M.H. Nanoencapsulation approach to improve antimicrobial and antioxidant activity of thyme essential oil in beef burgers during refrigerated storage. *Food Bioprocess Technol.* **2016**, *9*, 1187–1201. [**Google Scholar**] [**CrossRef**]

Guimarães, M.; Correia, S.; Briuglia, M.L.; Niosi, F.; Lamprou, D.A. Microfluidic manufacturing of phospholipid nanoparticles: Stability, encapsulation efficacy, and drug release. *Int. J. Pharm.* **2017**, *512*, 91–99. [**Google Scholar**] [**CrossRef**] [**PubMed**]

Hadian, M.; Rajaei, A.; Mohsenifar, A.; Tabatabaei, M. Encapsulation of rosmarinus officinalis essential oils in chitosan-benzoic acid nanogel with enhanced antibacterial activity in beef cutlet against *Salmonella typhimurium* during refrigerated storage. *LWT* **2017**, *84*, 394–401. [**Google Scholar**] [**CrossRef**]

Hadrup, N.; Lam, H.R. Oral toxicity of silver ions, silver nanoparticles and colloidal silver—a review. *Regul. Toxicol. Pharmacol.* **2014**, *68*, 1–7. [**Google Scholar**] [**CrossRef**] [**PubMed**]

Hasani-Javanmardi, M.; Fallah, A.A.; Abbasvali, M. Effect of safflower oil nanoemulsion and cumin essential oil combined with oxygen absorber packaging on the quality and shelf-life of refrigerated lamb loins. *LWT* **2021**, *147*, 111557. [**Google Scholar**] [**CrossRef**]

He, X.; Deng, H.; Hwang, H.-M. The current application of nanotechnology in food and agriculture. *J. Food Drug Anal.* **2019**, *27*, 1–21. [**Google Scholar**] [**CrossRef**]

He, X.; Hwang, H.M. Nanotechnology in food science: Functionality, applicability, and safety assessment. *J. Food Drug. Anal.* **2016**, *24*, 671–681.

Helmut Kaiser Consultancy. *Study: Nanotechnology in Food and Food Processing Industry Worldwide, 2006–2010–2015*; Helmut Kaiser Consultancy Group: Beijing, 2009.

Hosseini, S.N.; Pirsa, S.; Farzi, J. Biodegradable nano composite film based on modified starch-albumin/MgO; antibacterial, antioxidant and structural properties. *Polym. Test.* **2021**, *97*, 107182. [**Google Scholar**] [**CrossRef**]

Hussain, I.; Singh, N.B.; Singh, A.; Singh, H.; Singh, S.C. Green synthesis of nanoparticles and its potential application. *Biotechnol. Lett.* **2016**, *38*, 545–560. [**Google Scholar**] [**CrossRef**]

Ibrahim, D.; Kishawy, A.T.Y.; Khater, S.I.; Khalifa, E.; Ismail, A.; Mohammed, H.A.; Elnahriry, S.S.; Tolba, H.A.; Sherief, W.R.I.A.; Farag, M.F.M.; et al. Interactive effects of dietary quercetin nanoparticles on growth, flesh antioxidant capacity and transcription of cytokines and aeromonas hydrophila quorum sensing orchestrating genes in nile tilapia (*Oreochromis niloticus*). *Fish Shellfish Immunol.* **2021**, *119*, 1050–4648. [**Google Scholar**] [**CrossRef**]

Iravani, S. Green synthesis of metal nanoparticles using plants. *Green Chem.* **2011**, *13*, 2638–2650. [**Google Scholar**] [**CrossRef**]

Istiqola, A.; Syafiuddin, A. A review of silver nanoparticles in food packaging technologies: Regulation, methods, properties, migration, and future challenges. *J. Chin. Chem. Soc.* **2020**, *67*, 1942–1956. [**Google Scholar**]

Jadoun, S.; Arif, R.; Jangid, N.K.; Meena, R.K. Green synthesis of nanoparticles using plant extracts: A review. *Environ. Chem. Lett.* **2021**, *19*, 355–374. [**Google Scholar**] [**CrossRef**]

Jayasena, D.D.; Jo, C. Potential application of essential oils as natural antioxidants in meat and meat products: A review. *Food Rev. Int.* **2014**, *30*, 71–90. [**Google Scholar**] [**CrossRef**]

Jin, C.F.; Yuan, X.; Ge, W.W.; Hong, J.M.; Xin, X.Q. Synthesis of ZnO nanorods by solid state reaction at room temperature. *Nanotechnology*. **2003**, *14*, 667.

Joe, M.M.; Chauhan, P.S.; Bradeeba, K.; Shagol, C.; Sivakumaar, P.K.; Sa, T. Influence of sunflower oil based nanoemulsion (AUSN-4) on the shelf life and quality of Indo-Pacific king mackerel (*Scomberomorus guttatus*) steaks stored at 20 °C. *Food Control.* **2012**, *23*, 564–570. [**Google Scholar**] [**CrossRef**]

Joyner, J.R.; Kumar, D.V. Nanosensors and their applications in food analysis: A review. **2015,** *1*(4), 80–90.

Khan, I.; Saeed, K.; Khan, I. Nanoparticles: Properties, applications and toxicities. *Arab. J. Chem.* **2019**, *12*, 908–931. [**Google Scholar**] [**CrossRef**]

Khan, M.I.; Sahar, A.; Rahman, U. Nanotechnology in healthier meat processing. In *Novel Approaches of Nanotechnology in Food*; Grumezescu, A.M., Ed.; Elsevier Inc.: Amsterdam, The Netherlands, 2016; pp. 313–345. ISBN 9780128043080. [**Google Scholar**]

Kittler, S.; Greulich, C.; Diendorf, J.; Köller, M.; Epple, M. Toxicity of silver nanoparticles increases during storage because of slow dissolution under release of silver ions. *Chem. Mater.* **2010**, *22*, 4548–4554. [**Google Scholar**] [**CrossRef**]

Kumar, P.; Mahajan, P.; Kaur, R.; Gautam, S. Nanotechnology and its challenges in the food sector: A review. *Mater. Today Chem.* **2020a**, *17*, 100332. [**Google Scholar**] [**CrossRef**]

Kumar, R.; Lal, S. Synthesis of organic nanoparticles and their applications in drug delivery and food nanotechnology: A review. *J. Nanomater. Mol. Nanotechnol.* **2014**, *11*, 2. [**Google Scholar**]

Kumar, R.; Münstedt, H. Silver ion release from antimicrobial polyamide/silver composites. *Biomaterials* **2005**, *26*, 2081–2088. [**Google Scholar**] [**CrossRef**]

Kumar, S.; Mukherjee, A.; Dutta, J. Chitosan based nanocomposite films and coatings: Emerging antimicrobial food packaging alternatives. *Trends Food Sci. Technol.* **2020b**, *97*, 196–209. [**Google Scholar**] [**CrossRef**]

Kuuliala, L.; Pippuri, T.; Hultman, J.; Auvinen, S.M.; Kolppo, K.; Nieminen, T.; Karp, M.; Björkroth, J.; Kuusipalo, J.; Jääskeläinen, E. Preparation and antimicrobial characterization of silver-containing packaging materials for meat. *Food Packag. Shelf Life* **2015**, *6*, 53–60. [**Google Scholar**] [**CrossRef**]

Kuzma, J.; Romanchek, J.; Kokotovich, A. Upstream oversight assessment for agrifood nanotechnology: A case studies approach. *Risk Anal* **2008**, *28*(4), 1081–1098. https://doi.org/10.1111/j.1539-6924.2008.01071.x.

Lähde, A.; Koshevoy, I.; Karhunen, T.; Torvela, T.; Pakkanen, T.A.; Jokiniemi, J. Aerosol-assisted synthesis of gold nanoparticles. *J. Nanoparticle Res.* **2014**, *16*, 2716. [**Google Scholar**] [**CrossRef**]

Lamri, M.; Bhattacharya, T.; Boukid, F.; Chentir, I.; Dib, A.L.; Das, D.; Djenane, D.; Gagaoua, M. Nanotechnology as a processing and packaging tool to improve meat quality and safety. *Foods.* **2021**, *10*, 2633. https://doi.org/10.3390/foods10112633

Lech, O.; Pospiech, E.; Narine, S. Nanotechnologies in food and meat processing. *Acta Sci. Pol. Technol. Aliment.* **2010**, *9*, 401–412. [**Google Scholar**]

Lee, S.H.; Heng, D.; Ng, W.K.; Chan, H.-K.; Tan, R.B.H. Nano spray drying: A novel method for preparing protein nanoparticles for protein therapy. *Int. J. Pharm.* **2011**, *403*, 192–200. [**Google Scholar**] [**CrossRef**] [**PubMed**]

Li, X.; Quan, X.; Kutal, C. Synthesis and photocatalytic properties of quantum confined titanium dioxide nanoparticle. *Scr. Mater.* **2004**, *50*, 499–505.

Liu, T.; Liu, L. Fabrication and characterization of chitosan nanoemulsions loading thymol or thyme essential oil for the preservation of refrigerated pork. *Int. J. Biol. Macromol.* **2020**, *162*, 1509–1515. [**Google Scholar**] [**CrossRef**]

Liu, X.; Zhang, B.; Sohal, I.S.; Bello, D.; Chen, H. Chapter Eight—Is "nano safe to eat or not"? A review of the state-of-the art in soft engineered nanoparticle (sENP) formulation and delivery in Foods. In *Advances in Food and Nutrition Research*; Lim, L.-T.; Rogers, M., Eds.; Academic Press: Oxford, UK, 2019; Volume 88, pp. 299–335. [**Google Scholar**]

Liu, Y.; He, L.; Mustapha, A.; Li, H.; Hu, Z.Q.; Lin, M. Antibacterial activities of zinc oxide nanoparticles against *Escherichia coli* O157:H7. *J. Appl. Microbiol.* **2009**, *107*, 1193–1201. [**Google Scholar**] [**CrossRef**]

Long, M.; Wang, J.; Zhuang, H.; Zhang, Y.; Wu, H.; Zhang, J. Performance and mechanism of standard nano-TiO_2 (P-25) in photocatalytic disinfection of foodborne microorganisms–*Salmonella typhimurium* and *Listeria monocytogenes*. *Food Control.* **2014**, *39*, 68–74.

Loo, Y.Y.; Rukayadi, Y.; Nor-Khaizura, M.-A.-R.; Kuan, C.H.; Chieng, B.W.; Nishibuchi, M.; Radu, S. In vitro antimicrobial activity of green synthesized silver nanoparticles against selected gram-negative foodborne pathogens. *Front. Microbiol.* **2018**, *9*, 1555. [**Google Scholar**] [**CrossRef**]

Marchiore, N.G.; Manso, I.J.; Kaufmann, K.C.; Lemes, G.F.; Pizolli, A.P.d.O.; Droval, A.A.; Bracht, L.; Gonçalves, O.H.; Leimann, F.V. Migration evaluation of silver nanoparticles from antimicrobial edible coating to sausages. *LWT Food Sci. Technol.* **2017**, *76*, 203–208. [**Google Scholar**] [**CrossRef**]

McMillin, K.W. Advancements in meat packaging. *Meat Sci.* **2017**, *132*, 153–162. [**Google Scholar**]

Mirhosseini, M.; Arjmand, V. Reducing pathogens by using zinc oxide nanoparticles and acetic acid in sheep meat. *J. Food Prot.* **2014**, *77*, 1599–1604. [**Google Scholar**] [**CrossRef**]

Moraes-Lovison, M.; Marostegan, L.F.P.; Peres, M.S.; Menezes, I.F.; Ghiraldi, M.; Rodrigues, R.A.F.; Fernandes, A.M.; Pinho, S.C. Nanoemulsions encapsulating oregano essential oil: Production, stability, antibacterial activity and incorporation in chicken pâté. *LWT-Food Sci. Technol.* **2017**, *77*, 233–240. [**Google Scholar**] [**CrossRef**]

Morsy, M.K.; Elsabagh, R.; Trinetta, V. Evaluation of novel synergistic antimicrobial activity of nisin, lysozyme, EDTA nanoparticles, and/or ZnO nanoparticles to control foodborne pathogens on minced beef. *Food Control.* **2018**, *92*, 249–254. [**Google Scholar**] [**CrossRef**]

Nanda, P.K.; Das, A.K.; Dandapat, P.; Dhar, P.; Bandyopadhyay, S.; Dib, A.L.; Lorenzo, J.M.; Gagaoua, M. Nutritional aspects, flavour profile and health benefits of crab meat based novel food products and valorisation of processing waste to wealth: A review. *Trends Food Sci. Technol.* **2021**, *112*, 252–267. [**Google Scholar**] [**CrossRef**]

Neethirajan, S.; Jayas, D.S. Nanotechnology for the food and bioprocessing industries. *Food Bioprocess. Technol.* **2011**, *4*, 39–47.

Okoroh, D.; Aisida, S.; Asogwa, P. Synthesis and characterization of biopolymer capped zinc ferrite nanoparticles by a thermal treatment method. *IOSR-JAP.* **2018**, *10*, 52–56. [**Google Scholar**]

Oliver, S.; Wagh, H.; Liang, Y.; Yang, S.; Boyer, C. Enhancing the antimicrobial and antibiofilm effectiveness of silver nanoparticles prepared by green synthesis. *J. Mater. Chem. B.* **2018**, *6*, 4124–4138. [**Google Scholar**] [**CrossRef**]

Oymaci, P.; Altinkaya, S.A. Improvement of barrier and mechanical properties of whey protein isolate based food packaging films by incorporation of zein nanoparticles as a novel bionanocomposite. *Food Hydrocoll.* **2016**, *54*, 1–9. [**Google Scholar**] [**CrossRef**][**Green Version**]

Ozimek, L.; Pospiech, E.; Narine, S. Nanotechnologies in food and meat processing. *Acta Sci. Pol. Technol. Aliment.* **2010**, *9*, 401–412.

Pabast, M.; Shariatifar, N.; Beikzadeh, S.; Jahed, G. Effects of chitosan coatings incorporating with free or nano-encapsulated Satureja plant essential oil on quality characteristics of lamb meat. *Food Control.* **2018**, *91*, 185–192. [**Google Scholar**] [**CrossRef**]

Panea, B.; Ripoll, G.; González, J.; Fernández-Cuello, Á.; Albertí, P. Effect of nanocomposite packaging containing different proportions of ZnO and Ag on chicken breast meat quality. *J. Food Eng.* **2014**, *123*, 104–112.

Parveen, K.; Banse, V.; Ledwani, L. Green synthesis of nanoparticles: Their advantages and disadvantages. *AIP Conf. Proc.* **2016**, *1724*, 020048. [**Google Scholar**] [**CrossRef**]

Pateiro, M.; Munekata, P.E.S.; Sant'ana, A.S.; Domínguez, R.; Rodríguez-Lázaro, D.; Lorenzo, J.M. Application of essential oils as antimicrobial agents against spoilage and pathogenic microorganisms in meat products. *Int. J. Food Microbiol.* **2021**, *337*, 108966. [**Google Scholar**] [**CrossRef**] [**PubMed**]

Pinelli, J.J.; Helena de Abreu Martins, H.; Guimarães, A.S.; Isidoro, S.R.; Gonçalves, M.C.; Junqueira de Moraes, T.S.; Ramos, E.M.; Piccoli, R.H. Essential oil nanoemulsions for the control of Clostridium sporogenes in cooked meat product: An alternative? *LWT* **2021**, *143*, 111123. https://doi.org/10.1016/j.lwt.2021.111123

Purslow, P.P.; Warner, R.D.; Clarke, F.M.; Hughes, J.M. Variations in meat colour due to factors other than myoglobin chemistry; a synthesis of recent findings (invited review). *Meat Sci.* **2020**, *159*, 107941. [**Google Scholar**] [**CrossRef**]

Raghunath, A.; Perumal, E. Metal oxide nanoparticles as antimicrobial agents: A promise for the future. *Int. J. Antimicrob. Agents* **2017**, *49*, 137–152. [**Google Scholar**] [**CrossRef**]

Rai, M.; Ingle, A.P.; Gupta, I.; Pandit, R.; Paralikar, P.; Gade, A.; Chaud, M.V.; dos Santos, C.A. Smart nanopackaging for the enhancement of food shelf life. *Environ. Chem. Lett.* **2019**, *17*, 277–290. [**Google Scholar**] [**CrossRef**]

Rai, M.; Paralikar, P.; Jogee, P.; Agarkar, G.; Ingle, A.P.; Derita, M.; Zacchino, S. Synergistic antimicrobial potential of essential oils in combination with nanoparticles: Emerging trends and future perspectives. *Int. J. Pharm.* **2017**, *519*, 67–78. [**Google Scholar**] [**CrossRef**]

Ramachandraiah, K.; Han, S.G.; Chin, K.B. Nanotechnology in meat processing and packaging: Potential applications—a review. *Asian-Australas J. Anim. Sci.* **2015**, *28*, 290–302. [**Google Scholar**]

Ramachandraiah, K.; Hong, G.-P. Polymeric nanomaterials for the development of sustainable plant food value chains. *Food Biosci* **2021**, *41*, 100978. https://doi.org/10.1016/j.fbio.2021.100978

Ramos, K.; Ramos, L.; Gómez-Gómez, M.M. Simultaneous characterisation of silver nanoparticles and determination of dissolved silver in chicken meat subjected to in vitro human gastrointestinal digestion using single particle inductively coupled plasma mass spectrometry. *Food Chem.* **2017**, *221*, 822–828. [**Google Scholar**] [**CrossRef**]

Rane, A.V.; Kanny, K.; Abitha, V.K.; Thomas, S. Chapter 5—Methods for synthesis of nanoparticles and fabrication of nanocomposites. In *Synthesis of Inorganic Nanomaterials*; Mohan Bhagyaraj, S.; Oluwafemi, O.S.; Kalarikkal, N.; Thomas, S., Eds.; Woodhead Publishing: Duxford, UK, 2018; pp. 121–139. **[Google Scholar]**

Reverberi, A.; Salerno, M.; Fabiano, B. Inorganic nanoparticles synthesis by an aerosol-assisted wet chemical method. *Chem. Eng. Trans.* **2016**, *47*, 115–120. **[Google Scholar]**

Salas, G.; Costo, R.; Morales, M.d.P. Chapter 2—Synthesis of inorganic nanoparticles. In *Frontiers of Nanoscience*; de la Fuente, J.M.; Grazu, V., Eds.; Elsevier: Amsterdam, The Netherlands, 2012; Volume 4, pp. 35–79. **[Google Scholar]**

Salas, C.; Nypelö, T.; Rodriguez-Abreu, C.; Carrillo, C.; Rojas, O.J. Nanocellulose properties and applications in colloids and interfaces. *Curr. Opin. Colloid Interface Sci.* **2014**, *19*, 383–396. **[Google Scholar]**

Salgado, P.R.; Di Giorgio, L.; Musso, Y.S.; Mauri, A.N. Recent developments in smart food packaging focused on biobased and biodegradable polymers. *Front. Sustain. Food Syst.* **2021**, *5*, 125. **[Google Scholar] [CrossRef]**

Salminen, H.; Herrmann, K.; Weiss, J. Oil-in-water emulsions as a delivery system for n-3 fatty acids in meat products. *Meat Sci.* **2013**, *93*, 659–667.

Sanchez-Alvarado, D.I.; Guzmán-Pantoja, J.; Páramo-García, U.; Maciel-Cerda, A.; Martínez-Orozco, R.D.; Vera-Graziano, R. Morphological study of chitosan/poly (vinyl alcohol) nanofibers prepared by electrospinning, collected on reticulated vitreous carbon. *Int. J. Mol. Sci.* **2018**, *19*, 1718. **[Google Scholar] [CrossRef][Green Version]**

Shahverdi, A.R.; Fakhimi, A.; Shahverdi, H.R.; Minaian, S. Synthesis and effect of silver nanoparticles on the antibacterial activity of different antibiotics against Staphylococcus aureus and Escherichia coli. *Nanomedicine: NBM* **2007**, *3*, 168–171.

Shen, J.; Shafiq, M.; Ma, M.; Chen, H. Synthesis and surface engineering of inorganic nanomaterials based on microfluidic technology. *Nanomaterials* **2020**, *10*, 1177. **[Google Scholar] [CrossRef]**

Silvestre, C.; Duraccio, D.; Cimmino, S. Food packaging based on polymer nanomaterials. *Prog. Polym. Sci.* **2011**, *36*, 1766–1782.

Singh, J.; Dutta, T.; Kim, K.-H.; Rawat, M.; Samddar, P.; Kumar, P. 'Green' synthesis of metals and their oxide nanoparticles: Applications for environmental remediation. *J. Nanobiotechnology* **2018**, *16*, 84. **[Google Scholar] [CrossRef]**

Singh, M.; Manikandan, S.; Kumaraguru, A. Nanoparticles: A new technology with wide applications. *Res. J. Nanosci. Nanotechnol.* **2011**, *1*, 1–11. **[Google Scholar] [CrossRef]**

Singh, P.K.; Jairath, G.; Ahlawat, S.S. Nanotechnology: A future tool to improve quality and safety in meat industry. *J. Food Sci. Technol.* **2016**, *53*, 1739–1749. **[Google Scholar] [CrossRef]**

Singh, R.; Shedbalkar, U.U.; Wadhwani, S.A.; Chopade, B.A. Bacteriagenic silver nanoparticles: Synthesis, mechanism, and applications. *Appl. Microbiol. Biotechnol.* **2015**, *99*, 4579–4593. **[Google Scholar]**

Sinha, S.; Aman, A.K.; Singh, R.K.; Kr, N.; Shivani, K. Calcium oxide(CaO) nanomaterial (Kukutanda twak Bhasma) from egg shell: Green synthesis, physical properties and antimicrobial behaviour. *Mater. Today Proc.* **2021**, *43*, 3414–3419. **[Google Scholar] [CrossRef]**

Smolander, M.; Chaudhry, Q. Nanotechnologies in food packaging. In *Nanotechnologies in Food*; The Royal Society of Chemistry, 2010; pp. 86–101. https://doi.org/10.1039/9781847559883-00086

Störmer, A.; Bott, J.; Kemmer, D.; Franz, R. Critical review of the migration potential of nanoparticles in food contact plastics. *Trends Food Sci. Technol.* **2017**, *63*, 39–50. **[Google Scholar] [CrossRef]**

Suman, S.P.; Joseph, P. Myoglobin chemistry and meat color. *Annu. Rev. Food Sci. Technol.* **2013**, *4*, 79–99. **[Google Scholar] [CrossRef] [PubMed][Green Version]**

Sundar, S.; Kundu, J.; Kundu, S.C. Biopolymeric nanoparticles. *Sci. Technol. Adv. Mater.* **2010**, *11*, 014104. **[Google Scholar] [CrossRef]**

Taherimehr, M.; YousefniaPasha, H.; Tabatabaeekoloor, R.; Pesaranhajiabbas, E. Trends and challenges of biopolymer-based nanocomposites in food packaging. *Compr. Rev. Food Sci. Food Saf.* **2021**, 1–24. **[Google Scholar]**)

Tang, S.; Wang, Z.; Li, W.; Li, M.; Deng, Q.; Wang, Y.; Li, C.; Chu, P.K. Ecofriendly and biodegradable soybean protein isolate films incorporated with ZnO nanoparticles for food packaging. *ACS Appl. Bio Mater.* **2019**, *2*, 2202–2207. **[Google Scholar] [CrossRef]**

Teng, Z.; Luo, Y.; Wang, Q. Carboxymethyl chitosan–soy protein complex nanoparticles for the encapsulation and controlled release of vitamin D3. *Food Chem.* **2013**, *141*, 524–532. **[Google Scholar] [CrossRef]**

Tometri, S.S.; Ahmady, M.; Ariaii, P.; Soltani, M.S. Extraction and encapsulation of *Laurus nobilis* leaf extract with nanoliposome and its effect on oxidative, microbial, bacterial and sensory properties of minced beef. *J. Food Meas. Charact.* **2020**, *14*, 3333–3344. **[Google Scholar] [CrossRef]**

Tornuk, F.; Hancer, M.; Sagdic, O.; Yetim, H. LLDPE based food packaging incorporated with nanoclays grafted with bioactive compounds to extend shelf life of some meat products. *LWT Food Sci. Technol.* **2015**, *64*, 540–546.

Ulloa-Saavedra, A.; García-Betanzos, C.; Zambrano-Zaragoza, M.; Quintanar-Guerrero, D.; Mendoza-Elvira, S.; Velasco-Bejarano, B. Recent developments and applications of nanosystems in the preservation of meat and meat products. *Foods.* **2022**, *11*, 2150. https://doi.org/10.3390/foods11142150

Umaraw, P.; Munekata, P.E.S.; Verma, A.K.; Barba, F.J.; Singh, V.P.; Kumar, P.; Lorenzo, J.M. Edible films/coating with tailored properties for active packaging of meat, fish and derived products. *Trends Food Sci. Technol.* **2020**, *98*, 10–24. **[Google Scholar]**

Valdés, M.G.; González, A.C.V.; Calzón, J.A.G.; Díaz-García, M.E. Analytical nanotechnology for food analysis. *Microchim. Acta.* **2009**, *166*, 1–19.

Wang, J.; Zhuang, H.; Hinton, A.; Bowker, B.; Zhang, J. Photocatalytic disinfection of spoilage bacteria Pseudomonas fluorescens and Macrococcus caseolyticus by nano-TiO_2. *LWT-Food Sci. Technol.* **2014**, *59*, 1009–1017.

Wang, Q.; Zhang, L.; Ding, W. Eugenol nanocapsules embedded with gelatin-chitosan for chilled pork preservation. *Int. J. Biol. Macromol.* **2020**, *158*, 837–844. **[Google Scholar] [CrossRef]**

Weiss, J.; Gibis, M.; Schuh, V.; Salminen, H. Advances in ingredient and processing systems for meat and meat products. *Meat Sci.* **2010**, *86*, 196–213. **[Google Scholar] [CrossRef]**

Wesley, S.J.; Raja, P.; Raj, A.A.; Tiroutchelvamae, D. Review on-nanotechnology applications in food packaging and safety. *Int. J. Eng. Res.* **2014**, *3*, 645–651.

Wright, I.K.; Higginbotham, A.; Baker, S.M.; Donnelly, T.D. Generation of nanoparticles of controlled size using ultrasonic piezoelectric oscillators in solution. *ACS Appl.*

Mater. Interfaces. **2010**, *2*, 2360–2364. [**Google Scholar**] [**CrossRef**] [**PubMed**]

Yusop, S.M.; O'Sullivan, M.G.; Preuß, M.; Weber, H.; Kerry, J.F.; Kerry, J.P. Assessment of nanoparticle paprika oleoresin on marinating performance and sensory acceptance of poultry meat. *LWT Food Sci. Technol*. **2012**, *46*, 349–355. [**Google Scholar**] [**CrossRef**]

Zhang, H.; Liang, Y.; Li, X.; Kang, H. Effect of chitosan-gelatin coating containing nano-encapsulated tarragon essential oil on the preservation of pork slices. *Meat Sci*. **2020**, *166*, 108137. [**Google Scholar**] [**CrossRef**]

Zhang, L.; Liu, A.; Wang, W.; Ye, R.; Liu, Y.; Xiao, J.; Wang, K. Characterisation of microemulsion nanofilms based on Tilapia fish skin gelatine and ZnO nanoparticles incorporated with ginger essential oil: Meat packaging application. *Int. J. Food Sci. Technol*. **2017**, *52*, 1670–1679. [**Google Scholar**] [**CrossRef**]

Zhu, X.; Pathakoti, K.; Hwang, H.-M. Chapter 10—Green synthesis of titanium dioxide and zinc oxide nanoparticles and their usage for antimicrobial applications and environmental remediation. In *Green Synthesis, Characterization and Applications of Nanoparticles*; Shukla, A.K.; Iravani, S., Eds.; Elsevier: Amsterdam, The Netherlands, 2019; pp. 223–263. [**Google Scholar**] [**CrossRef**]

Zielińska, A.; Carreiró, F.; Oliveira, A.M.; Neves, A.; Pires, B.; Venkatesh, D.N.; Durazzo, A.; Lucarini, M.; Eder, P.; Silva, A.M.; et al. Polymeric nanoparticles: Production, characterization, toxicology and ecotoxicology. *Molecules* **2020**, *25*, 3731. [**Google Scholar**] [**CrossRef**]

8 Nanotechnology in Confectionery and Snacks Preservation

Davoud Habibzadeh[a]
[a]Food Science and Industry Engineering-Food Technology, Islamic Azad University-Tabriz Branch, Tabriz, East Azerbaijan, Iran

8.1 INTRODUCTION

The combination of various disciplines such as biological engineering, chemistry, mechanics and electronics has created an interdisciplinary specialized field called nanotechnology. This field can be used at the atomic/molecular/supramolecular levels to know, build and optimize the performance and quality of devices and systems. In fact, it can be said that the knowledge of nano and nanotechnology is the study of the application of very small objects (1 to 100 nanometers) and their use in all fields of science such as chemistry, biology, material engineering. Nano and nanotechnology are innovative sciences. The root of "nano" comes from the Greek language and the word "dwarf". The diameter of a human hair and virus is about 60,000 times larger than one nanometer. The thickness of an ordinary sheet of paper or newspaper is about 100,000 nanometers. One inch is 25,400,000 nanometers. On the other hand, the British Standard Institute (BSI), the science of nanotechnology, has introduced the use of nanoscale in all stages from design and production to application in devices and systems (1). Richard Feynman, an American physicist of the 20th century, first proposed the idea and concept of nanoscience and nanotechnology in a speech. In his speech, he talked about a process by which researchers could manipulate and direct the actions of distinct atoms and molecules. The term nanotechnology was popularized by Professor Norio Toniguchi 10 years later, when he was researching ultra-precise machines. Of course, nanotechnology was not particularly useful until scanning tunneling microscopes came to the market in 1981 and it became possible to observe individual atoms (2). Particles whose size has been reduced to the nanolevel have many differences in terms of physicochemical characteristics compared to the macro scale. Due to the larger surface-to-volume ratio, quantum features, physicochemical, biochemical functions, enzyme activation, mass transfer rate, permeability and catalytic behavior, nanoparticles are in better condition compared to larger particles. This feature of high surface-to-volume ratio in nanomaterials causes good features such as bioavailability, solubility, color, magnetism, optical properties, strength and diffusion. Mesoscopic materials possess properties that lie between the microscopic and macroscopic scales. Although the exact boundaries of the mesoscopic world are not well-defined, it can be roughly described as a regime involving systems with large numbers of atoms but still exhibiting quantum effects. Both mesoscopic and macroscopic objects share the feature of comprising many atoms. However, nanoparticles—due to their small size, significant surface-area-to-volume ratio, quantum behavior, high reactivity, and mechanical strength—are uniquely suited for use in a wide range of industrial applications (3). Nanomaterials, due to their suitable physicochemical and antimicrobial properties, have an assortment of applications, including being utilized in well-being care items, product preservation, the purification of water, food safety and food preservation (4, 5). The application of this science in food production and agriculture is relatively new compared to other sciences, such as pharmaceuticals (1). Type of nanomaterials are shown in Figure 8.1.

Alternative techniques introduced by nanotechnology are used in food processing to improve physicochemical quality and increase nutrient stability and bioavailability. The generation of new nourishment items with appropriate quality characteristics (sensory properties, taste, stability, texture, etc.) is the result of the use of nanotechnology in recent decades. Many studies have been conducted in the fields of assurance, quality, development of packaging, safety equipment and production of functional food products to investigate the participation of nanoparticles in them. On the other hand, researchers plan to use nanotechnology to treat obesity and diabetes (food-related diseases). They have designed appropriate nutritional regimens for the elderly population to ensure the maintainability of nourishment generation, and by way of nanoencapsulation and nutritional nanotherapy they can construct smart/intelligent frameworks for regular nutritive conveyance. Other nano applications include the use of preventive, additive agents, encapsulated edible compounds, bactericidal sensors, packaging composites, other nanostructures and additives that have an impact on food content or increase product long life, odor, contexture and pathogens (sign of food quality and safety) mentioned. Antimicrobial compounds, especially copper/copper oxide, silver, magnesium oxide, titanium dioxide, zinc oxide, carbon particles, mesoporous particles and graphene, are among bread materials that are used in food packaging. Features such as improved inhibitory features, mechanical and thermal resistance and biodegradability are among the advantages of nanotechnology over traditional polymer-based packaging methods. Nanotechnology can be used to detect food spoilage.

DOI: 10.1201/9781032677422-8

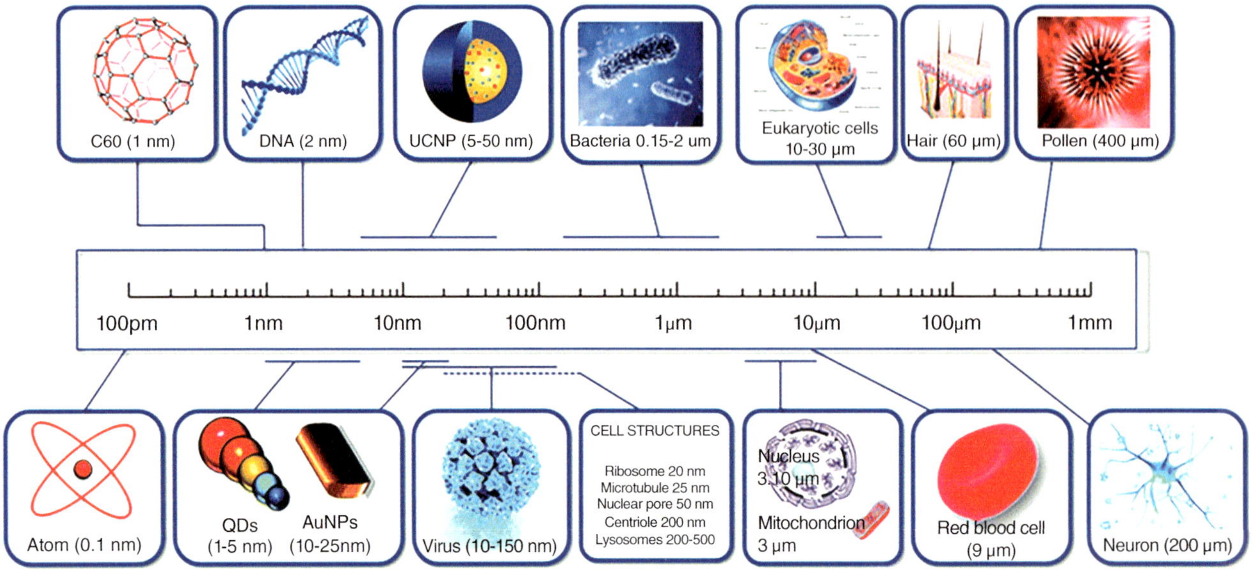

FIGURE 8.1 Type of nanomaterials (2).

Active food packaging with antimicrobial nanoparticles has a protective effect against dangerous and spoilage bacteria. On the other hand, they cause long-term shelf life and maintain quality during storage and customer satisfaction. Nanotechnology is used to enhance food safety by detecting infections and contamination during storage, packaging, processing, production and transport through nanosensors. One consequence of the use of modified nanoparticles in food is that people's exposure to oral nanoparticles increases as a result of their daily consumption of foods containing nanomaterials and particles that migrate from packaging. Studies show that the consumption of nanoparticles causes protein denaturation, DNA damage and biological consequences such as stimulation of oxidative stress responses. According to the mentioned cases, investigating the toxicity and food safety of nanoparticles by effective and appropriate techniques should be considered as an important issue. In this study, the use of nanoparticles for preservation, especially in confectionery and snacks, is investigated (2).

8.2 FOOD SNACKS

8.2.1 Definition of Food Snacks

Today, snacking is an important part of people's daily food culture. Snacks are eaten along with the main meal, especially between lunch and before late night. McCarthy first reviewed the history of snack food development in 2001. Over the centuries, snacking has become a common phenomenon in the world. Traditional snacks (fried snacks) are produced at home or on a small scale in many countries. Commercial production of potato chips and corn chips, which are considered salty snacks, has been carried out for the last 170 years and is a part of new foods. These delicious foods can be consumed during exercise, picnics or for informal parties along with a meal. Access to raw materials, primary and secondary materials and suitable technologies for processing are among the necessary conditions for producing a product with desirable qualitative and sensory characteristics. Most ingredients have the potential to be used to produce snack foods, while grains like complete grains, groundnuts, flours and pastry and meat products are more common materials. Using salt as a minor ingredient is mandatory, and ingredients such as flavors and flavor enhancers are used to encourage consumers to consume. Various methods are used to process snacks such as roasting, baking, frying and extrusion cooking, as shown in Figure 8.2. A mandatory step in the packaging process is the use of vacuum sealing and inert gases for flattened products, which not only improve the appearance but also enhance transportability and extend shelf life. These food products possess various characteristics influenced by public preferences, nutritional awareness, and health considerations. Snacks, in particular, add value to raw ingredients and provide enjoyment for consumers while serving as viable goods for trade, commerce, and export. Finally, snacks are nutritious and are associated with health issues, although consumers don't consider snacks to be nutritional goods. Like many foods, eating too many snacks can lead to health problems because it is accompanied by excess dietary energy intake and an excessive amount of fat, in particular fried or baked products, which are undesirable characteristics. The production of healthy products and limiting people's consumption of snacks are expected to resolve this problem. Snack ingredients, eaten among principal food, are an essential part of the way of life and culture. Although

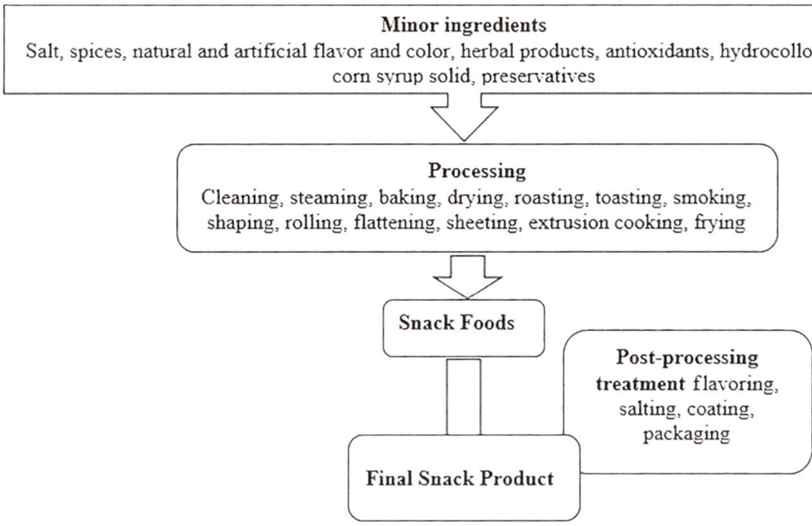

FIGURE 8.2 Snack product process (6).

snacks are typically eaten for enjoyment, they also offer numerous vitamins that supply nutritional energy. Consumers anticipate something exciting in snacks in comparison to primary food that provides no excitement. Eating behaviors have additionally modified because of the "much less cash, much less time" situation (6). See Figure 8.2.

8.2.2 Classification of Food Snacks

Snack are classified into four groups based on ingredients used, type of product process/technology and kind of food based totally on consumption or source (6).

Group 1, shown in Figure 8.3, (classified based on ingredients) includes flaked and puffed grains, legume- or paste-based snacks (such as fried products), salted nuts, potato-based chips, pastry items, as well as fried white meat, poultry, red meat, eggs, grains, legumes, and oilseeds (6).

FIGURE 8.3 Snack foods (6).

Group 2, shown in Figure 8.4, (based on the process technology used) includes spicy nuts; extruded snacks; batter-coated baked or fried vegetables, meat, and fish; as well as ready-to-cook snacks at the consumer level. These technologies involve processing such as puffing, extrusion, roasting, flavoring, coating, baking and frying.

Infrared/microwave/radio wave use and steaming Group 3, shown in Figure 8.5, (classification based on types of final product, such as ready to eat) includes new snack varieties featuring distinct forms such as bars, rings and wheels, and 3D snack and are part of the convenience food category. Other options in this group include extrusion-cooked snacks like corn and cheese balls, as well as potato fingers, burger (egg, meat, fish) snacks, chicken, coated vegetables and salty popcorn. These products are typically priced at a lower to medium range. Ready-made food items are typically more affordable compared to snacks that are ready to eat and have a longer shelf life.

Group 4, shown in Figure 8.6, (classification based on traditional snacks, such as where they come from or how well know they are) includes many conventional European cuisines that are already cooked and become well-linked snack items used in industries worldwide. Examples include snacks like potato chips and stuffed bread and tortillas originating from Central and South America. While potato chips are popular in England, pretzels are a traditional snack in France. Puffed rice and paratha are traditional foods commonly consumed in the Indian subcontinent, located in South Asia. Although many of these snacks originated in humble home kitchens, a variety of them have gained global recognition and economic importance (6).

8.2.3 Confectionery

Confectioneries are a group of foods that typically contain a high level of sugar and are commonly known as

FIGURE 8.4 Pastry-based snack products (6).

FIGURE 8.5 Snacks based on legumes (6).

FIGURE 8.6 Other snacks.

confectionery. Sweets is the practice of making desserts or treats with sugar, commonly in colorful hues. Sugar crafting is the practice of making sweet treats or zaranats (delicacy) using sugar as the main ingredient, typically in pastel colors. This word is derived from the ancient French word (Latin *confectio*) that means fastened together. Habitual sweets have been enjoyed since ancient times and have been consumed during different times from the Middle Ages to the present day. In fact, confectionery means delicious foods or foods that use sugar (as the main ingredient), coloring, flavoring, fruit or nuts in their formulation. In different parts of the world, including America, it is usually called candy. In Britain, it is called confectionery or boiled confectionery. Non-cocoa candies are mostly categorized broadly into two main groups based on their texture: hard candies and soft candies. The difference between these candies is that the sugar goes through specific steps during the crystallization process when boiling. Fondant, a type of sugar cooked until soft, serves as the foundation for many imaginative treats like cocoa cream. Initially, the main types of sweets in ancient Egypt, the Middle East and Asia were preserved fruits and honey-sweetened delicacies. Among the initial instances of utilizing sugary confections before the 14th century was to hide (coat) unpleasant drugs. These pasty confections made from dragon tree gum, white sugar and rose water were often sold by medieval doctors. Marzipan is one of the first confectioneries that still remain. Its composition includes crushed almonds or other nuts along with sugar and egg white, which has become a paste. During the medieval period, there were occasions when it was crafted into elaborate forms. Known in England in the 17th century, sugar plums (produced from cooked sugar) were unknown to the rest of the world until the 19th century. The showcasing of British boiled sweets at the 1851 national exhibition inspired other nations, particularly France, to create their own. By the mid-19th century, there were approximately 380 small-scale industries in the United States making pastilles, jujube paste and candies, but most of the high-quality candies were imported. The rise of modern machinery and the availability of sugar led to the growth of the industrial confectionery sector. In 2001, the United States experienced a surge in retail sales of chocolate, candy, and gum, with total sales estimated at $24 billion. That year also saw the launch of approximately 1,400 new candy brands. Nutriment products are typically rich in sugar content, high in calories and possess a delightful taste and aroma, making them easily digested and assimilated by the human body. The components of this mixture consist of berries, fruits, honey, syrup, sugar, flour (rye, barley, corn, soy), butter, milk, various fats, nuts, cocoa, acid, egg, gelatinizes and taste. These ingredients are transformed through heat and various types of machinery. The maximum nutritional level of confectionery can be attributed to their carbohydrate, fat and protein content; on the other hand, many confectioneries are enriched with vitamins (6).

8.2.4 Classification of Confectionery

Confectionery and chocolates are divided into three main groups (according to ingredients, production methods and final product). The first group includes various types of sweets like chocolates, candies, caramel, fruity confectionery such as halva, marmalade and other eastern sweets, coated candies and toffee. The second group includes varieties of baked products made from flour such as cakes pastries, wafers, cookies, crackers, galettes and shortbread. The third group includes dietary, therapeutic or special confectionery and chocolates; confectionery and chocolates are used as food for trips, mountain climbing and athletes because they keep their quality for a long time. This group is chemically different from ordinary confectionery. In some of them, sugars are replaced with sorbitol or xylitol (diabetic confectionery), in others, sources with hematogenous, iron source and complete protein are used (confectionery for anemic patients). It is also possible to use rare elements and materials such as laminaria algae as a source of enriched iodine and alginic acid for those who suffer from goiter, as well as food supplements for the elderly, in making confectionery. The use of compounds such as coffee in baking for children has been eliminated, so the quantity of cocoa available should be maintained at a low level. The use of low-volume but high-calorie micronutrient-enriched confectionery, such as formulated chocolate, has become common in specific military operations due to its ability to deliver concentrated energy in small portions (6).

8.3 APPLICATION AND IMPORTANCE OF NANOTECHNOLOGY IN FOOD PRODUCTION AND PRESERVATION

Nanoparticles can be used in the creation and development of packaging, practical food production, safety factors and food quality assurance. On the other hand, diseases such as obesity and diabetes (food-related diseases) are among the therapeutic goals of nanotechnology. Of course, this knowledge can be used to design suitable nutritional diets for the target communities of the elderly. The produced food is sustainable, and through nano diet and nanoencapsulation technology, it is possible to design a smart system with the aim of regular availability of nutrients for the body and creating enriched food. Preservatives, flavoring agents, encapsulation materials, antimicrobic detectors, packaging composites and nanoscale nanoparticles and additives used in food aim to influence lifetime product, odor, appearance, nutritional content, etc. And they can be used to identify food. On the other hand, they can be used to identify pathogens (a sign of food quality and safety). Nanotechnology and nanomaterials have been used in antimicrobic composites such as ZnO, TiO_2, MgO, Ag and Cu/CuO and carbon that have been labeled. Graphene and mesoporous particles are used in goods preservatives, nutriment preservation and packaging of food products. The advantages of nanotechnology compared to traditional packaging methods based on polymers include the following: 1) improved barrier properties, such as mechanical; 2) heat resistant and 3) biodegradability. Nanomaterials are used to detect food spoilage due to their increased antibacterial properties. Antimicrobial packaging, such as structured polymer films or encapsulation materials, act through the mechanism of limiting the growth phase of microbes and spreading active ingredients in the surrounding environment or onto food. Active packaging containing antimicrobial nanoparticles has the task of protecting food from dangerous and spoilage bacteria and provides the possibility of long-term storage and shelf life while maintaining freshness, quality and customer satisfaction. Also, active packaging with nanoparticles causes less access to oxygen and lighter and stronger packaging (7). In industries, during whole production stages, nanosensors are used with the aim of increasing food safety to identify infections or contamination. It should be noted that consuming foods modified with nanoparticles can increase the level of exposure to these substances in the human mouth. Foods fortified with nanoparticles as well as nanoparticles migrated from packaging are among the sources of aging exposed to nanoparticles in humans. Consuming nanoparticles can cause protein denaturation, stimulation of oxidative stress responses and DNA damage, followed by other related biological consequences. Other research shows that oral consumption of nanomaterials (solid nanoparticles) causes damage to the digestive system and secondary organs. The set of these factors causes careful focus on the utilization of nanoparticles in food. Due to this outcome, before using these particles in food, the degree of exposure to them should be determined and practical methods to check toxicity and safety should be considered (7). There are foods whose components are in nano dimensions and are different from artificially produced nanomaterials. Polymer/biopolymer (protein) nanoparticles, nanoemulsions, liposomes and nanocomposites are among the nanostructured materials with several types of nanocolloidal forms. These forms include the following: 1) nanoparticles with dimensions of 20 to 200 nm; they consist of biodegradable polymers with the purpose of being used in long-term treatments or antioxidant release. 2) Liposomes with dimensions of 100 to 400 nm are artificial vesicles of a round shape that contain more than two lipid layers. 3) 10 to 100 nm micelles are self-organized amphophilic (a chemical compound is hydrophilic or polar and lipophilic) surfactants stabilize particles capable of enclosing both lipophobic and lipophilic drugs. 4) Nano-size capsules in dimensions of 10 to 1000 nm, such as the encapsulation of significant amounts of drugs and nucleic acids such as siRNA, microRNA, shRNA and DNA. 5) Nanoconjugates are essentially polymers that have been chemically bonded with drug molecules. 6) Dendrimers ranging from 3 to 20 nm size are molecules that can be utilized to encapsulate drugs, imaging agents and targeting components either by encapsulation or covalent bonding. Nanostructures are created using cuisine ingredients through step-by-step methods that are uncomplicated and economical while having distinct physicochemical and biological characteristics (understanding natural and chemical reactions within food chain). A variety of nanomaterials (powders, rods, wires, tubes, clusters, thin films, particles and colloid) can be synthesized using various methods. The goal of nanotechnology is to improve the four main groups of applications in the following food products. The first group is the development of materials with new functions. The second group is micro- and nanoscale processing. The third group is product development. The fourth group is progress in enhancing food safety and biosecurity techniques and resources made through the utilization of nanostructured materials, which possess unique characteristics that facilitate the creation of innovative and efficient cuisine production including proceeding, guardianship, packaging and storing (7).

8.4 FOOD NANOSTRUCTURE MATERIALS IN PROCESSING

In food processing (initial resources to final goods) different and essential unit operations such as storage, packing, transport, dispersion and storage for the main purposes such as inactivating various pathogens/microorganisms and enzyme activity, removing substances, toxicity and nutrient enrichment (enrichment of nutritional supplements) are used. The result of many of these processes can be a possible improvement by applying nanotechnology, as outlined in Table 8.1. According to processing technicalities, packing, antimicrobial materials and cuisine ingredient can be a few of the prospective choice's applications of technology of nano into food, which perhaps is introduced as "direct" as "indirect" applications. Straight-line applications include participating in the accumulation of materials with a nano-level structure within the matrix of cuisine and also showing their presence and other uses such as antioxidants, flavors, preservatives, color agentic or bioactive compounds (polyphenols, omega 3, fatty acids, vitamins) and various types of compounds. Advances in nanotechnology in cuisine procedure exist mainly centered on enhancing the consistency of cuisine and enclosing cuisine enhancers or components, creating new flavors and controlling taste perception, improving the release flavors and enhancing the absorption of nutrients. Antioxidants, antimicrobials, vitamins, flavors, colors and various functional components, such as additives and preservatives, are commonly utilized by cuisine companies. Considering that this functional component may be damaged in the process of preparing, storing and consuming cuisine, measures must be taken to prevent their deterioration. The method of transfer (delivery) in effectiveness of food in the industries are impacted by nutrients and supplements, which are crucial factors. Regarding indirect applications among the various functions of the delivery method in transportation, the following can be mentioned. 1) Transfer practical materials to the desired location. 2) Protect active components in the delivery system against chemical or biological damage (destruction) in order to keep them active. 3) Manage the delivery of toxic substances in terms of the speed of release in addition to exact ecological conditions that result in their delivery. 4) Match plans for delivering

TABLE 8.1
How Different Kinds of Nano Techniques Are Utilized in Specific Areas of Different Activities Related to Cuisine Handling and Manufacturing

Food Technology Applications	Acting	Methods	Material or Particles with a Nanoscale Structure
Various carbohydrates and proteins are used in food systems, including hydrocolloids such as xanthan gum, guar gum, alginate, pectin, carrageenan, and starch. These compounds may be derived from sources like vegetable proteins, milk, and eggs	Emulsions nano shape	Contribute to improving the smoothness and firmness of ice cream	The process of preparing and producing cuisine (mixing, blending, encapsulating, adding ingredients, enhancing nutrition, strengthening, preserving cuisine components and various ways of processing), creating and manufacturing
Dammar resin, ester resin and sucrose acetate isobutyrate along with other additives, such as vegetable oil containing bromine, are commonly used in the production of certain products	–	This technique helps distribute and make nutrients in cuisine more available; it acts as a balancing agent and reduces creaming and sedimentation	–
Emulsions of nano shape	–	Creates additional cuisine items for personalized beverages, salads with dressing, sugar substitutes, infused oils and various packaged cuisine options	–
Nanoliposomes are created by combining nanoparticles made of zein fibers with gallic acid	Nanoscale encapsulation	Plants receive insecticides, fertilizers and vaccines while supplements, enzymes vitamins and other essential nutrients are provided to them through their cuisine	–

(Continued)

TABLE 8.1 (Continued)
How Different Kinds of Nano Techniques Are Utilized in Specific Areas of Different Activities Related to Cuisine Handling and Manufacturing

Food Technology Applications	Acting	Methods	Material or Particles with a Nanoscale Structure
Microcapsules made of colloidal particles		Improves the nutritional value of the cuisine by adding essential vitamins and minerals	–
Nano shape capsules	–	Enhancements have been made in the usefulness and bioavailability of product protection against oxidizing agents. Consistency and preservation of volatile components have also been improved The food product now traps scent and ugly components more effectively Controlled release can be induced by moisture, taste and pH	–
Nano shells (chocolate shake and nano shape tea)	–	Nanoclusters are employed in nanoencapsulation to enhance the flavor of the shake without the need for additional sugar	–
Ordinary promotion	–	Formerly, it was common practice to encapsulate vitamins or active biological components in beverages using nanotechnology	–
Pigment suspension	–	The technique is used to create paprika nano, apo carotene and beta-carotene emulsions	–
Archaeosomes are of an organism microscopic nature that consists of lipids found in archaeal membranes	–	Mechanism for distributing antioxidants	–
Nanocochleates are from nanomaterial that can utilize soy-based phospholipids	–	Aids in improving the quality of processed cuisine	–

products with the physical and chemical properties of the substance final goods (such as shelf life, taste, texture, and appearance) and other process components. Since materials with practical applications are very important, different ways of delivering products include biopolymer, matrices colloids and emulsions, and simple solutions should be considered to encapsulate them (7).

8.5 NANOSTRUCTURED MATERIALS OF FOOD IN PACKAGING AND STORAGE

Due to the fact that proper packaging can protect food from environmental and external factors such as temperature, humidity, microbiological infection and environmental gas mixtures, it is therefore one of the important elements in the food production process. There is a fundamental flaw in traditional food packaging and that is their permeable/porous nature. While there aren't packaging supplies that are completely impermeable being transformed into to H_2O gas and other environmental vapors, the use of nanotechnology in packaging can prevent the decomposition and loss of nutrients and, while creating safety, guarantee that food remains fresh for a prolonged period. Active packaging is essential for preserving cuisine freshness by creating an ineffective shield against environmental conditions and adapting to environmental changes. This protective mechanism is accomplished through the release of helpful substances like inhibiting the growth of microorganisms or substances that inhibits oxidation or through their ability to absorb gases. Such interactions that improve food stability can be seen in materials and technologies such as antimicrobials, oxygen scavengers and enzyme immobilization techniques. Controlled release packaging is another application of nanocomposites in active packaging, where they are employed to increase the intake of beneficial additives such as probiotics, vitamins and nutrients found in food with the aim of increasing some functional characteristics of conventional packaging of food products. It involves considerations such as strength, ability to block outside elements, resistance to heat, utilization of substance that inhibits its oxidation, potent bacteria-fighting substances, enzymes and extracts derived from vegetation; biological nanocomposite materials have been used for packaging design. The use of biodegradable or edible materials, nanocomposite materials and plant extracts has led to the development of environmentally conscious and eco-friendly packaging, which shows promise in minimizing the negative environment impact caused by synthetics. Due to the advanced functions of nanomaterial made up of metallic elements, nonmetals and their corresponding oxides, the application of nanotechnology in packaging of food is continuously advancing. Enhanced and intelligent packaging of food is among the packaging that uses nanocomposite materials. 1) Enhanced packaging for food items: Usage particles of nano in biological nanostructure substances increases puissance and stability characteristics like adaptability and gas barrier properties including stableness at carbon dioxide, oxygen and aromatic composites, as well as stability in different temperature and humidity conditions. 2) Smart food packaging: In addition to intelligent monitoring, this type of packaging actively interacts with the food contained in the package and preserves it. As a result, these packages are both active and smart. Smart packaging is not required to protect the cuisine; its main function is to track the unique attributes of package. Package of smart cuisine, combined with precise feedback and endorsement of the verified quality of packaged cuisine, serves as a safeguard opposed to fraud and fake products. It serves in the role of defense against being deceived and also as a visual condition indicator in adverse conditions such as temperature levels. Food packaging that is active helps to maintain the quality and freshness of cuisine using procedures created via intrinsic and acquired elements (the packaging material's inherent properties of antioxidants and antimicrobial properties) and enhances the longevity of cuisine products, thereby minimizing wastage. Utilizing nanoparticles into cuisine production brings many possibilities, such as enhancing the abilities of polymers (barriers whether physical characteristics properties) and presenting new properties (packaging, sensing and active or biologically active control) based on the inherent properties of nano-additives (7).

8.6 PRESERVATION

Individuals naturally have strong feelings when it comes to alterations in their diet. Historically, the incorporation of innovative technologies within the food industry has occasionally been met with pushback or animosity. There are various factors that can impact people's views on food, such as their apprehension of unfamiliar risk and their level of confidence in the safety of food. Regulations, along with various societal and psychological elements influenced by healthiness, environment, and methodical viewpoints, impact how impressive they are. The introduction and nanotechnology in the cuisine industry could potentially bring up these issues. Nevertheless, like previous technological advancements, new technologies have the potential to provide various advantages to individuals and the community. This study was initiated to explore the potential of nanotechnology in the cuisine industry and assess if it can contribute positively to this sector. Our objective is also to determine if appropriate measures are in place to guarantee consumer consciousness and safeguard against any potential risks posed by nanotechnology and explore and acknowledge the potential worries that the general population may hold regarding these emerging technologies. The field of nanotechnology allows researchers to control substances on increments of one billionth of a meter. At this minuscule level, substances could display unique characteristics like modified chemical responses or changed electronic, visual or magnetized attributes of these substances across various industries, including health care, construction and technology. Nanomaterials offer a variety of opportunities for the cuisine industry, with the potential to benefit both consumers and businesses. For instance, they can be utilized to produce cuisine products with reduced sugar, salt or fat content, while maintaining the original taste, as well as to develop innovative packaging solutions that enhance food preservation and provide information on freshness to consumers. At present, there are only a few food items that include nanomaterial, but this could increase in the coming years due to advancements in technology. To promote the responsible growth of nanotechnologies in the food industry and guarantee that any advantages for consumers and society are endorsed by the governments as needed, we offer a set of suggestions. Nanotechnology brings along both advantages and risks to consumers as a result of its unique properties. The diverse range of nanomaterials available means that some may be safe while others could be potentially hazardous. Our knowledge of how nanomaterials interact with the human body is not comprehensive enough to definitively forecast the impact on human health. Nanomaterials that are persistent raise specific worries as they do not degrade when consumed and could possibly leave the intestinal tract, circulate in the body and gather in cells. The lasting consequences remain unknown at this time. Regrettably, there is only a small body of research exploring the toxic effects of nanomaterials, particularly concerning the hazards associated with ingesting them. In order for regulatory agencies to properly evaluate product safety prior to being released on the market, it is essential to conduct this research. Our findings suggest that government and research councils do not prioritize research in these specific areas, especially considering the potential timing of the production of nanomaterial-containing products. Research funding organizations, especially, have not been proactive enough in promoting research in the critical areas of unpredictability that form the basis of risk evaluations for these substances. It is suggested that they take on a more proactive pattern in stimulating process (8). Nanoscience and nanotechnology have introduced fresh and inventive uses in the realm of production. Nanotechnology is a useful tool in various industries, particularly in food production and functional areas. With the emergence of any novel food processing technology, packaging of food substance or food product, it is essential to conduct additional research to prove the advantages of nanotechnology and engineered nanomaterials intended for food use while avoiding negative health impacts. Nanoemulsions offer several benefits compared to traditional emulsions. These advantages include superior optical clarity, enhanced physical stability against factors such as gravity-induced separation and droplets clumping and increased absorption of the encapsulated substances, rendering them ideal for use in the food industry. Nanoencapsulation represents the maximum amount desirable technology that has the possibility of entrapping bioactive chemicals. This review highlights the ongoing investigation in the field of nanomaterial of technology applications in cuisine and agricultural technology, which consists of nanosensors nanocomposites, nanoemulsion and nanoencapsulation and food packaging, in addition to suggestions of upcoming advancements in the expanding realm of agricultural nanotechnology. An introduction to the structure of nano elements and their present state usage and potential future advancements in the field of cuisine study is included (3).

8.7 SCIENTIFIC STUDIES ON NANOTECHNOLOGY IN THE PRESERVATIVE OF CONFECTIONERY AND SNACKS

8.7.1 Creation of a Frozen Vegetable Fortifier using Bio-Ingredients and Nanotechnology Containing Natural Healthy Protein Snacks

Pavlyuk et al. (2021) studied the creation of a frozen vegetable fortifier using bio-ingredients and nanotechnology containing natural healthy protein snacks. Their research showed nanotechnology was used to invent cryogen enrichments of vegetables (spices and carotene) with biological components (biologically active substances) and natural proteins for use in healthy snacks containing them. The aforementioned method is an innovative technique of conducting an in-depth analysis and transformation of natural resources (through cryo-processing and cryo-mechanical destruction) to obtain nano-vegetable additives, with an emphasis on preserving the biologically active substances of fresh vegetables, while transforming them into nanostructures and revealing their full biological potential. The mass ratio of biologically active substances to the total mass of the solution (mass fraction of biologically active substances) in the produced cryo-frozen materials is 3 to 2.3 more than the raw materials. As a result, a new generation of healthy, natural protein pastes was created with the help of nanomaterial from peas and soft cheese to prepare a breakfast snack. The protein content of other compounds was completely different and consisted of single monomers of α-amino acids with easy digestibility (50–55%) and particle size of 1 to 1.5 nm. Additives were produced using nanotechnology according to the application and nonfermentative catalyst, especially mechanical decomposition and mechanical degradation based on peas and in powder and pure form, which provided possibility through easily making digestible additives with nano dimensions. Utilization of technologies (cryo-processing and cryo-mechanical damage) and natural raw materials in the form of powder and nano extracts (spices and carotene) have been used, leading to the creation of a new formation of healthy protein, natural and nano-format snacks with the aim of achieving a health product with quality and no harmful additives. Due to the high content of β-carotene, phenolic compounds and tannins, about 30% of the daily protein requirement is obtained by eating every 100 grams of the product (9).

8.7.2 Dispersion of Microorganisms and the Antibacterial Properties of Nanoparticles (Silver Zinc Oxide and Silver Titanium Oxide) versus Microbes Found into Solid Pastry Snacks

Adebisi and Oyetayo (2019) studied dispersion of microorganisms and the antibacterial properties of nanoparticles (silver zinc oxide and silver titanium oxide) versus microbes found into solid pastry snacks. A review was conducted on the antimicrobial properties of silver zinc oxide and silver titanium oxide versus various microorganisms including *Aspergillus* (*fumigatus* and *niger*), *Staphylococcus* (*epidermidis* and *aureus*) and *Serratia marscens* from flour snacks (meat pie, egg and puff pastry) presented. Due to the lack of knowledge of most food vendors (street food) regarding fundamental principles of edible substance safety, the edible substances sold are the ones who were suspected of microscopic organism contamination. As a result, the health status of these people is one of the main reasons for reducing the safety and microbiological quality of road meal componence. The results from microscopic tests showed the pie made of meat had the maximum level microbial upload (80 to 150 billion colony-forming unit per gram) among flour snacks. Flour snacks gathered at night had a maximum bacterial load more so during the morning hours. A similar growth trend was observed in the number of mushrooms in morning and evening flour snacks. In general, the microbial load (aerobic, coliform, yeast and mold) of flour snacks varied ranging from 120 to 18000 colony-forming units per gram. Characteristics such as being resistant to high temperatures, drying out and exposure to radiation makes *Staphylococcus aureus* survive in some stages of food preparation. On the other hand, this microorganism is a living organism on human skin that can move from one place to another people to edibles. *Staphylococcus aureus* is a significant element food pathogen known for its capacity to generate heat resistant toxins and strong exotoxins. Strains of the *Bacillus* and *Aspergillus* species exist partly due to concern with ecological pollutants and common food poisoning, especially in floury foods, which survive due to their spores having resistance to thermal processes. They can survive heat treatment due to spores that are resistant to high temperatures. The used mercantile antibiotics showed different areas of inhibition (18–22 mm) against the microbial strains of flour snacks. Ampicillin, amoxicillin and septrin did not have an inhibitory effect on the strains of the aforementioned microorganisms, and antibiotics such as ciprofloxacin and pefloxacin showed a significant effect. Silver zinc oxide and silver titanium oxide showed important antibacterial properties strains (by inhibition zonal of 18–22 mm). The strains of epidermidis, *Staphylococcus aureus* and *Bacillus* species, were primarily sensitive in relation to the antibacterial properties of the aforementioned nanoparticles. Synthesized green nanoparticles (AgTiO and AgZnO) inhibited all microbial strains of flour snacks. ZnO NP had inhibitory effects equal to *Staphylococcus aureus*, and silver nanoparticles had antimicrobial activity (concentration-dependent) against *Escherichia coli* and *Pseudomonas aeruginosa* (10).

8.7.3 Impact of Using Films Made from Combination of Chitosan, Nano Titanium Dioxide, Thymol and Tween on Quality of Ready-To-Eat Cantaloupe Fruit

Qiao et al. (2019) examined the impact of using films made from combination of chitosan, nano titanium dioxide, thymol and tween on quality of ready-to-eat cantaloupe fruit. Increasing the awareness of consumers regarding the consumption of healthy food, the production of ready-made fruits and vegetables has become a vital matter. On the other hand, the increase in the price of fruit leads consumers to consume cut fruit instead of whole fruit. Changes related to microbial sensitivity, physicochemical and sensory stability due to washing, dewatering and packaging processes lead to a decrease in shelf life and overall quality. Practical methods such as the use of modified atmosphere, radiation, edible coating and film, along with ultraviolet light and ozone, are being utilized to increase the shelf life fruits such as cantaloupe. The application of nano metals is one of the effective methods in reducing the loss of nutrients in food, especially cut fruits. Adding nanoparticles to packaging materials improves physiological, chemical and transport characteristics. Chitosan can provide films with good properties despite the tensile stress and poor durability after wetting, stiffness and weakness in elongation. Nano titanium dioxide (TiO_2) can play the role of a photocatalyst and is used in various products across a broad-spectrum of food to cosmetics and medicines due to its nontoxicity. This material is used as a disinfectant and material in coatings that clean themselves. The effect of layers (coating) of chitosan, nano titanium dioxide, thymol and tween with antimicrobial agents on the quality of ready-to-eat cantaloupe fruit was studied with Kern's cold method. The leakage of ascorbic acid and fruit juice was not evident in the treated samples (chitosan/nano titanium dioxide [CH/TiO_2]) compared to the uncoated samples. On the other hand, in chitosan, nano titanium dioxide, thymol and tween covered fruits, the levels level of malondialdehyde and overall soluble solids were inhibited. The control fruit samples had more water related activities (AW) and polyphenol oxidase (PPO) than the fruits treated with the aforementioned nanoparticles. The final results showed that increasing duration of storage that is considered to be satisfactory quality of fruit from inside and outside of the cantaloupe can be affected by coating (CH/TiO_2/TT) (11).

8.7.4 Quality of Fresh-Cut Melon Influenced by Edible Coatings of Ag-Chitosan Nanocomposites

Ortiz-Duarte et al. (2019) studied the quality of fresh-cut melon influenced by edible coatings of Ag-chitosan nanocomposites. The effect of nano-form composites of Ag and chitosan (chitosan silver nanocomposites extracted from red claw crab) in edible quality coatings of freshly cut melon was studied during 13 days at 5°C. Covering melon treatments decreased the respiration rate of fruits compared to uncoated samples during shelf life and prevented softening during the storage period. In terms of sensory characteristics, these treatments received a higher score. After 9 to 10 days, the coated samples achieved equilibrium within packages that had CO_2 levels ranging from 12.6 to 16.2kPa and O_2 levels ranging from 2.3 to 3.7kPa. Nanocoating had little effect on sugar compounds consisting of fructose sucrose, glucose, mild edible acids such as malic and citric acid, soluble solids, color, pH and TA. On days 10 to 13, a decrease in microbial load (mesophils) was observed by 0.6 logarithmic units. Nanotechnology was used to prepare and develop vegetable protein supplements using chickpeas and use them in order to create a fresh lineup of sweets products. The use of heat-steam treatments focusing on using high quality ingredients and precise milling is a real innovation and using modern equipment in restaurant investments. The aforementioned technologies (steam processes and mechanical destruction) convert 40 to 70% of protein, starch, cellulose and pectin (biopolymers) in peas into single monomers with easy digestibility in nano dimensions. Peas have biologically active substances and nutrients such as 41–45% starch, 20–25% protein and equal amounts of glucose and fructose (2–25%). From a biological point of view, dry chickpea protein is complete and is 2 to 7.2 times more than the ideal protein limit considered by the World Health Organization. Only the amount of ethionine amino acid in it is less than the standard (3% less than the ideal amount). Using conventional technologies to make confectionery products, if their consumption is accompanied by stress, inactivity and inappropriate environmental conditions, causes cardiovascular diseases and eventually heart attack and stroke. The aforementioned proposed methods (deep processing) cause a better manifestation of the biological potential of raw materials. Formulation components such as vegetables and fruits like garlic, celery and ginger root, pumpkin, carrot, lime with skin and apple and their combination and protein supplements from peas, lead to developing an original line of candy products (waffle, sponge cake, dry waffle breakfast). In addition to significant amounts of protein, compared to older products, new confectionery products have high and different amounts of sugar, fat, protein, tannins, vitamin C, aromatic substances, phenolic compounds, β-carotene, etc., which can meet the daily needs of consumers. In addition, the mentioned products will have less harm in terms of constituent compounds (12).

8.7.5 Nanotechnology that are Being used to Produce Protein Plant Supplements from Chickpeas, Leading to the Creation of Innovative Confectionery Products

Pavlyuk et al. (2021) studied advancements in nanotechnology that are being used to produce protein plant supplements from chickpeas, leading to the creation of innovative confectionery products. Nanotechnology was used to prepare and develop vegetable protein supplements using chickpeas and use them to obtain a new line of candy treats using heat-steam treatments. The natural resources utilized and the significance of finely gritting them is a real innovation and uses modern equipment in restaurant investments. The aforementioned technologies (steam processes and mechanical destruction) convert 40 to 70% of protein, starch, cellulose and pectin (biopolymers) in peas into single monomers with easy digestibility in nano dimensions. Peas have biologically active substances and nutrients such as 41–45% starch, 20–25% protein and equal amounts of glucose and fructose (2–25%). From a biological point of view, dry chickpea protein is complete and is 2 to 7.2 times more than the ideal protein limit considered by the World Health Organization. Only the amount of ethionine amino acid in it is less than the standard (3% less than the ideal amount). Using conventional technologies to make confectionery products, if their consumption is accompanied by stress, inactivity and inappropriate environmental conditions, causes cardiovascular diseases and eventually heart attack and stroke. The aforementioned advanced processing methods enhance the biological potential of raw materials. Formulation components such as vegetables and fruits—including garlic, celery, ginger root, pumpkin, carrot, lime (with skin), and apple—along with protein supplements derived from peas, have led to the creation of a new generation of sweet treats such as waffles, sponge cakes, and dry waffles for breakfast.

Compared to older products, these new confectionery items contain significant amounts of protein as well as varying levels of sugar, fat, tannins, vitamin C, aromatic compounds, phenolic compounds, and β-carotene, all of which can help meet consumers' daily nutritional needs. Additionally, these products contain fewer harmful compounds in their composition (13).

8.7.6 Formulation Enriched in Nitrate, Betalains and Minerals Microencapsulated and Ready-To-Eat for Beetroot Soup that Offers Stability and Appeal

Da Trindade et al. (2023) studied a microencapsulated, ready-to-eat formulation for beetroot soup, enriched with nitrate, betalains, and minerals, which offers improved stability and sensory appeal. In recent years, the use of fruits and vegetables that have nutritional properties and bioactive compounds has been developed in ready-made food products. The use of new technologies such as encapsulation preserves and stabilizes functional or bioactive compounds. Food micro-grade particles cause the process of solidifying food suspensions through cooling processes during the lyophilization process. Easy transportation and supply, increasing the consistency of aromatic compounds sensitive to exposure to oxygen and high temperatures, reducing storage costs, preserving more bioactive compounds and nutrients are the advantages of using lyophilization. Encapsulation of already cooked beetroot soup (rich within nitrates, betalin, minerals and bioactive nutrients) by different encapsulating agents (starch or maltodextrin) was studied. The use of lyophilization and encapsulation preserves the functional compounds in the beet matrix (nitrate, minerals, saponins and betacyanins) in the ready-to-eat beetroot soup. Due to the physical barrier feature of the encapsulating agents against the impact of oxygen, light, heat and moisture, all pigments are preserved for at least 90 days after micro-encapsulating the beetroot soup. Different analyses showed that the encapsulated beetroot soup powder has sufficient microbiological safety and remained suitable for consumption even after one month. However, after two months of storage, visible growth of mold and yeast was detected in pure beetroot soup. The incidence of this microorganism was observed in all powdered soup products after 90 days (14).

8.7.7 The Use of Nanopackaging and Food Coating in Improving the Storage Life and Quality of the Seeds of Fresh Pistachios

Ahmadi et al. (2014) studied the use of nanopackaging and food coating in improving the storage life and quality of the seeds of fresh pistachios. Fresh pistachios are one of the most important agricultural products, and due to their very low durability, the drying process should be done on them. This issue reduces the nutritional value of this product and changes it from fresh fruit to dried fruit, which also requires paying a very high commercial tariff for export. Therefore, considering the importance of exporting fresh pistachios from an economic perspective standpoint, it holds significant importance to implement plans for increasing the expiration date of newly harvested pistachios and suitable packaging in this field. As a result of improper storage conditions or prolongation of storage period, the skin on the pistachio is cracked, and on the other hand, the color substances are removed from the pistachio skin, causing discoloration of the pistachio skin, ultimately leading to a decrease in its quality. One factor contributing to the short shelf life of fresh pistachios is their rapid respiration rate, which generates heat and can increase the temperature inside storage or processing environments. One way to enhance the long life of the goods is waxing or coating the fruits, which not only improves the appearance of the fruits, but also increases their duration storage and standard of excellence. The main benefit waxing is usually a reduction in fruit transpiration and respiration, and it is a method of improving and enhancing the storage of the fruits. These coatings will increase the quality of the products by changing and controlling the internal atmosphere of the fruits. One of the edible gels used for covering in recent years is aloe vera gel. Recently, the antioxidant, antipathogenic aloe vera gel characteristics have been introduced. Many of these properties are due to the presence of polysaccharides (galactan, mannan, araban and pectin substances) in mucilage gel. This covering material can be effective in maintaining the quality of the fruit due to its antioxidant properties, preventing weight loss and creating a barrier against the entry O_2 and Co_2. Essential oils and combinations of herbal nanocides have significant effects in enhancing the long life of goods. These nanoscale particles have a very high relative contact level that increases the contact with bacteria and fungi, so their antimicrobial and antifungal effects increase significantly. When nano silver is in contact with bacteria and fungi, it has adverse effects on the metabolism of their cells and causes the cell growth to stop. Nano silver prevents the division and growth of mold and bacteria that cause infection and odor. The mechanism of action of silver nanoparticles is that these nanoparticles cause the death of bacterial cells by deactivating the compounds of the thiol group present in the respiratory enzymes of bacterial cells. In this way, after reaching the outer layer of the cellular structure wall, and cellular membrane, silver nanoparticles destroy the respiratory system of the cell with the enzymes of the Ag+ chain of the bacteria in the form of interaction of respiratory ions. Also, during this process, silver ions in the cell wall break hydrogen bonds, activate oxygen production and destroy the microorganism. On the other hand, the bacterial cell loses its ability to multiply when it is under the DNA of silver ions, and it leads to the death of the bacterial cell. Packaging is also one of the effective methods that can be used to minimize the waste and waste of products and increase their durability and quality. The most important problem during the design of edible goods packaging is the existence of oxygen, which spoils the fat of food and destroys its color. The nature of plastics produced by nanotechnology is such that gases penetrate into the foodstuff in an insignificant amount. In this type of plastics, nanoparticles are arranged in a zigzag pattern; these particles are arranged in parallel layers and minimizes the penetration of oxygen. In other words, the path that oxygen must travel to enter the package becomes longer (15).

8.7.8 Using an Extract from Pomegranate Peel Nanoparticles to Potentially Decrease the Amount of Acrylamide Formed in Potato Chips when Deep-Fried in Sunflower Oil

Mekawi et al. (2019) studied using an extract from pomegranate peel nanoparticles to potentially decrease the amount of acrylamide formed in potato chips when deep-fried in sunflower oil. Frying and especially deep frying (complete immersion in hot oil) produces products with suitable characteristics (aroma, taste, texture and color) and ultimately the popularity of fried foods. The occurrence of simultaneous or sequential complex reactions caused by frying (Millard reactions, polymerization, hydrolysis and oxidation of lipids) are among the parameters of interest to researchers in the realm of cuisine and health industries. After oxidation of fatty acids, the resulting induced carbonyls can synthesize several reactive compounds with the help of protein residues. Many efforts have been made in connection with reducing the amount of acrylamide in foods since the discovery of the formation of acrylamide until today. Due to the toxicity of acrylamide, approaches to reducing the levels of this substance in thermally processed foods (above 120°C) have been made in recent European Union regulations. Considering the increasing consumption of fast foods, drinks (snacks) and foods like bread, fried cereals and potato chips, there may be a level of direct contact with acrylamide in society. The high amount of acrylamide in fried potatoes is attributed to the abundance of its precursor component present in the tubers, as well as the maximum heat applied during the cooking process. The absorbed oil in potato chips can reach a maximum of 40% compared to the concentration of acrylamide (525–722 μg/kg). Acrylamide levels in processed foods can be influenced by antioxidants (because of the capacity of various functional groups to interact with acrylamide precursors). Reducing the oxidation of lipids and the durability of oils in fried ingredients and products are among the positive effects of antioxidants. Extracts derived from plants like rosemary and pomegranate, which are abundant in phenolics, offer a viable natural option for synthetic antioxidants in preserving sunflower oil and decreasing acrylamide formation in fried potatoes. Due to the presence of maximum level of phenolics (ellagic tannin, gallic acid and ellagic acid), pomegranate skin is recommended to inhibit lipid oxidation in foods. Nanoparticle extract of lyophilized pomegranate skin has been used in deep frying to act as a powerful antioxidant, improving the oxidative stability of sunflower oil and decreasing the formation of acrylamide in potato chips. Considering the highest amount of 1674 mg/kg acrylamide into fried potato crisp after 20 fry cycles compared to the low initial amount of 192 mg/kg of acrylamide, lyophilized pomegranate peel nanoparticles extract as natural antioxidants because it can be employed to keep sunflower oil and reduce the production of acrylamide, increase safety, quality, durability and nutritional value (16).

8.7.9 Assessing the Impact of Chitosan–Pectin Nanoparticles on Fried Vegetables and Fruits Chips Exploring Both Technological and Biological Aspects

Sheir et al. (2022) studied assessing the impact of chitosan–pectin nanoparticles on fried vegetables and fruits chips exploring both technological and biological aspects. The study analyzed the impact of chitosan–pectin nanoparticles on the technological and biological properties of fried banana, potato and cassava chips. Fat frying, not deep frying, has been considered a cooking method since ancient times. Acceptability of consumers, crispy texture, quick and easy eating and increased useful life are the features of this method. This process causes discoloration, water loss, oil absorption, starch gelatinization and crust formation and increases hypertension, cancer, diabetes and heart disease. Because of these reasons, people want a healthier alternative to fried foods. The best step for proper changes (reducing energy, absorbing oil, moisture and increasing shelf life) in the frying process is a step before frying. The best way to make these changes is to use nanotechnology in food products. Treatment before the frying process (immersion in chitosan–pectin nanoparticle solution) for one minute effectively reduced the formation of acrylamide (49.62%), oil absorption (4.95%) and 1.98 mg/kg of malonaldehyde into bananas fried shape. The number of aforementioned properties was 54.59%, 29.76%, and 2.14 mg/kg in potato fried shape and cassava fried shape; also the sensory and textural properties were improved. The final product was given to albino mice for a three-month period, during which biological tests were conducted, liver tissue structure was examined and glycogen levels were displayed through histochemistry. No cases of death were observed in these animals during the mentioned period. As a result, methods like the aforementioned method can help food manufacturers in making healthier fried foods while maintaining better technological quality (17).

8.7.10 Evaluating the Impact of E171 (TiO_2) Food Additive in Colored Chocolate Candies in its Articulate/Nanoparticulate form through Analysis Using spICP-MS, TEM and Cellular in Vitro Models Both Prior to and Following Simulated Oral Ingestion

Espada-Bernabe et al. (2024) studied evaluating the impact of E171 (TiO_2) food additive in colored chocolate candies in its particulate/nanoparticulate form through analysis using spICP-MS, TEM and cellular in vitro models both prior to and following simulated oral ingestion. Titanium dioxide (E171) is one of the food additives and a whitening agent that is used in more than 900 food and confectionery products (low-fat dairy products, sauces, chewing gum, chocolate and candy). The use of this substance in food in EU countries is still being discussed and negotiated in connection with the risk assessment and the presence of nanometer titanium diacid in the labeling. Despite the proven risks of consumption and toxicological information, the European Food Safety Agency has banned the use of the substance, although it is still allowed in the United States, Great Britain and Asian countries. According to the formulation, type of product, route of exposure (swallowing, inhalation or skin absorption) and the amount of product consumption, it is difficult to determine the intake levels of titanium dioxide nanoparticles. On the other hand, the digestive system may have a different behavior against the properties of solubility, surface composition, state of accumulation and load of this additive compared to other additives in food. All of these factors cause differences in potential toxicity, bioavailability and biological fate compared to virgin titanium dioxide. Nanoencapsulated titanium oxide was used in confectionery products (colored chocolate candies) in order to investigate their behavior before and after oral consumption. Describing the physicochemical properties of particles using single-particle transmission electron microscopy and inductively coupled plasma-mass spectrometry techniques had been the first objective, and changing the physicochemical properties due to the laboratory digestion process in the digestive tract on particles was the second objective of this evaluation. The dimensions, density and proportion of particles measuring under 100 nanometers were studied in terms of particle stability over time. The results of mass spectrometry techniques showed the highest concentration in titanium content among coating colors in white candies. Transmission electron microscopic revealed the existence of nearly spherical particles measuring up to titanium dioxide (average size above 100 nm), while about 25% (particles below 100 nm) were found in all coatings. Due to the key role of the sample matrix in bioavailability, differences were observed between the study of the samples when comparing the watery extract of the covering to the entire cocoa sweet. Titanium dioxides bioavailability to be absorbed by the body as small particles is maintained. The research on cytotoxicity using Caco-2 cells indicated that cell viability was strongly reduced after 24 hours of exposure to GI extracts (18).

8.7.11 Characteristics Related to Physics and Chemistry, Along with the Ability to Prevent Antioxidant Chocolates Containing Specially Formulated Nano-Cinnamon Particles

Muhammad et al. (2018) researched characteristics related to physics and chemistry, along with the ability to prevent antioxidant chocolates containing specially formulated nano-cinnamon particles. From the past decades until now, producers have turned their attention to the use of useful bio-effective compounds in food. The main component of chocolate formulation (cocoa beans) contains polyphenols (as antioxidants of the body and their role in preventing metabolic syndromes of the body). These substances may be significantly reduced in chocolate production (fermentation and roasting) or transformed into their epimers. The goal of food industry experts in dealing with these problems, in the first phase, is to make changes in order to preserve cocoa compounds (bioactive compounds) during chocolate production stages. The second phase is the use of different phenolic compounds derived from plants (various herbs and spices, raspberry leaves, nettle leaves, mangrove extract, black, goji berries, skin of peanuts) to enrich and improve phenolic and phenolic activities. Chocolate is an oxidant. Cinnamon contains various polyphenols such as syringic acid, 4-hydroxy-m-anisic acid, vanillate, para-hydroxybenzoic acid, caffeic acid, para-hydroxybenzaldehyde, gallic acid, 4-hydroxycinnamic acid, proanthocyanidins, quercetin, ferulic acid, sinapic acid, and tannic acids. It also possesses antioxidant properties and volatile compounds like linalool and limonene, which contribute to its aroma and health benefits. In addition to the aforementioned effects, the enrichment of chocolate with cinnamon nanoparticles also has an effect on the characteristics of sensory tissues, rheological behavior and melting characteristics. The masking of unpleasant odors and tastes and control of bioactive emissions and biological activities, bioavailability is improved with the help of submicron encapsulation. Also, nano minerals play a role as flavor transfer agents and dietary antioxidant supplements. Since cinnamon extract is loaded to trap its aroma in lyophilized colloidal nanoparticles (nanoencapsulation method using precipitation method), it can be used without negative effects on the sensory characteristics of chocolate (19).

8.7.12 The Impact of Chitosan–Pectin Nanoparticle on Fried Vegetable and Fruits Chips on Both Technological and Biological Aspects

Sheir et al. (2022) studied assessing the impact of chitosan–pectin nanoparticle on fried vegetable and fruits chips on both technological and biological aspects. Various factors such as oxidation can cause the destruction of food and as a result reduce the quality and ultimately the nutritional value. The factors known as free radicals, which are produced during the mentioned process, pose a threat to people's health. Since artificial antioxidants may endanger the health of consumers, the use of natural antioxidants (without harmful effects, with therapeutic and nutritional effects) has gained many fans. Cumin seeds are traditionally used in the treatment of epilepsy, jaundice, diarrhea, indigestion and toothache due to their aromatic and special taste as a spice in the food of people from different parts of the world (Egypt, Africa, Middle East, India and China). High amounts of phenolic compounds in cumin significantly destroy free radicals, chelate carotenoid linoleic acids and reduce their activity. In Egypt, cloves (carnation flower or essential oil) are used. Cloves yield a significant amount (10–20%) of essential oil. The main component of this oil is eugenol, which exhibits potent antioxidant properties. The presence of unsaturated phenolic compounds, including eugenol, eugenol acetate and beta-caryophyllene, contributes to its ability to reduce free radicals. Its antiaging, anti-inflammatory and antioxidant properties are among its therapeutic effects. This particular vegetable oil is commonly utilized as a food enhancer at a sanctioned level of around 1500 prm. Inadequate durability under environmental modalities such as high temperatures, oxidation, exposure to light and low pH levels changes in organoleptic properties during normal storage, and low solubility in water are some reasons of limiting the application of either caraway or clove oil. Oxidation and dehydrogenation reactions (chemical or enzymatic) are among the spoilage factors oil (cumin/clove) within storage and proceeding. Various natural polymers such as serum protein concentrate, polylactic acid glycolic acid, maltodextrin, gum arabic and lecithin help prevent this defect and enhance its ability to dissolve in aqueous solution. The microencapsulated and nanoencapsulated formation oil (clove and cumin) by carboxymethyl cellulose was used in sweet and salty biscuits in order to evaluate antioxidant activity, increase long life and prevent oxidative deterioration while being kept (0 to 60 days). The DPPH values were the same in two batches in both of sweetish biscuits (cloves), and salty cookies (salt) by nano/micro capsule oils had a downward trend. This trend was present in the amount of power reduction in both samples. Nanoencapsulation in carboxymethyl cellulose provides long-term stabilization of the oil during storage. Sweet biscuit samples enriched with clove oil nanocapsules had higher amitase than other samples in terms of sensory characteristics (20).

8.7.13 The Creation of Emulsion Films Utilizing Bovine Gelatin, Nano-Chitin and Nano ZnO for the Purpose of Packaging Cake. Bakery Products (Bread, Cakes, Biscuits and Donuts) have Sources of Protein, Carbohydrates, Minerals and Vitamins, and Consuming Lipids is a Vital Component of Everyday Nutrition for Individuals

Sahraee et al. (2020) studied the creation of emulsion films utilizing bovine gelatin, nano-chitin and nano ZnO for the purpose of packaging cake. Bakery products (bread, cakes, biscuits and donuts) have sources of protein, carbohydrates, minerals and vitamins, and consuming lipids is a vital component of everyday nutrition for individuals. Due to their low price, general acceptance and nutritious properties, cakes are among the most popular among them. But there is a problem with cake storage and that is the growth of mold and oxidation of lipids. Using controlled active packaging, along with the incorporation of antioxidant and antimicrobial substances, can be beneficial in extending the freshness of bakery goods. Due to the characteristics of environmental compatibility in biodegradable packaging (organic waste materials), affordability and ease of transportability of the key ingredient portability of the intelligence components have been considered for the preservation of packaging materials. But their low resistance against water molecules is a big problem for this type of packaging material. Gelatin is among the top priorities sources films that can naturally decompose due to its film-forming properties, low price and ease of access. Through the use of nanomaterials (filling components), softeners, the utilization of cross-linking agents, and different film production processes, the inhibitory properties of biodegradable films can be improved or modified. On the one hand, it is possible to improve the physical properties of films through nanoparticles compatible with carbohydrate and protein polymers such as nanochitin, and on the other hand, antimicrobial properties are added to biopolymers. The ability to be safe for animals and humans, antimicrobial properties, high stability against process temperature and diffusion from packaging to food are among the most prominent properties of metal oxide nanoparticles (zinc, titanium, magnesium and calcium oxide). The use of biodegradable polymers can also reduce the use of preservatives in product formulations due to their high impermeable functions of N-chitin and N-ZnO were incorporated into gelatin nanocomposite layers consisting of sponge packaging cake and comparing durability and held up compared to cakes packed in plastic wrap from 0 to 28 days. The findings indicated that gelatin films alone do not effectively block high humidity water steam in food packaging. While comparing gelatin films that include N-chitin and N-ZnO to a gelatin film without additives showed that gelatin nanocomposites of cakes have better storage ability compared to pure gelatin films. Also, water vapor permeability was significantly reduced by adding zinc oxide nanoparticles and chitosan to the gelatin layer. The enhancement of gelatin layers' ability to block water steam is achieved by using N-chitin and N-ZnO to mobilize the filling properties and create a meandering route for water molecules. There was no significant in difference in the peroxide levels of both fresh and storage cakes that were wrapped in pure gelatin films on days 7 and 14. The good barrier property of gelatin layers may explain against cake packed in pure gelatin can resist ultraviolet rays, oxygen and moisture loss up to water activity level of 0.2–0.4, which itself reduces the rate of lipid oxidation to a minimum. While, due to the high humidity of the cakes, sufficient aw for the oxidation process, the production of the lipoxygenase enzyme, the growth of fungi, the low inhibitory properties of polyethylene protection from ultraviolet rays, and gas oxygen can produce hydroperoxide. The amount of peroxide in the cakes that were GE and GN sealed with remained relatively stable over 14 days of storage, as the barrier properties of both films were equally effective in protecting against water steam, O_2 and ultraviolet rays. The amount of peroxide in cakes stored in polyethylene films was found to be higher compared to other samples. Cakes stored in pure gelatin layers were left to dry for 3 days and experienced a decrease in moisture content from 7 to 14 days, resulting without fungous increase was observed on days 7, 14, 21 and 28. The reason for the lack of growth of fungi was caused by the low water activity levels in water cakes. Cakes packed in polyethylene had more fungal growth after 7, 14, 21 and 28 days than compared to those stored in GE and GNE. GNE covers contain N-chitin and N-ZnO, which help retain moisture in cakes and also have anti-fungoid characters that ability to enhance durability by inhibiting microbial development. The two-layer packaging can satisfy the water steam barricade character of gelatin emulsion wraps and strengthen the better anti-fungoid and physical characteristics of gelatin nanocomposite covers (21).

8.8 CONCLUSION

Due to the fact that nanotechnology has the potential to improve foods, it has gradually gained importance in the food industry. These changes exist from production to food packaging. Reduction of fat content, improvement of taste and texture of food and the encapsulation of nutrients (vitamins) are among the positive changes of using nanotechnology during the life of food storage.

REFERENCES

1. Robson A (2010) *Nanotechnologies and food: Volume II: Evidence*. Authority of the House of Lords, London, p. 1–372.
2. Bayda S, Adeel M, Tuccinardi T, Cordani M, Rizzolio F (2020) The history of nanoscience and nanotechnology: From chemical–physical applications to nanomedicine. *Molecules*, 25(1):112. 1–15 https://doi.org/10.3390/molecules25010112
3. Thiruvengadam M, Rajakumar G, Chung IM (2018) Nanotechnology: Current uses and future applications in the food industry. *3 Biotech* 8:74. 1–13 https://doi.org/10.1007/s13205-018-1104-7
4. Bajpai VK, Kamle M, Shukla S, Mahato DK, Chandra P, Hwang SK, Kumar P, HuhYS, Han YK (2018) Prospects of using nanotechnology for food preservation, safety, and security. *Journal of Food and Drug Analysis* 26:41. 201–1214 https://doi.org/10.1016/j.jfda.2018.06.011
5. Hutapea S, AL-Shawi SG, Chen TC, You X, Bokov D, Abdelbasset WK, Suksatan W (2022) Study on food preservation materials based on nano-particle reagents. *Food Science and Technology (Campinas)* 42. 1–12 https://doi.org/10.1590/fst.39721
6. Bhattacharya S (ed.) (2022) *Snack foods processing and technology*. Nikki Levy, Mysore.
7. Biswas R, Alam M, Sarkar A, Haque MdI, Hasan MdM, Hoque M (2022) Application of nanotechnology in food: Processing, preservation, packaging and safety assessment. *Heliyon* 8:11. 1–16 https://doi.org/10.1016/j.heliyon.2022.e11795
8. Robson A (2010) *Nanotechnologies and food: Volume I: Evidence*. Authority of the House of Lords, London, p. 1–112.
9. Pavlyuk R, Pogarskaya V, Yurieva O, Pogarskiy A, Maksymova N (2021) Devising nanotechnology for vegetable cryofozen enrichers with biocomponents and the natural

protein healthy snacks containing them. *Eastern-European Journal of Enterprise Technologies* 2:11(110). 15–22 https://doi.org/10.15587/1729-4061.2021.230165
10. Adebisi GD, Oyetayo VO (2019) Microbial distribution and antimicrobial effects of AgZnO and AgTiO nanoparticles (NPs) against microorganisms isolated from floury snacks sold in FUTA. *Journal of Microbiology and Experimentation* 7(2):67–70 https://doi.org/10.15406/jmen.2019.07.00244
11. Qiao G, Xiao Z, Ding W, Rok A (2019) Effect of chitosan/nano-titanium dioxide/thymol and tween Films on ready-to-eat cantaloupe fruit quality. *Coatings* 9(12):828. 1–10 https://doi.org/10.3390/coatings9120828
12. Ortiz-Duarte G, Perez-Cabrera LE, Artes-Hernandez F, Martínez-Hernandez GB (2019) Ag-chitosan nanocomposites in edible coatings affect the quality of fresh-cut melon. *Postharvest Biology and Technology* 174:174–184 https://doi.org/10.1016/j. postharvbio. 2018.09.021
13. Pavlyuk R, Pogarska V, Kotuyk T, Pogarskiy A, Balabai K (2020) Development of nanotechnology for processing chickpeas into protein plant supplements and their use to obtain a new generation of confectionery. *Eastern-European Journal of Enterprise Technologies* 6(11):108. 27–36 https://https://doi.org/10.15587/1729-4061.2020.217928
14. Da Trindade LR, Santos Baiao DD, Teixeira da Silva DV, Couto Almeida C, Petzold Pauli F, Francisco Ferreira V, Conte-Junior CA, Flosi Paschoalin VM (2023) Microencapsulated and ready to eat beetroot soup: A stable and attractive formulation enriched in nitrate, betalains and minerals foods. 12(7):1497. 2–26 https://doi.org/10.3390/foods12071497
15. Ahmadi Z, Mirdehghan SH, Hakmabadi H, Shamshiri MH (2014) The use of nano packaging and food coating in improving the storage life and quality of fresh pistachio seeds. *Journal of Horticultural Science* 27(4): 367–374. Agricultural Sciences and Industries, Mashhad.
16. Mekawi EM, Sharoba AM, Ramadan MF (2019) Reduction of acrylamide formation in potato chips during deep--frying in sunflower oil using pomegranate peel nanoparticles extract-. *Journal of Food Measurement and Characterization* 13(4):3298–3306 https://doi.org/10.1007/s11694-019-00252-y
17. Sheir MA, Ahmed RA, Farroh KY (2022) Evaluating the technological and biological effects of some fried vegetables and fruits chips treated with chitosan–pectin nanoparticles. *Egyptian Journal of Food Science* 50(1):33–45 https://doi.org/10.21608/ejfs.2022.111665.1117
18. Espada-Bernabe E, Moreno-Martin G, Gomez-Gomez B, Madrid Y (2024) Assesing the behaviour of particulate/nanoparticulate form of E171 (TiO2) food additive in colored chocolate candies before and after in vitro oral ingestion by spICP-MS, TEM and cellular in vitro models. *Food Chemistry* 1–8 https://doi.org/10.1016/j.foodchem.2023.137201
19. Muhammad DRA, Saputro AD, Rottiers H, Van de Walle D, Dewettinck K (2018) Physicochemical properties and antioxidant activities of chocolates enriched with engineered cinnamon nanoparticles. *European Food Research and Technology* 244:1185–1202 https://doi.org/10.1007/s00217-018-3035-2
20. Hussein AMS, Mahmoud KF, Kamil MM, Hegazy NA (2022) Effect the addition of Micro- and nano-capsule cumin and clove oils as antioxidants and anti-cancer on rancidity and shelf life in some biscuit products. *Egyptian Journal of Chemistry* 65(1):593–606 https://doi.org/10.21608/ejchem.2021.83661.4105
21. Sahraee S, Milani JM, Ghanbarzadeh B, Hamishehkar H (2020) Development of emulsion films based on bovine gelatin-nano chitin-nano ZnO for cake packaging. *Food Science Nutrition* 8:1303–1312 https://doi.org/10.1002/fsn3.1424

9 Nanotechnology in Beverage Preservation

J.O. Olusanya[a], *O.A. Afolabi*[a], *T.P. Mohan*[a], *and K. Kanny*[a]
[a]Composite Research Group (CRG) Department of Mechanical Engineering Durban University of Technology, Durban, South Africa

9.1 INTRODUCTION

The food and beverage industry places a high priority on beverage preservation to preserve the taste, safety, and shelf life of its various beverages. Preservation techniques are used to stop microorganisms from growing, stop chemical reactions, and reduce the influence of outside elements that can cause deterioration or spoiling (Tianli, Jiangbo et al. 2014). The concerns with oxidation and microbial contamination, nanocomposite materials that have nanoparticles embedded in packaging structures, are observed to have improved barrier properties against oxygen, moisture, and other gases (de Azeredo, Mattoso et al. 2011). Additionally, real-time beverage quality monitoring is now possible, thanks to the development of nanosensors, which offers important insights into freshness and safety parameters (Li and Gu 2023). By avoiding unfavorable interactions between the beverage and its packaging, the use of smart nanocoating on beverage containers helps to preserve product integrity (Motelica, Ficai et al. 2020). While there are many exciting opportunities for beverage preservation with nanotechnology, there are also many obstacles to overcome, including safety concerns, regulatory compliance, and consumer acceptability (Shafiq, Anjum et al. 2020). By introducing nanotechnology into the field of nutrition, new foods with improved solubility, thermal stability, and oral bioavailability have been designed and developed (Ayala-Fuentes and Chavez-Santoscoy 2021; Pateiro, Gómez et al. 2021). Within the field of food science, nanotechnology is used in a variety of ways. Its potential for enhancing food quality and safety is considerable. Nanotechnology plays a critical role in the field of food science, extending food's shelf life, enhancing food storage, monitoring and identifying pollutants, and incorporating antimicrobials or nutritional supplements into food (Sahoo, Vishwakarma et al. 2021).

Numerous studies verify that nanomaterials such as these can effectively increase food safety by improving food packaging effectiveness, shelf life, and nutritional value as additives without altering the flavor or physical properties of food products (Alfadul and Elneshwy 2010; Tianli, Jiangbo et al. 2014; Chaturvedi and Dave 2020; Pateiro, Gómez et al. 2021; Sridhar, Ponnuchamy et al. 2021). Nanomaterials have the potential to revolutionize the food industry with novel products and production methods, but they face a major obstacle when it comes to utilizing economical processing techniques to produce safe and palatable nano-delivery systems and efficient formulations for human consumption (Sampathkumar, Tan et al. 2020; Ashique, Kumar et al. 2023).

Nanoscale components can be found in a wide range of foods and drinks, and processing (dairy, for example) entails utilising naturally occurring nanoparticles (Shukla 2012; ur Rahman, Sahar et al. 2020). It is only recently that innovative applications have been researched for improved features and effective food and beverage delivery systems (Majid, Thakur et al. 2018; Köbnick, Velu et al. 2020; Muhamad, Nugroho et al. 2023). While food-grade and beverages nanotechnology products will probably be categorized as novel products and will need to undergo testing and approval, there are worries that accidentally released and ingested nanoparticles with unknown toxicity could occur, especially regarding food and beverage preservative materials (Powell and Colin 2008; Ranadheera, Prasanna et al. 2017). The acceptance of nanotechnology-based products by consumers will ultimately determine their success, so these concerns must be addressed. The food business is likely to be impacted by nanotechnology, both directly and indirectly, considering the recent surge in the broad availability of nanoproducts.

9.2 BEVERAGE PRESERVATION

Beverage preservation is a crucial process that guarantees that beverages stay tasty, safe, and pleasurable for customers (Aadil, Madni et al. 2019). To increase the shelf life of beverages and preserve their quality, a variety of techniques and technologies are used, such as (1) heat treatments, which are used to get rid of or lessen microorganisms in beverages, are pasteurization and sterilization. Products like milk and juices benefit from these processes by having their shelf lives extended and rotting prevented (Alcántara-Zavala, de Dios Figueroa-Cárdenas et al. 2019). (2) Cold storage, which through freezing and refrigeration, perishable beverages like dairy products, fresh juices, and some kinds of soft drinks are frequently kept in cold storage (Slavov, Denev et al. 2019). This allows delaying enzymatic

DOI: 10.1201/9781032677422-9

reactions and microbial growth. (3) Chemical preservatives such as sorbates and benzoates, when added to beverages, prevent molds, yeast, and bacteria growth. These additives are frequently found in fruit juices, soft drinks, and other nonalcoholic drinks (Vasilaki, Hatzikamari et al. 2019). (4) Filtration and microfiltration, by eliminating particles and microorganisms, improves clarity and lowers the chance of beverages getting spoilt (Reis, Madrona et al. 2019; Vieira, Balthazar et al. 2020). (5) Drinks can be preserved naturally without the use of artificial preservatives by using plant extracts or essential oils having antibacterial properties (Pandey and Negi 2018; Nemo and Bacha 2021). However, depending on the particulars of the beverage and the intended shelf life, a combination of these techniques can be used for effective beverage preservation. Technology developments and a growing focus on sustainable and natural preservation techniques are continuing to influence how the food and beverage industry handles beverage preservation.

9.3 ANTIMICROBIAL NANOPARTICLES

The term "antimicrobial nanoparticles" describes incredibly small particles with antimicrobial properties, frequently in the nanometer range (Wang, Hu et al. 2017). Due to their potential use in food packaging, medical devices, textiles, and beverage preservation, these nanoparticles have drawn attention (Nakazato, Kobayashi et al. 2017; Yang, Paulose et al. 2021; Garcia-Oliveira, Pereira et al. 2022). Utilizing antimicrobial nanoparticles as preservatives for beverages is a novel way to increase beverage shelf life without sacrificing quality or safety. The following are some of the ways antimicrobial nanoparticles could be useful in beverages preservation.

9.3.1 Application of Silver Nanoparticles in Beverage Packaging

The special antimicrobial qualities of silver nanoparticles (AgNPs) have drawn interest in the field of food and beverage packaging (Yang, Paulose et al. 2021; Amiri, Hadioui et al. 2022). By preventing bacteria and other microorganisms from growing, packaging materials containing silver nanoparticles can help preserve the beverages inside. Coatings for the interior surfaces of drink containers can be incorporated with silver nanoparticles, which produce a barrier that keeps bacteria from growing and stops contamination, prolonging the beverages' shelf lives (Yang, Paulose et al. 2021). It also has antimicrobial qualities that help stop pathogens and spoilage microorganisms from growing in beverages (Avila-Quezada and Espino-Solis 2019). Using silver nanoparticles in polymeric films for beverages packaging over time allow the release silver ions in the packaging films, which helps create an antimicrobial environment that helps prevent microbial contamination of the beverage while it is being stored (Amiri, Hadioui et al. 2022). Silver nanoparticles have the potential to offer UV protection by halting the deterioration of light-sensitive substances in beverages, like vitamins and flavorings, when added to packaging materials (Patil, Nille et al. 2021).

9.3.2 Inhibition of Spoilage Microorganisms

Inhibiting spoilage microorganisms can be achieved by adding antimicrobial nanoparticles to beverages, which can prolong product shelf life and preserve quality (Jamali, Assadpour et al. 2021). Antimicrobial nanoparticles help preserve beverages in several ways by preventing the growth of spoilage microorganisms. Unintentional fermentation by spoilage microorganisms can occur in some beverages, changing the flavor and level of carbonation (Malakar, Paul et al. 2020). Stopping this kind of fermentation, antimicrobial nanoparticles can maintain the stability of the beverage's composition.

9.3.3 Enhanced Safety and Shelf Life

Improved safety and longer shelf life are possible when antimicrobial nanoparticles are added to beverage preservation. Pathogenic microorganisms that pose health risks can be specifically targeted by nanoparticles' antimicrobial properties. The possibility of contamination with dangerous bacteria or viruses can be decreased by adding these nanoparticles to beverage formulations or packaging (Alfei, Marengo et al. 2020). Antimicrobial nanoparticles can aid in preventing temperature fluctuations from causing beverages to spoil. This is especially important for goods that might be stored and transported in environments with different temperatures (Sridhar, Ponnuchamy et al. 2021). The controlled release of antimicrobial nanoparticles prolonging and extending the shelf life of beverages, this controlled release protects against microbial growth over time (Garcia-Oliveira, Pereira et al. 2022).

9.4 NANOENCAPSULATION TECHNIQUES

Nanotechnology is a promising technology that is helping to solve issues related to food safety, food processing, and food packaging in the food industry as well as other sectors. Figure 9.1 shows the use of nanoparticles in the food industry across a range of sectors (Bajpai, Kamle et al. 2018). Enclosing active ingredients, like nutrients and flavors that are sensitive, inside nanoscale carriers is a novel technology known as nanoencapsulation (Mohebbi 2020). This method is especially useful for stabilizing, enabling controlled release, and shielding delicate compounds from the environment in the food and beverage sector. Using nanoencapsulation techniques, active ingredients, like antimicrobial nanoparticles, are contained within nanoscale carriers to enhance their overall efficacy, stability, and controlled release (Pateiro, Gómez et al. 2021). These methods can improve the effectiveness of antimicrobial nanoparticles in the context of beverage preservation (Prakash, Kujur et al. 2018).

9.4.1 Protection of Sensitive Flavors and Nutrients

There are commonly employed nanoencapsulation methods for safeguarding delicate flavors and nutrients in drinks including the following: (1) liposome encapsulation is

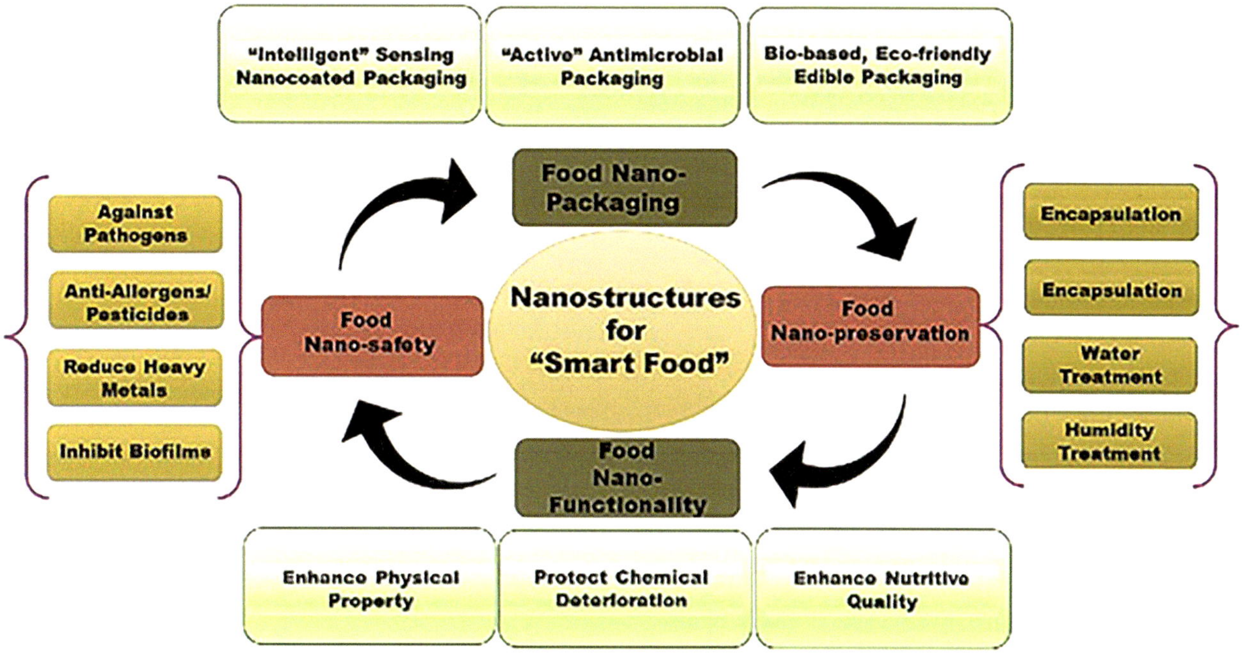

FIGURE 9.1 Schematic illustration of the use of nanoparticles in the food industry across a range of sectors. (Adopted from Bajpai, Kamle et al. 2018).

spheres with a lipid bilayer structure that aids in preventing oxidation, moisture, and degradation of delicate flavors and nutrients (Ishwarya and Anandharamakrishnan 2023); (2) nano-lipid carriers are lipid-based nanoparticles (NLCs) that create a barrier that keeps delicate flavors and nutrients safe from outside influences that might lower their quality (Abbasi, Sabahi et al. 2023); (3) polymeric nanoparticles provide both a controlled dispersion mechanism and environmental protection (Sridhar, Ponnuchamy et al. 2021); (4) nanoencapsulation in biopolymeric matrices (such as starch or gum arabic) provides protection and stability while allowing for the release of the encapsulated ingredients under specific conditions (Taheri and Jafari 2019); and (5) electro-spraying and spray drying is utilized to produce nanoparticles and microparticles with improved encapsulation effectiveness (Ghorbani and Aleman 2021).

9.4.2 Successful Application of Nanoencapsulation

The beverage industry has effectively used nanoencapsulation to improve the stability, bioavailability, and controlled release of different compounds (Paredes, Asensio et al. 2016). The use of antioxidants like vitamins C and E have been shielded from oxidation in beverages using nanoencapsulation. The antioxidant qualities of functional beverages are preserved, and their shelf life is increased by the encapsulation's ability to ward against oxidation and degradation (Hosseini, Tajiani et al. 2019). Another use of nanoencapsulation is to shield the fatty acids during beverages formulations by maintaining their stability and stopping the emergence of off-flavors (Muhamad, Zaidel et al. 2020). The use nanoencapsulation has been employed to protect catechins and other bioactive components present in tea (Dang, Gupta et al. 2015). This keeps tea's health-promoting qualities in ready-to-drink beverages intact. It has been established that live microorganisms, or probiotics, are beneficial to health when ingested (Soares, Almada et al. 2023). To maintain the viability of probiotics in beverages, nanoencapsulation is used to shield probiotics from environmental factors like oxygen and acidity (Reque and Brandelli 2021). Also, to keep the vitamins from degrading during processing and storage, the encapsulation improves the vitamins' solubility and stability in beverages (Raikos, Pirie et al. 2021).

9.5 NANOCOMPOSITE PACKAGING IN BEVERAGES PRESERVATION

Nanocomposite packaging has been investigated as a potential solution to enhance the safety, quality, and shelf life of beverages in the context of preservation (Sridhar, Ponnuchamy et al. 2021). Nanocomposites reduce the permeability of gases (such carbon dioxide and oxygen) and moisture, which enhances the barrier qualities of packing materials (Trinh, Chang et al. 2023). Since nanocomposite packaging has better barrier qualities, it helps prolong the shelf life of beverages by reducing the entry of outside elements that can cause deterioration, contamination, or spoiling. Addition of nanoparticles such as nanoparticles or nanoclay can enhance the mechanical strength and durability of packaging materials (Shekarchizadeh and Nazeri 2020). During handling and transit, this is especially helpful in avoiding physical damage. Nanocomposite materials can be used to filter out visible light and dangerous UV rays.

This is especially important for light-sensitive beverages because light exposure can cause some substances, like flavors and vitamins, to deteriorate (Roy, Ramakrishnan et al. 2023). It is possible to create active packaging systems out of nanocomposites that can communicate with the beverage environment. An extra degree of defense, for instance, can be provided by pH-sensitive nanocomposites, which can release antimicrobial agents in response to pH variations (Primožič, Knez et al. 2021).

9.6 NANOSENSORS FOR REAL-TIME BEVERAGE MONITORING

Nanosensors are becoming a practical instrument to guarantee food products' safety and quality. They are very good at detecting contaminants, pathogens, and early warning signs in food samples because of their great sensitivity, selectivity, and reaction time. The nanosensor software for detecting chemical pollutants such as pesticides, heavy metals, and mycotoxins is well secured. Nanomaterials enable ultrasensitive detection of harmful pollutants, even at trace levels, ensuring the safety and compliance of food items (Kumar, Guleria et al. 2017; Bauer-Estrada, Sandoval-Cuellar et al. 2023). A major and continuous public health concern is food safety. Poor food handling practices, a weakened food supply, and inadequate cleanliness all contribute to food safety concerns and increase the financial cost of foodborne illness.

FBDs are commonly linked to consumer disease, which entails high medical costs as well as a loss of revenue and output (Sarkar, Panigrahi et al. 2017).

9.6.1 Development and Application of Nanosensors

A promising technological advancement for enhancing food safety and quality is nanosensors. These sensors locate and assess the biological or chemical components of food samples using nanomaterials and biological identification components.

They provide various benefits, including rapid response, high sensitivity, and compactness. Numerous tasks, such as food manufacturing, processing, packaging, labelling, shipping, tracking, and maintaining high-quality meals, have involved the use of nanomaterials (Bauer-Estrada, Sandoval-Cuellar et al. 2023). In order to increase the capacity of green optical sensors to identify minute levels of pollution from diverse sources, nanosensors were also used.

Additionally, the use of nanomaterials has advanced the use of nanosensors in food safety, allowing for the quick and accurate identification of foodborne illnesses. Additionally, they have lightning-fast reaction times. Some of the applications of nanosensors include the following:

(i) s+They are used for quick and immediate interventions against food contamination and food waste.
(ii) They are used to find pollutants or pathogens in food samples.
(iii) They are used for critical identification of potential hazards.
(iv) They are used to ensure that food safety companies comply with regulatory standards.
(v) They are used to reduce false positive and negative output.
(vi) They ensure that the results of food processes are accurate and reliable.
(vii) They offer simple on-site monitoring and testing along the whole food supply chain.
(viii) They are used for real-time analysis in food chain supply.
(ix) They are used for spoilage and harmful substance detection.
(x) They are employed in the production, labelling, transportation, and inspection processes to guarantee the food's quality and safety.
(xi) They are used to monitor time-temperature and oxygen indicators for food freshness.

9.6.2 Monitoring Freshness and Safety Parameters

Food safety and hygienic practices assurance is a vital component of the food industry, ensuring that the food we eat is wholesome, safe, and free of contaminants.

Traditional approaches to evaluating food quality and safety may involve labor-intensive and time-consuming procedures that postpone the identification of potential threats and endanger the health of consumers. Scientists and scholars are searching for a potential solution in nanotechnology, specifically in the form of nanobiosensors, to tackle these problems.

In Figure 9.2, the concerns and possible solutions for nanosensors that are intended for food safety are explored (Yang and Duncan 2021).

To identify and assess particular biological or chemical compounds in food samples, nanosensors combine biological sensors and nanomaterials. These sensors are an attractive tool for improving food quality and safety because they have several advantages over conventional methods (Bashir, Bhat et al. 2022). Pathogenic bacteria, food pollutants, adulterants, vitamins, colors, fertilizers, pesticides, taste, and smell can all be detected using nanosensors. Food freshness can be determined using time-temperature and oxygen indicators. Using invisible nanobarcodes, product validity and brand protection may be verified (Ghosh, Raj et al. 2022). Food contamination is harmful to people's health as well. Toxins and adulterants can result in death, paralysis, vomiting, and diarrhea. Food safety, as a key and fundamental necessity of people, is a worldwide issue (Fuertes, Soto et al. 2016).

9.6.3 Real-Time Quality Control in Beverage Production

The use of innovative nanocomposite materials in food packaging has significantly altered the nutritional content

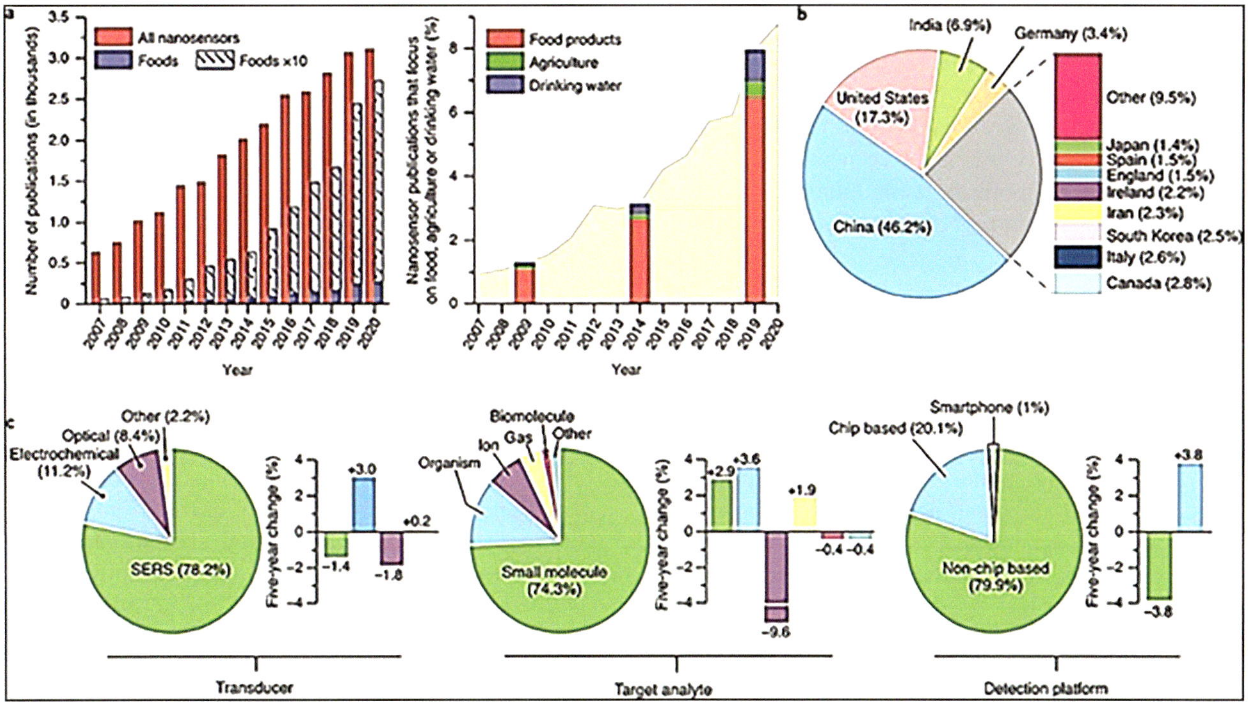

FIGURE 9.2 Graphs showing concerns and possible solutions for nanosensors that are intended for being used with food. (Adopted from Yang T. and Dunca T.V.) (Yang and Duncan 2021).

TABLE 9.1
Benefit of Nanotechnology in Beverages Packaging

Nanotechnology Applications	Their Benefits in Beverages and Food
Smart packaging	Enhanced flexibility, temperature stability, and moisture resistance in addition to increased food and drink shelf life.
Nanobiosensors	Beneficial in identifying food product spoilage and foreign particle contamination.
Nanoencapsulation	Targeted nutrition delivery and improved nutrient bioavailability.
Nano-biotracers	Good for tracking food particles over long distances when being shipped.
Nanoemulsions	Preserves the look and sensation for increased consumer appeal.
Nanocomposite	Protect the food component's structure and aid in maintaining the product's steady state, both of which extend the product's shell life.

of food. Because of their excellent antimicrobial qualities, mechanical strength, thermal stability, and barrier qualities that stop migration in food packaging systems, nanocomposite materials can help prolong the shelf life of perishable foods, cut down on food waste, and lessen the need for chemical preservatives.

Nanotechnology applications and their benefit in beverages and food are highlighted in Table 9.1. This helps to understand the possible effect of beverages preservation through nanotechnology applications.

To ensure food goods are safe and of high quality while being stored and distributed, it is essential to conduct a precise and forecast assessment of their preservation time (Ghosh, Raj et al. 2022). Therefore, for accurate shelf life monitoring, it is essential to comprehend the dynamics and linkages of these intricate matrices employing nanosensors. Nanosensors are extremely advanced detectors that monitor and control food quality, safety, and shelf life using the concepts of biotechnology and nanotechnology (Bashir, Bhat et al. 2022). The sensors are incorporated into devices or materials for packing, and they are used to continuously monitor vital parameters like temperature, humidity, concentration of gas, microbial activity, and accuracy.

In order to ensure the safety and freshness of packaged food goods, quick action and manufacturing control are made possible by the availability of nanobiosensors. Sensors use a variety of monitoring approaches based on optical, electrochemical, or nanomechanical principles to identify and assess certain materials of interest (Ghosh, Raj et al. 2022). Moreover, nanosensors help to improve intelligent package design by incorporating sensors into packaging or devices to assure food safety and quality. This enables better monitoring and control of material conditions, such as recognizing higher temperatures, decreased ventilation, or bacterial development and taking necessary measures to maintain ideal circumstances (Fuertes, Soto et al. 2016; Kumar, Guleria et al. 2017; Ghosh, Raj et al. 2022).

9.7 SMART NANOCOATINGS

In its many fields of application, nanocoating has soared to the top, both literally and figuratively as the best coating. One term used to describe nanocoating technology is "smart silis tech." It speaks of how the economy uses raw materials to improve performance while minimizing its negative effects on the environment (Fuertes, Soto et al. 2016).

9.7.1 Application on Beverage Containers

Some of the applications of smart nanocoating on beverage containers include the following:

(i) s+Scratch resistance: Excellent toughness makes nanocoatings perfect for usage on delicate surfaces like car paint, prescription eyeglasses, and scientific and medical equipment.
(ii) Self-cleansing: They aid in keeping surfaces clean, saving time and money on upkeep that might otherwise be required. For instance, nanocoating provides an excellent finish to the paint. "Stain-resistant" and "easy to clean."
(iii) Water repellent: Good visibility and safety are provided by nanocoating for translucent surfaces. Shields materials like metal, fabrics, wood, plastic, painted surfaces, and wood from water damage.
(iv) UV protection: Nanoparticles effectively prevent UV light that is harmful.
(v) Corrosion resistance: For a considerable amount of time, surfaces coated with nanoparticles show little to no visible deterioration.

9.7.2 Prevention of Interactions between Beverage and Packaging

Typical responsibilities of packaging include confining food, shielding it from deterioration processes (mostly brought on by outside factors including oxygen, light, and moisture) and giving consumers ingredient and nutritional information.

The selection of materials for packaging is critical for maintaining your food safe. Food packaging materials include plastics, paper, glass, and metals. Certified food-safe packaging bears an embossed "wine glass and fork" insignia. Interactions between packaging and food cause a loss of quality and the migration of harmful materials into the food. In order to improve beverage conservation—such as wine, juice, or beer—customer acceptability, and food safety, recent advancements in beverage packaging have focused on changing the structure of the material used for packaging and creating new active and/or intelligent systems that can interact with the product or its environment (Ramos, Valdés et al. 2015; Fuertes, Soto et al. 2016; Poeta, Liboà et al. 2023).

9.7.3 Maintenance of Product Integrity

Due to plastic's low cost and practical advantages, its use in beverage packaging has continued to grow (such as thermos-sealability, micro-wavability, optical characteristics, and infinite sizes and shapes) over conventional materials such as glass and tinplate. By utilising co-extrusion or lamination, plastic materials can be created as a composite of many plastics or as a single sheet. Combining materials often reduces the total amount of packing material required by allowing for the added benefit of each individual material's properties.

Plastics' variable permeability to light, gases, vapors, and low molecular weight chemicals is its primary drawback (Ramos, Valdés et al. 2015; Geueke, Groh et al. 2018). When using mechanically recycled plastic with food, it's important to be mindful of potential contaminants such as polymer and additive degradation, consumer misuse, waste disposal contamination, and environmental contaminants. Furthermore, similar-grade materials that are not fit for human consumption could find their way into the recycling process, raising the risk of contamination (Geueke, Groh et al. 2018). The safety of processed and packaged food products against harmful microorganisms like *Salmonella*, *Toxoplasma gondii*, *Listeria monocytogenes*, *Escherichia coli O157:H7*, *Campylobacter jejuni*, and *Staphylococcus aureus* is maintained by a number of factors (Sarkar, Panigrahi et al. 2017, Bashir, Bhat et al. 2022).

The United State Food and Drug Administration (USFDA), proposed some food labeling regulations which include the following:

- The nutrient panel includes the quantity being served size and nutritional information.
- The details panel offers the component list, directions for use, and more.
- The main display panel includes the manufacturer's name, item identification statement, and amount.
- The statements and warnings section states allergies and nutrient claims.

Other labeling ideas to ensure food maintenance and product integrity are as follows:

- Transparent packaging.
- Temperature sensitivity label.
- Color changing label.
- Communication sticker labels.

9.8 NANOFILTRATION AND ULTRAFILTRATION

Water can be filtered to remove particles by passing it through a substance that is porous.

This porous material could be a membrane wall made of different materials, or it could be something organic like sand, gravel, and clay. The process by which large particles settle before being filtered is called sedimentation.

The Conventional Water Treatment Coagulation and Filtration fact sheet provides general information on sedimentation and filtration (Martínez-Ballesta, Gil-Izquierdo et al. 2018).

Nanofiltration removes the same contaminants as ultrafiltration, as well as additional viruses, organic materials, and ions (such as calcium) that contribute to hard water. Other minerals, such as sodium chloride, can flow through nano-filters and remain in the water. Ultrafiltration eliminates bacteria, protozoa, and certain viruses from water. Nanofiltration eliminates these bacteria, in addition to most natural organic matter alongside certain minerals, particularly divalent ions, which generate hard water. Nanofiltration, however, does not remove the dissolved chemicals.

The size of the substances that can be eliminated during filtration is determined by the size of the pores in the filter. Particle filtration refers to traditional medium filtration, whereas the other forms are membrane filtrations (Nath, Dave et al. 2018; Van der Bruggen 2018).

9.8.1 Use of Nanoscale Pores in Membranes

Nanoporous membranes offer special qualities that set them apart from the many other porous polymers already on the market and can be beneficially used in the biomedical sciences. For example, because many significant filtrates have a size of less than 10 nm, nanoscale porosity makes biofiltration possible (Bernards and Desai 2010). Its function is to move tiny molecules through the pore by passive diffusion along a concentration gradient. Moreover, the transportation of large molecules and molecular structures demands energy (such proteins, RNA, and maybe ribosomes) through the nuclear membrane.

9.8.2 Removal of Impurities and Microorganisms

Various techniques of removing impurities and microorganisms including filtering, centrifugation, rectification, extraction, adsorption, or ion exchange are typically used to eliminate such contaminants. Water can be filtered naturally in several ways. Boiling water for a minimum of one minute is one way to eliminate most bacteria. Utilizing carbon filters or activated charcoal is an additional technique that can enhance flavor and eliminate contaminants (Azam, Akhtar et al. 2021). In addition, clear plastic water bottles can be left in direct sunlight for several hours to facilitate solar disinfection (SODIS). Sand, gravel, and cloth are examples of natural materials that can be used for effective filtering. However, for the best water purification, it's recommended to combine several techniques.

9.8.3 Improved Clarity and Safety of Beverages

Ways to improve clarity and safety of beverages include the following:

(i) Recognize the needs of the market. The market's desire for food goods fluctuates quickly.
(ii) Apply the forecasting technique.
(iii) Boost your inventory control.
(iv) Update and maintain equipment on a regular basis.
(v) Put a supply chain management system in place.
(vi) Perform a shelf-life test.
(vii) Run radiological and chemical tests.
(viii) Test for contaminants such as heavy metals or pesticides.
(ix) Monitor production.
(x) Use good manufacturing practice (GMP) and hygiene audits.

9.9 NANOCARRIERS FOR CONTROLLED DELIVERY

Currently, chemotherapeutic medications are delivered to precise regions using a number of state-of-the-art nanocarrier-mediated drug delivery techniques. Several noteworthy nanocarriers include polymeric nanoparticles, magnetic nanoparticles, liposomes, polymeric micelles, carbon nanotubes, dendrimers, solid lipid nanoparticles, and quantum dots. When creating nanocarriers, a variety of polymers are used; the active medicinal ingredients are either covalently attached to the polymeric matrix or physically entangled with it (Pudlarz and Szemraj 2018; Edis, Wang et al. 2021). The synthesized polymeric systems may resemble complex branched macromolecules, capsules, or amphiphilic molecules. The development of nanocarriers uses both natural and man-made polymers (Elzoghby, Samy et al. 2012; Lu, Zhang et al. 2021).

9.9.1 Encapsulation and Controlled Release of Preservatives and Bioactive Compounds

The dietary supplement industry's requirement to deliver active substances under precise conditions and maintain their benefits is driving up demand for encapsulation technologies. The food industry can use bioactive compound encapsulation to achieve the following goals:

(i) Maintain functional qualities.
(ii) Enhance the stability of compounds that are poorly soluble in pertinent (mostly aqueous) media.
(iii) Conceal offending flavors.
(iv) Boost the health benefits of food products (i.e., develop functional foods).
(v) Control the release of bioactive substances at a predetermined time and precise target.
(vi) Boost the bioavailability of substances that are bioactive.

Even though encapsulation has intriguing and special qualities that can benefit the food business, its use is still difficult since encapsulation systems must be developed utilizing GRAS (generally recognized as safe) components (Bernards and Desai 2010; Bourbon, Cerqueira et al. 2016; Martínez-Ballesta, Gil-Izquierdo et al. 2018).

9.9.2 Preservation of Beverage Quality throughout Its Life Cycle

Typically, thermal procedures like pasteurization and sterilization have been used to preserve beverages. In recent years, various nonthermal processing processes have been employed due to the detrimental effects of high process temperatures or timeframes on the nutritional value of beverages. Chemicals known as preservatives are included in beverages to prolong their shelf life by preventing microbial development and spoiling. They may be artificial or natural. Although some customers can voice concerns over their use, regulatory agencies make sure that preservatives that have been approved are safe when used in accordance with recommended dosages. These are a few typical preservatives for beverages: sorbates, ascorbic acid (vitamin C), benzoates, citric acid, grape seed extract, sulfites, rosemary, and green tea extract (Martínez-Ballesta, Gil-Izquierdo et al. 2018).

9.10 CHALLENGES AND CONSIDERATIONS

While many aspects of beverage packaging could be greatly improved by nanotechnology, there are several challenges and considerations that need to be taken into account. The following are some significant difficulties and factors to consider when using nanotechnology in beverage packaging:

(i) s+Assessing possible health hazards related to the application of nanomaterials in packaging.
(ii) Ensuring adherence to current laws and creating new regulations especially for packaging with nanomaterials.
(iii) Evaluating the cost-effectiveness of nanotechnology in beverage packaging, while critically taking into account the financial benefits to both producers and customers.
(iv) Overcoming technological and engineering obstacles is necessary to produce nanocomposite materials with consistent quality and properties on an industrial scale.
(v) Dispelling public fears and misconceptions by remembering it is imperative that the safety and advantages of nanotechnology in packaging be openly and transparently communicated.
(vi) Evaluating studies on compatibility are necessary for possible interactions that could impact the beverage's taste, odor, or chemical stability.
(vii) Ensuring the efficacy of these materials by imperatively comprehending their behavior over an extended duration, particularly when stored in different conditions.

Collaboration between researchers, industry stakeholders, regulatory bodies, and consumer advocacy groups is necessary to address these issues. Realizing the potential advantages of nanotechnology in beverage packaging while reducing related risks requires ongoing research, openness, and responsible innovation.

9.11 FUTURE PROSPECTS

One innovative area that shows promise is the use of nanotechnology in beverage packaging, which has the potential to improve product standards and safety. The application of nanotechnology in beverage packaging offers promising futures because of ongoing research and development aimed at finding solutions and creating new opportunities.

The development of intelligent and active packaging systems will heavily rely on nanotechnology and improve the bioavailability and controlled release of functional ingredients by developing more effective delivery methods (Martínez-Ballesta, Gil-Izquierdo et al. 2018). Real-time beverage quality monitoring could be facilitated by the use of nanosensors integrated into packaging materials (Chelliah, Wei et al. 2021). Antimicrobial nanocoating is expected to become more widely used. By doing this, the surface of the container may be protected from microbial growth, lowering the possibility of contamination and prolonging the shelf life of drinks (Mohammad and Ahmad 2024). Collaboration between researchers, industry participants, and regulatory agencies will be essential for the responsible development and implementation of nanocomposite packaging solutions in the beverage industry as nanotechnology advances. Future developments in the use of nanotechnology in beverage packaging are probably going to be fueled by the convergence of innovation, sustainability, and enhanced performance.

9.12 CONCLUSION

The use of nanotechnology in beverage packaging is a groundbreaking and creative strategy that has the potential to solve several issues and improve the general sustainability, safety, and quality of packaged beverages. Enhancing barrier qualities and regulating the release of functional ingredients are just two advantages of using nanomaterials and nanoencapsulation techniques.

Beverage packaging using nanocomposite materials offers better defenses against moisture, air, and other environmental elements, prolonging the shelf life of beverages. Improved beverage quality can be achieved through the controlled release of functional ingredients, nutrients, and flavors made possible by nanotechnology. Furthermore, it makes it easier to create intelligent and dynamic packaging systems that adapt to environmental changes by giving real-time data and monitoring features. Nanotechnology is becoming more scalable and economical due to advancements in manufacturing processes, which may lead to its widespread adoption. The increasing demand for sustainable and eco-friendly packaging solutions has prompted continued research into the creation of biodegradable nanocomposites.

One major concern is the safety of nanomaterials in applications when they come across food, such as beverages. Strict testing and risk analyses are required to guarantee consumer safety. To evaluate the possible health risks connected to ingesting these particles, research is required. To properly utilize nanotechnology as it develops, cooperation between scientists, business leaders, government agencies, and end users will be essential. The application of nanotechnology offers enhanced performance, sustainability, and customer satisfaction and has the potential to completely transform the beverage packaging sector. Future advancements in the application of nanotechnology in beverage packaging will most likely be driven by the convergence of improved performance, sustainability, and innovation.

REFERENCES

Aadil, R. M., G. M. Madni, U. Roobab, U. ur Rahman and X.-A. Zeng (2019). "Quality control in beverage production: An overview." *Quality control in the beverage industry*, Elsevier: 1–38.

Abbasi, S., A. Sabahi, N. Shahbazi, A. Karimi and H. Barkhordari (2023). "Nano lipid carrier in augmenting the bioavailability of functional bio-compounds." *Biointerface Research in Applied Chemistry* **13**(1): 1–14.

Alcántara-Zavala, A. E., J. de Dios Figueroa-Cárdenas, E. Morales-Sánchez, J. A. Aldrete-Tapia, S. M. Arvizu-Medrano and H. E. Martínez-Flores (2019). "Application of ohmic heating to extend shelf life and retain the physicochemical, microbiological, and sensory properties of pulque." *Food and Bioproducts Processing* **118**: 139–148.

Alfadul, S. and A. Elneshwy (2010). "Use of nanotechnology in food processing, packaging and safety–review." *African Journal of Food, Agriculture, Nutrition and Development* **10**(6).

Alfei, S., B. Marengo and G. Zuccari (2020). "Nanotechnology application in food packaging: A plethora of opportunities versus pending risks assessment and public concerns." *Food Research International* **137**: 109664.

Amiri, N., M. Hadioui, A. Chalifoux and K. J. Wilkinson (2022). "Characterizing silver nanoparticles in beverages and following their release from silver-containing food containers using sector field single particle inductively coupled plasma mass spectrometry (SP-ICP-MS)." *Measurement: Food* **8**: 100061.

Ashique, S., S. Kumar, E. Sirohi, A. Hussain, A. Farid, M. Faiyazuddin, N. Mishra and A. Garg (2023). "A comprehensive update on nanotechnology in functional food developments: Recent updates, challenges, and future perspectives." *Recent Patents on Nanotechnology* **19**: 241–256.

Avila-Quezada, G. D. and G. P. Espino-Solis (2019). "Silver nanoparticles offer effective control of pathogenic bacteria in a wide range of food products." *Pathogenic Bacteria*, chapter 12, Intec Publishers. DOI 10.5772/intechopen.78081. ISBN 978-1-78985-988-1

Ayala-Fuentes, J. C. and R. A. Chavez-Santoscoy (2021). "Nanotechnology as a key to enhance the benefits and improve the bioavailability of flavonoids in the food industry." *Foods* **10**(11): 2701.

Azam, K., S. Akhtar, Y. Y. Gong, M. N. Routledge, A. Ismail, C. A. Oliveira, S. Z. Iqbal and H. Ali (2021). "Evaluation of the impact of activated carbon-based filtration system on the concentration of aflatoxins and selected heavy metals in roasted coffee." *Food Control* **121**: 107583.

Bajpai, V. K., M. Kamle, S. Shukla, D. K. Mahato, P. Chandra, S. K. Hwang, P. Kumar, Y. S. Huh and Y.-K. Han (2018). "Prospects of using nanotechnology for food preservation, safety, and security." *Journal of Food and Drug Analysis* **26**(4): 1201–1214.

Bashir, O., S. A. Bhat, A. Basharat, M. Qamar, S. A. Qamar, M. Bilal and H. M. Iqbal (2022). "Nano-engineered materials for sensing food pollutants: Technological advancements and safety issues." *Chemosphere* **292**: 133320.

Bauer-Estrada, K., C. Sandoval-Cuellar, Y. Rojas-Muñoz and M. X. Quintanilla-Carvajal (2023). "The modulatory effect of encapsulated bioactives and probiotics on gut microbiota: Improving health status through functional food." *Food & Function* **14**(1): 32–55.

Bernards, D. A. and T. A. Desai (2010). "Nanoscale porosity in polymer films: Fabrication and therapeutic applications." *Soft Matter* **6**(8): 1621–1631.

Bourbon, A. I., M. A. Cerqueira and A. A. Vicente (2016). "Encapsulation and controlled release of bioactive compounds in lactoferrin-glycomacropeptide nanohydrogels: Curcumin and caffeine as model compounds." *Journal of Food Engineering* **180**: 110–119.

Chaturvedi, S. and P. N. Dave (2020). "Application of nanotechnology in foods and beverages." *Nanoengineering in the beverage industry*, Elsevier: 137–162.

Chelliah, R., S. Wei, E. B.-M. Daliri, M. Rubab, F. Elahi, S.-J. Yeon, K. H. Jo, P. Yan, S. Liu and D. H. Oh (2021). "Development of nanosensors based intelligent packaging systems: Food quality and medicine." *Nanomaterials* **11**(6): 1515.

Dang, S., S. Gupta, R. Bansal, J. Ali and R. Gabrani (2015). "Nano-encapsulation of a natural polyphenol, green tea catechins: Way to preserve its antioxidative potential." *Free Radicals in Human Health and Disease*, Springer Nature: 397–415.

de Azeredo, H. M. C., L. C. Mattoso and T. H. McHugh (2011). "Nanocomposites in food packaging–a review." *Advances in diverse industrial applications of nanocomposites*, InTech Open: 57–78.

Edis, Z., J. Wang, M. K. Waqas, M. Ijaz and M. Ijaz (2021). "Nanocarriers-mediated drug delivery systems for anticancer agents: An overview and perspectives." *International Journal of Nanomedicine*: 1313–1330.

Elzoghby, A. O., W. M. Samy and N. A. Elgindy (2012). "Protein-based nanocarriers as promising drug and gene delivery systems." *Journal of Controlled Release* **161**(1): 38–49.

Fuertes, G., I. Soto, R. Carrasco, M. Vargas, J. Sabattin and C. Lagos (2016). "Intelligent packaging systems: Sensors and nanosensors to monitor food quality and safety." *Journal of Sensors* **2016**.

Garcia-Oliveira, P., A. Pereira, M. Carpena, A. Carreira-Casais, M. Fraga-Corral, M. Prieto and J. Simal-Gandara (2022). "Application of releasing packaging in beverages." *Releasing systems in active food packaging: Preparation and applications*, Springer: 373–401.

Geueke, B., K. Groh and J. Muncke (2018). "Food packaging in the circular economy: Overview of chemical safety aspects for commonly used materials." *Journal of Cleaner Production* **193**: 491–505.

Ghorbani, M. and R. S. Aleman (2021). "Electro-spinning and electro-spraying as innovative approaches in developing of a suitable food vehicle for polyphenols-based functional

ingredients." *Bioactive compounds-biosynthesis, characterization and applications*, IntechOpen.
Ghosh, T., G. B. Raj and K. K. Dash (2022). "A comprehensive review on nanotechnology based sensors for monitoring quality and shelf life of food products." *Measurement: Food*: 100049.
Hosseini, H., Z. Tajiani and S. M. Jafari (2019). "Improving the shelf-life of food products by nano/micro-encapsulated ingredients." *Food quality and shelf life*, Elsevier: 159–200.
Ishwarya, S. P. and C. Anandharamakrishnan (2023). "Liposomal encapsulation of natural color, flavor, and additives for the food industry." *Liposomal encapsulation in food science and technology*, Elsevier: 145–167.
Jamali, S. N., E. Assadpour, J. Feng and S. M. Jafari (2021). "Natural antimicrobial-loaded nanoemulsions for the control of food spoilage/pathogenic microorganisms." *Advances in Colloid and Interface Science* **295**: 102504.
Köbnick, P., C. Velu and D. McFarlane (2020). "Preparing for industry 4.0: Digital business model innovation in the food and beverage industry." *International Journal of Mechatronics and Manufacturing Systems* **13**(1): 59–89.
Kumar, V., P. Guleria and S. K. Mehta (2017). "Nanosensors for food quality and safety assessment." *Environmental Chemistry Letters* **15**: 165–177.
Li, B. and Y. Gu (2023). "A machine learning method for the quality detection of base liquor and commercial liquor using multidimensional signals from an electronic nose." *Foods* **12**(7): 1508.
Lu, H., S. Zhang, J. Wang and Q. Chen (2021). "A review on polymer and lipid-based nanocarriers and its application to nano-pharmaceutical and food-based systems." *Frontiers in Nutrition* **8**: 783831.
Majid, I., M. Thakur and V. Nanda (2018). "Innovative and safe packaging technologies for food and beverages: Updated review." *Innovations in technologies for fermented food and beverage industries*, Springer: 257–287.
Malakar, S., S. K. Paul and K. J. Pou (2020). "Biotechnological interventions in beverage production." *Biotechnological progress and beverage consumption*, Elsevier: 1–37.
Martínez-Ballesta, M., Á. Gil-Izquierdo, C. García-Viguera and R. Domínguez-Perles (2018). "Nanoparticles and controlled delivery for bioactive compounds: Outlining challenges for new "smart-foods" for health." *Foods* **7**(5): 72.
Mohammad, Z. H. and F. Ahmad (2024). "Nanocoating and its application as antimicrobials in the food industry: A review." *International Journal of Biological Macromolecules* **254**: 127906.
Mohebbi, M. (2020). "Nanoencapsulation of flavors for beverage manufacturing." *Nanotechnology in the beverage industry*, Elsevier: 317–336.
Motelica, L., D. Ficai, O. C. Oprea, A. Ficai and E. Andronescu (2020). "Smart food packaging designed by nanotechnological and drug delivery approaches." *Coatings* **10**(9): 806.
Muhamad, I. I., D. N. A. Zaidel, Z. Hashim, N. A. Mohammad and N. F. A. Bakar (2020). "Improving the delivery system and bioavailability of beverages through nanoencapsulation." *Nanoengineering in the beverage industry*, Elsevier: 301–332.
Muhamad, W., H. Nugroho, S. Widaningsih and R. Hendriyanto (2023). "Designing a public API-based order delivery service system for the food and beverage industry." *Ingénierie des Systèmes d'Information* **28**(3).
Nakazato, G., R. K. Kobayashi, A. B. Seabra and N. Durán (2017). "Use of nanoparticles as a potential antimicrobial for food packaging." *Food preservation*, Elsevier: 413–447.
Nath, K., H. K. Dave and T. M. Patel (2018). "Revisiting the recent applications of nanofiltration in food processing industries: Progress and prognosis." *Trends in Food Science & Technology* **73**: 12–24.
Nemo, R. and K. Bacha (2021). "Natural preservative-based shelf-life enhancement of borde: A traditional Ethiopian low alcoholic beverage." *Journal of Food Processing and Preservation* **45**(11): e15968.
Pandey, A. and P. S. Negi (2018). "Use of natural preservatives for shelf life extension of fruit juices." *Fruit juices*, Elsevier: 571–605.
Paredes, A. J., C. M. Asensio, J. M. Llabot, D. A. Allemandi and S. D. Palma (2016). "Nanoencapsulation in the food industry: Manufacture, applications and characterization." *Journal of Food Bioengineering and Nanoprocessing* **1**: 56–70.
Pateiro, M., B. Gómez, P. E. Munekata, F. J. Barba, P. Putnik, D. B. Kovačević and J. M. Lorenzo (2021). "Nanoencapsulation of promising bioactive compounds to improve their absorption, stability, functionality and the appearance of the final food products." *Molecules* **26**(6): 1547.
Patil, A. S., O. S. Nille, G. B. Kolekar, D. Sohn and A. H. Gore (2021). "UV-shielding biopolymer@ nanocomposites for sustainable packaging applications." *Sustainable Packaging*: 177–202.
Poeta, E., A. Liboa, S. Mistrali, E. Núñez-Carmona and V. Sberveglieri (2023). "Nanotechnology and e-sensing for food chain quality and safety." *Sensors* **23**(20): 8429.
Powell, M. and M. Colin (2008). "Nanotechnology and food safety: Potential benefits, possible risks?" *CABI Reviews* **2008**: 16 pp.
Prakash, B., A. Kujur, A. Yadav, A. Kumar, P. P. Singh and N. Dubey (2018). "Nanoencapsulation: An efficient technology to boost the antimicrobial potential of plant essential oils in food system." *Food Control* **89**: 1–11.
Primožič, M., Ž. Knez and M. Leitgeb (2021). "(Bio) nanotechnology in food science—food packaging." *Nanomaterials* **11**(2): 292.
Pudlarz, A. and J. Szemraj (2018). "Nanoparticles as carriers of proteins, peptides and other therapeutic molecules." *Open Life Sciences* **13**(1): 285–298.
Raikos, V., L. P. Pirie, S. Gürel and H. E. Hayes (2021). "Encapsulation of vitamin E in yogurt-based beverage emulsions: Influence of bulk pasteurization and chilled storage on physicochemical stability and starter culture viability." *Molecules* **26**(6): 1504.
Ramos, M., A. Valdés, A. C. Mellinas and M. C. Garrigós (2015). "New trends in beverage packaging systems: A review." *Beverages* **1**(4): 248–272.
Ranadheera, C., P. Prasanna, J. Vidanarachchi, R. Mcconchie, N. Naumovski and D. Mellor (2017). "Nanotechnology in microbial food safety." *Nanotechnology applications in food*, Elsevier: 245–265.
Reis, M. H. M., G. S. Madrona, F. B. Ferreira, F. de Santana Magalhães, M. M. M. Bindes and V. L. Cardoso (2019). "Membrane filtration processes for the treatment of non-alcoholic beverages." *Engineering tools in the beverage industry*, Elsevier: 175–207.
Reque, P. M. and A. Brandelli (2021). "Encapsulation of probiotics and nutraceuticals: Applications in functional food industry." *Trends in Food Science & Technology* **114**: 1–10.
Roy, S., R. Ramakrishnan, G. Goksen, S. Singh and Ł. Łopusiewicz (2023). "Recent progress on UV-light barrier food packaging films–a systematic review." *Innovative Food Science & Emerging Technologies*: 103550.
Sahoo, M., S. Vishwakarma, C. Panigrahi and J. Kumar (2021). "Nanotechnology: Current applications and future scope in food." *Food Frontiers* **2**(1): 3–22.

Sampathkumar, K., K. X. Tan and S. C. J. Loo (2020). "Developing nano-delivery systems for agriculture and food applications with nature-derived polymers." *Iscience* **23**(5).

Sarkar, P., S. S. Panigrahi, E. Roy and P. Banerjee (2017). "Nanosensors in food safety." *Environmental Pollution* **16**: 185–207.

Shafiq, M., S. Anjum, C. Hano, I. Anjum and B. H. Abbasi (2020). "An overview of the applications of nanomaterials and nanodevices in the food industry." *Foods* **9**(2): 148.

Shekarchizadeh, H. and F. S. Nazeri (2020). "Active nanoenabled packaging for the beverage industry." *Nanotechnology in the beverage industry*, Elsevier: 587–607.

Shukla, K. (2012). "Nanotechnology and emerging trends in dairy foods: The inside story to food additives and ingredients." *International Journal of Nanoscience and Nanotechnology* **1**(1): 41–58.

Slavov, A. M., P. N. Denev, Z. R. Denkova, G. A. Kostov, R. S. Denkova-Kostova, R. M. Chochkov, I. N. Deseva and D. G. Teneva (2019). "Emerging cold pasteurization technologies to improve shelf life and ensure food quality." *Food quality and shelf life*, Elsevier: 55–123.

Soares, M. B., C. N. Almada, E. P. Pereira, B. M. Ferreira, C. F. Balthazar, N. Khorshidian, R. S. Rocha, D. Xavier-Santos, A. G. Cruz and C. S. Ranadheera (2023). "Sporeforming probiotic bacteria: Characteristics, health benefits, and technological aspects for their applications in foods and beverages." *Trends in Food Science & Technology* **138**: 453–469.

Sridhar, A., M. Ponnuchamy, P. S. Kumar and A. Kapoor (2021). "Food preservation techniques and nanotechnology for increased shelf life of fruits, vegetables, beverages and spices: A review." *Environmental Chemistry Letters* **19**: 1715–1735.

Taheri, A. and S. M. Jafari (2019). "Nanostructures of gums for encapsulation of food ingredients." *Biopolymer nanostructures for food encapsulation purposes*, Elsevier: 521–578.

Tianli, Y., Z. Jiangbo and Y. Yahong (2014). "Spoilage by Alicyclobacillus bacteria in juice and beverage products: Chemical, physical, and combined control methods." *Comprehensive Reviews in Food Science and Food Safety* **13**(5): 771–797.

Trinh, B. M., B. P. Chang and T. H. Mekonnen (2023). "The barrier properties of sustainable multiphase and multicomponent packaging materials: A review." *Progress in Materials Science*: 101071.

ur Rahman, U., A. Sahar, A. Ishaq and A. A. Khalil (2020). "Design of nanoparticles for future beverage industry." *Nanoengineering in the beverage industry*, Elsevier: 105–136.

Van der Bruggen, B. (2018). "Microfiltration, ultrafiltration, nanofiltration, reverse osmosis, and forward osmosis." *Fundamental modelling of membrane systems*, Elsevier: 25–70.

Vasilaki, A., M. Hatzikamari, A. Stagkos-Georgiadis, A. M. Goula and I. Mourtzinos (2019). "A natural approach in food preservation: Propolis extract as sorbate alternative in non-carbonated beverage." *Food chemistry* **298**: 125080.

Vieira, A. H., C. F. Balthazar, J. T. Guimaraes, R. S. Rocha, M. M. Pagani, E. A. Esmerino, M. C. Silva, R. S. Raices, R. V. Tonon and L. M. Cabral (2020). "Advantages of microfiltration processing of goat whey orange juice beverage." *Food Research International* **132**: 109060.

Wang, L., C. Hu and L. Shao (2017). "The antimicrobial activity of nanoparticles: Present situation and prospects for the future." *International Journal of Nanomedicine*: 1227–1249.

Yang, T. and T. V. Duncan (2021). "Challenges and potential solutions for nanosensors intended for use with foods." *Nature Nanotechnology* **16**(3): 251–265.

Yang, T., T. Paulose, B. W. Redan, J. C. Mabon and T. V. Duncan (2021). "Food and beverage ingredients induce the formation of silver nanoparticles in products stored within nanotechnology-enabled packaging." *ACS Applied Materials & Interfaces* **13**(1): 1398–1412.

10 Nanosensors and Biosensors for Real-Time Monitoring of Food Quality and Safety

Chiamaka Linda Mgbechidinma[a,b], Egong John Egong[c], Udoka Onyeka Okonta[d], Chibueze Anslem Mgbechidimma[e], Imo Yellow Sandy[f], and Otobong Donald Akan[f,g]

[a]School of Life Sciences, Centre for Cell and Development Biology and State Key Laboratory of Agrobiotechnology, School of Life Sciences, The Chinese University of Hong Kong, Shatin, New Territories, Hong Kong, China

[b]Department of Microbiology, University of Ibadan, Ibadan, Nigeria

[c]Faculty of Biological Sciences, University of Nigeria, Nsukka, Nigeria

[d]Faculty of Law and Business, University of Northampton, Northampton, United Kingdom

[e]Department of Computer Science, University of Ibadan, Ibadan, Nigeria

[f]Faculty of Biological Science, Akwa-Ibom State University, Uyo, Akwa-Ibom State, Nigeria

[g]Food Science and Engineering College, Central South University of Forestry and Technology, Changsha, Hunan, China

10.1 INTRODUCTION

Eating nutrient-dense food is vital for human life and optimal function; therefore, food safety should be recognized as a fundamental human right (Fung et al., 2018; Olovo et al., 2021). Although the hygiene standards in food production and processing (via the implementation of Hazard Analysis and Critical Control Point [HACCP], Good Manufacturing Practice [GMP], and Good Hygiene Practice [GHP]) have increased, there is still persistent contamination and unsafe foods (Zand et al., 2021). According to recent World Health Organization (WHO) data, food contamination significantly affects food safety. It results in monetary losses, illness inducement, adulteration, and waste and makes one in ten persons ill (Kraemer, 2016; Majer-Baranyi et al., 2023). While it might have a mild impact in countries with excess food supplies, food contamination can significantly harm food availability in countries with insufficient food production (Grace, 2015). Modern agricultural and industrial food operations frequently employ chemicals, such as fertilizers, pesticides, antibiotics, hormones, colourants, preservatives, antioxidants, etc., to boost competitiveness and profit margins. However, overuse of these chemicals has been considered a severe threat to human health (Nie & Nie, 2019). Numerous food-related risk factors can affect both food and customers. The contaminant effect could be single, additive, synergistic, or antagonistic, depending on the type and dose of exposure (DeGirolamo et al., 2022; Khan & Rahman, 2021).

Food security is significantly affected by the mass production of agricultural produce and goods, the growing number of imported foods, and new eating and consumption habits (Majer-Baranyi et al., 2023). There are a lot of inputs and exposures involved in the food sector due to the large number of people that work in this sector; this naturally raises the number of possible contamination chances from sowing seeds to processing food to selling it (Awuchi, 2023). However, to ensure food sufficiency and security, countries must not only revamp food production factors but also invest in factors that reduce food contamination (Abah et al., 2010; Kraemer, 2016). Fresh foods are susceptible to the actions of microorganisms, enzymes, and oxidation during storage (Poeta et al., 2023). Food quality and safety encompasses several attributes related to sensory traits, shelf life, freshness, and microbiological and technological parameters of food. Food safety procedures should detect the nutritional value and risk factors (such as the presence of bacteria, parasites, viruses, toxins, allergens, pesticide and veterinary drug residues, illegal food additives, restricted food ingredients, and heavy metals) in foods and generally ensure that food and food products meet labeling requirements and nutritional standards.

Conventional quality and safety monitoring methods are reliable but have limitations. They are not readily accessible to the everyday consumer, need a laboratory and qualified staff, and take several days to produce data and interventions. Therefore, emerging technologies that are more rapid, sensitive, portable, and specific are replacing conventional

DOI: 10.1201/9781032677422-10

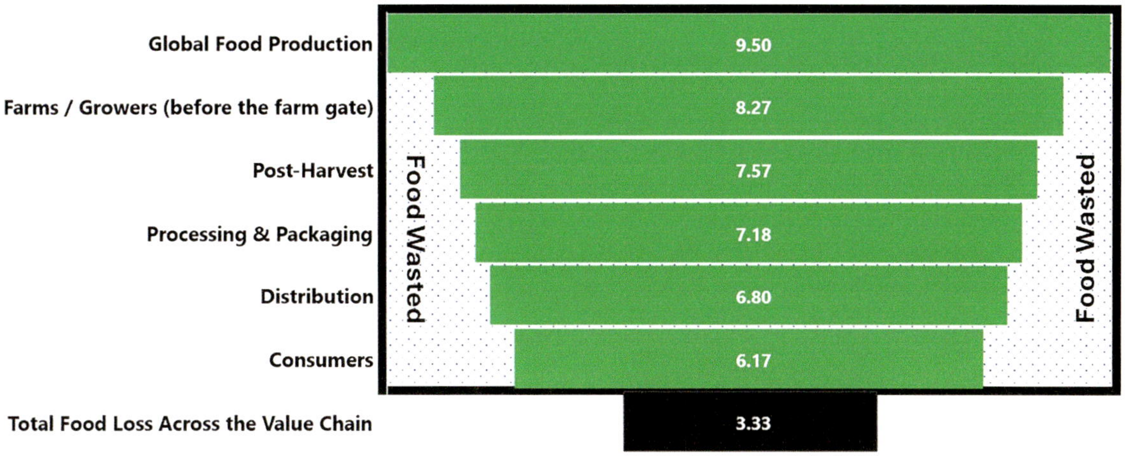

FIGURE 10.1 Average annual global food production and utilization per value chain segment (billion tonnes).

methods and reducing food wastage. Developments in biosensors and nanosensors technologies have allowed for the development of quick, reliable, affordable, sensitive, specialized, adaptable, and user-friendly devices. However, there is a need to sort out the downsides of these emerging technologies, such as reproducibility, drift, and the transfer of properties from one lot of sensors to another (Khan & Rahman, 2021).

10.2 THE NEED FOR FOOD QUALITY AND SAFETY MONITORING (FQSM)

Consuming contaminated food items comes with attendant disease outbreaks cum economic disasters, as unsafe foods are known vehicles for disease etiologies. A third of the world's food supply is lost within the food value chain, according to research by the Food and Agriculture Organization (FAO), which has sparked a conversation about food waste caused by improper handling and contamination problems (Gustavsson et al., 2011; Mgbechidinma, Akan, et al., 2022). Checks on food wastage will help combat hunger, raise income, and improve food security in the world's poorest countries. Figure 10.1 shows the global average food produced and utilized in billion tonnes annually per value chain segment. Data was collectively culled from recent UNDP and FAO data (FAO, 2021; UNDP, 2021). Food supply chains are global in reach; therefore, it's critical to use the most effective techniques for food quality and safety monitoring and to establish sustainability standards.

Stakeholders should strive to incorporate scientific principles in the handling, storage, and preparation of food items, and implement safety monitoring measures to prevent food contamination and losses (Awuchi, 2023). Technological developments in food and industry have helped food quality and safety monitoring (FQSM). However, food waste from contamination appears to be lower in industrialized nations due to the implementation of technological advancements across the food chain. On the other hand, losses are huge in developing countries due to poor storage issues, logistic nightmares, non-adoption of advanced safety technologies, and poor refrigeration due to incessant power outages (Bellemare et al., 2017; Grace, 2015; Tamasiga et al., 2022). Comparatively, consumers in developed countries budget about 10% of their income for food, while those in developing countries budget about 85% (Bellemare et al., 2017; Tamasiga et al., 2022). Food degrades mainly due to progressive natural biochemical and microbiological processes that changes its taste, colour, texture, aroma, and appearance. These changes are visible, especially during food storage, even though food contamination occurs in all aspects of food preparation. Traditional approaches (chemical, physical, microbiological, nutritional, and sensory criteria) used to evaluate food quality features and changes are based on the tested food's qualitative and quantitative properties (Rajput et al., 2019; Singham et al., 2015). While traditional methods have demonstrated efficacy in ensuring and assessing food safety, they are not without limitations. Traditional methods have limitations, including the need for skilled staff to obtain and analyze precise datasets, labour-intensive processes, laboratory-based procedures, significant resource requirements, and the risk of contamination (Kabiraz et al., 2023; Younes et al., 2023).

10.3 CONVENTIONAL FOOD QUALITY AND SAFETY MONITORING METHODS

The limitations associated with conventional food quality and safety monitoring methods notwithstanding have relative effectiveness and convenience. Some of these have long-standing uses. Typical examples are the human sense organs, molecular recombinant technologies, and immunoassays. Some commonly outlined conventional food quality and safety monitoring methods are listed next (Table 10.1).

10.3.1 Human Sense Organs

For time immemorial, man assessed food quality using multiple sense organs. Enzymatic and non-enzymatic redox reactions largely account for food deterioration—causing changes in colour, aroma, mouthfeel, texture, and taste. These could be easily verifiable by the senses of sight, touch, taste, and perception (Khan & Rahman, 2021). The human sensory experiences aid the delivery of pleasure and acceptability of food. When humans are used as testing machines, there will be a need to control errors due to psychological factors; therefore, trained hands are suggested (Kabiraz et al., 2023; Larmond, 1973). Commercial food companies and food quality assessment programs still rely on human organoleptic sensory parameters to evaluate, describe, and improve their products' characteristics (Larmond, 1973). Experts evaluate food quality attributes and control inputs during food processing and preparation using both quantitative (objective, discriminatory used for quality control) or qualitative (subjective or hedonic rating test) methods (Khan & Rahman, 2021).

However, there are limitations to their use since our senses are subjective (meaning there is no one standard for defining food safety following different people's senses of sight, smell, and taste). Secondly, our sense organs are not sensitive to contaminants such as toxins, chemicals, microbes, and even microbial spores in foods. Third, our sense organs are influenced by our moods, preconceived expectations and notions about food, and personal health (Khan & Rahman, 2021).

10.3.2 Microbiological Culture Analysis

Due to their ubiquity, microorganisms are crucial to the numerous processes that turn food products from "seeds to staples" (Erkmen & Bozoglu, 2016). Food components are susceptible to the diverse metabolic activities of self-perpetuating autochthonous and allochthonous microorganisms. The characteristics of these microbes finally crystallize when they exceed their growth threshold or reach their endpoints. Certain microbes play either a preservative, disease-causing, or spoiling implicative role; in any case, the presence of microorganisms in food can drastically affect its shelf life or safety ratings (Erkmen & Bozoglu, 2016; Marco et al., 2021). Aside from visible changes in colour, texture, flavour, and nutrient content due to the presence of vegetative bacteria and their enzymatic activities, microbial spores and toxins remain a concern in food safety and quality (André et al., 2017; Das et al., 2020).

The spores and toxins of microbes can still cause safety risks even after the vegetative cells are eliminated from the food. Conventional methods for evaluating the presence of microbes in food samples involve the "gold standard" plating methods (Zand et al., 2021). Techniques involve inoculating minute amounts of the food samples in enriched, differential, or selective media. These media allow for the growth of particular microbes with characteristic features. Preliminary isolation could take 1–3 days, while further microbial confirmation may require 7–10 days (Hahm et al., 2015). Notwithstanding, cultural methods could have or show (i) inadequate sensitivity, rapidity, and effectiveness; (ii) a high workload and time requirement; (iii) difficulties in differentiating between viable and non-viable microbial cells; or (iv) detect only cultivable microbes (Zand et al., 2021). Aside from these limitations, the traditional cultural method is still relatively reliable, accurate, and cost-effective.

10.3.3 Microbial Morphological-Based Methods

Aside from cultural methods, microbes in food can be identified and classified using their visual characteristics. Alternative to culturing, optical equipment like microscopes detect microbial properties, including size, shape, and other physical qualities. Some commonly used techniques are (i) the direct epifluorescent filter technique (DEFT), which uses a microscope and special filters to observe the fluorescence of specific dyes in microorganisms. The DEFT technique uses a combination of fluorescent dyes, a fluorescent microscope, and a special filter to visualize and detect microorganisms in food samples, especially water. This technique detects and differentiates microorganisms using emitting colours when exposed to fluorescent microscope light (Zand et al., 2021). (ii) Flow cytometry uses a laser to detect and classify microorganisms based on their size, shape, and other physical features. It is a thorough-put technique that analyzes the physical properties of many microorganisms as they move past a sensor (Zand et al., 2021). (iii) Solid-phase laser cytometry (SPC) uses a laser to detect and classify microorganisms based on the physical properties of the food sample (Zand et al., 2021). The laser detects microorganisms labeled with fluorescent antibodies that bind to specific antigens. It detects even small (or large) amounts of microorganisms on solid food surfaces in a concise time frame. These visual techniques have also been investigated for microbial identification in food samples. They complement molecular-based techniques that identify microorganisms through nucleic acid sequencing or antigen-antibody reactions.

10.3.4 PCR-Based Methods and Immunological Techniques

Food experts can effectively detect the presence of microbes and their toxins in food samples using molecular methods like recombinant technologies and antibody-antigen interaction methods (Kabiraz et al., 2023). These molecular methods can be conducted within drastically reduced time frames and work through identifying target pathogenic or spoilage microbes' nucleic acid sequences. At the same time, immunoassay with enzymatic analysis detects both microbes and expressed toxins (Lan et al., 2022; Smits et al., 2023). A small food sample is dispensed into a microplate well and incubated for a predetermined amount of time. The sample is, thereafter, rinsed to remove any unbound components. Then, a signal-generating molecule is added to the

TABLE 10.1
Features of Traditional Food Quality and Safety Monitoring Methods

	Methods	Working Principle	Food Type/Component	References
1	**Human sense organs** (evaluate the appearance, smell, taste, texture, and acceptability of food)	Use multiple (most times trained) people to evaluate the organoleptic characteristics of food to get a more accurate assessment.	Fruits and vegetables by looking for mould, discoloration, or slimy textures. Dairy products like milk and cheese can be evaluated by their appearance and smell to detect spoilage. Canned goods can also be monitored by their appearance, smell, and taste.	(Khan & Rahman, 2021; Molnár, 2011)
2	**Microbiological culture analysis** (certain bacteria, moulds, and yeast may either be naturally present in a food (natural flora) or introduced during handling, storage, or processing)	Analysing food samples to check for the presence of spoilage of pathogenic microorganisms.	Wide variety of fresh and processed foods.	(Erkmen & Bozoglu, 2016; Marco et al., 2021)
3	**Microbial Morphological-Based Methods**	Evaluating food safety by identifying the presence of certain microorganisms using their distinguishing shapes and structures.	Preserved or fermented, such as yogurt, cheese, sauerkraut, and pickles.	(Zand et al., 2021)
4	**PCR-Based Methods**	Targeting bacterial DNA sequences in food samples through polymerase chain reaction helps identify the presence or absence of specific bacteria.		(Ndraha et al., 2023)
	Real-Time Polymerase Chain Reaction (RT-PCR)	Commonly used analytical techniques for detecting specific DNA sequences present in food-borne pathogens and allergens.	Detects viruses in vegetables and fruits, foodborne pathogens (*Listeria*, *Campylobacter*, *Salmonella*, *E. coli*, etc.) in meat, dairy, and poultry products, and GMOs in processed foods.	(Salihah et al., 2016)
	Randomly Amplified Polymorphic DNA Polymerase Chain Reaction (RAPD-PCR)	Analytical technique that uses random primers to amplify specific DNA sequences of GMOs and foodborne pathogens (such as *E. coli* O157:H7, etc.) in food and food products.	Rapid detection of pathogens in fresh food produce; used to analyze fruits, vegetables, and processed foods prone to contamination.	(Atienzar & Jha, 2006)
5	**Munoanalytical Methods**	Uses monoclonal antibodies to detect the presence of specific compounds in a food sample.	Detects a wide range of compounds in food, including toxins, allergens, and foodborne pathogens.	(Smits et al., 2023)
	Enzyme-linked Immunoanalytical Methods (ELISA, EIA)	Commonly detects antigen-antibody complexes in samples using an enzymatic reaction's sensitivity. The test produces coloured compounds that are read by an ELISA reader, which measures the absorbance (intensity) of the colours at a specific wavelength. The absorbance value corresponds to the number of pathogenic microorganisms present in the food sample.	Pathogens such as *Salmonella enterica*, *Bacillus cereus*, *B. thuringiensis*, *B. licheniformis*, *B. subtilis*, *B. perfringens*, and *Staphylococcus aureus* in rice, dairy products, meat, eggs, fruits, and vegetables and *E. coli* O157:H7 in milk. Amine oxides can be used to determine the concentration of biogenic amides by monitoring oxygen consumption or hydrogen peroxide (H_2O_2) production in a reaction mixture containing the amide and the amine oxide.	(Kabiraz et al., 2023; Majer-Baranyi et al., 2023)
	Lateral-Flow Immunoassay	Commonly used to detect pathogens by fluorescent antibody where a fluorescent reporter molecule (fluorescein isothiocyanate) is labelled with an antibody.	Directly and rapidly detects pathogens from food samples.	(Kabiraz et al., 2023)

(Continued)

TABLE 10.1 (*Continued*)
Features of Traditional Food Quality and Safety Monitoring Methods

	Methods	Working Principle	Food Type/Component	References
	Immunomagnetic Separation	Antibodies are attached to magnetic beads, and the antibodies bind to the proteins or toxins in the food sample. The magnetic beads are then passed through a magnet, which causes the proteins or toxins to be separated from the rest of the food sample.	Commonly used to separate and identify pathogens (their proteins or toxins) from food.	(Kabiraz et al., 2023)
6	**Chromatography**	Differences in size, polarity, and solubility are exploited to identify food components. Compounds with different properties will travel through the column at different rates, allowing them to be identified and quantified.		(Cortes-Herrera et al., 2019; M. Mishra, 2022)
	Gas Chromatography (GC)	Employs a gas as the mobile phase and a column covered with a stationary phase. It separates volatile compounds by considering their boiling points and interactions with the stationary phase.	Volatiles and semi-volatiles samples (mineral oil hydrocarbons (MOHs), plasticizers, and herbicides, pesticides in foods (vegetables, edible oils, and meat) and beverages, antibiotics, packaging materials in processed foods (strawberries, milk, and other dairy products).	(M. Mishra, 2022; Nolvachai et al., 2023)
	Liquid Chromatography (LC)	LC is a versatile technique that can analyze various ions and polar molecules. Provides information about the composition, concentration, and identity of these compounds, which are related to the nutritional, sensory, functional, and safety aspects of food products.	Only small molecules, non- or weak-volatile compounds, and thermally unstable compounds. Polyphenols (fruits, vegetables, and some beverages), triacylglycerols (vegetable and animal fats), and carotenoids in citrus essential oils.	(Martín-pozo et al., 2022)
	High-Performance Liquid Chromatography (HPLC)	A liquid mobile phase is pumped through a column packed with a stationary phase. As the mobile phase moves through the column, the compounds in the food mixture will interact with the stationary phase differently based on their chemical properties. This results in different compounds being eluted (separated) at different times, which allows them to be identified and quantified.	Biogenic amines (histamine, cadaverine, putrescine, tyramine, spermidine, and spermine) in fish and fish products.	(Nie & Nie, 2019; Zare & Ghazali, 2016)
	Mass Spectrometry (MS)	Used to identify and quantify food compounds by measuring their mass-to-charge ratio. Compounds are ionized by bombarding them with electrons, and these ions can then be separated based on their mass-to-charge ratio using a magnetic field.	Fruits and vegetables, meat, and dairy products can be contaminated with a variety of harmful substances, including mycotoxins (produced by moulds), bacterial toxins, plant-based toxins, marine biotoxins (produced by algae or marine organisms), heavy metals, antibiotics, and pesticides.	(Majer-Baranyi et al., 2023)
7	**Coupled Methods**	Multiple techniques can be used in combination to provide a more comprehensive and accurate analysis of a food sample. This technique is known as "coupled analysis" and can help to detect different types of contaminants or pathogens in a sample.		

(*Continued*)

TABLE 10.1 (*Continued*)
Features of Traditional Food Quality and Safety Monitoring Methods

Methods	Working Principle	Food Type/Component	References
HPLC-MS and GC-MS	This coupled technique uses both retention time and the mass-to-charge ratio of the compounds in the food.	Detects non-biological food components (additives, allergens, toxins, preservatives) and contaminants.	(Nie & Nie, 2019; Zare & Ghazali, 2016)
Liquid Chromatography-Mass Spectrometry (LC-MS)	Conventional method to detect trace quantity of contaminants and can distinguish compounds with similar molecular weight.	Pesticide residues on fruits and vegetables, mercury, and polychlorinated biphenyls (even at low levels) in fish and seafood.	(Smaoui et al., 2020)
Gas Chromatography Ion mobility spectroscopy (GC-IMS)	Ion mobility spectrometry (IMS) is used to detect volatile organic compounds (VOCs) in foods using ions in a gas buffer under an electric field. It differentiates ions using their dimensions, structure, and charge, which are associated with their collision cross section (CCS).	Detects the volatile components of both raw and processed foods.	(Cui et al., 2024; Mgbechidinma, Zheng, et al., 2022; Qian et al., 2021)
Atomic and Molecular Spectroscopy	Analyzes the emission and absorption of light in food atoms and molecules. Used to detect and identify organic compounds (such as toxins or pesticides) and heavy metals in food samples.	Detects more recalcitrant contaminants like Pb in infant formula, and DDT in vegetables and fruits.	(Zaukuu et al., 2022)
Nuclear Magnetic Resonance (NMR) Spectroscopy	Food analyzing technique that uses magnetic properties of the nuclei to detect food contaminants. NMR spectroscopy measures the resonance frequencies of nuclei in a magnetic field which helps in contaminant identification to the specific level.	Detects multiple target microbial hybridizing to pathogen's DNA or protein in the same device chamber that runs assays using nucleic acid, antibodies, and other biomarkers.	(Fung et al., 2018)

microplate well, which will bind to the captured microbe or toxin. The intensity of the signal (as either fluorescence or colourimetry) is proportional to the presence of the target microbe or toxin. These techniques are useful in detecting a large range of microorganisms and toxins and, when combined, offer even increased detection accuracy.

PCR methods use a DNA-specific enzyme (polymerase) to amplify and detect specific DNA sequences of microorganisms. This method is very sensitive and can detect minute amounts of DNA. Advancement with qRT-PCR has significantly improved the accuracy and specificity of PCR-based detections by measuring fluorescence (targeted) amplified microbial sequences (Ndraha et al., 2023). Immunoassays use antibodies to detect specific proteins or toxins produced by microorganisms. ELIZA, a plate-based technique, is highly precise, sensitive, easy to use, cost-effective, and applicable for large-scale analysis (Smits et al., 2023). However, the immunoassay methods sustain a major setback in that there is a relatively low antibody affinity towards the microbes and other contaminants that do interfere (Younes et al., 2023).

10.3.5 Chromatographic and Spectroscopic Techniques

Good manufacturing practices (GMPs) and the hazard analysis and critical control point (HACCP) are geared to prevent, reduce, control, mitigate, or eliminate foodborne hazards along the food processing chain (Malik et al., 2010). The ability to identify and measure food components, elucidate nutritional requirements and novel dietary components, enhance food quality, and assess organic contaminants in foods have all contributed to human growth (Lehotay & Hajslova, 2002). High-performance liquid chromatography (HPLC) and gas chromatography (GC) are two analytical methods frequently used to detect and identify food components (Nie & Nie, 2019). These methods combine absorption, transmission, and emission of electromagnetic light radiation with added food materials based on the radiation wavelength (M. Mishra, 2022). Prepared food components are separated based on their interactions with a stationary phase, which can then be detected by an attached device. These methods are highly sensitive and specific in analyzing non-biological components of foods, such as additives, allergens, toxins, preservatives, and other food contaminants (-Park et al., 2021). One advantage of the chromatographic methods over the molecular methods is their versatility, rapidity, and the availability of various stationary phases.

Chromatographic techniques break down complex mixtures into individual compounds. For example, HPLC separates small molecules at non-physiological pH levels using a pump that delivers the mobile phase, injector, degasser, sampler, column, detector, and data system (Cortes-Herrera et al., 2019; Younes et al., 2023). However, liquid chromatography demands high solvents, which could pose toxicity issues and reduce the qualitative power of the technique (Younes et al., 2023). On the other hand, GC can effectively analyze volatile compounds in food samples with a gas supply with regulators, autosamplers, inlets, and detectors for data recording and processing (-Park et al., 2021). GC is a useful method for analyzing volatile, semi-volatile, nonpolar, and semi-polar compounds (Lehotay & Hajslova, 2002). The different components of food have unique adsorption/desorption properties when they interact with a stationary and a mobile phase in a chromatographic process. This property allows the components to be separated and easily detected and measured by suitable detectors and data-handling system (Nie & Nie, 2019). Spectroscopic techniques, such as mass spectrometry and nuclear magnetic resonance spectroscopy, are used to identify and quantify the components of food samples that have been separated using chromatographic techniques such as gas chromatography and liquid chromatography (Cavdaroglu & Ozen, 2023; Hong et al., 2023). For example, a sample can be analyzed by high-performance liquid chromatography (HPLC) to separate the different components and then by ultraviolet-visible (UV-Vis) spectroscopy to identify and quantify those components. There are other spectroscopy methods-UV-Vis (ultraviolet and visible), NIR (near-infrared), and MIR (Mid-infrared) that differ in molecular and atomic transition based on the absorption and release of radiation within their respective frequency ranges.

10.4 EMERGING FOOD QUALITY AND SAFETY MONITORING TECHNOLOGIES

More effective analytes and technical toolkits have recently been used by specialists to identify nutritious and harmful food ingredients quickly. While traditional methods of food analysis are still commonly used due to their relatively low cost and efficacy, they have several drawbacks that limit their usefulness, including lengthy time requirements, high risk of contamination, and a need for laboratory-based procedures (Kabiraz et al., 2023; Younes et al., 2023). Biosensors and nanosensors are emerging and advanced food quality and safety monitoring technologies that give more rapid and accurate results. They are more sensitive and portable and can be used in the field, curbing the transportation of samples to a laboratory. Also, they are less expensive to operate and maintain. They can be used for real-time monitoring of food production processes, allowing for rapid and accurate detection of contaminants and intervention (Majer-Baranyi et al., 2023; G. K. Mishra et al., 2018; Shan et al., 2023). The signals produced could be optical, such as colourimetric, fluorescence, chemiluminescence, surface plasmon resonance, or electrical, like voltammetry, impedance, capacitance, magneto-electric, or piezoelectric signals (Y. Kumar & Narsaiah, 2020). A few companies have even utilized sensors in intelligent packaging (Fuertes et al., 2021). These packages are integrated with sensors that generate signals that tell "how wholesome or safe" a packaged food is. However, most operable monitoring devices—also referred to as point-of-care (POC) devices—are operated by computer programs that are compatible with them and evaluate the signals they produce. These devices are installed with sensors that utilize materials such as paper, chips, threads, cuvettes, glass slides, and nanomaterials as output units (Y. Kumar & Narsaiah, 2020).

A biosensor monitoring device uses biological-recognition molecules (an aptamer, antibody, peptide nucleic acid [PNA], synthetic molecularly imprinted polymers [MIPs], nanobodies, or enzyme) to detect and measure a specific analyte in a food sample (Polat et al., 2022). The bio-recognition elements generate a physiological or biological signal that can be converted by a transducer and measured and recorded by a computer (Polat et al., 2022). However, the development of these monitoring devices could be costly. Conversely, even at the nanoscale level, a food sample's chemical and biological analytes can be efficiently detected and measured by a nanosensor using nanomaterials (nanoparticles or nanowires) (Fuertes et al., 2021; Poeta et al., 2023). Nanomaterials are highly effective sensors for detecting natural and synthetic food components because of their ability to selectively amplify signals, generate significant responses, exhibit antimicrobial properties, and maintain their structural stability. These qualities make nanomaterial-based sensors durable and efficient for food analysis (-Dehdashtian et al., 2023; Poeta et al., 2023).

10.4.1 Advanced Microorganisms, Toxins, and Adulterants Detections

10.4.1.1 Detection of Microbial Presence

Microorganisms that are either native to food products (autochthonous) or introduced during production, handling, or storage (allochthonous) can persist in the food and contribute to deterioration, disease, or transformation (fermentation). These microbes can propagate on their own and cause changes in the food, resulting in spoilage or desirable effects such as fermentation (Antia et al., 2018; Mgbechidinma et al., 2020; Xing et al., 2023). The main way to identify entire cells or their genetic material in food samples is to perform advanced microbial detection. Nanoparticles have been used to identify bacteria with remarkable sensitivity and accuracy by reacting with the genetic material of the target species. Magnetic iron oxide nanoparticles identified the milk-borne pathogenic bacterium *Listeria monocytogenes* by isolating its DNA fragments (Amagliani et al., 2004). Here, researchers relied on the paramagnetic nanoparticle-based isolation of the bacterial DNA directly from milk and subsequent PCR with selective primers for the listeriolysin O (hlyA) gene. Other scientists have developed nanoparticles that detect microbes in standard cultures and complex food samples (Inbaraj & Chen, 2015). Ren et al. (2021) used a magnetic focus lateral sensor (mLFS) platform to identify pathogenic *E. coli* K-12 MG1655, *Bacillus subtilis*, and *Pseudomonas aeruginosa* using their 16S ribosomal RNA (16S rRNA). This was highly sensitive and gave out a defining colour that could be seen with the naked eye.

Also, Satoh et al. (2022) used the light-scattering properties of oligonucleotide-functionalized gold nanoparticle probes (Au-nanoprobes) to identify unamplified RNA of enterohemorrhagic *Escherichia coli* (EHEC) O157:H7. This assay allowed for the sensitive quantification of bacterial 16S rRNA and mRNA without reverse transcription and cDNA amplification within 10 minutes of RNA extraction. Another study combined multiple biotin-labeled DNA probes and lateral-flow nucleic acid biosensors to detect *Staphylococcus aureus* (Zheng et al., 2019). The images of the lateral flow detection results obtained using a smartphone were transmitted to a computer via Wi-Fi or Bluetooth for quantitative processing by ImageJ.

10.4.1.2 Detection of Microbial Toxins

Lyophilic and stable toxins from vegetative microbial cells can persist after being produced by eliminated cells in contaminated food samples, causing food safety hazards. Repeated exposure to even smaller doses of these microbial toxins can eventually lead to harmful effects as they accumulate (Rajkovic, 2014). Some of these are depsipeptides, cereulide, and beauvericin. A study that developed an ultrasensitive electrochemiluminescence-based sensor detected palytoxin (PlTX), a marine toxin in seafood (Zamolo et al., 2012). The high specificity of anti-PlTX was achieved by the good conductive properties of carbon nanotubes and the excellent sensitivity of the luminescence-based transducer. Aflatoxins in foods contaminated with *Aspergillus flavus* and *Aspergillus parasiticus* could pose serious health risks (Ahmad et al., 2014). A study developed a rapid and sensitive immunochromatographic strip made of gold nanoparticle (AuNP) for detecting aflatoxin B1 (AFB1) in suspicious fungi-contaminated food samples (Sojinrin et al., 2019). AFB1 antibody was conjugated to BSA in a 96-well plate using the N-hydroxysuccinimide ester method. This caused a dramatic hue shift, from red to purple-blue, with a detection limit of 2 ng/mL. Another study coupled superparamagnetic beads and gold nanoprobes to detect aflatoxin M1 (AFM) in milk using dynamic light scattering (DLS) (Zhang et al., 2013). The concentration of unattached nanoprobes was positively proportional to the concentration of AFM in the sample using DLS. These nanoparticle-based detection methods provide valuable tools for ensuring food safety by detecting and monitoring the presence of harmful contaminants.

10.4.1.3 Detection of Harmful Agro-Residues

Sensor monitoring technologies have applications for detecting pesticides, fertilizers, and other harmful agro-residues in foods. Although these substances are purposefully released into the environment to aid in the production of more food, a sizable portion persist, and as non-biodegradable materials, enter the food chain and pose health hazards (Okeke et al., 2022; Tseng et al., 2020). A study examined using gold nanoparticles' (AuNPs) special optical feature to create a clever, portable, easy-to-use, and highly sensitive sensor for assessing pesticide levels in various matrices (Tseng et al., 2020). A team used highly passivated commercial CdSe/ZnS core-shell quantum dots (QDs) to detect phosphorothiolate pesticides in tomato fruits (Yang et al., 2018). Thiophosphorus hydrolyzates from the alkaline hydrolysis of phosphorothiolates markedly diminished the fluorescence emission of the QDs by an electron transfer mechanism. To detect the pesticide methyl parathion (MP) in water samples, a team created a unique electrode consisting of copper (Cu), tin dioxide-coupled multiwalled carbon nanotubes (SnO_2-cMWCNTs), and acetylcholinesterase (AChE). It is highly sensitive, biocompatible, and nontoxic (Dhull, 2018).

10.4.1.4 Detection of Food Integrity

Ensuring that food is fresh, safe, and free of tampering or contamination (adulteration) is vital for food security, which refers to the reliable access to sufficient and nutritious food for all people (Haji et al., 2023; Momtaz et al., 2023). Food integrity refers to the overall quality of food and includes factors such as safety, authenticity, and sustainability. Recent work with Kudzu roots and flowers showed that different growth factors and origins could influence the flavonoid content of different Kudzu plant samples (Zhu et al., 2023). Works on food integrity shed light on food origins and integral contents. Works conducting real-time (food integrity) detections using suitable sensors have been reported: luminescent oxygen nanosensors prepared by addressable staining of poly(styrene-block-vinylpyrrolidone) nanobeads with metal-ligand complexes lose their luminescence within the presence of oxygen (Borisov & Klimant, 2009). The authors could sense and monitor oxygen during glucose oxidation in meat products. Another work described using a nontargeted approach to determine the optimal duration of postharvest cold storage of oranges (Perez-Marin, 2020). A study used a portable short-wave NIR spectroscopic technique and chemometric data analysis to report/evaluate coffee samples' integrity (Boadu et al., 2023). Another study used zinc/aluminum-layered double hydroxide-quinclorac (Zn/Al-LDH-QC) as a modifier of multiwalled carbon nanotubes (MWCNT) paste electrode to check for the presence of bisphenol A (BPA) in babies' bottle and mineral water samples (Zainul et al., 2019).

10.4.2 Artificial Intelligence and Food Sensing

Currently, devices that can sense and analyze food based on smell, image, and taste are in use. With these devices, terms like "electronic olfaction," "electronic eye," and "electronic gustation," bio-mimicked human nose, eyes, and tongue are now commonplace in the food security field (V. Kumar et al., 2017; Munekata et al., 2023). Electronic olfaction, computer vision, and electronic gustation are often used with other technologies, such as bar codes, to create sophisticated food-sensing systems. Bar codes are information encoding that employ patterned lines and spaces to monitor food goods along the supply chain and provide details about their ingredients, provenance, and other attributes. However, conventional two-dimensional barcodes are easily detectable and thus susceptible to damage, alteration, and counterfeiting. In contrast, nanoparticle-based invisible barcodes offer enhanced security and are difficult to manipulate (Shikha et al., 2017). Many encoding elements have been used to encode nanostructures with varying geometries to create a wide range of barcodes.

TABLE 10.2
Some Examples of Artificial Intelligence Employed in Food Quality and Safety Monitoring

	AI-Based Tools	Working Principle	Examples	References
1	**Electronic nose** (electronic olfaction; e-nose)	Combines a gas or chemical sensor array with machine learning algorithms. The gas or chemical sensor arrays differentiate between different volatile organic compounds (VOCs) in the air.	Differentiated between chilled and frozen-thawed chicken meat using high-performance fuzzy K-nearest neighbours (F-KNN) algorithm. Electronic sensors classified the meat products into five classes according to their shelf life.	(Mirzaee-Ghaleh et al., 2020)
			A portable electronic equipped with a smartphone was developed to identify the Chinese dry-cured ham of three grades. A hybrid filter-wrapper method optimized the feature set.	(Qian et al., 2021)
			An E-nose detected and distinguished the volatile flavour of the sugar-smoked foods produced by lipid oxidation and caramelization reactions.	(Chang et al., 2021)
			E-nose (solid-phase micro-extraction coupled with gas chromatography-mass spectrometry and high-performance liquid chromatography) was deployed to evaluate concentration change in volatile organic compounds (VOCs) and biogenic amines (BAs) in smoked bacon during 45-day refrigerated storage.	(Li et al., 2018)
			A particular electronic nose (equipped with 10 separate metal oxide sensors) recognised variations in the volatile production of tomato ripening states. It differentiated and classified tomato samples according to their maturity states.	(Gómez et al., 2006)

(Continued)

TABLE 10.2 (*Continued*)
Some Examples of Artificial Intelligence Employed in Food Quality and Safety Monitoring

	AI-Based Tools	Working Principle	Examples	References
2	**Electronic Tongue** (electronic gustation; e-tongue)	Combines an array of taste sensors with machine learning algorithms. These sensors differentiate between tastes, such as sweet, salty, umami, sour, and bitter. It is composed of several types of lipid/polymer membranes that transform taste information into electrical signals that can be read by a computer and used to identify and classify different tastes.	The electrochemical responses of sensors to aminic compounds (ammonia and putrescine) in beef. The multisensor system discriminated and classified beef samples according to their storage time.	(Apetrei & Apetrei, 2016)
			Electronic nose prototype, based on the FIGARO semiconductor sensors, classified Polish honey types-acacia flower, linden flower, rape, buckwheat, and honeydew.	(Dymerski et al., 2014)
			A duplex bioelectronic tongue (DBT) based on graphene field-effect transistors that were functionalized with heterodimeric human umami taste and sweet taste receptor nanovesicles.	(Ahn et al., 2016)
			A potentiometric lab-made E-tongue (40 lipid sensor membranes) differentiated four common food bacteria contaminants (including two Gram-positive-*Enterococcus faecalis, Staphylococcus aureus,* and two Gram-negative *Escherichia coli, Pseudomonas aeruginosa*).	(Ghrissi et al., 2021)
			The e-tongue discriminated among paneer cheese samples containing different levels of capsaicin (1.875, 3.75, 7.5, 15, and 30 ppm).	(Schlossareck & Ross, 2019)
3	**Electronic eyes** (computer vision; e-eyes)	Combines a camera with image processing algorithms. The algorithm analyses the food item in terms of colour, shape, texture, defects or contamination that may be present, and type of food. This curbs the subjective deviation of human vision.	A computer vision system (CVS) and spectral information from the near-infrared (NIR) region identified and classified Chickens with wooden breasts by linear and nonlinear algorithms.	(Geronimo et al., 2019)
			Computer vision system (CVS) measured pork loin under industry speed requirements (National Pork Board standards). The CVS modelling with support vector machine prediction accuracy was 92.5% for the measured pork colour score and 75.0% for the measured pork marbling score.	(Sun et al., 2018)
			A machine vision system measured colour changes over 15 days of iced-stored Gulf sturgeon fillets.	(Oliveira & Balaban, 2006)
			An AI-based method properly recognized and quantified enzootic pneumonia-like lesions on digital images captured from slaughtered pigs under routine abattoir conditions without interfering with the slaughter chain routine.	(Bonicelli et al., 2021)
			A prototype system assessed and identified footpad dermatitis in broiler chickens.	(Vanderhasselt et al., 2012)

Some notable nanostructures are nanobeads, nanoclusters, nanowires, nanorods, nanotubes, nanostars, nanodiscs, nanopillars, nanoscale metal-organic frameworks (NMOF), and sequence barcodes (Shikha et al., 2017).

Most reports on using these systems give a prediction accuracy of 80 to 96%; even at this high score, many applications are combined for increased accuracy (Tan & Xu, 2020). The application of these systems shows versatility in characterizing and discriminating preservation techniques, adulteration techniques, grading quality, monitoring shelf life-related quality, and rapidly identifying the presence of pathogenic microbes (and their toxins) and other harmful compounds accurately (Munekata et al., 2023). Table 10.2 contains some of the artificial intelligence applications in food safety.

10.5 CONCLUSION

Because so many people are engaged in food production and processing these days, it can be challenging to guarantee the authenticity and integrity of some food (from cultivation to consumption point). In addition to wastage and illness problems, contamination with food has had a detrimental impact on the economy. Food security and safety issues are fundamental to the survival of the growing global population. However, weighing the setbacks with the application of conventional methods, it is easy to accept the benefits advanced food quality and safety monitoring methods offer rapid, accurate, simple (although technologically complex), on-site applications, non-destructive detection, portable, convenient, high sensitivity, and relatively cheap.

Biosensors and nanosensors food quality and safety monitoring measures have found applications even in intelligent food packaging, aiding buyers and consumers to consume only safe food items off the shelves. Additionally, considering the short shelf life of certain food items, developing rapid analytic systems that can show great specificity for food biomarkers (including complex food matrix) is advantageous economically and health wise.

REFERENCES

Abah, J., Ishaq, M. N., & Wada, A. C. (2010). The Role of Biotechnology in Ensuring Food Security and Sustainable Agriculture. *African Journal of Biotechnology*, *9*(52), 8896–8900.

Ahmad, M. M., Ahmad, M., Ali, A., Hamid, R., Javed, S., & Abdin, M. Z. (2014). Detection of *Aspergillus flavus* and *Aspergillus parasiticus* from Aflatoxin-Contaminated Peanuts and their Differentiation Using PCR-RFLP. *Annals of Microbiology*, *64*, 1597–1605. https://doi.org/10.1007/s13213-014-0803-5

Ahn, S. R., An, J. H., Song, H. S., Park, J. W., Lee, S. H., Kim, J. hyun, Jang, J., & Park, T. hyun. (2016). Duplex Bioelectronic Tongue for Sensing Umami and Sweet Tastes Based on Human Taste Receptor Nanovesicles. *ACS Nano*, *10*, 7287–7296. https://doi.org/10.1021/acsnano.6b02547

Amagliani, G., Brandi, G., Omiccioli, E., Casiere, A., Bruce, I. J., & Magnani, M. (2004). Direct Detection of Listeria Monocytogenes from Milk by Magnetic Based DNA Isolation and PCR. *Food Microbiology*, *21*, 597–603. https://doi.org/10.1016/j.fm.2003.10.008

André, S., Vallaeys, T., & Planchon, S. (2017). Spore-Forming Bacteria Responsible for Food Spoilage. *Research in Microbiology*, *168*(4), 379–387. https://doi.org/10.1016/j.resmic.2016.10.003

Antia, U. E., Akan, O. D., Stephen, N. U., Eno-Ibanga, C. K., & Akpan, N. G. (2018). Isolation and Screening of Yeast Isolates Indigenous Palm Wine for Ethanol Production. *Philippine Journal of Science*, *147*(3), 411–417.

Apetrei, I. M., & Apetrei, C. (2016). Application of Voltammetric e-Tongue for the Detection of Ammonia and Putrescine in Beef Products. *Sensors and Actuators, B: Chemical*, *234*, 371–379. https://doi.org/10.1016/j.snb.2016.05.005

Atienzar, F. A., & Jha, A. N. (2006). The Random Amplified Polymorphic DNA (RAPD) Assay and Related Techniques Applied to Genotoxicity and Carcinogenesis Studies: A Critical Review. *Mutation Research*, *613*, 76–102. https://doi.org/10.1016/j.mrrev.2006.06.001

Awuchi, C. G. (2023). HACCP, Quality, and Food Safety Management in Food and Agricultural Systems. *Cogent Food & Agriculture*, *9*(1), 1–30. https://doi.org/10.1080/23311932.2023.2176280

Bellemare, M. F., Çakır, M., Peterson, H. H., Novak, L., & Rudi, J. (2017). On the Measurement of Food Waste*. *American Journal of Agricultural Economics*, 1148–1158.

Boadu, V. G., Teye, E., Amuah, C. L. Y., Lamptey, F. P., & Sam-amoah, L. K. (2023). Portable NIR Spectroscopic Application for Coffee Integrity and Detection of Adulteration with Coffee Husk. *Processes*, *11*(1140), 1–13.

Bonicelli, L., Trachtman, A. R., Rosamilia, A., Liuzzo, G., Hattab, J., Alcaraz, E. M., Negro, E. Del, Vincenzi, S., Dondona, A. C., Calderara, S., & Marruchella, G. (2021). Training Convolutional Neural Networks to Score Pneumonia in Slaughtered Pigs. *Animals*, *11*(3290), 1–13. https://doi.org/10.3390/ani11113290

Borisov, S. M., & Klimant, I. (2009). Luminescent Nanobeads for Optical Sensing and Imaging of Dissolved Oxygen. *Microchimica Acta*, *164*, 7–15. https://doi.org/10.1007/s00604-008-0047-9

Cavdaroglu, C., & Ozen, B. (2023). Applications of UV-visible, Fluorescence and Mid-Infrared Spectroscopic Methods Combined with Chemometrics fir Authentication of Apple Vinegar. *Foods*, *12*(1139), 1–13. https://doi.org/10.3390/foods12061139

Chang, H., Wang, Y., Xia, Q., Pan, D., He, J., Zhang, H., & Cao, J. (2021). Characterization of the Physicochemical Changes and Volatile Compound Fingerprinting During the Chicken Sugar-smoking Process. *Poultry Science*, *100*(1), 377–387. https://doi.org/10.1016/j.psj.2020.09.059

Cortes-Herrera, C., Artavia, G., Leiva, A., & Granados-Chinchilla, F. (2019). Liquid Chromatography Analysis of Common Nutritional Components, in Feed and Food. *Foods*, *8*(1), 1–63. https://doi.org/10.3390/foods8010001

Cui, F., Liu, M., Li, X., Wang, D., Ma, F., Yu, L., Hu, C., Li, P., & Zhang, L. (2024). Gas Chromatography Ion Mobility Spectroscopy: A Rapid and Effective Tool for Monitoring Oil Oxidation. *Food Research International*, *176*(113842), 1–7. https://doi.org/10.1016/j.foodres.2023.113842

Das, S., Chaudhari, A. K., Singh, A., Deepika, Singh, V. K., Dwivedy, A. K., & Dubey, N. K. (2020). Foodborne Microbial Toxins and Their Inhibition by Plant-Based Chemicals. In *Functional and Preservative Properties of Phytochemicals*. https://doi.org/10.1016/B978-0-12-818593-3.00006-3

DeGirolamo, A., Lippolis, V., & Pascale, M. (2022). Overview of Recent Liquid Chromatography Mass Spectrometry-Based Methods for Natural Toxins Detection in Food products. *Toxins*, *14*(328), 1–33.

Dehdashtian, S., Wang, S., Murray, T. A., Chegeni, M., Rostamnia, S., & Fattahi, N. (2023). Determination of Vanillin in Different Food Samples by Using SMM/Au@ZIF--67 Electrochemical Sensor-. *Scientific Reports*, *13*(17907), 1–11. https://doi.org/10.1038/s41598-023-45342-6

Dhull, V. (2018). Fabrication of AChE/SnO2-cMWCNTs/Cu Nanocomposite-Based Sensor Electrode for Detection of Methyl Parathion in Water. *International Journal of Analytical Chemistry*, *2018*(2874059), 1–8. https://doi.org/10.1155/2018/2874059

Dymerski, T., Gebicki, J., Wardencki, W., & Namieśnik, J. (2014). Application of an Electronic Nose Instrument to Fast Classification of Polish Honey Types. *Sensors (Switzerland)*, *14*(6), 10709–10724. https://doi.org/10.3390/s140610709

Erkmen, O., & Bozoglu, F. T. (2016). Types of Microorganisms in Foods. In *Food Microbiology: Principles into Practice* (1st Edition, pp. 35–80). John Wiley and Sons, Ltd. https://doi.org/10.1002/9781119237860.ch3

FAO. (2021). World Food and Agriculture-Statistical Yearbook 2021. In *FAO Statistics* (Vol. 274). Food and Agriculture Organisation of the United Nations. https://doi.org/10.4060/cb4477en

Fuertes, G., Soto, I., Carrasco, R., Vargas, M., Sabattin, J., & Lagos, C. (2021). Intelligent Packaging Systems: Sensors and Nanosensors to Monitor Food Quality and Safety. *Journal of Sensors*, *2016*, 1–8. https://doi.org/http://dx.doi.org/10.1155/2016/4046061

Fung, F., Wang, H. S., & Menon, S. (2018). Food Safety in the 21st Century. *Biomedical Journal*, *41*(2), 88–95. https://doi.org/10.1016/j.bj.2018.03.003

Geronimo, B. C., Mastelini, S. M., Carvalho, R. H., Barbon Júnior, S., Barbin, D. F., Shimokomaki, M., & Ida, E. I. (2019). Computer Vision System and Near-infrared Spectroscopy for Identification and Classification of Chicken with Wooden Breast, and Physicochemical and Technological Characterization. *Infrared Physics and Technology*, *96*, 303–310. https://doi.org/10.1016/j.infrared.2018.11.036

Ghrissi, H., Veloso, A. C. A., Marx, M. G. I., Dias, T., & Peres, A. M. (2021). A Potentiometric Electronic Tongue as a Discrimination Tool of Water-Food Indicator/Contamination Bacteria. *Chemosensors*, *9*(143), 1–15. https://doi.org/doi.org/10.3390/chemosensors9060143

Gómez, A. H., Hu, G., Wang, J., & Pereira, A. G. (2006). Evaluation of Tomato Maturity by Electronic Nose. *Computers and Electronics in Agriculture*, *54*(1), 44–52. https://doi.org/10.1016/j.compag.2006.07.002

Grace, D. (2015). Food Safety in Low and Middle Income Countries. *International Journal of Environmental Research and Public Health*, *12*, 10490–10507. https://doi.org/10.3390/ijerph120910490

Gustavsson, J., Cederberg, C., Sonesson, U., Van-Otterdijk, R., & Meybeck, A. (2011). *Global Food Losses and Food Waste: Extent, Cause and Prevention*. Food and Agriculture Organisation of the United Nations.

Hahm, B. K., Kim, H., Singh, A. K., & Bhunia, A. K. (2015). Pathogen Enrichment Device (PED) Enables One-Step Growth, Enrichment and Separation of Pathogen from Food Matrices for Detection Using Bioanalytical Platforms. *Journal of Microbiological Methods*, *117*, 64–73. https://doi.org/10.1016/j.mimet.2015.07.016

Haji, A., Desalegn, K., & Hassen, H. (2023). Selected Food Items Adulteration, Their Impacts on Public Health, and Detection Methods: A Review. *Food Science and Nutrition*, *11*, 7534–7545. https://doi.org/10.1002/fsn3.3732

Hong, Y., Birse, N., Quinn, B., Li, Y., Jia, W., Mccarron, P., Wu, D., Da-Silva, G. R., Vanhaecke, L., Ruth, S. Van, & Elliott, C. T. (2023). Data Fusion and Multivariate Analysis for Food Authenticity Analysis. *Nature Communications*, *14*(3309). https://doi.org/10.1038/s41467-023-38382-z

Inbaraj, B. S., & Chen, B. H. (2015). Nanomaterial-Based Sensors for Detection of Foodborne Bacterial Pathogens and Toxins as Well as Pork Adulteration in Meat Products. *Journal of Food and Drug Analysis*, *24*(1), 15–28. https://doi.org/10.1016/j.jfda.2015.05.001

Kabiraz, M. P., Majumdar, P. R., Mahmud, M. M. C., Bhowmik, S., & Ali, A. (2023). Conventional and Advanced Detection Techniques of Foodborne Pathogens: A Comprehensive Review. *Heliyon*, *9*(4), 1–24. https://doi.org/10.1016/j.heliyon.2023.e15482

Khan, M. S., & Rahman, M. S. (2021). Techniques to Measure Food Safety and Quality: Microbial, Chemical, and Sensory. In M. S. Khan & M. S. Rahman (Eds.), *Techniques to Measure Food Safety and Quality: Microbial, Chemical, and Sensory* (Issue September 2021). Springer Nature. https://doi.org/10.1007/978-3-030-68636-9

Kraemer, K. H. (2016). *The Critical Role of Food Safety in Ensuring Food Security* (D. Crean & A. Ayalew, Eds.; pp. 313–325). Karger Publishers.

Kumar, V., Guleria, P., & Mehta, S. K. (2017). Nanosensors for Food Quality and Safety Assessment. *Environmental Chemistry Letters*, *15*(2), 165–177. https://doi.org/10.1007/s10311-017-0616-4

Kumar, Y., & Narsaiah, K. (2020). Rapid Point-of-care Testing Methods/Devices for Meat Species Identification: A Review. *Comprehensive Reviews in Food Science and Food Safety*, 1–24. https://doi.org/10.1111/1541-4337.12674

Lan, J., Yang, S., Wang, Y., Guo, N., Liu, X., Zhu, K., Gong, P., & Lv, S. (2022). Evaluation of Microbial Contamination in Cold Dishes and Prevalence of Foodborne Pathogens in Jilin Province. *Journal of Food Protection*, *85*(5), 728–734. https://doi.org/10.4315/JFP-21-328

Larmond, E. (1973). *Methods for Sensory Evaluation of Food*. Agriculture and Agri-Food Canada.

Lehotay, S. J., & Hajslova, J. (2002). Application of Gas Chromatography in Food Analysis. *Trends in Analytical Chemistry*, *9–10*, 686–697.

Li, X., Zhu, J., Li, C., Ye, H., Wang, Z., Wu, X., & Xu, B. (2018). Evolution of Volatile Compounds and Spoilage Bacteria in Smoked Bacon During Refrigeration Using an E-nose and GC-MS Combined with Partial Least Square Regression. *Molecules*, *23*(3286), 1–20. https://doi.org/10.3390/molecules23123286

Majer-Baranyi, K., Székács, A., & Adányi, N. (2023). Application of Electrochemical Biosensors for Determination of Food Spoilage. *Biosensors*, *13*(4), 1–26. https://doi.org/10.3390/bios13040456

Malik, A. K., Blasco, C., & Picó, Y. (2010). Liquid Chromatography—Mass Spectrometry in Food Safety. *Journal of Chromatography*, *1217*, 4018–4040. https://doi.org/10.1016/j.chroma.2010.03.015

Marco, M. L., Sanders, M. E., Gänzle, M., Arrieta, M. C., Cotter, P. D., De Vuyst, L., Hill, C., Holzapfel, W., Lebeer, S., Merenstein, D., Reid, G., Wolfe, B. E., & Hutkins, R. (2021). The International Scientific Association for Probiotics and Prebiotics (ISAPP) Consensus Statement on Fermented Foods. *Nature Reviews Gastroenterology and Hepatology*, *18*(3), 196–208. https://doi.org/10.1038/s41575-020-00390-5

Martín-pozo, L., Arena, K., Cacciola, F., Dugo, P., & Mondello, L. (2022). Comprehensive Two-Dimensional Liquid Chromatography in Food Analysis. Is Any Sample Preparation Necessary? *Green Analytical Chemistry*, *3*(100025), 1–9. https://doi.org/10.1016/j.greeac.2022.100025

Mgbechidinma, C. L., Adegoke, C. O., & Ogunbanwo, S. T. (2020). Lactic Acid Bacteria as Bioactive Potential against Selected Resistance Canada Species and Pathogenic Bacteria. *International Journal of Pharmaceutical and Biological Science Archive PubMed*, *8*(2), 19–32. www.ijpba.in

Mgbechidinma, C. L., Akan, O. D., Zhang, C., Huang, M., Linus, N., Zhu, H., & Wakil, S. M. (2022). Integration of Green Economy Concepts for Sustainable Biosurfactant Production—a Review. *Bioresource Technology*, *364*. https://doi.org/10.1016/j.biortech.2022.128021

Mgbechidinma, C. L., Zheng, G., Baguya, E. B., Zhou, H., Okon, S. U., & Zhang, C. (2022). Fatty Acid Composition and Nutritional Analysis of Waste Crude Fish Oil Obtained by Optimized Milder Extraction Methods. *Environmental Engineering Research, In Press*. https://doi.org/10.4491/eer.2022.034

Mirzaee-Ghaleh, E., Taheri-Garavand, A., Ayari, F., & Lozano, J. (2020). Identification of Fresh-Chilled and Frozen-Thawed Chicken Meat and Estimation of Their Shelf Life Using an E-Nose Machine Coupled Fuzzy KNN. *Food Analytical Methods*, *13*(3), 678–689. https://doi.org/10.1007/s12161-019-01682-6

Mishra, G. K., Barfidokht, A., Tehrani, F., & Mishra, R. K. (2018). Food Safety Analysis Using Electrochemical Biosensors. *Foods*, *7*(9), 1–11. https://doi.org/10.3390/foods7090141

Mishra, M. (2022). Spectroscopic Techniques for the Analysis of Food Quality, Chemistry, and Function. In A. K. Shukla (Ed.), *Advanced Spectroscopic Techniques for Food Quality* (Issue 32). The Royal Society of Chemistry. https://doi.org/10.1039/9781839165849–00001

Molnár, P. (2011). Food Quality Indices. *Food Quality and Standards, II*, 32.

Momtaz, M., Bubli, S. Y., & Khan, M. S. (2023). Mechanisms and Health Aspects of Food Adulteration: A Comprehensive Review. *Foods, 12*(199), 1–25.

Munekata, P. E. S., Finardi, S., Souza, C. K. De, Meinert, C., Pateiro, M., Hoffmann, T. G., Bertoli, L., Kumar, M., & Lorenzo, J. M. (2023). Applications of Electronic Nose, Electronic Eye and Electronic Tongue in Quality, Safety and Shelf Life of Meat and Meat Products: A Review. *Sensors, 23*(672), 1–27. https://doi.org/10.3390/ s23020672

Ndraha, N., Lin, H. Y., Wang, C. Y., Hsiao, H. I., & Lin, H. J. (2023). Rapid Detection Methods for Foodborne Pathogens based on Nucleic Acid Amplification: Recent Advances, Remaining Challenges, and Possible Opportunities. *Food Chemistry: Molecular Sciences, 7*(September), 100183. https://doi.org/10.1016/j.fochms.2023.100183

Nie, Q., & Nie, S. (2019). High-Performance Liquid Chromatography for Food Quality Evaluation. In *Evaluation Technologies for Food Quality*. Elsevier Inc. https://doi.org/10.1016/B978-0-12-814217-2.00013-5

Nolvachai, Y., Amaral, M. S. S., & Marriott, P. J. (2023). Foods and Contaminants Analysis Using Multidimensional Gas Chromatography: An Update of Recent Studies, Technology, and Applications. *Analytical Chemistry, 95*, 238–263. https://doi.org/10.1021/acs.analchem.2c04680

Okeke, E. S., Okoye, C. O., Atakpa, E. O., Ita, R. E., Nyaruaba, R., Mgbechidinma, C. L., & Akan, O. D. (2022). Microplastics in Agroecosystems-Impacts on Ecosystem Functions and Food Chain. *Resources, Conservation and Recycling, 177*(105961), 1–17. https://doi.org/10.1016/j.resconrec.2021.105961

Oliveira, A. C. M., & Balaban, M. O. (2006). Comparison of a Colorimeter with a Machine Vision System in Measuring Colour of Gulf of Mexico Sturgeon Fillets. *Applied Engineering in Agriculture, 22*(1997), 583–587.

Olovo, C. V., Udoekong, N. S., & Akan, O. D. (2021). Precision Nutrition, Diet and Gut-Microbiota in Obesity. *Journal of Biotechnology & Bioresearch, 2*(5), 4–6.

Park, M. H., Kim, C. J., Lee, J. Y., Kim, I. S., & Kim, S. K. (2021). Development and Validation of a Gas Chromatography Method for the Determination of β--Caryophyllene in Clove Extract and its Application-. *Scientific Reports, 11*(13853), 1–8. https://doi.org/10.1038/s41598-021-93306-5

Perez-Marin, L. (2020). Innovation for Food Integrity Assessment and Fraud Detection Using NIRS as a Nontargeted Method: Towards Intelligent Product and Process Control. *NIR News, 31*(7–8), 5.8. https://doi.org/10.1177/0960336020978715

Poeta, E., Liboa, A., Mistrali, S., Núñez-Carmona, E., & Sberveglieri, V. (2023). Nanotechnology and E-Sensing for Food Chain Quality and Safety. *Sensors, 23*(8429), 1–31. https://doi.org/10.3390/s23208429

Polat, E. O., Cetin, M. M., Tabak, A. F., Guven, E. B., Uysal, B. O., Arsan, T., Kabbani, A., Hamed, H., & Gul, S. B. (2022). Transducer Technologies for Biosensors and Their Wearable. *Biosensors, 12*(385), 1–32. https://doi.org/10.3390/bios12060385

Qian, K., Bao, Y., Zhu, J., Wang, J., & Wei, Z. (2021). Development of a Portable Electronic Nose Based on a Hybrid Filter-Wrapper Method for Identifying the Chinese Dry-Cured Ham of Different Grades. *Journal of Food Engineering, 290*(110250), 1–13. https://doi.org/10.1016/j.jfoodeng.2020.110250

Rajkovic, A. (2014). Microbial Toxins and Low Level of Foodborne Exposure. *Trends in Food Science & Technology, 38*(2), 149–157. https://doi.org/10.1016/j.tifs.2014.04.006

Rajput, H., Rehal, J., Goswami, D., & Mandge, M. H. (2019) Publisher: Apple Academic Press, New York, 299–320. Methods for Food Analysis and Quality Control. In *State-of-the-Art Technologies in Food Science*. Apple Academic Press, New York, 299–320.

Ren, W., Ahmad, S., & Irudayaraj, J. (2021). 16S rRNA Monitoring Point-of-Care Magnetic Focus Lateral Flow Sensor. *ACS Omega, 6*, 11095–11102. https://doi.org/10.1021/acsomega.1c01307

Salihah, N. T., Hossain, M. M., & Lubis, H. (2016). Trends and Advances in Food Analysis by Real-Time Polymerase Chain Reaction. *Journal of Food Science and Technology, 53*(5), 2196–2209. https://doi.org/10.1007/s13197-016-2205-0

Satoh, H., Nakajima, M., Hirano, R., & Nakaya, Y. (2022). Highly Sensitive and Homogeneous Detection of Unamplified RNA Based on the Light Scattering Properties of Gold Nanoparticle Probes. *Biosensors and Bioelectronics: X, 12*(100249), 1–4. https://doi.org/10.1016/j.biosx.2022.100249

Schlossareck, C., & Ross, C. F. (2019). Electronic Tongue and Consumer Sensory Evaluation of Spicy Paneer Cheese. *Journal of Food Science, 84*(6), 1563–1569. https://doi.org/10.1111/1750-3841.14604

Shan, Y., Lu, Y. N., Yi, W., Wang, B., Li, J., Guo, J., Li, W., Yin, Y., Wang, S., & Liu, F. (2023). On-Site Food Safety Detection: Opportunities, Advancements, and Prospects. *Biosensors and Bioelectronics: X, 14*, 1–12. https://doi.org/10.1016/j.biosx.2023.100350

Shikha, S., Salafi, T., Cheng, J., & Zhang, Y. (2017). Versatile Design and Sythesis of Nano-Barcodes. *Chemical Society Reviews, 46*, 7054–7093. https://doi.org/10.1039/c7cs00271h

Singham, P., Birwal, P., & Yadav, B. (2015). Importance of Objective and Subjective Measurement of Food Quality and Their Inter-Relationship. *Journal of Food Processing & Technology, 06*(09), 1–8. https://doi.org/10.4172/2157–7110.1000488

Smaoui, S., Ben Braïek, O., & Ben Hlima, H. (2020). Mycotoxins Analysis in Cereals and Related Foodstuffs by Liquid Chromatography-Tandem Mass Spectrometry Techniques. *Journal of Food Quality, 2020*, 1–23. https://doi.org/10.1155/2020/8888117

Smits, N. G. E., De Dominicis, E., Koops, A. J., Kraan, R., Saner, S., Van Der Fels-Klerx, H. J., & Hoek-van den Hil, E. (2023). Comparison of Commercial Allergen ELISA Kits for Egg Detection in Food Matrices. *Heliyon, 9*(10), 1–13. https://doi.org/10.1016/j.heliyon.2023.e19687

Sojinrin, T., Liu, K., Wang, K., Cui, D., Byrne, H. J., Curtin, J. F., & Tian, F. (2019). Developing Gold Nanoparticles-Conjugated Aflatoxin B1 Antifungal Strips. *Molecular Sciences, 20*(6260), 1–16. https://doi.org/10.3390/ijms20246260

Sun, X., Young, J., Liu, J., & Newman, D. (2018). Prediction of Pork Loin Quality Using Online Computer Vision System and Artificial Intelligence Model. *Meat Science, 140*, 72–77. https://doi.org/10.1016/j.meatsci.2018.03.005

Tamasiga, P., Miri, T., Onyeaka, H., & Hart, A. (2022). Food Waste and Circular Economy: Challenges and Opportunities. *Sustainability, 14*(9896). https://doi.org/ 10.3390/su14169896

Tan, J., & Xu, J. (2020). Artificial Intelligence in Agriculture Applications of Electronic Nose (e-Nose) and Electronic Tongue (e-Tongue) in Food Quality-Related Properties Determination: A Review. *Artificial Intelligence in Agriculture*, *4*, 104–115. https://doi.org/10.1016/j.aiia.2020.06.003

Tseng, W., Hsieh, M.-M., Chen, C.-H., Chiu, T.-C., & Tseng, W.-L. (2020). Functionalized Gold Nanoparticles for Sensing of Pesticides: A Review. *Journal of Food and Drug Analysis*, 28(4):521–538. doi: 10.38212/2224-6614.1092.

UNDP. (2021). *Food Waste Index Report 2021*. United Nations Environment Porgramme.

Vanderhasselt, R. F., Sprenger, M., Duchateau, L., & Tuyttens, F. A. M. (2012). Automated Assessment of Footpad Dermatitis in Broiler Chickens at the Slaughter-Line: Evaluation and Correspondence with Human Expert Scores. *Poultry Science*, *92*(1), 12–18. https://doi.org/10.3382/ps.2012-02153

Xing, Y., Huang, M., Olovo, C. V., Mgbechidinma, C. L., Yang, Y., Liu, J., Li, B., Zhu, M., Yu, K., Zhu, H., Yao, X., Bo, L., & Akan, O. D. (2023). Traditional Fermented Foods: Challenges, Sources, and Health Benefits of Fatty Acids. *Fermentation*, *9*(2), 1–23. https://doi.org/10.3390/fermentation9020110

Yang, L., Zhang, X., & Jiang, L. (2018). Determination of Organophosphorus Pesticides in Fortified Tomatoes by Fluorescence Quenching of Cadmium Selenium—Zinc Sulfide Quantum Dots Determination of Organophosphorus Pesticides in Fortified Tomatoes by Fluorescence Quenching of Cadmium Selenium –. *Analytical Letters*, *0*(0), 1–16. https://doi.org/10.1080/00032719.2018.1490311

Younes, N., Yassine, H. M., Kourentzi, K., Tang, P., Litvinov, D., Willson, R. C., Abu-Raddad, L. J., & Nasrallah, G. K. (2023). A Review of Rapid Food Safety Testing: Using Lateral Flow Assay Platform to Detect Foodborne Pathogens. *Critical Reviews in Food Science and Nutrition*, *0*(0), 1–23. https://doi.org/10.1080/10408398.2023.2217921

Zainul, R., Azis, N. A., Isa, I. M., Hashim, N., Ahmad, M. S., Saidin, M. I., & Mukdasai, S. (2019). Zinc/Aluminium-Quinclorac Layered Nanocompsite Modified Multi-Walled Carbon Nanotube Paste Electrode for Electrochemical Determination of Bisphenol A. *Sensors*, *19*(941), 1–15. https://doi.org/10.3390/s19040941

Zamolo, V. A., Valenti, G., Venturelli, E., Chaloin, O., Marcaccio, M., Boscolo, S., Castagnola, V., Sosa, S., Berti, F., Fontanive, G., Poli, M., Tubaro, A., Bianco, A., Paolucci, F., & Prato, M. (2012). Highly Sensitive Electrochemiluminescent Nanobiosensor for the Detection of Palytoxin. *American Chemical Society*, *6*(9), 7989–7997. https://doi.org/10.1021/nn302573c

Zand, E., Froehling, A., Schoenher, C., Zunabovic-Pichler, M., Schlueter, O., & Jaeger, H. (2021). Potential of Flow Cytometric Approaches for Rapid Microbial Detection and Chracterization in the Food industry—a Review. *Foods*, *10*(12), 1–43. https://doi.org/10.3390/foods10123112

Zare, D., & Ghazali, H. M. (2016). Assessing the Quality of Sardine Based on Biogenic Amines Using a Fuzzy Logic Model. *Food Chemistry*, *221*, 936–943. https://doi.org/10.1016/j.foodchem.2016.11.071

Zaukuu, J. L. Z., Benes, E., Bázár, G., Kovács, Z., & Fodor, M. (2022). Agricultural Potentials of Molecular Spectroscopy and Advances for Food Authentication: An Overview. *Processes*, *10*(2), 1–42. https://doi.org/10.3390/pr10020214

Zhang, Z., Lin, M., Zhang, S., & Vardhanabhuti, B. (2013). Detection of Aflatoxin M1 in Milk by Dynamic Light Scattering Coupled with Superparamagnetic Beads and Gold Nanoprobes. *Journal of Agricultural and Food Chemistry*, *61*, 4520–4525. https://doi.org/dx.doi.org/10.1021/jf400043z

Zheng, X., Wang, Y., Bu, S., Chen, Z., & Wan, J. (2019). Point-of-care Detection of 16S rRNA of *Staphylococcus aureus* Based on Multiple Biotin-labeled DNA Probes. *Molecular and Cellular Probes*, *47*(101427), 1–6. https://doi.org/10.1016/j.mcp.2019.101427

Zhu, H., Xing, Y., Akan, O. D., & Yang, T. (2023). Ultrafine Comminution-Assisted Ultrasonic-microwave Synergistic Extraction of *Pueraria mirifica* (Kudzu Flower and Root) Flavonoids. *Heliyon*, *9*(11), 1–12. https://doi.org/10.1016/j.heliyon.2023.e21137

11 Advancing Novel Applications of Nanotechnology in the Food Industry

Anele Sithole[a] *and Shalini Singh*[a]
[a]Department of Operations and Quality Management Durban University of Technology, Durban, South Africa

11.1 INTRODUCTION

Organisations experience many obstacles that hinder productivity. In the twenty-first century these obstacles arise mainly from climate change, the rapid growth of the global population, diseases and the risks associated with potential pandemic outbreaks as seen with the coronavirus, constant droughts in agricultural zones and concomitant food waste and food insecurity (Sreelakshmi et al. 2021). Unfortunately, the food industry has identified threats to food sustainability and food security (Lata, Bhardwaj and Garg 2022). Another school of thought indicated that the food industry is one of the sectors that has created a negative footprint in the impact of climate change. One can argue that this is because the food industry is under great pressure to supply the rapidly growing population while complying with the constantly changing customer and regulatory requirements. According to recent customer perceptions, there is a high demand for fresh, natural food free from chemical additives and preservatives. Consequently, the food industry is also exploring techniques to review current formulations to produce healthier food products derived from green processes that adhere to clean label solutions. Similarly, there is a need for advanced technologies that will allow the production of high yields while extensively minimising the environmental pollution that has been claimed to have been caused by the Industrial Revolution (Lata, Bhardwaj and Garg 2022). Hence, more sound strategies have been documented and implemented successfully while others are still at the foundation phase or are being explored to address the aforementioned obstacles (Haris et al. 2023). There is agreement that there is an urgent need for innovative sustainable resolutions to ensure continued increasing food production or yields and improved food availability to serve the world's growing population (Brar, Bhardwaj and Chauhan 2023; Lata et al. 2022; Zhang et al. 2023a).

Thus, there is a high demand for robust technologies to be employed in food processing industries to produce fresh and safe products with high quality. Interestingly, the food industry supports the use of diverse nanotechnologies in research and development to combat the challenges encountered and are thus allowing mass production of food within a short period while providing long-term economic benefits serving specific customer needs with increased customer awareness, informed decision-making and strong consumer perspectives towards food safety and quality (Haris et al. 2023). The application of nanotechnology is mooted as a novel frontier that can make a significant impact on improving the nutritional value and functional food characteristics (Brar, Bhardwaj and Chauhan 2023; Hadidi et al. 2023). Furthermore, it can be agreed that nanoparticles enhance the organoleptic properties of food such as colour, texture, taste and odour (Haris et al. 2023). Other views indicate that NPs have antimicrobial properties that can be used for preservative effects and extend product shelf life without altering food organoleptic attributes (Lata, Bhardwaj and Garg 2022). Moreover, the application of nanosensors and biosensors in food manufacturing has been proposed as a critical control point to either detect or eliminate food safety hazards (Karthik and Jayendra 2022).

Despite these positive attributes, there are implications associated with the application of nanotechnology in the food industry. This is due to the potential risks and limited regulatory controls associated with nanotechnology in the food industry (Haris et al. 2023; Hadidi et al. 2023). Despite the concerns linked with the use of NPs, food organisations have adopted nanotechnology to improve their products. Hence, this chapter used suitable case studies of successful nanotechnology applications within the food industry by systematically reviewing the literature. Furthermore, appropriate keywords and specific exclusive and inclusive criteria were used to maintain data credibility and dependability.

11.2 EVOLVING HISTORY OF NANOTECHNOLOGY

In 1959, Richard Phillips Feynman, a theoretical physicist who is well-known for expanding the knowledge of quantum mechanics and electrodynamics, laid a foundation for the establishment of the concept of nanotechnology (Moradi et al. 2022; Vijayalakshmi, Sachini and Kirtan 2017). Nowadays, NPs can be derived from various sources using different methods either naturally, physically or synthetically (Karthik and Jayendra 2022). Indeed, some NPs

DOI: 10.1201/9781032677422-11

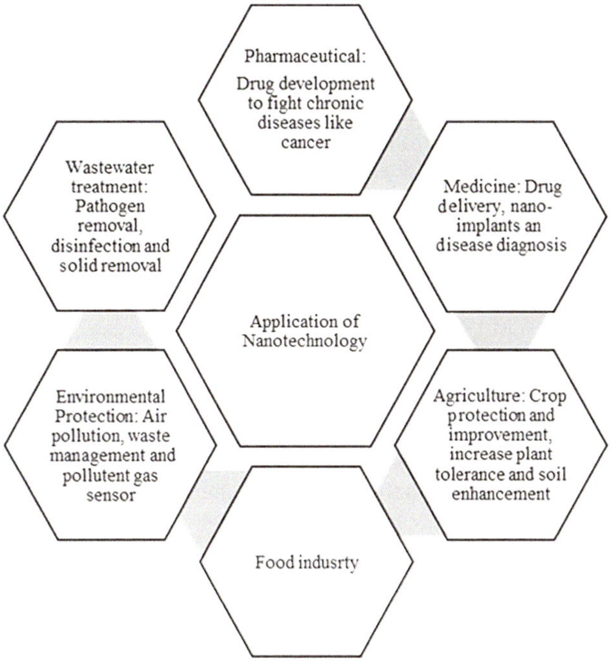

FIGURE 11.1 Application of nanotechnology. (Adapted by authors).

can occur naturally; as a result, they can be found everywhere, in ocean spray, volcanic ash, dust and fine sand. Nanoparticles can also be obtained naturally from plants and animals. To exemplify, inorganic materials such as cobalt, silver, copper, nickel and zinc are derived from biotic plants like *Helianthus annus*, *Medicago sativa* and *Brassica juncea* while others are chemically synthesised or derived from microorganisms. Furthermore, different types of systems are used for the biosynthesis of metals, and they include *Aspergillus flavus*, *Phanerochaete chrysosporium*, *Aspergillus sp*, *Aspergillus furnigatus* and *Fusarium oxysporium* (Karthik and Jayendra 2022). Nowadays, numerous microorganisms are used to develop green nanoparticles. On the other hand, synthetic NPs are developed by manipulating the atomic molecules of the material to create novel products that serve new opportunities. Nanoparticles can be classified into various categories depending on their sizes, structure, shapes and properties. Metal-based NPs are commonly used when compared to other types of NP. They include titanium dioxide NPs, gold NPs, silver NPs, zinc oxide NPs and other metal oxides (Chand et al. 2020). Figure 11.1 outlines different applications of nanotechnology in various sectors.

11.2.1 Application of Nanotechnology in the Food Industry

There is consensus in the literature that nanotechnology is a revolutionary approach in the food industry. In the same context, Figure 11.2 presents various applications of nanotechnology in the food sector ranging from the

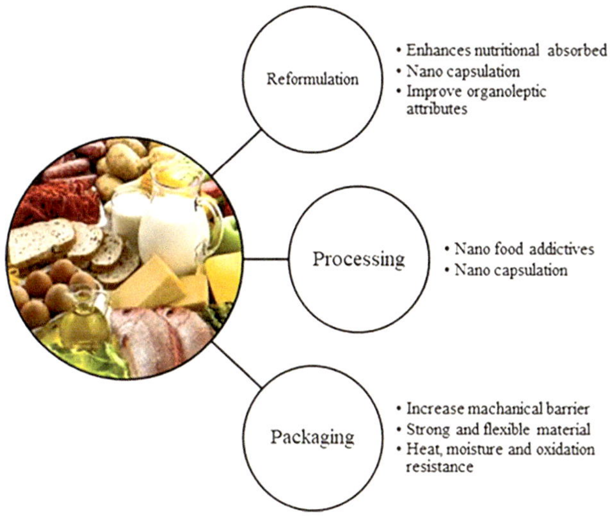

FIGURE 11.2 Potential application of nanotechnology in the food industry. (Adapted by authors from Moradi et al. 2022).

reformulation of materials, optimisation of food processing, enhancement of packaging materials and improvement of food safety and security.

11.2.2 Reformulation of Materials

In the present era, the potential of nanotechnology to improve modern agriculture has led to the advancement of the food industry (Moradi et al. 2022).

11.2.2.1 Enhances Nutritional Absorption and Raw Materials Availability

Extensive research has been conducted on the application of nanotechnology and its impact on the improvement of nutritional content and nutritional uptake in crops and plant harvests. The continuous improvement of nutritional content in the food industry is directly related to upstreaming industries including agriculture, processing engineering, manufacturing, packaging, marketing and supply chain (Brar, Bhardwaj and Chauhan 2023). Nevertheless, there is a strong relationship between the food industry and the agricultural sector. Even though the food industry is divided into the eat-out industry and the manufacturing industry, the connection between these two industries is still high and interdependent. Statistics have cautioned a rapidly growing global population, concomitantly requiring the agri-food sector to ensure food security and availability (Lata, Bhardwaj and Garg 2022; Brar, Bhardwaj and Chauhan 2023; Zhang et al. 2023b). Another element to consider is that various developing countries are highly reliant on the agri-food sector for their economic prosperity. The yield within the agriculture sector continues to pose a threat due to soil erosion, poor groundwater and soil contamination, growth of weeds, river eutrophication and climate change. The application of chemicals such as pesticides, fertilisers, insecticides and herbicides for production improvement also harmed groundwater and soil contamination contributing to significant adverse effects on living organisms. The use of nano-based biosensors for early detection of the presence of chemical contaminants and pathogens and the sufficiency of nutrients can be useful to enhance the quantity and quality of the agri-produce (Lata, Bhardwaj and Garg 2022).

For instance, various techniques have been developed and adopted within the domain of nanotechnology to upsurge nutrient absorption by crops, develop genetic improvement of crops and animals, control mechanisms for crops and animal disease, identify crops that are resistant to climate change and reduce the application of hazardous pesticides or antibiotics (Lata, Bhardwaj and Garg 2022; Haris et al. 2023; Brar, Bhardwaj and Chauhan 2023). Additionally, nano-fertiliser is perceived as a promising alternative compared to traditional fertilizers. This is because nano-fertilisers are known to improve the micro and macro elements of soil and enhance nutrients while preventing water pollution (Alhalili 2023). Furthermore, the application of nanotechnology in pathogen control has been observed as a highly effective and cost-effective approach (Zhang et al. 2023a). A different school of thought indicated that the use of micronutrients in a nano-size can be utilised to improve the yield of crops. Therefore, the application of nanotechnology can create crop sustainability, improve yield, minimise input cost and improve profitability in the agricultural sector (Lata, Bhardwaj and Garg 2022; Brar, Bhardwaj and Chauhan 2023). It is reasonable to assume that the increasing yield in agriculture can have a positive impact on the food industry. In light that the agricultural sector supplies raw materials to the food industry, the food quality standard will increase and improve compliance the customer and regulatory requirements, the cost of food will become affordable and increase food availability, improve food security, enhance nutritional content and remove export barriers (Lata, Bhardwaj and Garg 2022; Zhang et al. 2023a). Indeed, in developing countries, the application of nanotechnology is a solution to obtain sustainability in the food system (Brar, Bhardwaj and Chauhan 2023). Another element to consider is nanoencapsulated fertilisers which serve a similar purpose such as increasing nutrient uptake (Brar, Bhardwaj and Chauhan 2023).

The application of nano-fertiliser at the agricultural level helps in improving the nutritional content of the harvested crops. Therefore, increasing the nutrients of crops at the agricultural level directly increases the nutritional content of fresh produce. By doing so, the nutritional content of raw materials and food products increases (Zhang et al. 2023a). Furthermore, nanocapsules, nanoparticles and nanofiber can be used to acquire desired traits or transport foreign material, for example, deoxyribonucleic acid to change genes. Even though changing the plant's genomes is complicated, de novo synthesis is a strategy that is employed for designing and creating new desired gene analogs (Lata, Bhardwaj and Garg 2022; Karthik and Jeyendra 2022). Genetical engineering continues to play a significant role in silencing undesired endogenic genes and substituting or adding desired foreign genes to improve crops. Typically, genetically modified crops consist of one or a combination of desirable traits like insect resistance, disease resistance, abiotic stress tolerance herbicide tolerance and nutritional improvement (Awad-Allah 2023).

Given that consumers are being more health conscious and more informed, there is a growing global demand for plant-based food. Successful applications of NPs used in agriculture have been outlined in Table 11.1.

Various NPs are used in agriculture to improve harvest, nutritional content and produce quality. For example, a plant growth experiment performed by Awad-Allah 2023 using silica nanoparticles (SNP/nano-SiO2) at the nanoscale ranging from 20–40 nm indicated that SNP increased wheat seed germination by 95.5%, and this was

TABLE 11.1
Various Applications of NP in Agri-Food Industry

Application	Nanoparticles	Characterisation of NPs	Outcome	Reference
Wheat *Triticum aestivim*	Silicon/silica NPs	Nano size: 20–40 nm	Wheat seed germination increased by 95.5%.	Alhalili 2023
Rice *Oryza sativa*	Selenium and silicon NPs	Concentration ranging from 5, 10 and 20 mg L−1	Hazardous heavy metals content (Cd and Pb) decreased by 62 and 52%. Foliar application of both Se and Si NPs increase grain yield, quality and rice biomass.	Kumar et al. 2020
Bok choy *Brassica chinensis*	Selenium NP	Nano size: 40–60 nm	Enhance nutritional content of bok choy, increase in yield and led to 29% investment increase.	Wang et al. 2022
Rapeseed *Brassica napus L*	Silicon NP	Shape: Spherical shapes Nano size: 14–35 nm Concentration: 250 mg/kg	Improved the plant growth parameters and increased the antioxidant enzymes for plant protection and reduced the absorption of Cd.	Rai et al. 2018
Beans or legumes *Phaseolus vulgaris*	Silicon nanoparticles	Concentration: 5 mmol/L	Bio-Si-NPs improved legumes yield and reduced its contamination level from heavy metals.	Ahari et al. 2022

(*Continued*)

TABLE 11.1 (*Continued*)
Various Applications of NP in Agri-Food Industry

Application	Nanoparticles	Characterisation of NPs	Outcome	Reference
Rice	ZnO NP	Concentration: 40 ppm	Foliar spraying using ZnO NP improves the organoleptic properties for rice grain, eventually, the rice yield increased as well as the nutritional content of the rice.	Elshayb et al. 2021
Tomatoes *Solanum lycopersicum L.*	Copper Nanoparticles	With concentration ranging from 50, 125, 250, 500 mg/L Size: 50 nm proforated twice a day	CuNP enhances the quality of tomatoes, especially their firmness, increases the content of bio-compounds and decreases their enzymatic activities.	López-Vargas et al. 2018

achieved by providing plants with adequate nutrients and suitable pH. It was concluded that SiO_2 is an essential element for plant growth, and at the nanoscale, SiO_2 increases stress tolerance of the wheat crops noting that the agricultural sector faces episodes of droughts and unpredicted climate change. Even though numerous scholars perceive this metalloid as an essential element for plant growth, others disagree as they state that SiO_2 is the non-essential element for plant development. Nevertheless, there is tangible data in the literature indicating that SiO_2 has a great potential for plant growth, particularly under environmental stress conditions to resist the negative effects contributed by nonliving things like salt, droughts and heavy metals (Chand et al. 2020).

Likewise, cadmium (Cd), lead (Pb), arsenic (As), copper (Cu) and zinc (Zn) are heavy metals that are normally absorbed by the plant and have been identified as a hazardous contaminant that pose a threat to the agricultural ecosystem and food safety (Kumar et al. 2020; Wang et al. 2022; Rai et al. 2018; Chand et al. 2020). The main source of these heavy metals is the illegal disposal of untreated industrial wastewater in oceans, reservoirs, lakes, lagoons, rivers and other water bodies (Kumar et al. 2020; Rai et al. 2018). Hence, another plant growth trial was conducted in southern China using selenium and silicon nanoparticles to minimise Cd and Pb content in brown rice. The interchanging ratio of both nanoparticles was used ranging from 5, 10 and 20 mg/L. The result indicated that the content of Cd and Pb was reduced from 62% and 52%, respectively. The foliar application for both Se and Si NPs increased grain yield, harvest quality and rice biomass (Kumar et al. 2020; Rai et al. 2018).

A similar observation was noted in research performed to find the effectiveness of SeNPs on bok choy. The result demonstrated that a negatively charged SeNP with a size ranging from 40–60 nm can enhance the nutritional content of bok choy, increasing yield and leading to a 29% investment increase. Therefore, it was deduced that the use of NPs with suitable size and concentration can lead to the improvement of crop productivity, improve the nutritional content and minimise undesired heavy metals in harvested produce. Consistent results were noted in various studies of the effectiveness of NPs in improving yield production, harvest quality, nutritional content and reducing heavy metals content (Lata, Bhardwaj and Garg 2022; Li, Xu and Ma 2023). The NPs listed in Table 11.1 show promising capability to solve the worldwide obstacles faced by this sector, and that includes fulfilling the food demand of a rapidly growing population, enhancing food security, complying with changing regulatory requirements and improving food quality and food safety.

11.2.3 The Application of Nanotechnology in Food Processing

The application of nanotechnology is regarded as a novel solution in the food processing industry. Several studies have explored various applications of nanotechnology from nanoencapsulation, food additives, antiaging and others (Brar, Bhardwaj and Chauhan 2023; El-Saadony et al. 2020). The following section will explore nanoencapsulation and food additives because of their great interest.

11.2.3.1 Exploring Nanoencapsulation

In general, encapsulation is a technique that is commonly applied in the food industry and other industries such as pharmaceutical industries. It includes preservation and/or protection of sensitive, volatile and bioactive substances from thermal and biochemical degradation (El-Saadony et al. 2020; Awad-Allah 2023). Historically, encapsulation was established six decades ago as a technique for encapsulating the gaseous, liquid and solid compounds (Dar et al. 2020). The application of coats preserves the critical compounds and delivers them to the target site even under undesired conditions (El-Saadony et al. 2020). The coating material varies, for instance it can be composed of monolayers or different layers of coats. The selection and design of encapsulated materials depends based on the nature of the internal compounds, coating techniques and end products (Awad-Allah 2023). However, there are challenges associated with the use of encapsulated material derived from natural sources (such as polysaccharides, proteins, gums and lipids) as they are known to be sensitive to oxygen and ultraviolet as well as being susceptible to chemical degradation during storage (Dar et al. 2020). Hence, there is growing interest in the application of nanoencapsulations due to their ability to efficiently encapsulate, add barrier protection and maintain the integrity of the substance until it is released to its target area (El-Saadony et al. 2020; Awad-Allah 2023). Nowadays, nanoencapsulation is applied for preserving phenolic compounds, antimicrobial agents, antioxidants, vitamins and minerals, essential oils, food colourants and flavours to deliver to them to the target area in the food system (El-Saadony et al. 2020). However, this chapter will explore the advances in the application of nanoencapsulation of bioactive compounds and food colourants and flavours. This is because nanoencapsulation of bioactive compounds and food colourants and flavours is a topic of interest even if it was introduced just a few years ago (Dar et al. 2020).

11.2.3.2 Nanoencapsulation of Bioactive Compounds

Compared with the previous consumer trends, the modern consumer has shifted from ready-to-eat food products to natural foods that are good for health. Moreover, the aquatic medium rosemary (*Rosemarinus Officinalis*) provides a source of substances that can improve health, such as polyunsaturated fats, bioactive peptides, antimicrobials, antioxidants and photosynthetic pigments (Ngasotter, Xavier and Meitei 2023). Nevertheless, currently, bioactive compounds are a topic of interest and have gained popularity from consumers, food organisations, investors and researchers due to their phenomenal health benefits, such as anticancer, antioxidants, antiaging and anti-inflammatory properties for human health (Ngasotter, Xavier and Meitei 2023; Hosseini, Ramezanzade and McClements 2021; Premjit et al. 2022; Tahir et al. 2021). These compounds are essential for human health, and the human body cannot secrete them. Furthermore, there are various sources of bioactive compounds; however, plants are the most common and appropriate source of bioactive compounds and other functional foods. Interestingly, these bioactive compounds are derived from various parts of the plant including roots, stems, gums, leaves, flower peels, tubers and rhizomes, notwithstanding that the concentration of bioactive compounds may vary between different parts (Hosseini, Ramezanzade and McClements 2021). Nowadays, the bioactive compounds that have gained much interest include essential oils, vitamins, fatty acids, polyphenols, carotenoids, flavonoids, alkaloids, terpenes and bioactive peptides (Hosseini, Ramezanzade and McClements 2021; Saifullaha et al. 2019).

Despite the benefit of bioactive compounds present (for example, saponins, phytosterols and polyphenols), concerns have been reported about its undesired sensory attributes (acrid and bitter taste), making it difficult to incorporate with food materials (Hosseini, Ramezanzade and McClements 2021). Moreover, additional limitations of bioactive compounds include thermal instability, photodegradation, low water solubility, undesired food matrix interactions, low bioaccessibility and bioavailability as well as highly volatility. Significantly, the functional properties of bioactive compounds degrade during processing and digestion because they are sensitive (Hosseini, Ramezanzade and McClement 2021; Premjit et al. 2022; Tahir et al. 2021).

During food processing, there is a high likelihood that anthocyanins and vitamin C may be inactivated. Additionally, most nutraceuticals and bioactive compounds are degraded when moving across the gastrointestinal tract due to their sensitivity to other environmental factors like temperature, acidic pH, enzymatic degradation and oxygen (Hosseini, Ramezanzade and McClement 2021; Premjit et al. 2022; Tahir et al. 2021; Lata, Bhardwaj and Garg 2022). In addition, the gastrointestinal system can degrade these substances before they are absorbed by the body (Hosseini, Ramezanzade and McClement 2021; Saifullaha et al. 2019; Lata, Bhardwaj and Garg 2022; Sheybani et al. 2023). Moreover, if there is excessive production of oxygen or an imbalance of oxygen in the cells or products a phenomenon called oxidative stress occurs. Consequently, the oxidative stress causes phenolic compounds to undergo oxidation and produce free radicals. The presence of free radicals in food products decreases the shelf life and quality of the food product (Ngasotter, Xavier and Meitei 2023; Tahir et al. 2021; Lata, Bhardwaj and Garg 2022). In addition, the consumption of food high in free radicals can cause diseases in human beings including certain cancers (Primjit et al. 2022).

Consequently, nanoencapsulation is a novel approach that is applied for conserving bioactive compounds and maintaining their functional properties in supplements, nutraceuticals and food by delivering these micronutrients to targeted areas of human bodies (Sahani and Sharma 2021). Numerous scholars indicated that nanoparticles are considered the best option for improved bioavailability, protection efficiency, prevention of undesirable flavour masking, sustainable encapsulation and controlled release to the target area compared to traditional encapsulating materials (Hadidi et al. 2023; Karthik and Jeyendra 2022; Moradi et al. 2022).

However, there is widespread agreement with the challenges associated with the ingestion of these bioactive compounds effectively, and this is because of their low bioaccessibility, chemical or thermal instability, low lipid and aqueous solubilities and low pH sensitivity (Ngasotter, Xavier and Meitei 2023; Primjit et al. 2022; Tahir et al. 2021). Additionally, these bioactive compounds are normally administrated orally as food matrix and contain some food constituents that may hinder their stability and bioaccessibility. As a result, it is important to preserve the integrity of these bioactive compounds by using appropriate nanostructures. Currently, there is an emerging application of plant proteins as nanoencapsulation agents for protection, ensuring delivery to a targeted site and enhancing the shelf life of bioactivity. Furthermore, plant proteins add stability, bioavailability, and solubility to bioactive compounds even though different delivery systems are used. Additionally, the integration of plant protein together with polysaccharides was found to form an effective barrier mechanism for food bioactive (Hosseini, Ramezanzade and McClement

2021). Even though numerous research articles are based on the encapsulation of bioactive compounds, encapsulation can be performed to protect antimicrobial and antioxidant agents either in food or biopolymer matrix. However, there is limited data on the use of edible coating and its safety (Hadidi et al. 2023). Seemingly, edible nanoencapsulating agents are derived from proteins, lipids and polysaccharides (Sheybani et al. 2023).

Several researchers suggest that polysaccharides are appropriate nanoencapsulation agents in food processing because they are highly soluble, flexible and modifiable and they can bind with other substances in their functional group. To exemplify, chitosan is a suitable food nanoencapsulation agent since it is biodegradable, biocompatible and nontoxic as well as other properties like emulsifying, antimicrobial, flocculent dietary, antioxidant and chelating characteristics (Saifullaha et al. 2019). Furthermore, chitosan poses remarkable properties that can be modified by binding certain ligand atoms allowing easy penetration to achieve an effective delivery system. However, there are deficiencies associated with the use of chitosan as it is poorly soluble in water (El-Saadony et al. 2020). Henceforth, the topical nanoencapsulation technology uses lipid-based colloidal carriers like solid lipid nanoparticles (SLNs), nanostructured lipid carriers (NLCs), nanoemulsion, nano-pyrosomes and nanoliposomes because they are simple to produce in massive production (Premjit et al. 2022; Tahir et al. 2021; Lata, Bhardwaj and Garg 2022). NLCs have recently gained popularity globally because of their remarkable potential benefits and applications (Ngasotter, Xavier and Meitei 2023; Tahir et al. 2021). Additionally, NLCs have positive benefits like low toxicity, biocompatibility, improved stability, improved loading capacity and extended shelf life when compared to other traditional lipid-based nanocarriers (Ngasotter, Xavier and Meitei 2023).

Alternatively, the use of a nanoliposome delivery system has been the topic of interest for the encapsulation of all kinds of bioactive agents such as hydrophobic, amphiphilic and hydrophilic molecules, and it is regarded as a safe food application. This is because their structure is comprised of concentrical spheres of amphiphilic lipids containing an aqueous centre (Shahparast et al. 2019). Then again, it is challenging to produce liposomes at a larger scale due to their low stability. Despite this, heat is applied to scale up and strengthen liposomes. Consequently, there is limited data on the successful application of liposomes. According to some studies the researchers and new product development institutes are intensively exploring adequate techniques to maintain and improve the efficacy of food bioactive and nutraceuticals.

Table 11.2 shows that there are various encapsulating agents and nanoencapsulating techniques that are adopted in the food industry to preserve nutraceuticals and bioactive compounds from undesired environmental conditions and processes. Numerous testing trials discovered that by stabilising these compounds the shelf life of food products was prolonged. For instance, PUFA is used to fortify food products because they play a vital role in human health for brain and cardiovascular function optimisation. However, PUFAs is a poor water-soluble compound and are prone to undergo oxidation during food production and food storage, resulting in the degradation of food organoleptic attributes, deterioration and the formation of toxic oxidative products

TABLE 11.2
The Potential of Encapsulation Agents in Food Applications

Encapsulating Agents	Method	Outcome	Reference
Nanostructured lipid carrier integrated with fish oil (omega-3) as well as α-tocopherol (antioxidant).	Hot ultrasonication methods	The oxidation of polyunsaturated fatty acids (PUFAs) was inhibited as a result of the development of toxic hydroperoxides, and off-flavour was prevented. Furthermore, the shelf life was improved. The pharmacokinetic behaviour of PUFA improved.	Maleki, Woltering and Mozafari 2022; Tahir et al. 2021; Lata, Bhardwaj and Garg 2022
Nanostructured lipid carrier integrated α-tocopherol for coating camelina oil.	Hot homogenization method	The results demonstrated a decrease in radical scavenging, oxidative stability total phenol content and nevertheless an increase in peroxide value.	Ngasotter, Xavier and Meitei 2023
Chitosan NP (ChNps), TPP in conjunction with cellulose nanocrystals (CNCs), for folic acid entrapment.	Ionic gelation	More than 90% nanoencapsulation efficiency was achieved.	Shahparast et al. 2019
A hybrid of ovalbumin (OVA) with high methoxyl pectin (PEC) was used for the encapsulation of Vitamin D_3 (VD_3).	Nanoemulsions	96.37% of VD_3 encapsulation efficiency was obtained and the stability of stored VD_3 was remarkably enhanced after encapsulation.	Rafiee and Jafari 2019
Potato protein nanocomplex was used to coat vitamin D as their low water solubility limits their integration and are susceptible to oxidation.	Protein nanocomplex	The shelf life of vitamin D was extended.	Yimaztekin et al. 2019

(Rafiee and Jafari 2019). Hence, a study was conducted for 40 days to investigate the prevention of oxidation of PUFAs in food products. The results indicated that the oxidation of PUFAs was successfully inhibited, and the shelf life of food products was extended. Preventing the oxidation of PUFA hindered the development of oxidative products such as hydroperoxides and off-flavours (Tahir et al. 2021). Based on these observations, it was deduced that applying nanostructured lipid carriers to preserve PUFAs from reacting with oxygen ions inhibited the process of oxidation. Furthermore, similar observations were confirmed by Lata, Bhardwaj and Garg 2022 and Maleki, Woltering and Mozafari 2022.

Another field experiment was performed on the application of NLC to preserve camelina oil. According to the existing literature, camelina oil is rich in omega fatty acids and vitamin D3 and helps lower heart rates, enhance exercise stamina and improve responsiveness to insulin. A study examined the effectiveness of NCLs incorporated with α-tocopherol for stabilisation of camelina oil from oxidation. This study was performed within 90 days at 110 °C. Several analytical tests were performed as per International Standards for Organizations (ISO) number 3960, 6885 and 6886 to assess the influence of oxidation in camelina oil during storage conditions. The results demonstrated a decrease in radical scavenging, oxidative stability, total phenol content and an increase in peroxide value (PV) (Ngasotter, Xavier and Meitei 2023). Furthermore, it can be deduced that at high storage temperatures and increasing time, PV increases (Poornima et al. 2022).

Another element to consider is the application of animal-derived polysaccharides such as chitosan NPs for nanocarrier synthesis of essential compounds that are considered safe and eco-friendly. This is due to their biodegradability biocompatibility and polycation characteristics (Shahparast et al. 2019; Xiao et al. 2019). Folic acid is a sensitive bioactive compound, and it is normally transported by chitosan NPs to expand its stability and bioavailability. In contrast, there are limitations associated with the application of chitosan NPs alone as a nanoencapsulation agent, such as low pH sensitivity and polydispersity (Sekhon 2014). Hence, the use of chitosan NPs is normally incorporated with other polysaccharides to improve their encapsulation efficiency. For instance, to achieve better results Pooraima et al. (2022) incorporated chitosan NPs, thiamine pyrophosphate (TPP) and cellulose nanocrystals (CNC) to preserve and release folic acid to a target site. As a result, they deduced that the use of hybrid nanosystems created great potential for the encapsulation, stability and targeted distribution of delicate essential compounds through adequate encapsulation efficiency exceeding 90%.

Lastly, the effectiveness of applying the integration protein and polysaccharide delivery system vitamin D_3 (VD_3) was studied. VB_3 is an essential nutritional substance that is crucial for human health. According to Travares and Norena (2020), a deficiency in VD_3 can result in bone medical conditions such as osteoporosis, muscle weakness, growth retardation and fragile bones. Furthermore, VD_3 is regarded as one of the fat-soluble vitamins, prone to oxidation if exposed to the air, sensitive to light and has low bioavailability properties. Hence, there is a need to establish an adequate transporting system to preserve VD_3 in food and to deliver it to its target site. By doing so, the VB_3 intake was enhanced. Another study was conducted to explore the effectiveness of hybrid OVA together with PEC for encapsulation of VD_3 using nanoemulsion. The results showed 96.37% of VD_3 encapsulation efficiency was obtained and the stability of stored VD_3 remarkably enhanced after nanoencapsulation (Rafiee and Jafari 2019). In agreement, protein-polysaccharide NPs provide several benefits for the advancement of nanoencapsulation effectiveness, bioavailability and accurate delivery of nutraceuticals to the target site. Furthermore, various pectin–protein delivery systems are currently being applied for the encapsulation of nutraceutical substances including the controlled release of orange peel oil that is utilized in the food sector as a flavouring agent (Shahparast et al. 2019).

On close analysis, nano-delivery systems and nanoencapsulating agents have proved to add numerous advantages in the food industry focusing on nutraceuticals and bioactive compounds. Moreover, it adds advantages such as inhibiting oxidation and extending the shelf life of fatty acids (Blanco-Llamero 2022; Shahparast et al. 2019; Rafiee and Jafari 2019), increasing the stability of camelina oil (Sheybani et al. 2023), produced >90% of folic acid encapsulation efficiency (Pooraima et al. 2022; Okamoto-Schalch 2020) and >96.3% of vitamin D3.

11.2.3.3 Nanoencapsulation of Flavours and Its Application in the Food Industry

In general, flavouring agents are known as unstable compounds and are usually applied in various industries such as confectionery, dairy, beverage, frozen products and snacks (Khan, Saeed and Khan 2019). These compounds play a significant role in human sensory attributes and have an impact on the product's acceptability to the public. Nevertheless, these compounds are in the liquid form making them suspectable to evaporation and downgrade fast as it is difficult to preserve it through undesired conditioning and interacting with other ingredients during processing and storage (Khan, Saeed and Khan 2019; Alhalili 2023). Furthermore, food organisations are worried about the stability of the flavours as it is easily affected by the enzymatic action, acid, salt and moisture (Poornima et al. 2022). Moreover, flavouring compounds possess organic molecules with light molecular mass and making them highly volatile and vulnerable to moisture, air, light and heat (Khan, Saeed and Khan 2019; Poornima et al. 2022). Lastly, with an exception to the external factors, there are other internal factors affecting the release of flavours.

11.2.3.3.1 Understanding Flavour Release and The Mechanism

There are various ways to describe the rate of flavour release, but a study published by Saifullah et al. (2019) describes it as a transfer of flavour particles from one area to another at a certain time. This concept needs to be significantly considered in the development of food products because flavour can be associated with food quality and taste. The flavour molecule together with air molecules spread out and end up contacting the receptor cells found in the membranous tissue called olfactory epithelium, which is normally located in the nasal cavity. The nerves will send a message to the brain that will detect the type of flavour and identify its taste and food quality. Alternatively, during chewing, the flavouring molecules migrate from the food to the mouth cavity, combine with air, enter into the nasal cavity and interrelate odour proteins to active olfactory transduction. With that being said, numerous factors affecting the release of flavour have been identified. These include kinetic factors (mass transfer and diffusion), thermodynamic factors, weight and size, food composition (protein, lipid, carbohydrate), concentration of flavour molecules and oral processing of food (saliva flow, chewing rate and swallow frequency). In this context, the chemical composition and molecular weight of the cover material affect its attributes. For instance, lighter flavoured molecules increase the diffusion of the flavour/aroma. Furthermore, food with high-fat content restrains the flavour, as a result, heat is required in order to release flavour molecules. On the other hand, too much heat can alter or inactivate the flavour molecule before it serves it purpose (Poornima et al. 2022).

11.2.3.3.2 The Application of Encapsulation in the Food Industry

Encapsulation is employed in order to decrease the degradation and enhance the flavour stability of the flavouring agent. It also improves the oxidative and thermal stability of the flavouring agent during processing (Khan, Saeed and Khan 2019). The core material is encapsulated using an active ingredient in order to preserve it, and this forms a shielding coat for masking unpleasant taste. Additionally, encapsulation enhances the intended function of the flavouring agent (Poornima et al. 2022). There are two techniques that are commonly used for encapsulating flavouring agents: nanoencapsulation and microencapsulation (Poornima et al. 2022). These techniques provide enhanced product integrity and serve by protecting the product from undesired flavours, enhancing the encapsulation efficiency and increasing bioavailability, loading capability and durability (Poornima et al. 2022).

Several studies have supported that nanoencapsulation offers superior functional properties compared to microencapsulation (Khan, Saeed and Khan 2019; Boutillier, Fourmentin and Laperche 2020) because of its high surface-to-volume ratio and its reduced size. Its dynamic physiochemical properties allow it to prevent flavouring molecules from interacting with compounds during processing, improving its sensorial properties and solubility. Furthermore, nanoencapsulation also protects flavouring substances from chemical light and degradation. Hence, nanoencapsulation is currently regarded as the greatest application in masking or preserving aroma and flavour compounds (Poornima et al. 2022). It also preserves the flavouring agents from oxidation and thermal processes and improves their food delivery system (Khan, Saeed and Khan 2019). Nevertheless, some authors strongly disagree with the application of nanoencapsulation, and a recent study reported that microencapsulation technology showed high efficiency in encapsulation compared to nanoencapsulation. Microencapsulation is also reported to be more adequate in controlling the release of flavouring substance; it is easy to handle and exhibits high industrial capability. For instance, a study was conducted on flavoured ice cream. Carnauba wax (microencapsulation) was used to coat peppermint essential oil used to add a flavour to the ice cream and was found acceptable (Poornima et al. 2022). Trained panellists assessed the taste of the peppermint ice cream models and ranked them low at 7 while the control sample ranked at 8 on a scale of 1–10, indicating that there is room for improving this strategy (Blanco-Llamero et al. 2022). However, sometimes the product acceptability depends on the nature of the food that consists of the flavour substance (Poornima et al. 2022). This alternative perception stipulates that the physicochemical, techno-function and biological characteristics of micro or nanoencapsulation systems may perhaps depend on the carrier materials, conditions of encapsulation, nature of the flavouring agent and encapsulation technology. Hence, Table 11.3 outlines successful case studies and the advancement of flavour nanoencapsulation and encapsulation efficiency (Rafiee and Jafari 2019).

11.2.3.3.3 Explicit Discussion of Nanoencapsulation

The studies in Table 11.3 confirm that nanoencapsulation of flavour results in the effective stability of aromatic compounds and a controlled release of flavour. For instance, limonene is the flavour substance that is normally found in citrus fruit peels such as lemon, orange and others. Apart from limonene being used as a flavouring agent, it is also used as an anticancer substance. Furthermore, it is a more reliable substitute for synthetic preservatives. However, limonene is highly volatile, has a high rate of reaction, a hydrophobic structure and is sensitive to oxidation. A study was conducted to explore the application of nanoencapsulation as the mitigation of these aforesaid challenges. Ninety-two percent of limonene was encapsulated using the amylose nanostructures. This was achieved by integrating sonication and thermal means. Successful encapsulation (stability and control released) of limonene was noted, although 5% of flavour burst was noted under acid conditions (Simbine et al. 2019).

Vanilla is a flavour compound or aromatic aldehyde linked with the functional group that is widely used in various cakes from muffins to sponge cakes and dessert cakes. Moreover, it also displays some bioactive properties such as antioxidant, antimicrobial and antitumor properties. However, it is sensitive to light and heat; hence more studies have been conducted to find sound strategic techniques (nanoencapsulation) to maintain its original function even

TABLE 11.3
The Application of Flavour Nanoencapsulation in the Food Industry

Flavour Substances	Type of Encapsulation	Process Description	Major Findings	References
92% Limonene	Nanoencapsulation	Limonene was encapsulated using the amylose nanostructures. This was achieved by integrating sonication and thermal means.	Successful encapsulation of limonene was noted, although 5% of flavour burst was noted under acidic conditions.	Simbine et al. 2019
Vanilla	Nanoencapsulation	Acceptable sensory evaluation was conducted on a muffin incorporated with galactan-flavoured emulsions (GFE) vanilla and control normal vanilla cake.	Sensory evaluation data demonstrated high acceptability GFE vanilla in relation to the controlled vanilla muffin.	Ahari et al. 2022
Oregano essential oil (OEO)	Nanoencapsulation	OEO was nonencapsulated using chia mucilage and gelatine with a ratio of 1:2.	90% of encapsulation efficiency of oregano essential oil was established.	Ahari et al. 2022
Watermelon flavour	Nanoencapsulation	Comparison of watermelon encapsulated with γ-cyclodextrin and pure watermelon flavour.	Superior effective control release of the flavour was noted in a sample that was encapsulated.	Rafiee and Jafari 2019
Peppermint flavour	Nanoencapsulation	Gum arabic was used for encapsulation of peppermint flavour.	The shelf life of flavour was extended and stabilised.	Simbine et al. 2019

if it is exposed to heat during processing (baking). A creative study using acceptable sensory evaluation was conducted on a muffin incorporated with galactan-flavoured emulsions vanilla and control normal vanilla cake. The sensory evaluation data demonstrated the high acceptability of galactan-flavoured emulsions (GFE) vanilla in relation to the controlled vanilla muffin. Therefore, it was deduced that nanoencapsulation can be a suitable technique to adopt in order to stabilise the function of vanilla in muffins (Wu et al. 2022).

Another study was performed to evaluate to use of γ-cyclodextrin (γ-CD) to nanoencapsulate watermelon flavour. Two samples of watermelon flavours were used; Sample 1 was encapsulated γ-CD, and Sample 2 was controlled (nanoencapsulation). The results indicated that there was a greater control release of the flavour and high efficiency in Sample 1 (Rafiee and Jafari 2019). The nanoencapsulation of ginger essential oil using cyclodextrin was successful in a similar study performed by Taghpour et al. 2024. Therefore, it was deduced that nanoencapsulation of flavour and aromatic compound extends the shelf life of flavour controls the release and preserves flavour during processing.

11.2.4 Improvement of Organoleptic Properties Using Nanofood Additives

For the past few decades, the rapidly increasing global population and economic hardships in developing or underdeveloped countries have raised the demand for the optimisation of food production that is cheaper while fulfilling regulatory and customer demands. Microbiological limits, chemical limits and sensory characteristics are customary requirements that are required by the regulator. Thus, the use of food additives is regarded as a solution for the rapidly growing population and economic hindrances. At present, food additives have opened new opportunities in the food and food processing industry. To better describe food additives, it is any substance that is intentionally added to food to enhance its organoleptic attributes, preservation, and other qualities (Kim et al. 2019). Food additives are divided into two categories that include synthetic and natural food additives. Like any other novel approaches, synthetic food additives have curved a market niche in the food industry over natural food additives because they are cheaper, are easier to manufacture and have a long shelf life (Kim et al. 2019). Conversely, natural food additives are regarded as propelling the emerging novel compounds that are accepted by consumers (Wang et al. 2022; Jurić et al. 2022). Figure 11.3 shows several applications of food additives in the food industry (Kim et al. 2019).

Despite different views on the types of food additives, there is agreement amongst scholars on the sensory pleasure and financial opportunities of food additives. Importantly, concerns have been raised by relevant stakeholders about the potential risk associated with the application of synthetic food additives in the food industry. However, studies and regulatory bodies have emphasised the use of food-grade and acceptable limits in food products. As a result, organisations continue to enjoy the benefits linked with the use of food additives such as antioxidants emulsifiers, stabilizers, colourants, preservatives thickeners sweeteners and gelling agents (Kim et al. 2019; Bhat et al. 2023; Mingming et al. 2014).

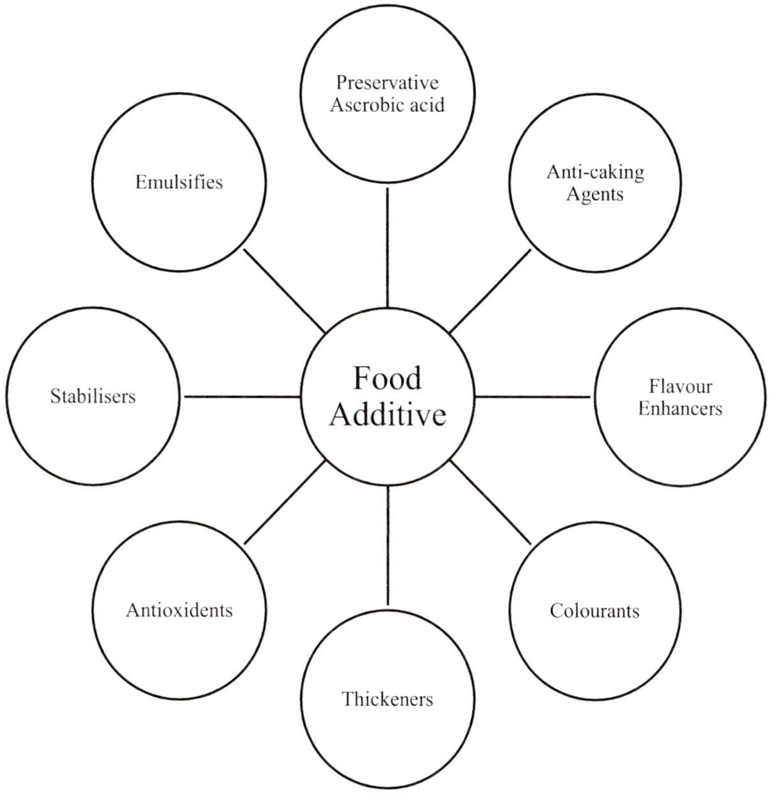

FIGURE 11.3 This demonstration illustrates various functions of food additives in the food industry. (Adapted by authors from Kim et al. 2019).

11.2.4.1 Introduction to Nanofood Additives

The use of nanomaterials for the enhancement of food additives is widely researched and discussed. Numerous food products containing nano-additives are available on the market. Additionally, these nano-additives are intentionally added as ingredients, used for food processing and infused in the packaging materials to improve their original function (Mingming et al. 2014). Moreover, these nano-additives can be used to trigger a reaction and/or improve the bioavailability of functioning substances (Bhat et al. 2023). Given this, colour is one of the vital food organoleptic attributes because it attracts consumers; it is also linked with flavour, nutritional value and food safety. For instance, if the fruit is fully ripened it gives the impression that it has high nutritional value. Secondly, if the bread is a grey to greenish colour it indicates that the bread is no longer safe for consumption. Therefore, it can be deduced that colour can have a direct influence on the consumer's perspectives and advertising of the product in the market. Furthermore, food additives are also used as a food nutrition enhancer to fortify degraded nutrients throughout food processing (Akbari-Alavijeh, Shaddel and Jafari 2020). Therefore, they can act to combat food insecurity and malnutrition and encourage nutritional content (Kim et al. 2019; Mingming et al. 2014). Numerous studies have outlined various nano-additives in the food industry to enhance sensory properties and extend the shelf life and nutritional content of the product (Mingming et al. 2014).

11.2.4.2 Nanofood Additives and Their Application

It is safe to say that the application of nanotechnology in this domain has brought numerous benefits that could have not been achieved if these nanomaterials were not available or explored. Despite the benefits associated with nanofood additives in the food industry, access to utilise nanofood additives has the potential to cause risks to human health (Mingming et al. 2014). However, according to the existing literature, there are inconsistent findings on the safety of nanoparticles, either directly or indirectly incorporated into food products. Therefore, there is a need for a comprehensive exploration of NPs (Mingming et al. 2014). Furthermore, NPs can be administered into the human body in many ways and distributed via blood streams to cells resulting in mitochondrial dysfunction, altering cell membrane permeability, generating excessive reactive oxygen and resulting in health risks such as chronic diseases (Bhat et al. 2023). Previous literature provided studies on the application of nano-additives in the food industry and their success (Mingming et al. 2014). Therefore, it makes sense to explore and review the different types of nanofood additives, methods to produce them and outcomes of those studies.

11.2.4.2.1 Pigment Properties of Titanium Dioxide and Its Application as Food Ingredient

Titanium dioxide (E171 or TiO_2) has gained popularity over the past decades due to its whitening or brightening properties that are extensively applied in numerous food products and beverages (Alhalili 2023). Various studies have reported that TiO_2 NPs are present in several food products for the enhancement of organoleptic properties and product quality improvement (Sahoo et al. 2021; Wang et al. 2022). According to several reports, TiO has been identified in confectionaries like candies, chocolate and chewing gums (Alhalili 2023; Premjit et al. 2022). Furthermore, TiO_2 is also used in non-white food like soups, dried vegetables, sauces, pastries, cakes and cheeses to enhance the opacity and whiteness of these products. Moreover, it also applies to these commodities to prevent their discoloration (Alhalili 2023). Lastly, adequate brightening or whitening colour is attained with a size diameter ranging from 200 to 300 nm because of their light-scattering activities (Primjit et al. 2022). It has been reported that the presence of TiO_2 does not alter the nutritional content of food products nor change flavour.

However, recently emerged evidence has caused public and scholars concerns. The accumulation of TiO_2 has been noted in internal organs like kidneys, lungs and lungs tissues and brain tissues. It has been mooted in recent reports that this accumulation contributes to carcinogenic and genotoxic effects (El-Wakil et al. 2015; Kavitake et al. 2020; Premjit et al. 2022). According to an animal study performed for 90 days, it was found that the daily intake of E171 can cause cardiovascular effects in mice (Premjit et al. 2022). Consequently, and very controversially, the European Food Safety Authority responded to those reports stating that insufficient evidence was revealed indicating that TiO_2 is harmful to human beings (Kavitake et al. 2020). Also, some in vivo and in vitro studies were found to be unreliable as they did not use validated and reliable (standardised) toxicological testing techniques (Kavitake et al. 2020; Premjit et al. 2022). At the same time, some studies used excessive daily dosage in their animal studies, which was extremely above the acceptable daily dosage of 2.25 mg/kg authorised by the (Kavitake et al. 2020). Although the EFSA criticised these studies, the safety of E171 was reviewed (Pateiro et al. 2021). In this respect, the EFSA published a scientific opinion recognising the existing uncertainty concerning the safety of consuming E171 because the EFSA concluded that this food additive is no longer safe, and it can induce genotoxicity effects to the consumers and their progeny (English et al. 2023). On the other hand, the United Kingdom's Committee on Toxicity and the Government of Canada examined the evidence and data presented by various scholars concerning the consumption of E171 and concluded that there is insufficient scientific evidence proving that E171 can potentially harm the consumer (Ganasan et al. 2022).

Nevertheless, in the beginning of 2022, the European Commission together with the members of the states suggested that imported food containing E171 must be banned, and later that year all food products containing E171 was banned. Furthermore, it is worthy to note that France had banned the use of E171 earlier for similar reasons. In parallel, the USFDA allows the use of TiO_2 provided it does not exceed 1% of the overall weight of the foodstuff (English et al. 2023).

The uncertainty with regards to the safety of consuming E171, followed by the ban on the import or production of food products with E171 as an ingredient led to a shift in research area from the successful applications to cautioning consumers on the use of food with E171 due to fear of the possible toxicity and adverse health effects that E171 might cause. Given the aforementioned concerns, this book chapter could not find any updated case studies of the successful application of E171 as a pigment agent. Despite these concerns, according to Bucher et al. (2024), E171 is still applied in several countries around the globe, and various studies have reported detecting it in numerous confectionaries even though it is not mentioned in the product ingredients list. As much as E171 is explicitly discussed in this section, modern customer demands have shifted from the direct use of synthetic food additives as ingredients to food additives used as a fillers in food packaging. This is due to its potential to cause health risks. Hence, there is limited data on the successful application of food additives, especially synthetic ones.

11.2.5 Nanofood Additives That Are Usually Incorporated into Food Packaging Materials

Food additives are also incorporated in food packaging. AgNPs, Zn NPs and Cu NPs are normally incorporated with various modern packaging materials or as coating agents to either minimise or inhibit the growth of pathogenic or spoilage microorganisms (Ganje et al. 2019; EFSA 2021; Simbine et al. 2019; Vos et al. 2020). By doing so, it provides added efficiency for the prevention of undesired microbial growth for extended periods. Even though the traditional method of adding direct food preservatives in foodstuff is a common practice in the food industry, it has presented hindrances in the field, like altering the original taste of food products, good compatibility between preservatives and microorganisms, and lack of selection as they inactivate a broad spectrum of microorganism, even the desired microbes (Simbine et al. 2019). There is agreement among the scholars that among the current NPs, metal-based NPs have drawn attention as a topic of great interest presently. Furthermore, the incorporation of NPs in food packaging has been explored for the past decade, and successful applications have been achieved and shared publicly as shown in Table 11.4.

11.2.5.1 Antibacterial Activities of TiO_2 NPs and Their Application in Food Packaging

TiO_2 is described as an inorganic NP and has been reported to provide constant chemical characteristics and can be used

TABLE 11.4
Successful Applications of Nanofood Additives

Food Additive	Food Application	Results	Reference
Titanium dioxide (TiO_2)	Combination of chitosan, CNC, and TiO_2 NPs in fresh-cut apples.	The quality of fresh sliced apples was preserved for up to 11 days. Ethylene production decreased.	Elshayb et al. 2021
TiO_2	Biodegradable nano-reinforced packaging with Zn-doped TiO_2 to pack plums.	The shelf life of plums improved from 5 days to 21 days.	-Wen et al. 2021
AgNPs	AgNP absorbent pads.	Effective against both *S. Aureus* and *E. Coli*.	Ahari and Lahijani 2021
ZnO and AgNPs	Polymer film embedded with either ZnO and AgNPs against *Lactobacillus* normally found in orange juice stored at 4 °C for 112 days.	The observations illustrated an AgNP highest antimicrobial effect if equated to ZnO NPs. However, some organoleptic properties were compromised.	Zhang et al. 2023b
ZnO NPs	Chitosan biodegradable polymer infused with ZnO NPs were packed chicken meat.	The presence of ZnO enhanced the barrier properties of the polymer, the total bacteria count decreased, the organoleptic properties improved, and the shelf life of chicken increased from 5 days to 7 days.	Wanga et al. 2020a
ZnO NPs	The film was combining ZnO NPs, chitosan and nanoliposomes was used to pack beef meat and evaluated against pure chitosan packaging.	The film showed good mechanical characteristics, high antibacterial activities, high surface hydrophobicity and effective antioxidant effects.	Sahani and Sharma 2021
ZnO NPs	ZnO NPs were tested against both Gram-positive and Gram-negative.	It was found effective against both Gram-negative and positive microorganisms.	Saifullaha et al. 2019
ZnO NPs	Poly vinyl alcohol (PVA) and carbon nanotubes polymer doped with ZnO NPs were evaluated against the pure PVA.	The nanocomposite film has shown more antimicrobial activities and barrier properties compared to pure PVA that was used to pack chicken meat.	Saifullaha et al. 2019
AuNPs	AuNPs embedded in PVA packaging film were used to pack a banana.	This film was found effective against *E. coli* and the shelf life of bananas were extended.	Elshayb et al. 2021
AuNPs	Gallic acid coupled with AuNPs.	Antimicrobials and scavenging effects were noted.	Ahari and Lahijani 2021
Iron NPs	Iron sachet and vacuum technology were used for the preservation of lasagna pasta.	The growth of spoilage microorganisms was minimised, and the shelf life of lasagna pasta was extended from 5 days to 30 days.	Hernández-Nava et al. 2020
CuO + ZnO	CuO + ZnO infused in chitosan-based nanocomposite polymer was used to pack guava.	The antimicrobial and antioxidant effects were noted, and as a result, the quality of guavas was maintained for over a week whereas the quality of guavas packed with traditional packaging was degraded.	Premjit et al. 2022
Chitosan and ZnO	250 and 500 mg/L of chitosan and ZnO coatings	Extended the shelf life and improved the quality of fresh pistachios to 35–40 days. Upon sensory evaluation, crispness and palatability was noted in nanocomposite unlike chitosan only or none-chitosan sample.	Wanga et al. 2020a

for antibacterial activities because it is cheap. Hydrogen gas, carbon dioxide and ethylene are undesired components that stimulate spoilage in food. The blend of TiO_2 NPs with polymeric film minimises these components, prevents undesired microorganism that grows on the product surface and inhibits meat deterioration. For instance, a study confirmed that if TiO_2 is infused in polyvinyl chloride (PVC) it can extend the shelf life of a product (Yilmaztekin et al. 2019). A recent study conducted on sliced apples used bio-nanocomposite film consisting of a mixture of TiO_2, cellulose nanocrystal (CNC) and chitosan to pack sliced apples. Typical, ethylene degradation was decreased by 70% and the shelf life of slices of apples was extended to 11 days. Apart from scavenging activities, barrier, thermal and mechanical properties were improved (Gmoshinski, Bagryantseva and Khotimchenko 2019). Similar observations were noted in other studies (Premjit et al. 2022).

In the same context, a study was conducted using biodegradable nano-reinforced packaging with Zn-doped TiO_2 to pack plums. This improved its shelf life to 21 days. It was noted that as the concentration of Zn-doped TiO_2 increased to 200 µg/ml the maximum zone of inhibition increased. This observation varies per bacteria, for example, *S. aureus* (9.03 mm), *B. pumilus* (22.03 mm), *P. oleovorans* (40.03 mm) and *E. coli* (17.03 mm) (-Wen et al. 2021). Several studies confirmed and supported the antibacterial and antifungal effects of TiO_2 NPs against a broad spectrum of microorganisms. Additionally, TiO_2 NPs are also well-known for strong oxidising properties catalysing the generation of harmful free radicals like superoxide-anion and hydroxyl radicals resulting in the inactivation of microorganisms (-Wen et al. 2021). Therefore, it was supported that the ability of TiO_2 to use photolytic and oxidising properties to inactivate bacteria was effective.

11.2.5.2 Antibacterial Activities of AgNPs and Their Application in Food Packaging

Many studies stated that AgNPs cannot react with the microorganism chemically but serve as a catalyst in suffocating the microbial cells until it is denatured (Hernández-Nava et al. 2020). Hence, the microorganisms are not resistant to them, and the possibility of genetic mutation is unlikely to occur. A limitation associated with the use of AgNPs is that it is one of the most expensive metals. Therefore, the cost of producing AgNPs is expensive. Although AgNPs are expensive to produce, they are commonly used due to their remarkable chemical and physical properties. Apart from the cost of producing AgNPs, other factors make AgNPs a study of interest, especially in food packaging, because it effortlessly fuses in various packaging films, making them versatile enough to be applied to numerous applications (Bhat et al. 2023).

Indeed, several studies have confirmed that the infusion of AgNPs in a pullulan edible film adds benefits to deli meat such as turkey (Umair et al. 2023; Wanga et al. 2020a). Another experimental study was conducted to examine the effectiveness of the AgNP absorbent pad's antibacterial activities against *S. aureus* and *E. coli* in poultry meat. Similarly, the incorporation of low-density polyethylene together with ZnO and AgNP was found effective against *Listeria monocytogenes* and *Pseudomonas aeruginosa* on chicken breast meat. Another stimulating study was performed to assess the inhibition effect of a polymer film embedded with ZnO and AgNPs against *Lactobacillus plantarum* normally found in orange juice stored at 4 °C for 112 days. The observations illustrated that AgNPs produced the highest antimicrobial activities compared to ZnO NPs (Wu et al. 2022). Even though antimicrobial activities were observed in a film incorporated with both AgNPs and ZnO NPs in other past studies, organoleptic properties of meat such as odour and colour were altered (Hosseini, Ramezanzade and McClements 2021). In the same framework, a blend of AgNPs and ethylene-vinyl alcohol matrix was developed to selectively inactivate *Listeria innocua* (Maleki, Woltering and Mozafari 2022). The results indicated that this blend was effective against *Listeria innocua*. Evidently, several studies have confirmed the antibacterial effect of ZnO if incorporated with nanocomposite film on both negative and positive bacteria.

11.2.5.3 Antibacterial Activities of ZnO NPs and Their Application Food Packaging

The use of ZnO NPs for inhibition of microbial growth and cross-contamination is an emerging method in this domain. Studies have shown that ZnO NPs are an antimicrobial agent that is effective against the growth of pathogenic microbial such as *Escherichia coli*, *Klebsiella pneumonia*, *Staphylococcus aureus*, *Bacillus subtilis*, *Candida albicans*, *Bacillus megaterium*, *Aspergillus Niger*, *Pseudomonas vulgaris* and so on (Karthik and Jayendra 2022; Wanga, Chang and Dong 2020; Sahani and Sharma 2021). Secondly, ZnO NPs also show good ultraviolet (UV) protection properties in food products that are UV light sensitive (Saifullaha et al. 2019). Thirdly, ZnO NPs are resistant to adverse conditions like extreme pH values, high temperatures and more. Fourthly, this nanoparticle is regarded as a safe substance for consumption either in bulk or nano-form by the United States Food and Drug Administration. Lastly, ZnO NPs also serve as a biosensor in detecting biological and chemical contaminants in water and food products during storage (Wang et al. 2022). A study was conducted to investigate the antimicrobial activeness of a polymer mixing PVA, carbon nanotubes and ZnO NPs against pure PVA using chicken meat. The content of ZnO ranged from 0, 0.3, 0.6, and 0.9%, and the result indicated that the nanocomposite film with 0.9% showed more antimicrobial activities and barrier properties compared to other samples that were used to pack chicken meat (Wanga et al. 2020a). Another novel active packaging was developed by Rodríguez-Ibarra et al. (2022). This nanocomposite had the ability to release antimicrobial compounds, and it was composed of PVA, gelatine (PVA/G) and ZnO NPs. These scholars evaluated the efficiency of ZnO NPs antibacterial properties, and its impact in on the sensory properties of raw shrimps after 12 days in the refrigerator. The sensory properties and total microbial count of raw shrimps that were stored in a refrigerator for 6 days packed with pure PVA/G compared with the shrimp's film blended with ZnO NPs. A decrease in microbial growth was determined in the raw shrimp packed with active packaging. Also, the organoleptic attributes of shrimps improved. Furthermore, similar results were noted in other related studies (Wang et al. 2022; Sahani and Sharma 2021).

11.2.5.4 Antibacterial Activities of AuNPs and Their Application in Packaging

One more food additive to consider is AuNP, also known as E175. Like any other food additive, Au can be utilized in a nano or microform for the adornment of cakes, confectionery, chocolate, liquors and others. In addition, nowadays, AuNPs are infused in food packaging for its properties such as antioxidants, antimicrobials and barrier properties (Ganje et al. 2019). With an exception to the normal properties, AuNPs have other benefits such as being less corrosive, inert and nontoxic (Elshayb et al. 2021). For instance, AuNPs have demonstrated a visible potential to enhance the properties of food packaging relative to the barrier properties and antimicrobial and scavenging activities. Given the active packaging in the food sector, the infusion of AuNPs into packaging film poses exceptional benefits. Hence, numerous studies investigated the effectiveness of polymeric nanocomposite infused with AuNPs. A study was conducted to validate the application of AuNPs embedded in PVA packaging film for antibacterial activities against *Escherichia coli* causing spoilage in bananas. In this study, the antimicrobial activity of a nanocomposite film was observed, and the shelf life of the banana was qualitatively extended (Elshayb et al. 2021). Another creative study was performed to examine the effects of gallic acid coupled with AuNPs as well as its ability to minimise the degradation of organoleptic attributes on food products because of oxidation. Additionally, this study explored the ability of gallic acid coupled with AuNPs to scavenge free radicals and antimicrobial activities (Akbari-Alavijeh, Shaddel and Jafari 2020). The results demonstrated that gallic acid-AuNPs are the most effective and safe food packaging material in the food packaging sector (Ganje et al. 2019).

11.2.5.5 Antibacterial Activities of Iron NPs and Their Application in Packaging

Like the AuNP application, iron powder sachets are mostly used as oxygen scavengers. Additionally, the combination of sodium chloride together with iron powder is used in modern sachets as scavengers. To further clarify this concept, once the iron powder sachets are sealed and stored at ambient temperature, the relative humidity (RH) of the surroundings will increase the moisture content of the product. The increase in moisture content is a trigger for the oxidation of iron-to-iron oxide, while sodium chloride acts as a catalyst for this reaction to occur even if the relative humidity (RH) is lower than 65%. For this reaction to occur, a RH must be above 65%. Once iron is oxidised, there will be no oxygen left for organisms to use (1 g of iron absorbs 300 cm^3 of oxygen). Furthermore, iron is an important micronutrient required by microorganisms to live. Normally, microorganisms will bind with iron for nutritional benefit. However, in this case, the microorganism will bind with iron-oxide instead of iron and cause oxidative stress to the microorganisms. For example, an iron sachet together with vacuum technology was used for the effective preservation of lasagna pasta. The results indicated that the growth of spoilage microorganisms (such as mould, yeast and coliforms) decreased to acceptable levels for a period of 30 days (Hernández-Nava et al. 2020). It was then concluded that the use of iron powder in conjunction with vacuum technology is effective. Another school of thought suggests that iron NPs can be incorporated into a polymeric film for oxygen-scavenging purposes. This was observed in sausages packaged with packaging nanocomposite infused with iron NPs packaging for 4 days. According to the results, it was noted that as the temperature increased, the process of oxygen scavenging increased (Hernández-Nava et al. 2020).

11.2.5.6 Antibacterial Activities of CuONPs and Their Application in Packaging

Significantly, CuNPs are antimicrobial agents that are universally available and reasonably cheaper compared to silver (Simbine et al. 2019; Vos et al. 2020). Metal oxide NPs have been confirmed to pose prominent applications with antimicrobial effects if blended with a nanocomposite film. A study was conducted to prolong the shelf life by a week of packed guavas using both CuO extracted from *Urtica dioica* leaf and ZnO together with chitosan-based nanocomposite polymer. The results indicated that CuO and ZnO exhibited antimicrobial and antioxidant effects, but CuO was found to be more effective. Furthermore, the study showed a great improvement in guava's appearance within 1 week, compared to unpackaged control fruits (Premjit et al. 2022).

11.3 CONCLUSION

This chapter showed that for the past decades, nanotechnology has gained popularity due to its dynamic physiochemical properties. In addition, the use of nanoparticles in the food industry is increasing, and its applications have gained widespread interest and created many conflicting opinions among peers. The small size and high surface area of NPs increase their rate of reaction, and it is easy to manipulate them to form desirable end-products. Consequently, nanotechnology is mooted as a novel solution to humanities' challenges and offers opportunities for food security from farm to fork. Its application in improving the quality, nutritional content and raw material availability has become apparent. The use of nanoparticles like silica NPs, selenium NPs and copper NPs increased seed germination and yield, nutritional content and quality of plant-based food products. Interestingly, typical chemical toxins and pathogenic contaminants found in crops were reduced to an acceptable level. Therefore, this resulted in food security and increases in profit both in the agricultural and food sectors. Unfortunately, concern about the safety of the application of nano-enabled food additives as ingredients has led to modern consumers being cautious and shifting from food with approved food additives to food products that are free from food additives. Consequently, there is limited data on the successful application of the direct use of food additives.

Nevertheless, a remarkable outcome has been noted in the nanoencapsulation of bioactive compounds and flavour substances. It can be asserted that nanoencapsulation can preserve or protect bioactive compounds and flavour substances from undesirable environmental conditions such as acid, light, high temperature and oxidation during processing or storage till it reaches the target site.

Based on the evidence compiled, active packaging material incorporated with metal oxide nanoparticles proved to improve the safety and quality of foodstuffs. It can also be asserted that the blend of polymeric material with selected metal oxide NPs can also enhance the shelf life of food products.

REFERENCES

Ahari H, Fakhrabadipour M, Paidari S et al (2022). Role of AuNPs in active food packaging improvement: A review. *Molecules* 27: 8027. https://doi.org/10.3390/molecules27228027.

Ahari H, Lahijani LK (2021) Migration of silver and copper nanoparticles from food coating. *Coatings*. https://doi.org/10.3390/coatings11040380.

Akbari-Alavijeh S, Shaddel R, Jafari SM (2020) Encapsulation of food bioactives and nutraceuticals by various chitosan-based nanocarriers. *Food Hydrocoll*. https://doi.org/10.1016/j.foodhyd.2020.105774.

Alhalili Z (2023) Metal oxides nanoparticles: General structural description, chemical, physical, and biological synthesis methods, role in pesticides and heavy metal removal through wastewater treatment. *Molecules*. https://doi.org/10.3390/molecules28073086.

Awad-Allah EFA (2023) Effectiveness of silica nanoparticle application as plant nano-nutrition: A review. *J Plant Nutr* 46(11): 2763–2776. https://doi.org/10.1080/01904167.2022.2160735.

Bhat MI, Shahi NC, Lohani UC et al (2023) Cellulose nanocrystals reinforced chitosan/titanium dioxide bionanocomposite with enhanced interfacial compatibility: Fabrication, characterization, and application in fresh-cut apple slices. *Intern J Bio Macromolecules*. https://doi.org/10.1016/j.ijbiomac.2023.127498.

Blanco-Llamero C, Fonseca J, Durazzo A et al (2022) Nutraceuticals and food-grade lipid nanoparticles: From natural sources to a circular bioeconomy approach. *Foods*. https://doi.org/10.3390/foods11152318.

Boutillier S, Fourmentin S, Laperche B (2020) Food additives and the future of health: An analysis of the ongoing controversy on titanium dioxide. *Futures* 122. https://doi.org/10.1016/j.futures.2020.102598.

Brar PS, Bhardwaj G, Chauhan A (2023) A review on understanding the efficient source of balanced crop nutrition through nanotechnology in agriculture. *J Plant Nutri* 46(13): 3221–3231. https://doi.org/10.1080/01904167.2023.2179923

Bucher G, Hadri HE, Asensio O (2024) Large-scale screening of E171 food additive (TiO_2) on the French market from 2018 to 2022: Occurrence and particle size distribution in various food categories. *Food Cont*. https://doi.org/10.1016/j.foodcont.2023.110102.

Chand K, Cao D, Fouad DE et al (2020) Photocatalytic and antimicrobial activity of biosynthesized silver and titanium dioxide nanoparticles: A comparative study. *J Molecular Liq* 316. https://doi.org/10.1016/j.molliq.2020.113821

Dar AH, Rashid N, Majid I et al (2020) Nanotechnology interventions in aquaculture and seafood preservation. *Crit Rev Food Sci Nutri* 60(11): 1912–1921. https://doi.org/10.1080/10408398.2019.1617232.

EFSA (2021) Physicochemical characterization of nanoparticles in food additives in the context of risk identification. https://doi.org/10.2903/sp.efsa.2021.EN-6678.

El-Saadony MT, Abd El-Hack ME, Taha AE (2020) Ecofriendly synthesis and insecticidal application of copper nanoparticles against the storage pest *Tribolium castaneum*. *Nanomaterials*: 10 130–147. https://doi.org/10.3390/nano10030587.

Elshayb OM, Farroh KY, Amin HE et al (2021) Green synthesis of zinc oxide nanoparticles: Fortification for rice grain yield and nutrients uptake enhancement. *Molecules*. https://doi.org/10.3390/molecules26030584.

El-Wakil NA, Hassana EA, Abou-Zeid RE et al (2015) Development of wheat gluten/nanocellulose/titanium dioxide nanocomposites for active food packaging. *Carbohydr Polym* :124 337–346. https://doi.org/10.1016/j.carbpol.2015.01.076.

English M, Okagu OD, Stephens K et al (2023) Flavour encapsulation: A comparative analysis of relevant techniques, physiochemical characterisation, stability, and food applications. *Front Nutri*. https://doi.org/10.3389/fnut.2023.1019211.

Ganasan E, Yusoff H, Azmi A et al (2022) Food additives for the synthesis of metal nanoparticles: A review. *Enviro Chem Letters* :21 525–538. https://doi.org/10.1007/s10311-022-01473-2.

Ganje M, Jafari S, Tramadon AM et al (2019) *Food Hydrocoll* :95 186–194. https://doi.org/10.1016/j.foodhyd.2019.04.045.

Gmoshinski IV, Bagryantseva OV, Khotimchenko SA (2019) Toxicological and hygienic assessment of titanium dioxide nanoparticles as a component of E171 food additive (review of the literature and metahanalysis). *Health Risk Analysis*. https://doi.org/10.21668/health.risk/2019.2.17.eng.

Hadidi M, Tan C, Assadpour E et al (2023) Emerging plant proteins as nanocarriers of bioactive compounds. *J Cont Release*. https://doi.org/10.1016/j.jconrel.2023.01.069.

Haris M, Hussain T, Mohamed HI et al (2023) Nanotechnology—a new frontier of nano-farming in agricultural and food production and its development. *Sci Total Enviro*. https://doi.org/10.1016/j.scitotenv.2022.159639

Hernández-Nava R, López-Malo A, Palou E et al (2020) Encapsulation of oregano essential oil (*Origanum vulgare*) by complex coacervation between gelatin and chia mucilage and its properties after spray drying. *Food Hydrocoll*. https://doi.org/10.1016/j.foodhyd.2020.106077.

Hosseini SF, Ramezanzade L, McClements DJ (2021) Recent advances in nanoencapsulation of hydrophobic marine bioactives: Bioavailability, safety, and sensory attributes of nano-fortified functional foods. *Trends in Food Sci & Tech* :109 322–339. https://doi.org/10.1016/j.tifs.2021.01.045.

Jurić S, Jurić M, Król-Kilińska Z et al (2022) Sources, stability, encapsulation and application of natural pigments in foods. *Food Rev Int* 38(8): 1735–1790. https://doi.org/10.1080/87559129.2020.1837862.

Karthik SB, Jayendra KA (2022) Generation methods, stability, detection techniques, and applications of bulk nanobubbles in agro-food industries: A review and future perspective. *Cri Rev Food Sci Nut*. https://doi.org/10.1080/10408398.

Kavitake D, Kalahasti K, Devi et al (2020) Galactan exopolysaccharide-based flavour emulsions and their application in improving the texture and sensorial properties of muffin. *Bioact Carbohydr Diet Fibre*. https://doi.org/10.1016/j.bcdf.2020.100248.

Khan I, Saeed K, Khan I (2019) Nanoparticles: Properties, applications and toxicities. *Arab J Chem* :12 908–931. https://doi.org/10.1016/j.arabjc.2017.05.011

Kim EHJ, Paredes D, Motoi L et al (2019) Dynamic flavor perception of encapsulated flavors in a soft chewable matrix. *Food Res Int* :123 241–250. https://doi.org/10.1016/j.foodres.2019.04.038.

Kumar K, Gambhir G, Dass A et al (2020) Genetically modifed crops: Current status and future prospects. https://doi.org/10.1007/s00425-020-03372-8.

Lata S, Bhardwaj S, Garg R (2022) Nanomaterials for sensing and biosensing: Applications in agri-food diagnostics. *Int J Enviro Analyt Chem*. https://doi.org/10.1080/03067319.2022.2115895.

Li Y, Xu R, Ma C (2023) Potential functions of engineered nanomaterials in cadmium remediation in soil-plant system: A review. *Enviro Pol*. https://doi.org/10.1016/j.envpol.2023.122340.

López-Vargas ER, Ortega-Ortíz H, Cadenas-Pliego G (2018) Foliar application of copper nanoparticles increases the fruit quality and the content of bioactive compounds in tomatoes. *App Sci*. https://doi.org/10.3390/app8071020.

Maleki G, Woltering EJ, Mozafari MR (2022) Applications of chitosan-based carrier as an encapsulating agent in food industry. *Trends Food Sci Tech* :120 88–99. https://doi.org/10.1016/j.tifs.2022.01.001.

Mingming Guo M, Jin TZ, Wang L, et al (2014) Antimicrobial films and coatings for inactivation of Listeria innocua on ready-to-eat deli turkey meat. *Food Cont* :40 64–70. https://doi.org/10.1016/j.foodcont.2013.11.018.

Moradi M, Razavi R, Omer A et al (2022) Interactions between nanoparticle-based food additives and other food ingredients: A review of current knowledge. *Trends Food Sci Tech* :120 75–87. https://doi.org/10.1016/j.tifs.2022.01.012.

Ngasotter S, Xavier KA, Meitei MM (2023) Crustacean shell waste derived chitin and chitin nanomaterials for application in agriculture, food, and health—a review. *Carbohydrate Poly Tech Appl*. https://doi.org/10.1016/j.carpta.2023.100349.

Pateiro M, Gómez B, Munekata PES et al (2021) Nanoencapsulation of promising bioactive compounds to improve their absorption, stability, functionality and the appearance of the final food products. *Molecules*. https://doi.org/10.3390/molecules2606154.

Poornima VB, Katiyar R, Banerjee S, Patti A, Arora A (2022) Polysaccharide-based nanostructures as nutraceutical carriers from: *Handbook of Nanotechnology in Nutraceutical*.

Premjit Y, Pandhi S, Kumar A et al (2022) Current trends in flavor encapsulation: A comprehensive review of emerging encapsulation techniques, flavour release, and mathematical modelling. *Food Res Int* 151. https://doi.org/10.1016/j.foodres.2021.110879.

Rafiee Z, Jafari SM (2019) Application of lipid nanocarrier for the food industry. *Bioact Mol Food* 623–665. https://doi.org/10.1007/978-3-319-78030-6_93.

Rai M, Ingle AP, Pandit R et al (2018) Copper and copper nanoparticles: Role in management of insect-pests and pathogenic microbes. *Nanotechnol Rev* 7(4): 303–315. https://doi.org/10.1515/ntrev-2018-0031.

Rodríguez-Ibarra C, Medina-Reyes E, Deciga-Alcaraz A et al (2022) Food grade titanium dioxide accumulation leads to cellular alterations in colon cells after removal of a 24-hour exposure. https://doi.org/10.1016/j.tox.2022.153280.

Sahani S, Sharma YC (2021) Advancements in applications of nanotechnology in global food industry. *Food Chem* 342. https://doi.org/10.1016/j.foodchem.2020.128318

Sahoo M, Vishwakarma S, Panigrahi C et al (2021) Nanotechnology: Current applications and future scope in food. *Food Front* 2(1): 3–22. https://doi.org/10.1002/fft2.58.

Saifullaha MD, Shishirc MRI, Ferdowsi R et al (2019) Micro and nano encapsulation, retention and controlled release of flavor and aroma compounds: A critical review. *Trends Food Sci Tech* :86 230–251. https://doi.org/10.1016/j.tifs.2019.02.030.

Sekhon BS (2014) Nanotechnology in agri-food production: An overview. *Nanotechnol Sci Appl*: 31–53. https://doi.org/10.2147/NSA.S39406.

Shahparast Y, Eskandani M, Rajaei A et al (2019) Preparation, physicochemical characterization and oxidative stability of omega-3 fish oil/α-tocopherol-co-loaded nanostructured lipidic carriers. *Adv Pharm Bulletin* 9(3). https://doi.org/10.15171/apb.2019.046.

Sheybani F, Rashidi L, Nateghi L et al (2023) Application of nano-structured lipid carriers containing α-tocopherol for oxidative stability enhancement of camelina oil. *Indust Crops Prod* 202. https://doi.org/10.1016/j.indcrop.2023.117007.

Simbine EO, Rodrigues LC, Lapa-Guimaraes J et al (2019) Application of silver nanoparticles in food packages: A review. *Food Sci Tech Campinas* 39(4): 793–802. https://doi.org/10.1590/fst.36318.

Sreelakshmi B, Induja S, Adarsh PP et al (2021) Drought stress amelioration in plants using green synthesised iron oxide nanoparticles. *Mat Today Proc* :41 723–727. https://doi.org/10.1016/j.matpr.2020.05.801

Taghipour S, Nia AE, Hokmabadi H et al (2024) Physicochemical and quality characters of fresh pistachio (*Pistacia vera* L.) cultivars in response to chitosan/ZnO nanocomposite coating. *Food Chem* 435. https://doi.org/10.1016/j.foodchem.2023.137136.

Tahir A, Ahmad RS, Imran M et al (2021) Recent approaches for utilization of food components as nano-encapsulation: A review. *Int J Food Prop* 24(1): 1074–1096. https://doi.org/10.1080/10942912.2021.1953067.

Travares L, Norena CP (2020) Encapsulation of ginger essential oil using complex coacervation method: Coacervate formation, rheological property, and physicochemical characterization. *Food Bioproc Tech* :13 1405–1420. https://doi.org/10.1007/s11947-020-02480-3.

Umair M, Sultana T, Xun S et al (2023) Advances in the application of functional nanomaterial and cold plasma for the fresh-keeping active packaging of meat. *Food Sci Nutr*. https://doi.org/10.1002/fsn3.3540.

Vijayalakshmi S, Sachin C, Kirtan T (2017) Nanotechnology: A growing need for agriculture and food sectors. *Integra Ferroelectrics* 185(1): 73–81. https://doi.org/10.1080/10584587.2017.1370341

Vos SD, Waegeneersm N, Verleysen E et al (2020) Physicochemical characterisation of the fraction of silver (nano) particles in pristine food additive E174 and in E174-containing confectionery. *Food Addi & Contam* 37(11): 1831–1846. https://doi.org/10.1080/19440049.2020.1809719.

Wang H, Iqbal A, Murtaza A et al (2022) A review of discoloration in fruits and vegetables: Formation mechanisms and Inhibition. 39(9): 6478–6499. https://doi.org/10.1080/87559129.2022.2119997.

Wanga C, Chang T, Dong S (2020a) Biopolymer films based on chitosan/potato protein/linseed oil/ZnO NPs to maintain the storage quality of raw meat. *Food Chem*. https://doi.org/10.1016/j.foodchem.2020.127375.

Wen YH, Tsou CH, Guzman MR et al (2021) ntibacterial nanocomposite flms of poly (vinyl alcohol) modifed with zinc oxide--doped multiwalled carbon nanotubes as food packaging-. *Poly Bull* :79 3847–3866.

Wu L, Zhang C, Long Y et al (2022) Food additives: From functions to analytical methods. *Crit Rev Food Sci Nutr* 62(30): 8497–8517. https://doi.org/10.1080/10408398.2021.1929823.

Xiao Z, Hou W, Kang Y et al (2019) Encapsulation and sustained release properties of watermelon flavor and its characteristic aroma compounds from γ-cyclodextrin inclusion complexes. *Food Hydrocolloids*. https://doi.org/10.1016/j.foodhyd.2019.105202.

Yilmaztekin M, Lević S, Kalušević A et al (2019) Characterisation of peppermint (*Mentha piperita* L.) essential oil encapsulates. *J Microencapsul* :36 109–119. https://doi.org/10.1080/02652048.2019.1607596.

Zhang T, Li L, Chunta S et al (2023a) Enhanced oral bioavailability from food protein nanoparticles: A mini review. *J Contr Release* 54(6): 146–154. https://doi.org/10.1016/j.jconrel.2022.12.043.

Zhang ZH, Wang S, Cheng L et al (2023b) Micro-nano-bubble technology and its applications in food industry: A critical review. *Food Rev Int* 39(7): 4213–4235. https://doi.org/10.1080/87559129.2021.2023172.

12 Nanotechnology in Food Packaging and Preservation

Veerapandi Loganathan[a], Abhinash P[b], and Lekhashri Vijayan[c]
[a,b]Department of Food Technology, Nehru Institute of Technology, Kaliyapuram, Coimbatore, Tamil Nadu, India
[c] Department of Food Technology, Karunya Institute of Technology and Sciences, Coimbatore, Tamil Nadu, India

12.1 INTRODUCTION

Nanotechnology is a multidisciplinary area incorporating principles from biochemical and various engineering fields (de Francisco and García-Estepa 2018). The primary objective is on systems, components, and subsatellites within the size range of 1–100 nanometers, where the physicochemical properties diverge significantly from those of larger-scale materials made from the same components (Chellaram et al. 2014). NPs, viewed as minuscule entities exhibiting singular qualities and performance, represent a distinct level of engagement in nanotechnology (Ozimek et al. 2010). Compared to bigger particles with similar composition, their small size results in a greater area of contact and more substantial transfer of mass, which is explained by their behavior as single units with enhanced synthetic and biological activities, kinetic actions, metabolic stimulation, and atomic attributes (Avella et al. 2007).

Nanomaterials are classified according to molecular proportions, properties, and structures, providing a systematic framework for understanding and manipulating these materials. Nanoparticles possess beneficial functions such as dispersion, accessibility, optical science, hue, power, electromagnetism, and kinetics (Sahoo et al. 2021). The implications of nanotechnology extend across diverse applications, ranging from medicine and drug delivery to electronics and materials science. The unique characteristics of nanomaterials make them appropriate for pioneering technologies as well as products. However, the advancement of nanotechnology also raises ethical considerations and potential challenges, necessitating a balanced approach to harness its benefits while mitigating risks. In essence, nanotechnology is the convergence of various scientific disciplines for exploring and manipulating materials at the nanoscale, opening avenues for groundbreaking advancements across industries. The manipulation of materials within the 1–100 nanometer range unlocks unique properties that differentiate them from their macro counterparts. Nanoparticles, which function as singular entities, display enhanced attributes owing to their size, contributing to the development of novel technologies (Ariyarathna et al. 2017; Omerović et al. 2021). This characteristic makes nanomaterials particularly promising for useful functions in the drug and microelectronics fields (Hoseinnejad et al. 2018; Sharma et al. 2020). While nanotechnology holds immense promise, it is imperative to approach its development with a mindful consideration of ethical implications and potential risks. The unique properties of nanomaterials, while beneficial, also pose challenges that require careful navigation. Achieving equilibrium between innovation and conscientious development is essential for harnessing the complete capabilities of nanotechnology (Nile and Kai 2021; Wang et al. 2021). The manipulation of materials at the nanoscale, in conjunction with the multidisciplinary character of nanotechnology, has created new avenues for scientific and technological advancement. Nanoparticles, with their distinctive characteristics, redefine possibilities in various fields. The systematic categorization of nanomaterials enhances our understanding, paving the way for innovative applications. However, responsible development and ethical considerations must accompany these advancements to ensure a sustainable and beneficial future for nanotechnology (Mohammadian et al. 2020).

According to Patra et al. (2018), the application of nanotechnology has turned into a crucial element in the food processing industry because it offers novel techniques to enhance physical and chemical attributes and increase dietary retention and absorption. NPs have become extremely desirable in a variety of industries due to their unique mesoscopic qualities, which consist of a substantial area, enhanced reaction, microscopic dimension, resilience, nuclear impact, and flexibility (Ramakrishnan et al. 2021). The development of the food and agriculture sectors mirrors the advancements witnessed in medication delivery and pharmaceuticals, with nanotechnology playing a pivotal role in this transformation (Lopez-Rubio et al. 2006). Investigations of the synthesis, classification, functions, and evaluation of nanoparticles have increased considerably because of the uniqueness of nanoparticles, which can be distinguished from larger particles by variations in their physical and biological characteristics. In recent years, this investigation has encouraged creativity and modernized the entire agricultural products business (Bouwmeester et al. 2018; Siddiqi et al. 2018).

DOI: 10.1201/9781032677422-12

The utilization of nanomaterials in dietary processes has contributed to the invention of an extensive variety of creative goods that have improved chewiness, flavor, aroma, and consistency (Anandharamakrishnan and Parthasarathi 2019). Researchers and industry groups alike have underscored the vast applicability of nanotechnology in almost every aspect of the food industry, spanning agriculture (Pathakoti et al. 2017; Sahoo et al. 2021). The viability of using nanoparticles to enhance hygiene procedures, maintain the quality of food, improve packing methods, and even develop dietary supplements with distinctive characteristics and improved dietary value has become the focus of several studies (Noruzi 2016; Singh 2016; Wyser et al. 2016; Ariyarathna et al. 2017).

Nano therapy offers itself as a potential approach in the field of dietary intake, facilitating the discovery of innovative products through nutritional enhancement and computer programs for controlling the transportation of nutrients (Sahoo et al. 2021). This intersection of the food industry and nanotechnology not only opens doors to unprecedented advancements in food quality but also holds the capacity to tackle principal issues in food safety, nutrition, and sustainability. The ongoing research and application of nanotechnology in this context promises to reshape the landscape of the agri-food sector, offering solutions to contemporary issues and paving the way for a more efficient and sustainable future. To affect nutritional composition and optimize product characteristics, including longevity, odor, and appearance, the food and beverage industries utilize increasing amounts of antioxidants, coloring substances, encapsulated ingredients, microbial detectors, sealing compounds, and other particles. In addition, according to Bott et al. (2014), these technologies are essential for recognizing pathogenic bacteria in food and providing data concerning the quality of food specifications.

In numerous sectors involving food manufacturing, including storage, chemical additives, and packaging, the field of nanotechnology has been utilized when combined with antimicrobial compounds (Pathakoti et al. 2014; Omerović et al. 2021). Enhanced tensile limitations, thermal stability, and decomposition are only a few of the positive aspects that nanotechnology offers above conventional techniques for packaging (Dekkers et al. 2016). Because of their increased antibiotic abilities, nanoparticles have the potential to be used with nanosensors for detecting the degradation of food (McClements and Xiao 2012). Antimicrobial packaging's objective is to hinder the reproduction of microbes on the outer layer of a food item by dispersing chemical components to the product itself or to its surroundings. These components are typically encapsulated in sheets or rigid polymers (Drago et al. 2020). The application of antibiotic microscopic particles in functional packaging has two benefits: it increases the mechanical strength, reduces the weight, and restricts the oxygen accessibility of the packaging while simultaneously shielding food from harmful bacteria, enhancing the shelf life, and preserving freshness (Hoseinnejad et al. 2018; Sharma et al. 2019). This strategy is in line with the more general objectives of effective and sustainable food packaging.

Applications of nanotechnology in the food sector are diverse and impactful. A paradigm change in food processing and packaging has been brought about by the introduction of antibacterial agents and nanomaterials, which enhance sensory attributes. The application of nanoscale technology helps create more resilient, lightweight, and ecologically friendly packaging solutions in addition to extending the shelf life of products. As this field of study progresses, the potential for nanotechnology to revolutionize the food industry remains significant, promising continued improvements in food quality, safety, and sustainability (Committee et al. 2021). Throughout the manufacturing of food and transport networks, production, preparation, wrapping, and storage facilities are available. It represents a single approach in which the field of nanotechnology is used in the food industry to promote food safety (Nile and Kai 2021; Wang et al. 2021). Nevertheless, the use of customized particles in foods raises questions regarding the gastrointestinal contact of humans with nanoparticles (Cao et al. 2016).

Human exposure to these tiny particles can occur primarily through the ingestion of meals enhanced with nanoparticles and through the possible migration of nanoparticles from packaging materials (Wang et al. 2018). Many investigations have related the consumption of nanomaterials to hazardous results, which include harm to genetic material (Lu et al. 2015), denaturation of amino acids (Hong et al. 2017), and the initiation of inflammatory reactions (Khanna et al. 2015). In particular, several studies on the direct consumption of nanomaterials, metallic nanoparticles in the intestine and subsequent harm to tissues have been reported (Cao et al. 2016). This highlights the significance of observing the consumption of nanomaterials in food products. Given these concerns, it is imperative to closely monitor the extent of human exposure to nanoparticles and develop effective techniques for assessing toxicity and ensuring safety (Wyser et al. 2016; Sahoo et al. 2021). Addressing these practical challenges is crucial before the implementation of comprehensive monitoring systems for nanoparticles in food becomes a feasible reality. Finding a balance between nanotechnology's potential advantages in enhancing food security and the imperative to mitigate any associated health risks is essential for responsible and sustainable integration. Rigorous research endeavors to establish regulatory measures are necessary to ensure that nanotechnology contributes to food safety without compromising human health.

The critical function of nanotechnology in strengthening the fortification is evident through the strategic utilization of nanosensors at different food supply chain phases. This encompasses the entire life cycle, ranging from manufacturing and processing to packaging, storage, and transportation. Despite these advancements, the introduction of modified nanoparticles into food products raises valid concerns regarding human exposure, particularly through oral ingestion (Dudefoi et al. 2018). The consumption of foods imbued with nanoparticles may migrate from packaging materials, which could provide significant routes for human exposure. Numerous studies have shed light on the potential adverse effects associated with nanoparticle ingestion, including protein denaturation, oxidative stress responses, DNA damage, and other physiological consequences. Of particular concern is the association between harmful effects on the gastrointestinal tract and other organs and oral exposure to nanomaterials, particularly solid nanoparticles (Xu et al. 2013; Ojha and Kumar 2018).

These considerations highlight the imperative of scrutinizing the application of nanoparticles in food products. To address these challenges, it is crucial to rigorously monitor the extent of human exposure to nanoparticles and develop effective methodologies for assessing toxicity and ensuring the overall safety of nanoparticles in food. These efforts must precede the establishment of comprehensive monitoring systems for nanoparticles in the food industry (Bashir et al. 2022). Achieving a harmonious equilibrium between the potential advantages of nanotechnology in fortifying food security and the need to mitigate associated health risks is paramount for the responsible and sustainable integration of nanotechnology in the food sector. Therefore, a concerted focus on robust research initiatives and the formulation of effective regulatory measures is essential to ensure that nanotechnology contributes to food safety without compromising human health (Bajpai et al. 2018; Brand et al. 2020).

In this chapter, we discuss a thorough investigation and highlight key findings concerning the utilization of nanoparticles in diverse segments of the food industry. Our primary focus was on the practical application and enhancement of nanomaterials within the domain of active or intelligent food packaging. This entailed a comprehensive exploration of the various nanomaterials employed in the packaging of food products, their unique characteristics, and their role in preserving the freshness and overall quality of food products. Additionally, our study investigated the crucial aspect of safety, encompassing evaluations relevant to the health risks of nanomaterials under dietary conditions. We also addressed the inherent challenges associated with the use of nanoparticles in the food industry.

12.2 NANOMATERIALS USED IN FOOD PACKAGING

The process of preparing the key ingredients for the final product indicates that a variety of unit operations are involved in food processing. The primary objectives of handling, sustainability, preservation, transportation, delivery, and storage are the sterilization and elimination of hazardous materials, the deactivation of numerous infectious microorganisms and chemical reactions, the enhancement of vitamins and minerals, and further processing. Employing technological advances that involve nanomaterials may improve the efficiency of processes. The use of nanotechnology in food products falls into one of two categories: "dependent" or "independent", which involve shipment, antibacterial substances, and food constituents (Figure 12.1). These applications depend on process technologies (Dhara and Nayak 2022).

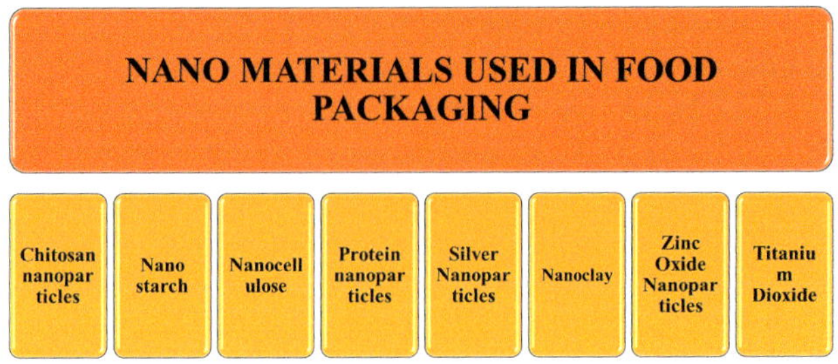

FIGURE 12.1 Different types of nanomaterials used in food packaging.

Nanostructured materials have diverse applications in food processing, both directly and indirectly. In terms of direct utilization, they are integrated into the food matrix itself, serving purposes such as indicating their presence and enhancing various components. Additionally, these materials contribute to the sensory aspects of food, participating in texture improvement, flavor regulation, and the encapsulation of additives. On the indirect front, nanostructured particles are harnessed through smart packaging techniques, ensuring effective hydration of lipids. Moreover, expertly nanostructured catalysts contribute to improve the food processing. The primary goals of the development of nanotechnology in food processing are to improve food quality, introduce new flavors, and maximize nutritional bioavailability. A scientific foundation underlies these advancements, with references to studies by Moraru et al. (2009), McClements and Xiao (2012), and McClements et al. (2017) highlighting the depth of research in this transformative field (Abbas et al. 2009). Nanotechnology plays a crucial part in advancing and improving food processing and production by employing diverse nano techniques and applications (Chau et al. 2007). Given the significance of functional materials in delivery methods, various encapsulation techniques have been implemented (Graveland-Bikker and De Kruif 2006).

Incorporating nanoparticles into food products represents a means to provide nutrients effectively, potentially achieving greater nutrient absorption. Additionally, incorporating nanoparticles into foods enhances their absorption into the body and increases their lifespan. Because of their unique characteristics, particles, fluids, and loaded particles prolong the duration of storage of the product. The top-down approach uses methods including dry processing, high-pressure homogenization, micro fluidization, and ultrasonic emulsification to prepare nanoparticles (Pathakoti et al. 2013). A vast range of food products have been produced differently because of the application of nanomaterials (Kentish et al. 2008).

Nanotechnology has revolutionized the food sector by introducing innovative techniques and applications to enhance various aspects of food processing and production. From the encapsulation of functional elements to the preparation of nanoparticles using advanced methods, nanotechnology offers a versatile toolkit for enhancing the food product's nutritional value, shelf life, and quality. Consumers benefit from the customization of nanoencapsulated foods to meet their specific preferences, marking a significant stride in the intersection of technology and nutrition (Kumar and Kirupavathy 2022).

12.2.1 Chitosan Nanoparticles

Chitosan, a polysaccharide derived from chitin found in the cell walls of crustaceans, arthropod exoskeletons, and fungi, is becoming increasingly popular as a possible nanoparticle for packaging uses. Because of its environmentally friendly profile, nontoxic nature, and remarkable antimicrobial properties, chitosan holds promise for enhancing the attributes of packaging materials (Divya and Jisha 2018). The incorporation of chitosan into biodegradable polymers, like polylactic acid films, enhances the qualities of gas and moisture barriers (Yanat and Schroën 2021). Its incorporation into bio composite films extends benefits beyond physical and mechanical enhancements, encompassing elevated barrier properties and heightened antimicrobial efficacy. The films' antibacterial qualities are further enhanced by the chelation effect and ionic interactions between chitosan nanoparticles and bacterial cell surfaces, which are crucial in limiting nutrition transport and causing cell death.

Furthermore, chitosan nanoparticles positively impact moisture impermeability in bio composite films. The hydrophobic property of chitosan, the creation of hydrogen bonds, and the covalent bonding between the chitosan nanoparticles and the biopolymer are the causes of this beneficial effect. These interactions collectively mitigate the rate of moisture diffusion, reinforcing the resistance of the films to moisture penetration (Garavand et al. 2022). In summary, chitosan has emerged as a versatile and environmentally friendly component of packaging materials. Its antimicrobial properties, combined with its ability to enhance gas and moisture barrier properties in biodegradable polymer films, position chitosan nanoparticles as an asset in the creation of efficient and environmentally friendly packaging. The intricate interplay between chitosan and biopolymers contributes to an overall enhancement of physical, mechanical, and barrier characteristics in bio composite films (Istiqola and Syafiuddin 2020).

12.2.2 Nano Starch

Starch, a complex polysaccharide comprising two polymers, amylose and amylopectin, undergoes a transformation into nano starch through a process involving the disintegration of starch granules via a range of physical and chemical treatments (Morán et al. 2021). Characterized by having at least one dimension smaller than 300 nm and a heightened surface area per unit volume, these starch nanoparticles, also known as starch nanocrystals, play a crucial role in enhancing various properties of composites (Morán et al. 2021). When used as nanofillers in composite materials, starch nanoparticles contribute to notable improvements in strength, flexibility, biodegradability, water impermeability, thermal characteristics, and barrier properties (Dularia et al. 2019; Yu et al. 2021). This utilization of starch nanoparticles as nanofillers is particularly significant because of their positive influence on the overall performance of composite materials.

The integration of starch nanoparticles into composites results in heightened strength and flexibility, resulting in a substantial enhancement in the material properties. Moreover, the increased biodegradability aligns with the contemporary emphasis on sustainable and environmentally friendly materials. The improved resistance to water permeability enhances the suitability of these composites for applications requiring moisture resistance. Additionally, the incorporation of starch nanoparticles positively affects the thermal and barrier properties of composites, expanding their utility across diverse industries. The application of starch nanoparticles as nanofillers is an effective strategy for optimizing the properties of composite materials, meeting a broad spectrum of performance requirements (Oliver-Ortega et al. 2021a, 2021b).

12.2.3 Nanocellulose

Cellulose, an abundant polysaccharide found widely in plant cell walls, serves as a fundamental polymer contributing to the structural strength of plants. The derivative known as nanocellulose takes on a nanostructured form, presenting as nanocrystals or nanofibers with dimensions of a few micrometers in length and diameters less than 100 nm (Sharma et al. 2019). The extraction of nanocellulose from plants involves mechanical and chemical processes. It's remarkable hydrogen bonding capability is a key feature, enabling the creation of robust and impermeable surfaces resistant to molecular penetration, thus establishing excellent barrier properties (Ferrer et al. 2017). In various industries, particularly paper and composites, nanocellulose plays a crucial role in enhancing material properties. Its application contributes to improved strength, mechanical characteristics, uniformity, and biodegradability (Naseer et al. 2018). However, the incorporation of nanocellulose into biopolymers poses a notable challenge due to inherent interfacial differences between the two constituents. Nevertheless, achieving optimal dispersion of nanocellulose can lead to significant enhancements in barrier, thermal, and mechanical properties (Arrieta et al. 2016). The industrial utilization of nanocellulose underscores its transformative potential, leveraging its unique ability to enhance material properties across diverse applications. Whether reinforcing the strength of paper or fortifying composites, nanocellulose stands out as a valuable component, characterized by its nanostructured form and exceptional hydrogen bonding capabilities. Despite the complexities associated with integrating nanocellulose into biopolymers, the anticipated benefits in terms of improved properties make it an enticing avenue for exploration and application in various industries, paving the way for advanced and sustainable materials (Kim et al. 2022).

12.2.4 Protein Nanoparticles

In food packaging, nanoparticles derived from proteins play a pivotal role in augmenting the strength and barrier properties of materials, particularly in terms of water resistance (Zubair and Ullah 2020). A noteworthy example is the utilization of peanut protein nanoparticles, which have demonstrated the capacity to enhance strength, temperature resistance, and moisture barrier properties. When incorporated into protein-starch-based bio composites, these nanoparticles contribute to an overall improvement in material characteristics (Li et al. 2015). Similarly, the integration of zein nanoparticles has been found to enhance both the mechanical strength and moisture barrier properties of films made from whey protein isolate (Oymaci and Altinkaya 2016). The application of protein-based nanoparticles in food packaging addresses key considerations related to material durability and protection against external factors. The incorporation of peanut protein nanoparticles into bio composites has a positive impact on various properties, including strength, resistance to temperature variations, and barriers against moisture. Likewise, the introduction of zein nanoparticles into whey protein isolate-based films results in improved mechanical strength and enhanced resistance to moisture, reinforcing the overall performance of the material. These findings underscore the versatility and potential uses in advancing the functional properties of packaging materials, contributing to their effectiveness in preserving food products and ensuring the integrity of the packaged contents (Biswal and Misra 2020).

12.2.5 Silver Nanoparticles

Silver particles have been employed for many years as antimicrobial agents to safeguard food and beverages. Historically, silver has served a dual role, not only as a container material for water and wine but also in the form of silver vessels mounted beneath bottles to preserve liquids such as milk and water (Li et al. 2017). As a stable metal element, silver poses no significant threat to biological systems when it is used in quantities compliant with the standards set by the Food and Drug Administration (FDA) (Kraśniewska et al. 2020). The utilization of silver nanoparticles is based on their extensive biocidal properties and low volatility, encompassing a broad spectrum of antimicrobial, anti-yeast, antifungal, and antiviral characteristics. Microscale AgNPs contribute to their efficacy while remaining minimally toxic to eukaryotic cells (McClements and Xiao 2017; Abbas et al. 2019). Silver NPs are particularly effective against Gram-negative bacteria because they can easily traverse their cell walls. Through binding to microbial surfaces, these nanoparticles induce the degradation of lipopolysaccharides, resulting in the formation of pits within the cell membrane. Additionally, silver NPs disrupt the ribosomal activity of various microbes, hindering the production of crucial enzymes (Chaudhary et al. 2020).

In contemporary applications, the integration of silver NPs into plastic polymers for packaging materials is prevalent due to their ability to resist microbial growth. Various methods have been employed, such as incorporating silver ions into porous zeolites (Mohr et al. 2019). Research suggests that when silver and zinc oxide nanoparticles are incorporated into low-density polyethylene (LDPE), they effectively enhance the preservation and prolong the shelf life of fruits and vegetables by inhibiting ethylene (Kraśniewska et al. 2020). These active silver nanocomposites, coupled with heat treatment properties, have proven to be highly efficient at exhibiting antimicrobial effects. In recent years, silver has demonstrated efficacy against antibiotic-resistant strains of *Staphylococcus aureus*, gaining popularity as an effective and infrequently utilized clinical intervention (Sungur et al. 2020). The multifaceted antimicrobial properties of silver nanoparticles continue to make them valuable components in various applications, especially in food preservation and packaging.

The historical use of silver vessels in preserving liquids underscores the enduring recognition of silver's antimicrobial properties. As a stable metal with a long history of safe application, silver has a modern role in the form of nanoparticles, contributing to a diverse array of antimicrobial functions. The enhanced surface area of silver nanoparticles compared to their microscale counterparts amplifies their efficacy, ensuring potent antimicrobial activity while maintaining minimal toxicity toward eukaryotic cells (Baig et al. 2021). A significant contemporary application of silver nanoparticles lies in their incorporation into plastic polymers for packaging materials, capitalizing on their ability to inhibit microbial growth. This integration employs various methods, including embedding silver ions into porous zeolite, demonstrating the adaptability of silver nanoparticles in diverse contexts. Notably, studies have revealed their effectiveness, in addition to zinc oxide nanoparticles, in preserving and extending the shelf life of perishable items by absorbing and breaking down the ethylene hormone in low-density polyethylene (Arsene et al. 2021a, 2021b).

Beyond their conventional uses, silver nanoparticles have proven particularly adept at addressing antibiotic-resistant strains of *Staphylococcus aureus*, emerging as a compelling solution with limited clinical incidence. The multifaceted antimicrobial properties of silver nanoparticles underscore their versatility and efficacy in contemporary applications, especially in the critical domains of food preservation and packaging. As technological advancements continue, the enduring role of silver in combating microbial threats remains a testament to its historical significance and ongoing relevance in modern antimicrobial strategies (Primozi et al. 2021).

12.2.6 Nanoclay

The utilization of nanoclays in polymer nanocomposites represents a noteworthy advancement in the market, particularly as it has emerged as the most widely employed nanomaterial in the domain of food packaging (Venkatasubbu et al. 2016). Nanoclay plays a pivotal role in augmenting the physical properties of plastic materials, imparting improved resistance to moisture and gases. The tactoid structure, with clay platelet clusters exhibiting a fence-like arrangement due to their limited compatibility with polymers, is particularly noteworthy within nanocomposites (Bandyopadhyay and Ray 2019). Nanoparticles containing minimal amounts of clay, combined with biodegradable materials, serve as essential fillers in biobased polymers, enhancing barrier properties, resistance to creep, and the mechanical potential of biopolymers (Yadav et al. 2012).

The application of nanoclays in food packaging and medical uses dates to the 1990s. An example in this context is Bayer's development of a nanoclay-engineered polyamide film, commercially known as Durethan (Zhuang et al. 2016). Durethan has found widespread use in beverage packaging due to its exceptional barrier characteristics (Varghese et al. 2019). In recent years, the considerable reduction in the use of plastic bottles in the beer and soda industries, attributed to their suboptimal gas barrier properties, has prompted the adoption of a polymer matrix incorporating nanoclays to address and overcome these inherent challenges.

12.2.7 Zinc Oxide Nanoparticles

Zinc oxide nanoparticles (ZnO NPs) have demonstrated robust inhibitory effects against a wide range of bacterial species, with their effectiveness increasing as the particle size decreases (Manzoor et al. 2018). Notably, ZnO NPs require visible light for activation, and their applications span diverse fields, including modified atmospheric packaging (MAP), cosmetics, medication delivery, and medical devices. Particularly noteworthy is the frequent use of ZnO NPs in food packaging due to their cost-effectiveness and mild toxicity towards eukaryotic organisms, as they can be used in conjunction with nanoparticles such as AgNPs (Arrieta et al. 2016).

Upon exposure to ultraviolet light, ZnO NPs exhibit substantial generation of hydrogen peroxide (H_2O_2), resulting in oxidative stress. Its antimicrobial properties extend to various pathogenic microbes, bacteria, and germs, especially under the influence of UV rays (Habte et al. 2019). Previous studies have suggested that Zn+ ions play a role in inhibiting bacterial growth, showing bacteriostatic effects (Noshirvani et al. 2018a, 2018b). The direct interaction of ZnO with the cell wall of microorganisms is reported to compromise the integrity of bacterial cells, triggering the release of antimicrobial ions such as Zn2+ and the generation of reactive oxygen species (ROS) (Soleimani Zohr Shiri et al. 2019). Comparative analyses underscore the superior performance of ZnO against the growth inhibition of bacteria (Shukla et al. 2019; Arsene et al. 2021a). In dental applications, 80% of ZnO has been proven to be effective at inhibiting *S. sobrinus* biofilms, demonstrating antibacterial effects on S. enteritidis monocyte genes. Furthermore, when integrated into polybutylene succinate (PBS) films along with essential oils, ZnO strongly eliminates the undesirable odors associated with the oils, resulting in enhanced antibacterial effects. Moreover, in the preservation of fresh-cut apples, ZnO NPs contribute to a reduction in decay by catalyzing the oxidation of ethylene into H_2O and CO_2 (Salarbashi et al. 2021).

12.2.8 Titanium Dioxide

Titanium dioxide, an early and crucially significant metal oxide nanomaterial, is distinguished by its remarkable thermostability. Its crystal size places it into three primary phases: anatase, rutile, and brookite (Arfat et al. 2017). Praised for being cost-effective, nontoxic, and photostable, titanium dioxide possesses the unique ability to modify biodegradable films. Approval by the United States Food and Drug Administration (USFDA) for the use of nanomaterials (NMs) in food supplements and materials in contact with food dates back to 1996 (Wang et al. 2018). In food packaging, titanium dioxide is extensively utilized as a dye and stabilizer. Its snowy nanoparticles efficiently scatter visible light, serving their purpose effectively. This nanoparticle contributes to heightened brightness, white light, and overall potency when applied to coated objects. The dairy sector has discovered that introducing titanium dioxide nanoparticles in pasteurizing dairy products after processing, followed by extrusion and packaging in recyclable plastic and polypropylene bottles, significantly minimizes the negative impacts of ultraviolet radiation on the quality of milk (Tabatabaei et al. 2018). Additionally, the extremely small layer of titanium dioxide that is found on packaged food products remains completely inactive in all kinds of foods and food components (Dularia et al. 2019).

Titanium nitride (TiN) serves as an example of a popular titanium-based product used in the production of plastics for food packaging. Beyond its role as a whitener or photocatalyst, TiN exhibits antimicrobial properties, features self-cleaning structures, engages in water filtration, and is utilized for air and water purification, contributing to economic and environmental goals (Rostamzad et al. 2016). Its antimicrobial efficacy arises from the oxidation of the unsaturated poly phospholipid component of microbial cell membranes, establishing its role as an effective biocidal agent against pathogens. The multifaceted applications of titanium dioxide underscore its significance in various industries, particularly as a versatile and reliable component in food packaging and related sectors.

The exceptional photocatalytic properties of titanium dioxide make it highly responsive, allowing the binding of biological molecules (Khalaj et al. 2016; Trifol et al. 2016; Jeevahan and Chandrasekaran 2019). The potency of its nanoparticle façade strengthens, playing a pivotal role in interactions with polymers and fillers. Titanium dioxide is utilized in the fabrication of various nanomaterials with different nanostructures, all of which exhibit antibacterial effects upon exposure to infrared rays (Shankar et al. 2016; Yang et al. 2016; Slavin et al. 2017; Fasihnia et al. 2018; Yousuf et al. 2018; Dash et al. 2019). Although the precise mechanism governing its biocidal activity is incompletely understood, titanium dioxide may initiate oxidative stress on both the outer and inner bacterial cell membranes. This process leads to alterations in coenzyme A activity and eventual DNA damage facilitated by hydroxyl radicals (Ramesh and Radhakrishnan 2019).

Exploration of the impact of titanium dioxide nanoparticles on various microorganisms has revealed their significant effects on antibiotic activity. Particularly noteworthy is the application of titanium dioxide, especially after the photoactivation of ethylene-vinyl alcohol (EVOH) films, which have demonstrated efficacy against these microorganisms. Furthermore, the use of titanium dioxide reduced spoilage in fruits such as apples, tomatoes, and lemons against *Penicillium expansum*. This reduction indicated that growth suppression was achieved through the photocatalytic activity of titanium dioxide NPs under light exposure (Zhang et al. 2019; Ghaani et al. 2016). These diverse and promising applications of titanium dioxide in various antimicrobial contexts, particularly in the domains of food packaging and preservation.

12.3 BARRIER PROPERTIES ENHANCEMENT WITH NANOTECHNOLOGY

Respiration, a vital physiological process, plays a significant role in the postharvest senescence of fresh fruits and vegetables (FVs), as it involves the consumption of substances (Jiang et al. 2013). To address this, nanotechnology has been integrated into the packaging industry to regulate CO_2/O_2 levels, offering innovative solutions to mitigate respiration in postharvest FVs, particularly those with high respiration rates such as broccoli, sweet corn, spinach, mushrooms, and berries.

In a study by Tirgar et al. (2018), a nanopackaging system was developed as an ethylene absorber that effectively oxidizes ethylene into CO_2. This process indirectly inhibited the respiratory rate of packaged banana fruits. Similarly, Gholami et al. (2017) introduced a modified atmospheric package utilizing nanocomposites. This packaging exhibited exceptional oxygen barrier properties while allowing controlled permeability to CO_2 under both ambient and refrigerated conditions. Notably, nanopackaging, through the oxygenolysis of ethylene, led to the accumulation of CO_2, creating a preservative atmosphere. This atmospheric condition proved effective in mitigating the deterioration of postharvest FVs. Additionally, nanopackaging, with its high-water vapor barrier properties, contributed to reducing moisture loss and delaying the cap opening of mushrooms during storage (Gholami et al. 2017). Ultimately, nano-based modified atmosphere packaging has emerged as a promising approach to effectively control microbial growth and physiological activity in postharvest FVs.

12.4 ACTIVE PACKAGING SYSTEMS

Approximately 25% of food consumed by humans is waste, and studies suggest that a massive portion, ranging from 59% to 65% of this food waste is avoidable (Gupta et al. 2021). Food, broadly classified into plant-based (fruits, vegetables, cereals) and animal-based (meat) categories, is prone to spoilage due to various factors (Gupta et al. 2021; Kumar et al. 2022). Current techniques employed to prevent spoilage and minimize waste in the supply chain have proven insufficient and often have adverse environmental impacts (Pandiselvam et al. 2019; Gupta et al. 2021).

The staggering amount of avoidable food waste highlights the need for more effective strategies for managing food throughout the supply chain. The challenges are multifaceted, encompassing issues related to the intrinsic nature of diverse food products, environmental considerations, and the limitations of existing preservation techniques. In response to these challenges, researchers have turned their attention to the potential of advancements in food packaging technology (Kumar and Thakur 2020). The classification of food into plant-based and animal-based categories underscores the diverse nature of food products susceptible to spoilage. Factors contributing to spoilage range from microbial contamination to undesirable environmental conditions, necessitating comprehensive solutions that extend beyond conventional preservation methods. Current practices, while offering some level of protection, fall short in achieving the necessary balance between preservation efficacy and environmental sustainability (Blanke 2014).

In the pursuit of more effective and sustainable solutions, the focus has shifted toward enhancing packaging attributes. The integration of innovative materials, such as metal-organic frameworks (MOFs), represents a promising avenue in this endeavor. These advanced materials, with their unique properties, offer the potential to revolutionize the way we approach food packaging. The exploration of MOFs in this context is a testament to ongoing efforts to address the intricate challenges posed by food spoilage and waste (Figure 12.2).

12.4.1 Biodegradable Active Packaging Coatings

The use of recyclable packaging has increased beyond the use of chemical substitutes because of increasing wellness and environmental concerns. Bioplastics synthesized from natural sources, including proteins, lipids, and

FIGURE 12.2 Active packaging systems in foods.

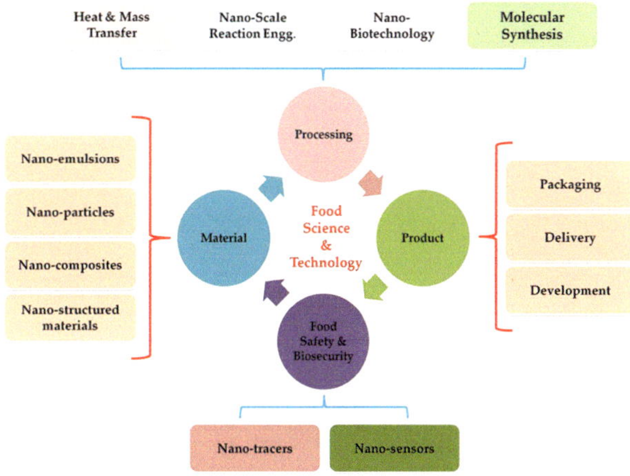

FIGURE 12.3 Application of nanotechnology in different food industries.

polysaccharides, are frequently employed as foundational materials for biodegradable packaging. Notably, these materials are characterized by recyclability, swift decomposition, and nontoxicity (Zhang et al. 2014). A promising approach to environmentally friendly food packaging involves the incorporation of herbal compounds, which have become recognized for their extensive phytochemical characteristics and organic sources. Leveraging these extracts allows the development of active materials that can improve the overall worth of the product in addition to increasing its lifespan. The application methods for these innovative materials involve directly coating a food surface through various techniques. Efforts to address these limitations involve the formulation of blends using two different biodegradable plastics or the incorporation of additives tailored to specific modifications and improvements (Singh 2018). This strategic approach aims to enhance the overall performance and applicability of biodegradable packaging, making it more competitive with traditional synthetic materials while aligning with sustainability goals.

When beneficial substances derived from aromatic compounds and botanical modifications are incorporated into organic coating materials, their oxygen limitation and antimicrobial abilities are enhanced. According to Agarwal et al. (2023), investigations are also being conducted on the incorporation of nanomaterials in biological product packaging to enhance the qualities, notably the longevity, of food. Because they are recyclable, amino acids and polysaccharides play essential roles in the synthesis of films as well as coatings. Among the primary macromolecules that contribute to the development of barriers are carbohydrates, cellulose, salt, gum, etc. (Avramescu et al. 2020).

Proteins are categorized as typical polymers composed of amino acids from various sources used for packing purposes. Due to their abundance and excellent film-forming properties, proteins are extensively utilized. Protein-based films exhibit remarkable barrier effects against gases (O_2 and CO_2) and unstable complexes (Bora et al. 2020; Namratha et al. 2020). Films are prepared with materials that are hydrophobic, such as fatty substances, resin, and hydrocarbons, to enhance their hydration sensitivity (Avramescu et al. 2020). This multifaceted approach involving various natural materials shows the versatility and potential of bio-based films/coatings for improving packaging characteristics in the food industry (Figure 12.3).

12.5 NANOMATERIALS-BASED ACTIVE PACKAGING SYSTEMS

Smart packaging for food is a novel technology that has commercial applications in storing, transporting, and distributing specific food products (Kumar and Thakur 2020). This approach aims to effectively inhibit decay caused by pathogenic bacteria by maintaining appropriate concentrations on the surfaces of food items. Edible films and coatings, which are integral components of active packaging systems, are primarily utilized for prolonging the lifespan of food (Bora et al. 2020). Inhibiting properties in active packaging remove unwanted components from the food surroundings, such as moisture, harmful gases, or odors. On the other hand, emitters release desired components, such as antimicrobials and antioxidants, into packaged foods. The choice of the type of packaging substance changes depending on the product's mechanical activities, demanding specific planning for maximizing longevity (Kumar and Thakur 2020). The resulting films developed from these natural extracts were assessed for their contributions to their bioactive and functional characteristics. This innovative approach represents a dynamic and evolving field with the potential to revolutionize how food items are preserved, ensuring both safety and extended shelf life.

Active packaging, empowered by nanotechnology, plays a pivotal role in elevating the standards of food quality, safety, and shelf life by employing active carrier components that interact with internal and external factors, initiating positive responses (Figure 12.4). The application of

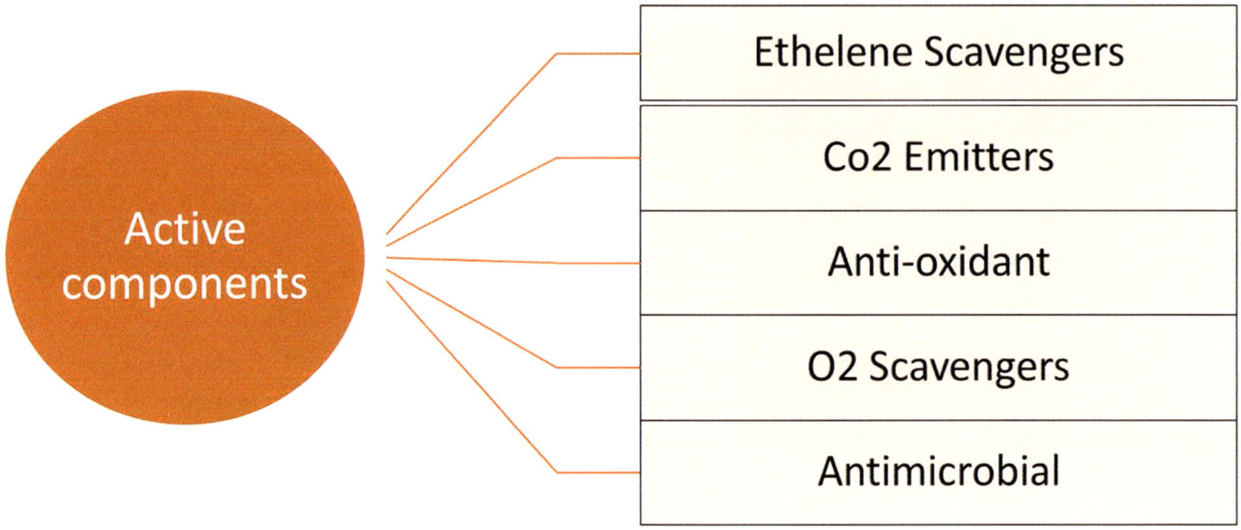

FIGURE 12.4 Active packaging systems.

nanocomposites and coating materials is a pivotal strategy for advancing food packaging. Active packaging is realized by introducing various compounds into food packaging materials to regulate microbial proliferation. While organic compounds such as essential oils, organic acids, and bacteriocins exhibit antimicrobial properties suitable for antimicrobial packaging (Perumal et al. 2018), their reliance on high temperatures and pressures during food processing renders them less ideal for certain applications. In contrast, inorganic nanoparticles employed in antimicrobial food packaging show heightened stability under extreme conditions and demonstrate remarkably potent antibacterial activity even at low concentrations. In what is termed active packaging, this approach hinders bacterial development on the surface of food products, thereby prolonging their shelf life. Films containing various active compounds, such as metal nanoparticles, play a significant role in improving the shelf life and microbial quality of food items. Commonly used metal NPs, such as chitosan, copper, gold, silver, and metal oxide NPs such as TiO and ZnO, possess antibacterial properties and prove highly effective across a spectrum of food applications. The utilization of these nanoparticles is particularly advantageous due to their stability and efficiency against various food spoilage strains, along with their ability to oxidize ethylene or scavenge oxygen. Active packaging, underpinned by nanotechnology and incorporating these innovative materials, has emerged as a promising approach for transforming the landscape of food preservation and storage (Noshirvani et al. 2018a).

12.6 NANOTECHNOLOGY FOR FRESHNESS AND QUALITY MAINTENANCE

Nanotechnology has ushered in a transformative era for the food industry, playing a pivotal role in elevating food safety, extending shelf life, and refining both packaging and food quality. NPs have emerged as potent agents for shielding food from chemical deterioration. Nanomaterials, characterized by reduced reactivity and functioning as carriers of antioxidants, have been harnessed to avert undesirable chemical reactions across a spectrum of food contexts. The encapsulating micronutrients within polymeric nanoparticles have been investigated for controlled release in acidic environments, offering protection against browning and oxidation in fresh produce. Furthermore, nanomaterials have been instrumental in enhancing the functional properties of food by integrating nutraceuticals to mitigate cholesterol levels. Encapsulation methods ensure the stability and controlled release of active compounds, such as epigallocatechin gallate, thereby enhancing antioxidant performance and extending the shelf life of fatty food products (Sharma et al. 2019).

In terms of augmenting the physical properties of both packaging and food materials, the incorporation of engineered nanomaterials has yielded promising outcomes. The development of polymer nanocomposites, which exhibit UV radiation protection and high flame resistance, exemplifies the achievements made in this domain. Notably, nanoparticles such as TiO_2 have gained approval as food additive colorants, while SiO_2 has been employed as an anti-caking agent and a carrier of scents in both edible and non-edible materials. The stabilization of β-carotene through encapsulation within solid lipid nanoparticles has improved the physicochemical stability of the compound, resulting in significant improvements in food aesthetics and stability made possible by nanotechnology (Mehrad et al. 2018a, 2018b). Nanotechnology has played a pivotal role in ensuring food safety by facilitating rapid and precise detection of foodborne pathogens. Nanobiosensors have been employed for detecting various foodborne pathogens. These biosensing methods enable the early identification of pathogens through advanced technologies such as surface-enhanced

Raman scattering (SERS) and lateral-flow immunological test strips. To address the challenge of heavy metal contamination, a range of nanomaterials have been employed for the efficient removal of heavy metals from food products and wastewater. NPs, including silica-modified magnetite and carbon dots, developed as effective adsorbents for heavy metal ions, demonstrating their potential for environmental remediation and pollution control. Moreover, the use of nanotechnology has proven beneficial for inhibiting biofilm formation and preventing bacterial contamination in food processing industries. Nanoparticles such as nano silver and nickel oxide exhibit antimicrobial properties, effectively inhibiting biofilm formation in various bacterial strains, including *Bacillus subtilis* and *Klebsiella pneumoniae*.

These materials significantly contribute to enhancing food safety, prolonging shelf life, and preventing chemical and biological contamination. Despite potential concerns, the promise of nanotechnology to revolutionize food functionality and safety remains robust, with ongoing research and development expected to uncover more advanced applications and solutions (Singh et al. 2017). Nanotechnology stands poised to redefine the landscape of the food industry, offering unprecedented opportunities for innovation and advancement.

12.7 SAFETY AND REGULATORY CONSIDERATIONS

Many goods that use nanotechnology have raised concerns regarding health risks to the general public. The incorporation of intentionally developed nanomaterials in the dietary chain increases the risk of toxic chemicals, resulting in desirable problems for users and impacts on the environment, although numerous food ingredients have existing nanomaterials. The physical and permeability properties of both traditional and organic packaging substances have been enhanced by the inclusion of nanoparticles in food packaging. According to Dudefoi et al. (2018), it additionally supports the production of biologically active and intelligent packaging that utilizes substances that absorb or release chemical substances in the food that has been sealed.

Nanostructured substances possess beneficial properties such as antioxidant substances and antimicrobial agents and can additionally trigger an increase in the production of reactive oxygen species, which may harm tissues or induce oxidative damage. According to Lyu et al. (2017), nanotoxicity may trigger cell death, toxicity to cells, damage to genomes, unregulated proliferation of cells, alterations to cell movement, adverse effects, and the growth of tumors. Numerous factors, including the nature of the nanomaterials utilized, the container composition, the rate of migration, and how quickly specific foods are consumed, can influence nanotoxicity (McClements and Xiao 2017). To guarantee the safe and responsible integration of nanotechnology in food packaging, careful thought and assessment are needed, weighing its advantages against hazards to the environment and public health. The advancement of nanotechnology has introduced products with potential implications for public well-being and eco-friendliness. Usage of NPs into packaging objects to enhance material properties and introduce active or smart functionalities. However, concerns arise regarding potential health risks due to inadequate packaging performance, leading to the transfer of nanomaterials into food. Despite the positive applications of nanostructured materials, their potential to generate reactive oxygen species and induce nanotoxicity necessitates careful evaluation. Factors such as nanoparticle type, packaging matrix composition, migration extent, and food absorption rates contribute to the overall impact.

The evidence indicating health risks linked to the absorption of nanoparticles is steadily increasing, and the use of nano-based food products is raising concerns due to their adverse impact on health, including risks associated with excessive intake, bioaccumulation, and potential overactivity (López-Rubio et al. 2019). The pathways of penetration, further accumulation, and the process of nanoparticle transport inside an organism must be considered when analyzing the possible hazards associated with nanostructured substances (Chau et al. 2007). Growing exposure to certain chemicals throughout epidermal or pulmonary penetration poses major security risks, leading to a thorough examination of the associated hazards, particularly toxicity and mortality (Pathakoti et al. 2013). The consumption of nanomaterials originates through the intestines, which accelerates their movement across the immune system before they reach the bloodstream (Radoi et al. 2008). The emergence of potential health risks associated with nanoparticle absorption underscores the imperative need for comprehensive research and risk assessments. These efforts are crucial for ensuring the safety of nano-based food products and understanding their implications for human health.

The utilization of titanium oxide and its release can have repercussions on both human health and the environment, increasing concerns about potential harm (Zhang et al. 2014). There are only a limited number of detailed studies on the adverse effects of foodborne nanoparticles on illness. Initially, in the investigation of a nanomaterial that has been suggested for potential use throughout the nutrition network, a number of fundamental issues need to be resolved. The process of significant material evaluation of a substance to be considered in the category of nanomaterial. The distinctive qualities or effects of a nanomaterial are intricately linked to its nanoscale characteristics, and if a nanomaterial loses these properties due to processes such as degradation or disintegration, it is expected to behave similarly to its non-nanomaterial counterpart, if one exists. Moreover, other conceivable risks tied to nanomaterials in the food/feed chain include issues of food contamination and their effects on food components (Cha and Chinnan 2004). It is imperative to recognize and address these complex challenges early in the evaluation process to gain a comprehensive understanding of potential risks and to develop effective strategies for mitigating these risks in the context of food safety and human health. Nanotechnology has had a transformative impact on the food industry, revolutionizing aspects such as food safety, shelf life extension, and packaging improvements. NPs have emerged as effective agents for safeguarding food against chemical deterioration, offering enhanced antioxidant capabilities to prevent undesirable reactions within various food matrices (Branton et al. 2008). Employing nanomaterials with reduced reactivity as carriers for antioxidants has been pivotal in preventing undesired chemical reactions.

In addition to improving the physical properties of packaging and food materials, the incorporation of nanomaterials has shown promising results in developing polymer nanocomposites with UV radiation protection and high flame resistance. Certain nanoparticles, such as TiO_2, have gained approval as food additive colorants, while SiO_2 has been found to be useful as an anti-caking agent and scent carrier in both edible and non-edible items. However, the potential benefits of nanotechnology in the food industry are significant and ensuring its safe and responsible implementation is paramount. Establishing safety requirements, permissible ranges, and limitations for nanoparticles in food applications is essential, as is adhering to environmental regulations for their disposal. This framework will facilitate ongoing research and advancements while ensuring the responsible and sustainable use of nanotechnology in food.

12.8 NANOTECHNOLOGY FOR INTELLIGENT PACKAGING

Food quality is tracked and assessed by intelligent packaging. It evaluates the sensory aspects, specifically focusing on typical issues accompanying the decomposition of food, including temperature variations, changes in oxygen levels due to microbial growth, and the generation of gases and acids resulting from food fermentation processes. Intelligent packaging addresses these concerns by managing temperature, utilizing pH indicators, and incorporating radio and biosensors to detect food quality and contamination levels (Ansari et al. 2010).

Smart or intelligent packaging goes beyond conventional methods by providing consumers with real-time information about packed foods. Its capacity to sense, detect, and record changes allows it to provide information on temperature, pH, and oxygen levels. The temperature variations may result in more water activity, undesired phase changes, structural modifications, accelerated degradation rates, or product chill injuries, reducing the shelf life of a product (Branton et al. 2008). An e-nose, functioning as an electronic nose (e-nose), effectively serves this purpose (Lopez-Rubio et al. 2006). A sophisticated electronic nose that uses peptide receptors has been developed to assess raw seafood quality in real time. This innovative technology specifically targets the detection of trimethylamine, a volatile compound associated with food spoilage (Zhou et al. 2009). An ultraviolet-based nano gas sensor exhibiting colorless properties under UV light and turning blue in the presence of oxygen has been developed. Innovations in nanotechnology have yielded an electronic nose with peptide receptors, revolutionizing the real-time assessment of raw seafood quality. The focus on detecting trimethylamine, a spoilage-associated volatile compound, highlights the precision of this technology. Additionally, the e-nose extends its ability to identify volatile sulfides in meat products through the application of silver and/or copper nanofilms, providing crucial insights into meat quality. Gas sensors, which are pivotal in both qualitative and quantitative detection, monitor microbial populations by measuring emitted gas levels (Radoi et al. 2008). The application of antibodies specific to certain pathogens on nanoparticle-conjugated membranes results in a visible color change upon their presence. In addition to these functionalities, intelligent packaging offers additional advantages to manufacturing companies, integrating anti-counterfeit procedures to offer security and safeguard their innovations and products. The enhanced bioavailability of bioactive compounds like vitamin E, vitamin D, beta-carotene, lutein, coenzyme Q10, lycopene, astaxanthin, curcumin, polyunsaturated fatty acids, and essential oils has been achieved using nano-sized delivery systems like nanoemulsions, nano dispersions, and solid lipid nanoparticles. Iron, calcium, and selenium inorganic nanoparticles have demonstrated better bioavailability than their bulk equivalents. Using nano-delivery technologies can help alleviate a significant issue with the use of antioxidants in food, which is their intrinsic poor solubility, structural instability, and poor tissue accumulation (Wang et al. 2009).

12.9 CHALLENGES AND FUTURE PERSPECTIVES

Scientific advances have been made possible by the food industry's quick adoption of nanotechnology. Although there has been significant advancement in the incorporation of nanotechnology in different facets of food production, the development of nanostructures has not experienced a parallel surge. This divergence is significant, given the vast potential that nanotechnology holds for innovating food commodities and processes. However, a series of barriers stand in the way of harnessing these promises effectively. Discussions about the efficacy of substances integrated with nanoparticles that were developed for consumption by individuals have become significant challenges. As highlighted by Agarwal et al. (2023), these concerns intensify with the advancement of studies in food nanotechnology. As such, it is now essential to thoroughly investigate any risks to human health prior to the commercialization of nanofood products. Despite the evident benefits, comprehensive guidelines to assess the food's nanoparticles level of safety have yet to be established. Central to these challenges is the necessity of developing cost-effective and safe edible delivery systems (Gholami et al. 2017). At the nanoscale, materials exhibit distinct behaviors, necessitating a deeper understanding of their effects and toxicities. Addressing these concerns is paramount for enhancing safety standards in the pragmatic use of nanomaterials.

Addressing the issues related to consumer risk requires an understanding of the migratory dynamics of nanoparticles (Jokar et al. 2017). The complexities are made more complex by the ways in which different physiochemical traits affect absorption in the gastrointestinal tract. (McClements et al. 2017). Beyond the scope of human health concerns, the environmental implications of nanotechnology must be addressed. NPs, synthesized by various chemical techniques, are non-eco-friendly by-products that contribute to environmental pollution. The far-reaching consequences of nanoparticles penetrating cells and organs, as observed by Das et al. (2011), necessitate a comprehensive risk assessment strategy (Bajpai et al. 2018).

Additionally, the synthesis of antimicrobial nanoparticles presents additional challenges. Das et al. (2011) emphasized the importance of numerous studies. regulatory agencies like the European Commission and the Food and Drug Administration (FDA) play pivotal roles in navigating these challenges. Consequently, a nuanced approach that integrates public demand, popularity, and a thorough understanding among the possible dangers is indispensable for the continued development, production, and consumption of nano-based food products.

12.10 CONCLUSION

The integration of nanotechnology in the food industry has accompanied a transformative era, offering numerous opportunities to enhance food safety, quality, and sustainability. NPs have demonstrated remarkable potential in revolutionizing various aspects of the food sector, including food preservation, packaging, and nutritional fortification. The unique physicochemical and biological characteristics of nanomaterials have spurred extensive research into their synthesis, categorization, applications, and evaluation, driving innovation and reshaping the agri-food sector in recent decades. One of the most significant contributions of nanotechnology to the food industry is its role in preserving food against chemical deterioration and extending shelf life. NPs, characterized by reduced reactivity, have been harnessed as carriers of antioxidants, effectively preventing undesirable chemical reactions within diverse food matrices. The encapsulation of bioactive substances within polymeric nanoparticles has shown promise in facilitating controlled release in acidic environments, thereby protecting freshness from oxidation and maintaining its nutritional integrity. Furthermore, nanomaterials have been instrumental in enhancing the functional properties of food, integrating nutraceuticals to mitigate cholesterol levels and prolong the shelf life of fatty food products. In addition to improving the physical properties of packaging and food materials, the incorporation of nanomaterials has shown promising results in developing polymer nanocomposites with UV radiation protection and high flame resistance. These advancements in packaging technology hold immense potential for addressing the multifaceted challenges associated with food spoilage and waste, offering greater effectiveness and sustainability.

However, the promising applications of nanotechnology in the food industry are accompanied by concerns regarding human exposure to nanoparticles through oral consumption and potential environmental impacts. Studies have associated the ingestion of nanoparticles with adverse effects, including protein denaturation, oxidative stress responses, DNA damage, and gastrointestinal and secondary organ damage. These findings emphasize the critical need for rigorous analysis of the application of nanoparticles in food products and the development of effective methodologies for assessing toxicity and ensuring overall safety. To address these challenges, it is imperative to establish comprehensive monitoring systems for nanoparticles in the food industry and achieve an equilibrium between the potential advantages of nanotechnology in fortifying food security and the need to mitigate associated health risks. This requires an intensive focus on healthy research initiatives and the formulation of effective regulatory measures to ensure that nanotechnology contributes to food safety without compromising human health.

In conclusion, the utilization of nanoparticles in diverse segments of the food industry presents both opportunities and challenges. While nanotechnology holds immense promise for enhancing food safety, quality, and sustainability, its responsible integration requires a balanced approach that prioritizes thorough research, strict safety assessments, and the establishment of effective regulatory frameworks. By addressing these considerations, the food industry can harness the full potential of nanotechnology while ensuring the well-being of consumers and the environment.

REFERENCES

Abbas KA, Saleh AM, Mohamed A, MohdAzhan N (2009) The recent advances in nanotechnology and its applications in food processing: A review. *J Food Agric Environ* :714–17

Abbas M, Buntinx M, Deferme W, Peeters R (2019) (Bio) polymer/ZnO nanocomposites for packaging applications: A review of gas barrier and mechanical properties. *Nanomaterials* 9:1494

Agarwal A, Shaida B, Rastogi M, Singh NB (2023) Food packaging materials with special reference to biopolymers-properties and applications. *Chem Afr* :6117–144

Anandharamakrishnan C, Parthasarathi S (2019) *Food nanotechnology: Principles and applications*. CRC Press

Ansari AA, Alhoshan M, Alsalhi MS, Aldwayyan AS (2010) Prospects of nanotechnology in clinical immunodiagnostics. *Sensors* :106535–6581

Arfat YA, Ahmed J, Hiremath N, et al (2017) Thermo-mechanical, rheological, structural and antimicrobial properties of bionanocomposite films based on fish skin gelatin and silver-copper nanoparticles. *Food Hydrocoll* :62191–202

Ariyarathna IR, Rajakaruna R, Karunaratne DN (2017) The rise of inorganic nanomaterial implementation in food applications. *Food Control* :77251–259

Arrieta MP, Fortunati E, Burgos N, et al (2016) Nanocellulose-based polymeric blends for food packaging applications. In: *Multifunctional polymeric nanocomposites based on cellulosic reinforcements*. pp. 205–252.

Avella M, Bruno G, Errico ME, et al (2007) Innovative packaging for minimally processed fruits. *Packag Technol Sci Int J* :20325–335

Avramescu SM, Butean C, Popa CV, et al (2020) Edible and functionalized films/coatings—performances and perspectives. *Coatings* 10:687

Baig N, Kammakakam I, Falath W (2021) Nanomaterials: A review of synthesis methods, properties, recent progress, and challenges. *Mater Adv* 2:1821–1871

Bajpai VK, Kamle M, Shukla S, et al (2018) Prospects of using nanotechnology for food preservation, safety, and security. *J Food Drug Anal* 26:1201–1214

Bandyopadhyay J, Ray SS (2019) Are nanoclay-containing polymer composites safe for food packaging applications?—An overview. *J Appl Polym Sci* 136:47214

Bashir O, Bhat SA, Basharat A, et al (2022) Nano-engineered materials for sensing food pollutants: Technological advancements and safety issues. *Chemosphere* 292:133320

Biswal AK, Misra PK (2020) Biosynthesis and characterization of silver nanoparticles for prospective application in food packaging and biomedical fields. *Mater Chem Phys* 250:123014

Blanke MM (2014) Reducing ethylene levels along the food supply chain: A key to reducing food waste? *J Sci Food Agric* 94:2357–2361

Bora H, Kamle M, Mahato DK, et al (2020) Citrus essential oils (CEOs) and their applications in food: An overview. *Plants* 9:357

Bott J, Störmer A, Franz R (2014) A comprehensive study into the migration potential of nano silver particles from food contact polyolefins. In: *Chemistry of food, food supplements, and food contact materials: From production to plate*. ACS Publications, pp. 51–70

Bouwmeester H, van der Zande M, Jepson MA (2018) Effects of food-borne nanomaterials on gastrointestinal tissues and microbiota. *Wiley Interdiscip Rev Nanomed Nanobiotechnol* 10:e1481

Brand W, Van Kesteren PCE, Oomen AG (2020) Potential health risks of nanomaterials in food: A methodology to identify signals and prioritise risks.

Branton D, Deamer DW, Marziali A, et al (2008) The potential and challenges of nanopore sequencing. *Nat Biotechnol* 26:1146–1153

Cao Y, Li J, Liu F, et al (2016) Consideration of interaction between nanoparticles and food components for the safety assessment of nanoparticles following oral exposure: A review. *Environ Toxicol Pharmacol* :46206–210

Cha DS, Chinnan MS (2004) Biopolymer-based antimicrobial packaging: A review. *Crit Rev Food Sci Nutr* :44223–237

Chau C-F, Wu S-H, Yen G-C (2007) The development of regulations for food nanotechnology. *Trends Food Sci Technol* :18269–280

Chaudhary P, Fatima F, Kumar A (2020) Relevance of nanomaterials in food packaging and its advanced future prospects. *J Inorg Organomet Polym Mater* :305180–5192

Chellaram C, Murugaboopathi G, John AA, et al (2014) Significance of nanotechnology in the food industry. *APCBEE Procedia* :8109–113

Committee ES, More S, Bampidis V, et al (2021) Guidance on technical requirements for regulated food and feed product applications to establish the presence of small particles including nanoparticles. *EFSA Journal* 19:e06769

Das S, Jagan L, Isiah R, et al (2011) Nanotechnology in oncology: Characterization and in vitro release kinetics of cisplatin-loaded albumin nanoparticles: Implications in anticancer drug delivery. *Indian J Pharmacol* 43:409

Dash KK, Ali NA, Das D, Mohanta D (2019) Thorough evaluation of sweet potato starch and lemon-waste pectin based-edible films with nano-titania inclusions for food packaging applications. *Int J Biol Macromol* :139449–458

de Francisco EV, García-Estepa RM (2018) Nanotechnology in the agrofood industry. *J Food Eng* :2381–11

Dekkers S, Oomen AG, Bleeker EAJ, et al (2016) Towards a nanospecific approach for risk assessment. *Regul Toxicol Pharmacol* :8046–59

Dhara AK, Nayak AK (2022) Biological macromolecules: Sources, properties, and functions. In: *Biological macromolecules*. Elsevier, pp. 3–22

Divya K, Jisha MS (2018) Chitosan nanoparticles preparation and applications. *Environ Chem Lett* :16101–112

Drago E, Campardelli R, Pettinato M, Perego P (2020) Innovations in smart packaging concepts for food: An extensive review. *Foods* 9:1628

Dudefoi W, Villares A, Peyron S, et al (2018) Nanoscience and nanotechnologies for biobased materials, packaging and food applications: New opportunities and concerns. *Innov Food Sci Emerg Technol* :46107–121

Dularia C, Sinhmar A, Thory R, et al (2019) Development of starch nanoparticles based composite films from non-conventional source-water chestnut (*Trapa bispinosa*). *Int J Biol Macromol* 136:1161–1168

Fasihnia SH, Peighambardoust SH, Peighambardoust SJ (2018) Nanocomposite films containing organoclay nanoparticles as an antimicrobial (active) packaging for potential food application. *J Food Process Preserv* 42:e13488

Ferrer A, Pal L, Hubbe M (2017) Nanocellulose in packaging: Advances in barrier layer technologies. *Ind Crops Prod* :95574–582

Garavand F, Cacciotti I, Vahedikia N, et al (2022) A comprehensive review on the nanocomposites loaded with chitosan nanoparticles for food packaging. *Crit Rev Food Sci Nutr* 62:1383–1416

Ghaani M, Cozzolino CA, Castelli G, Farris S (2016) An overview of the intelligent packaging technologies in the food sector. *Trends Food Sci Technol* :511–11

Gholami R, Ahmadi E, Farris S (2017) Shelf-life extension of white mushrooms (*Agaricus bisporus*) by low temperatures conditioning, modified atmosphere, and nanocomposite packaging material. *Food Packag Shelf Life* :1488–95

Graveland-Bikker JF, De Kruif CG (2006) Unique milk protein-based nanotubes: Food and nanotechnology meet. *Trends Food Sci Technol* :17196–203

Gupta V, Yadav R, Tanwar R, Gaikwad KK (2021) κ-Carrageenan-based bio-nanocomposite film reinforced with cellulose nanocrystals derived from amla pomace for food packaging. *Biomass Convers Biorefin* 1–10

Habte L, Shiferaw N, Mulatu D, et al (2019) Synthesis of nano-calcium oxide from waste eggshell by sol-gel method. *Sustainability* 11:3196

Hong F, Wu N, Zhou Y, et al (2017) Gastric toxicity involving alterations of gastritis-related protein expression in mice following long-term exposure to nano TiO2. *Food Res Int* :9538–45

Hoseinnejad M, Jafari SM, Katouzian I (2018) Inorganic and metal nanoparticles and their antimicrobial activity in food packaging applications. *Crit Rev Microbiol* :44161–181

Istiqola A, Syafiuddin A (2020) A review of silver nanoparticles in food packaging technologies: Regulation, methods, properties, migration, and future challenges. *J Chin Chem Soc* 67:1942–1956

Jeevahan J, Chandrasekaran M (2019) Nanoedible films for food packaging: A review. *J Mater Sci* :5412290–12318

Jiang T, Feng L, Wang Y (2013) Effect of alginate/nano-Ag coating on microbial and physicochemical characteristics of shiitake mushroom (*Lentinus edodes*) during cold storage. *Food Chem* :141954–960

Jokar M, Pedersen GA, Loeschner K (2017) Six open questions about the migration of engineered nano-objects from polymer-based food-contact materials: A review. *Food Addit Contam Part A* :34434–450

Kentish S, Wooster TJ, Ashokkumar M, et al (2008) The use of ultrasonics for nanoemulsion preparation. *Innov Food Sci Emerg Technol* :9170–175

Khalaj M-J, Ahmadi H, Lesankhosh R, Khalaj G (2016) Study of physical and mechanical properties of polypropylene nanocomposites for food packaging application: Nano-clay modified with iron nanoparticles. *Trends Food Sci Technol* :5141–48

Khanna P, Ong C, Bay BH, Baeg GH (2015) Nanotoxicity: An interplay of oxidative stress, inflammation and cell death. *Nanomaterials* 5:1163–1180

Kim I, Viswanathan K, Kasi G, et al (2022) ZnO nanostructures in active antibacterial food packaging: Preparation methods, antimicrobial mechanisms, safety issues, future prospects, and challenges. *Food Rev Int* :38537–565

Kraśniewska K, Galus S, Gniewosz M (2020) Biopolymers-based materials containing silver nanoparticles as active packaging for food applications–a review. *Int J Mol Sci* 21:698

Kumar L, Ramakanth D, Akhila K, Gaikwad KK (2022) Edible films and coatings for food packaging applications: A review. *Environ Chem Lett* 1–26

Kumar MIS, Kirupavathy SS (2022) Investigations on the capacitive behaviour of hydrothermally synthesised cadmium meta niobate incorporated reduced graphene oxide hybrid nanocomposite electrode material. *J Mater Sci: Mater Electron* 33:2428–2449

Kumar S, Thakur KS (2020) Active packaging technology to retain storage quality of pear cv. "Bartlett" during shelf-life periods under ambient holding after periodic cold storage. *Packag Technol Sci* :33239–254

Li W, Zhang C, Chi H, et al (2017) Development of antimicrobial packaging film made from poly (lactic acid) incorporating titanium dioxide and silver nanoparticles. *Molecules* 22:1170

Li X, Ji N, Qiu C, et al (2015) The effect of peanut protein nanoparticles on characteristics of protein- and starch-based nanocomposite films: A comparative study. *Ind Crops Prod* :77565–574

López-Rubio A, Fabra MJ, Martínez-Sanz M (2019) Food packaging based on nanomaterials. *Nanomaterials* 9:1224

Lopez-Rubio A, Gavara R, Lagarón JM (2006) Bioactive packaging: Turning foods into healthier foods through biomaterials. *Trends Food Sci Technol* :17567–575

Lu C-F, Yuan X-Y, Li L-Z, et al (2015) Combined exposure to nano-silica and lead induced potentiation of oxidative stress and DNA damage in human lung epithelial cells. *Ecotoxicol Environ Saf* :122537–544

Lyu H, Gao B, He F, et al (2017) Ball-milled carbon nanomaterials for energy and environmental applications. *ACS Sustain Chem Eng* :59568–9585

Manzoor M, Rafiq A, Ikram M, et al (2018) Structural, optical, and magnetic study of Ni-doped TiO 2 nanoparticles synthesized by sol–gel method. *Int Nano Lett* 8:1–8

McClements DJ, Xiao H (2012) Potential biological fate of ingested nanoemulsions: Influence of particle characteristics. *Food Funct* :3202–220

McClements DJ, Xiao H (2017) Is nano safe in foods? Establishing the factors impacting the gastrointestinal fate and toxicity of organic and inorganic food-grade nanoparticles. *NPJ Sci Food* 1:6

McClements DJ, Xiao H, Demokritou P (2017) Physicochemical and colloidal aspects of food matrix effects on gastrointestinal fate of ingested inorganic nanoparticles. *Adv Colloid Interface Sci* :246165–180

Mohammadian M, Waly MI, Moghadam M, et al (2020) Nanostructured food proteins as efficient systems for the encapsulation of bioactive compounds. *Food Sci Hum Well* :9199–213

Mohr LC, Capelezzo AP, Baretta C, et al (2019) Titanium dioxide nanoparticles applied as ultraviolet radiation blocker in the polylactic acid bidegradable polymer. *Polym Test* 77:105867

Morán D, Gutiérrez G, Blanco-López MC, et al (2021) Synthesis of starch nanoparticles and their applications for bioactive compound encapsulation. *Appl Sci* 11:4547

Moraru C, Huang Q, Takhistov P, et al (2009) Food nanotechnology: Current developments and future prospects. In: *Global issues in food science and technology* 369–399

Namratha S, Sreejit V, Preetha R (2020) Fabrication and evaluation of physicochemical properties of probiotic edible film based on pectin–alginate–casein composite. *Int J Food Sci Technol* 55:1497–1505

Naseer B, Srivastava G, Qadri OS, et al (2018) Importance and health hazards of nanoparticles used in the food industry. *Nanotechnol Rev* :7623–641

Nile SH, Kai G (2021) Recent clinical trials on natural products and traditional Chinese medicine combating the COVID–19. *Indian J Microbiol* :6110–15

Noruzi M (2016) Electrospunnanofibres in agriculture and the food industry: A review. *J Sci Food Agric* :964663–4678

Ojha S, Kumar B (2018) A review on nanotechnology-based innovations in diagnosis and treatment of multiple sclerosis. *J Cell Immunother* :456–64

Oliver-Ortega H, Tresserras J, Julian F, et al (2021a) Nanocomposites materials of PLA Reinforced with nanoclays using a masterbatch technology: A study of the mechanical performance and its sustainability. *Polymers (Basel)* 13:2133

Oliver-Ortega H, Vandemoortele V, Bala A, et al (2021b) Nanoclay effect into the biodegradation and processability of poly (lactic acid) nanocomposites for food packaging. *Polymers (Basel)* 13:2741

Omerović N, Djisalov M, Živojević K, et al (2021) Antimicrobial nanoparticles and biodegradable polymer composites for active food packaging applications. *Compr Rev Food Sci Food Saf* 20:2428–2454

Oymaci P, Altinkaya SA (2016) Improvement of barrier and mechanical properties of whey protein isolate based food packaging films by incorporation of zein nanoparticles as a novel bionanocomposite. *Food Hydrocoll* 54:1–9

Ozimek L, Pospiech E, Narine S (2010) Nanotechnologies in food and meat processing. *Acta Sci Pol Technol Aliment* :9401–412

Pandiselvam R, Subhashini S, Banuu Priya EP, et al (2019) Ozone based food preservation: A promising green technology for enhanced food safety. *Ozone Sci Eng* :4117–34

Pathakoti K, Huang M-J, Watts JD, et al (2014) Using experimental data of Escherichia coli to develop a QSAR model for predicting the photo-induced cytotoxicity of metal oxide nanoparticles. *J Photochem Photobiol B* :130234–240

Pathakoti K, Manubolu M, Hwang H-M (2017) Nanostructures: Current uses and future applications in food science. *J Food Drug Anal* :25245–253

Pathakoti K, Morrow S, Han C, et al (2013) Photoinactivation of Escherichia coli by sulfur-doped and nitrogen–fluorine-codoped TiO2 nanoparticles under solar simulated light and visible light irradiation. *Environ Sci Technol* :479988–9996

Patra JK, Shin H-S, Paramithiotis S (2018) Application of nanotechnology in food science and food microbiology. *Front Microbiol* 9:714

Perumal AB, Sellamuthu PS, Nambiar RB, Sadiku ER (2018) Development of polyvinyl alcohol/chitosan bio-nanocomposite films reinforced with cellulose nanocrystals isolated from rice straw. *Appl Surf Sci* :449591–602

Primozi CM, Knez Z, Leitgeb M (2021) Bionanotechnology in food science food packaging

Radoi A, Targa M, Prieto-Simon B, Marty J-L (2008) Enzyme-linked immunosorbent assay (ELISA) based on superparamagnetic nanoparticles for aflatoxin M1 detection. *Talanta* :77138–143

Ramakrishnan S, Mahawer SK, Prasad M, et al (2021) Nanomaterials in integrated methods for soil remediation (biological/physiological combination processes). In: *Nanomaterials for soil remediation*. Elsevier, pp. 445–462

Ramesh S, Radhakrishnan P (2019) Cellulose nanoparticles from agro-industrial waste for the development of active packaging. *Appl Surf Sci* 484:1274–1281

Rostamzad H, Paighambari SY, Shabanpour B, et al (2016) Improvement of fish protein film with nanoclay and transglutaminase for food packaging. *Food Packag Shelf Life* 7:1–7

Sahoo M, Vishwakarma S, Panigrahi C, Kumar J (2021) Nanotechnology: Current applications and future scope in food. *Food Front* :23–22

Salarbashi D, Tafaghodi M, Bazzaz BSF, et al (2021) pH-sensitive soluble soybean polysaccharide/SiO2 incorporated with curcumin for intelligent packaging applications. *Food Sci Nutr* 9:2169–2179

Shankar S, Jaiswal L, Rhim J-W (2016) Gelatin-based nanocomposite films: Potential use in antimicrobial active packaging. In: *Antimicrobial food packaging*. Elsevier, pp. 339–348

Sharma A, Thakur M, Bhattacharya M, et al (2019) Commercial application of cellulose nano-composites–a review. *Biotechnol Rep* 21: e00316

Sharma R, Jafari SM, Sharma S (2020) Antimicrobial bio-nanocomposites and their potential applications in food packaging. *Food Control* 112:107086

Shukla P, Chaurasia P, Younis K, et al (2019) Nanotechnology in sustainable agriculture: Studies from seed priming to post-harvest management. *Nanotechnol Environ Eng* :41–15

Siddiqi KS, Husen A, Rao RAK (2018) A review on biosynthesis of silver nanoparticles and their biocidal properties. *J Nanobiotechnol* :161–28

Singh H (2016) Nanotechnology applications in functional foods; opportunities and challenges. *Prev Nutr Food Sci* 21:1

Singh R (2018) Metal organic frameworks for drug delivery. In: *Applications of nanocomposite materials in drug delivery*. Elsevier, pp. 605–617

Singh T, Shukla S, Kumar P, et al (2017) Application of nanotechnology in food science: Perception and overview. *Front Microbiol* 8:1501

Slavin YN, Asnis J, Hńfeli UO, Bach H (2017) Metal nanoparticles: Understanding the mechanisms behind antibacterial activity. *J Nanobiotechnol* :151–20

Soleimani Zohr Shiri M, Henderson W, Mucalo MR (2019) A review of the lesser-studied microemulsion-based synthesis methodologies used for preparing nanoparticle systems of the noble metals, Os, Re, Ir and Rh. *Materials* 12:1896

Sungur Ş, Kaya P, Koroglu M (2020) Determination of titanium dioxide nanoparticles used in various foods. *Food Addit Contam Part B* :13260–267

Tabatabaei RH, Jafari SM, Mirzaei H, et al (2018) Preparation and characterization of nano-SiO2 reinforced gelatin-k-carrageenan biocomposites. *Int J Biol Macromol* 111:1091–1099

Tirgar A, Han D, Steckl AJ (2018) Absorption of ethylene on membranes containing potassium permanganate loaded into alumina-nanoparticle-incorporated alumina/carbon nanofibers. *J Agric Food Chem* :665635–5643

Trifol J, Plackett D, Sillard C, et al (2016) Hybrid poly (lactic acid)/nanocellulose/nanoclay composites with synergistically enhanced barrier properties and improved thermomechanical resistance. *Polym Int* :65988–995

Varghese RJ, Parani S, Thomas S, et al (2019) Introduction to nanomaterials: Synthesis and applications. In: *Nanomaterials for solar cell applications*. Elsevier, pp. 75–95

Venkatasubbu GD, Baskar R, Anusuya T, et al (2016) Toxicity mechanism of titanium dioxide and zinc oxide nanoparticles against food pathogens. *Colloids Surf B Biointerfaces* :148600–606

Wang J, Gardner DJ, Stark NM, et al (2018) Moisture and oxygen barrier properties of cellulose nanomaterial-based films. *ACS Sustain Chem Eng* :649–70

Wang L, Chen W, Xu D, et al (2009) Simple, rapid, sensitive, and versatile SWNT– paper sensor for environmental toxin detection competitive with ELISA. *Nano Lett* :94147–4152

Wang M, Li S, Chen Z, et al (2021) Safety assessment of nanoparticles in food: Current status and prospective. *Nano Today* 39:101169

Wyser Y, Adams M, Avella M, et al (2016) Outlook and challenges of nanotechnologies for food packaging. *Packag Technol Sci* :29615–648

Xu Q, Kambhampati SP, Kannan RM (2013) Nanotechnology approaches for ocular drug delivery. *Middle East Afr J Ophthalmol* 20:26

Yadav TP, Yadav RM, Singh DP (2012) Mechanical milling: A top-down approach for the synthesis of nanomaterials and nanocomposites. *Nanosci Nanotechnol* :222–48

Yanat M, Schroën K (2021) Preparation methods and applications of chitosan nanoparticles; with an outlook toward reinforcement of biodegradable packaging. *React Funct Polym* 161:104849

Yang W, Owczarek JS, Fortunati E, et al (2016) Antioxidant and antibacterial lignin nanoparticles in polyvinyl alcohol/chitosan films for active packaging. *Ind Crops Prod* :94800–811

Yousuf B, Qadri OS, Srivastava AK (2018) Recent developments in shelf-life extension of fresh-cut fruits and vegetables by application of different edible coatings: A review. *Lwt* :89198–209

Yu M, Ji N, Wang Y, et al (2021) Starch-based nanoparticles: Stimuli responsiveness, toxicity, and interactions with food components. *Compr Rev Food Sci Food Saf* 20:1075–1100

Zhang X, Liu Y, Yong H, et al (2019) Development of multifunctional food packaging films based on chitosan, TiO2 nanoparticles and anthocyanin-rich black plum peel extract. *Food Hydrocoll* :9480–92

Zhang Z-H, Zhang Q-Q, Feng S, et al (2014) Fluorinated metal–organic frameworks of 1, 4-bis (1, 2, 4-triazol-1-yl-methyl)-2, 3, 5, 6-tetrafluorobenzene: Synergistic interactions of ligand isomerism and counteranions. *Dalton Trans* :43646–655

Zhou Y, Pan F-G, Li Y-S, et al (2009) Colloidal gold probe-based immunochromatographic assay for the rapid detection of brevetoxins in fishery product samples. *Biosens Bioelectron* 24:2744–2747

Zhuang S, Lee ES, Lei L, et al (2016) Synthesis of nitrogen-doped graphene catalyst by high-energy wet ball milling for electrochemical systems. *Int J Energy Res* 40:2136–2149

Zubair M, Ullah A (2020) Recent advances in protein derived bionanocomposites for food packaging applications. *Crit Rev Food Sci Nutr* :60406–434

13 Nanoparticle and Nanostructure for Enhanced Food Preservation

Meghraj Suryawanshi[a,b,f], *Pankaj Nerkar*[c], *Aarti P. Nikam*[d], *Shrinivas Bhumrela*[e], *and Taufik Mulla*[b]
[a]School of Pharmaceutical Sciences, Jaipur National University, Jaipur, Rajasthan, India
[b]Department of Pharmaceutics Sandip Institute of Pharmaceutical Sciences (SIPS), Affiliated To Savitribai Phule Pune University (SPPU, Pune), Nashik, Maharashtra, India
[c]SES'S R. C. Patel Institute of Pharmaceutical Education and Research, Shirpur, Maharashtra, India
[d]MGV,S SPH College of Pharmacy Malegaon, Nashik, Maharashtra, India
[e]Latur College of Pharmacy, Banasthali University, Latur, Maharashtra, India
[f]All-Well Nutritech LLP, Dharangaon, Maharashtra, India

13.1 INTRODUCTION

Nanotechnology (NT) involves the controlling of matter at the molecular and atomic levels, and nanomaterials (NM) encompass both newly developed and old-style products intentionally structured at the nanoscale (Zhang et al., 2020; Yaseen et al., 2020). "NT" refers to the creation, manufacturing, and use of functional structures at the nanoscale size that have at least one distinguishing dimension, where 1 nm equals 10^{-9} meters. This groundbreaking technology has provided humanity with tools capable of profoundly transforming our lives by manipulating materials and processes on atomic and molecular levels (Tatli Seven et al., 2018).

Being a multidisciplinary area, NT has the possibility toward bringing about significant changes in numerous domains on account of its extensive scope. It has found applications in diverse industries, with some sectors already achieving notable success and others still undergoing research and development. NT is revolutionizing agri-food, aviation, electronics, chemicals, health care, manufacturing, energy, environment, and transport (Qiu et al., 2022; Ruiz-Rico et al., 2018).

In the last decade, NT research has advanced significantly, with major corporations now actively developing novel nanoscale materials for numerous applications. Among the industries, the food sector has been relatively slow in embracing this technology, largely on account of the public's inclination toward natural foods, which has hindered the adoption of food technology, containing NT. Despite being a less recognized aspect in the field of food, scientists and industry leaders assert that NT holds potential applications across nearly each characteristic of the food industry (Melini & Melini, 2018; Nazir & Azad, 2019).

Key areas where NT could take part as crucial character in the food industry include food processing, focusing on improving the texture as well as quality of food, as well as the production of gel-like materials. Furthermore, NT can make a substantial contribution to FP by protecting nutrients and foods from pathogens and harmful gases and by using nanosensors to discover these substances. Within these segments, the adulteration and assessment of packaged food and safety against UV rays are predominantly vital, through NT increasingly finding application in FP. This propensity is likely because food does not directly include NPs, preserving the food's natural structure (Kowalewski, 2008; Kumar et al., 2020; Molajou et al., 2021).

The widespread usage of plastics in contemporary society has raised environmental concerns, with recycling proving to be a predominantly challenging aspect. Plastics, being neither recyclable nor biodegradable, present a noteworthy hazard to the atmosphere. There is an increasing interest in tackling these issues in exploring biodegradable materials' application in food packaging on account of their considerable potential. These materials offer an auspicious substitute toward non-biodegradable plastics, helping mitigate the environmental impact. However, it's worth noting that biodegradable materials exhibit lower resistance to the intrusion of water and gas and possess inferior mechanical characteristics. To address these limitations, one strategy involves combining biodegradable polymers with other beneficial materials, offering a potential solution to enhance their overall performance (Cacciatore et al., 2021; El-Saadony et al., 2021).

A solitary critical aspect of ensuring food safety is the packaging, and NT emerges as a valuable tool toward increasing the efficiency and quality of packaging materials. NT enables smart packaging that reacts to environmental changes and alerts consumers. By infusing nanoparticles (NPs) into plastics, packaging can better block oxygen (O_2), water (H_2O), and carbon dioxide (CO_2), enhancing food preservation. For example, clay NPs improve mechanical strength and obstacle properties, prolonging freshness (Boskovic et al., 2019). These innovative packages help preserve food freshness, color, flavor, and prevent microbial growth, extending SL (SL).

DOI: 10.1201/9781032677422-13

Nanotechnologies play a critical part in numerous features of food and agricultural systems; identify areas, for example, packaging materials, food safety, drug delivery, disease treatment, and advancements in the study of cells and molecules; and innovative materials in order to identify infections (Sekhon, 2010). Soon, we'll see NT's real-world effects, especially in dairy and food industries. It's utilized in nanoparticle distribution, advanced packaging like nanocomposite bottles, and enhancing biosecurity and food safety measures (Prakash et al., 2018; Sahoo et al., 2021).

Soon, NT will revolutionize food and dairy production, focusing on two main areas: extending SL and enhancing food additives. This includes technologies like nanocoatings to preserve freshness, smart packaging to detect spoilage, and the usage of nano-antimicrobials to prolong food lifespan. Additionally, nanoscale food additives will influence the appearance, taste, nutritional enhancement, and functionality of food, while also serving the crucial role of detecting pathogens (Pal et al., 2020).

NPs and nanoscale additives are utilized in food to enhance nutrition, flavor, and texture, while also increasing stability and looking for infections to guarantee the quality and safety of food (Bott et al., 2014). NT is widely employed in food preservation and additives, notably through antimicrobial agents like copper, silver, magnesium oxide, titanium dioxide, zinc oxide, carbon dots and mesoporous silica particles, and graphene, among others (Omerović et al., 2021; Pathakoti et al., 2017).

Compared to conventional packaging approaches, NT boasts numerous benefits and contains enhanced mechanical obstacle properties, heat resistance, and then biodegradability (De Azeredo, 2009). NMs, with their increased antibacterial properties, can be harnessed for detecting food deterioration through nanosensors (McClements & Xiao, 2012). Antimicrobial packaging, involving materials for encapsulation or structured polymeric films, distributes active substances toward or through the food to stop bacteria from growing on its surface (Drago et al., 2020). Active packaging, incorporating antimicrobial NPs, serves to protect food from harmful bacteria, extending the expiry period and maintaining superior freshness. These NPs are also integrated into dynamic packaging to enhance strength, reduce weight, and limit O_2 accessibility (Hoseinnejad et al., 2018; Omerović et al., 2021; Sharma et al., 2020).

NMs find widespread application in combating numerous microbial pathogens and contribute significantly due to their exceptional antibacterial properties, crop security, health care, food care, and water treatment, in addition to food retreat physicochemical functionality. Within the food industry (FI), nanostructured materials (NSMs) are currently employed as encapsulated food components, advanced packaging materials, and nanosensors (He et al., 2019; Kim et al., 2018; Mustafa & Andreescu, 2020).

By using nanosensors to detect contamination or illnesses in food throughout a variety of processes, including production, handling, wrapping, storing, and transportation, NT plays a critical role in improving food security within the food sector (Nile & Kai, 2021; Wang et al., 2021). However, concerns about oral exposure to individuals are raised by the usage of altered NPs in food (Cao et al., 2016). Humans come into contact with NPs in two main ways: either through eating food that has been enhanced with NPs or by ingesting NPs that have leaked from packing (Wang et al., 2021).

Numerous studies have linked nanoparticle ingestion to adverse effects, for example protein denaturation (Hong et al., 2017), activation of reactions to oxidative stress (Khanna et al., 2015), DNA impairment (Lu et al., 2015), and further biological repercussions. Numerous studies have linked the gastrointestinal tract (GIT) and subsequent organ damage to oral exposure to nanoparticles, primarily solid NPs (Cao et al., 2016). This underscores the importance of closely monitoring the application of NPs in food products. Therefore, establishing the level of exposure and developing effective techniques to investigate the toxicity and ensure the safety of NPs in food are crucial practical concerns that must be solved before comprehensive monitoring of NPs in food can be implemented (Wyser et al., 2016; Sahoo et al., 2021).

Considering the main emphasis on the fabrication and execution of nanocomposite-based active or smart FP, this chapter did a thorough analysis and highlighted pertinent investigation on NPs utilized in several food sectors. The focus was on biodegradable polymers and antibacterial nanofillers, although safety evaluations and toxicity studies pertaining to food NPs were also carried out.

This chapter examined and emphasized the relevant studies on nanoparticles utilized in numerous food sectors, with an emphasis on the development and use of active or smart nanoparticle-based nanocomposite filters. We focused on biodegradable polymers and antibacterial nanofillers, and we investigated the toxicity and safety evaluations related to NPs in food. The usage of NT in food processing and food preservation is shown in Figure 13.1.

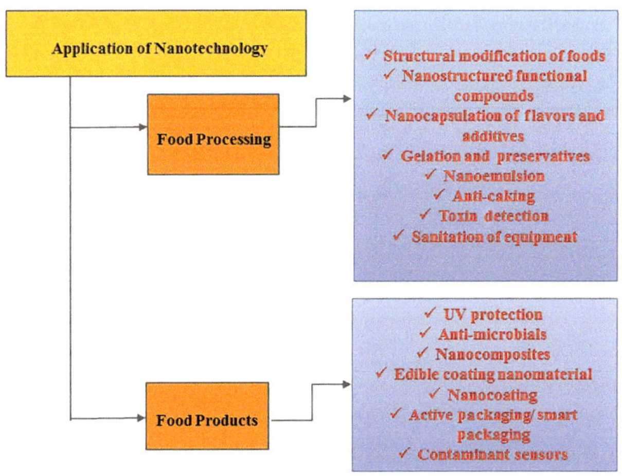

FIGURE 13.1 Application of NT in food preservation and packaging.

13.2 NANOSTRUCTURES FOR FOOD PRESERVATION

As food is gathered, it starts to exhibit signs of spoiling. One long-term answer to the issue of food waste is food preservation. This concept was developed by our forebears while they were attempting to preserve fresh and edible food in the past. Based on seasonal and climatic conditions, concepts like pasteurization, salting, and sun drying were introduced (Sridhar et al., 2021). Preservation gave people the ability to live in communities, prevented them from assassinating animals, and then promoted a leisure-based mindset by prolonging food's SL.

It was made possible by rapid industrial developments and the introduction of lean techniques to create methods like heat treatment, canning, and freezing that enhanced shelf-life extension by reducing microbes. However, as the population grew and consumer standards and needs for wholesome food increased, food safety and security became a top priority. With the intention of feeding everyone, the idea of food preservation thus expanded quickly. Inhibiting metabolic reactions and limiting bacterial or fungal penetration are the two main objectives of food preservation.

Large enterprises have started using some of the common old-style preservation techniques like drying, freezing, and heating. Nevertheless, it has been found that freezing and heat treatment techniques have some drawbacks, containing food's texture, shrinkage, and nutritional decline in addition to organic property loss, that result in a notable decline for the food product as a whole. Food preservation has found an efficient answer in the usage of additives, coatings, and other polyphenolic plant extracts in chemical and microbiological treatments conducted in the current century.

The engineering of matter in its atomic and molecular forms is the focus of the field of NT. NMs are both novel and conventional items that are extensively utilized in NT and have been deliberately nanostructured. Designing, producing, and implementing functional structures on the nanometer scale (1 nm = 10^{-9} m) with at least one of their distinctive dimensions is referred to as NT. Thanks to NT, people can now combine materials and control atomic and molecular processes, opening the door to profound changes in human life. This technology is an interdisciplinary approach with the ability to impact change in a multitude of disciplines on account of its broad breadth. Regarding food safety, one of the most crucial aspects is the packaging. Food safety can be ensured by using NT to enhance the effectiveness and caliber of packaging materials. Nanoparticle packages have the ability to react to their surroundings and alert operators to harmful compounds or air pollution (Malik et al., 2023). Plastics are utilized extensively in the food processing industry. They are not very successful at stopping the flow of CO_2, water, or O_2. Food packaging permeability can be increased by the usage of plastics that include NP packages containing clay NPs, for instance, have enhanced mechanical, thermal, and obstacle qualities that stop moisture, CO_2, and O_2 from passing through. Additionally, they elongate the food's SL, maintain its flavor and color, and stop germs from growing. Food additives and FP are two examples of the two types of nanofood applications that NT will be heavily utilizing in the manufacture of food and dairy products in the near future. FP consists of edible nanocoatings that enclose food products to stop the exchange of gases and moisture, smart packaging that recognizes when food is spoiled and releases nano-anti-microbes to extend the SL of the food, along nanoscale additives to food that can be used to change the look, taste, nutrition, functionality, and detection of pathogens (Sumihar et al., 2022).

13.2.1 Common Food Preservation Technologies

Following are some food preservation technologies that are widely utilized in industries.

1. Thermal treatment
2. Freezing
3. Ultrasound
4. Ozone treatment
5. Pulse electric field
6. NT

13.2.1.1 Thermal Treatment

The use of heat, or thermal treatment, is one of the most innovative methods for food preservation. The method has been well-established for many years in several food-related industries containing dairy, veggies and fruits, as well as bakeries. Foods are typically heated during the procedure to a temperature of 75°C–90°C or higher and held for 25 to 30 seconds. Foods that have been heated lessen the germs. However, a wealth of study has also found that there are reductions in the food matrix, flavor changes, energy waste, and nutrient losses. In an effort to upsurge the SL, optimization studies of the effects of heat exposure have been conducted repeatedly over the years. Lately, minor alterations and adaptations to earlier technology have helped to considerable improvements using a mix of thermal and electrical techniques. The development of the food sector has been significantly impacted by numerous procedures like microwave heating, ohmic heating, and electro plasmolysis (Leong & Oey, 2022).

13.2.1.2 Freezing Treatment

The preservation of the nutritional value and flavor of dairy products and spices, along with green vegetables, has been achieved through the widespread use of cooling and freezing techniques. Common freezing techniques include air blast, cryogenic, direct contact, and immersion freezing. Dehydration freezing, electromagnetic disturbance, high pressure, and ultrasound assistance are examples of sophisticated approaches. Heat transfer is the primary mechanism involved in the freezing and cooling processes. A cooling agreement is reached when heat energy from the meal in addition to its packaging container is transferred to the neighboring air. Thus, the rate of freezing or cooling is significantly influenced by thermal diffusivity and conductivity. Recent years have seen a significant increase in interest in preservation techniques due to the introduction of food products that are ready to consume and satisfy consumer demands. Food safety is always maintained when food is stored in an appropriate container and maintained at a low temperature. Although freezing and cooling work well enough on their own, there are some notable problems with chilling times, ice crystal formation that occurs at different rates, storage costs, and specific conditions (Zhu et al., 2021).

13.2.1.3 Ultrasound Treatment

Utilizing high frequency and intensity sound waves, food items are subjected to ultrasound therapy. The cost-effective technology is selected over more sophisticated instruments on account of it is less complicated to use and has lower equipment needs. Ultrasound is a versatile tool that finds usage in a widespread assortment of industries, including the food, medical, and health care sectors. For both liquid and powdered meals, bath and probe sonicators can be utilized in this procedure. The target solution is subjected to ultrasonic radiation during the ultrasound procedure. Particles shatter and diffuse into the solvent as a result of this action's disruption of the solution's solid constituents. It should be noted that maintaining the procedure's strength over time is necessary. This is because increasing intensity causes intramolecular tensions, which in turn cause the particle–particle bond to break and allow solvent to enter the molecules between the particles in a process called cavitation. Higher capillary effect, enhanced swelling capacity, penetration, and disruption of cells are all necessary for further ultrasonic extraction augmentation (Bachu et al., 2021).

13.2.1.4 Ozone Treatment

One of the O_2 allotropes is ozone. When O_2 splits the molecule is formed into a single O_2 or nascent O_2 in the occurrence of light or UV radiation. As an efficient antibacterial and antiviral agent, the compound has a higher oxidation potential (2.07V) and breaks down quickly into O_2 molecules when compared to chemical preservatives like chlorine (1.35V), hydrogen peroxide (1.78V), and hypochlorous acid (1.79V). In addition, on account of the fact that ozone can be produced instantaneously, it eliminates the need to stockpile dangerous chemicals. In addition, in contrast to heat treatment, which prioritizes SL, less energy is utilized. In recent years, the Food and Drug Administration (FDA) has designated ozone as generally recognized as safe (GRAS). This has compelled the food processing and preservation industries to make difficult decisions to ensure standards and product safety; unlike chlorine, it breaks down with relatively little filtrate once combined with solid meals or beverages (Botondi et al., 2021).

13.2.1.5 Pulse Electric Field

A more sophisticated method of pre-drying food that processes food with a smaller time of residence is pulse electric field technology. The method became well-known due to its continuous functioning as well as minimal electric field demand (0.5–5 kV/cm). Because the method works at low temperatures over 40°C, it can be used in addition to or instead of thermal drying to improve food drying. As part of the pulse electric field methodology, the meal (fruit, vegetable, dairy, or beverage of any sort) is placed between two electrodes. The next step is to apply a high-voltage pulse for a short duration of time (50 kV/cm). The main idea is to combine electroporation and electro-permeabilization. The food's cell membrane matrix is broken by the electric field, improving its nutritional value, safety, and SL. The temperature, polarity, treatment duration, pulse width, frequency, and field strength all influence the pulse electric field (Ranjha et al., 2021).

13.2.1.6 Nanotechnology Application in Food Preservation

NT has emerged as a major innovation with enormous promise for advancing sustainability. Disciplines in applied science like food technology, environmental engineering, physics, biology, medicine, and materials processing are all integrated. To put it simply, NT is the study of materials or NPs with one or more dimensions of less than 100 nm. The technology is chosen on account of its numerous qualities, which include high surface area, precision action on active areas, target specificity, and gradual release action. NT's success is due to its anticipated results, zero emission of pollutants, energy conservation, and minimal spatial requirements. In addition to these key success characteristics, NT has demonstrated several applications in the food industry, agriculture, and environmental domains concerning risk assessment, toxicology, and safety (Diallo et al., 2013).

A lot of fresh produce is susceptible to ethylene, O_2, and water permeability, which can lower food quality. Food packaging is therefore essential to solving this problem. The best options have turned out to be polymer-based composites and NPs. Food preservation has lately showed promise with the usage of a biopolymer or a natural polymer and placing it over the edible surface. While the usage of NPs in smart packaging is still in its early stages, it has advanced quickly over time since it provides a sustainable and safe method (Gaikwad et al., 2020). Table 13.1 depicts advanced techniques in food preservation.

13.2.2 Nanocoatings in Food Packaging

A consistent layer made by repeatedly covering a substrate with a substance is referred to as a coating. Ultra-thin layers (less than 1–100 nm thick) are grown upward as nanocoatings on surfaces. Unlike paints, which smooth out flaws and improve surface quality, nanocoatings do not change the surface's topography or resist wear and abrasion. They are used to impart specific chemical or physical function(s) to a surface as gas-obstacle coatings, hydrophilic/hydrophobic or oleophobic properties, increased corrosion resistance, and enhanced insulating or conductive capabilities. One or more molecular layers of varying thickness can be combined to create nanocoatings. Certain nanocoatings are made of polymers, either synthesized beforehand or in situ. Coatings that have many beneficial purposes. For example, regulated release of active substances, self-healing properties, or

TABLE 13.1
Advanced Techniques in Food Preservation

Sr No.	Food Preservation Techniques		Mechanism Involved
1	**Thermal Treatment**	1. Electro plasmolysis	It includes using a strong electric current to effectively destroy the cell membrane matrix of numerous food ingredients.
		2. Microwave heating (post pasteurization)	It involves the food's temperature rising as a result of electromagnetic energy absorption.
		3. Ohmic heating	Food is heated with the application of electric current.
2	**Freezing Treatment**	1. High-pressure freezing	It involves rapidly forming tiny ice crystals when high pressure is applied to freeze water below 0°C.
		2. Ultrasound-assisted freezing	It involves sound waves traveling through the food.
		3. Radioactive freezing	Because water is dielectric and dipolar, radio waves cause a force that causes the water molecule to spin, which leads to the formation of an ice cluster.
		4. Dehydration freezing/ osmo-dehydro freezing	Food is frozen right away after being dehydrated (water removed).
3	**Ultrasound Treatment**	1. Ultrasonication	It entails the use of sound waves with high frequencies and intensities that are absorbed by food items.
		2. Manosonication	It involves applying pressure and ultrasonic waves simultaneously.
		3. Thermo-sonication	It involves applying heat and ultrasound in conjunction. It is used in pasteurization or sterilization processes.
		4. Manothermo-sonication	It involves applying pressure, ultrasound, and heat in conjunction.
4	**Ozone Treatment**	Compared to chemical preservatives, ozone has a high oxidation potential (2.07 V) and breaks down quickly into O_2 molecules, which renders it a potent antibacterial and antiviral substance.	
5	**Pulse Electric Field**	An intense voltage pulse (50 kV/cm) is applied to the food—fruit, vegetable, milk, or any juice—for a brief length of time after it is sandwiched between two electrodes. The fundamental idea is to combine electroporation and electropermeabilization.	

thermal insulation, are referred to as "smart coatings." Very small amounts of coatings are enhanced by additive substance. Many different features, like flow characteristics, weather resistance, gloss, surface tension, UV protection, and structure, can be altered by them. Depending on the intended usage, functional coatings based on NT employ a variety of NMs: for example, organic acids, proteins, vegetable oils, bacteriocins, and zinc oxide, titanium dioxide, and silver for antimicrobial coatings; and certain natural bioactive substances, for example, polysaccharides, herbs, and spices, can be used for sanitary surfaces or food packaging (Vasile, 2018).

One or more molecular layers of varying thickness can be combined to create nanocoatings. Food packaging uses a wide range of nanocoatings, including vacuum-deposited aluminum films on plastic; coating glass bottles and jars for food and drink; coatings on the inside that are sandwiched between layers of laminated multilayered film for packaging as well as positioned outdoors in packets; edible films and coatings that serve as lipid, gas, and moisture barriers on a variety of meals; and so forth (Farooq et al., 2022).

Whereas the active agent in non-migratory technologies is supposed to remain stable in the packing matrix, in coating methods that employ embedded agents for a controlled release, the active agent is meant to migrate to the packed item. Certain techniques for creating nanocoatings are interchangeable with those for creating nanocomposites, while others are tailored specifically for this usage. Certain applications of nanocoatings called for sophisticated methodologies and specialized tools, like chemical vapor deposition or plasma spray techniques. Within a packed product, the active ingredients are anticipated to move and carry out their distinct antibacterial, antioxidant, biocatalytic, or nutraceutical actions. Polymeric materials can contain active chemicals by techniques, for example extrusion, mixing, pip-coating, or solution casting (Bastarrachea et al., 2015).

13.2.3 Nanocoating Procedures

The following are some techniques that are utilized in nanocoating in industry.

13.2.3.1 Solution Casting

This procedure comprises combining the intended active component with the polymer and dissolving it in the proper solvent. After that, the solution is poured onto a neutral surface, forming films with specific functionalities like antimicrobial, antioxidant, or pharmaceutical properties through solvent evaporation. While primarily utilized on a laboratory scale, it can be adapted for industrial usage with roll-to-roll and wet-coating stations. But its practical and economic application is limited, particularly when it comes to polymers found in food packaging that need elevated temperatures as well as certain organic solvents for dissolution. This can jeopardize the consistency and efficiency of the incorporated active ingredients. Therefore, solution casting is predominantly applied using biodegradable plastics that dissolve at lower amount in solvents, for instance, poly(lactic acid), poly(butylene adipate-co-terephthalate), and derivatives of cellulose (Yuhong et al., 2011).

13.2.3.2 Extrusion

In this approach, active compounds are combined with polymeric materials while in a molten state, typically utilizing an extruder with a single or twin screws where heat facilitates their integration. This process yields a mixture from which film can be created. However, it faces significant drawbacks on account of the many inherent active chemicals that lack temperature stability, as indicated earlier. This instability can lead to the disappearance of active compounds via deterioration or evaporation during the heat transfer processes involved. Achieving an undeviating dispersion of the active agent inside the matrix during these operations is challenging. The active compound's release is influenced by numerous characteristics of the packaging material, containing its affinity with the matrix, morphology, porosity, and the temperature of the surrounding medium. Multi-laminated systems, in which a neighboring layer acts as an impediment and encapsulates a layer holding the active composite, are used to control the release rate. But these change the bulk characteristics, such as heat conductivity and tensile strength, thus coating techniques are the better choice (Yuhong et al., 2011).

13.2.3.3 Sol-Gel Procedure

This method is widely employed for producing coatings based on nanotechnology. A thick colloidal sol is applied onto a surface through coating by dip, spray, or spin techniques. The breadth of the resulting layers ranges from a few nanometers to between 0.5 and 3 micrometers. This process operates at low temperatures and within a short timeframe, consuming minimal energy. Importantly, it preserves the arrangement and operation of biological complexes or natural aggregates. The gel-like nanocoating can show a porous three-dimensional network. (Yuhong et al., 2011).

13.2.3.4 Wet Techniques

Wet approaches involve the application of caustic solutions that are in direct contact with polymer substrates. These liquids include hydrogen peroxide and sulfuric acid that have been dissolved, graft copolymerization, potassium permanganate, sodium hydroxide and sulfuric acid combined, chromic acid, and nitric acid. Among these techniques, graft copolymerization has garnered significant scientific and practical interest (Yuhong et al., 2011).

13.2.3.5 Photografting

The photografting method involves a process of free radical photografting onto a surface made of polymers, which is exposed to UV radiation with a wavelength of 315–400 nm. This is facilitated by the presence of monomers usually containing ketone groups and photoinitiators (e.g., benzophenone, polymers, curcumin). During photografting, subsequently grafting a polymer chain with reactive functional groups, the active agent can either be integrated straightly or become immobilized later (like acrylic acid) (Yuhong et al., 2011).

13.2.3.6 Biological Techniques

Pre-adsorption of proteins, medications, enzyme immobilization, cell seeding, and pre-clotting are examples of biological procedures (Yuhong et al., 2011).

13.2.3.7 Chemical Vapor Deposition (CVD)

In a characteristic CVD procedure, the substrate comes into contact with one or more vaporous precursors. These precursors produce the intended deposit by reacting and/or disintegrating on the substrate surface. DLC, or diamond-like carbon, possesses garnered attention as a coating material for industrial purposes on account of its exceptional properties, containing elevated hardness, resilience to wear, chemical inertness, and biocompatibility. There's a growing demand for polymeric materials, predominantly highly functionalized ones, for numerous applications including biomaterials to environmentally friendly substances. To enhance the functionality and biocompatibility of polymers, DLC can be efficiently nanocoated onto polymeric surfaces, typically through CVD techniques. This process can lead to the development of a novel polymer/DLC nanocomposite type appropriate for solicitations in food packaging and biomedical devices (El-Ghoul et al., 2021).

13.2.3.8 Layer-by-Layer Assembly

It showed that deposition is a process used to create coatings for active packaging by incorporating active substances either within the structure of the material or between layers of polyelectrolytes. Depositing onto a polymeric-activated substrate is accomplished by covalently connecting a polyelectrolyte in a sequence of alternating layers, such as proteins, polysaccharides, or synthetic polymers, or by the reciprocal attraction of polyelectrolytes with opposite charges. Polyelectrolyte solutions can be sprayed onto the substrate or the substrate can be submerged in them to achieve deposition. To guarantee complete protonation or deprotonation and maximum charge, the pH of the solutions is adjusted to optimize the deposition process. Interactions between layers can take the form of covalent bonds created by cross-linkers, electrostatic interactions, or both. A saturation state that is determined by the system's stability and thickness sets a limit on the number of deposited layers (Sarfraz et al., 2020).

13.2.4 Nanocomposites

Nanocomposites refer to biopolymers in addition to nanoreinforcements/supports for example cellulose microfibrils, metallic nanoparticles, carbon nanotubes, and nanoclays, which enhance the biological and structural characteristics required for food packaging. When nanofillers are included into composites, extra functionalities, for example antimicrobial effects, O_2 or ethylene removal, and the inclusion of nanosensors, ensure food safety. Adding nanofillers to polymers with a high composite material's barrier performance against the diffusion of permeable molecules improves with a higher surface area-to-thickness ratio. This improvement is explained by the fact that, in comparison to pure polymer composites, the route that gas and vapor molecules must take through the nanocomposite is more complicated. Nanocomposites are materials where at least one component has nanoscale dimensions (1 nm=10^{-9} m). They display greater dimensional stability, increased thermal expansion, and synergistic flame retardant qualities in comparison to micro- or macro-composites, along with superior mechanical, optical, and obstacle properties.

Composites consist of a matrix and a reinforcing agent, two separate materials, each possessing notably many properties, both chemical and physical. The medium serves as the continuous phase, which might be made of ceramic, metal, or polymer. It functions to shield the fibers from environmental factors and abrasion while ensuring their proper alignment. Acting as a reinforcement agent performs several functions in the scattered or discontinuous phase, for example improving the matrix's mechanical and physical qualities. Table 13.2 illustrated numerous types of reinforcement agents, composites, and nanocomposite materials.

13.2.5 Incorporation in Packaging Material

To create polymer nanocomposites effectively, the preparation techniques must ensure thorough dispersion of the nanofiller within the matrix, leading to either structures that are exfoliated or intercalated. Partial exfoliation is an interim step among intercalation and exfoliation, and ordered and disordered exfoliations are further classifications for exfoliated nanocomposites. Differences in the morphology of polymer nanocomposites are directly related to variations in their mechanical and physical properties (Pielichowska, 2022). There are two primary approaches for producing nanostructures and subsequently polymer coatings and nanocomposites: the "top-down" as well as the ex situ synthesis method "bottom-up" or onsite chemical synthesis method (e.g., casting, spin coating, and in situ polymerization). In the top-down method, nanostructures are made by mechanically disintegrating (for example, milling) previously prepared material, or NPs are first created separately and then dispersed into a polymer melt or monomer solution, which is then polymerized. The bottom-up method, on the other hand, entails creating nanostructures inside the monomer polymerizing solution or polymer matrix by means of a chemical process that produces NPs by the thermal, photolytic, or chemical breakdown of precursors. Numerous bottom-up techniques such as sol-gel processing, chemical syntheses, and CVD, along with top-down methods like sonication and mechanical milling, enable the fabrication of nanocomposites. These techniques determine the concentration and distribution of nanofillers within polymers, examples include nylon 6, poly(vinyl acetate), and in situ synthesis of polyurethanes (PU) in dimethyl formamide (DMF). In situ polymerization processes, like emulsion polymerization of styrene in water, are extensively employed for producing diverse nanocomposites (Wypij et al., 2023).

TABLE 13.2
Numerous Types of Reinforcement Agents, Composites, and Nanocomposite Materials

Sr. No.	Particular	Types/Classes
1	Reinforcement agents	1. Particle reinforced 2. Short fibre reinforced 3. Continuous fibre reinforced elements
2	Composites	1. Laminated composites 2. Fibrous composites 3. Particulate composites
3	Nanocomposite materials	1. Polymer matrix nanocomposites 2. Metal matrix nanocomposites 3. Ceramic matrix nanocomposites.

13.2.6 Release of the Active Substances

Numerous strategies and techniques for functionalizing nanostructures are employed to enhance encapsulation efficiency, typically ranging from 10% to 75%. The efficacy of these methods depends on the properties of the bioactive material and the choice of encapsulating polymers like starch, poly-ε-caprolactone, polylactic acid, zein, and pectin. The release kinetics of the active ingredient within functionalized nanostructures, embedded in an edible polymer matrix, are influenced by the selection of wall material, encapsulating substance, and support matrix in edible coating production. In addition, many challenges that the bioactive ingredient needs to overcome in order to interact with the food surface and achieve its main goal of prolonging the period of commercialization while maintaining freshness and functional qualities are taken into account. As a result, to examine kinetics and their connection to longer shelf lives, it is essential to investigate and understand the release kinetics of functionalized bioactive chemicals in nanostructures (González-Reza et al., 2018).

Edible coatings offer precise control over bioactive compound release, considering factors like temperature, humidity, and pH. Mechanisms for controlled release include compound solubility, permeability through the polymer, and nanostructure degradation. Various release models, such as zero-order and first-order kinetics, regulate release kinetics. Zero-order kinetics (Mt = k0t + M0) ensure constant release, maintaining compound concentration for prolonged food quality. Conversely, first-order kinetics (Log Mt = log M0 – k1t/2.303) adapt release to changing concentrations during storage, optimizing compound release from the coating (González-Reza et al., 2018).

13.3 ENHANCED FOOD SAFETY AND QUALITY

13.3.1 Impermeable Polymer Nanocomposites

Integrating at least two distinct materials—each with unique properties of its own—creates composite materials. The qualities of the parts are carried over into the finished product. One example comprises reinforced concrete, which combines the strength of concrete with the flexibility of iron. The use of nanocomposites in place of conventional polymer packaging is one area of NT that is developing quickly. A scattered phase of NPs and an ongoing polymeric phase make up nanocomposites; a material is referred to as a nanocomposite if it contains NPs in its dispersed phase. High surface-to-volume ratios, flexibility without sacrificing strength or abrasion resistance, and advantageous optical qualities with transparency associated with particle size are all displayed by nanocomposites. Numerous NPs, including clay and metals, along with additional substances, serve as reinforcements in these composites. NPs typically account for less than 10% of the overall weight in polymer nanocomposites, enhancing polymer strength and reducing weight. Nanobiocomposites applied in FP not only preserve and extend SL but also offer environmental benefits. Conventional packaging materials often contribute to pollution on account of their non-recyclable nature, whereas the gas-shielding capability of FP has recently improved with the usage of nano-bio-composites. Table 13.3 provides a summary of diverse nanoparticle applications in FP. The prevalence of biodegradable polymers reinforced with eco-friendly nanofillers is increasing in FP, contributing to environmental sustainability (Hutapea et al., 2021).

13.3.2 Active Packaging

In contrast to old-style FP, active packaging has the ability to respond by releasing beneficial compounds, for example antimicrobial and antioxidant agents or eliminating harmful elements. This reactivity leads to improved food stability. The process of active packing involves the incorporation of particular molecules that can either attract or repel particles from the surrounding atmosphere of the packed food. Currently, antibacterial applications are the main use of FP's active polymer nanoparticles. A range of bioactive compounds can be added to food packaging materials by techniques such as encapsulation, NM capping, or other nano techniques, in order to enhance their usefulness and efficiency.

13.3.3 Antimicrobial Packaging Systems

These are becoming increasingly popular on account of their ability to regulate the development of infections and spoiling agents in food. These bundles are intended for release preservatives once the food inside starts to spoil. In FP, antimicrobial methods are used, for example silver NPs. In addition to being poisonous to some microbes, silver has advantages including decreased volatility and stability at high temperatures. Adherence to the cell surface, lipopolysaccharide degradation, cell membrane cavity development, and diffusion into bacterial cells, DNA impairment, as well as the discharge of antibacterial silver ions, are the mechanisms behind the antimicrobial activity of these packages (Hutapea et al., 2021).

13.3.4 Inhibitors of Gas and Moisture

The presence of O_2 in FP leads to the degradation of fats and oils and, therefore, the proliferation of microorganisms. Additionally, O_2 accelerates processes within FP that result in issues like discoloration, changes in texture, loss of sharpness, unpleasant taste, and other undesirable effects. Ethylene, a gas connected through the ripening of fruits and vegetables, is an additional element affecting the SL of these products. Lowering the amount of ethylene in these products' packaging is crucial to extending their SL. NT offers a solution by producing inhibitors for O_2 and ethylene, preventing unfavorable changes in packaged foods or significantly slowing down the degradation of their quality (Hutapea et al., 2021).

TABLE 13.3
NP Utilized in Food Packing

Types of NP	Matrix	Application
Silver Oxide	Apple slice	Slows down the growth of microorganisms causing spoilage.
Titanium Oxide	Strawberry, Chinese jujubes	Diminishes the rate of ripening, browning, decay, and senescence.
Zinc Oxide	Orange juice	Effectively decreases the counts of mold, yeast, *Salmonella*, and *Lactobacillus Plantarum* without altering superiority factors.
Silver	Beef-meat exudates, fresh-cut melon, poultry meat, orange juice, asparagus	Slows down the growth of aerobic psychotropic bacteria, yeasts, and molds; exhibits antimicrobial effects against *E. coli* and *S. aureus*.

13.3.5 Smart Packaging

Smart packaging refers to a system that can interact with consumers, providing a more dynamic approach to storing products and mitigating the impact of environmental influences on food. Effective control of O_2 besides moisture is crucial for preserving the superiority of packaged food. Smart packaging containers address challenges related to O_2 and moisture levels by incorporating compounds that can absorb or release these elements as needed. The ability of intelligent packaging materials to absorb O_2 is a crucial component that keeps the food's flavor and freshness intact by inhibiting the growth of aerobic bacteria.

Another characteristic of smart packaging is its ability to facilitate the passage of CO_2. This feature is predominantly beneficial in preventing the growth of harmful bacteria in items, for example cheese, poultry, and red meat. The purpose of smart packaging is to detect microbiological or biochemical alterations in food products containing the formation of pathogens. Sensors within smart packaging detect the freshness of the contents, focusing on assessing and understanding the product's status to ensure quality (freshness, ripeness, and stability) and safety.

Manufacturers, retailers, and customers can all receive information on the product via intelligent packaging, which is capable of monitoring it. Food sensors embedded in packaging can identify toxins, infections, and chemicals; their main purpose is to provide data regarding the quality of the goods and the conditions within the box. Biosensors, containing those utilizing NT through the monitoring method are aided by chemical alteration, sensitization of packaging labels, smart chips, or nano cells.

The development of smart packaging, utilizing NT, aims to enhance storage life and enable the transportation of products over longer distances. Tracking and disseminating data about food quality is the main goal of smart packaging. Ripeness markers, RFID recognition, and time-temperature indicators (TTIs) are a few examples of system components. Smart gadgets are either integrated into the packing materials or affixed to the exterior or interior of the box (Hutapea et al., 2021).

13.3.6 Edible Films and Coating

The use of biodegradable materials has received renewed attention due to contaminants arising from biopolymers, and throughout the last 20 years, research on biodegradable materials made of proteins and carbohydrates has greatly increased. A good substitute are synthetic polymers made from petroleum sources. Biodegradable packaging comes in two edible forms films and food coatings that can be ingested accompanied by the food. Edible films, produced in thin layers are utilized as synthetic polymers in packaging (Díaz-Montes & Castro-Muñoz, 2021). Unlike the common belief, edible coatings adhere to the food, becoming an integral part of the product through techniques, for example waxing, spraying, or immersion. Oral films and coatings offer distinct benefits compared to synthetic polymers containing biodegradability, effective inhibition of respiratory gas exchange for controlling respiration of fruits and vegetables, inhibition of the transfer of delicious and odorous compounds, and defense against mechanical harm (Hutapea et al., 2021).

Packaged goods have antimicrobial coatings and films intended for oral consumption. It is essential that the food's organoleptic qualities be unaffected by the makeup of these films or coatings. While some flavoring compounds may be intentionally added to create a specific taste or mask undesirable flavors, the preference is for tasteless food films and coatings to minimize any impact on the food's flavor. Antimicrobial films consist of three main components—film-forming biopolymers, emollients, and antimicrobials—all of which must be tasteless. Collagen and gelatin films are bland alternatives, while films derived from isolates of peanut or soy proteins may taste slightly like beans. Emollients like polyethylene glycol, glycerol, and sorbitol are frequently used; glycerol and sorbitol have a nice sweetness, and polyethylene glycol is flavorless. Although antimicrobials may have their own distinct flavor, their minimal presence makes their effect on taste in food films insignificant (Hutapea et al., 2021).

Table 13.4 describes how NT techniques are used in the food sector for packaging and preservation, and Table 13.5

TABLE 13.4
Application of Nano Techniques for FP and Preservation in Food Sector

Techn Ques and Material	Activity	Application in Food Technology	Reference
NP: Categories in Inorganic and Metal NP			
Silver Based	Enhanced obstacle and mechanical properties; reduced yellowness, improved transparency, and heat stability; increased antioxidant activity; antibacterial effectiveness against both Gram-positive and Gram-negative bacteria.	Utilizing active packaging to elongate the SL of food and regulate the presence of pathogenic and spoilage microorganisms and bacteria.	(Biswas et al., 2022)
Zinc Oxide	Potent antibacterial substance; exposure to UV-A radiation did not impact the mechanical properties of the produced nanomaterial; materials that scavenge activated O_2 are employed to hinder O_2 movement within packaging containers.	The emphasis on packaging for food preservation underscores its antimicrobial effects and is employed to elongate the SL of fresh food items while preventing the sticking together of packaged foods.	(Biswas et al., 2022)

(Continued)

TABLE 13.4 (*Continued*)
Application of Nano Techniques for FP and Preservation in Food Sector

Techn Ques and Material	Activity	Application in Food Technology	Reference
Copper Based	Used to prevent the growth of viruses, bacteria, and fungi; due to their large surface area, they can interact with cell membranes to intensify the antibacterial effect; they also possess antimicrobial qualities, UV and heat resistance, and water vapor permeation.	Utilizing active packaging to extend the SL of food and regulate the presence of pathogenic and spoilage microorganisms and bacteria.	(Biswas et al., 2022)
Titanium Dioxide	Provides numerous benefits, for example cost-effectiveness, nontoxicity, and photostability; increasingly recognized as a superior photocatalyst for cost-effective applications in power generation (water splitting), antibacterial applications, cleansing of air and water, and surfaces that clean themselves; demonstrates antibacterial qualities; enhances the polymer nanocomposites' mechanical characteristics; utilized as food whiteners in dairy, cheese, and other numerous others.		(Biswas et al., 2022)
Silicon Dioxide	Absorbs water molecules in food to exhibit hygroscopic properties; minimizes moisture leakage; serves as food coloring, drying, and anti-caking agent; has a typical particle size, a large surface area, stability, biocompatibility, low toxicity, low heat conductivity, and good insulation.	Outspreading the SL of food and controlling the amount of germs and harmful microorganisms that cause spoiling.	(Biswas et al., 2022)
Nanoclay and Silicate	Elevated levels of total volatiles, organic acids, antioxidant activity, and antibacterial properties.		(López-Rubio et al., 2019)
Polymer-Based NP			
PVA	Enhance the mechanical properties linked to its appropriate structure, along with hydrophilic attributes, for example solvent resistance, mechanical performance, biocompatibility, and high hydrophilicity; exhibit improved antibacterial effects, with no cytotoxic impact, and ensure cell survival of over 90%.	Utilizing active packaging to extend the SL of food and regulate the presence of pathogenic and spoilage microorganisms and bacteria.	(Biswas et al., 2022)
PLA	Demonstrates notable qualities, for example exceptional mechanical capabilities, renewability, crystallinity, biodegradability, and process ability.		(Biswas et al., 2022)
Starch	When multiwalled carbon nanotubes are added, the mechanical properties are greatly enhanced, resulting in a decrease in the transmission of water vapor and moisture absorption. The antibacterial action of nanocellulose reduced both Gram-positive and Gram-negative pathogens, and its crystallinity index, was lower than that of micro-crystalline cellulose (MCC).		(Biswas et al., 2022)
Cellulose	When multiwalled carbon nanotubes are added, the mechanical properties are greatly enhanced, resulting in a decrease in the transmission of water vapor and moisture absorption. The antibacterial action of nanocellulose reduced both Gram-positive and Gram-negative pathogens, and its crystallinity index was lower than that of MCC.		(López-Rubio et al., 2019)
Chitosan	Incorporated with nanocapsules containing epicatechin gallate and assessed their antioxidant properties; incorporated into packaging films; significantly diminished microbiological deterioration on the fruit's contact surface.		(Biswas et al., 2022)
Zein	Enhanced mechanical strength and water moisture obstacle properties without affecting bacterial growth was markedly inhibited by decreased hydrophilicity, fractional free volume, and film elongation; exhibited increased tensile strength, decreased elasticity, and an initial surge in tensile strength.		(Biswas et al., 2022)
Whey Protein	The permeability of films to water vapor has been decreased; the water resistance and obstacle properties of the films have been improved; there is a decrease in transparency.		(López-Rubio et al., 2019)

TABLE 13.5
Application as Nanocomposites

Techniques and Material	Activity	Application in Food Technology	Reference
Nanocomposites ZnO_2, pediocin, and silver coating	Degradation of lipopolysaccharides; inflicts irreversible damage to bacterial DNA; aids in combating microorganisms.	Enhanced composition of food packaging with unique attributes (antimicrobial agent).	(Sundaramoorthy & Nagarajan, 2022)
Polymer and NP (nanoclay)	Gas obstacles are employed to minimize the escape of CO_2 from bottles containing carbonated beverages.		(Sundaramoorthy & Nagarajan, 2022)
Nanolaminates (nanoencapsulation)	It's used as a coating for baked items, cheeses, veggies, and fruits.		(Biswas et al., 2022)
Garlic oil nano-composites coated with PEG	Eliminate the bugs that often contaminate packaged foods in retail establishments.		(Biswas et al., 2022)
DS13 Top Screen and Guard IN Fresh	Remove ethylene molecules to aid in the ripening of fruits and vegetables.		(Biswas et al., 2022)
Nanocor	To prevent CO_2 leakage from beverage containers, it is employed in the manufacturing of plastic beer bottles.		(Biswas et al., 2022)
Durethan (polyamide)	Imparts stiffness to paper packaging jars for fruit juice.		(Biswas et al., 2022)
Indicator of the entire temperature history	Detecting fluctuations in temperature in frozen food items; monitoring changes in temperature over a period.	Intelligent food packaging (Smart FP) for extending SL and managing the identification and control of pathogenic and spoilage bacteria.	(Hu & Miao, 2022)

gives a summary of the numerous uses of nanocomposites (Hutapea et al., 2021).

13.4 CHALLENGES AND CONSIDERATIONS

13.4.1 Health Risk, Toxicity, and Safety Assessment with Certain Aspects of NPs in Food

The continual advancements in NT have resulted in the creation of numerous products, posing a serious risk to the general public's health. Although nanostructures are present in many food components naturally, the addition of synthetic NPs to the food chain raises questions about the possibility that food may become contaminated, endangering human health and having an adverse effect on the environment. NPs used in FP is intended in the direction of advance the mechanical and obstacle characteristics of conventional and biobased materials. Additionally, they bring in active and smart features, whereby the materials used for packaging are engineered to have parts that can release or absorb chemicals into, onto, or out of the food that is being packed, or that can convey vital information about how the food should be used (Dudefoi et al., 2018).

However, inadequate packaging performance may lead to exposure pathways and carry significant health hazards, mostly related to toxicity and ecotoxicity, since food can absorb particle nanoparticles from the container. Reactive oxygen species (ROS) produced by NSMs are expected to negate their beneficial effects as antioxidants and antimicrobials by inducing oxidative stress or cell damage. Damage to DNA, unchecked cell growth, altered cell motility, cytotoxicity, apoptosis, and the enlargement of malignancy are all possible outcomes of nanotoxicity (Fu et al., 2014). The degree of migration, the type of packing matrix, the toxicity of the nanoparticle, and the rate at which a particular meal is absorbed are some of the variables that determine how these health hazards may manifest (Biswas et al., 2022).

Growing evidence indicates health risks associated with nanoparticle absorption, as nano-based food products may negatively impact health on account of excessive intake, bioaccumulation, and overactivity (Biswas et al., 2022). Depending on their invading following accumulation, and movement within the body, NSMs can be evaluated for potential dangers (Biswas et al., 2022). Higher exposure levels to these chemicals through skin or respiratory absorption may cause major safety issues that call for additional study and risk assessment, namely in the area of long-term cytotoxicity and toxicity (Biswas et al., 2022).

Despite research indicating that the elements used in NM packaging may leak into food and are ingested by customers, there is not much data available regarding the toxicity of these materials (Echegoyen & Nerín, 2013). Numerous human organs, such as the liver, small intestine, lungs, spleen, stomach, kidneys, including main distribution organs, are susceptible to NP accumulation. ZnO NP can lead to issues such as liver toxicity, renal difficulties, and lung damage with a single oral dosage. Nanoparticles

TABLE 13.6
Risks and Effects on Health of Utilizing NT in Food Systems

Application	Harmful Materials	Nanoemulsion	Risk Factors	References
FP	Carbohydrates, protein, and lipids are all surfactants	Organic NSMs that are indigestible	Cellular disruption and accumulation in biological systems.	(Biswas et al., 2022)
	Nanotubes of carbon	Inorganic NSMs that are indigestible	Toxic to the skin and lungs.	
	Silver NP	Inorganic NSMs that are indigestible	Augmenting the production of reactive O_2 species (ROS) while reducing the body's ATP levels.	(Biswas et al., 2022)
Food Preservation	Carbohydrates, protein, and lipids are all surfactants	Organic NSMs that are indigestible	Overweight/obesity.	
	Silver NP	Inorganic NSMs that are indigestible	Carcinogenic, genotoxic, and cytotoxic are terms employed to characterize substances with the potential to induce cancer.	

that are easily able to pass through biological barriers and enter the bloodstream can be consumed through the GIT (Esmaeillou et al., 2013; Biswas et al., 2022). Table 13.6 highlights the impact of NT on the biological framework with respect to food systems. The usage of titanium oxide as well as the release of it raises several concerns that may have an impact on both the environment and human health (Yang et al., 2014).

13.5 PRESENT AND FUTURE ASPECTS OF NT IN FOOD APPLICATION

There have been notable advancements in the application of nanotechnology in nutrition science and research. By monitoring, tracking, and documenting, NT is essential for identifying contaminants, bacteria, and pesticides as well as for guaranteeing the safety of food. By adding carbon nanotubes into packaging materials, it is now possible to track dangerous food deterioration, germs, and proteins. Moreover, FP materials may one day become smart devices through the use of carbon nanotubes. The use of NM has been linked to toxicity problems and possible risks, which need to be addressed in further research. Because these nanoparticles' characteristics change over time, it is still unknown how they will affect people, animals, and the environment. Moreover, these nanoparticles possess the capability to transcend biological barriers, such as the blood-brain barrier, to gain entry into many cells and organs.

It has been established that NT has potential uses in the food and pharmaceutical industries. Many facets of food production, packing, and storage involve the use of NT. There are 400–500 nano-based FP applications and about 150–600 commercially produced nano-based food products (Chuah et al., 2014). By adding nanoemulsions, companies like Unilever and Nestle have produced ice cream with lower fat content while retaining the fatty flavors and texture (Renton, 2006). Commercially, bioactive substances including vitamins and fatty acids that have been nanoencapsulated are available for use in food processing and preservation. German start-up Aqua Nova has created micro micelles (30 nm) that can be used to preserve fatty acids and vitamins C and E.

Despite possessing numerous beneficial properties, NT-based food products are still awaiting full commercialization. This delay is attributed to numerous factors, for example limited knowledge about the benefits and safety of nanoproducts, the unfamiliarity of consumers with the technology, and the absence of appropriate regulatory guidelines. Nevertheless, it is projected that the global NT market will witness significant growth, reaching USD 112.49 billion during the period of 2019–2023 (Global Food NT Market, 2023). This data prompts an examination of consumer perceptions regarding nanofood and emphasizes the need for in-depth studies on the safety of NT. Table 13.7 provides a list of commercially available nanoproducts in food, showcasing their advantageous properties.

13.6 FUTURE PERSPECTIVES

In the area of food science and research, NT has accomplished notable advancements. By monitoring, tracking, and recording, it contributes to the overall safety of food by playing a critical role in the detection of contaminants, pathogens, and pesticides. Because carbon nanotubes may be incorporated into packaging materials, they have become particularly attractive for their use in tracking bacteria, proteins, and dangerous food deterioration. Furthermore, these nanotubes may be able to turn food packaging materials into smart gadgets in the future. Current research has shown possible risks and toxicity problems related to NM, which call for their resolution. The uncertain impact of nanoscale particles on the atmosphere, livestock, and human health is attributed in the direction of deviations in their properties over time. Moreover, these NPs may traverse biological obstacles, for example the blood-brain obstacle, to access diverse cells and organs.

TABLE 13.7
Commercially Available Nano-Based Products for Food Applications

Food Application Type	Product Name	Nanomaterials	Company Name	Properties	Reference
Preservatives	Aquasol	Nanocapsules	Aquanova	Improve the preservative impact and enhance the absorption of nutritional content.	(Kirdar, 2012)
Additives	Bioral™	Nano-chelates	Bio-delivery sciences international	Improve the bioavailability of Omega 3 fatty acids in cakes, muffins, pasta, soups, cookies, and cereals.	(Kirdar, 2012)
Functional Beverages	Oat chocolate and oat vanilla drink	Iron NP	Toddler health	Increase the reactivity as well as bioavailability of micronutrient NPs.	(Kirdar, 2012)
Contact Materials	Teapot treated with AgNP	Silver NP	Song Sing NT Co.	Exhibit antimicrobial activity.	(Alfadul & Elneshwy, 2010)
Contact Materials	Ag-Nano cutting board	Silver NP	A Do global	Exhibit antimicrobial activity.	(Kirdar, 2012)
FP Materials	Fresherlonger™	Silver NP	Sharper Image Company, USA	Exhibit antimicrobial activity. Food remains fresh 3 to 4 times longer.	(Bumbudsanpharoke & Ko, 2015)
FP Materials	Sina Antibacterial Food Storages	Silver NP polypropylene and silicon	Dai Dong Tien Corporation, Vietnam	Eliminate the bad smells; prevent the microorganism's growth; improve the SL of food.	(Bumbudsanpharoke & Ko, 2015)

13.7 CONCLUSION

Nanotechnology is a comparatively novel expertise that has the potential to develop food safety and quality. The identification of foodborne pathogens and the creation of sophisticated packaging are its main uses. The packaging industry has been greatly impacted by the molecular-level change of material structures, which has made large-scale industrial production possible. However, before being widely adopted in the food processing (FP) industry, a quantity of important concerns needs to be resolute. Given the possibility that NPs could infiltrate food through FP containers, food safety is the most important factor to take into account. Because NPs can pass across cell membranes, more research is needed to determine the long-term health impacts of several NPs interacting with food. Notably, tests and research results show that nanocoatings significantly increase the SL of meat and other items. As a practical replacement for conventional materials, nanocoatings stop bacterial development and lessen the waste that microorganisms produce. Many of the features of NM are influenced by their surface-to-volume ratio, which is an important factor. Because of this high ratio, nanoparticles can interact with biological molecules and microbes more successfully, demonstrating their antibacterial qualities. As a result, it is expected that adding nanoparticles to FP will improve the qualities of the packaging barrier, requiring less raw material and producing less waste. One possible strategy for preventing gas and pathogen permeability in FP is to use composite polymers. Because of their great impermeability and lightweight design, clay polymer nanocomposites are well-known for their affordability and extended SL. Strong antibacterial qualities of silver and copper nanocomposites protect food from a wide range of microorganisms, greatly extending its SL. By identifying gases, smells, chemical pollutants, and pathogens within FP without causing damage, the incorporation of nanosensors into smart packaging adds value. It's important to remember, though that research on the health risks posed by breathing in NPs and ultrafine particles is still underway. It is established that there are worries regarding the bioaccumulation of NM in food and humans, for example nanosilver that is extracted from nano packing or plants and animals. The use of NM as insecticides, microbicides, and activation catalysts presents unknown risks, underscoring the necessity of stringent risk management procedures when handling food. Even with NT's possible advantages, developing a secure and long-lasting food supply is still a difficult task. As NT is implemented, public education about the health, safety, and environmental effects of NT in the food chain is crucial.

REFERENCES

Alfadul SM, Elneshwy AA (2010) Usage of NT in food processing, packaging and safety–review. *Afr J Food Agric Nutr Dev* 10(6):2719–2739. https://doi.org/10.4314/ajfand.v10i6.58068.

Bachu VS, Kedda J, Suk I, Green JJ, Tyler B (2021) High-intensity focutilized ultrasound: A review of mechanisms and clinical applications. *Ann Biomed Eng* 49(9):1975–1991. https://doi.org/10.1007/s10439-021-02833-9.

Bastarrachea LJ, Wong DE, Roman MJ, Lin Z, Goddard JM (2015) Active packaging coatings. *Coat* :5771–791. https://doi.org/10.3390/coatings5040771.

Biswas R, Alam M, Sarkar A, Haque MI, Hasan MM, Hoque M (2022) Application of NT in food: Processing, preservation, packaging and safety assessment. *Heliyon* 8:e11795. https://doi.org/10.1016/j.heliyon.2022.e11795.

Boskovic M, Glisic M, Djordjevic J, Baltic MZ (2019) NT and plant extracts as a future control strategy for meat and milk products. In Prasad R (Ed.), *Plant nanobionics*. Springer, Cham 201–253. https://doi.org/10.1007/978-3-030-12496-0_10.

Botondi R, Barone M, Grasso C (2021) A review into the effectiveness of ozone technology for improving the safety and preserving the quality of fresh-cut fruits and vegetables. *Foods* 10(4):748. https://doi.org/10.3390/foods10040748.

Bott J, Störmer A, Franz RA (2014) comprehensive study into the migration potential of nano silver particles from food contact polyolefins. In *Chemistry of food, food supplements, and food contact materials: From production to plate*. American Chemical Society 51–70. https://doi.org/10.1021/bk-2014-1159.ch005.

Bumbudsanpharoke N, Ko S (2015) Nano-food packaging: An overview of market, migration research, and safety regulations. *J Food Sci* 80(5):R910–R923. https://doi.org/10.1111/1750-3841.12861.

Cacciatore FA, Brandelli A, Malheiros PD (2021) Combining natural antimicrobials and NT for disinfecting food surfaces and control microbial biofilm formation. *Crit Rev Food Sci Nutr* 61(22):3771–3782. https://doi.org/10.1080/10408398.2020.1806782.

Cao Y, Li J, Liu F, Li X, Jiang Q, Cheng S, Gu Y (2016) Consideration of interaction between NP and food components for the safety assessment of NP following oral exposure: A review. *Environ Toxicol Pharmacol* :46206–210. https://doi.org/10.1016/j.etap.2016.07.023.

Chuah LH, Roberts CJ, Billa N, Abdullah S, Rosli R (2014) Cellular uptake and anticancer effects of mucoadhesive curcumin-containing chitosan NP. *Colloids Surf B* :116228–236. https://doi.org/10.1016/j.colsurfb.2014.01.007.

De Azeredo HM (2009) Nanocomposites for FP applications. *Food Res Int* 42(9):1240–1253. https://doi.org/10.1016/j.foodres.2009.03.019.

Diallo MS, Fromer NA, Jhon MS (2013) NT for sustainable development: Retrospective and outlook. *J Nanopart Res* 15:2044. https://doi.org/10.1007/s11051-013-2044-0.

Díaz-Montes E, Castro-Muñoz R (2021) Edible films and coatings as food-quality preservers: An overview. *Foods* 10(2):249. https://doi.org/10.3390/foods10020249.

Drago E, Campardelli R, Pettinato M, Perego P (2020) Innovations in smart packaging concepts for food: An extensive review. *Foods* 9(11):1628. https://doi.org/10.3390/foods9111628.

Dudefoi W, Villares A, Peyron S, Moreau C, Ropers MH, Gontard N, Cathala B (2018) Nanoscience and nanotechnologies for biobased materials, packaging and food applications: New opportunities and concerns. *Innovat Food Sci Emerg Technol* :46107–121. https://doi.org/10.1016/j.ifset.2017.09.007.

Echegoyen Y, Nerín C (2013) Nanoparticle release from nanosilver antimicrobial food containers. *Food Chem Toxicol* :6216–22. https://doi.org/10.1016/j.fct.2013.08.014.

El-Ghoul Y, Alminderej FM, Alsubaie FM, Alrasheed R, Almousa NH (2021) Recent advances in functional polymer materials for energy, water, and biomedical applications: A review. *Polym (Basel)* 13(24):4327. https://doi.org/10.3390/polym13244327.

El-Saadony MT, Sitohy MZ, Ramadan MF, Saad AM (2021) Green NT for preserving and enriching yogurt with biologically available iron (II). *Innov Food Sci Emerg Technol* 69:102645. https://doi.org/10.1016/j.ifset.2021.102645.

Esmaeillou M, Moharamnejad M, Hsankhani R, Tehrani AA, Maadi H (2013) Toxicity of ZnO NP in healthy adult mice. *Environ Toxicol Pharmacol* 35(1):67–71. https://doi.org/10.1016/j.etap.2012.11.003.

Farooq SA, Raina A, Mohan S, Arvind Singh R, Jayalakshmi S, Irfan Ul Haq M (2022) Nanostructured coatings: Review on processing techniques, corrosion behaviour and tribological performance. *Nanomater (Basel)* 12(8):1323. https://doi.org/10.3390/nano12081323.

Fu PP, Xia Q, Hwang HM, Ray PC, Yu H (2014) Mechanisms of nanotoxicity: Generation of reactive O2 species. *J Food Drug Anal* 22(1):64–75. https://doi.org/10.1016/j.jfda.2014.01.005.

Gaikwad KK, Singh S, Negi YS (2020) Ethylene scavengers for active packaging of fresh food produces. *Environ Chem Lett* :18269–284. https://doi.org/10.1007/s10311-019-00938-1.

Global Food NT Market (2019–2023). *Growth analysis and forecast*. https://www.businesswire.com/news/home/20190117005434/en/Global-FoodNT-Market-2019–2023-Growth-Analysis-and-Forecast-Technavio (accessed on 22 December 2023).

González-Reza RM, García-Betanzos CI, Sánchez-Valdes LI, Quintanar-Guerrero D, Cornejo-Villegas MA, Zambrano-Zaragoza ML (2018) The functionalization of nanostructures and their potential applications in edible coatings. *Coat* 8:160. https://doi.org/10.3390/coatings8050160.

He X, Deng H, Hwang HM (2019) The current application of NT in food and agriculture. *J Food Drug Anal* 27(1):1–21. https://doi.org/10.1016/j.jfda.2018.12.002.

Hong F, Wu N, Zhou Y, Ji L, Chen T, Wang L (2017) Gastric toxicity involving alterations of gastritis-related protein expression in mice following long-term exposure to nano TiO2. *Food Res Int* :9538–45. https://doi.org/10.1016/j.foodres.2017.02.013.

Hoseinnejad M, Jafari SM, Katouzian I (2018) Inorganic and metal NP and their antimicrobial activity in FP applications. *Crit Rev Microbiol* 44(2):161–181. https://doi.org/10.1080/1040841X.2017.1332001.

Hu X, Miao M (2022) Fabrication of intelligent packaging systems using nano-indicators and sensors. In: *Nanotechnol. Enhanced food packaging* 183–211. https://doi.org/10.1002/9783527827718.ch8

Hutapea S, Ghazi Al-Shawi S, Chen TC, You X, Bokov D, Abdelbasset WK, Suksatan W (2021) Study on food preservation materials based on nano-particle reagents. *Food Sci Technol* 42:e39721. https://doi.org/10.1590/fst.39721.

Khanna P, Ong C, Bay BH, Baeg GH (2015) Nanotoxicity: Interplay of oxidative stress, inflammation and cell death. *Nanomater* 5(3):1163–1180. https://doi.org/10.3390/nano5031163.

Kim DY, Kadam A, Shinde S, Saratale RG, Patra J, Ghodake G (2018) Recent developments in NT transforming the agricultural sector: A transition replete with opportunities. *J Sci Food Agric* 98(3):849–864. https://doi.org/10.1002/jsfa.8749.

Kirdar SS (2012) A survey on chemical, biochemical and microbiological characteristics of a old-style dairy product in Mediterranean region: Kes. *J Anim Vet Adv* 11(3):330–334.

Kowalewski Z (2008) Actions for small water retention undertaken in Poland. *J Water Land Dev* 12(1):155–167. https://doi.org/10.2478/v10025-009-0012-y.

Kumar A, Singh P, Gupta V, Prakash B (2020) Application of NT to boost the functional and preservative properties of essential oils. In: Prakash B (Ed.), *Functional and preservative properties of phytochemicals*. Elsevier, Amsterdam 241–246. https://doi.org/10.1016/B978-0-12-818593-3.00008-7.

Leong SY, Oey I (2022) Application of novel thermal technology in foods processing. *Foods* 11(1):125. https://doi.org/10.3390/foods11010125.

López-Rubio A, Fabra MJ, Martínez-Sanz M (2019) Food packaging based on NM. *Nanomater* 9(9):1224. https://doi.org/10.3390/nano9091224.

Lu CF, Yuan XY, Li LZ, Zhou W, Zhao J, Wang YM, Peng SQ (2015) Combined exposure to nano-silica and lead induced potentiation of oxidative stress and DNA damage in human lung epithelial cells. *Ecotoxicol Environ Saf* :122537–544. https://doi.org/10.1016/j.ecoenv.2015.09.030.

Malik S, Muhammad K, Waheed Y. (2023) NT: A revolution in modern industry. *Mol* 9;28(2):661. https://doi.org/10.3390/molecules28020661.

McClements DJ, Xiao H (2012) Potential biological fate of ingested nanoemulsions: Influence of particle characteristics. *Food Funct* 3(3):202–220.

Melini V, Melini F (2018) Strategies to extend bread and GF bread shelf-life: From sourdough to antimicrobial active packaging and NT. *Fermentation* 4(1):9. https://doi.org/10.3390/fermentation4010009.

Molajou A, Afshar A, Khosravi M, Soleimanian E, Vahabzadeh M, Variani HA (2021) A new paradigm of water, food, and energy nexus. *Environ Sci Pollut Res* 1–10. https://doi.org/10.1007/s11356-021-13034-1.

Mustafa F, Andreescu S (2020) NT-based approaches for food sensing and packaging applications. *RSC Adv* 10(33):LL19309–19336. https://doi.org/10.1039/D0RA01084G.

Nazir S, Azaz Ahmad Azad ZR (2019) Food NT: An emerging technology in food processing and preservation. In: Malik A, Erginkaya Z, Erten H (Eds.), *Health and safety aspects of food processing technologies*. Springer, Cham 567–576. https://doi.org/10.1007/978-3-030-24903-8_20.

Nile SH, Kai G (2021) Recent clinical trials on natural products and old-style Chinese medicine combating the COVID-19. *Indian J Microbiol* 61(1):10–15. https://doi.org/10.1007/s12088-020-00919-x.

Omerović N, Djisalov M, Živojević K, Mladenović M, Vunduk J, Milenković I, Knežević NŽ, Gadjanski I, Vidić J (2021) Antimicrobial NP and biodegradable polymer composites for active FP applications. *Compr Rev Food Sci Food Saf* 20(3):2428–2454. https://doi.org/10.1111/1541-4337.12727.

Pal A, Das R, Kumar M, Roy DN (2020) Application of nanoscience and NT in advancement of safety issues of foods. In Verma ML (Ed.), *Biotechnological approaches in food adulterants*. CRC Press, Boca Raton 282–307.

Pathakoti K, Manubolu M, Hwang HM (2017) Nanostructures: Current usages and future applications in food science. *Food Drug Anal* 25(2):245–253. https://doi.org/10.1016/j.jfda.2017.02.004.

Pielichowska K. (2022) Polymer nanocomposites: Preparation, characterisation and applications. *Nanomater (Basel)* 12(11):1900. https://doi.org/10.3390/nano12111900.

Prakash B, Kujur A, Yadav A, Kumar A, Singh PP, Dubey NK (2018). Nanoencapsulation: An efficient technology to boost the antimicrobial potential of plant essential oils in food system. *Food Control* :891–11. https://doi.org/10.1016/j.foodcont.2018.01.018.

Qiu L, Zhang M, Bhandari B, Yang C (2022) SL extension of aquatic products by applying NT: A review. *Crit Rev Food Sci Nutr* 62(6):1521–1535. https://doi.org/10.1080/10408398.2020.1844139.

Ranjha MMAN, Kanwal R, Shafique B, Arshad RN, Irfan S, Kieliszek M, Kowalczewski PŁ, Irfan M, Khalid MZ, Roobab U, Aadil RM (2021) A critical review on pulsed electric field: A novel technology for the extraction of phytoconstituents. *Mol* 26(16):4893. https://doi.org/10.3390/molecules26164893.

Renton A (2006) Welcome to the world of nanofoods. *Guardian Unlimited, UK*. http://observer.guardian.co.uk/foodmonthly/futureoffood/story/0,1971266,00.html (accessed on 22 December 2023).

Ruiz-Rico M, Pérez-Esteve É, de la Torre C, Jiménez-Belenguer AI, Quiles A, Marcos MD, Martínez-Máñez R, Barat JM (2018) Improving the antimicrobial power of low-effective antimicrobial molecules through NT. *J Food Sci* 83(8):2140–2147. https://doi.org/10.1111/1750-3841.14211.

Sahoo M, Vishwakarma S, Panigrahi C, Kumar J (2021) NT: Current applications and future scope in food. *Food Frontiers* 2(1):3–22. https://doi.org/10.1002/fft2.58.

Sarfraz J, Gulin-Sarfraz T, Nilsen-Nygaard J, Pettersen MK (2020) Nanocomposites for food packaging applications: An overview. *Nanomater (Basel)* 11(1):10. https://doi.org/10.3390/nano11010010.

Sekhon BS (2010) Food NT–an overview. *Nanotechnol Sci Appl* :31–15. https://doi.org/10.2147/nsa.s12187498.

Sharma R, Jafari SM, Sharma S (2020) Antimicrobial bio-nanocomposites and their potential applications in FP. *Food Control* 112:107086. https://doi.org/10.1016/j.foodcont.2020.107086.

Sridhar A, Ponnuchamy M, Kumar PS et al. (2021) Food preservation techniques and NT for increased SL of fruits, vegetables, beverages and spices: A review. *Environ Chem Lett* 19:1715–1735. https://doi.org/10.1007/s10311-020-01126-2.

Sumihar H, Sarmad G, Tzu-Chia C, Xiang Y, Dmitry B, Walid KA, Wanich S (2022) Study on food preservation materials based on nano-particle reagents. *Food Sci Technol* 42:e39721. https://doi.org/10.1590/fst.39721.

Sundaramoorthy NS, Nagarajan S (2022) Can NP help in the battle against drug-resistant bacterial infections in "post-antibiotic era"? In: Kumar V, Shriram V, Paul A, Thakur M. (Eds.), *Antimicrobial resistance*. Springer, Singapore 175–213. https://doi.org/10.1007/978-981-16-3120-7_7.

Tatli Seven P, Seven I, Gul Baykalir B, Iflazoglu Mutlu S, Salem AZ (2018) NT and nano-propolis in animal production and health: An overview. *Ital J Anim Sci* 17(4):921–930. https://doi.org/10.1080/1828051X.2018.1448726.

Vasile C (2018) Polymeric nanocomposites and nanocoatings for food packaging: A review. *Materials (Basel)* 11(10):1834. https://doi.org/10.3390/ma11101834.

Wang M, Li S, Chen Z, Zhu J, Hao W, Jia G, Chen W, Zheng Y, Qu W, Liu Y (2021) Safety assessment of NP in food: Current status and prospective. *Nano Today* 39:101169. https://doi.org/10.1016/j.nantod.2021.101169.

Wypij M, Trzcińska-Wencel J, Golińska P, Avila-Quezada GD, Ingle AP, Rai M (2023) The strategic applications of natural polymer nanocomposites in food packaging and agriculture: Chances, challenges, and consumers' perception. *Front Chem* 10:1106230. https://doi.org/10.3389/fchem.2022.1106230.

Wyser Y, Adams M, Avella M, Carlander D, Garcia L, Pieper G, Rennen M, Schuermans J, Weiss J (2016) Outlook and challenges of nanotechnologies for FP. *Packag Technol Sci* 29 (12):615–648. https://doi.org/10.1002/pts.2221.

Yang Y, Doudrick K, Bi X, Hristovski K, Herckes P, Westerhoff P, Kaegi R (2014) Characterization of food-grade titanium dioxide: The presence of nanosized particles. *Environ Sci Technol* 48(11):6391–6400. https://doi.org/10.1021/es500436x.

Yaseen R, Kotp YH, Eissa D (2020) The impact of production of silver NP using soil fungi and its applications for reducing irrigation water salinity. *J Water Land Dev* :46216–228. https://doi.org/10.24425/jwld.2020.134216.

Yuhong Ma, Lianying Liu, Wantai Yang (2011) Photo-induced living/controlled surface radical grafting polymerization and its application in fabricating 3-D micro-architectures on the surface of flat/particulate organic substrates. *Polym* 52(19):4159–4173. https://doi.org/10.1016/j.polymer.2011.07.027.

Zhang J, Hu S, Du Y, Cao D, Wang G, Yuan Z (2020) Improved food additive analysis by ever-increasing NT. *J Food Drug Anal* 28(4):623–641. https://doi.org/10.38212/2224-6614.1152.

Zhu Z, Li T, Sun DW (2021) Pressure-related cooling and freezing techniques for the food industry: Fundamentals and applications. *Crit Rev Food Sci Nutr* 61(17):2793–2808. https://doi.org/10.1080/10408398.2020.1841729.

14 Chemistry of Carbon-Based Nanomaterials and Their Composites in Food Packaging and Preservation

Chinmay Mittal[a], *Anirudh Singh Bhathiwal*[b], *Vishaka Chauhan*[a], *and Anjaneyulu Bendi*[c]

[a]School of Sciences, Geeta University, Panipat, Haryana, India

[b]Department of Chemistry, MMV, Banaras Hindu University, Varanasi, Uttar Pradesh, India

[c]Innovation and Translational Research Hub (iTRH) & Department of Chemistry, Presidency University, Rajanukunte, Itgalpura, Bangalore, Karnataka, India

14.1 INTRODUCTION

The intricate confluence of food packaging and preservation has long posed a formidable challenge, demanding innovative solutions to ensure consumables' safety, quality, and endurance. Against the backdrop of escalating global food demand and the increasing intricacies of supply chains, the imperative for packaging solutions that protract shelf life, forestall spoilage, and unequivocally ensure food safety has reached unprecedented significance. In this milieu, nanotechnology has emerged as an unparalleled catalyst for transformative change, with nanoparticles assuming a pivotal role in navigating the nuanced intricacies of food packaging and preservation. The preservation of food, spanning the continuum from farm to fork, represents a nuanced and multidimensional challenge shaped by myriad adversarial factors. The threats of microbial contamination, oxidation, moisture infiltration, and temperature oscillations collectively conspire to erode the quality and safety of food products. Traditional preservation methodologies, though displaying efficacy to a certain extent, often need to improve in adequately addressing the evolving threats that imperil the integrity of packaged foods. Consequently, the pursuit of innovative and sustainable solutions has propelled researchers toward an exploration of the distinctive attributes of nanoparticles and their transformative capacity to revolutionize food packaging (Ahire et al., 2014; Alexandre and Dubois, 2000; Althues et al., 2007; An et al., 2008; Anthonissen et al., 2010).

Nanoparticles, occupying a size range between 1 and 100 nanometres, exhibit exceptional characteristics, distinguishing them from bulk materials. Their expansive surface area, distinctive optical properties, and heightened reactivity make nanoparticles eminently suitable for food packaging and preservation applications. A paramount challenge addressed by nanoparticles pertains to the prevention of microbial contamination, an omnipresent threat to the safety of packaged foods (Chowdhury et al., 2020; Cutter, 2006; Damm et al., 2007; Danielli et al., 2008; Rezić et al., 2017).

The extensively researched antimicrobial properties of specific nanoparticles, notably their proficiency in inhibiting the growth of bacteria and fungi, have been harnessed by integrating these nanoparticles into packaging materials. The resultant protective barrier actively deters and eradicates harmful microorganisms, substantially prolonging perishable goods' shelf life. This proactive stance toward microbial control represents a paradigm shift from conventional preservatives, offering a more nuanced, targeted, and sustainable solution against foodborne pathogens.

Moving beyond microbial challenges, the menace of oxidation looms large over the quality of packaged foods. Notably, lipid oxidation precipitates off-flavors, rancidity, and a decline in nutritional value. Nanoparticles endowed with antioxidant properties, such as titanium dioxide and zinc oxide, seamlessly integrate into packaging materials to neutralize free radicals and impede oxidation. This dual function elevates the food's sensory attributes and contributes significantly to preserving its nutritional content.

Efficacious food packaging must also contend with moisture control, given that moisture ingress can precipitate textural alterations, mold proliferation, and an overall diminution in product quality. Carefully incorporated nanoparticles give packaging materials moisture-resistant properties that keep out water vapor and protect the food within. This attribute proves particularly advantageous in preserving the crispness of snacks, the freshness of baked goods, and the overall palatability of a diverse range of food items (Cheng et al., 2006; Chawengkijwanich and Hayata, 2008; Barlow and Morgan, 2013; Cushen et al., 2012; Choi et al., 2010; Cuq et al., 1995; Xu et al., 2017).

Temperature fluctuations during storage and transportation are further complicating the task of food preservation. Nanotechnology presents a sophisticated solution by developing innovative packaging materials with built-in sensors

DOI: 10.1201/9781032677422-14

and indicators. These sensors, tuned to detect temperature variations, provide real-time information about the conditions to which the packaged food is exposed, facilitate the maintenance of optimal storage conditions, and ensure that consumers are apprised of the safety and quality attributes of the products they procure.

Carbon-based nanomaterials and other cutting-edge technologies have a symbiotic relationship that is becoming more apparent as the field of nanotechnology as a whole takes shape. Carbon-based nanomaterials, which include carbon nanotubes (CNTs), graphene, fullerenes, and carbon nanoparticles, exhibit features that make them a highly adaptable and dynamic class with the potential to be strategically utilized to improve the safety and quality of food packaging. This exposition thoroughly explores the chemistry of carbon-based nanomaterials and their composites, delving meticulously into their diverse applications in food packaging and preservation.

Carbon nanotubes, characterized by cylindrical structures composed of rolled-up graphene sheets, boast exceptional strength and electrical conductivity. These characteristics make them especially appealing for use in intelligent packaging, where sensors that are seamlessly integrated are used to check freshness and identify any impurities. Furthermore, carbon nanotubes' large surface area and high aspect ratio offer chances for enhanced barrier qualities, halting the entry of gases and moisture that may otherwise degrade the quality of food that has been packed.

Graphene, a hexagonally latticed atomic monolayer of carbon, is well-known for its exceptional mechanical strength and thermal conductivity. These attributes position graphene as a compelling candidate for developing packaging materials adept at temperature regulation and protection against external factors. The two-dimensional nature of graphene also facilitates the creation of thin, lightweight packaging with augmented strength, thus reducing environmental impact.

Fullerenes manifest as spherical carbon molecules with distinctive cage-like structures colloquially known as buckyballs or buckytubes. Endowed with antioxidant properties, fullerenes assume significance in preserving food by scavenging free radicals that precipitate spoilage. The strategic incorporation of fullerenes into packaging materials holds the promise of augmenting the oxidative stability of food products, thereby extending their shelf life and preserving nutritional quality.

Carbon nanoparticles, offering a versatile platform for synthesizing composite materials with tailored properties, epitomize the synergy between carbon-based nanomaterials and other nanotechnologies. By judiciously amalgamating carbon nanoparticles with polymers or alternative nanomaterials, researchers can engineer multifunctional composites featuring enhanced mechanical strength, flexibility, and barrier properties. These bespoke composites, customized to address specific challenges associated with diverse food products, represent a pioneering frontier in the field. The narrative surrounding the chemistry of carbon-based nanomaterials and their composites stands as a compelling discourse within food packaging and preservation. The distinctive properties of carbon nanotubes, graphene, fullerenes, and carbon nanoparticles proffer an expansive spectrum of opportunities for confronting the multifaceted challenges related to food quality and safety preservation. As the trajectory of research in this realm advances, the integration of carbon-based nanomaterials into food packaging emerges as a transformative force, promising to redefine the paradigms governing food storage, transportation, and consumption. This paradigm shift, in turn, ensures a more sustainable and secure global food supply chain (Appendini and Hotchkiss, 2002; Arvanitoyannis and Bosnea, 2004; Cockburn et al., 2012; Damm et al., 2008; Asharani et al., 2011; Aymonier et al., 2002).

14.2 CHEMISTRY OF GRAPHENE OXIDE-BASED MATERIALS

Huang et al. synthesized nanocomposite films by mixing graphene oxide and zinc oxide nanoparticles with polylactic acid (PLA) (Figure 14.1) using the solution-blending method. Both polylactic acid and GO-ZnO particles were dissolved in chloroform separately and later thoroughly mixed at higher temperatures for a few hours. Now, the resultant mixture was cast on coating machines and dried off to form the target blend films (Huang et al., 2015).

The added GO-ZnO nanoparticles raised the tensile strength from 60.0 to 72 MPa, and the glass transition temperature and UV shielding capabilities also significantly increased, according to the data. With activities of 99.2% and 97.6%, respectively, the films' antibacterial activity against *S. aureus* and *E. coli* were examined, and the results indicated that the films were most active when exposed to light.

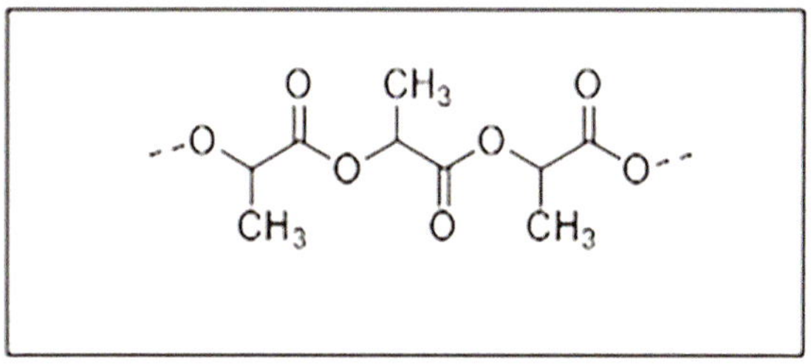

FIGURE 14.1 Chemical structure of polylactic acid.

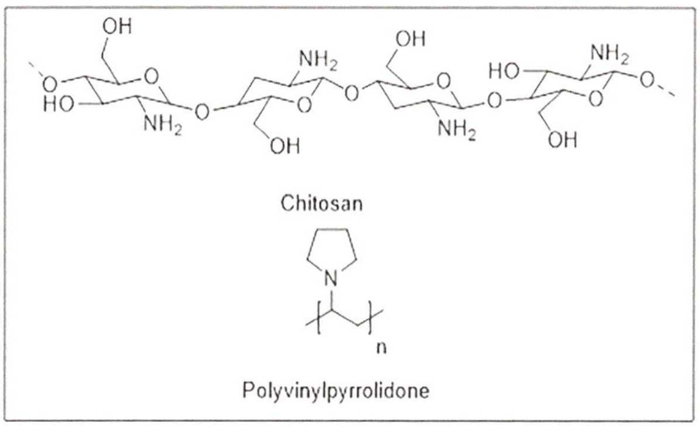

FIGURE 14.2 Structures of chitosan and polyvinyl pyrrolidone.

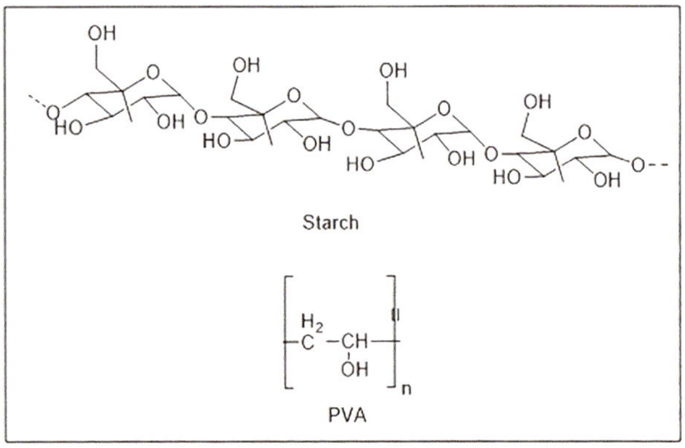

FIGURE 14.3 Structures of starch and PVA.

Mahmoudi et al. developed composite films by combining chitosan (CS) films with polyvinylpyrrolidone (PVP) (Figure 14.2), and graphene oxide (GO). To synthesize the target blend film, a chitosan solution was first prepared using acetic acid, and then an aqueous solution of polyvinylpyrrolidone was also formed. Next the synthesized solutions were mixed in different proportions. Finally, graphene oxide suspension was slowly added to the mixture in such a way that graphene oxide contained about 1, 2, and 3 wt.% (Mahmoudi et al., 2016).

Examining the films for several functional characteristics, it was found that the PVP addition increased the water vapor permeability, opacity, and flexibility of the resultant films. Because of the inclusion of graphene oxide and chitosan, the mix films exhibited significant antibacterial activity.

In a similar approach, Usman et al. formed nanocomposites by incorporating graphene oxide with polyvinyl alcohol (PVA), starch (Figure 14.3), and silver. To synthesize the target film, solutions of PVA and GO were formed separately using deionized water, which were later thoroughly mixed. Subsequently, silver nitrate and starch were added to the mixture, and the resulting solution was applied to cast plates to create the films (Usman et al., 2016).

After being evaluated for a number of characteristics, the films' young modulus and tensile strength values improved by 145% and 26.81%, respectively. Antimicrobial activity also showed a notable uptick, with a zone of inhibition values for *E. coli* and *S. aureus* rising to 17 and 20.5 mm, respectively.

In order to reduce food spoilage due to microorganisms, Xu et al. synthesized blend films using graphene oxide, chitosan, and titanium oxide. Similar methods of formation were used, where graphene oxide solutions were thoroughly mixed with chitosan solutions, and later on, colloidal titanium dioxide was added to form three kinds of solutions with different ratios.

They were tested against *Bacillus subtilis* and *Aspergillus niger*, and the results showed that film with a ratio of 1:20:4 (GO:CS:TiO_2) exhibited the highest activity.

Arfat et al. investigated the antimicrobial activities of graphene oxide nanosheets incorporated with polylactic acid and clove essential oils (Figure 14.4) (CLO). The target films were prepared using the solvent casting method, in which PLA along with polyethylene glycol was added in the presence of dichloromethane. The suspension of GO nanosheets was prepared using dichloromethane. Both solutions were mixed and sonicated for a few minutes. The mixture was then fully mixed for a few more minutes after the addition of CLO, and the mixture was then poured onto glass petri dishes to produce the sheets or films (Arfat et al., 2018).

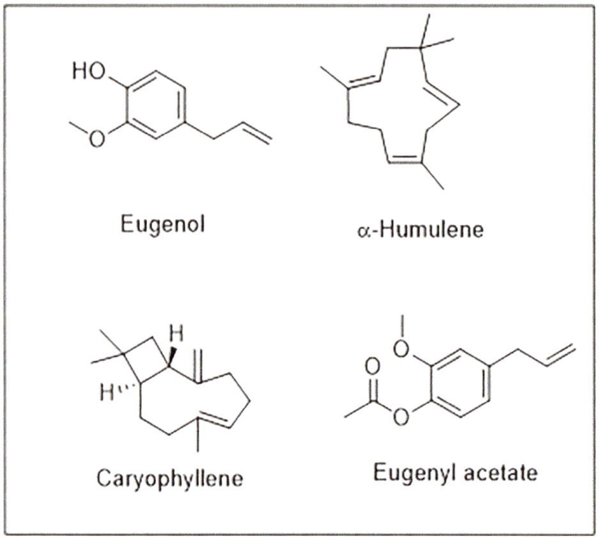

FIGURE 14.4 Chemical components of clove essential oils.

Examining the sheets for different functional qualities revealed that the presence of clove oils increased the films' flexibility while graphene oxide had a substantial impact on the thermo-mechanical and barrier properties. Significant antimicrobial action against *S. aureus* and *E. coli* was also demonstrated by the films.

Sharma et al. developed PVA-based films by incorporating starch and laponite RD. Further, the sheets were coated with graphene oxide to improve the water barrier activities of the sheets. The blend films were developed using melt mixing and solution mixing methods (C. Sharma et al., 2017).

Lin et al. tested the antibacterial activity of composite films, which were prepared by adding thyme oil to polyvinyl alcohol and GO. After making the aqueous solution, glycerine, Tween-80, graphene oxide, and polyvinyl alcohol were well combined. After a little while, the mixture was heated to a higher temperature, and thyme oil was added. The mixture was continuously mixed. After drying out and casting the solution onto plates, it was finally formed into sheets or films (Lin et al., 2020).

After testing, the composite film with 1% oregano oil (Figure 14.5) was shown to have better qualities. They were examined for antimicrobial activities against *E. coli* and *L. monocytogenes*, where they showed promising antibacterial properties with zone of inhibition values up to 11.7 and 12.6 mm, respectively. The films also showed 25.9% and 20.760

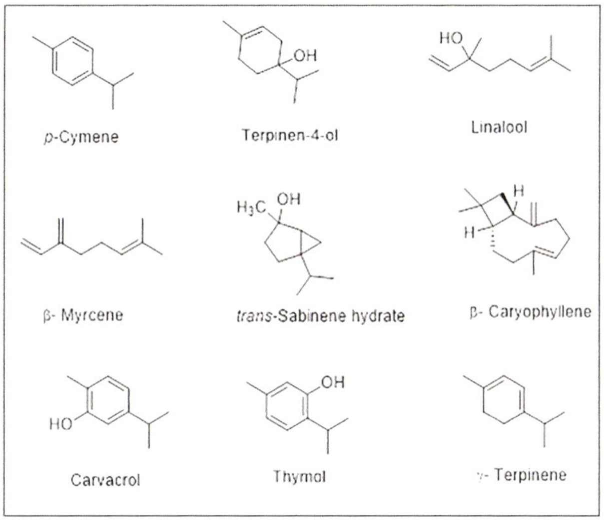

FIGURE 14.5 Chemical components of oregano oil.

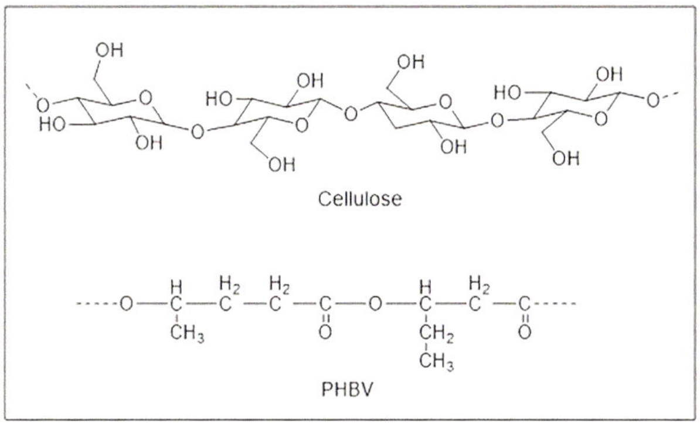

FIGURE 14.6 Structures of cellulose and PHBV.

MPa of tensile fracture elongation and tensile strength, respectively.

Poly (3-hydroxybutyrate-co-3-hydroxyvalerate) (PHBV) (Figure 14.6) was combined with graphene oxide-cellulose nanocrystal hybrids by the solution casting process to create high-grade food packaging materials by Fang, Li, et al. Upon screening the synthesized films for several functional parameters, it was observed that the elongation to break and tensile strength had increased, with respective values of 52.1% and 170.2%. Similarly, the films displayed excellent barrier properties and antibacterial activities (Li et al., 2019).

Shen et al. formed blend films by integrating polylactic acid with a hybrid of silver and graphene oxide using polymerization and mechanical techniques. To synthesize the blend film, a solution of silver-graphene oxide was prepared using lactic acid, which was distilled at 100°C for a couple of hours. To the solution, stannous octoate and toluene sulfonic acid were added, and the mixture was stirred thoroughly for a few more hours under a nitrogen atmosphere (Shen et al., 2019).

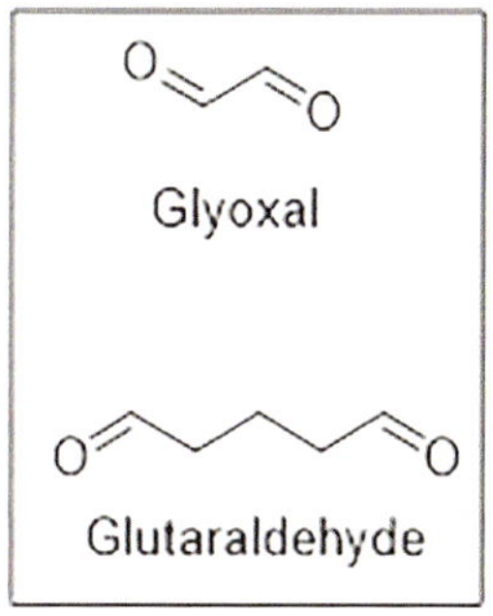

FIGURE 14.7 Structures of glyoxal and glutaraldehyde.

Due to the presence of a silver-graphene oxide hybrid in the film, the flexural strength increased many folds and reached values of 942K Pa, whereas a similar increase was observed when they were tested against *S. aureus* and *E. coli*, with values reaching up to 99%.

Barra et al. synthesized bio-nanocomposites by blending reduced graphene oxide with chitosan to achieve packaging materials with better electrical conductivity. Firstly, the graphene oxide was treated with caffeic acid to form the reduced form of graphene oxide, which was then mixed in distilled water. Now, to it, chitosan and acetic acid were added, along with glycerol as a plasticizer. Lastly, the mixture solution was put onto acrylic plates to form the sheets (Barra et al., 2019).

The resulting bio-nanocomposite's water solubility and tensile strength were significantly improved. The antioxidant effect enhanced in the presence of decreased graphene oxide, with inhibition up to 80%. The blend films showed the best electrical conductivity, with values of 0.7 S/m in the plane.

Another type of blend film using reduced graphene oxide was synthesized by Pal et al. where polylactic acid polymer was integrated with rGO and cellulose. For better dispersion of cellulose and reduced graphene oxide, they were oxidized with TEMPO radical, followed by a reaction with polyethylene glycol (Pal et al., 2019).

The synthesized films showed considerable improvement in various functional properties, with a reduction in water vapor permeability of up to 20%, while the antioxidant activity increased significantly, with values reaching up to 18%.

A similar composite film of polyvinyl alcohol and graphene oxide with gold nanoparticles (AuNPs) was synthesized by Chowdhury et al. using two different cross-linking agents, i.e., glyoxal or glutaraldehyde (Figure 14.7) (GA). First, two PVA-based films were prepared using an aqueous solution of polyvinyl alcohol and the aforementioned cross-linking agents, which led to the formation of nanocomposites. Lastly, gold nanoparticles were added to both films to form the target nanocomposite (Chowdhury et al., 2020).

Pourmoslemi et al. developed a packaging film using polyvinyl alcohol along with graphene oxide, silver nanoparticles, and tragacanth gum (Figure 14.8) (TG). In the aqueous solution of PVA, glycine and Tween-80 were added along with silver oxide nanoparticles, graphene oxide, and TG solution. To form films or sheets, the prepared solution was placed on glass plates and dried at higher temperatures for a whole day (Pourmoslemi et al., 2021).

The created films' antibacterial activities were assessed against *S. aureus* and *E. coli*. The outcomes demonstrated the films' outstanding antibacterial qualities, with the zone of inhibition values of around 39.8 and 41.2 mm, respectively.

Rivera-Briso et al. fabricated food packaging films with the help of carbon nanofibers and graphene oxide (Rivera-Briso et al., 2020).

A weight percentage of 1% GO, or CNFs nanoparticles, was used to make the films. Enhancements in wettability, compression performance, and thermal behavior were similarly achieved by both nanomaterials. When nanomaterials were added to the polymer matrix, both the nanocomposites developed antibacterial activity that is sufficient to stop certain *S. aureus* bacterial growth. Compared to CNFs-used nanocomposites, GO-used nanocomposites demonstrated greater proliferative activity versus time, cell adhesion, and antibacterial activity. However, both nanocomposite films have shown great potential for biomedical applications.

S. Sharma et al. formed nanocomposites based on GO for better and longer storage of fruits (S. Sharma et al., 2018).

Graphene oxide was supposed to produce salicylaldehyde, a preservative when acid synthesized in the overripened fruits cleaves acid-labile hydrazone. Two-step activation was used to optimize loading and get over the problem of GO decrease by hydrazine. The finished composite extends the shelf life of the fruit and demonstrates effective preservative release in response to overripe fruit juice. In both bacterial strains, the antibacterial activity of the composite was shown to be greater than that of free SA. It has been discovered that *E. coli* is more vulnerable to the composite than *P. syringae*. In comparison to the free preservative, the composite exhibits lower toxicity and offers more potential for reuse.

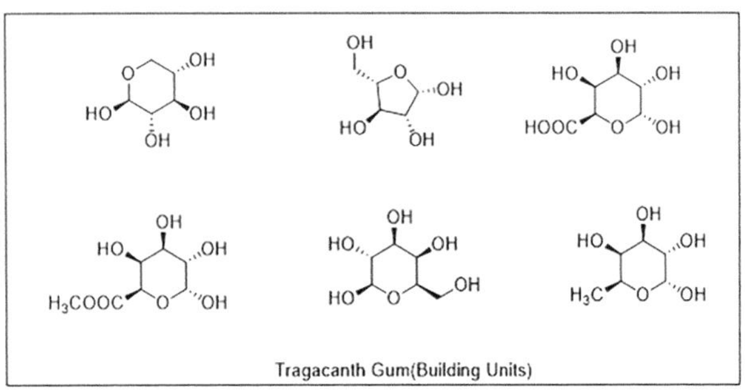

FIGURE 14.8 Structures of tragacanth gum units.

Gouvêa et al. fabricated nanocomposites for active food packaging (Gouvêa et al., 2018).

The hybrid materials were prepared by mixing reduced graphene oxide, zinc oxide, and glycerol-plasticized poly(3-hydroxybutyrate-co-3-hydroxyvalerate) by melt extrusion. The modulus and Tg values were maintained in agreement with the extruded polymer matrix thanks to the inclusion of the composite in the plasticized matrix. A bactericidal action against the Gram –ve bacterium *E. coli* was shown upon direct contact between the hybrid's surface and the bacteria cells. Migration experiments to food simulants revealed that the amount of ZnO that migrated was less than the maximum recommended limit for Zn^{2+}, except for the acetic acid solution.

Zhang et al. fabricated nanofiber composites for enhanced food packaging applications (Zhang et al., 2019).

The ultrasonic processing and electrospinning methods were used to create the nanofiber composites. The composite nanofibrous membrane performed best in terms of water vapor permeability when it was sonicated for 30 minutes and had a ZnO-GO mass ratio of 7:3. The composite nanofibrous membrane's ability to block UV light was improved. Additionally, the nanocomposite films demonstrated potent antibacterial action against foodborne pathogenic bacteria of both gram +ve and gram –ve varieties, with the impact being comparatively more pronounced against gram +ve bacteria. To extend the shelf life of packaged food, the created nanocomposite films offer a great deal of promise as environmentally friendly, antimicrobial material for packaging with ultraviolet barrier qualities.

Naskar et al. fabricated biocompatible nanocomposites based on metal oxides for enhanced food packaging applications (Naskar et al., 2018).

The nanocomposites were synthesized using a solution method. Antibiofilm activity against *Pseudomonas aeruginosa* and *Staphylococcus aureus* microorganisms has been studied in order to confirm the efficacy of the nanocomposites in eradicating the bacterial cells. The results indicate that nanocomposites based on concentration of silver nitrate, administered at a specific dose amount, exhibit about 95% prevention of biofilm development and eradication of pre-existing biofilm. Agar-based film's texture study reveals increased film tensile strength. Even after being stored for an extended amount of—roughly 90 days—the film still had good antibacterial activity. Therefore, it can be concluded that nanocomposites can be a very promising candidate for biomedical and food packaging applications.

Dong et al. synthesized hybrid fibrous nanocomposite films for preservation of peppers (Dong et al., 2021).

The nanofibrous films were made by electrostatic spinning with the help of ultrasonics. Addition of nanoparticles to nanofibrous films to improve their tensile strength, elongation at break, and water barrier properties. In the

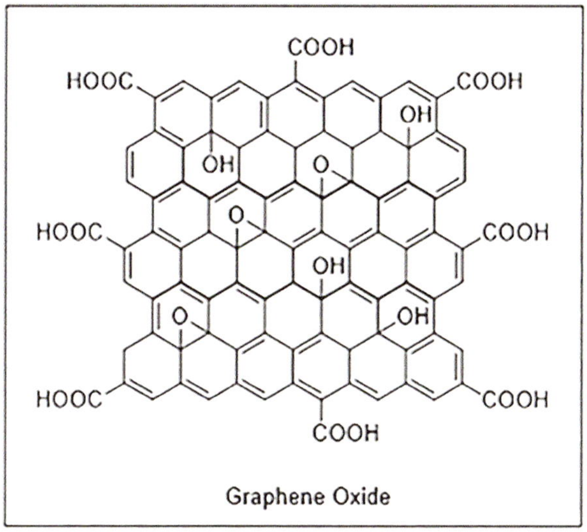

FIGURE 14.9 Structure of graphene oxide.

antibacterial experiment, the nanofibrous films' inhibition rates against *Escherichia coli* and *Staphylococcus aureus* reached 94% and 93%, respectively, after being exposed to UV light for a whole day. The findings demonstrated the superior ability of nanocomposite fibrous films to preserve green peppers' sensory qualities, slow down their rate of deterioration, and increase their shelf life.

Wu et al. fabricated nanocomposite films for better and enhanced food packaging applications (Wu et al., 2019).

Casting was the procedure used to create the nanocomposite films. Tensile strength first rose and subsequently declined with the increasing quantity of nanoparticles, followed by an increase in thickness. The tensile strength peaked at a concentration of 2 mg/ml of nanoparticles. The film's ability to withstand moisture was improved by the concentration of nanoparticles growing. The findings indicate that while adding nanoparticles does not alter the film's composition, it can modify the matrix agglomeration process and enhance the film's thermal stability. Nanoparticle-enhanced composite film preparation enhances its characteristics and broadens the use of nanomaterials in biodegradable composite film research.

The antibacterial and antifungal activity of graphene oxide (Figure 14.9) was examined by Chen et al. against *F. graminearum*, *F. oxysporum*, *P. syringae*, and *X. campestris pv. Udulosa*. The modified Hummer process was used to create the graphene oxide. Graphite particles were oxidized with strong sulfuric acid to produce graphite oxide, which was subsequently dissociated in deionized water and sonicated for several hours to generate sheets of graphene oxide. The synthesized graphene oxides were able to restrict the growth of bacteria by 90% and that of fungal pathogens by up to 90% (Chen et al., 2014).

Venkateshaiah et al. combined graphene oxide with gum arabic, gum karaya, and kondagogu gum to create three distinct nanocomposites. The solution casting process was used to create the target blend films, which involved dissolving graphene oxide in deionized water and stirring for a short while. Later, 1% weight or volume of various tree gums (Figure 14.10) was added along with glycerol. The solution was again sonicated for a few minutes to form a homogenous mixture. Now, the homogenous mixture formed was cast onto plates to form the hybrid films (Venkateshaiah et al., 2020).

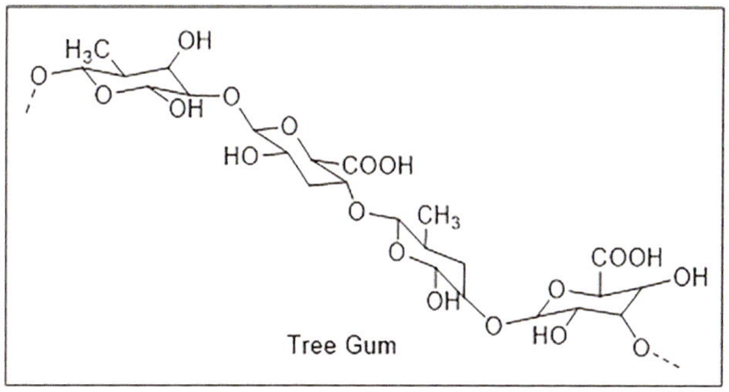

FIGURE 14.10 Chemical structure of tree gum.

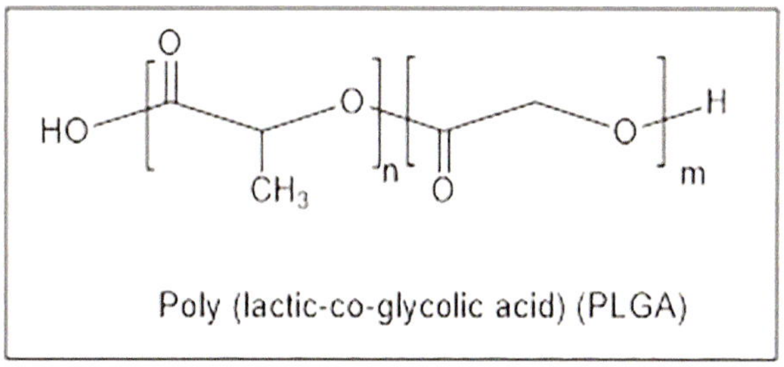

FIGURE 14.11 Chemical structure of PLGA.

When the hybrid films were examined for various functional properties, the results showed that they acted as excellent oxygen gas barriers and significantly decreased water vapor transmittance.

14.3 CHEMISTRY OF CARBON NANOTUBE-BASED MATERIALS

Aslan et al. combined poly (lactic-co-glycolic acid) (PLGA) (Figure 14.11) with single-walled carbon nanotubes (SWNTs) to enhance the antibacterial qualities of pure PLGA polymer. To create the target films, PLGA was gradually added to a solution of SWNTs made with chloroform, and the mixture was then constantly sonicated for a few minutes. The synthetic film's ability to inhibit *E. coli* and *S. epidermidis* bacteria was assessed. The findings demonstrated that the inclusion of single-walled carbon nanotubes greatly enhanced the antibacterial activity, as seen by the ability to eradicate 98% of the bacterial population in just one hour (Aslan et al., 2010).

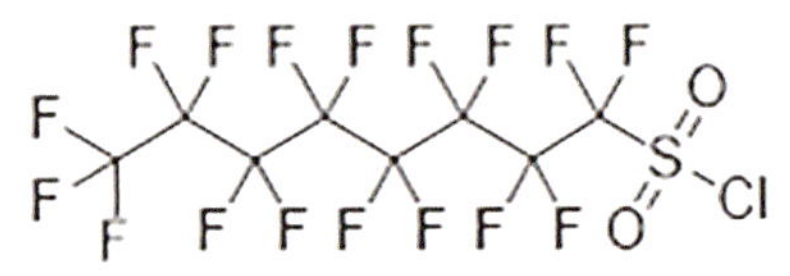

FIGURE 14.12 The chemical structure of PSC.

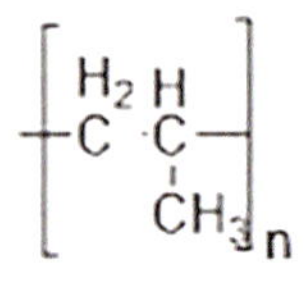

FIGURE 14.13 Structure of polypropylene.

Song et al. synthesized a nano-hybrid membrane having multiwalled carbon nanotubes integrated with chitosan and a silicon couple agent, which were further modified using perfluorooctanesulfonyl fluoride (PFOSF). The synthesized bio-nanocomposites (Figure 14.12) were evaluated for their antibacterial properties against *S. aureus* and *E. coli*, and the results stated that they showed excellent antibacterial activity, with growth inhibition values lying in the range of 26–89% (Song et al., 2015).

To develop a better-preserving material for strawberries, Liu et al. synthesized nanocomposites of fibers by using polylactic acid with carbon nanotubes and chitosan (CS). A solution of chitosan and carbon nanotubes (CNT) was prepared using acetic acid and concentrated sulfuric acid/nitric acid, respectively. Further, the CNT solution was diluted using deionized water, which was then dried at higher temperatures, followed by the addition of dimethylformamide. Now, different amounts of chitosan were added to the PLA/CNT solution to form the nanocomposite. The mixture solution was then converted to fibers using electrospinning techniques (Liu et al., 2019).

When the composite fiber was evaluated for various functional properties, the results showed that the fiber with 7% chitosan was most efficient as it exhibited better antimicrobial activity, with a bacteriostatic rate reaching up to 90%.

Mohamed et al. created four chitosan derivatives: derivative 1 was created by reacting the chitosan with benzaldehyde; derivative 2 was created by adding epichlorojydrine; and derivative 3 was created by reacting the chitosan with aminobenzhydrazide. To obtain the derivative 4, the benzaldehyde group was removed from the derivative 3. Further, by using the derivative 4, multiwalled carbon composites were prepared and examined for their antimicrobial activities. The results came out to be promising with zone of inhibition values ranging from approximately 13.6–22.9 mm (Mohamed and Abd El-Ghany, 2018).

Azizi et al. formed nanocomposites using polypropylene (Figure 14.13) (PP), polylactic acid, and multiwalled carbon nanotubes using the melt-compound method. When they underwent additional screening for different functional characteristics, the results revealed that the tensile strength and modulus of composites containing 2% weight of carbon nanotubes rose by 95% and 60%, respectively. Regarding

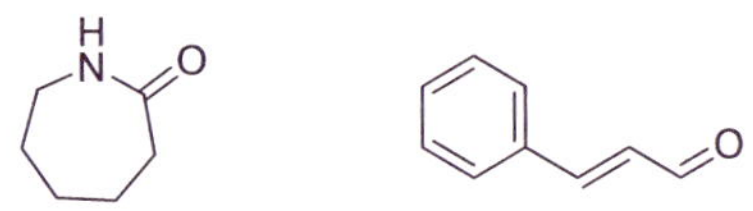

FIGURE 14.14 Structures of caprolactum and cinnamaldehyde.

electrical conductivity, a comparable pattern was noted, with improvements up to nine times greater than those reported with PP/PLA mixes (Azizi et al., 2019).

Cui et al. used the solvent evaporation technique to integrate polylactic acid, poly-ε-caprolactone (PCL), carbon nanotubes, and cinnamon aldehyde (CIN) (Figure 14.14) to produce antibacterial films. Upon examination, the barrier characteristics of the films remained unchanged, but the addition of carbon nanotubes and cinnamaldehyde resulted in a decrease in the films' UV resistance. When the glass transition temperature dropped to 44.8°C and the elongation at break rose, they were able to strengthen the films. Additionally, the blended films have outstanding antimicrobial properties due to their role as effective transporters of cinnamaldehyde (Cui et al., 2020).

Marques et al. fabricated carbon nanotube composites for enhanced food packaging applications (Marques et al., 2021).

Changes in film structure, such as lump formation and decreased roughness, were brought about by the inclusion of complex and carbon nanotubes, respectively. Enhancing the antibacterial agent's retention by its complexation with *β*-CD and CNT application is meant to promote controlled diffusion of the agent and hence increase the film's preservation impact. However, care must be taken to avoid damaging surface modifications in the films when contemplating the usage of *β*-CD inclusion complexes. Both bacteria's development was halted by the active coatings, which also damaged the cell membrane. It was confirmed that in all films examined, the minimum inhibitory concentrations (MIC) of *S. choleraesuis* were less than those of *L. innocua*.

Youssef et al. fabricated nanocomposite food packaging films from carbon nanotubes (Youssef et al., 2019).

Functionalized multiwall carbon nanotubes were integrated into cross-linked poly (vinyl alcohol) nanocomposite films at varying concentrations. Ultimately, a powder form of hybrid nanocomposite polymers was obtained by using citric acid as a cross-linking agent, which was then immediately transformed into nanocomposite films. The antimicrobial properties were assessed against a variety of bacterial and fungal species, and it was shown that the inhibition zone was significantly increased at large functionalized nanocomposites loadings in the hybrid polymer nanocomposites, resulting in considerable antibacterial activity.

14.4 CHEMISTRY OF GRAPHENE-BASED MATERIALS

Montes et al. produced graphene-based nanocomposites, which were then used to make films by the hot-pressing technique (Montes et al., 2018).

The nanocomposites were mixed in two separate ratios and added to the polylactic acid using the melt blending process, loading it up to a total of 1 weight percent. A study of the mechanical parameters of the GR/CNC (ratio 50/50) film over PLA-T showed improvements of 11% and 8%, respectively, in the film's tensile strength and Young's modulus. The thermal properties also improved with the inclusion of both nanofillers. In a similar vein, the films' oxygen transmission rate (OTR) improved by 23% due to improvements in gas barrier characteristics. *A. niger* was used to test the antifungal qualities, and the hybrid films showed better antifungal activity.

Demitri et al. synthesized films based on nanocomposites for better food packaging applications (Demitri et al., 2016).

Chitosan reacted with cinnamaldehyde at various doses to create the Schiff base required for chitosan functionalization. The chitosan films functionalized with cinnamaldehyde exhibited enhanced mechanical capabilities as a consequence of the incorporation of nanometric graphene stacks. Through the use of in vitro studies against *R. stolonifer*, the fungicidal impact was effectively assessed. These experiments demonstrated that the concentration of cinnamaldehyde boosted the inhibitory effect.

Kumar et al. fabricated graphene-based nanocomposite for more food packaging applications (Kumar et al., 2016).

For improved food packaging films, several PCL composites were made utilizing the solvent precipitation approach.

When evenly distributed nanocomposites were utilized, PCL's modulus increased by 77%, and electrical conductivity increased by four orders of magnitude to 10^{-7} S/cm at 5 weight percent filler—significantly more than PCL's gains when rGO or AgNP alone was employed. In contrast to the highly hazardous nanocomposites lacking rGO, human mesenchymal stem cells were unaffected by nanocomposites or by nanocomposites without Ag. The significant antibacterial effect of the nanocomposites was attributed to a combination of mechanical breaking induced by the RGO in the hybrid nanoparticles and silver ion release from the AgNPs. In comparison to plain PCL, the bacterial count decreased by 37% at 5 weight percent of rGO. Bacterial count decrease was greatest in nanocomposites without rGO. The nanocomposites without rGO revealed an 81% decline. Good antibacterial activity was also demonstrated by the hybrid nanocomposites added to PCL, with 59% greater inhibition than on pure PCL. Thus, it may very well be a good option for use in food packaging.

Goh et al. synthesized graphene-based nanocomposites food packaging films (Goh et al., 2016).

The rGO core barrier's hydrophobicity and compact lamellar microstructure are responsible for its exceptional barrier qualities. Because of its sandwiched construction, the nanocomposite film offers improved processability and manageability. Simulations utilizing the nanocomposite food packaging film suggest that the shelf life of edible oil and potato chips—which are prone to moisture and oxygen—may be increased by a minimum of eight times. Together, the protective encapsulation and barrier considerably minimize the water vapor permeability by 87.6%. These characteristics collectively have shown the great potential of a nanocomposite film with a sandwich design for food packaging applications.

Torres-Giner et al. synthesized nanocomposites for enhanced food packaging applications (Torres-Giner et al., 2018).

Graphene nanoplatelets were created by synthesizing graphite powder and then electrospinning them into fibers in the range of 0.1–2% weight percent. The electro-spun fiber mats with 0.5 weight percent graphene had the best nanoplatelets dispersion, according to Raman studies. To create continuous and contact transparent layers, the fiber mat was ultimately post-treated at 158°C. The resultant films took on a grey hue and grew in opaqueness as the graphene concentration rose, yet the contact transparency remained intact. The ability of electrospinning to create materials containing graphene nanoparticles and having a high electrical conductivity, which may be used for "smart" labels or tags in intelligent packaging applications.

14.5 CONCLUSION

The global market is shifting toward products that are less harmful to the environment. Regarding food applications, especially packaging materials, carbon-based nanomaterials, and their composites are seen to be the most excellent alternatives to synthetic materials. The chapter gives a general review of the uses of graphene, carbon nanotubes, and nanomaterials based on graphene oxide in food preservation and packaging. The research aims to create nanocomposites with suitable cross-linkages, enhancing the film's functional qualities, such as its mechanical potency and water resistance.

14.6 FUTURE ASPECT

Humans depend on food, and because of this, the environment in which we live is continuously being contaminated using different kinds of materials in its packing and preservation. Therefore, instead of using synthetic items that are more harmful, the food industry needs to move to packaging materials that are less toxic to prevent pollution. Eco-friendly and cost-effective carbon-based nanomaterials and composites have shown great promise in various food applications.

ACKNOWLEDGMENT

The authors would like to express their heartfelt gratitude to the management of SGT University, Gurugram, Haryana, India, Banaras Hindu University, Varanasi, Uttar Pradesh, India, and Presidency University, Bangalore, Karnataka, India for providing the resources to finish this chapter on time.

Conflict of Interest

The authors declare no conflict of interest.

REFERENCES

Ahire, J. J., & Dicks, L. M. T. (2014). Nisin incorporated with 2,3-dihydroxybenzoic acid in nanofibers inhibits biofilm formation by a methicillin-resistant strain of Staphylococcus aureus. *Probiotics and Antimicrobial Proteins*, *7*, 52–59.

Alexandre, M., & Dubois, P. (2000). Polymer-layered silicate nanocomposites: Preparation, properties, and uses of a new class of materials. *Materials Science and Engineering*, *28*, 1–63.

Althues, H., Henle, J., & Kaskel, S., 2007. Functional inorganic nanofillers for transparent polymers. *Chemical Society Reviews*, *36*(9), 1454–1465.

An, J., Zhang, M., Wang, S., & Tang, J. (2008). Physical, chemical and microbiological changes in stored green asparagus spears as affected by coating of silver nanoparticles-PVP. *LWT - Food Science and Technology*, *41*(6), 1100–1107.

Anthonissen, R., Toussaint, O., Vankoningsloo, S., Saout, C., Alfaro-Moreno, E., Hoet, P., Gonzalez, L., & Dubruel, P. (2010). Eco-, geno- and human toxicology of bio-active nanoparticles for biomedical applications. *Toxicology*, *269*, 170–181.

Appendini, P., & Hotchkiss, J. H. (2002). Review of antimicrobial food packaging. *Innovative Food Science and Emerging Technologies*, *3*, 113–126.

Arfat, Y. A., Ahmed, J., Ejaz, M., & Mullah, M. (2018). Polylactide/graphene oxide nanosheets/clove essential oil composite films for potential food packaging applications. *International Journal of Biological Macromolecules*, *107*(PartA), 194–203. https://doi.org/10.1016/j.ijbiomac.2017.08.156.

Arvanitoyannis, I. S., & Bosnea, L. (2004). Migration of substances from food packaging materials to foods. *Critical Reviews in Food Science and Nutrition*, *44*, 63–76.

Asharani, P. V., Lianwu, Y., Gong, Z., & Valiyaveettil, S. (2011). Toxicity of silver nanoparticles on zebrafish models. *Nanotoxicology*, *5*, 43–54.

Aslan, S., Loebick, C. Z., Kang, S., Elimelech, M., Pfefferle, L. D., & Van Tassel, P. R. (2010). Antimicrobial biomaterials based on carbon nanotubes dispersed in poly(lactic-co-glycolic acid). *Nanoscale*, *2*(9), 1789–1794. https://doi.org/10.1039/c0nr00329h.

Aymonier, C., Schlotterbeck, U., Antonietti, L., Zacharias, P., Thomann, R., Tiller, J. C., & Mecking, S. (2002). *Chemical Communications*, *24*, 3018–3019.

Azizi, S., Azizi, M., & Sabetzadeh, M. (2019). The role of multiwalled carbon nanotubes in the mechanical, thermal, rheological, and electrical properties of pp/pla/mwcnts nanocomposites. *Journal of Composites Science*, *3*(3). https://doi.org/10.3390/jcs3030064.

Barlow, C. Y., & Morgan, D. C. (2013). Polymer film packaging for food, an environmental assessment. *Resources, Conservation and Recycling*, *78*, 74–80.

Barra, A., Ferreira, N. M., Martins, M. A., Lazar, O., Pantazi, A., Jderu, A. A., Neumayer, S. M., Rodriguez, B. J., Enăchescu, M., Ferreira, P., & Nunes, C. (2019). Eco-friendly preparation of electrically conductive chitosan—reduced graphene oxide flexible bionanocomposites for food packaging and biological applications. *Composites Science and Technology*, *173*, 53–60. https://doi.org/10.1016/j.compscitech.2019.01.027.

Chawengkijwanich, C., & Hayata, Y. (2008). Development of TiO_2 powder-coated food packaging film and its ability to inactivate Escherichia coli in vitro and in actual tests. *International Journal of Food Microbiology*, *123*(3), 288–292.

Chen, J., Peng, H., Wang, X., Shao, F., Yuan, Z., & Han, H. (2014). Graphene oxide exhibits broad-spectrum antimicrobial activity against bacterial phytopathogens and fungal conidia by intertwining and membrane perturbation. *Nanoscale*, *6*(3), 1879–1889. https://doi.org/10.1039/c3nr04941h.

Cheng, Q., Li, C., Pavlinek, V., Saha, P., & Wang, H. (2006). Surface-modified antibacterial TiO2/Ag+ nanoparticles: Preparation and properties. *Applied Surface Science*, *252*, 4154–4160.

Choi, O., Yu, C. P., Fernández, G. E., & Hu, Z. (2010). Interactions of nanosilver with Escherichia coli cells in planktonic and biofilm cultures. *Water Research*, *44*, 6095–6103.

Chowdhury, S., Teoh, Y. L., Ong, K. M., Rafflisman Zaidi, N. S., & Mah, S. K. (2020). Poly(vinyl) alcohol crosslinked composite packaging film containing gold nanoparticles on shelf life extension of banana. *Food Packaging and Shelf Life*, *24*. https://doi.org/10.1016/j.fpsl.2020.100463.

Cockburn, A., Bradford, R., Buck, N., Constable, A., Edwards, G., Haber, B., Hepburn, P., Howlett, J., Kampers, E., Klem, C., Radomski, M., Stamm, H., Wijnhoven, S., & Wildemann, T. (2012). Approaches to the safety assessment of engineered nanomaterials in food. *Food and Chemical Toxicology*, *50*, 2224–2242.

Cui, R., Jiang, K., Yuan, M., Cao, J., Lin, L., Tang, Z., & Qin, Y. (2020). Antimicrobial film based on polylactic acid and carbon nanotube for controlled cinnamaldehyde release. *Journal of Materials Research and Technology*, *9*(5), 10130–10138. https://doi.org/10.1016/j.jmrt.2020.07.016

Cuq, B., Gontard, N., & Guilbert, S. (1995). Edible films and coatings as active layers. In: M. Rooney (Ed.), *Active Food Packaging*. Blackie Academic and Professional, Glasgow, pp. 11–142.

Cushen, M., Kerry, J., Morris, M., Cruz-Romero, M., & Cummins E. (2012). Nanotechnologies in the food industry—recent developments, risks and regulation. *Trends in Food Science and Technology*, *24*, 30–46.

Cutter, C. N. (2006). Opportunities for bio-based packaging technologies to improve the quality and safety of fresh and further processed muscle foods. *Meat Science*, *74*, 131–142.

Damm, C., Münstedt, H., & Rösch, A. (2007). Long-term antimicrobial polyamide 6/silver nanocomposites. *Journal of Materials Science*, *42*(15), 6067–6073.

Damm, C., Münstedt, H., & Rösch, A. (2008). The antimicrobial efficacy of polyamide 6/silver-nano- and microcomposites. *Materials Chemistry and Physics*, *108*, 61–66.

Danielli, D., et al. (2008). Active and intelligent food packaging: Legal aspects and safety concerns. *Trends in Food Science and Technology*, *19*(1), 103–112.

Demitri, C., De Benedictis, V. M., Madaghiele, M., Corcione, C. E., & Maffezzoli, A. (2016). Nanostructured active chitosan-based films for food packaging applications: Effect of graphene stacks on mechanical properties. *Measurement: Journal of the International Measurement Confederation*, *90*, 418–423. https://doi.org/10.1016/j.measurement.2016.05.012.

Dong, X., Liang, X., Zhou, Y., Bao, K., Sameen, D. E., Ahmed, S., Dai, J., Qin, W., & Liu, Y. (2021). Preparation of polylactic acid/TiO2/GO nano-fibrous films and their preservation effect on green peppers. *International Journal of Biological Macromolecules*, *177*, 135–148. https://doi.org/10.1016/j.ijbiomac.2021.02.125.

Goh, K., Heising, J. K., Yuan, Y., Karahan, H. E., Wei, L., Zhai, S., Koh, J. X., Htin, N. M., Zhang, F., Wang, R., Fane, A. G., Dekker, M., Dehghani, F., & Chen, Y. (2016). Sandwich-architectured poly(lactic acid)- graphene composite food packaging films. *ACS Applied Materials and Interfaces*, *8*(15), 9994–10004. https://doi.org/10.1021/acsami.6b02498

Gouvêa, R. F., Del Aguila, E. M., Paschoalin, V. M. F., & Andrade, C. T. (2018). Extruded hybrids based on poly(3-hydroxybutyrate-co-3-hydroxyvalerate) and reduced graphene oxide composite for active food packaging. *Food Packaging and Shelf Life*, *16*, 77–85. https://doi.org/10.1016/j.fpsl.2018.02.002

Huang, Y., Wang, T., Zhao, X., Wang, X., Zhou, L., Yang, Y., Liao, F., & Ju, Y. (2015). Poly(lactic acid)/graphene oxide-ZnO nanocomposite films with good mechanical, dynamic mechanical, anti-UV and antibacterial properties. *Journal of Chemical Technology and Biotechnology*, *90*(9), 1677–1684. https://doi.org/10.1002/jctb.4476.

Kumar, S., Raj, S., Jain, S., & Chatterjee, K. (2016). Multifunctional biodegradable polymer nanocomposite incorporating graphene-silver hybrid for biomedical applications. *Materials and Design*, *108*, 319–332. https://doi.org/10.1016/j.matdes.2016.06.107.

Li, F., Yu, H. Y., Wang, Y. Y., Zhou, Y., Zhang, H., Yao, J. M., Abdalkarim, S. Y. H., & Tam, K. C. (2019). Natural biodegradable poly(3-hydroxybutyrate-co-3-hydroxyvalerate) nanocomposites with multifunctional cellulose nanocrystals/graphene oxide hybrids for high-performance food packaging. *Journal of Agricultural and Food Chemistry*, *67*(39), 10954–10967. https://doi.org/10.1021/acs.jafc.9b03110

Lin, D., Wu, Z., Huang, Y., Wu, J., Li, C., Qin, W., Wu, D., Li, S., Chen, H., & Zhang, Q. (2020). Physical, mechanical, structural and antibacterial properties of polyvinyl alcohol/oregano oil/graphene oxide composite films. *Journal of Polymers and the Environment*, *28*(2), 638–646. https://doi.org/10.1007/s10924-019-01627-4

Liu, Y., Wang, S., Lan, W., & Qin, W. (2019). Fabrication of polylactic acid/carbon nanotubes/chitosan composite fibers by electrospinning for strawberry preservation. *International Journal of Biological Macromolecules*, *121*, 1329–1336. https://doi.org/10.1016/j.ijbiomac.2018.09.042

Mahmoudi, N., Ostadhossein, F., & Simchi, A. (2016). Physicochemical and antibacterial properties of chitosan-polyvinylpyrrolidone films containing self-organized graphene oxide nanolayers. *Journal of Applied Polymer Science*, *133*(11). https://doi.org/10.1002/app.43194

Marques, C. S., Dias, M. V., Soares, N. de F. F., Borges, S. V., de Oliveira, I. R. N., Pires, A. C. dos S., Medeiros, E. A. A., & Alves, E. (2021). Ultrastructural and antimicrobial impacts of allyl isothiocyanate incorporated in cellulose, β-cyclodextrin, and carbon nanotubes nanocomposites. *Journal of Vinyl and Additive Technology*, *27*(4), 795–805. https://doi.org/10.1002/vnl.21850.

Mohamed, N. A., & Abd El-Ghany, N. A. (2018). Novel amino-hydrazide cross-linked chitosan filled with multi-walled carbon nanotubes as antimicrobial agents. *International Journal of Biological Macromolecules*, *115*, 651–662. https://doi.org/10.1016/j.ijbiomac.2018.04.101.

Montes, S., Etxeberria, A., Mocholi, V., Rekondo, A., Grande, H., & Labidi, J. (2018). Effect of combining cellulose nanocrystals and graphene nanoplatelets on the properties of poly(lactic acid) based films. *Express Polymer Letters*, *12*(6), 543–555. https://doi.org/10.3144/expresspolymlett.2018.45.

Naskar, A., Khan, H., Sarkar, R., Kumar, S., Halder, D., & Jana, S. (2018). Anti-biofilm activity and food packaging application of room temperature solution process-based polyethylene glycol capped Ag-ZnO-graphene nanocomposite. *Materials Science and Engineering C*, *91*, 743–753. https://doi.org/10.1016/j.msec.2018.06.009.

Pal, N., Banerjee, S., Roy, P., & Pal, K. (2019). Reduced graphene oxide and PEG-grafted TEMPO-oxidized cellulose nanocrystal reinforced poly-lactic acid nanocomposite film for biomedical application. *Materials Science and Engineering C*, *104*. https://doi.org/10.1016/j.msec.2019.109956.

Pourmoslemi, S., Shokouhi, S., & Mahjub, R. (2021). Investigation of antibacterial activity of polyvinyl alcohol packaging films composed of silver oxide nanoparticles, graphene oxide and tragacanth gum using Box–Behnken design. *Packaging Technology and Science*, *34*(10), 613–622. https://doi.org/10.1002/pts.2599.

Rezić, I., Haramina, T., & Rezić, T. (2017). Metal nanoparticles and carbon nanotubes—perfect antimicrobial nano-fillers in polymer-based food packaging materialsa. In *Food Packaging*. Elsevier, pp. 497–532. https://doi.org/10.1016/b978-0-12-804302-8.00015-7.

Rivera-Briso, A. L., Aachmann, F. L., Moreno-Manzano, V., & Serrano-Aroca, Á. (2020). Graphene oxide nanosheets versus carbon nanofibers: Enhancement of physical and biological properties of poly(3-hydroxybutyrate-co-3-hydroxyvalerate) films for biomedical applications. *International Journal of Biological Macromolecules*, *143*, 1000–1008. https://doi.org/10.1016/j.ijbiomac.2019.10.034.

Sharma, C., Manepalli, P. H., Thatte, A., Thomas, S., Kalarikkal, N., & Alavi, S. (2017). Biodegradable starch/PVOH/laponite RD-based bionanocomposite films coated with graphene oxide: Preparation and performance characterization for food packaging applications. *Colloid and Polymer Science*, *295*(9), 1695–1708. https://doi.org/10.1007/s00396-017-4114-9.

Sharma, S., Biswal, B. K., Kumari, D., Bindra, P., Kumar, S., Stobdan, T., & Shanmugam, V. (2018). Ecofriendly fruit switches: Graphene oxide-based wrapper for programmed fruit preservative delivery to extend shelf life. *ACS Applied Materials and Interfaces*, *10*(22), 18478–18488. https://doi.org/10.1021/acsami.8b02048.

Shen, X. J., Yang, S., Shen, J. X., Ma, J. L., Wu, Y. Q., Zeng, X. L., & Fu, S. Y. (2019). Improved mechanical and antibacterial properties of silver-graphene oxide hybrid/polylactid acid composites by in-situ polymerization. *Industrial Crops and Products*, *130*, 571–579. https://doi.org/10.1016/j.indcrop.2019.01.018.

Song, K., Gao, A., Cheng, X., & Xie, K. (2015). Preparation of the superhydrophobic nano-hybrid membrane containing carbon nanotube based on chitosan and its antibacterial activity. *Carbohydrate Polymers*, *130*, 381–387. https://doi.org/10.1016/j.carbpol.2015.05.023.

Torres-Giner, S., Echegoyen, Y., Teruel-Juanes, R., Badia, J. D., Ribes-Greus, A., & Lagaron, J. M. (2018). Electrospun poly(ethylene-co-vinyl alcohol)/graphene nanoplatelets composites of interest in intelligent food packaging applications. *Nanomaterials*, *8*(10). https://doi.org/10.3390/nano8100745.

Usman, A., Hussain, Z., Riaz, A., & Khan, A. N. (2016). Enhanced mechanical, thermal and antimicrobial properties of poly(vinyl alcohol)/graphene oxide/starch/silver nanocomposites films. *Carbohydrate Polymers*, *153*, 592–599. https://doi.org/10.1016/j.carbpol.2016.08.026.

Venkateshaiah, A., Cheong, J. Y., Habel, C., Wacławek, S., Lederer, T., Černík, M., Kim, I. D., Padil, V. V. T., & Agarwal, S. (2020). Tree gum-graphene oxide nanocomposite films as gas barriers. *ACS Applied Nano Materials*, *3*(1), 633–640. https://doi.org/10.1021/acsanm.9b02166.

Wu, Z., Huang, Y., Xiao, L., Lin, D., Yang, Y., Wang, H., Yang, Y., Wu, D., Chen, H., Zhang, Q., Qin, W., & Pu, S. (2019). Physical properties and structural characterization of starch/polyvinyl alcohol/graphene oxide composite films. *International Journal of Biological Macromolecules*, *123*, 569–575. https://doi.org/10.1016/j.ijbiomac.2018.11.071.

Xu, W., Xie, W., Huang, X., Chen, X., Huang, N., Wang, X., & Liu, J. (2017). The graphene oxide and chitosan biopolymer loads TiO2 for antibacterial and preservative research. *Food Chemistry*, *221*, 267–277. https://doi.org/10.1016/j.foodchem.2016.10.054.

Youssef, A. M., El-Naggar, M. E., Malhat, F. M., & El Sharkawi, H. M. (2019). Efficient removal of pesticides and heavy metals from wastewater and the antimicrobial activity of f-MWCNTs/PVA nanocomposite film. *Journal of Cleaner Production*, *206*, 315–325. https://doi.org/10.1016/j.jclepro.2018.09.163.

Zhang, R., Wang, Y., Ma, D., Ahmed, S., Qin, W., & Liu, Y. (2019). Effects of ultrasonication duration and graphene oxide and nano-zinc oxide contents on the properties of polyvinyl alcohol nanocomposites. *Ultrasonics Sonochemistry*, *59*. https://doi.org/10.1016/j.ultsonch.2019.104731.

15 Nanotechnology in Food Supply Chain Management

Shazia Akhter[a], *Jonaid Ahmad Malik*[b], *Syed Zameer Hussain*[a], *and Aaruba Maqbool Sofi*[a]

[a]Division of Food Science and Technology, Sheri-e-Kashmir University of Agricultural Sciences and Technology, Srinagar Kashmir, India

[b]Department of Biomedical Engineering, Indian Institute of Technology, Ropar, Punjab, India

15.1 INTRODUCTION

The word "nanotechnology" was introduced by Norio Taniguchi in the year 1974, and Richard Feynman introduced the foundational concepts in 1959 [1, 2]. Nanotechnology encompasses exploring structures, devices, and substances through construction, characterization, production, and control processes, where at least one dimension falls within the 1 to 100 nanometer range [3, 4]. The emergence of properties at the nanoscale occurs, showing significant disparities from the characteristics of larger-scale materials formed from identical components when the particle size goes below this critical threshold [3]. Nanoparticles demonstrate elevated catalytic efficacy due to their expanded surface area and reduced dimensions. These modifications result in unique physicochemical attributes, including improved adsorption capabilities and antimicrobial potency, originating from their increased surface-to-volume ratio and small-scale [5].

A supply chain can be broadly perceived as the dynamic interface among multiple entities actively facilitating the movement of goods and services to their ultimate recipients [6]. The food supply chain encompasses many participants, from farmers and manufacturers to processors, transporters, retailers, and consumers. Ensuring safety measures and efficient operations at each stage is imperative to deliver secure and quality products to the end consumer. In the food industry, customer satisfaction and trust are the main motivators for success. The food journey, from farm to fork, involves various critical stages, including selection, cleaning, processing, manufacturing, packaging, and storage. This intricate supply chain necessitates transparency for detecting foodborne illnesses, fostering consumer confidence, ensuring safety, and obtaining approval from regulatory authorities [2]. Food supply chain management entails organizing logistics and key features to boost responsiveness to customer requirements and maximize delivery efficiency while maintaining safety standards. It signifies the integrated administration of human-operated services and automated procedures within this multifaceted system [7].

Nanotechnology is a groundbreaking technology that has recently changed the food industry significantly. Increasing concerns about food quality and safety fuel a rising demand for nanoparticle-based materials in the food supply chain. Nanotechnology is applied in developing consumer products like additives, food packaging, and preservation methods, with nanometer-scale food additives becoming more common [8]. Nanotechnology significantly impacts food production by enhancing efficiency and sustainability. In production processes, nanomaterials streamline operations, reduce resource usage, and ensure quality. Moving to processing, nanoparticles improve nutrient delivery and flavor, elevating overall product quality. Nanotechnology enhances freshness and safety in packaging, extending shelf life and meeting the stringent demands of modern food supply chains.

15.2 LIMITATIONS OF TRADITIONAL PACKAGING

The packaging industry has relied on traditional techniques to safeguard and transport products for decades. Nevertheless, these methods exhibit limitations in the realm of advanced packaging. With ongoing technological advancements, conventional/traditional packaging approaches are proving less efficient and increasingly expensive.

15.2.1 Limited Protection

The packaging material should be able to prevent the penetration of water, grease, oxygen (O_2) microorganisms, carbon dioxide (CO_2), aromatic compounds, and oils [9]. However, traditional packaging techniques such as paper and cardboard-based packaging demonstrate less effective barrier properties, allowing the permeation of gases, water vapors, and liquids. Consequently, transportation can cause damage to products, resulting in expensive returns and harm to the business's reputation [10].

DOI: 10.1201/9781032677422-15

15.2.2 Limited Customization and Limited Sustainability

Customization poses a challenge for traditional packaging methods, as most packaging materials are tailored for generic products, making it difficult to find suitable sizes and shapes. This can result in inefficiencies, high costs, and wasted space during the packaging process. Customized packaging provides a more stable, secure, and protective enclosure, which is particularly beneficial for products with distinctive shapes, features, or susceptibility to wear and tear [11]. Traditional/conventional packaging techniques frequently lack sustainability, with many packaging materials derived from non-renewable resources and lacking recyclability. This elevates waste production and exerts a detrimental effect on the environment [12].

15.2.3 Need for Advanced Packaging Techniques

To address the limitations of traditional/conventional packaging methods, advanced packaging techniques must focus on quality, safety, shelf life, and usability [13]. Examples of various advanced packaging techniques are as follows: a) 3D printing has the potential to significantly contribute to the commercial utilization of environmentally friendly materials in packaging, thus leveraging its inherent adaptability and scalability [14]; b) with nanotechnology, by incorporating nanoparticles, the attributes of packaging polymers, including their robustness, adaptability, resistance, longevity, and ability to be recycled, can be improved [15]; c) sustainable packaging utilizes materials that are renewable and recyclable to reduce both environmental impact and waste [16].

15.3 ROLE OF NANOTECHNOLOGY IN MODERN FOOD SUPPLY

Nanotechnology has permeated nearly every sector, astonishing the world with its many potential applications across various fields. Nanostructured materials are preferred over their larger-scale counterparts due to their distinctive chemical and physical attributes, enhancing performance [17]. Nanotechnology rapidly introduces innovative advancement opportunities within the food industry [18]. Applications of nanotechnology within the realm of food encompass the encapsulation and precise delivery of substances, enhancing taste, embedding antimicrobial particles into packaging materials, extending product longevity, and identifying impurities [19]. Modern food supply chains benefit from the key role of nanotechnology, which boosts food safety by providing improved visibility, traceability, and safeguards against counterfeiting. Specifically, invisible nanobarcodes applied directly onto food products and packaging enable efficient tracing, tracking, and brand protection, offering a technological edge in maintaining secure and transparent supply chains [20].

15.4 CHALLENGES OF IMPLEMENTING NANOTECHNOLOGY IN THE FOOD INDUSTRY

Developing edible delivery systems poses a substantial challenge that requires cost-effective processing methods and the incorporation of formulations that are effective for human ingestion and ensure safety. Upholding food integrity requires tackling issues associated with the transfer of nanoparticles from packaging materials into food products. This multifaceted challenge necessitates a comprehensive approach that encompasses the optimization of processing techniques, formulation enhancement, and rigorous safety measures. Achieving success in this endeavor is pivotal for advancing the field of edible delivery systems and aligning technological innovation with efficacy and safety standards.

Additionally, a sophisticated assessment of the risks linked to nanoparticle migration is vital for instilling confidence in the application of these technologies in food packaging. Ongoing research and advancements in this area will undoubtedly be key to developing sustainable edible delivery systems that adapt to the evolving needs of the food industry and consumer preferences, ensuring both safety and efficacy [21].

Evaluating the potential health and toxicity risks of nanoparticles requires examining their environmental pathways, including exposure, transformation, movement, and fate. It involves selecting more suitable biomarkers and tests for the risk assessment framework. Adopting a holistic approach supports a seamless transition from experimental research to industrial application [22].

The gradual integration of technology into commercialization and industrialization raises questions about the cost-benefit balance of advanced technologies, particularly in the agriculture, water, and energy sectors. There are concerns about safety and public perception regarding the use of nanoparticles in food supply chains. Challenges in nanoparticle life-cycle processes include creating structures enabled by nanotechnology, large-scale sustainable production of nanoparticles, extracting them from composites, as well as issues related to their decomposition and recovery. Concerns also emerge in engineering nanoparticles for both food and environmental applications [23]. The Food and Drug Administration has determined that studies evaluating the effectiveness, safety, health risks, or regulation status of nanotechnology products should encompass any distinctive features and characteristics introduced by nanotechnology. The European Commission has consolidated comprehensive legislation in this direction, specifically addressing advancement in technology involving nanoparticles and active and smart (intelligent) materials. This guarantees the recognition of technological advancements based on nanostructured materials with properties that markedly differ from those observed at larger scales [3].

Research calls should aim to narrow the delay between nanotechnology research and the eventual arrival of finished products in the market, simultaneously enhancing economic sustainability, fostering the competitiveness of smaller enterprises, stimulating specialized employment opportunities, and boosting overall quality of life [24]. The nanomaterials utilized in the production of nano-engineered foods and packaging materials are currently not subjected to assessments as novel chemicals. Industries, at present, follow established safety assessment protocols to ensure these materials' safety and regulatory compliance. However, despite these precautions, there is still an incomplete understanding of the complex interactions between nanoparticles and living cells and the resulting effects on both the industry and consumers. This knowledge gap underscores the ongoing need for extensive research and exploration to address potential risks, optimize safety measures, and facilitate the responsible advancement of nano-based technologies in food and packaging. As the field evolves, we must stay vigilant and proactive in advancing our understanding of the possible impacts on health, the environment, and the overall safety of nanomaterial applications in the food industry [25]. Recognizing the pros and cons inherent in each technology, it is imperative to acknowledge nanotechnology's potential negative environmental impacts. Further research is essential to harness its benefits, making it a suitable and necessary next step.

15.5 ADVANCEMENTS IN NANOTECHNOLOGY TOOLS IN THE FOOD INDUSTRY

Nowadays, there is a growing emphasis on enhancing the features of food packaging, characterized by superior properties, increased functionality, reduced weight, and biodegradability. This emerging trend is facilitated by the application of nanotechnology, which allows for the creation of innovative nanostructured materials endowed with exceptional qualities. Innovative packaging solutions (active packaging and intelligent packaging) have been developed to improve the quality and safety of food, as well as to streamline consumer interaction [26].

15.5.1 Active Packaging

A packaging system with active functionality involves intentionally introducing elements that either take in or emit substances to and from the food in the package. This concept indicates that active packaging systems might incorporate emitters and absorbents [26]. Recent progress in nanotechnology has accelerated the creation of functional materials with outstanding performances.

15.5.1.1 Antimicrobial Packaging

Active packaging with antimicrobial properties is designed to decrease, limit, or retard the growth of microbes in food or packaging materials. By incorporating antimicrobial agents, this type of packaging prolongs the lag phase and decreases the growth phase of microbial development, effectively curbing the overall proliferation of microorganisms [27]. Antimicrobial packaging emerges as a crucial tool in managing foodborne diseases by reducing or impeding microbial growth. It also offers continuous defense for packaged food against microbial dangers during storage and transit, enhancing product safety and protecting consumer health. This type of packaging typically incorporates antimicrobial agents into the packaging materials, which can be organic (such as enzymes, essential oils, and chitosan) or inorganic (such as titanium dioxide and silver). Nanotechnology breakthroughs enable the creation of distinctive antimicrobial compounds that can be precisely released. Notably, nano-antimicrobial agents demonstrate superior efficacy compared to bulk materials due to the enhanced penetration facilitated by their smaller size through bacterial cell walls [26]. A significant focus is on improving packaging with antimicrobial properties by incorporating different nanomaterials, including titanium dioxide, copper oxide, silver, and zinc oxide [26].

15.5.1.2 Gas Scavenger

Gas scavengers are engineered to eliminate unwanted substances such as oxygen and ethylene from food packaging. Active scavenging systems are essential for maintaining the safety and integrity of food. Oxygen, for example, can hasten food spoilage by triggering processes like lipid oxidation, protein and vitamin degradation, and the accelerated growth of microorganisms within the packaging [28]. Oxygen scavengers have the potential to establish oxygen-free environments within the headspace of packages possessing moderate barrier properties [27]. Utilizing nanotechnology presents an effective approach to improve the performance of oxygen scavengers. Nanoscale scavengers provide absorption mechanisms with increased surface area to volume ratios, optimizing their oxygen absorption capabilities [26].

15.5.1.3 Gas Emitter

Research is increasingly investigating the use of gas generators in active packaging, allowing precise dispensing of substances to improve the safety and integrity of fresh food. For example, research indicates that carbon dioxide enhances food preservation and prolongs longevity. This is due to its ability to form carbonic acid, which can compromise bacterial cell membranes and reduce enzyme function [26]. Research on applying nanotechnology to gas emitter systems is currently limited.

15.5.2 Intelligent Packaging

Over time, alterations occur in the headspace composition within food packages. Instruments capable of detecting, measuring, and communicating variations in the package's atmospheric conditions, temperature fluctuations during transportation and storage, and the microbiological status of the food offer crucial insights to end consumers and producers/marketers. This information helps assess the efficacy of preservation strategies employed throughout the marketing chain [29]. Indicators are categorized into three types: gas indicators, spoilage and freshness indicators, and

TABLE 15.1
Nanomaterials Used in Different Active and Intelligent Packaging System

Active and Intelligent Packaging Systems	Nanomaterials Present	References
Antimicrobial packaging (AP)*	Chitosan, titanium dioxide (TiO_2), zinc oxide (ZnO), copper oxide (CuO), silver nanomaterials.	[5, 26]
Gas indicators (AP)	Montmorillonite nanoparticles, carbon-supported Zn nanostructures, bimetallic Zn-Fe nanoparticles, titanium dioxide nanoparticles, nano-silica.	[35–37]
Gas emitter (AP)	Electrospun nanofibers.	[38]
Gas indicator (IP)*	Titanium dioxide nanoparticles (TiO_2), Ag-loaded titanium dioxide nanotube.	[39, 40]
Spoilage and freshness indicators (IP)	Chitin nanofibers, oxidized-chitin nanocrystals, nanocellulose, bacterial cellulose nanofibers.	[26]
Time-temperature indicators (IP)	Electrospun chitosan nanofibers, gelatin, and gold nanoparticles.	[26]

*AP: Active packaging and IP: Intelligent packaging

time-temperature indicators. The synergy between intelligent packaging and nanotechnology leads to enhanced packaging performance, including better mechanical and physicochemical properties and more efficient indicator functions [26].

15.5.2.1 Gas Indicators

A label adhered to the package surface is designed to indicate variations in the gas composition within the package. This indicator type generally shows carbon dioxide, oxygen, and ethylene levels. To efficiently monitor conditions, these indicators need to remain in continuous contact with the product and the internal environment of the package [30]. These are crucial for monitoring food freshness. Traditionally based on redox dyes, oxygen indicators evolve with UV-activated options like titanium dioxide nanotubes. Meanwhile, CO_2 indicators, utilizing solid support and pH-sensitive dyes, offer direct monitoring of food quality, demonstrating rapid color changes in response to CO_2 concentration [26, 30].

15.5.2.2 Spoilage and Freshness Indicators

Indicators for freshness and spoilage offer insights into the quality of a food product, revealing information about microbial growth or chemical alterations. Visual data regarding the quality of microbial products is derived from the interaction between by-products of microbial growth and embedded indicators in the packaging [27]. Researchers are enhancing freshness indicators with improved solid supports and responsive dyes. A freshness indicator that senses pH was developed using a combination of chitosan, chitin, and anthocyanin nanofibers. This indicator shifts color when exposed to volatile basic nitrogen, signaling the freshness of fish. Eco-friendly methods involve natural nanomaterials such as nanocellulose sourced from sugarcane bagasse [31] and bacterial cellulose nanofibers [32]. Natural pigments like red cabbage anthocyanins show pH-responsive properties, but immobilization challenges persist [33].

15.5.2.3 Time-Temperature Indicators

Temperature is a critical environmental factor influencing food items' physical, chemical, and microbial deterioration rates. Time-temperature indicators (TTIs) provide an important function in packaging by alerting consumers to temperature variations that could affect the quality and safety of the food [27].

TTIs, classified as full or partial history indicators, benefit temperature-sensitive foods like chilled and frozen products. They are vital in evaluating fish quality and have been applied to dairy and frozen foods. Recent developments include room temperature-storable, air-activated TTIs, addressing storage challenges. TTIs find applications in chill management, quality, and shelf life indicators, aiding in predicting spoilage through mathematical models [35]. Recent advancements incorporate nanocomposites to enhance TTI performance. Enzymatic TTIs, though advantageous for visibility, face challenges like enzymatic instability. Innovative approaches include using electrospun nanofibers for enzyme immobilization and novel self-healing nanofibers that address conventional TTI limitations, offering flexibility, damage tolerance, and sensitivity, as well as unveiling warning signs through color changes in response to temperature shifts [26].

The incorporation of nanotechnology has transformed food packaging, resulting in the creation of active and intelligent packaging solutions. These innovations, encompassing antimicrobial solutions, gas scavengers, and intelligent indicators, enhance food safety, quality, and consumer communication. Table 15.1 represents nanomaterials present in different active and intelligent packaging systems.

15.6 VARIOUS NANOMATERIALS USED IN THE FOOD SUPPLY CHAIN

Nanotechnology is driving transformative change in the food supply chain by introducing a variety of innovative nanomaterials. Table 15.2 sheds light on pivotal categories,

TABLE 15.2
Various Types of Nanomaterials Used in the Food Supply Chain

Category	Examples	Applications
Inorganic nanomaterials	Engineered nanomaterials (ENMs) of silver, iron, calcium, magnesium, selenium, and titanium dioxide	• Growing use of nanosilver in consumer products, including food and health food, water surfaces of foods, and packaging materials • Antimicrobial, anti-odorant, and health supplement applications of nanosilver • Nano-silica is utilized in food contact surfaces and packaging materials • Nano-calcium salts for chewing gums and health supplements • Nanoselenium marketed as an additive with proclaimed health benefits • Nano-iron as a health supplement and water treatment
Surface functionalized materials	Nanoclays (Montmorillonite/Bentonite)	• Enhancement of gas barrier properties in food packaging • Binding with polymer matrix for improved mechanical strength
Organic Nanomaterials	Benzoic acid, ascorbic acid Vitamins (A, E), citric acid, isoflavones, proteins, fat, sugar molecules, omega-3 fatty acids, neutraceuticals	• Increased uptake and absorption in food/feed products • Improved bioavailability of nutrients compared to bulk equivalents

Source: [18]

representative examples, and their applications, delineating how nanotechnology reshapes packaging, nutritional uptake, and preservation within the food industry. Explore the diverse roles of inorganic, surface functionalized, and organic nanomaterials, unraveling their contributions to the dynamic landscape of the food supply chain.

15.7 NANOSENSORS IN FOOD SAFETY AND QUALITY CONTROL

Nanosensors are crucial for ensuring food safety and quality by consistently tracking the presence of microbes, contaminants, and toxins throughout the entire food production process. Their application extends to food packaging, where nanosensors play a key role in real-time monitoring, ensuring the maintenance of product quality from production through to consumption.

15.7.1 Safeguards Food Quality through Pathogen Detection

Identifying pathogenic bacteria in food materials typically requires recognizing either the genetic material of the bacteria or the entire bacterial cell. Using nanoparticles for DNA isolation and bacterial detection has shown to be quicker and more precise compared to traditional methods [41]. Ruengruglikit et al. developed a food packaging system incorporating a range of nanosensors. These sensors rapidly sense gases released by food, resulting in a visible shift in color that signals the food's quality [41]. Cheng et al. utilized adenosine triphosphate bioluminescence in conjunction with biofunctional magnetic nanoparticles, having a diameter of approximately 20 nm, to showcase the faster detection of *Escherichia coli* in food [42]. Fu et al. developed a biosensor for detecting *Salmonella* in food samples. This cutting-edge sensor uses fluorescent dye particles bound to anti-*Salmonella* antibodies placed on a gold/silicon nanorod array. The clear visibility of the dye particles on the sensor allows for quick identification of *Salmonella* bacteria contamination in the sample. This biosensor provides fast and efficient detection of *Salmonella* in food [43]. Other nanomaterials that detect food pathogens are functionalized gold nanoparticles, nanostructured zinc oxide, silver core and gold shell glycol nanoparticles, and carbon nanotubes [44].

15.7.2 Precise Detection of Toxins Present in Food

Biosensors provide an efficient method for identifying toxins in food samples. Hazardous substances like aflatoxins, found in food tainted by *Aspergillus flavus* and *Aspergillus parasiticus*, can be detected using gold nanoprobes. These gold nanoprobes are widely utilized to detect aflatoxin M1, specifically in milk samples. Applying stable and highly effective magnetic nanocarriers, specifically surface-active maghemite nanoparticles, demonstrated the elimination of toxins such as mycotoxins [45]. Additionally, Mohammad and Jafari highlighted the effectiveness of biosensors utilizing nanomaterials for identifying food spoilage and spotting cases of tampering [46]. Gold nanoparticles, graphene oxide, quantum dots, and dendrimers possess outstanding optical and electronic characteristics, making them essential components in fluorescent sensing devices for detecting mycotoxins. Custom-designed oligonucleotides with high specificity, known as aptamers, are a key element. The technique of fluorescence resonance energy transfer represents a significant advancement in identifying Fumonisin B1, Ochratoxin A, and Aflatoxin M1 in food [47].

15.7.3 Detection of Pesticides in Food

Incorporating innovative elements like aptamers, molecularly imprinted polymers, and various materials, nanobiosensors provide heightened sensitivity and selectivity. Examples include a Glutamic acid-functionalized graphene quantum dots-based electrochemical aptasensor for detecting neonicotinoid insecticide residues and a tuned fluorescent nanosensor for herbicide detection with a detection limit of 0.2 ppm [47]. Colorimetric and fluorometric nanosensors incorporating gold nanoparticles are useful in identifying organophosphorus and carbamate pesticides [8].

15.7.4 Detection of Antibiotics in Food

Nanobiosensors are crucial in detecting antibiotics in food products and addressing health risks associated with residues. Various methods include fluorescence, colorimetry, and electrochemical sensing, utilizing materials like gold nanoparticles and graphene oxide. These techniques offer accurate and sensitive detection of antibiotics such as Penicillin G, Ampicillin, sulfadimethoxine, chloramphenicol, and tetracyclines in diverse food samples [47].

15.8 PRACTICAL APPLICATIONS OF NANOTECHNOLOGY IN FOOD SUPPLY

15.8.1 Food Production

Nanotechnology can improve agricultural output, encompassing the development of nanoformulations for agrochemicals to facilitate the targeted application of pesticides and fertilizers for enhanced crop yield. Additionally, nanosensors and nanobiosensors find utility in safeguarding crops by identifying diseases and spotting traces of pesticides. Nanotechnology's reach encompasses genetic manipulation in plants via nanodevices and its applications in breeding, animal health, poultry, farming, and and postharvest handling [48]. Nanotechnology plays a significant role in revolutionizing various aspects of food production, contributing to enhanced efficiency, safety, and the development of innovative products. Here are the practical applications of nanotechnology in food production.

15.8.1.1 Nanotechnology in Agri-Production

15.8.1.1.1 Precision Agriculture

Precision farming employs separate controllers for each growth factor, including nutrition, light, and temperature. Satellite systems oversee information regarding planting and harvesting times, empowering farmers to make well-informed choices and avoid adverse weather conditions. These systems assist farmers in determining the most favorable times for planting and harvesting to optimize yields. Additionally, they manage the efficient application of fertilizers, irrigation, lighting, and temperature. Nanotechnology plays a vital role by utilizing sensitive nuclear links within GPS controllers to ensure precise control and monitoring [49].

15.8.1.1.2 Nanofertilizers

Nanofertilizers possess distinctive characteristics such as exceptionally high absorption rates, enhanced crop yields, increased photosynthetic activity, and substantial enlargement of leaf surface areas. Applying nanofertilizers results in enhanced elemental efficiency, minimizing adverse impacts associated with excessive fertilizer use, mitigating soil toxicity, and decreasing the frequency of fertilizer applications [50].

15.8.1.2 Nanotechnology in Animal Production

Enhancing the feed utilization and nutritional intake in livestock, mitigating challenges posed by illnesses affecting animals and converting their biological residues, and addressing environmental issues are some ways nanotechnology finds application in animal husbandry to create value-added products [50, 51].

15.8.1.3 Nanotechnology in Aquaculture

Nanotechnology benefits aquaculture by promoting faster growth in carp and sturgeon through iron nanoparticles. Nanoselenium enhances weight and antioxidant status in crucian carp, outperforming organic selenomethionine. Nano-level nutraceuticals positively impact fish growth, but direct use of silver nanoparticles harms trout. A silver-coated water filter prevents fungal infections in indoor rainbow trout farming [50, 52].

15.8.1.4 Nanotechnology in Poultry Production

Nanoparticles enable the rapid production of poultry products with enhanced safety measures. Nanobiotic-silver, known for its antimicrobial properties in poultry nutrition, can influence microbial populations without causing resistance. Additionally, it may boost anabolic activities. Diagnosis kits based on gold nanoparticles are effective in identifying poultry affected by the influenza virus [50, 53].

15.8.2 Food Processing

Progress in nanotechnology in food processing focuses on improving food texture, the encapsulation of additives, enhanced nutritional availability, and precision control over flavor release. Nanotechnology is utilized to bring innovation and enhance a broad spectrum of food items and commodities throughout the entire food processing and production process, employing diverse nano techniques and their applications. The following are a few tangible examples of how nanotechnology is practically applied in processing.

15.8.2.1 Improving Food Texture, Taste, and Flavor

Nanotechnology offers diverse possibilities for enhancing food quality and flavor. Nanoencapsulation methods are widely employed to optimize the release and retention of flavors, ensuring a balanced culinary experience. The widespread use of nanoemulsions for delivering fat-soluble bioactive compounds is due to their capacity to use natural food ingredients through uncomplicated production methods. Moreover, these nanoemulsions can be specifically created to improve water dispersion and bioaccessibility. Food items have historically used metallic oxides, such as silicon dioxide and titanium dioxide, as flow agents or colorants.

Silicon dioxide derived nanoparticles find widespread use in the food industry as mediators for tastes or aromas in various food items [54]. The Nanoclusters™ technique has been utilized to produce various products, such as Slim Chocolate Shake. This product uses cocoa-coated silica nanoparticles to enhance the flavor of chocolate [55].

15.8.2.2 Safeguard Nutrient Integrity

Certain highly nutritious ingredients face degradation during processing. For instance, easily oxidizable omega-3 polyunsaturated fatty acids (PUFAs) necessitate stabilization and protection strategies innovatively crafted casein nanostructures, demonstrating exceptional protection against docosahexaenoic acid oxidation. Using soy as a source, Bio-Delivery Sciences International unveiled the Bioral™ nanocochleate technology, which ensures GRAS (generally recognized as safe) status and safeguards micronutrients during manufacturing. Nanoencapsulation of natural dipeptide antioxidants and coenzyme Q10 showcases the potential of nanotechnology in enhancing therapeutic efficacy and bioavailability for improved consumer health. Biodegradable polymeric nanostructures further contribute to stability (PLGA-based with quaternary ammonium salt didodecyldimethyl ammonium bromide) [55].

15.8.2.3 Enhance Nutritional Boost in Processed Foods

The utilization of nanomaterials within food products is growing drastically, especially when it comes to adding healthy ingredients. The Medical Institute of U.S. National Academy of Sciences describes functional foods as those that provide healthcare benefits above and beyond their conventional nutritional composition. Nanoencapsulation, a process involving active ingredients within nano-sized capsules, is increasingly applied. Cereals, dairy products, bread, and beverages are added to vitamins, minerals, probiotics, antioxidants, bioactive proteins, and phytosterols. Notably, several of these active components are incorporated into food in nanomaterials, usually with diameters between a few hundred and thousand nanometers [56].

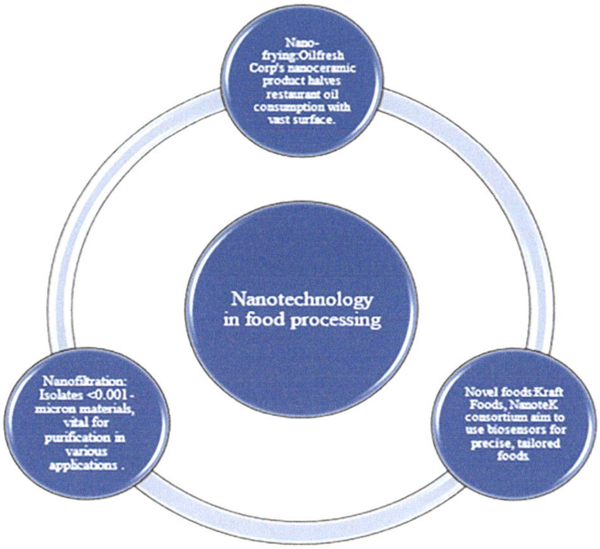

FIGURE 15.1 Additional applications of nanotechnology (food processing). (*Source*: [55, 58]).

Additionally, utilizing nanoparticles in food processing aims to enhance its nutritional quality, creating healthier foods with reduced levels of fat, sugar, and salt. This approach also seeks to mitigate food-related diseases. Nutraceuticals such as beta-carotene, lycopene, and plant sterols can be integrated at the nanoscale into food products, which are commonly present in foods targeted at health-conscious consumers aiming to prevent elevated cholesterol levels [57]. Further applications of nanotechnology in the food production sector are illustrated in Figure 15.1.

15.8.3 Food Packaging

To uphold the safety and standards of food products throughout the food chain, proper storage facilities with modern packaging materials are crucial. A crucial requirement is to prevent the contamination of food during storage. An inventive packaging solution involves integrating active materials, effectively enhancing and preserving the packaged product's sensory quality while improving its microbiological safety [5]. Over 400 nanopackaging products are currently available in the market, making up around 25 percent of the total packaging market [56, 58]. Packaging films incorporate nanomaterials to provide properties like antimicrobial features, antioxidants, improved barriers, and other characteristics. Certain characteristics of nanomaterials have been previously addressed in active and intelligent packaging systems. This integration helps minimize the exchange of gases with the external storage environment. Implementing such nano-based packaging systems enhances food products' functionality and ensures their safety [5]. Here are some notable practical applications of nanotechnology in packaging.

15.8.3.1 Barrier Enhancement

Nanotechnology has shown significant potential in enhancing barriers in food packaging, contributing to improved shelf life, safety, and quality. Nanocomposite materials in food packaging enhance packaging performance and provide intelligent and active functionalities [59] as already discussed. Nanostructured materials like silver-based, copper-based, titanium dioxide, polymer-based, polysaccharide-based, and protein-based act as improved barriers in food packaging. Introducing nano-sized fillers in a uniform polymer film has dual impacts on the barrier properties. The initial mechanism entails creating a convoluted route for the diffusion of gases. As these additives consist of impermeable inorganic crystals, air molecules are compelled to navigate around them, deviating from a direct, perpendicular channel through the film surface. As a result, when fillers are present, the overall route for gas movement through the film is lengthened [3, 60]. Roilo et al. incorporate TiO_2 nanoparticles in pure cellulose nanofibril coatings for food packaging, enhancing gas barrier properties and limiting the permeation of oxygen (O_2), carbon dioxide (CO_2), and nitrogen (N_2). Additionally, TiO_2 reduces the diffusivity of small penetrants such as helium and deuterium, contributing to improved barrier performance.

15.8.3.2 Antimicrobial Properties

The evolution of antimicrobial packaging offers a promising method to hinder, minimize, or delay microbial growth. Nowadays, the prevailing trend involves incorporating antimicrobial-based natural ingredients into packaging. Some substances such as benzoic, sorbic, propionic, acetic, ascorbic acids, enzymes, bacteriocins, and ethanol have recently been integrated into biobased food packaging nanocomposite films to initiate them with antimicrobial properties [61]. These substances can alter the transport and permeability of membranes, ultimately lowering the pH inside cells [62]. In a similar vein, nanoparticles composed of inorganic materials and metal oxides are seen as powerful agents for combating microbes in food packaging because of their exceptional stability and robust antimicrobial characteristics. They have the ability to curb the proliferation of microbes [63].

15.8.3.3 Traceability and Authentication

In the intricate landscape of the food supply chain, nanotechnology is emerging as a disruptive factor in the complex food supply chain, especially in the domain of food packaging authenticity and visibility. These communication devices guarantee authenticity, count counterfeiting, and prevent food product theft [20]. These tools, commonly known as data carrier devices, encompass widely used technologies, including radio frequency identification (RFID) tags and barcoding system, frequently utilized in the food industry for effective implementation [64].

15.8.3.3.1 RIFD Tags

RFID technology uses radio waves for object identification and data collection, including product details and environmental parameters like temperature. In the food industry, RFID is employed for supply chain management, temperature monitoring, and safety [65]. The utilization of RFID technology extends beyond individual items, allowing for the evaluation of every node within the whole supply chain process [20]. The system comprises a tag and a reader, with active tags using batteries and passive tags powered by magnetic induction or electromagnetic waves. Frequencies of 100 kHz to 30 MHz are common in RFID devices, beneficial for ensuring proper temperature in food transportation [65].

15.8.3.3.2 Barcodes

Barcodes have become integral in packaging, facilitating retail checkout and distribution. The Universal Product Code (UPC) is a common barcode system with a 12-digit representation of information. However, due to limited storage capacity, UPC barcodes offer minimal details, primarily the manufacturer or item number. Advancements introduce two-dimensional data matrix barcodes like PDF417, overcoming storage limitations. These barcodes have the potential to hold large amounts of data, which makes it possible to include additional details like recipe instructions, nutritional facts, or a link to the manufacturer's website for more product information [20].

15.9 FUTURE TRENDS AND INNOVATIONS

As the world grapples with the complexities of modern supply chain management of food (SCM), the integration of nanotechnology emerges as a beacon of innovation, offering transformative solutions and assisting in an entirely novel period of accuracy and efficiency. Nanotechnology promises transformative developments in addressing the food industry's logistical and quality control challenges. A notable stride involves the creation of sensors for monitoring chemical, physical, and biological properties, as well as detecting pathogens and toxins. Innovations also target pathogen control, emphasizing improved food safety and waste reduction. In water treatment, ongoing advancements focus on membrane and sorbent technologies, particularly in decentralized applications. Additionally, the emergence of materials designed for precise agrochemical delivery underscores the versatile applications of technology across sectors [23].

Delving into nanotechnology's impact on the food and beverage sector reveals many applications, especially in nutraceuticals and functional food formulation. Recent investigations delve into novel functionalities and efficient delivery mechanisms. Nanotechnology introduces groundbreaking possibilities, enabling foods to adapt color, flavor, or nutrients to individual preferences. Filters can modify flavors based on molecular shape, while packaging gains the ability to detect spoilage. Understanding food materials and processing at the nanoscale proves crucial for innovating enhanced food products [66]. Nanotechnology is poised to revolutionize the visibility of both farming as well as food products with advancements in artificial intelligence, the Internet of Things, blockchain, and innovative nanotechnological devices. Integration of nanosensors, nanobiosensors, and nanoparticle-based aptasensors in traceability systems holds significant promise for monitoring food safety, detecting contaminants, and ensuring a more secure and efficient food supply chain in the future [20].

15.10 CONCLUSION

Implementing nanotechnology within the food sector, particularly packaging, marks a paradigm shift with transformative advancements. Traditional packaging methods face protection, customization, and sustainability limitations, necessitating nanotechnological solutions.

Nanotechnology's pivotal role in modern food supply chains is evident through its contributions to enhancing food safety, traceability, and counterfeiting protection. Active and intelligent packaging systems utilizing nanomaterials like chitosan, titanium dioxide, copper oxide, and silver nanoparticles address challenges in antimicrobial efficacy, gas scavenging, and intelligent monitoring of freshness and temperature.

However, challenges persist in implementing nanotechnology, requiring solutions for cost-effective processing, nanoparticle migration concerns, and comprehensive safety assessments. Regulatory bodies emphasize the need for thorough evaluations to manage nanotechnology products' unique characteristics and potential risks.

The evolving landscape of nanotechnology tools within the food sector, including antimicrobial packaging (AM), gas scavengers, and intelligent indicators, highlights the instrumental role of nanomaterials like electrospun nanofibers and chitin nanofibers.

While nanotechnology holds vast potential for revolutionizing food supply chains, continued research, responsible implementation, and collaboration between researchers, industry stakeholders, and regulatory bodies are imperative. This concerted effort will ensure the realization of nanotechnology's benefits, mitigating potential risks and contributing toward a safer, more effective, and economically viable future in the food sector.

REFERENCES

1. J.K. Momin, C. Jayakumar, J.B. Prajapati, Potential of nanotechnology in functional foods, *Emirates J. Food Agric.* 25 (2013) 10–19. https://doi.org/10.9755/ejfa.v25i1.9368.
2. B.A. Kehinde, A. Panghal, P. Bishnoi, N. Chhikara, Role of nanotechnology in food supply chain, *Nanotechnol. Appl. Food Saf. Qual. Monit.* (2022) 415–434. https://doi.org/10.1016/B978-0-323-85791-8.00010-0.
3. R. Biswas, M. Alam, A. Sarkar, M.I. Haque, M.M. Hasan, M. Hoque, Application of nanotechnology in food: Processing, preservation, packaging and safety assessment, *Heliyon.* 8 (2022) e11795. https://doi.org/10.1016/j.heliyon.2022.e11795.
4. N. Gaur, S. Kumari, R. Chaudhary, S. Bajpai, Nanotechnology for food and bioprocess industry, *Nanomater. Bioreact. Bioprocess. Appl.* (2023) 291–305. https://doi.org/10.1016/B978-0-323-91782-7.00003-5.
5. Sharma, A., Nagarajan, J., Gopalakrishnan, K., Bodana, V., Singh, A., Prabhakar, P. K., ... & Kumar, R. (2023). Nanotechnology applications and implications in food industry. In Nanotechnology applications for food safety and quality monitoring (pp. 171–182). Academic Press.
6. S. Chen, S. Brahma, J. Mackay, C. Cao, B. Aliakbarian, The role of smart packaging system in food supply chain, *J. Food Sci.* 85 (2020) 517–525. https://doi.org/10.1111/1750-3841.15046.
7. N. Mahalik, K. Kim, *The role of information technology developments in food supply chain integration and monitoring*, Elsevier Ltd, 2016. https://doi.org/10.1016/B978-1-78242-447-5.00002-2.
8. K. Arunachalam, V.K. Nathan, N. Sharma, S. Rajeshkumar, L. Marwaha, R. Subbaiah, G. Mukupa, A.J.A. Ranjitsingh, A. Mboshi, M. Siankuku, Nanotechnology applications in food and bioprocess industries, *Nanomater. Bioreact. Bioprocess. Appl.* (2023) 335–364. https://doi.org/10.1016/B978-0-323-91782-7.00005-9.
9. A. Sangroniz, J.-B. Zhu, X. Tang, A. Etxeberria, E.Y.-X. Chen, H. Sardon, Packaging materials with desired mechanical and barrier properties and full chemical recyclability, *Nat. Commun.* 10 (2019) 3559. https://doi.org/10.1038/s41467-019-11525-x.
10. M. Mujtaba, J. Lipponen, M. Ojanen, S. Puttonen, H. Vaittinen, Trends and challenges in the development of bio-based barrier coating materials for paper/cardboard food packaging; a review, *Sci. Total Environ.* 851 (2022) 158328. https://doi.org/10.1016/j.scitotenv.2022.158328.
11. O.O. Oloyede, S. Lignou, Sustainable paper-based packaging: A consumer's perspective, *Foods.* 10 (2021). https://doi.org/10.3390/foods10051035.
12. P. Tyagi, K.S. Salem, M.A. Hubbe, L. Pal, Advances in barrier coatings and film technologies for achieving sustainable packaging of food products—a review, *Trends Food Sci. Technol.* 115 (2021) 461–485. https://doi.org/10.1016/j.tifs.2021.06.036.
13. D. Schaefer, W.M. Cheung, Smart packaging: Opportunities and challenges, *Procedia CIRP.* 72 (2018) 1022–1027. https://doi.org/10.1016/j.procir.2018.03.240.
14. C.T. Tracey, A.L. Predeina, E.F. Krivoshapkina, E. Kumacheva, A 3D printing approach to intelligent food packaging, *Trends Food Sci. Technol.* 127 (2022) 87–98. https://doi.org/10.1016/j.tifs.2022.05.003.
15. A. Ashfaq, N. Khursheed, S. Fatima, Z. Anjum, K. Younis, Application of nanotechnology in food packaging: Pros and Cons, *J. Agric. Food Res.* 7 (2022) 100270. https://doi.org/10.1016/j.jafr.2022.100270.
16. P. Jain, D.M. Hudnurkar, Sustainable packaging in the FMCG industry, *Clean. Responsible Consum.* 7 (2022) 100075. https://doi.org/10.1016/j.clrc.2022.100075.
17. R.K. Gupta, P. Guha, P.P. Srivastav, Natural polymers in bio-degradable/edible film: A review on environmental concerns, cold plasma technology and nanotechnology application on food packaging- a recent trends, *Food Chem. Adv.* 1 (2022) 100135. https://doi.org/10.1016/j.focha.2022.100135.
18. B.S. Sekhon, NSA_8677_food-nanotechnology—an-overview (2024). https://doi.org/10.2147/nsa.s12187498.
19. C. Chellaram, G. Murugaboopathi, A.A. John, R. Sivakumar, S. Ganesan, S. Krithika, G. Priya, Significance of nanotechnology in food industry, *APCBEE Procedia.* 8 (2014) 109–113. https://doi.org/10.1016/j.apcbee.2014.03.010.
20. Aggarwal, D., & Idrishi, R. (2023). Nanotechnology applications for food traceability. In Nanotechnology Applications for Food Safety and Quality Monitoring (pp. 457–472). Academic Press.
21. M. Sahoo, S. Vishwakarma, C. Panigrahi, J. Kumar, Nanotechnology: Current applications and future scope in food, *Food Front.* 2 (2021) 3–22. https://doi.org/10.1002/fft2.58.

22. S. Liu, Y. Lu, W. Chen, Bridge knowledge gaps in environmental health and safety for sustainable development of nano-industries, *Nano Today.* 23 (2018) 11–15. https://doi.org/10.1016/j.nantod.2018.09.002.
23. S.M. Rodrigues, P. Demokritou, N. Dokoozlian, C.O. Hendren, B. Karn, M.S. Mauter, O.A. Sadik, M. Safarpour, J.M. Unrine, J. Viers, P. Welle, J.C. White, M.R. Wiesner, G. V. Lowry, Nanotechnology for sustainable food production: Promising opportunities and scientific challenges, *Environ. Sci. Nano.* 4 (2017) 767–781. https://doi.org/10.1039/c6en00573j.
24. M.Â. Cerqueira, A.C. Pinheiro, O.L. Ramos, H. Silva, A.I. Bourbon, A.A. Vicente, Chapter two—Advances in food nanotechnology, in: R.B.T.-E.N. in F.S. Busquets (Ed.), *Micro Nano Technol.*, Elsevier, Boston, 2017: pp. 11–38. https://doi.org/10.1016/B978-0-323-42980-1.00002-9.
25. S. Neethirajan, D.S. Jayas, Nanotechnology for the food and bioprocessing industries, *Food Bioprocess Technol.* 4 (2011) 39–47. https://doi.org/10.1007/s11947-010-0328-2.
26. S. Huang, V. Tyagi, B. Bhattacharya, Y. Wang, Nanotechnology: A new approach to advanced food packaging, *Nanotechnol. Appl. Food Saf. Qual. Monit.* (2022) 241–264. https://doi.org/10.1016/B978-0-323-85791-8.00003-3.
27. Panja, P., Mani, A., & Thakur, P. K. (2018). Current status of active and intelligent packaging in food technologies. Trends and prospects in postharvest management of horticultural crops, 299–320.
28. Z. Kordjazi, A. Ajji, Development of TiO_2 catalyzed HTPB based oxygen scavenging films for food packaging applications, *Food Control.* 121 (2021) 107639. https://doi.org/10.1016/j.foodcont.2020.107639.
29. D.A.P. de Abreu, J.M. Cruz, P.P. Losada, Active and intelligent packaging for the food industry, *Food Rev. Int.* 28 (2012) 146–187. https://doi.org/10.1080/87559129.2011.595022.
30. M. Soltani Firouz, K. Mohi-Alden, M. Omid, A critical review on intelligent and active packaging in the food industry: Research and development, *Food Res. Int.* 141 (2021) 110113. https://doi.org/10.1016/j.foodres.2021.110113.
31. P. Lu, Y. Yang, R. Liu, X. Liu, J. Ma, M. Wu, S. Wang, Preparation of sugarcane bagasse nanocellulose hydrogel as a colourimetric freshness indicator for intelligent food packaging, *Carbohydr. Polym.* 249 (2020) 116831. https://doi.org/10.1016/j.carbpol.2020.116831.
32. M. Moradi, H. Tajik, H. Almasi, M. Forough, P. Ezati, A novel pH-sensing indicator based on bacterial cellulose nanofibers and black carrot anthocyanins for monitoring fish freshness, *Carbohydr. Polym.* 222 (2019) 115030. https://doi.org/10.1016/j.carbpol.2019.115030.
33. S. Kalpana, S.R. Priyadarshini, M. Maria Leena, J.A. Moses, C. Anandharamakrishnan, Intelligent packaging: Trends and applications in food systems, *Trends Food Sci. Technol.* 93 (2019) 145–157. https://doi.org/10.1016/j.tifs.2019.09.008.
34. B. Gomes, P.J. Ferreira, S. Carvalho, Zinc nanostructures for oxygen scavenging, *Nanoscale.* 9 (2017) 5254–5262.
35. A. Castro, I. Carvalho, L. Marques, P.J. Ferreira, A. Cavaleiro, S. Carvalho, S. Calderon V., Galvanic oxidation of bimetallic Zn-Fe nanoparticles for oxygen scavenging, *Appl. Surf. Sci.* 537 (2021) 147896. https://doi.org/10.1016/j.apsusc.2020.147896.
36. A. Ebrahimi, M. Zabihzadeh Khajavi, A.M. Mortazavian, H. Asilian-Mahabadi, S. Rafiee, M. Farhoodi, S. Ahmadi, Preparation of novel nano–based films impregnated by potassium permanganate as ethylene scavengers: An optimization study, *Polym. Test.* 93 (2021) 106934. https://doi.org/10.1016/j.polymertesting.2020.106934.
37. C. Zhang, Y. Li, P. Wang, H. Zhang, Electrospinning of nanofibers: Potentials and perspectives for active food packaging, *Compr. Rev. Food Sci. Food Saf.* 19 (2020) 479–502. https://doi.org/10.1111/1541-4337.12536.
38. Z. Gao, L. Liu, Z. Tian, Z. Feng, B. Jiang, W. Wang, Fast-response flexible photochromic gels for self-erasing rewritable media and colorimetric oxygen indicator applications, *ACS Appl. Mater. Interfaces.* 10 (2018) 33423–33433. https://doi.org/10.1021/acsami.8b09825.
39. J. Wen, S. Huang, L. Jia, F. Ding, H. Li, L. Chen, X. Liu, Visible colorimetric oxygen indicator based on Ag-loaded TiO_2 nanotubes for quick response and real-time monitoring of the integrity of modified atmosphere packaging, *Adv. Mater. Technol.* 4 (2019) 1–7. https://doi.org/10.1002/admt.201900121.
40. V. Kumar, P. Guleria, S.K. Mehta, Nanosensors for food quality and safety assessment, *Environ. Chem. Lett.* 15 (2017) 165–177. https://doi.org/10.1007/s10311-017-0616-4.
41. C. Ruengruglikit, H.C. Kim, R.D. Miller, Q.R. Huang, Fabrication of nanoporous oligonucleotide microarray for pathogen detection and identification, in: *Abstr. Pap. Am. Chem. Soc.*, Amer Chemical Soc, Washington, DC, 2004: pp. U464–U464.
42. Y. Cheng, Y. Liu, J. Huang, K. Li, W. Zhang, Y. Xian, L. Jin, Combining biofunctional magnetic nanoparticles and ATP bioluminescence for rapid detection of *Escherichia coli*, *Talanta.* 77 (2009) 1332–1336.
43. J. Fu, B. Park, G. Siragusa, L. Jones, R. Tripp, Y. Zhao, Y.-J. Cho, An Au/Si hetero-nanorod-based biosensor for Salmonella detection, *Nanotechnology.* 19 (2008) 155502.
44. B. Kuswandi, D. Futra, L.Y. Heng, *Nanosensors for the detection of food contaminants*, Elsevier Inc., 2017. https://doi.org/10.1016/B978-0-12-811942-6.00015-7.
45. M. Magro, D. Baratella, E. Bonaiuto, J. de A. Roger, F. Vianello, New perspectives on biomedical applications of iron oxide nanoparticles, *Curr. Med. Chem.* 25 (2017) 540–555. https://doi.org/10.2174/0929867324666170616102922.
46. Z. Mohammadi, S.M. Jafari, Detection of food spoilage and adulteration by novel nanomaterial-based sensors, *Adv. Colloid Interface Sci.* 286 (2020) 102297. https://doi.org/10.1016/j.cis.2020.102297.
47. K. Pushparaj, W.C. Liu, A. Meyyazhagan, A. Orlacchio, M. Pappusamy, C. Vadivalagan, A.A. Robert, V.A. Arumugam, H. Kamyab, J.J. Klemeš, T. Khademi, M. Mesbah, S. Chelliapan, B. Balasubramanian, Nano- from nature to nurture: A comprehensive review on facets, trends, perspectives and sustainability of nanotechnology in the food sector, energy. 240 (2022). https://doi.org/10.1016/j.energy.2021.122732.
48. I.M. Chung, G. Rajakumar, T. Gomathi, S.K. Park, S.H. Kim, M. Thiruvengadam, Nanotechnology for human food: Advances and perspective, *Front. Life Sci.* 10 (2017) 63–72. https://doi.org/10.1080/21553769.2017.1365775.
49. S.R. Mousavi, M. Rezaei, Nanotechnology in agriculture and food production, *J. Appl. Env. Biol Sci.* 1 (2011) 414–419.
50. B. Singh Sekhon, Nanotechnology in agri-food production: An overview, *Nanotechnol. Sci. Appl.* 7 (2014) 31–53. https://doi.org/10.2147/NSA.S39406.
51. H. Chen, R. Yada, Nanotechnologies in agriculture: New tools for sustainable development, *Trends Food Sci. Technol.* 22 (2011) 585–594. https://doi.org/10.1016/j.tifs.2011.09.004.

52. X. Zhou, Y. Wang, Q. Gu, W. Li, Effects of different dietary selenium sources (selenium nanoparticle and selenomethionine) on growth performance, muscle composition and glutathione peroxidase enzyme activity of crucian carp (*Carassius auratus* gibelio), *Aquaculture*. 291 (2009) 78–81. https://doi.org/10.1016/j.aquaculture.2009.03.007.
53. T. Emami, R. Madani, S.M. Rezayat, F. Golchinfar, S. Sarkar, Applying of gold nanoparticle to avoid diffusion of the conserved peptide of avian influenza nonstructural protein from membrane in Western blot, *J. Appl. Poult. Res*. 21 (2012) 563–566. https://doi.org/10.3382/japr.2011-00456.
54. T. Singh, S. Shukla, P. Kumar, V. Wahla, V.K. Bajpai, Application of nanotechnology in food science: Perception and overview, *Front. Microbiol*. 8 (2017) 1–7. https://doi.org/10.3389/fmicb.2017.01501.
55. Dasgupta, N., Ranjan, S., Mundekkad, D., Ramalingam, C., Shanker, R., & Kumar, A. (2015). Nanotechnology in agro-food: from field to plate. Food Research International, 69, 381–400.
56. S.M. Alfadul, A.A. Elneshwy, Use of nanotechnology in food processing, packaging and safety–review, *African J. Food, Agric. Nutr. Dev*. 10 (2010).
57. L. Cruz-Lopes, M. Macena, R.P.F. Guiné, Application of nanotechnologies along the food supply chain, *Open Agric*. 6 (2021) 749–760. https://doi.org/10.1515/opag-2021-0052.
58. B. Park, Current and future applications of nanotechnology, *Issues Environ. Sci. Technol*. 24 (2007) 1–18. https://doi.org/10.1039/9781847557766-00001.
59. C. Sharma, R. Dhiman, N. Rokana, H. Panwar, Nanotechnology: An untapped resource for food packaging, *Front. Microbiol*. 8 (2017). https://doi.org/10.3389/fmicb.2017.01735.
60. D. Roilo, C.A. Maestri, M. Scarpa, P. Bettotti, R. Checchetto, Gas barrier and optical properties of cellulose nanofiber coatings with dispersed TiO_2 nanoparticles, *Surf. Coatings Technol*. 343 (2018) 131–137. https://doi.org/10.1016/j.surfcoat.2017.10.015.
61. A.E. Kapetanakou, P.N. Skandamis, Applications of active packaging for increasing microbial stability in foods: Natural volatile antimicrobial compounds, *Curr. Opin. Food Sci*. 12 (2016) 1–12. https://doi.org/10.1016/j.cofs.2016.06.001.
62. P. Jha, Effect of plasticizer and antimicrobial agents on functional properties of bionanocomposite films based on corn starch-chitosan for food packaging applications, *Int. J. Biol. Macromol*. 160 (2020) 571–582. https://doi.org/10.1016/j.ijbiomac.2020.05.242.
63. P.R. Salgado, L. Di Giorgio, Y.S. Musso, A.N. Mauri, Recent developments in smart food packaging focused on biobased and biodegradable polymers, *Front. Sustain. Food Syst*. 5 (2021) 1–30. https://doi.org/10.3389/fsufs.2021.630393.
64. A. Mirza Alizadeh, M. Masoomian, M. Shakooie, M. Zabihzadeh Khajavi, M. Farhoodi, Trends and applications of intelligent packaging in dairy products: A review, *Crit. Rev. Food Sci. Nutr*. 62 (2021) 383–397. https://doi.org/10.1080/10408398.2020.1817847.
65 J. Muthukumar, S. Kumar, N. Arunachalam, L. Kumar, R. Chidambaram, Quality assurance of packaged food using nanotechnology (2022). https://doi.org/10.1016/B978-0-323-85791-8.00019-7.
66. R. Ravichandran, Nanotechnology applications in food and food processing: Innovative green approaches, opportunities and uncertainties for global market, *Int. J. Green Nanotechnol. Phys. Chem*. 1 (2010) P72–P96. https://doi.org/10.1080/19430871003684440.

16 Nanoemulsion of Essential Oils against Foodborne Pathogens and Stored Product Insect Pests

N.U. Visakh[a], M. Safna[a], Anu Thomas[a], Athira Shajin[a], and Berin Pathrose[a]
[a]Department of Agricultural Entomology, College of Agriculture, Kerala Agricultural University, Thrissur, Kerala, India

16.1 INTRODUCTION

Preserving food integrity during storage and ensuring food safety within stored grain insect infestation and foodborne diseases is crucial in contemporary agricultural and food systems (Fu and Ying 2014). The challenges posed by these threats require a comprehensive and research-driven approach to minimize risks. The detrimental effects caused by stored grain insect pests on food supplies and nutritional quality emphasize the need for strategic interventions. The prevalence of foodborne diseases emphasizes the significance of strict protection measures in both food manufacturing and storage (Sahoo et al. 2022). To tackle these challenges, the investigations must thoroughly explore the intricate biological nuances of insect infestations and the microbial origins of foodborne illnesses. Innovations in storage methodologies, encompassing controlled atmospheres and hermetic storage, emerge as promising avenues for mitigating insect activity and microbial contamination (Kamano 2022). Additionally, incorporating sustainable practices into storage protocols aligns with broader environmental and ethical considerations.

Securing food storage against insect pests and foodborne pathogens remains critical for food safety and public health. Stored food pests, ranging from insects to rodents, substantially threaten the integrity of stored food items, causing economic losses and jeopardizing nutritional quality. Simultaneously, the risk of foodborne pathogens introduces an additional layer of complexity, potentially leading to severe public health consequences (Stejskal et al. 2015). Effective management of food storage pests involves a multidisciplinary approach, incorporating entomology, ecology, and integrated pest management tactics. Implementing preventive measures, like ensuring adequate sanitation, airtight storage, and temperature control, is crucial in mitigating the proliferation of pests. Furthermore, ongoing research into novel, environmentally sustainable methods for pest control is imperative (Windsor et al. 2015).

The challenge of foodborne pathogens demands rigorous adherence to hygienic practices throughout the food chain. Contamination may happen at different points, from production to storage and distribution, necessitating robust monitoring and control measures. Advanced technologies, including molecular diagnostics and traceability systems, play a significant role in identifying and mitigating the risks associated with foodborne pathogens (Sahoo et al. 2022). The persistent challenges related to food storage pests and foodborne microbes have spurred exploration into alternative strategies, notably using chemical pesticides and botanicals. Chemical pesticides have conventionally served as a primary means to combat pests and microbes in stored food. However, concerns over their environmental impact, development of resistance, and potential residual impacts on human health have resulted in reconsidering alternative approaches (Sharma et al. 2020).

Botanicals from plants present a promising avenue as natural alternatives to chemical pesticides. Compounds extracted from various botanical sources exhibit inherent insecticidal and antimicrobial properties (Ilangovan et al. 2018). Noteworthy examples include neem oil, pyrethrum, and essential oils such as citronella and peppermint (Stoytcheva 2011). These botanicals demonstrate efficacy in pest control and offer environmentally sustainable solutions, aligning with contemporary demands for ecologically sound pest management. Moreover, botanicals often possess multifaceted bioactive compounds, acting on diverse biochemical pathways in pests and microbes. This complexity hinders the development of resistance, a persistent concern with chemical pesticides. However, challenges such as formulation stability and standardized application protocols necessitate ongoing research to optimize the efficacy of botanicals in diverse storage conditions.

Aromatic plant-derived essential oils have emerged as promising botanicals in food storage, exhibiting notable efficacy in managing storage pests and foodborne microbes. The complex composition of essential oils, comprising volatile compounds terpenes and phenolics, contributes to their inherent insecticidal and antimicrobial properties (Chouhan et al. 2017). When it comes to storage pests, essential oils disrupt the physiological processes of insects, impeding their development and reproduction. Their effectiveness is attributed to a diverse array of bioactive compounds that target the nervous system, respiratory system, and cuticle of insects. Importantly, essential oils offer an environmentally

DOI: 10.1201/9781032677422-16

friendly alternative to conventional chemical pesticides, mitigating concerns regarding ecological impact and resistance development (Ayilara et al. 2023). Simultaneously, essential oils demonstrate pronounced antimicrobial effectiveness against pathogens found in food. These oils interfere with microbial cell membranes, inhibit enzyme activity, and disrupt cellular structures, thereby limiting the growth and survival of harmful microorganisms (Maurya et al. 2021).

The multidimensional characteristics of essential oils make them particularly appealing, as their mechanisms of action make it challenging for pests and microbes to develop resistance. While essential oils show great promise, challenges such as stability, solubility, and standardized application methods necessitate ongoing research to optimize their use in diverse storage conditions (Rout et al. 2022). Nevertheless, their potential as botanicals underscores a paradigm shift toward sustainable and biologically sound approaches for managing storage pests and ensuring food safety in storage facilities (Nawaz et al. 2019).

Nanotechnology has ushered in a transformative paradigm in formulating essential oils, introducing innovative solutions for combatting storage pests and foodborne pathogens. Nano-based formulations, including nanoemulsions and nanoparticles, leverage nanomaterials' unique properties to improve essential oils' stability, miscibility, and efficacy. Nanoemulsions, a notable application of nanotechnology, involve the creation of stable, nanoscale emulsions that encapsulate imperative oil components (Barradas and de Holanda e Silva 2020). This process addresses long-standing challenges, extending the longevity of essential oils and facilitating their controlled release. The reduced particle size achieved through nanoemulsification increases surface area, improving dispersibility and penetration into pests and pathogens (Mustafa and Hussein 2020).

This chapter aims to explore nanotechnology's innovative applications in the development of nanoemulsions containing essential oils to manage storage pests and foodborne pathogens. It seeks to elucidate the unique properties of nanoemulsions in enhancing the stability, solubility, and controlled release of essential oil components, offering a promising avenue for effective pest and pathogen control. Furthermore, the chapter aims to critically evaluate the potential ecological implications and safety considerations associated with the utilization of these nanoemulsion-based formulations in revolutionizing pest and pathogen management within the context of food storage.

16.1.1 Essential Oils

For centuries, essential oils have been employed as a noninvasive and natural way to improve overall well-being. These oils, derived from various plant sources, contain fragrant compounds with unique chemical compositions that offer a wide range of therapeutic benefits. From aiding in relaxation to alleviating symptoms of certain ailments, essential oils have found their way into various industries, including perfumery, cosmetics, and even the food and beverage industry. Their aromatic and therapeutic properties make them a versatile and valuable resource for promoting physical and psychological health (Bunse et al. 2022).

With the increasing awareness of consumers, there is a growing trend toward natural additives in the food industry, as opposed to synthetic chemical additives. Essential oils and bioactive compounds extracted from aromatic plants have become popular choices for food preservation and flavor enhancement (Christaki et al. 2012). These natural alternatives serve as industrial raw materials for functional food production and offer antimicrobial properties, making them ideal for prolonging the shelf life of food items (Gyawali and Ibrahim 2014). Furthermore, the unique composition of essential oils makes them valuable for the perfume and cosmetic industries. Not only do they serve as natural fragrances, but they also offer preservation benefits for products such as perfumes, makeup, and various personal care items. As the demand for natural and sustainable ingredients continues to rise, essential oils have become a preferred choice for scenting and preserving various consumer goods (Guzmán and Lucia 2021).

In addition to their various applications in different industries, essential oils exhibit a wide range of bioactivities due to their chemical composition. These bioactivities include antibacterial, antimicrobial, insecticidal, anti-inflammatory, anticancer, antifungal, antiulcerogenic, antianxiety, and antidiabetic actions (Aisha et al. 2024; Ankitha et al. 2024; Zielińska-Błajet and Feder-Kubis 2020; Paul et al. 2024; Visakh et al. 2023a; Visakh et al. 2022a). The extensive use of essential oils in perfumery, cosmetics, and agrofood industries is primarily attributed to their aromatic properties, which enhance the sensory experience of products. Furthermore, the encapsulation of active compounds, including essential oils, has gained significant attention in the food, cosmetic, and pharmaceutical industries due to their potential benefits (Zhu et al. 2021).

Moreover, the consumer shift toward natural additives in the food sector has propelled the utilization of bioactive compounds derived from aromatic plants and essential oils. These natural alternatives contribute to the production of functional foods and play a crucial role in food products' protection and flavor enhancement. The significance of essential oils in various industries, their biological activities, and the consumer-driven demand for natural additives emphasize their potential as versatile and valuable resources (Tomić et al. 2023). As such, their applications extend beyond traditional uses, making them an integral component in developing and enhancing products across different sectors. The chemical characterization and composition of essential oils play a crucial role in determining their biological activities and potential applications. For example, monoterpenes are the main compounds found in essential oils and are responsible for their fragrance and many of their biological effects (Ribeiro-Santos et al. 2018).

16.1.2 Nanoemulsions

An emulsion can be considered as two liquids mixed together that cannot dissolve in each other, with one liquid existing as small droplets suspended within the other. The emulsifying agent stabilizes the interface, preventing droplets from aggregating and separating. Common examples of emulsion systems include oil suspended in water and water suspended in oil systems, in which oil or water is scattered as droplets of microscopic size in the other liquid (Tadros 2013). Emulsions create a biphasic liquid system, where the internal phase, consisting of dispersed droplets, is distinct from the continuous external phase enveloping it. This distinctive separation into internal and external phases characterizes the structure of emulsions (Patel and Joshi 2012; Naseema et al. 2021; Wilson et al. 2022). Physical conditions and thermodynamical properties of emulsion determine the size range of droplets. Based on this, macroemulsions, microemulsions, and nanoemulsions are used (Gadhave 2014).

Nanoemulsions represent a multiphase colloidal dispersion of water and oil, kinetically stable yet thermodynamically unstable, and are sustained by surfactant and co-surfactant molecules. The size of the droplet in these colloidal systems averages in the range of 100 and 500 nm (Patel and Joshi 2012; Wilson et al. 2022). These are also called ultrafine emulsions and submicron emulsions (Hussan 2011). The droplets that are in Brownian motion impart kinetic stability of the nanoemulsion, resulting in comparatively slower phase separation (Anton and Vandamme 2011; Friberg et al. 2011; McClements 2012; Kumar et al. 2021). Since the mean size of the droplet is below the visible spectrum wavelength, nanoemulsions exhibit a translucent, milky, or transparent appearance (Peng et al. 2019; Sayyar and Jafarizadeh-Malmiri 2020; Kumar et al. 2021).

16.1.2.1 Types of Nanoemulsions

Depending upon the size of the droplet, nanoemulsions are of two types: the translucent/transparent type with a diameter ranging from 50–200 nm and the milky type with a diameter of up to 500 nm (Izquierdo et al. 2001). Nanoemulsions, categorized based on the proportions of water and oil, encompass three primary types. The first category consists of oil suspended in water (O/W) nanoemulsions, with oil as the dispersion phase and water as the continuous phase (Kuntworbe et al. 2012). The second category of emulsion is the water suspended in oil (W/O) type nanoemulsions, which is distinguished by the presence of a dispersed aqueous phase in oil. The last category is bicontinuous nanoemulsions, in which both phases are interlinked, creating a continuous network. This emulsion exhibits the properties and characteristics of both emulsions (Sutradhar and Amin 2013; Bhosale et al. 2014).

16.1.2.2 Composition of Nanoemulsions

The major components that make up a nanoemulsion include an oleaginous phase or the nonpolar oil phase, surfactants, co-surfactants, and a polar phase of water (Kumar et al. 2021). Factors that intrinsically determine the stability and characteristics of the emulsion are refractive index, viscosity, density, tension between the interfaces, polarity, and water solubility (Table 16.1).

The choice of the surfactant remains the critical factor for synthesizing nanoemulsions (Wilson et al. 2022). Surfactants act by reducing the surface tension, which helps prevent the coalescence of particles (Kabalnov 1998). When the surfactant fails to play its role, the co-surfactant addition takes up its function, giving fluidity properties (Sadeq 2020). Factors like pH, electrolytes, and ionic content determine the size and stability (Tadros et al. 2004).

TABLE 16.1
Composition of Nanoemulsions

Components	Examples	Reference
Oil	Lauric acid, linoleic acid, oleic acid	(Halnor et al. 2018)
	Esters of fatty acid	
Surfactant	Cationic—Benzalkonium chloride, amines, cetrimonium bromide, quaternary ammonium compounds	(Halnor et al. 2018; Mondal et al. 2022)
	Anionic—Sodium lauryl sulfate, carboxylate groups, lignosulfonate, sulfonates	
	Non-ionic—Sorbitan alkyl esters, hydrogenated castor oil, polyoxyl 40, polyoxyethylene lauryl ether, polysorbate 20, polyoxyl 40 stearate, polyethoxylated tallow amine, Cetostearyl alcohol ethoxylate, Sorbitan monooleate	
Co-Surfactant	Glycols of short chain	(Jaiswal et al. 2015)
	Alcohols of medium chain	
	Alcohols of short chain	
Aqueous Phase	Plain water, N, N-methyl oleate, Ringer's solution, D-limonene, phosphate buffered saline	(Sarker et al. 2015)

16.1.2.3 Nanoemulsion Preparation

The process of emulsification and the appropriate proportions of all the key components of nanoemulsion are critical factors in forming a nanoemulsion. A nanoemulsion's desirable characteristics and stability will only be achieved with the proper balance of these elements. A considerable amount of energy input is necessary to produce nanoemulsions, which could be exploited from the chemical potential of the components or externally provided through mechanical means (Shafiq et al. 2007). Nanoemulsions could be prepared mainly with two different methods, one with the use of mechanical equipment, which uses a lumpsum of energy like high-pressure homogenization (Maali and Mosavian 2013; Gadhave 2014; Jasmina et al. 2017; Pathak 2017). The high-energy approach involves the droplets' deformation followed by the droplets' disruption into lower size configurations. The second type utilizes only low energy, like spontaneous emulsification and emulsion inversion points, and exploits the inherent chemical energy of the component materials (Gadhave 2014; Yukuyama et al. 2016). The energy consumed during the production period and the time required for its production determines the volume of the nanoemulsion being synthesized. The end product scaling up is comparatively more straightforward in the latter as uniform emulsification occurs across the whole mixture (Anton et al. 2008).

16.1.2.4 High-Energy Approaches

i. **High-Pressure Homogenization:** This is employed in producing nanoemulsions of particle size that start at 10 nm and extend to 100 nm (Flanagan and Singh 2006). It involves passing a blend of oil and aqueous phases through a narrow orifice under significant pressure, typically ranging from 500 to 5000 psi, which results in extremely fine particles due to hydraulic shear and high turbulence (Bhatt and Madhav 2011; Bhosale et al. 2014). Increasing the pressure leads to a reduction in particle size, while shorter homogenization cycles result in the production of smaller particles (Patel and Joshi 2012).
ii. **Ultrasonication:** The operating principle is the cavitation mechanism. The choice of the surfactant dictates droplet size. It is low cost and surfactant-flexible (Koroleva and Yurtov 2012). The size distributions of ultrasonicated nanoemulsions are broader and bimodal (Jafari et al. 2007). Ultrasonication outperforms more expensive high-energy technologies in terms of particle size reduction and material processing (Anton et al. 2008). While following this method for synthesizing nanoemulsions, the dispersed phase particle size decreases with the duration of ultrasonic homogenization, the concentration of surfactant, and power levels (Kentish et al. 2008; Kumar et al. 2015; Halnor et al. 2018).
iii. **Microfluidization:** A pump of high pressure (500 to 20000 psi) positive displacement is exploited to force the two immiscible liquids through microchannels, which are small channels in the interaction chamber. Further, it flows into an impingement area, which results in very fine particles (Bouchemal et al. 2004; Jasmina et al. 2017). The advantage of this method includes controlled-size droplets and the fact that multiple emulsions can be prepared (Koroleva and Yurtov 2012; Jasmina et al. 2017).

16.1.2.5 Low-Energy Approaches

i. **Emulsion Inversion Point:** When water is poured into a premixture of an appropriate oil and surfactant, W/O nanoemulsion forms. The inversion of the phase takes place catastrophically. Upon increasing the amount of water above a critical point results in the inversion of phase through the formation of bicontinuous structures. The rate of water added, along with the swirling speed, determines the droplet size of the nanoemulsion (Sokolov 2014).
ii. **Spontaneous Emulsification:** This is alternatively termed self-emulsification or emulsification by solvent diffusion. Nanoemulsion formation occurs based on dispersion and condensation mechanisms (Anton et al. 2008). When mixed with the surfactant and gently swirled with the gradual addition of water over a constant temperature, the oil results in its formation (Jasmina et al. 2017; Sneha and Kumar 2022). Surfactant structure and its concentration, interfacial tension, and interfacial and bulk viscosity determine the process's spontaneity. The execution of this process is feasible at room temperature (Bouchemal et al. 2004). It is among the cost-effective and readily scalable procedures (Koroleva and Yurtov 2012).
iii. **Phase Inversion Method:** The amalgamation of two immiscible phases with the addition of surfactant at elevated temperatures ends up forming a water suspended in oil emulsion, which transitions into an oil suspended in water emulsion upon subsequent cooling (Friberg et al. 2011; Wilson et al. 2022). Phase inversion composition results from adequately transiting phases that are generated with the composition adjustments while maintaining a consistent temperature. Alternatively, it can be produced by adjusting the temperature and ensuring a consistent makeup, the phase inversion temperature method (Shinoda and Saito 1968; Yukuyama et al. 2016). The concentration of surfactant, balance between the hydrophilic–lipophilic property of surfactant, ambient temperature, and phase transition temperature difference and water-to-oil ratio stabilizes this process (Tadros 2013; Kumar et al. 2019). This is widely used in

industrial-scale nanoemulsion production. PIT is founded on temperature-driven shifts in the solubility of non-ionic surfactants, specifically those of the polyoxyethylene type (Shinoda and Saito 1968). PIC involves manipulating the equilibrium that exists in the system while maintaining a consistent temperature (Koroleva and Yurtov 2012; Maali and Mosavian 2013; Jasmina et al. 2017).

iv. **Membrane Emulsification:** This method produces emulsions with a narrow range of size distribution, utilizing only minimal surfactants (Sanguansri and Augustin 2006).

16.1.3 Nanoemulsion Characterization

As the nanoemulsions lack thermodynamic stability, the synthesis method determines what properties the nanoemulsion will exhibit. The appropriate choice of the proportion of surfactants and co-surfactants and their properties also impact nanoemulsion characters and their utility (Morales et al. 2003).

16.1.3.1 Droplet Size Analysis, Polydispersity Index

Dynamic light scatterer, a light-scattering particle size analyzer that works by utilizing laser light diffusion, could be used in analyzing the particle size distribution (Bhosale et al. 2014). This device can also be used to evaluate the polydispersity index and the average droplet sizes. The distribution of size will be narrow when the PDI value is in the lower range (Jaiswal et al. 2015; Jadhav et al. 2020). Spectroscopy by the photon correlation can also be utilized for the same purpose (Flanagan and Singh 2006).

16.1.3.2 Turbidity

Turbidity of nanoemulsions can be assessed using an ultraviolet-visible light spectrophotometer by calculating the percent of light transmitted by undiluted emulsions at the wavelength of 600 nm, with distilled water as the standard (Liew et al. 2020).

16.1.3.3 Zeta Potential

The load on the droplet surface and physical stability in the long term can be quantified with an instrument called ZetaPALS/Zetasizer (Bali et al. 2010). The generation of surface charges also serves as a physical barrier, which produces repulsion among the droplets, thus preventing coalescence (Yilmaz and Borchert 2005). If the zeta potential value is below –30 mV, it demonstrates higher physical stability. If the nanoemulsion is stable, then the zeta potential will be more negative, and its net droplet charge will be higher (Jadhav et al. 2020).

16.1.3.4 Study of Phase Behavior

This study can determine the dispersibility and the phase of nanoemulsion of the product produced by the PIT technique method (Sharma et al. 2010). The O/W or W/O phase established by the developed nanoemulsion can be determined through measurements using a conductometer (Rao and Shao 2008).

16.1.3.5 Conductance Measurement

In a conductometer, two electrodes are connected to a lamp and completely immersed within the emulsion. An oil dispersed in water type emulsion is indicated when water, acting as a good conductor, allows the current to flow, causing the lamp to glow. Conversely, electricity is not conducted in a W/O type emulsion, where the external phase is oil and the lamp remains unlit (Jaiswal et al. 2015).

16.1.3.6 Viscosity

Measurement of viscosity can be done using an Ostwald viscometer or Brookfield rotary-type viscometer, which can be determined at different temperatures and at varied rates of shear (Jaiswal et al. 2015; Heydari et al. 2020).

16.1.3.7 Interfacial Tension

Measurement of interfacial tension with a spinning-drop apparatus helps study nanoemulsions' properties and formation. If the values are ultra-low, it means the presence of a surfactant phase that is in equilibrium with respect to oil and aqueous phases (Singh and Vingkar 2008).

16.1.3.8 Fluorescence Test

The oils with fluorescence properties can be tested by this method. Nanoemulsion, when viewed under UV light, shows fluorescence of the whole field if it's a W/O type and patchy fluorescence if it's an O/W (Jaiswal et al. 2015).

16.1.3.9 Transmission Electron Microscopy (TEM)

The morphology and internal structure can be effectively studied with TEM (Akhter et al. 2008). Surfactant aggregates could be detected using the high-resolution images, which cannot be detected with the particle size analysis alone without changing the original state of the product (Klang et al. 2012).

16.1.3.10 Scanning Electron Microscopy (SEM)

Characteristics of the surface of particles and topography can be determined by SEM and provides crucial information on the surface features of the sample. Cryo Sem is its variation, which analyses the sample by rapid freezing (Klang et al. 2012).

16.2 APPLICATION OF NANOEMULSION TECHNOLOGY AGAINST FOOD MICROBIAL CONTAMINANTS

Food industries face a significant threat from food contamination, which can occur due to various factors such as bacteria, fungi, viruses, synthetic chemicals, and preservatives, eventually leading to food infection and posing risks to human health. Food pathogens, including bacterial pathogens like *Escherichia coli*, *Bacillus sublitis*, *Klebsiella pneumon*iae, *Bacillus cereus*, *Listeria* *monocytogenes*, *Pseudomonas aeruginosa*, *Staphylococcus aureus*, *Salmonella typhimurium*, and *Acinetobacter baumannii* are contaminants in food. Food pathogens also comprise kingdom fungi such as *Alternaria* sp., *Aspergillus* sp., *Fusarium* sp., *Rhizopus* sp., *Penicillium* sp., etc. (Das et al. 2020). In several developing countries, 25–30% of food loss is attributed to contamination (Bondi et al. 2017), with microbes producing thermostable toxins that persist even under high temperatures (Rajkovic et al. 2020). Bacterial toxins include proteins (exotoxins), while lipopolysaccharides constitute endotoxins. Fungi contribute to contamination through sporulation and the production of mycotoxins, *viz*., aflatoxin, which may cause adverse effects on health (Aldars-Garcia et al. 2016). Mycotoxins, in turn, affect food quality and decline consumer commodity preference. It may also reduce shelf life and make the commodity unfit for consumption (Deepika et al. 2021).

Essential oils, synthesized by aromatic plants and medicinal crops, serve as natural preservatives with antimicrobial capabilities (Albaqami et al. 2022; Narayanankutty et al. 2022; Kuttithodi et al. 2023; Job et al. 2024). The United States Food and Drug Administration classified essential oils as environmentally friendly and under the "generally recognized as safe" category. Studies, such as the one by Bhavaniramya et al. (2019), underscore their effectiveness and safety, avoiding toxicity in mammals.

Essential oils contain terpenes, polyphenols, carotenoids, phenylpropanoids, and terpenoids (Winska et al. 2019). The broad antimicrobial properties of essential oils have been documented extensively across various stored foods (Kumar et al. 2020). These plant-based antimicrobials offer a safe and eco-friendly alternative, yet they come with challenges like high volatility, limited solubility, and instability. As these oils are highly volatile, less stable, and have inadequate solubility, it is challenging to utilize them extensively in modern food preservation (He et al. 2020).

The advent of nanoencapsulation addressed the challenges faced in the practical efficacy of essential oil in food preservation. Nanoencapsulation emerges as a promising solution for enhancing the practical utility of essential oils. Essential oil nanoemulsion revolutionizes the food industry by promising prolonged shelf life and maintaining food quality.

Additionally, nanoemulsions have greater surface area, higher stability, and enhanced bioavailability than pure oil (Zhang et al. 2021). Nanoemulsion of essential oils enhanced the antimicrobial efficacy with proper dispersion and direct and prolonged interaction at target sites of food pathogens (Mushtaq et al. 2023). Essential oil nanoemulsion has more impact in the food sector than pure oils. Nanoemulsions provide room for the direct action of essential oil on cell membranes for a prolonged period, thereby increasing microbial inhibition efficacy (Donsì and Ferrari 2016). Interaction with cell membranes eventually leads to increased permeability and apoptosis. Growing demand for clean and healthy food yearns for non-synthetic preservatives, making nanoemulsion of essential oils promising candidates in modern food industries (Zamuz et al. 2021).

16.2.1 Advantages of Essential Oil Nanoemulsions

Microbial deterioration is always a nightmare for farmers and food industries because it adversely affects the quality standard of food commodities and shortens their durability. Essential oil nanoemulsions are reassuring as they are non-synthetic and attract consumer preference. Additionally, it adds flavor and odor at proper concentrations. Nanoemulsions facilitate food preservation by directly incorporating into the food, penetrating permeable food surfaces, or layering on it.

16.2.2 Nanoemulsion of Essential Oils against Foodborne Microbes

The components of essential oil have a significant function in inhibiting microbes in stored food. The primary way of inhibition is infusion into cell membranes and triggering apoptosis (Pateiro et al. 2021). Nanoemulsion facilitates the dispersion and lengthens the action time on pathogens, thus increasing the efficacy. The phospholipid layer is the foundation of the cell membrane, where essential oil and its low molecular weight constituents attack fatty acids and integral proteins, leading to greater plasma membrane permeability. The imbalance in cellular ions due to leakage through a highly permeable membrane disrupts the electron transport chain and topples the equilibrium of their metabolic activity (Tang et al. 2020). Subsequently, the firmness of the plasma membrane loses and disturbs microbial cells' normal metabolism and ATP pool draining (Rao et al. 2019). Prolonged interaction with essential oil will eventually leak away most of the cell content, which causes cell death through apoptosis or necrosis (Ju et al. 2019). Nanoemulsions also interfere with the communication signal phenomenon of quorum sensing in bacteria, thus inhibiting the formation of biofilms (Maurya et al. 2021a).

16.2.3 Mechanism of Essential Oil Action on Bacteria

Essential oils encompass a diverse blend of active organic compounds, categorized into terpenes, terpenoids, and phenylpropanoids (Pateiro et al. 2021). Eugenol and thymol are more effective in inhibiting microbes than terpenes and ketones (Pathania et al. 2018). Nanoencapsulation amplifies the effectiveness of action more than free oils. Nanoemulsion of essential oils impacts the phospholipid bilayer of bacteria, leading to increased permeability, damage to membrane proteins, and cellular leakage. Nanoencapsulated essential oils disrupt the phospholipid layer within bacteria, impairing membrane proteins, increasing permeability, and causing subsequent leakage of cellular constituents (Yang et al. 2021). This disruption affects ions such as Na+, Mg2+, K+, and genetic constituents like DNA, RNA, and proteins, ultimately disintegrating bacterial cells (Maurya et al. 2021b).

16.2.4 Essential Oil Nanoemulsions against Bacterial Foodborne Pathogens

The variance in the plasma membrane composition of Gram-positive and Gram-negative bacteria is primarily influencing bacterial inhibition by nanoemulsions of essential oils. Ghosh et al. (2014) observed that applying concentrated eugenol oil nanoemulsion resulted in no viability just one minute after application. At 10% concentration, 50% of the *S. aureus* population was eliminated within 30 minutes. Li et al. (2018) also observed substantial growth inhibition of *E. coli* at 0.156 µl/ml, *B. subtilis* at 0.312 µl/ml, and *S. aureus* at 0.312 µl/ml notably higher than pure oil. According to Moraes-Lovison et al. (2017), the minimum inhibitory concentrations (MICs) of oregano nanoemulsion were 0.60 and 0.56 mg/ml for *E. coli* and *S. aureus*, respectively, which is necessary to preserve chicken pate. As *E. coli* is a Gram-negative bacteria, minimum bactericidal concentrations (MBC) increased during the 30 days of storage, which is attributable to the structural difference in the cell membrane. Bhargava et al. 2015 conducted a study where fresh lettuce, artificially contaminated with *S. typhimurium*, *E. coli*, and *L. monocytogenes* was immersed in Origanum nanoemulsions 0.1% for 1 minute reduced populations approximately into 2.31, 3.05, and 3.44 log CFU/g. The SEM result confirmed a reduction in population by revealing cell wall disruption.

16.2.5 Evaluating the Antibacterial Activity of Essential Oil Nanoemulsion

Nanoemulsion of various essential oils impacts permeability enhancement, membrane protein damage, and cellular leakage. There are many essential oil-based nanoemulsions utilized for the evaluation of antibacterial activity (Table 16.2). Similarly, Paudel et al. (2019) found that cinnamon oil nanoemulsions reduced populations of *L. monocytogenes* into 7.7 log CFU/g and *Salmonella* populations into 5.5 log CFU/g in contaminated honeydew and cantaloupe. Moghimi et al. (2016) discovered that the minimum inhibitory concentration of salvia essential oil nanoemulsion against *S. typ*hi and *E. coli* was four times lower than pure oil, attributed to extensive disintegration of the plasma membrane, which surges the oozing of cellular constituents. Additionally, Hadian et al., 2017 reported that *Rosmarinus officinalis* nanoemulsion at 2 mg/g significantly reduced the contamination of *Salmonella* sp. during refrigeration of beef cutlet, compared to unbound oil. He et al. (2021) revealed potent inhibition of nanoemulsion prepared from *Thymus* defenses against *E. coli* at 0.25 mg/mL, showing greater effectiveness than pure oil when applied to contaminated cherry tomatoes without causing any changes in their color or firmness. Hassanshahian et al. (2020) discovered higher inhibition with *Alhagi maurorum* essential oil nanoemulsion against various bacterial strains, with *E. coli* at 1.75 mg/ml (MIC) exhibiting twice the effectiveness of the oil. Zhang et al. (2017) observed that a blended clove-cinnamon oil nanoemulsion resulted in inhibition zone diameters of 25.50, 22.63, 21.87, and 14.17 against bacterial strains *S. aureus*, *E. coli*, *S. typhimurium*, and *B. subtilis*, respectively, highlighting significant antimicrobial action compared to using pure oil.

Furthermore, incorporating essential oil nanoemulsion into mushroom sauce did not alter taste and flavor, demonstrating considerable potential in the food sector. Moradi et al. 2023 observed inhibition of microbial growth in mayonnaise with *Cuminum cyminum* L. essential oil nanoemulsion against two bacterial strains, *E. coli* with 6 mg/ml MBC, whereas the MBC value for *Staphylococcus aureus* was 3 mg/mL, which shows that *E. coli* was the most resistant bacteria as compared to *S. aureus*. Shehabeldine et al. (2023) observed that clove oil nanoemulsion significantly inhibited quorum sensing and biofilm formation of *S. aureus* compared with pure oil. Jo et al. (2015) observed that trans-cinnamaldehyde nanoemulsion inhibited *Salmonella typhimurium*, *Staphylococcus aureus*, with an inhibition zone of 30 and 37 mm diameter at 8% concentration in watermelon juice. Venkadesaperumal et al. (2016) indicated that the MIC of cumin nanoemulsion was 60 mg/ml for *S. typhimurium*, *E. coli, and C. violaceum*. In contrast, the nanoemulsion of fennel oil exhibited 60 mg/ml MIC against *C. violaceum*. Gharenaghadeh et al. (2017) found that *Salvia multicaulis* essential oil nanoemulsion significantly inhibited *B. cereus* (625 mg/ml MIC), *Enterococcus faecalis* (375.5 mg/ml MIC), and *Klebsiella pneumoniae* (1250 mg/ml MIC). Moreover, inhibition by *Moraxella catarrhalis* nanoemulsion (1250 mg/ml MIC) was significantly higher than free oil. He also observed that the effect of nanoemulsion is higher on Gram-positive bacteria. Donsì et al. (2014) noticed that carvacrol nanoemulsions have a higher diffusion rate in zucchini and cooked meat sausages artificially inoculated with *E. coli* compared to unbound oil. According to Liew et al. (2020), among three nanoemulsions of essential oil, calamansi manifested the highest bacterial inhibition of 9.98, 7.71, and 8.34 mm (inhibition zone diameter) against *S. aureus*, *Salmonella* spp., and *E. coli*, respectively.

16.2.6 Mechanism of Essential Oil Action on Fungi Pathogens

Various bioactive molecules in essential oil synergistically inhibit fungal pathogens by interfering with cell wall biosynthesis. Their lipophilic nature specifically targets the phospholipid component of the plasma membrane of the cell, thus modifying cell integrity and permeability and facilitating the movement of cellular constituents through the membrane. This process paves the way for the loss of cellular components such as Ca2+, K+, and Mg2+, disrupting cell integrity. Consequently, it inhibits the mitochondrial electron transport system, the proton pump, ultimately culminating in cell disruption (Singh et al. 2020). Thus, essential oils are potential microbial inhibitors against food pathogens.

TABLE 16.2
Essential Oil Nanoemulsions against Food Pathogenic Bacteria

Essential Oil	Other Components	Inhibited Bacteria	Emulsification Technique	ConcentRation	References
Eugenol	Tween 80	*S.aureus*	Ultrasonic cavitation	0.3%	(Gosh et al. 2014)
Citrus medica L. *var. sarcodactylis*	Cremophor EL	*B. sublitis* *S. aureus* *E. coli*	Spontaneous emulsification	0.312 µL/mL 0.312 µL/mL 0.156 µL/mL	(Li et al. 2018)
Oregano	Cremophor RH 40 Span 80	*S. aureus* *E. coli*	Phase inversion temperature method (PIT)	0.56 mg 0.60 mg (MIC)	(Moraes-Lovison et al. 2017)
Origanum spp.	Tween 80	*E. coli* *L. monocytogenes* *S. typhimurium*	Ultrasonication	0.1%	(Bhargava et al. 2015)
Cinnamon	Tween 80	*Salmonella spp.* *L. monocytogenes*	Ultrasonication	0.5%	(Paudel et al. 2019)
Sage (*Salvia officinalis*)	Tween 80, Span 80	*E. coli* *S. typhi*	Ultrasonication	0.0078–32 mg/ml	(Moghimi et al. 2016)
Rosmarinus officinalis	Tween 80, chitosan (CS) benzoic acid (BA)	*S. typhimurium*	Sonication	2 mg/g	(Hadian et al. 2017)
Thyme (*Thymus daenensis*)	cetylpyridinium chloride	*E. coli*	Ultrasonication	0.25 mg/ml	(He et al. 2021)
Alhagi maurorum	Tween 80	*K. pneumoniae* *S. aureus* E. *coli* *A. baumannii* *B. cereus* *P. aeruginosa*	Ultrasonication	100 mg/ml	(Hassanshahian et al. 2020)
Cloves/cinnamon	Tween 80 ethanol	*E. coli* *B. subtilis* *S. typhimurium* *S. aureus*	Ultrasonication	100 µl	(Zhang et al. 2017)
Cuminum cyminum L.	Tween 80	*S. aureus* *E. coli*	Ultrasonication	3 mg/ml 6 mg/ml	(Moradi et al. 2023)
trans-cinnamaldehyde	Tween 20	*S. typhimurium* *S. aureus*	High-speed homogenization and high-pressure homogenization	0.8%	(Jo et al. 2015)
Spice oil cumin (*C. cyminum*), pepper (*Piper nigrum*) fennel (*Foeniculum vulgare*)	Tween80	*C. violaceum CV026* *S. typhimurium* *E. coli* *K. pneumoniae*	Ultrasonication	10–200 µl	(Venkadesaperumal et al. 2016)
Salvia multicaulis	Span 80 Tween 80	*K. pneumoniae* *E. faecalis* *M. catarrhalis* *B. cereus*	Ultrasonication	19.53–10,000 µg/ml	(Gharenaghadeh et al. 2017)
Carvacrol	Tween 20 Monoolein Fluid soy lecithin	*E. coli*	High-pressure homogenization	0.2 g/100 g	(Donsì et al. 2014)
Kaffir lime (*Citrus hystrix*) Lime (*Citrus aurantifolia*), Calamansi lime (*Citrofortunella microcarpa*)	Tween 80	*E. coli*, *Salmonella spp* *S. aureus*	Homogenization	0.01 ml	(Liew et al. 2020)

16.2.7 Nanoemulsions of Essential Oils against Fungi Pathogens

Nanoemulsions possess significantly smaller sized (10 to 100 nm) droplets than sole emulsions, which typically range from 100 nm to 100 µm. The smaller size contributes to their enhanced stability. Smaller droplet size leads to active interaction at target sites of fungi, such as cell membranes, and acts for a prolonged time, effectively modifying membrane permeability. According to Sugumar et al. (2016), nanoemulsion of concentrated orange oil eradicated all yeast cells within 48 hours by using orange nanoemulsion. When diluted tenfold, 67% of viable cells were present after 4 hours, 10% after 12 hours, and no viable cells after 48 hours. Pongsumpun et al. (2019) discovered that the inhibition zones against *Aspergillus. niger*, *Penicillium* sp., *Colletotrichum. gleosporioides*, and *Rhizopus arrhizus* measuring 23.74, 19.27, 16.34, and 22.12 mm, respectively, compared to pure cinnamon essential oil, which showed zones smaller than 12.5 mm. This enhancement was attributed to the essential oil's limited solubility, reducing its interaction with fungal cell walls.

16.2.7.1 Evaluating Antifungal Activity of Nanoemulsion Oils

There are many essential oil-based nanoemulsions utilized for the evaluation of antifungal activity (Table 16.3.). Yousef et al. found that 0.2% of cinnamon nanoemulsions exhibited the highest inhibition of grey mold caused by *Botrytis cinerea* (5.43% infection). At a concentration of 0.2%, it showed the highest decline in rottenness (13.50% infection). Wan et al. 2019, observed that thyme oil nanoemulsion (EC_{90}=7.61 mg/g) showed the highest microbial inhibition as compared to the peppermint oil nanoemulsion (EC_{90}= 23.672 mg/g) against *Fusarium graminearum*. Singh et al. (2020) found that clove oil nanoemulsion exhibited minimum inhibition of *Fusarium proliferatum* (MIC) at 9 µl, while the fungal concentration was minimum (MFC) at 50 µl nanoemulsion in maize. These findings indicated effective inhibition of fungal growth under storage conditions. According to Jiang et al. (2023), a rise in the concentration of nanoemulsified hope oil (14.00 mg/g) correspondingly increased mycelial growth inhibition (55.37%) in a population of *F. graminearum*. de Castro e Silva et al. (2019) observed an inhibition zone of 25 mm for *P. citrinum* and 28 mm for *A. flavus* by nanoemulsion of neem oil at 3% with infected soybean seeds, which assures the compatibility of neem oil nanoemulsion for preserving stored agricultural commodities. Shehabeldine et al. (2023) observed that clove oil nanoemulsion shows significant inhibition against fungi, *Candida albicans*, *Cryptococcus neoformans*, *A. flavus*, *A. fumigatus*, and *Aspergillus brasiliensis* with MICs 12.5, 3.12, 1.56, 1.56, and 0.78 mg/mL, respectively.

TABLE 16.3
Essential Oil Based Nanoemulsions against Food Pathogenic Fungi

Essential Oil	Other Components	Inhibited Microorganisms	Emulsification Technique	Conc.	References
Orange oil *Citrus sinensis*	Tween 80	*Saccharomyces cerevisiae*	Ultrasonication	2 µl/ml	(Sugumar et al. 2016)
Cinnamon	Tween 80	*A. niger*, *R. arrhizus*, *Penicillium* sp. *C. gleosporioides*	Ultrasonication	1–3%	(Pongsumpun et al. 2019)
Thyme	Tween 80	*Fusarium graminearum*	Microfludizer and optimized homogenization	7.61 mg/g (EC_{90})	(Wan et al. 2019)
Cinnamon zeylanicum L.	Tween 80 Lecithin Span80	*Rhizopus stolonifera* *Botrytis cinerea*	Ultrasonication	0.2%	(Yousef et al. 2019)
Clove oil	Tween 80	*Fusarium proliferatum* sp.	Wash out method and ultrasonication	9 µl (MIC)	(Singh et al. 2020)
Hop oil	Tween 80	*F. graminearum* 10–124–1	Microfludizer and optimized homogenization	14.00 mg/g	(Jiang et al. 2023)
Neem oil	Tween 20	*Aspergillus flavus* *Penicillium citrinum*	Ultrasonication	3%	(de Castro e Silva et al. 2019)
Clove oil	Tween 80	*Aspergillus fumigatus* *Aspergillus flavus* *Cryptococcus neoformans* *Candida albicans* *Aspergillus brasiliensis*	Ultrasonication	1.56 mg/ml 1.56 mg/ml 3.12 mg/ml 12.5 mg/ml 0.78 mg/ml	(Shehabeldine et al. 2023)

16.3 NANOEMULSIONS OF ESSENTIAL OILS FOR THE MANAGEMENT OF STORED PRODUCT INSECTS

An escalation in global population growth amplifies the demand for food and agricultural commodities. Despite increased food production, substantial quantities of these products are lost annually due to stored pests and other bio-agents. Rajendran and Gunasekaran (2002) reported losses of 5% to 10% in temperate regions and 20% to 30% or more in tropics due to stored cereals infestation, diminishing both quantity and quality. Moreover, such infestations foster mold growth, including toxin-producing species (Magan et al. 2003). Chemical pesticides and fumigants, long relied upon for controlling stored insect pests in various storage systems, are facing escalating restrictions due to their environmental and organismal hazards (United States, Office of Mission to Planet Earth 1995; Danse et al. 1984).

Reports worldwide indicate the emergence of resistance to fumigants and chemical pesticides among stored grain insects (Wakil et al. 2021; Feroz et al. 2020). In response to the potential adverse impacts of synthetic substances, there is a growing market demand for botanicals and bio-based materials. These botanicals, especially essential oils derived from plants, play a significant role in eco-friendly management strategies against various stored grain pests (Visakh et al. 2023b; Visakh et al. 2022b; Sulhath et al. 2024). Ju et al. (2018) highlighted the US Food and Drug Administration's recognition of plant-derived aromatic volatile secondary metabolites as "generally recognized as safe" (GRAS) substances. These compounds, which include terpenoids, alcohols, carbonyl compounds, phenolics, and aliphatic compounds, show promise in controlling mites, nematodes, insects, and more. However, challenges such as water insolubility, instability, and volatility hinder their effective utilization across diverse environments. These concerns are addressed by nanoemulsion-based formulations, which have proven to be superior to bulk equivalents (essential oils and simple emulsions) regarding water miscibility, stability, and volatility.

16.3.1 Toxicity Assessment of Essential Oil-Based Nanoemulsions against Stored Grain Pests

The nanoemulsion formulations of essential oils hold promise for augmenting the potency of botanical insecticides slated for commercial application. By incorporating adjuvants and surfactants, the effectiveness of nanoemulsion formulations can be further enhanced. Noteworthy for their high solubility in water and ability to dissolve lipophilic and hydrophilic compounds, these formulations diminish the necessity for inert and active materials, presenting a more economically viable option. Moreover, they exhibit remarkable stability across a broad temperature spectrum ranging from 10 to 55°C (Pavoni et al. 2020).

Exploration into essential oil-based nanoemulsions underscores their potential as insect growth regulators, fumigants, antifeedants, and repellents against stored insect pests. The envisioned delivery systems for nanoemulsions seek to bolster the physicochemical attributes of essential oils by curbing volatility, bolstering stability, and shielding them from environmental influences (Nenaah 2014). The adoption of essential oil nanoformulations proves pragmatic for safeguarding handling equipment, production facilities, warehouses, and machinery from stored product insect pests. Notably, cold aerosol and gel formulations of essential oil-based nano-insecticides emerge as effective means for managing pests in stored products, offering promise for integration into sustainable and biorational strategies (Palermo et al. 2021).

Among the prevalent stored insect pests, *Sitophilus oryzae L.* (rice weevil) (Coleoptera: Curculionidae) and *Tribolium castaneum*, the red flour beetle Herbst (Coleoptera: Tenebrionidae), are identified as the most common species (Guru-Pirasanna-Pandi et al. 2018). Additionally, other frequently encountered stored pests comprise the *Trogoderma granarium* (Dermestidae), *Oryzaephilus surinamensis* (Cucujidae), *Rhyzopertha dominica* (Bostrychidae), *Lasioderma serricorne* (Ptinidae), *Callosobruchus* spp. (Bostrychidae), *Stegobium paniceum* (Ptinidae), *Sitotroga cereallela* (Gelechidae), *Ephestia cautella* (Phycitidae), and *Corcyra cephalonica* (Galleridae). Many stored product insects have been evaluated for their bio-efficacy of nanoemulsion of essential oils (Table 16.4).

16.3.1.1 Toxicity of Essential Oil-Based Nanoemulsions against Tenebrionidae

Tribolium spp. stands out as one of the primary secondary and generalist pests within the Tenebrionidae family of the Coleoptera order. The detrimental impact inflicted by these pests encompasses both qualitative (such as nutritional and industrial value) and quantitative (like grain weight) losses (Gad et al. 2020). Draz et al. 2022 observed a notable increase in the toxicity of anise (*Pimpinella anisum*) and thyme (*Thymus vulgaris*) essential oils, up by 1.50 and 1.47 times, respectively, when formulated into nanoemulsions. Utilizing a nanoemulsion formulation of *Mentha longifolia* (L.) Huds. (Lamiaceae) at a concentration of one thousand parts per million (10% w/w essential oil-based) exhibited notable efficacy as a wheat protectant against *T. castaneum* and *T. confusum*, resulting in mortality rates of 93.3% and 96.7%, respectively (Kavallieratos et al. 2022). Palermo et al. (2021) underscored the effectiveness of cold aerosol and gel formulations of essential oil-based nano-insecticides derived from various plants such as artemisia (*Artemisia vulgaris*), garlic (*Allium sativum*), anise (*Pimpinella anisum*), lavender (*Lavandula angustifolia*), fennel (*Foeniculum vulgare*), sage (*Salvia officinalis*), mint (*Mentha piperita*), and rosemary (*Rosmarinus officinalis*) in controlling *T. confusum* adults infesting stored products. Employing a low-energy, solvent-free, sustainable process, Lima et al. (2021) introduced a novel class of potential nanopesticides derived from the essential oil of *Baccharis reticularia* (Asteraceae), which exhibited repellent activity at 1.1 µg/cm2 against *T. castaneum*, with limonene and α-pinene identified as the key repellent constituents. Fumigation using nanoemulsions of sweet orange (*Citrus sinensis*) essential oil in cold aerosol form displayed a sevenfold increase in insecticidal activity compared to fumigation against adults of *T. confusum* (Giunti et al. 2019). Therefore, applying cold aerosol-based nanoemulsion treatments in manufacturing spaces, storage facilities, and equipment presents a promising approach to sanitation. Nanoemulsions of *Achillea* essential oils exhibited significant fumigant activity targeting both adults and larvae of *T. castaneum*, with LC50 values ranging from 8.8 to 21.3 µL/L and 11.0 to 27.5 µL/L, respectively, after 96 hours of exposure (Nenaah 2014).

16.3.1.2 Toxicity of Essential Oil-Based Nanoemulsions against Curculionidae

Sitophilus oryzae is a major pest in storage and the resistance and tolerance problems with the synthetic pesticides now in use hinder their ability to manage the pest effectively (Kavallieratos et al. 2022). Hence, the effectiveness of essential oil-based nanoemulsions from *Hazomalania voyronii* (Jum.) Capuron, *Mentha longifolia* (L.) Huds., and *Carlina acaulis* L. were examined by Kavallieratos et al. (2023) against *S. oryzae*. Essential oil-based nanoemulsions of *C. acaulis* were most effective, with 94.4% mortality of adults of *S. oryzae* in barley one week after exposure at 1000 ppm, followed by essential oil-based nanoemulsions of *H. voyronii* (83.3%).

Draz et al. (2022) reported the increased toxicity of anise, *Pimpinella anisum*, and thyme, *Thymus vulgaris*, oils as nanoemulsion formulations to *S. oryzae* (1.41 to 1.48 times) as compared to their pure essential oils. The nanoemulsion of *Pimpinella anisum* showed significantly higher toxicity to *S. oryzae* with LC50 values of 3858.88 mg/L, which also reduced the emergence of *S. oryzae* by 94.64%. Formulated nanoemulsions (LC_{50} 0.45 µL/cm^2) outperformed eucalyptus oil (0.795 µL/cm^2) against *Sitophilus oryzae* and were found to be safer substitutes for hazardous chemical pesticides for controlling stored grain pests (Adak et al. 2019). Mesbah et al. (2023) demonstrated higher toxicity of *Pelargonium graveolens* nanoemulsion to adults of *S. oryzae* with an LC_{50} value of 2.29 ppm cm^{-2} compared to the free essential oil with an LC_{50} value of 67.662. However, the nanoemulsion formulation reduced the germination of wheat.

For the nanoemulsion formulation made from *Myristica fragrans* and *Jatropha curcas*, the contact approach showed the greatest toxicity to *Sitophilus zeamais* with an LC_{50} of 0.6% and 0.69% after 72 hrs of treatment among the other application procedures. The seed dressing came next with an LC_{50} of 0.74% and 0.80%, and fumigation produced the lowest results with 1.51% and 0.89% LC_{50} values, respectively (Nuryanti et al. 2023). Nanoemulsions formulated using ultrasonic emulsification and Tween 20 as a surfactant of Jojobe (*Simmondasia chinensis*) oil was more effective (LC_{50} = 0.31ml/kg wheat) compared with the bulk oil with an LC_{50} value of 3.12 ml/kg wheat. Hence, it was concluded that the formulated nanoemulsions can be employed to manage rice weevil, *S. oryzae* (Sh et al. 2015).

16.3.1.3 Toxicity of Essential Oil-Based Nanoemulsions against Cucujidae

Essential oil based nanoemulsions increased the effectiveness of pure citronella oil by more than two times (LC_{50} = 20.3 µL/L), in controlling *O. surinamensis* (Gharsan et al. 2022). Dose-dependent response of essential oil-based nanoemulsion of *Citrus sinensis* in the form of cold aerosol showed five folds higher insecticidal activity than fumigation in adults of *Cryptolestes ferrugineus* with an LD_{50} value of 36.79 ppm (Giunti et al. 2019). The aforementioned studies suggest essential oil-based nanoemulsion as a promising substitute for chemical pesticides in controlling stored pests belonging to Cucujidae.

16.3.1.4 Toxicity of Essential Oil-Based Nanoemulsions against Bruchidae

The nanoemulsion procured from aerial parts of *Deverra tortuosa* showed contact, fumigant, and residual bioactivity against cowpea weevil, *C. maculatus* with LC_{50} of 10.3 µg/cm^2, and 6.9 µl/l, respectively (Almadiy et al. 2022). *Mentha longifolia* formulation exhibited increased fumigant toxicity against *C. maculatus*, effectively eliminating 50% of the egg population in 4.7 days at a dosage of 9 µL/L (Louni et al. 2019). Contact and fumigant toxicity of two nanoemulsion formulations of *Eucalyptus globulus* using gum arabic and Span-80 as surfactants successfully managed cowpea weevil adults (Ya-Ali et al. 2020).

16.3.1.5 Toxicity of Essential Oil-Based Nanoemulsions against Bostrychidae

Rhyzopertha dominica (F.), commonly known as lesser grain borer, thrives on substances, including stored medications, legumes, mud plaster, leather stuffing, and packaging materials. Essential oil based nanoemulsion (<200 nm) of sweet orange (*Citrus sinensis*), mint (*Mentha x piperita*), and fennel (*Foeniculum vulgare*) showed good repellence to adults of *R. dominica* adults and decreased the activity after subsequent exposures (Giunti et al. 2021).

16.4 CONCLUSION

The chapter discusses the potential of nanotechnology to enhance the stability and efficacy of biopesticides, which will produce more effective and biocompatible biopesticides. It offers prospects for these formulations in integrated pest management in addition to an updated literature approach. The chapter draws attention to the necessity of tight and effective cooperation between various societal players (such as research institutes, colleges, businesses, and government regulatory bodies) to fix and expand these biopesticide-based products for the agricultural mark.

In addressing storage pests and foodborne pathogens, nanoemulsion-based formulations of essential oils exhibit

TABLE 16.4
Major Constituents and Targeted Insects of Nanoemulsion Formulation of Essential Oils

Nanoemulsions of Essential Oils	Major Chemical Compound	Targeted Test Insects	Reference
Pimpinella anisum	Trans-a nethole (93.9%) and estragole (2.4%)	*S. oryzae, T. castaneum*	Draz et al. 2022
Thymus vulgaris	Camphene (17.19%), camphor (38.54%), 1, 8-cineole (5.44%), α-pinene (9.35%), β-pinene (3.90%), and borneol (4.91%)		
Citronella	Citronellol (9.49%), linalool (9.46%), and citronellal (46.95%)	*O. surinamensis*	Gharsan et al. 2022
Achillea species	*Cis*-ascaridol (33.8%), *P*-cymene (22.4%), piperitone (5.4%), 1,8-cineole (6.3%), camphor (8.6%), and carvenone oxide (4.1%)	*T. castaneum*	Nenaah 2014
Citrus sinensis	*R*-limonene (93.35%), α-pinene (1.14%), β-myrcene (3.38%), linalool (0.54%), and sabinene (0.50%)	*T. confusum* *C. ferrugineus*	Giunti et al. 2019
Baccharis reticularia	Limonene (14.6%), bicyclogermacrene (12.5%), and β-myrcene (12.6%)	*T. castaneum*	Lima et al. 2021
Citrus sinensis	*R*-limonene (93.25%), α-pinene (1.14%), and β-myrcene (3.37%)	*R. dominica*	Giunti et al. 2021
Mentha x piperita	Menthol (36.72%), isomenthone (13.56%), menthone (18.23%), eucalyptol (4.73%), and menthyl acetate (5.79%)		
Foeniculum vulgare	Estragole (3.15%), (*E*)-anethole (43.81%), limonene (23.98%), α-phellandrene (3.33%), and fenchone (10.14%),		
Pimpinella anisum	(E)-anethole (87.11%), foeniculin (2.01%), estragol (3.98%), and linalool (1.94%)	*T. confusum*	Palermo et al. 2021
Foenicum vulgare	Estragol (2.10%) and phenylpropanoids (E)-anethole (42.13%)		
Artemisia vulgaris	Oxygenated monoterpene (84.52%), *a*-thujone (28.42%), chrysantenyl acetate (4.97%), *b*-thujone (13.35%), camphor (22.21%), and chrysanthenone (5.97%)		
Lavandula angustifolia	Linalool (38.12%), camphor (7.01%), linalyl acetate (36.31%), and eucalyptol (5.01%)		
Mentha piperita	(-)-menthol (37.67%), limonene (7.73%), isomenthone (8.75%), (-)-neomenthol (6.32%), and menthone (22.10%)		
Rosmarinus officinalis	Eucalyptol (52.72%), cymene (4.01%), α pinene (11.28%), ortho- and *b*-pinene (4.01%), camphor (12.04%)		
Salvia officinalis	Eucalyptol (20.07%), *a*-pinene (21.20%), camphor (12.85%), camphene (4.67%), and *a*-thujone (16.23%)		
Allium sativum	Diallyl disulfide (30.09%), diallyl tetrasulfide (13.17%), diallyl trisulfide (22.70%), and diallyl sulfide (11.09%)		
Carlina acaulis	Carlina oxide (94.6%) and benzaldehyde (3.1%)	*S. oryzae*	Kavallieratos et al. 2023
Mentha longifolia	Myrcene (10.8%), piperitenone oxide (61.1%), limonene (5.3%), and 1,8-cineole (5.3%)		
Hazomalania voyronii	Perilla aldehyde (43.0%),1,8-cineole (33.2%), and limonene (13.0%)		
Eucalyptus spp	Eucalyptol (64.80%), β-pinene (8.19%), α-Phellandrene (3.88%), γ-Terpinene (5.91%), α-Pinene (11.17%), terpinen-4-ol (0.72%), and α-terpineol (1.01%)	*S. oryzae, T. castaneum*	Adak et al. 2019
Pelargonium graveolens	Citronellol (29.90%), geranyl acetate (4.52%), isomenthone (5.44%), 10-epi- γ -eudesmol (8.27%), trans -geraniol (18.03%), γ -cadinene (2.89%), linalool (5.13%), gemacrene D (2.05%), geranyl tiglate (2.50%), and geranyl butyrate (2.53%)	*S. oryzae*	Mesbah et al. 2023
Myristica fragrans	Sabinene (17.2%), myristicin (9.1%), eugenol (16.6%), α-pinene (5.4%), caryophyllene (8.8%), β-pinene (6.4%), limonene (5.0%)	*S. zeamais*	Nuryanti et al. 2023
Jatropha curcas	Phytol (47.42%), 7- benzyl-8-theophylline (12.53%), 1- butoxyperethyl-homotetrasilsesquioxane (36.11%), β-elemol (13.45%), γ-gurjunene (13.99%), and guaiene (13.13%)		
Deverra tortuosa	Dillapiole (41.6%), myristicin (5.1%), elemicin, (7.3%), and sabinene (4.2%)	*C. maculatus*	Almadiy et al. 2022
Mentha longifolia	Pulegone (26.07%), piperitone (11.88%), piperitone oxide (19.72%), cis-piperitone oxide (6.35%), 1,8-cineole (8.21%), and borneol (5.96%)	*E. kuhniella*	Louni et al. 2018
Mentha longifolia	Pulegone (26.07%), cis-piperitone oxide (6.35%), piperitone (11.88%), piperitone oxide (19.72%), 1,8-cineole (8.21%), and borneol (5.96%)	*C. maculates*	Louni et al. 2019
Eucalyptus globulus	Eucalyptol (43.79%), limonene (14.27%)	*C. maculates*	Ya-Ali et al. 2020
Simmondasia chinensis	Alpha-linolenic acid (37.1%), nervonic acid (13.8%)	*S. oryzae*	Sh et al. 2015

heightened bioactivity. The nanoscale emulsions enhance the delivery of essential oil components, ensuring sustained and targeted release. This augmentation of insecticidal properties disrupts physiological processes in pests, impeding their proliferation effectively. Similarly, in the context of foodborne pathogens, nanotechnology-enabled formulations of essential oils, including nanoemulsions, provide enhanced antimicrobial effects. These nanoemulsions improve the bioavailability of active compounds, facilitating greater interaction with microbial cell membranes and inhibiting their growth more effectively. Despite the immense promise of nanotechnology-enabled formulations, ongoing research is imperative to elucidate potential ecological implications and ensure the safety of these nano-based solutions. Integrating nanoemulsions, along with other nanotechnological applications, into essential oil formulations represents a cutting-edge approach with substantial potential to revolutionize pest and pathogen management in food storage.

REFERENCES

Aisha K, Visakh NU, Pathrose B, Mori N, Baeshen RS, Shawer R (2019) Nanoemulsion of eucalyptus oil: An alternative to synthetic pesticides against two major storage insects (*Sitophilus oryzae* (L.) and *Tribolium castaneum* (Herbst)) of rice. *Industrial Crops and Products* 143:111849. https://doi.org/10.1016/j.indcrop.2019.111849

Aisha K, Visakh NU, Pathrose B, Mori N, Baeshen RS, Shawer R (2024) Extraction, chemical composition and insecticidal activities of *Lantana camara* Linn. leaf essential oils against *Tribolium castaneum*, *Lasioderma serricorne* and *Callosobruchus chinensis*. *Molecules* 29:344. https://doi.org/10.3390/molecules29020344

Akhter S, Jain GK, Ahmad FJ, Khar RK, Jain N, Khan ZI, Talegaonkar S, (2008) Investigation of nanoemulsion system for transdermal delivery of domperidone: Ex-vivo and in vivo studies. *Current Nanoscience* :4381–390. https://doi.org/10.2174/157341308786306071

Albaqami JJ, Hamdi H, Narayanankutty A, Visakh NU, Sasidharan A, Kuttithodi AM, Famurewa AC, Pathrose B (2022) Chemical composition and biological activities of the lEaf Essential oils of *Curcuma longa*, *Curcuma aromatica* and *Curcuma angustifolia*. *Antibiotics* 11:1547. https://doi.org/10.3390/antibiotics11111547

Aldars-Garcia L, Ramos AJ, Sanchis V, Marin S (2016) Modeling postharvest mycotoxins in foods: Recent research. *Current Opinion in Food Science* :1146–50. https://doi.org/10.1016/j.cofs.2016.09.005

Almadiy AA, Nenaah GE, Albogami BZ (2022) Bioactivity of *Deverra tortuosa* essential oil, its nanoemulsion, and phenylpropanoids against the cowpea weevil, a stored grain pest with eco-toxicological evaluations. *Environmental Science and Pollution Research* :2965112–65127. https://doi.org/10.1007/s11356-022-20404-w

Ankitha TA, Visakh NU, Pathrose B, Mori N, Baeshen RS, Shawer R (2024) Phytochemical characterization of *Callistemon lanceolatus* leaf essential oils and their application as sustainable stored grain protectants against major storage insect pests. *Sustainability* 16:1055. https://doi.org/10.3390/su16031055

Anton N, Benoit JP, Saulnier P (2008) Design and production of nanoparticles formulated from nano-emulsion templates—a review. *Journal of Controlled Release* :128185–199. https://doi.org/10.1016/J.JCONREL.2008.02.007

Anton N, Vandamme TF (2011) Nano-emulsions and micro-emulsions: Clarifications of the critical differences. *Pharmaceutical Research* :28978–985. https://doi.org/10.1007/S11095-010-0309-1/METRICS

Ayilara MS, Adeleke BS, Akinola SA, Fayose CA, Adeyemi UT, Gbadegesin LA, Omole RK, Johnson RM, Uthman QO, Babalola OO (2023) Biopesticides as a promising alternative to synthetic pesticides: A case for microbial pesticides, phytopesticides, and nanobiopesticides. *Frontiers in Microbiology* 14. https://doi.org/10.3389/fmicb.2023.1040901

Bali V, Ali M, Ali J (2010) Study of surfactant combinations and development of a novel nanoemulsion for minimizing variations in bioavailability of ezetimibe. *Colloids and Surfaces B: Biointerfaces* :76410–420. https://doi.org/10.1016/J.COLSURFB.2009.11.021

Barradas TN, de Holanda e Silva KG (2020) Nanoemulsions of essential oils to improve solubility, stability and permeability: A review. *Environmental Chemistry Letters* 19:1153–1171. https://doi.org/10.1007/s10311-020-01142-2

Bhargava K, Conti DS, da Rocha SR, Zhang Y (2015) Application of an oregano oil nano-emulsion to control foodborne bacteria on fresh lettuce. *Food Microbiology* :4769–73. https://doi.org/10.1016/j.fm.2014.11.007

Bhatt P, Madhav S (2011) A detailed review on nanoemulsion drug delivery system. *International Journal of Pharmaceutical Sciences and Research* 2.

Bhavaniramya S, Vishnupriya S, Al-Aboody MS, Vijayakumar R, Baskaran D (2019) Role of essential oils in food safety: Antimicrobial and antioxidant applications. *Grain & Oil Science and Technology* :249–55. https://doi.org/10.1016/j.gaost.2019.03.001

Bhosale RR, Osmani RA, Ghodake PP, Shaikh SM, Chavan SR, (2014) Corresponding author: Nanoemulsion: A review on novel profusion in advanced drug delivery. *Indian Journal of Pharmaceutical and Biological Research* 2:122. https://doi.org/10.30750/ijpbr.2.1.19

Bondi M, Lauková A, de Niederhausern S, Messi P, Papadopoulou C (2017) Natural preservatives to improve food quality and safety. *Journal of Food Quality* 2017:1090932. https://doi.org/10.1155/2017/1090932

Bouchemal K, Briançon S, Perrier E, Fessi H (2004) Nano-emulsion formulation using spontaneous emulsification: Solvent, oil and surfactant optimization. *International Journal of Pharmaceutics* :280241–251. https://doi.org/10.1016/J.IJPHARM.2004.05.016

Bunse M, Daniels R, Gründemann C, Heilmann J, Kammerer DR, Keusgen M, Lindequist U, Melzig MF, Morlock GE, Schulz H, Schweiggert R (2022) Essential oils as multicomponent mixtures and their potential for human health and well-being. *Frontiers in Pharmacology* 13:956541. https://doi.org/10.3389/fphar.2022.956541

Chouhan S, Sharma K, Guleria S (2017) Antimicrobial activity of some essential oils—present status and future perspectives. *Medicines* 4:58. https://doi.org/10.3390/medicines4030058

Christaki E, Bonos E, Giannenas I, Florou-Paneri P (2012) Aromatic plants as a source of bioactive compounds. *Agriculture* :2228–243. https://doi.org/10.3390/agriculture2030228

Danse LHJC, Van Velsen FL, Van Der Heijden CA (1984) Methylbromide: Carcinogenic effects in the rat forestomach. *Toxicology and Applied Pharmacology* 72(2): 262–271.

Das S, Singh VK, Dwivedy AK, Chaudhari AK, Upadhyay N, Singh A (2020) Fabrication, characterization and practical efficacy of *Myristica fragrans* essential oil nanoemulsion delivery system against postharvest biodeterioration. *Ecotoxicology and Environmental Safety* 189:110. https://doi.org/10.1016/j.ecoenv.2019.110000

de Castro e Silva P, Pereira LAS, de Rezende ÉM, dos Reis MV, Lago AMT, Carvalho GR (2019) Production and efficacy of neem nano-emulsion in the control of *Aspergillus flavus* and *Penicillium citrinum* in soybean seeds. *European Journal of Plant Pathology* 155(4):1105–1116.

Deepika, Chaudhari AK, Singh A, Das S, Dubey NK (2021) Nanoencapsulated *Petroselinum crispum* essential oil: Characterization and practical efficacy against fungal and aflatoxin contamination of stored chia seeds. *Food Bioscience* 42:101117. https://doi.org/10.1016/j.fbio.2021.101117

Donsì F, Cuomo A, Marchese E, Ferrari G (2014) Infusion of essential oils for food stabilization: Unraveling the role of nano-emulsion-based delivery systems on mass transfer and antimicrobial activity. *Innovative Food Science & Emerging Technologies* :22212–220.

Draz KA, Tabikha RM, Eldosouky MI, Darwish AA, Abdelnasser M (2022) Biotoxicity of essential oils and their nanoemulsions against the coleopteran stored product insect pests Sitophilus oryzae L. and Tribolium castaneum herbst. *International Journal of Pest Management* 70(4): 729–743. https://doi.org/10.1080/09670874.2022.2036862

Donsì F, Ferrari G (2016) Essential oil nano-emulsions as antimicrobial agents in food. *Journal of Biotechnology* :233106–120. https://doi.org/10.1016/j.jbiotec.2016.07.005

Feroz A, Shakoori FR, Riaz T, Shakoori AR (2020) Development of resistance in stored grain pest, *Trogoderma granarium* (Everts), against deltamethrin and its effective control by synergistic toxicity of bifenthrin and chlorpyrifos. *Journal of Stored Products Research* 88: 101673.

Flanagan J, Singh H (2006) Microemulsions: A potential delivery system for bioactives in food. *Critical Reviews in Food Science and Nutrition* :46221–237. https://doi.org/10.1080/10408690590956710

Friberg SE, Corkery RW, Blute IA (2011) Phase inversion temperature (PIT) emulsification process. *Journal of Chemical & Engineering Data* :564282–4290. https://doi.org/10.1021/JE101179S

Fu X, Ying Y (2014) Food safety evaluation based on near infrared spectroscopy and imaging: A review. *Critical Reviews in Food Science and Nutrition* 56:1913–1924. https://doi.org/10.1080/10408398.2013.807418

Gad HA, Al-Anany MS, Abdelgaleil SA (2020) Enhancement of the efficacy of spinosad for the control of *Sitophilus oryzae* by combined application with diatomaceous earth and *Trichoderma harzianum*. *Journal of Stored Products Research* 88: 101663.

Gadhave AD (2014) Nanoemulsions: formation, stability and applications. International Journal for Research in Science & Advanced Technologies. 23(2):38–43.

Gharenaghadeh S, Karimi N, Forghani S, Nourazarian M, Gharehnaghadeh S, Kafil HS (2017) Application of *Salvia multicaulis* essential oil-containing nano-emulsion against food-borne pathogens. *Food Bioscience* :19128–133.

Gharsan FN, Kamel WM, Alghamdi TS, Alghamdi AA, Althagafi AO, Aljassim FJ, Al-Ghamdi SN (2022) Toxicity of citronella essential oil and its nanoemulsion against the saw-toothed grain beetle *Oryzaephilus surinamensis* (Coleoptera: Silvanidae). *Industrial Crops and Products* 184: 15024. https://doi.org/10.1016/j.indcrop.2022.115024

Ghosh V, Mukherjee A, Chandrasekaran N (2014) Eugenol-loaded antimicrobial nano-emulsion preserves fruit juice against microbial spoilage. *Colloids and Surfaces B: Biointerfaces* :114392–397.

Giunti G, Campolo O, Laudani F, Zappalà L, Palmeri V (2021) Bioactivity of essential oil-based nano-biopesticides toward *Rhyzopertha dominica* (Coleoptera: Bostrichidae). *Industrial Crops and Products* 162: 113257.

Giunti G, Palermo D, Laudani F, Algeri GM, Campolo O, Palmeri V (2019) Repellence and acute toxicity of a nano-emulsion of sweet orange essential oil toward two major stored grain insect pests. *Industrial Crops and Products* 142: 111869. https://doi.org/10.1016/j.indcrop.2019.111869

Guru-Pirasanna-Pandi G, Adak T, Gowda B, Patil N, Annamalai M, Jena M (2018) Toxicological effect of the underutilized plant, *Cleistanthus collinus* leaf extracts against two major stored grain pests, the rice weevil, *Sitophilus oryzae*, and red flour beetle *Tribolium castaneum*. *Ecotoxicology and Environmental Safety* :154 92–99.

Guzmán E, Lucia A (2021) Essential oils and their individual components in cosmetic products. *Cosmetics* 8:114. https://doi.org/10.3390/cosmetics8040114

Gyawali R, Ibrahim SA (2014) Natural products as antimicrobial agents. *Food Control* :46412–429. https://doi.org/10.1016/j.foodcont.2014.05.047

Hadian M, Rajaei A, Mohsenifar A, Tabatabaei M (2017) Encapsulation of *Rosmarinus officinalis* essential oils in chitosan-benzoic acid nanogel with enhanced antibacterial activity in beef cutlet against *Salmonella typhimurium* during refrigerated storage. *Lebensmittel-Wissenschaft + Technologie* :84394–401. https://https://doi.org/10.1016/j.lwt.2017.05.075

Halnor V, Pande V, Borawake D, Nagare H (2018) Nanoemulsion: A novel platform for drug delivery system nanoemulsion: A novel platform for drug delivery system. *Journal of Materials Science & Nanotechnology* 6:104.

Hassanshahian M, Saadatfar A, Masoumipour F (2020) Formulation and characterization of nano-emulsion from *Alhagi maurorum* essential oil and study of its antimicrobial, antibiofilm, and plasmid curing activity against antibiotic-resistant pathogenic bacteria. *Journal of Environmental Health Science and Engineering* 18:1015–1027. https://doi.org/10.1007/s40201-020-00523-7

He F, Wang W, Wu M, Fang Y, Wang S, Yang Y, Ye C, Xiang F (2020) Antioxidant and antibacterial activities of essential oil from Atractylodes lancea rhizomes. *Industrial Crops and Products* 153:112552. https://doi.org/10.1016/j.indcrop.2020.112552

He Q, Guo M, Jin TZ, Arabi SA, Liu D (2021) Ultrasound improves the decontamination effect of thyme essential oil nano-emulsions against *Escherichia coli* O157: H7 on cherry tomatoes. *International Journal of Food Microbiology* 337:108936. https://https://doi.org/10.1016/j.ijfoodmicro.2020.108936

Heydari M, Amirjani A, Bagheri M, Sharifian I, Sabahi Q, et al (2020) Eco-friendly pesticide based on peppermint oil nanoemulsion: Preparation, physicochemical properties, and its aphicidal activity against cotton aphid. *Environmental Science and Pollution Research* :276667–6679. https://doi.org/10.1007/S11356-019-07332-Y/METRICS

Hussan K (2011) Nanoemulsion as a novel transdermal drug delivery system. *International Journal of Pharmaceutical Sciences and Research* 2:1938–1946.

Ilangovan M, Guna V, Hu C, Nagananda GS, Reddy N, et al (2018) *Curcuma longa* L. plant residue as a source for natural cellulose fibers with antimicrobial activity. *Industrial Crops and Products* :112556–560. https://doi.org/10.1016/j.indcrop.2017.12.042

Izquierdo P, Esquena J, Tadros TF, Dederen C, Garcia MJ, Azemar N, Solans C, et al (2001) Formation and stability of nano-emulsions prepared using the phase inversion temperature method. *Langmuir* :1826–30. https://doi.org/10.1021/LA010808C

Jadhav RP, Koli VW, Kamble AB, Bhutkar MA (2020) A review on nanoemulsion. *Asian Journal of Research in Pharmaceutical Science* :10103–108. https://doi.org/10.5958/2231-5659.2020.00020.X

Jafari SM, He Y, Bhandari B (2007) Production of sub-micron emulsions by ultrasound and microfluidization techniques. *Journal of Food Engineering* :82478–488. https://doi.org/10.1016/J.JFOODENG.2007.03.007

Jaiswal M, Dudhe R, Sharma PK (2015) Nanoemulsion: An advanced mode of drug delivery system. *3 Biotech* :5123–127. https://doi.org/10.1007/S13205-014-0214-0/METRICS

Jasmina H, Džana O, Alisa E, Edina V, Ognjenka R, et al (2017) Preparation of nanoemulsions by high-energy and lowenergy emulsification methods. *IFMBE Proc* :62317–322. https://doi.org/10.1007/978-981-10-4166-2_48/COVER

Jiang H, Zhong S, Schwarz P, Chen B, Rao J (2023) Antifungal activity, mycotoxin inhibitory efficacy, and mode of action of hop essential oil nano-emulsion against *Fusarium graminearum*. *Food Chemistry* 400:134016.

Jo YJ, Chun JY, Kwon YJ, Min SG, Hong GP, Choi MJ (2015) Physical and antimicrobial properties of trans-cinnamaldehyde nanoemulsions in water melon juice. *LWT-Food Science and Technology* 60:444–451. https://doi.org/10.1016/j.lwt.2014.09.041

Job JT, Visakh NU, Pathrose B, Alfarhan A, Rajagopal R, Thayyullathil J, Thejass P, Ramesh V, Narayanankutty A, et al (2024) Chemical composition and biological activities of the essential oil from *Citrus reticulata* blanco peels collected from agrowastes. *Chemistry & Biodiversity* 21:e202301223. https://doi.org/10.1002/cbdv.202301223

Ju J, Chen X, Xie Y, Yu H, Guo Y, Cheng Y, Qian H, Yao W (2019) Application of essential oil as a sustained release preparation in food packaging. *Trends in Food Science & Technology* :9222–32. https://doi.org/10.1016/j.tifs.2019.08.005

Ju J, Xu X, Xie Y, Guo Y, Cheng Y, Qian H, Yao W, et al (2018) Inhibitory effects of cinnamon and clove essential oils on mold growth on baked foods. *Food Chemistry* :240850–855. https://doi.org/10.1016/j.foodchem.2017.07.120

Kabalnov A (1998) Thermodynamic and theoretical aspects of emulsions and their stability. *Current Opinion in Colloid & Interface Science* :3270–275. https://doi.org/10.1016/S1359-0294(98)80071-X

Kamano HM (2022) Efficacy of plasma technology in eliminating fungi and aflatoxins in maize in Makueni and Baringo counties, Kenya. In: *erepository.uonbi.ac.ke*. http://erepository.uonbi.ac.ke/handle/11295/161529. Accessed 16 Dec 2023

Kavallieratos NG, Bonacucina G, Nika EP, Benelli G (2023) The type of grain counts: Effectiveness of three essential oil-based nanoemulsions against *Sitophilus oryzae*. *Plants* 12(4):813. https://doi.org/10.3390/plants12040813

Kavallieratos NG, Nika EP, Skourti A, Xefteri DN, Maggi F (2022) Piperitenone oxide-rich *Mentha longifolia* essential oil and its nanoemulsion to manage different developmental stages of insect and mite pests attacking stored wheat. *Industrial Crops and Products* 178:114600.

Kentish S, Wooster TJ, Ashokkumar M, Balachandran S, Mawson R, Simons L, et al (2008) The use of ultrasonics for nanoemulsion preparation. *Innovative Food Science & Emerging Technologies* :9170–175. https://doi.org/10.1016/J.IFSET.2007.07.005

Klang V, Matsko NB, Valenta C, Hofer F (2012) Electron microscopy of nanoemulsions: An essential tool for characterisation and stability assessment. *Micron* :4385–103.

Koroleva MY, Yurtov E V (2012) Nanoemulsions: The properties, methods of preparation and promising applications. *Russian ChemicalReviews*:8121–43.https://doi.org/10.1070/RC2012V081N01ABEH004219/XML

Sharma N, Bansal M, Visht S, Sharma PK, Kulkarni GT (2010) Nanoemulsion: A new concept of delivery system. Chronicles of Young Scientists. 1(2):2-6.

Kumar A, Li S, Cheng CM, Lee D (2015) Recent developments in phase inversion emulsification. *Industrial & Engineering Chemistry Research* :548375–8396. https://doi.org/10.1021/ACS.IECR.5B01122/ASSET/IMAGES/MEDIUM/IE-2015-01122T_0021.GIF

Kumar A, Singh PP, Gupta V, Prakash B (2020) Assessing the antifungal and aflatoxin B1 inhibitory efficacy of nanoencapsulated antifungal formulation based on a combination of *Ocimum spp.* essential oils. *International Journal of Food Microbiology* 330:108766. https://doi.org/10.1016/j.ijfoodmicro.2020.108766

Kumar M, Bishnoi RS, Shukla AK, Jain CP (2019) Techniques for formulation of nanoemulsion drug delivery system: A review. *Preventive Nutrition and Food Science* 24:225. https://doi.org/10.3746/PNF.2019.24.3.225

Kumar N, Verma A, Mandal A (2021) Formation, characteristics and oil industry applications of nanoemulsions: A review. *Journal of Petroleum Science and Engineering* 206:109042. https://doi.org/10.1016/J.PETROL.2021.109042

Kuntworbe N, Martini N, Shaw J, Al-Kassas R (2012) Malaria intervention policies and pharmaceutical nanotechnology as a potential tool for malaria management. *Drug Development Research* :73167–184. https://doi.org/10.1002/DDR.21010

Kuttithodi AM, Narayanankutty A, Visakh NU, Job JT, Pathrose B, Olatunji OJ, Alfarhan A, Ramesh V, et al (2023) Chemical composition of the *Cinnamomum malabatrum* leaf essential oil and analysis of its antioxidant, enzyme inhibitory and antibacterial activities. 12:940–940. https://doi.org/10.3390/antibiotics12050940

Li ZH, Cai M, Liu YS, Sun PL (2018) Development of finger citron (*Citrus medica* L. var. sarcodactylis) essential oil loaded nano-emulsion and its antimicrobial activity. *Food Control* :94317–323.

Liew SN, Utra U, Alias AK, Tan TB, Tan CP, Yussof NS (2020) Physical, morphological and antibacterial properties of lime essential oil nano-emulsions prepared via spontaneous emulsification method. *LWT—Food Science and Technology* 128. https://doi.org/10.1016/j.lwt.2020.109388

Lima L A, Ferreira-Sá PS, Garcia Jr. MD, Pereira VLP, Carvalho JCT, Rocha L, Cruz RA (2021) Nano-emulsions of the essential oil of *Baccharis reticularia* and its constituents as eco-friendly repellents against *Tribolium castaneum*. *Industrial Crops and Products* 162:113282. https://doi.org/10.1016/j.indcrop.2021.113282

Louni M, Shakarami J, Negahban M (2018) Insecticidal efficacy of nanoemulsion containing *Mentha longifolia* essential oil against *Ephestia kuehniella* (Lepidoptera: Pyralidae). *Journal of Crop Protection* 7(2):171–182.

Louni M, Shakarami J, Negahban M (2019) Study on insecticidal properties of nanoemulsion *Mentha longifolia* L. (Lamiaceae) essential oil against *Callosobruchus maculatus* (Fabricius)(Coleoptera: Chrysomelidae). *Journal of Entomological Society of Iran* 39(2): 151–163.

Maali A, Mosavian MTH (2013) Preparation and application of nanoemulsions in the last decade (2000–2010). *Journal of*

Dispersion Science and Technology :3492–105. https://doi.org/10.1080/01932691.2011.648498

Magan N, Hope R, Cairns V, Aldred D (2003) Post-harvest fungal ecology: Impact of fungal growth and mycotoxin accumulation in stored grain. In *Epidemiology of Mycotoxin Producing Fungi: Under the Aegis of COST Action 835 'Agriculturally Important Toxigenic Fungi 1998–2003', EU project (QLK 1-CT-1998-01380)*: 723–730.

Maurya A, Prasad J, Das S, Dwivedy AK (2021a) Essential oils and their application in food safety. *Frontiers in Sustainable Food Systems* 5. https://doi.org/10.3389/fsufs.2021.653420C:\Users\LENOVO\Downloads\Query Integration (1)\Query Integration\24064-0309-Ref Mismatch Report.docx - LStERROR_8

Maurya A, Singh VK, Das S, Prasad J, Kedia A, Upadhyay N, Dwivedy AK (2021b) Essential oil nano-emulsion as an eco-friendly and safe preservative: Bioefficacy against microbial food deterioration and toxin secretion, mode of action, and future opportunities. *Frontiers in Microbiology* 12:751062.

McClements DJ (2012) Nanoemulsions versus microemulsions: Terminology, differences, and similarities. *Soft Matter* 8:1719–1729. https://doi.org/10.1039/C2SM06903B

Mesbah HA, Massoud MA, Aajel MS, Abdelsalam NR, Adel MR (2023) Nano-formulation of *Pelargonium graveolens* essential oil: Physicochemical characterization and its bioactivity against rice weevil *Sitophilus oryzae* on stored wheat grain. *Egyptian Journal of Chemistry* 66(2): 459–469. https://doi.org/10.21608/ejchem.2022.119611.5375

Moghimi R, Aliahmadi A, McClements DJ, Rafati H (2016) Investigations of the effectiveness of nano-emulsions from sage oil as antibacterial agents on some foodborne pathogens. *Lebensmittel-Wissenschaft + Technologie* :7169–76. https://doi.org/10.1016/j.lwt.2016.03.018

Mondal P, Laishram R, Sarkar P, Kumar R, Karmakar R, Hazra DK, Banerjee K, Pal K, Choudhury A, et al (2022) Plant essential oil-based nanoemulsions: A novel asset in the crop protection arsenal. *Agricultural Nanobiotechnology: Biogenic Nanoparticles, Nanofertilizers and Nanoscale Biocontrol Agents* 325–353. https://doi.org/10.1016/B978-0-323-91908-1.00007-9C:\Users\LENOVO\Downloads\Query Integration (1)\Query Integration\24064-0309-Ref Mismatch Report.docx - LStERROR_9

Moradi A, Davati N, Emamifar A (2023) Effects of *Cuminum cyminum* L. essential oil and its nano-emulsion on oxidative stability and microbial growth in mayonnaise during storage. Food Science & Nutrition. https://doi.org/10.1002/fsn3.3457

Moraes-Lovison M, Marostegan LF, Peres MS, Menezes IF, Ghiraldi M, Rodrigues RA, Pinho SC (2017) Nanoemulsions encapsulating oregano essential oil: Production, stability, antibacterial activity and incorporation in chicken pâté. *LWT* :77233–240.

Morales D, Gutiérrez JM, García-Celma MJ, Solans YC (2003) A study of the relation between bicontinuous microemulsions and oil/water nano-emulsion formation. *Langmuir* :197196–7200. https://doi.org/10.1021/LA0300737

Mushtaq A, Mohd Wani S, Malik AR, Gull A, Ramniwas S, Ahmad Nayik G, Ercisli S, Alina Marc R, Ullah R, Bari A (2023) Recent insights into nano-emulsions: Their preparation, properties and applications. *Food Chemistry* X 18:100684. https://doi.org/10.1016/j.fochx.2023.100684

Mustafa IF, Hussein MZ (2020) Synthesis and technology of nanoemulsion-based pesticide formulation. *Nanomaterials* 10:1608. https://doi.org/10.3390/nano10081608

Narayanankutty A, Visakh NU, Sasidharan A, Pathrose B, Olatunji OJ, Al-Ansari A, Alfarhan A, Ramesh V, et al (2022) Chemical composition, antioxidant, anti-bacterial, and anti-cancer activities of essential oils extracted from *Citrus limetta* risso peel waste remains after commercial use. *Molecules* 27:8329. https://doi.org/10.3390/molecules27238329

Naseema A, Kovooru L, Behera AK, Kumar KP, Srivastava P, et al (2021) A critical review of synthesis procedures, applications and future potential of nanoemulsions. *Advances in Colloid and Interface Science* 287:102318. https://doi.org/10.1016/J.CIS.2020.102318

Nawaz A, Muhammad Sufyan, Muhammad Dildar Gogi, Muhammad Wajid Javed (2019) *Sustainable management of insect-pests*. Springer eBooks 287–335. https://doi.org/10.1007/978-3-030-23169-9_10

Nenaah GE (2014) Chemical composition, toxicity, and growth inhibitory activities of essential oils of three Achillea species and their nano-emulsions against *Tribolium castaneum* (Herbst). *Industrial Crops and Products* :53252–260. https://doi.org/10.1016/j.indcrop.2013.12.042

Nuryanti SP, Bulbarti L, Dulbari D, Sutrisno H, Sudrajat D, Yuriansyah Y, Priyadi P, Rahmadi R, Rochman F, Sari EY, Maharani JS (2023) Activity of nanoemulsion botanical insecticides from *Myristica fragrans* and *Jatropha curcas* essential oil against *Sitophilus zeamais*. *Biodiversitas Journal of Biological Diversity* 24(10).

Palermo D, Giunti G, Laudani F, Palmeri V, Campolo O (2021) Essential oil-based nano-biopesticides: Formulation and bioactivity against the confused flour beetle *Tribolium confusum*. *Sustainability* 13(17): 9746. https://doi.org/10.3390/su13179746

Pateiro M, Munekata PE, Sant'Ana AS, Domínguez R, Rodríguez-Lázaro D, Lorenzo JM (2021) Application of essential oils as antimicrobial agents against spoilage and pathogenic microorganisms in meat products. *International Journal of Food Microbiology* 337:108966. https://doi.org/10.1016/j.ijfoodmicro.2020.108966

Patel RP, Joshi JR (2012) An overview on nanoemulsion: A novel approach. *International Journal of Pharmaceutical Science and Research* 3.

Pathak M (2017) Nanoemulsions and their stability for enhancing functional properties of food ingredients. *Nanotechnology Applications in Food: Flavor, Stability, Nutrition and Safety* 87–106. https://doi.org/10.1016/B978-0-12-811942-6.00005-4

Pathania R, Kaushik R, Khan MA (2018) Essential oil nano-emulsions and their antimicrobial and food applications. *Current Research in Nutrition and Food Science* :6626–643. https://doi.org/10.12944/CRNFSJ.6.3.05

Paudel SK, Bhargava K, Kotturi H (2019) Antimicrobial activity of cinnamon oil nano-emulsion against *Listeria monocytogenes* and *Salmonella spp.* on melons. *Lebensmittel-Wissenschaft + Technologie* :111682–687. https://doi.org/10.1016/j.lwt.2019.05.087

Paul A, Visakh NU, Pathrose B, Mori N, Baeshen RS, Shawer R, et al (2024) Exploring the chemical characterization and insecticidal activities of *Curcuma angustifolia* roxb. leaf essential oils against three major stored product insects. *Saudi Journal of Biological Sciences* :31103986–103986. https://doi.org/10.1016/j.sjbs.2024.103986

Pavoni L, Perinelli DR, Bonacucina G, Cespi M, Palmieri GF (2020) An overview of micro- and nanoemulsions as vehicles for essential oils: Formulation, preparation and stability. *Nanomaterials* 10(1):135.

Peng F, Ke Y, Zhao Y, Hu X, Zhao X, et al (2019) The influence of organically intercalated montmorillonites on the interfacial tension and structure of oil-in-water nanoemulsions. *RSC Advances* :913378–13385. https://doi.org/10.1039/C8RA10595B

Pongsumpun P, Iwamoto S, Siripatrawan U (2020) Response surface methodology for optimization of cinnamon essential oil nano-emulsion with improved stability and antifungal activity. *Ultrasonics Sonochemistry* 60:104604. https://doi.org/10.1016/j.ultsonch.2019.05.021

Rajendran S, Gunasekaran N (2002) The response of phosphine-resistant lesser grain borer *Rhyzopertha dominica* and rice weevil *Sitophilus oryzae* in mixed-age cultures to varying concentrations of phosphine. *Pest Management Science: Formerly Pesticide Science* 58:277–281. https://doi.org/10.1002/ps.446

Rajkovic A, Jovanovic J, Monteiro S, Decleer M, Andjelkovic M, Foubert A (2020) Detection of toxins involved in foodborne diseases caused by Gram-positive bacteria. *Comprehensive Reviews in Food Science and Food Safety* 19:1605–1657. https://doi.org/10.1111/1541-4337.12571.

Rao J, Chen B, McClements DJ (2019) Improving the efficacy of essential oils as antimicrobials in foods: Mechanisms of action. *Annual Review of Food Science and Technology* : 10365–387. https://doi.org/10.1146/annurev-food-032818-121727

Rao S, Shao J (2008) Self-nanoemulsifying drug delivery systems (SNEDDS) for oral delivery of protein drugs: I. Formulation development. *International Journal of Pharmaceutics* :3622–9. https://doi.org/10.1016/J.IJPHARM.2008.05.018

Ribeiro-Santos R, Andrade M, Sanches-Silva A, de Melo NR (2018) Essential oils for food application: Natural substances with established biological activities. *Food and Bioprocess Technology* 11:43–71.

Rout S, Tambe S, Deshmukh RK, Mali S, Cruz J, Srivastav PP, Amin PD, Gaikwad KK, de Aguiar Andrade EH, de Oliveira MS, et al (2022) Recent trends in the application of essential oils: The next generation of food preservation and food packaging. *Trends in Food Science & Technology* :129421–439. https://doi.org/10.1016/j.tifs.2022.10.012

Sadeq ZA (2020) Review on nanoemulsion: Preparation and evaluation. *International Journal of Drug Delivery Technology* :10187–189.

Sahoo M, Panigrahi C, Aradwad P (2022) Management strategies emphasizing advanced food processing approaches to mitigate food borne zoonotic pathogens in food system. *Food Frontiers*. https://doi.org/10.1002/fft2.153

Sanguansri P, Augustin MA (2006) Nanoscale materials development—a food industry perspective. *Trends in Food Science and Technology* :17547–556. https://doi.org/10.1016/J.TIFS.2006.04.010

Sarker A, Shimu IJ, Tuhin RH, Raju AA (2015) Nanoemulsion: An excellent mode for delivery of poorly soluble drug through different routes. *Journal of Chemical and Pharmaceutical Research* :7966–976.

Sayyar Z, Jafarizadeh-Malmiri H (2020) Process intensification for curcumin nanodispersion preparation using subcritical water—optimization and characterization. *Chemical Engineering and Processing—Process Intensification* 153: 107938. https://doi.org/10.1016/J.CEP.2020.107938

Abouelkassem. Sh, Abdelrazeik AB, Rakha OM (2015) Nanoemulsion of jojoba oil, preparation, characterization and insecticidal activity against *Sitophilus oryzae* (coleoptera: Curculionidae) on wheat. *International Journal of Agricultural Innovation and Research* :472–75.

Shafiq S, Shakeel F, Talegaonkar S, Ahmad FJ, Khar RK, Ali M, et al (2007) Development and bioavailability assessment of ramipril nanoemulsion formulation. *European Journal of Pharmaceutics and Biopharmaceutics* :66227–243. https://doi.org/10.1016/J.EJPB.2006.10.014

Sharma A, Shukla A, Attri K, Kumar M, Kumar P, Suttee A, Singh G, Barnwal RP, Singla N, et al (2020) Global trends in pesticides: A looming threat and viable alternatives. *Ecotoxicology and Environmental Safety* 201:110812. https://doi.org/10.1016/j.ecoenv.2020.110812

Shehabeldine AM, Doghish AS, El-Dakroury WA, Hassanin MM, Al-Askar AA, AbdElgawad H, Hashem AH (2023) Antimicrobial, antibiofilm, and anticancer activities of *Syzygium aromaticum* essential oil nano-emulsion. *Molecules* 28(15):5812.

Shinoda K, Saito H (1968) The effect of temperature on the phase equilibria and the types of dispersions of the ternary system composed of water, cyclohexane, and nonionic surfactant. *Journal of Colloid and Interface Science* :2670–74. https://doi.org/10.1016/0021-9797(68)90273-7

Singh A, Chaudhari AK, Das S, Singh VK, Dwivedy AK, Shivalingam RK (2020) Assessment of preservative potential of *Bunium persicum* (Boiss) essential oil against fungal and aflatoxin contamination of stored masticatories and improvement in efficacy through encapsulation into chitosan nanomatrix. *Environmental Science and Pollution Research* :2727635–27650. https://doi.org/10.1007/s11356-020-08872-4

Singh KK, Vingkar SK (2008) Formulation, antimalarial activity and biodistribution of oral lipid nanoemulsion of primaquine. *Interrnational Journal of Pharmaceutics* :347136–143. https://doi.org/10.1016/J.IJPHARM.2007.06.035

Sneha K, Kumar A (2022) Nanoemulsions: Techniques for the preparation and the recent advances in their food applications. *Innovative Food Science & Emerging Technologies* 76:102914. https://doi.org/10.1016/J.IFSET.2021.102914

Sokolov YV (2014) Nanoemulsion formation by low-energy methods: A review. *News of Pharmacy* 3:16–19.

Stejskal V, Hubert J, Aulicky R, Kucerova Z (2015) Overview of present and past and pest-associated risks in stored food and feed products: European perspective. *Journal of Stored Products Research* :64122–132. https://doi.org/10.1016/j.jspr.2014.12.006

Stoytcheva M (2011) *Pesticides in the modern world: Pesticides use and management*. IntechOpen, London. https://doi.org/10.5772/17170

Sugumar S, Singh S, Mukherjee A, Chandrasekaran N (2016) Nano-emulsion of orange oil with non-ionic surfactant produced emulsion using ultrasonication technique: Evaluating against food spoilage yeast. *Applied Nanoscience* :6113–120.

Sulhath TAA, Visakh NU, Pathrose B, George SB (2024) Investigating the insecticidal properties of essential oils extracted from wild turmeric (*Curcuma aromatica* salisb) leaves waste against three key stored product pests. *Sustainable Chemistry and Pharmacy* :38101482–101482. https://doi.org/10.1016/j.scp.2024.101482

Sutradhar KB, Amin L (2013) Nanoemulsions: Increasing possibilities in drug delivery. *European Journal of Nanomedicine* : 597–110. https://doi.org/10.1515/EJNM-2013-0001/MACHINE READABLECITATION/RIS

Tadros T, Izquierdo P, Esquena J, Solans C (2004) Formation and stability of nano-emulsions. *Advances in Colloid and Interface Science* 108–109:303–318. https://doi.org/10.1016/J.CIS.2003.10.023

Tadros TF (2013) *Encyclopedia of colloid and interface science*. Springer Reference, Heidelberg.

Tang C, Chen J, Zhang L, Zhang R, Zhang S, Ye S, Zhao Z, Yang D (2020) Exploring the antibacterial mechanism of essential oils by membrane permeability, apoptosis and biofilm formation combination with proteomics analysis against methicillin-resistant *Staphylococcus aureus*. *International Journal of Medical Microbiology* 310:151435. https://doi.org/10.1016/j.ijmm.2020.151435

Tomić A, Šovljanski O, Erceg T (2023) Insight on incorporation of essential oils as antimicrobial substances in biopolymer-based active packaging. *Antibiotics* 12:1473. https://doi.org/10.3390/antibiotics12091473

United States, Office of Mission to Planet Earth (1995) *Scientific assessment of ozone depletion 1994* (No. 37). World Meteorological Organization.

Venkadesaperumal G, Rucha S, Sundar K, Shetty PH (2016) Anti-quorum sensing activity of spice oil nano-emulsions against food-borne pathogens. *LWT-Food Science and Technology* :66225–231.

Visakh NU, Pathrose B, Chellappan M, Ranjith MT, Sindhu PV, Mathew D, et al (2023a) Extraction and chemical characterisation of agro-waste from turmeric leaves as a source of bioactive essential oils with insecticidal and antioxidant activities. *Waste Management* :1691–10. https://doi.org/10.1016/j.wasman.2023.06.030

Visakh NU, Pathrose B, Chellappan M, Ranjith MT, Sindhu PV, Mathew D, et al (2022a) Chemical characterisation, insecticidal and antioxidant activities of essential oils from four *Citrus spp.* fruit peel waste. *Food Bioscience* 50:102163. https://doi.org/10.1016/j.fbio.2022.102163

Visakh NU, Pathrose B, Narayanankutty A (2023b) Characterization of secondary metabolites from the leaves of curry leaf (*Murraya koenigii* L.) essential oils with insecticidal activities against stored product insects. *Biocatalysis and Agricultural Biotechnology* 54:102973. https://doi.org/10.1016/j.bcab.2023.102973

Visakh NU, Pathrose B, Narayanankutty A, Alfarhan A, Ramesh V, et al (2022b) Utilization of pomelo (*Citrus maxima*) peel waste into bioactive essential oils: Chemical composition and insecticidal properties. *Insects* 13:480. https://doi.org/10.3390/insects13050480

Wakil W, Kavallieratos NG, Usman M, Gulzar S, El-Shafie HA, et al (2021) Detection of phosphine resistance in field populations of four key stored-grain insect pests in Pakistan. *Insects* 12(4):288. https://doi.org/10.3390/insects12040288

Wan J, Zhong S, Schwarz P, Chen B, Rao J (2019) Physical properties, antifungal and mycotoxin inhibitory activities of five essential oil nano-emulsions: Impact of oil compositions and processing parameters. *Food Chemistry* :291199–206.

Wilson RJ, Li Y, Yang G, Zhao CX (2022) Nanoemulsions for drug delivery. *Particuology* :6485–97. https://doi.org/10.1016/j.partic.2021.05.009

Windsor P, Pinniger D, Bacon L, Child B, Harris K, et al (2015) *Integrated pest management for collections: Proceedings of 2011: A pest Odyssey, 10 years later*. English Heritage.

Winska K, Maczka W, Łyczko J, Grabarczyk M, Czubaszek A, Szumny A (2019) ESSENTIAL oils as antimicrobial agents—myth or real alternative? *Molecules* 24:2130. https://doi.org/10.3390/molecules24112130

Ya-Ali P, Yarahmadi F, Mehrnia MA (2020) Efficacies of two nano-formulations of Tasmanian blue gum essential oil to control *Callosobruchus maculatus*. *Journal of Economic Entomology* 113(3): 1555–1562. https://doi.org/10.1093/jee/toaa069

Yang K, Liu A, Hu A, Li J, Zen Z, Liu Y (2021) Preparation and characterization of cinnamon essential oil nanocapsules and comparison of volatile components and antibacterial ability of cinnamon essential oil before and after encapsulation. *Food Control* 123:107783. https://doi.org/10.1016/j.foodcont.2020.107783

Yilmaz E, Borchert HH (2005) Design of a phytosphingosine-containing, positively-charged nanoemulsion as a colloidal carrier system for dermal application of ceramides. *European Journal of Pharmaceutics and Biopharmaceutics* :6091–98. https://doi.org/10.1016/J.EJPB.2004.11.009

Yousef N, Niloufar M, Elena P (2019) Antipathogenic effects of emulsion and nano-emulsion of cinnamon essential oil against Rhizopus rot and grey mold on strawberry fruits. *Foods and Raw Materials* :7210–216. https://https://doi.org/10.21603/2308-4057-2019-1-210-216

Yukuyama MN, Ghisleni DDM, Pinto TJA, Bou-Chacra NA (2016) Nanoemulsion: Process selection and application in cosmetics—a review. *International Journal of Cosmetic Science* :3813–24. https://doi.org/10.1111/ICS.12260

Zamuz S, Munekata PES, Meiselman HL, Zhang W, Xing L, Lorenzo JM (2021) Consumer and market demand for sustainable food products. In: *Sustainable production technology in food*. Elsevier, pp. 23–35. https://doi.org/10.1016/B978-0-12-821233-2.00008-3

Zhang S, Zhang M, Fang Z, Liu Y (2017) Preparation and characterization of blended cloves/cinnamon essential oil nano-emulsions. *Lebensmittel-Wissenschaft + Technologie* :75316–322. https://doi.org/10.1016/j.lwt.2016.08.046

Zhang Y, Liao Y, Tang Q, Lin J, Huang, P (2021) Biomimetic nanoemulsion for synergistic photodynamic-immunotherapy against hypoxic breast tumor. *Angewandte Chemie International Edition* 60:10647–10653. https://doi.org/10.1002/anie.202015590

Zhu Y, Li C, Cui H, Lin L (2021) Encapsulation strategies to enhance the antibacterial properties of essential oils in food system. *Food Control* 123:107856. https://doi.org/10.1016/j.foodcont.2020.107856

Zielińska-Błajet M, Feder-Kubis J (2020) Monoterpenes and their derivatives—recent development in biological and medical applications. *International Journal of Molecular Sciences* 21:7078. https://doi.org/10.3390/ijms21197078C:\Users\LENOVO\Downloads\Query Integration (1)\Query Integration\24064-0309-Ref Mismatch Report.docx - LStERROR_16

17 Nanotechnology-Driven Delivery Systems in Revolutionizing Food Preservation

Ghania Qindeel[a] *and Muhammad Shahzaib*[b]

[a]Department of Industrial Biotechnology, Government College University, Lahore, Punjab, Pakistan

[b]Institute of Tropical Agriculture and Forestry, Hainan University, Haikou, Hainan, People's Republic of China

17.1 INTRODUCTION

Nanotechnology is the study of to design, development, characterization, and application of material structures and systems with unique features resulting from the extremely small size of the nanometer, on a scale of 1–100 nm. One important feature of nanomaterials is that their structural components are less than 1 micrometer in at least one dimension (Dutta and Goswami 2020).

In the previous few years, food nanotechnology has become a booming industry. This increase is motivated by the assumption that these materials, with their large surface area to volume ratio, can improve the bioavailability of active compounds, enable controlled or target release, enhance sensory aspects, and so on. Numerous nanotechnology projects in the late 1990s and the advent of food-grade additives ideal for nanoparticle production have led to the majority of the growth in the food nanotechnology domain since the year 2000 (Ahamed et al. 2014). The food industry's nanotechnology product market is approaching $1 billion now, with over $20 billion in potential growth over the next decade. The majority of this growth is attributable to nanoparticle coatings for packaging technologies, health-promoting products, and drinks (Alfadul and Elneshwy 2010). Recently, researchers have started looking for ways to use nanomaterials in the food and drink business. According to Neethirajan and Jayas (2011), there are an estimated 400 companies worldwide working on nanotechnology's potential uses in the food and packaging industries. Nanomaterials in food can be naturally occurring or artificially added. Nanomaterials that have been purposefully added to food might come from naturally occurring or manufactured sources (Figure 17.1). Casein micelles in milk and pectin nanostructures in fruit both contain nanoscale components, with dimensions ranging from 300 to 400 nm and polymer chain lengths of 100 to 400 nm, respectively (Neethirajan and Jayas 2011).

Additionally, nanoparticles can encapsulate vitamins or minerals, making them a viable carrier for these nutrients. Nutrients can now be delivered to food, drink, and supplement products through nanomaterials, all without altering their flavor or visual appeal. According to the study, nanocapsules can transport vitamins from the human stomach into the bloodstream (Nguyen et al. 2014). But, insecticides and food contact packaging both pose risks of nanomaterial migration into food, which could lead to accidental contamination. Nanotechnology, which provides a new lightweight material with stronger packaging barriers, allows for the protection of food quality during transportation, the extension of the freshness of fruits and vegetables during storage, and the preservation of meats or poultry against viruses. Nanoparticles, a novel alternative additive source for increasing the polymeric properties of packaging materials, have recently found their way into various types of food contact containers and packaging (Belluco et al.). To give just a few examples, certain nanomaterials can improve barrier, mechanical, and heat resistance qualities. By following standard procedures, polymer blends can have their barrier and mechanical properties improved (McClements et al. 2016).

The gap between food waste and methods for preserving food has not been adequately investigated. Nanotechnology has several potential uses in the food industry, including the creation of new flavorings, additives, and supplements as well as the enhancement of food quality and safety (Singh Sekhon 2014). Packaging materials with improved thermal and mechanical qualities as well as increased safety can be made using nanotechnology in the food sector. A good example of this is the usage of nanosensors in food packaging systems, which notify buyers when their food has gone bad. One further usage of nanotechnology is in the production of nutritious meals. Because of the great variety of nanostructures that are available and their distinctive characteristics, foods, and packaging materials that incorporate them have the potential to enhance the nutritional content of foods (Sekhon 2010).

Surely, the food business is changing due to nanotechnology. First, nanostructures have the potential to enhance food quality. Second, they can be used for bioactive fortification. Third, they can be used to control the release of bioactive compounds through nanocarrier encapsulation. Fourth,

DOI: 10.1201/9781032677422-17

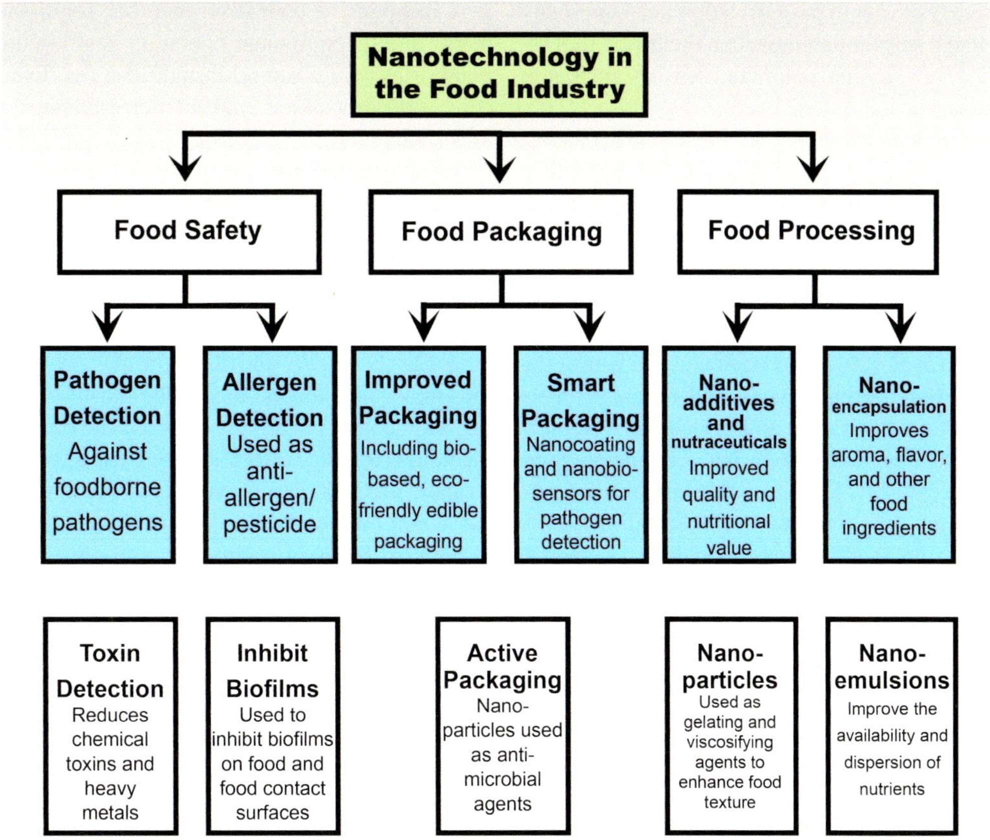

FIGURE 17.1 Application of nanotechnology in the food industry.

they can alter the structures and textures of food. Finally, intelligent packaging systems can detect and neutralize biochemical, microbiological, and chemical changes. When it comes to making food and making it healthier, there are a lot of different nanomaterials that can be used (Kumar 2015). Protein nanoparticles with desired functional characteristics can be assembled in food substances because of protein solubility, which is why they find application in food product manufacturing (Berezin 2015). Current food nanosensing and nanostructured ingredient applications are the mainstays of nanotechnology's impact on the industry. Food formulation and food packaging are examples of nanostructured food ingredients. Food nano detection allows for more accurate evaluations of safety and quality used to achieve more favorable outcomes. Nanopolymers are gradually displacing more traditional materials used for food packaging (Cho and Jones 2019).

The food business may be able to produce innovative functional foods because of the cutting-edge properties of nanotechnology. The creation of novel engineering techniques based on nanoparticles to regulate the distribution of nutrients and bioactive substances in functional foods is another potential application for nanotechnology. The food industry may find new uses for nanotechnology in the development of nanopackages, nanosensors, nanocarriers, nanotubes, and nanocapsules (Sothornvit 2019).

With an emphasis on the toxicity and health concerns associated with the intake, this chapter showcases current advancements in nanobiotechnology aimed at improving the safety and/or quality of food. The main points covered in this chapter are the most recent developments and trends in food preservation methods, as well as how these methods are applied to various foodstuffs, including fruits, vegetables, liquids, and spices. Traditional and cutting-edge technological developments in the aforementioned domains are covered extensively in the chapter. Nanotechnology and its potential uses in the food, agricultural, and packaging industries along with future trends also receive extensive attention

17.2 PREVALENT FOOD PRESERVATION TECHNOLOGIES

17.2.1 Thermal Treatment

Among the more recent methods for preserving food, heat or thermal treatment stands out. The method has been successfully used for a long time in many different areas of the food industry, including dairy, fruits and vegetables, and baking. In most cases, this method calls for heating items to a temperature of 75–90 °C or more for 25–30 seconds. Studies that looked at the preservation enhancement of apple juice beverages through pasteurization and heat treatment of maize had a substantial impact on the flavor, digestibility, glycemic index, aroma, color, and sensory attributes (Singh et al. 2017).

Some microorganisms are killed off when food is heated. On the other hand, a large body of research has found that there are nutritional losses, energy wastages, flavor modifications, and a decrease in the food matrix. Two investigations found that treatment differentiating light and dark honey affected their physicochemical characteristics, antioxidant activity, and nutritional variances (Ahmad et al. 2017).

Codina-Torrella et al. (2017) and de Souza and Kawaguti (2021) found that liquid foods, juices, and beverages also cause responses of gelatinization and browning. Research into the effects of heat on food storage has been ongoing for some time, and recent years have seen substantial advancements made by combining electrical and thermal methods, thanks to tweaks and modifications to older technology. The food industry has been greatly influenced by several methods that have led to innovations, such as electroplasmolysis, ohmic heating, and microwave heating of meals (Sridhar et al. 2021).

17.2.2 Freezing

The sensory features and nutritional qualities of leafy greens, spices, and dairy products can be preserved by cooling and freezing the food (El-Sayed and Youssef 2019). Freezing methods include air blast, cryogenic, immersion, and direct contact, as well as high pressure, ultrasound, electromagnetic disturbance, and dehydration. Direct contact and air blast freezing are commonly used methods. Heat transmission is the key mechanism that allows for the cooling and freezing processes (Chopra, Sodhi, and Lücker 2021).

Cooling is the process by which the thermal energy contained in food and its packaging is transferred to the environment. Cooling or freezing rates are, thus, significantly affected by thermal diffusivity and thermal conductivity. The advent of ready-to-eat foods that specifically target consumer demands has led to a surge in interest in storage technology in recent years. Properly packaged foods kept at a cold temperature will always keep food safe by preventing microbes from entering. Although freezing and cooling are effective on their own, there are several factors to consider, including the amount of time it takes to chill, the uneven rate of ice crystal formation, the cost of storage, and the environment. Thermodynamic and fluid flow patterns were investigated with computational fluid dynamics and three-dimensional mathematical models with various food recipes. This allowed researchers to find ways to reduce the impact of the problem (Gokularaman et al. 2017).

17.2.3 Ultrasound

The process of ultrasound treatment entails penetrating food ingredients with sound waves of very high frequency and intensity. The efficiency technology was selected since it is inexpensive compared to other sophisticated instruments and its equipment operation is straightforward. The many uses of ultrasound in industries as diverse as health care, food processing, and medicine attest to the technology's adaptability (Lagnika, Zhang, and Mothibe 2013). Ultrasonic radiation is used in this technique to pass through the desired solution. The solid particles in the solution are disrupted and eventually break apart as a result of this action, which allows the solvent to diffuse into them. Bear in mind that you should not vary the intensity of the procedure. This is because cavitation, a phenomenon described by Fu et al. (2020) and Khan et al. (2020), occurs when intramolecular forces break the particle–particle bonding and solvent penetrates between the molecules. Factors including capillary effect, swelling capacity, enhanced penetration, and cell disruption can further improve ultrasound extraction (Wu et al. 2018). This is far better as compared to the challenges in traditional food preservation methods (Table 17.1).

17.3 PRINCIPLES OF NANOTECHNOLOGY DELIVERY SYSTEMS

17.3.1 Nanomaterials Used in Food Preservation

In today's food market, packaging is crucial. Yet the most important part of food preservation is making sure the quality stays high from manufacturing to consumption. Food packaging has a lot of room to improve its efficiency through the application of nanotechnology. Nanoparticles are produced and used extensively in the industry due to their ability to encapsulate active compounds and enhance functionality, stability, and bioavailability (Jahangirian et al. 2017). The desirable properties and practical uses of nanofillers make them ideal for use in food packaging. Nanofillers have many uses in biosensors, including protecting food from the elements, enhancing the functionality of packaging materials, and opening up new avenues for innovation in the food packaging market (Benoit et al. 2014).

Nanotechnology opens up a world of possibilities for enhancing the nutritional content and flavor of food. The duodenum and stomach have an extremely acidic environment and enzymes make short work of most bioactive molecules, such as proteins, carbs, lipids, and vitamins. The major reasons for recycling nanoencapsulation technologies are to keep a culinary equilibrium and to gradually release tastes. Anthocyanins and other highly reactive and unstable plant pigments are encased in nanoparticles. To improve the photostability and thermal stability of recombinant soybean kernels, chemicals such as cyanidin-3-O-glucoside can be enclosed within their core cavities (Wang, Hu, and Siahaan 2016). The presence of ferritin, a subunit of H-2, improves photostability and thermal stability (rH-2). The food industry rarely uses rutin, even though it is a dietary flavonoid with several important pharmacological characteristics, because of how poorly it dissolves in water. As compared

TABLE 17.1
Challenges in Traditional Food Preservation Methods

Preservation Method	Challenges	References
Salting and Drying	—Loss of nutrients, especially vitamins and minerals—Potential for microbial contamination during processing and storage—High salt content can be detrimental to health—Energy-intensive process, particularly for drying	(Brandelli, Pinilla, and Lopes 2019)
Pickling and Fermentation	—Inconsistent quality and safety due to variations in fermentation conditions—Risk of contamination with undesirable microorganisms—Potential for over-fermentation leading to off-flavors and textures— Limited shelf life compared to some modern methods	(Rattanapanone et al. 2001)
Freezing	—Formation of ice crystals can damage food texture and appearance—Loss of nutrients, particularly vitamin C—Requires specialized equipment and energy for freezing and storage—Limited shelf life for some foods	(Bahrami et al. 2019a)
Sugaring and Canning	—High sugar content can be detrimental to health—Potential for botulism if canning process is not properly controlled—Loss of nutrients, particularly vitamins and antioxidants—May not be suitable for all types of foods	(Yildirim, Mavi, and Kara 2001)
Smoking	—Formation of polycyclic aromatic hydrocarbons (PAHs) during the smoking process, which are potential carcinogens—Risk of contamination with foodborne pathogens—Unpleasant smoke flavor may not be appealing to all consumers—Limited shelf life compared to some modern methods	(Tongo, Ogbeide, and Ezemonye 2017)

to unattached rutin, ferritin-confined rutin had better heat, UV radiation, and hardness properties after being encased in ferritin nanocages. Because of their usefulness in boosting bioavailability and water dispersion, nanoemulsions are increasingly being used to produce bioactive chemicals that are lipid-soluble (Malik, Muhammad, and Waheed 2023).

In addition to enabling them to fight that difficult environment, bioactive multifarious encapsulation also allows them to willingly adapt to food commodities. Utilizing non-capsulated bioactive chemicals, which are less soluble, makes achieving this phenomenon very difficult. To enhance the absorption of drugs, vitamins, and delicate micronutrients, tiny edible capsules coated with nanoparticles are being used in common foods (Hongcai Zhang and Chen 2019).

17.3.2 Properties of Nanomaterials

Food product safety is a major concern in the food industry. Over 45% of packaged and processed foods are vulnerable to contamination and deterioration, per a recent survey. Many problems about the safety and quality of food products have been resolved thanks in part to nanotechnology. Currently, a variety of nanomaterials and nanodevices, including temperature-time indicators, liposomal nanovesicles, polymeric nanoparticles, and nano-loaded emulsions, are used to improve food quality by extending shelf life, detecting an allergen, the presence of chemicals and heavy metals in food, and sensing freshness (Savolainen et al. 2010). Various nanomaterials can be coupled with various nanosensors and nano tracers to detect a broad variety of contaminants and degradants that impact food quality. According to Youssef et al. (2019), these include immunomagnetic liposomes, silicon nanorods, magnetic beads, quantum dots, palladium nanoparticles, single- and multiwalled carbon nanotubes, and gold nanoparticles. Due to its speed and accuracy, radio frequency identification (RFID) is a relatively new technology that has shown significant promise in several applications relating to supply chain management and food engineering. By tracking the origin of pollutants in different food products, RFID technology may also benefit food companies in terms of safety and security. This chapter quickly discusses the many nanomaterials and devices that are used to ensure food safety and quality (Nielsen 1967).

17.3.3 Nanomaterials to Prevent the Formation of Biofilms

Because biofilms are made of densely adhering bacterial cells and an impenetrable polymeric extracellular matrix, they present a challenge to the food business. They arise from the Van der Waals forces that bind free-floating bacteria to a substrate surface, causing biofouling, bio corrosion, and interference with food processing. Many different kinds of nanomaterials help prevent the formation of biofilms in food products. For example, filter membranes composed of nanofibers are highly effective in preventing the formation of biofilms in many bacterial strains that are resistant to many drugs (Fernández-Pan et al. 2010). Furthermore, numerous scientific publications have extensively shown the use of various nanoparticles, including silver oxide, in antifungal and antibacterial creation. In an industrial setting, *Bacillus subtilis* fermentation results in high-value products, but it also prevents food products from being processed by forming biofilms. Ranmadugala et al. assessed the efficacy of superparamagnetic iron oxide nanoparticles, both coated and naked, against *B. subtilis*. Similarly, *K. pneumonia* and *S. aureus* were unable to produce biofilms when gold nanoparticles conjugated with chlorhexidine were present (Espitia et al. 2012).

17.3.4 Nanoparticles and Nanostructures

Novel properties of nanoparticles open up a world of possibilities for the food industry. Adding different types of functional nanostructures to food as building blocks can offer it new functionalities and creative structures. Nanoemulsions, nanoliposomes, nanofibers, and nanoparticles make up these. Several of these structures have been described by Weiss, along with their present and future uses in the food industry. Nanomaterials made of organic and inorganic substances are employed in food preparation, according to the available data. Inorganic, surface functionalized, and organic engineered nanomaterials (ENMs) are the three main categories of ENMs that can be found in nanofood products. Some examples of inorganic nanomaterials that have potential use in food, food additives, packaging, or storage include enantiomeric nanomaterials (ENs) of transition metals, alkaline earth metals, nonmetals, and silicates (Buzea and Pacheco 2017). Titanium dioxide is another EN that has the potential to be useful in the food industry. Food packaging is where metal oxide ENs are most commonly used. Nanosilver is being used in an increasing range of consumer goods, including water, food contact surfaces, health food, and packaging materials. While other ENs have found use in a variety of industries, nanosilver's antibacterial, anti-odorant, and (so-called) health supplement uses have far overtaken them. Food packaging and other surfaces that come into touch with food are two areas where amorphous nano-silica finds use. Nanoselenium has been introduced as a green tea beverage, which is said to have many health benefits due to enhanced selenium absorption (Rauscher et al. 2019).

Nano-calcium salts are the subject of patent applications that aim to incorporate them into gum. Additionally, some supplements contain nanomagnesium and nano-calcium salts. To purify polluted water, you can take nano-iron, a nutritional supplement. The theory goes that it can remove organic pollutants and microbiological illnesses from water. Customers will be able to cut back on salt use thanks to the widespread coverage of a small amount of nano salt on food surfaces. Researchers are looking at nano salt, a soluble nanomaterial. Some of the food products that are being developed at Wageningen University in Holland using nanotechnology include nano milk with a cola flavor and nano mayonnaise that has less fat. Thanks to nanotechnology, "smart" packaging could extend the shelf life of food and allow it to be transported farther (Dube et al. 2010).

The use of "smart" packaging with nanosensors and antibacterial activators will allow supermarkets to keep food for increasingly longer durations before selling it. Foods will have a longer shelf life thanks to smart packaging that can detect when they are deteriorating and unleash nano-antimicrobials. One possible use for nanosensors is as electronic barcodes, which might be embedded in food as tiny, almost invisible chips. It is critical to find and describe nano-delivery mechanisms in food to understand their potential benefits and risks. Methods for the nano-delivery of food based on lipids, proteins, and polysaccharides have been well-documented. The analytical tools available today can detect and characterize these food product delivery systems (Jemil et al. 2014). There are now three main types of analytical methods: separation procedures, imaging techniques, and characterization procedures. There have been articles discussing the most recent advances in nanosensing-based analytical nanotechnology and how they pertain to the food business and food analysis. A new method of feature selection is being used by MS-based electronic nose applications. To prove how well it works, we used two MS e-nose databases. Some have claimed that an electronic nose can pick up on fruit aromas (Pathakoti, Manubolu, and Hwang 2017).

17.3.5 Nanosensors as Emerging Devices in the Food Industry

Nanosensors are bioanalytical devices that employ an integrated system design to leverage a wide range of nanostructured materials and biological receptors. The food industry significantly relies on nanosensors, which have recently received widespread attention due to their rapid detection effectiveness, cost-efficiency, and integrity. Nanosensors have exceptional sensitivity and specificity, making them appropriate for integration with a wide range of analytes. These devices have an impressive surface-to-volume ratio and exceptional electric and optical properties as a result of conjugation with a variety of nanomaterials, including carbon nanotubes, semiconductor nanoparticles, nanorods, nanowires, nano biofilms, nanofibers, and quantum dots (Drusch and Mannino 2009).

17.3.6 Nanobarcodes for Product Authenticity

Worldwide, two-dimensional barcodes are used to visually authenticate products; yet they are easily broken, forged, and manipulated. To address these problems, distinct invisible barcodes based on nanoparticles and nanodisks have been created recently to verify the legitimacy of different food products. A Raman microscope can be used to scan the nanodisk barcode, which can take the shape of a silver nanodisk code, a linear gold nanodisk array, or a silver–gold heterodimer nanodisk (Shi et al. 2019). Nanodisks can be further enhanced by functionalizing and combining them with metal composites. The future of food safety and quality assurance is in the development of effective nanobarcode systems powered by nanotechnology (Hongcai Zhang and Chen 2019).

17.4 NANOTECHNOLOGY APPLICATIONS IN FOOD PRESERVATION

17.4.1 Antimicrobial Nanomaterials for Combating Foodborne Pathogens

Major threats to food safety and quality come from the increasing demand for "fresh" foods that are minimally processed, easy to prepare, and ready to eat, as well as from the globalization of food trade and delivery. A quest for novel approaches to inhibiting microbial development in foods while preserving their quality, freshness, and safety has been prompted by recent outbreaks of foodborne microbial diseases (Chlumsky et al. 2021). Using packaging to add a quality and safety buffer is one possibility. Materials that inhibit the growth of microbes might be a part of the food packaging of the future. Foods could have their shelf life extended and the risk of viruses reduced with the help of these packaging solutions. There may be further food contact applications for antimicrobial polymers (Khiralla and El-Deeb 2015). There is antimicrobial active packaging. The goal of active packaging is to achieve a certain result by interacting with the product or the area between the package and the food system. In a similar vein, antimicrobial food packaging neutralizes or slows the proliferation of any microbes that might be in the food or packaging (Mousavi Khaneghah, Hashemi, and Limbo 2018).

17.4.1.1 Types of Antimicrobial Packaging

There are various forms that antimicrobial packaging can take:

1. Including sachets or pads that contain volatile antimicrobial agents in the delivered product.
2. Incorporating antimicrobial agents, both volatile and non-volatile, into polymers (Bahrami et al. 2019).
3. Placing antimicrobial coatings or adsorbents onto surfaces made of polymers.
4. Attaching, electrostatically or covalently, antimicrobials to polymers.
5. Employing polymers with inherent antibacterial properties (Appendini and Hotchkiss 2002).

17.4.1.2 Applications of Antimicrobial Packaging in Foods

According to Appendini and Hotchkiss (1997), there are various food-related uses for antimicrobial polymers, one of which is packaging.

- First, an option is to decrease the rate of growth of certain microbes when the package comes into direct touch with solid items (such as meats, cheese, etc.) or liquids (such as milk or meat exudates) to increase the product's shelf life and ensure its safety.
- Second, there is the possibility that antimicrobial packaging materials can sanitize or self-sterilize. Antimicrobial packaging materials like these make it easier to treat materials without contaminating processed goods and drastically cut down on the chance of recontamination. One possible alternative to peroxide treatment in aseptic packaging is self-sterilizing packaging (Al-Tayyar, Youssef, and Al-hindi 2020).
- Third, this has the potential to turn foods and drinks into self-sterilizing ones, at least in theory. Fruit juices and other items with a high acidity level can benefit greatly from this. Equipment utilized in the food processing industry might potentially have their surfaces coated with antimicrobial polymers, allowing them to self-sanitize as needed. Equipment for personal hygiene, such as filler gaskets, conveyors, gloves, and clothing, is one example (Movassaghian, Merkel, and Torchilin 2015).

Within antimicrobial packaging, the content of the food as well as the target microorganisms must be taken into account. The pace of development and physiological state of the targeted microorganisms, as well as the antimicrobial's method of action, chemical composition, range of activity, and manner of incorporation into polymers, are all important considerations (Anvar, Ahari, and Ataee 2021).

17.4.2 Antioxidant Nanomaterials for Extending Shelf Life and Reducing Food Spoilage

Using barriers in food is an effective strategy to minimize gas exchange, minimize moisture loss, and lessen the harm that oxidizing agents and bacteria can do. Packaging typically serves this purpose for processed foods, but when it comes to fresh food, plastic packaging may not always be the best option because it cannot control vapor exchange, block light from entering the food, or stop microorganisms or live cell enzymes from causing deterioration (Rahaiee et al. 2015). Adding functional molecules or particles that can produce qualities like antioxidant and antibacterial while acting as a selective gas barrier is one way to enhance the performance of a traditional package (Table 17.2). As a result, the food industry has found that using edible films (EFs) to preserve fresh, little processed, and even cooked meals, is a viable option (Mohammadi, Hashemi, and Hosseini 2016). The antioxidant impact of coatings made of pectin, gelatin, and beeswax and plasticized with sorbitol and glycerol was investigated by Bermudez et al. The coating mixes were functionalized using hydroxytyrosol and 3,4-dihydroxyphenylglycol, two antioxidants found in olives. The beef was coated and kept at 4 °C for 7 days. The analysis of reactive chemicals for thiobarbituric acid (TBARS) revealed that the meat oxidation processes were greatly suppressed in the samples covered with films

TABLE 17.2
A Comparison of Nanotechnology-Based Food Preservation Methods

Preservation Method	Mechanism of Action	Advantages	Disadvantages	References
Nanoparticle-based antimicrobial coatings	Inhibition of microbial growth by direct contact with nanoparticles	Broad-spectrum antimicrobial activity	Potential migration of nanoparticles into food	(Sridhar et al. 2021)
Nanocomposite packaging materials	Enhanced barrier properties against oxygen, moisture, and microorganisms	Improved shelf life and food quality	Higher cost compared to traditional methods	(Sokamté Tégang et al. 2022)
Cold plasma technology	Generation of reactive oxygen species (ROS) that damage microbial cells	Environmentally friendly	Limited data on long-term safety	(Dong et al. 2015)
High-pressure processing (HPP)	Inactivation of microorganisms by high-pressure	Minimal impact on food quality	High initial investment	(Garcia-Gonzalez et al. 2007)
Pulsed electric field (PEF)	Disruption of microbial cell membranes by electric pulses	Rapid processing time	Potential for nutrient loss	(Saulis 2010)

containing antioxidants, as compared to the control samples that did not include antioxidants (Yun et al. 2013). A coating based on chitosan and gelatin was created by Xiong et al. and functionalized with nisin and grape seed extract, two distinct antioxidants. For 20 days, the combinations were used as a coating over fresh pork meat and stored at 4 °C. The meat's degradation was assessed using a variety of metrics, including color change, protein oxidation, lipid oxidation, and microbiological viability. Furthermore, samples functionalized with nisin did not exhibit any antibacterial or antioxidant properties (Leyva-Porras et al. 2021).

17.5 CASE STUDIES OF NANOTECHNOLOGY-DRIVEN FOOD PRESERVATION TECHNOLOGIES

17.5.1 Nanoencapsulation

Nanoencapsulation is a technique that uses techniques like nanocomposite, nanoemulsification, and nanostructuration to pack substances into very small packages and then add functionality to those packages, including controlled release of the core. Producing functional meals with enhanced functionality and stability while safeguarding bioactive components like proteins, lipids, vitamins, and antioxidants is now within reach, all thanks to this approach. Both the various technologies developed for synthesizing nanocapsules and their actual application have been documented (Awang and Zakaria 2019). For usage in nutraceutical products, researchers have developed a novel, patent-pending process for nanoencapsulating a variety of bioactive and active components. It was shown that these nanocapsules disintegrate and are absorbed as conventional foods once they have delivered their active ingredients (Sogvar et al. 2016). The design philosophy of a new food delivery system and advancements in controlled release and encapsulation technologies have both been reported. By reducing the amount of active chemicals needed, formulators can save a significant amount of money through the use of nanoencapsulation (Ezhilarasi et al. 2013).

Scientists looked at nanotechnological methods for encapsulating and regulating the release of active ingredients in food. The molecular complex of octenyl succinic anhydride and polylysine shows promise as a surfactant or emulsifier in addition to being a possible encapsulant for nutraceuticals or pharmaceuticals. These molecules could also have antimicrobial properties. Curcumin-containing micelles were produced by hydrophobically modified starch. The use of lipid-based nanoencapsulation technology has several benefits, including increasing the bioavailability and solubility of antioxidants, stabilizing them both in laboratory settings and in living organisms, and preventing their accidental interaction with other food components. Food and nutraceuticals can be protected and distributed using one of three main lipid-based nanoencapsulation systems: nanoliposomes, nanocochleates, or archaeosomes. Nanoliposome technology has opened up many new possibilities for food technologists, such as controlled release and encapsulation of food products and increased bioavailability, stability, and shelf life of sensitive compounds (Mozafari et al. 2008).

Nanoliposomes have been studied for their potential as nutrient carriers, enzymes, food additives, and antimicrobials. The desired stability and encapsulation quality were achieved in the 80 coenzyme Q10 nanoliposomes that were produced. Colloidosomes are tiny vesicles that can form a hollow hull by themselves. Smaller than human cells by a factor of 10 make up these entities. Scientists are considering adding vitamins, drugs, and fat blockers to colloidosomes. The presentation focused on a method for the synthesis of water-soluble nanoparticles containing controlled-functional beta-carotene. Despite beta-carotene's natural hydrophobicity, researchers have developed a nanostructured lipid carrier that enhances the compound's solubility and stability in beverages (Weiss et al. 2009). Problems with the controlled release of taste components and efficient delivery of nutritious nutrients are plaguing the food industry. Nanoencapsulation technology, though, has the potential to solve these problems. To protect lipids from degradation, the prolamine zein binds and enrobes them in maize endosperm. It has also been shown that zein can produce periodic patterns when it adsorbs fatty acids. Cooperatively constructed nanoscale layers of zein and fatty acid sheets are of special significance (Chellaram et al. 2014).

To make aqueous nanodispersions with a lot of water-insoluble actives, soy lecithin is the main structural component. Among these functional components are nutraceuticals that do not dissolve in water, vitamins that are soluble in fat, and flavors. The encapsulated actives have better stability and bioavailability since they dissolve well into water-based formulations. A different study indicated that compared to regular powder formulations, nanodispersions increased intestinal cell absorption of CoQ10 by a factor of seven. Nanocochleates are nanocoiled particles that show promise for increasing the nutritional content of processed foods and for stabilizing and protecting a variety of micronutrients (Pendergast and Hoek 2011).

17.5.2 Nanoemulsions

Food packaging equipment and food itself can be decontaminated with the use of nanoemulsions. One common instance is a nanomicelle-based item asserted to have glycerin that is naturally occurring. In addition to cleaning silverware and produce of any remaining oil or grime, it also gets rid of pesticide residues. The food sector has recently taken an interest in nanoemulsions because of their exceptional clarity. These allow for the seamless incorporation of nanoemulsified bioactives and flavors into beverages, preserving their original look. Nanoemulsions can prevent the growth of many different types of foodborne bacteria, including Gram-negative ones. You can use them to reduce surface contamination of chicken skin and decontaminate food processing industries. When treated with nanoemulsion, *Salmonella typhimurium* colonies no longer proliferate. Microencapsulated fish oil's physicochemical characteristics suggest that sugar beetroot pectin might work as a suitable alternative to gum arabic and milk proteins for encapsulating functional food ingredients. The nanoemulsions demonstrated significant potential for several uses, including in the beverage industry (Lee et al.).

The polyelectrolytes, like chitosan and alginate, could give rise to a wide variety of nanoemulsions, including single-layer, double-layer, and triple-layer varieties. Solid lipid nanoparticles, made by controlled crystallization of food nanoemulsions, can transport bioactives such as carotenoids and lycopene. Among the several advantages of solid lipid nanoparticles are their long-term stability, high concentration of functional molecules, suitability for large-scale manufacture without organic solvents, and the capacity to be spray-dried into powder. As a hunger stimulant, megestrol acetate oral suspension (MAOS) can help AIDS patients with cachexia. Megace® (MAOS) and Megace® ES (MA-ES), both formulated with nanocrystals, are available for purchase from Bristol-Myers Squibb in New York, USA. In fasting participants, MA-ES had higher bioavailability and absorption than MAOS; hence, it is recommended that patients with cachectic symptoms, who are reducing their caloric intake, be administered megestrol acetate in this form (Meng et al. 2008).

17.5.3 Nanocomposites in Food Packaging

Novel food packaging could be created using nanotechnology. Nanocomposites have many potential uses in food packaging, including improving mechanical strength, decreasing weight, increasing heat resistance, and creating a more effective barrier against oxygen, carbon dioxide, moisture, ultraviolet radiation, and volatile chemicals. Research on the most common types of nanoparticles and their potential uses in food packaging systems was reviewed (Hong et al. 2020). Plastics are enhanced compared to their traditional alternatives by adding fine nanoparticles, which are 100 nm or smaller. Thermoplastic polymers with nanoscale inclusions ranging from 2–8% by weight are known as polymer nanocomposites (Sanchez-Garcia, Lopez-Rubio, and Lagaron 2010). Nanoclays, carbon nanoparticles, metal oxides, and polymeric resins are all examples of nanoscale inclusions. When compared to their macroscale equivalents, nanocomposites exhibit radically different properties due to their extraordinarily high surface-to-volume ratio, which makes them highly reactive (Tamayo et al. 2016).

An enormous opportunity exists in the active food packaging market for nanocomposite materials derived from natural biopolymers that possess biofunctional characteristics. Natural biopolymer-based films and nanocomposites have recently made strides in their preparation, and these materials show promise for usage in packaging. Carbon nanotubes are one example of a new nanotechnology advancement that has the potential to enhance the mechanical qualities of food packaging (Hai Zhang et al. 2017). Their potential for strong antibacterial actions has only just been uncovered. When exposed to clusters of carbon nanotubes, *E. coli* bacteria are quickly killed. Recently, polymer-clay nanocomposite has become more popular as a material for food packaging due to its exceptional mechanical, thermal, and barrier properties (Moustafa et al. 2019). Preparation, characterization, characteristics, current advancements, and prospects have all been highlighted as scientists have explored the possibility of using these polymer composites as innovative food packaging materials (Zafar et al. 2016).

17.5.4 Nanocoatings and Films

According to reports, the coatings can compete with more conventional active packaging technologies, including oxygen scavengers, in terms of their ability to hold carbon dioxide and prevent the passage of oxygen. Examples include gas barriers made of plasma arc deposited amorphous carbon inside PET bottles and nanocoatings. Coatings as thin as 5 nm on the nanoscale, which are edible, are now within reach, thanks to nanotechnology. Edible nanocoatings can prolong the shelf life of processed foods even after they are opened (Sridhar et al. 2021). These coatings could protect foods from moisture and gas exchange, add flavor and color, and even include antioxidants, enzymes, and anti-browning agents. They could also be applied to confections, baked goods, fast food, fruit and vegetables, and meat. Cellulose nanofibers can greatly enhance the qualities of mango puree edible films. You can directly apply an edible antibacterial nanocoating to baked items (Olivas and Barbosa-Cánovas 2005).

17.5.5 Nanosensors/Nanobiosensors

Nanobiosensors have several potential applications in agriculture, including the detection of a broad range of fertilizers, herbicides, pesticides, insecticides, pathogens, moisture, and soil pH. When used responsibly, they can contribute to sustainable farming practices and increase crop yields. With the assistance of smart sensors, precision farming has the potential to boost agricultural productivity. This technology allows farmers to make greater use of fertilizer, reduce inputs, and manage their time and surroundings more efficiently (Ong et al. 2014). Through precision farming, nanosensors and nano-based smart delivery systems have the potential to aid in the efficient use of water, fertilizers, and pesticides, among other agricultural natural resources. Plant viruses and other agricultural diseases, as well as soil nutrient levels, can be detected by nanosensors scattered around the field (Acosta 2009). Quickly gauging pollution levels in the environment is possible with gas sensors and nano-smart dust, which utilizes small wireless sensors and transponders total. The quality of agricultural products might likewise be tracked using nanobarcodes and nano-processing (Gallocchio, Belluco, and Ricci 2015). Novel nanosensors, better mechanical and barrier qualities in new packaging materials, and efficient and targeted nutrient delivery systems are just a few of the many food industry applications of functionalized nanostructured materials. To get the food business to accept nanotechnology, we need a better knowledge of the pros and cons of the technology based on solid scientific evidence (Shafiq et al. 2020).

Atomic force spectroscopy is becoming more useful for the detection of herbicides that block enzymes as biosensors based on particular interactions are developed. Bionanosensors also enable faster and more precise detection of viruses and bacteria, which improves food safety for consumers (Jain et al. 2018). The environmental, agricultural, and food industries are anticipated to be influenced by nanosensors. There have been five Nanotechnology Signature Initiatives sponsored by National Nanotechnology Initiative agencies, the most recent being "Sensors for Nanotechnology and Nanotechnology for Health and Safety: Enhancing and Preserving the Natural World". Quicker diagnoses and treatments are possible with the use of portable nanodevices that can identify insects, diseases, pathogens, toxins, and pollutants with remarkable speed (Dimitrijevic et al. 2015).

17.5.6 Nanocomposite and Nanolaminates for Nanopackaging

The primary functions of food packaging are environmental protection, product containment, customer nutrition information provision, food shelf life extension, and food quality enhancement. These days, nanotechnology is being used more and more in food packaging. According to Coles, McDowell, and Kirwan (2003), the nanopackaging industry is projected to grow from a $660 million business in 2006 to a $30 billion one in the following decade. By altering the permeability to gases or water vapor, nanoparticles might enhance the mechanical and thermal resistance qualities of food packaging, extending its shelf life. Polymer silicate nanocomposites, for instance, enhance the gas barrier, mechanical strength, and heat resistance characteristics of food packaging compared to polymers alone. An organic substance (such as a protein, peptide, or lipid) interacting with an inorganic substance (like calcium carbonate) to create a more durable material was the inspiration for the first nanocomposite (Youssef et al. 2019). A material for packaging that uses potato starch and calcium carbonate is one example. Fast food packaging can be made using this biodegradable and thermally stable foam instead of polystyrene. By combining starch with tucinin whiskers and amorphous poly (beta hydroxyoctanoate), nanocomposites were created in 2002. Thermoplastic starch (TPS)/clay nanocomposite, made by Choi and colleagues (2003), outperformed pure TPS in terms of tensile strength and water vapor transmission rate. A nanocomposite barrier material was utilized to coat plastic films like PET, which outperforms films coated with silica and alumina, is clear, and aesthetically pleasing for food packaging (Sanchez-Garcia, Lopez-Rubio, and Lagaron 2010).

Two or more layers of nanomaterials that are physically or chemically bound to each other form nanolaminates, which find a useful application in the food sector. Food items like fruit, vegetables, meats, chocolate, chocolates, bakery goods, and French fries can all benefit from the usage of nanolaminates in the production of edible coatings and films, in addition to foaming. According to several studies. Among the many possible applications for these films and coatings are enhancing the textural characteristics of meals, protecting them from gases, moisture, lipids, and oils, and delivering functional agents including flavors, colors, antioxidants, minerals, and antimicrobials Chlumsky et al. 2021). Lipids, polysaccharides, and proteins are now components of these films and coatings. The incredibly thin and delicate character of nanolaminates makes them ideal for usage as a coating material on food surfaces (Youssef et al. 2019).

17.6 DIFFERENT ASPECTS AND ROLES OF NANOTECHNOLOGY IN THE FOOD INDUSTRY

Many parts of food science and the food industry have been revolutionized by the efficient advancements in nanotechnology, which have increased investment and market share. The present uses of nanotechnology in food processing, packaging, and preserving nutrients and freshness are the focus of this article (Jha Indira Gandhi Krishi Vishwavidyalaya et al.) The food industry has been revolutionized by nanotechnology in numerous ways.

Nanoencapsulation undeniably enhances the product's nutritional value. Chemical processing of food ingredients at the nanoscale not only increases nutritional value but also improves scent, texture, and flavor (Sahoo et al. 2021). Nanoparticles outperform more traditional methods of product encapsulation in terms of both encapsulation properties and release efficiency. Food products can be shielded from heat and moisture with nanomaterial encapsulations, which also regulate the active release of components and how they interact with each other in the food matrix. Some examples of such vehicles include nanoemulsions, which increase oral bioavailability, and liposomes, which transport hydrophobic compounds (Oehlke et al. 2014). Nutrients can be delivered to cells more efficiently using nanocochleates, and they don't affect the color or flavor of food in any way. On the other hand, dilated micelles self-assembling nanoscale liquid structures smaller than 30 nm are utilized as carriers. They have been utilized for the distribution of beverages and nutraceuticals (Salgado et al. 2018).

17.7 APPLICATION OF NANOTECHNOLOGY IN FOOD PRESERVATION

What we call the word "nanofood", means the food that has been grown, made, processed, or packaged utilizing nanotechnology in some way, or that has nanomaterials added to it. Nanotechnology is quickly making its way into every facet of the food production process, leaving the lab behind. Nanotechnology stands as a technological foundation. More powerful food coloring, flavoring, nutritional additions, and antibacterial components for food packaging are only a few of the new possibilities presented by the characteristics of modern nanomaterials (Mohammad et al. 2022) There are numerous current uses for nanotechnology in food processing and packaging industries, as well as in agriculture for the controlled release of fertilizers and herbicides. In addition, nanotechnologies have several applications in fields such as biotechnology (today called nanobiotechnology), insecticides, veterinary medicine, animal feed, biocides, and food hygiene. The food business, along with packaging and storage applications, makes commercial use of

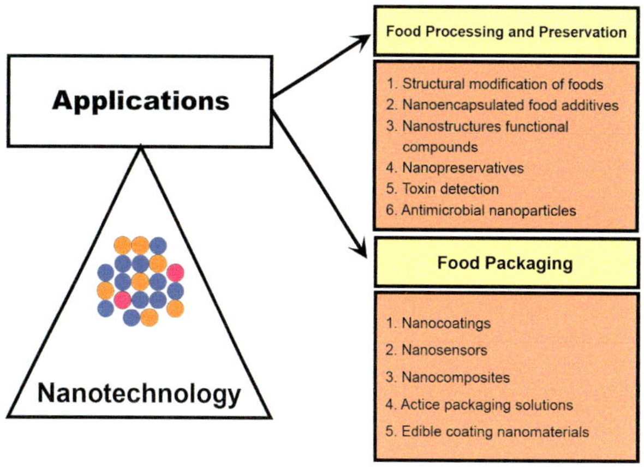

FIGURE 17.2 Nanotechnology applications in food processing, preservation, and packaging.

nanotechnologies. There are an estimated 150–600 commercially accessible nanofoods, with 400–500 potential uses for nanofood packaging. All areas of food manufacturing, packaging, and monitoring could benefit from nanotechnology (V. K. Bajpai et al. 2018). Companies that make functional foods like chips, ice cream, chocolate, and soft drinks often claim that their products are healthier because they have cut back on calories, fat, or carbs while upping their protein, fiber, or vitamin content (Figure 17.2) (Ijaz et al. 2020).

Many nanomaterials have many industrial-scale uses in the food industry. Particles of titanium dioxide less than 100 nm are commonly utilized as an antibacterial agent and food additive in containers and packaging. In addition to their usage as a nutritional supplement and antibacterial agent in food packaging, storage containers, cutting boards, and freezers, silver nanoparticles have several other potential applications. Nutritional additions and antibacterial agents in food packaging, zinc, and zinc oxide have multiple uses (L. Chen, Remondetto, and Subirade 2006). Food additives and packaging materials often include carbon and silicon dioxide, both of which have particle sizes of a few hundred nanometers. To enhance food analysis, biosensors made of platinum and gold nanowires are utilized. To maintain dominance in the food processing industry and provide consumers with fresh, authentic, convenient, and flavorful food products, technology is essential. The objective is to enhance food quality while simultaneously extending the shelf life and freshness of the product. One sector that might be impacted by nanotechnology's revolutionary potential is the food business (Pradhan et al. 2015).

The ability to manipulate substances at the molecular level has allowed for the creation of novel food textures, flavors, and sensations while also erasing the lines between several conventional areas of food science (Liu, Zhang, and Bhandari 2020). Although nanotechnology's uses in the food industry are still in their infancy, experts anticipate a meteoric rise in their popularity in the years to come. An investigation into the potential use of nanotechnology in

TABLE 17.3
Overview of Applications of Nanomaterials in the Food Industry

Application	Nanomaterials Used	Benefits	Potential Risks	References
Food Packaging	Silver nanoparticles (AgNPs), titanium dioxide (TiO_2), zinc oxide (ZnO), nanoclays	—Improved barrier properties against oxygen, moisture, and microorganisms—Enhanced mechanical strength—Reduced spoilage and longer shelf life—Active and intelligent packaging for monitoring food quality	—Potential migration of nanoparticles into food—Uncertain long-term health impacts—Environmental risks of nanoparticles	(Liu, Zhang, and Bhandari 2020)
Food Additives	Encapsulation of flavors, vitamins, and nutraceuticals	—Controlled release of flavors and nutrients—Enhanced bioavailability of nutrients—Improved stability and shelf life— Development of new functional foods	—Potential toxicity of nanocarriers—Uncertain impact on gut microbiome—Regulatory hurdles	(Chaudhry et al. 2008)
Food Processing	Nanozymes, nanosensors	—Improved efficiency and selectivity in enzymatic reactions—Detection of contaminants and pathogens in food—Real-time monitoring of food quality	—Risk of unintended reactions with food components—Potential for false-positive or negative results	(H. Chen, Seiber, and Hotze 2014)
Food Safety	Antimicrobial nanomaterials, biosensors	—Inhibition of foodborne pathogens—Detection of foodborne contaminants and toxins—Improved food safety and quality	—Potential development of resistant bacteria—Unintended interactions with other food components—Uncertain environmental impact	(Tarhan 2020)
Nutritional Supplements	Nanocapsules, nanofibers	—Enhanced bioavailability of nutrients—Improved delivery to target cells—Reduced dosage requirements—Development of personalized nutrition solutions	—Potential toxicity of nanocarriers—Uncertain impact on gut microbiome—Regulatory hurdles	(L. Chen, Remondetto, and Subirade 2006)
Food Texture and Sensory Properties	Nanocellulose, nanoemulsions	—Improved texture and mouthfeel—Enhanced flavor and aroma—Reduced fat and sugar content—Development of healthier and more appealing food products	—Potential allergenicity of nanomaterials—Unintended interactions with other food components—Sensory changes may not be universally accepted	(Rashidi and Khosravi-Darani 2011)

food or food packaging has reportedly been underway by a number of the world's largest food firms. Advancements in this field have already led to a wide variety of practical applications, such as enhanced food flavor, color, texture, and consistency; enhanced nutrient absorption and bioavailability; innovative materials for food packaging that improve mechanical strength, barrier effectiveness, and antibacterial capabilities; and nanosensors for tracking food's whereabouts and tracking its condition as it travels through storage and transportation (Table 17.3) (Chellaram et al. 2014).

Although nanotechnology's application to the food sector is in its early stages, great strides have been achieved in this area in recent years. Up until now, most innovations have been on improving the bioavailability of nutritious components, creating novel tastes and sensations, managing the release of flavors, encapsulating food components or additives, or changing their texture. New materials with enhanced mechanical, barrier, and antibacterial qualities have been developed as a result of these discoveries and are now being used in food packaging (L. Chen, Remondetto, and Subirade 2006). The following are some of the most common and up-to-date examples of nanotechnology's potential uses in the food industry. Nanotechnology has almost no recognized uses in food or food packaging in the United Kingdom, according to this study. The majority of these uses are in other countries, including the United States, Australia, New Zealand, South Korea, Taiwan, China, and Israel. The data collected for this study did not reveal the presence of any food company in the United Kingdom or the European Union that has acknowledged or is known to be utilizing nanofood additives or components. The sole two products that don't fall into this category are lycopene, a synthetic version of the tomato carotenoid, and NovaSOL, a nanomicelle-based carrier system developed by Aquanova Germany to introduce antioxidants and supplements into food and drink. Several solubilisates have been developed using the NovaSOL method, which relies on "product micelles" that have a diameter of approximately 30 nm. Some examples of their uses include dietary supplements and functional food ingredients, food additives such as citric acid, ascorbic acid, and benzoic acid and vitamins A and E, lipoic acid, soybean isoflavones, and coenzyme (Cleveland et al. 2009).

17.8 TYPES OF NANOTECHNOLOGY-BASED FOOD PACKAGING

There are two primary categories into which food packaging falls: smart packaging and active packaging. Consumers can be alerted to product contamination or the presence of pathogens with smart packaging that uses nanoparticle-based sensors. In contrast, active packaging includes antibacterial nanoparticles as well as agents that regulate moisture, including oxygen scavengers, CO_2 emitters, and scavengers (Robles-García et al. 2016).

17.8.1 Active Packaging Systems

The development of active packaging technologies is mostly driven by the goals of food storage, but in general, active packaging improves packaged food and extends its shelf life. Parts of this packaging system are specifically engineered to either release antioxidants or antimicrobials into or from the packaged food or to absorb oxygen (Arroyo et al. 2019). An effective way to increase the shelf life and quality of packaged food is to combine a polymer with active substances such as nanoparticles, antimicrobials, water vapor, and oxygen absorbers, among others. Various active packaging applications have made considerable use of metal oxide nanoparticles and various metals, including zinc, gold, silver, titanium dioxide, silicon oxide, etc. These nanoparticles can react with organics in food through gradual migration or direct contact (Augustin and Sanguansri 2009). Due to their well-established antibacterial efficacy against pathogenic strains, silver nanoparticles are among the most commonly utilized nanoparticles. On top of that, they block the growth of several fungal infections and viruses, including monkeypox (Bowman and Hodge 2006). In addition to inhibiting respiratory chain enzymes, silver nanoparticles have been found to enhance reactive oxygen species (ROS) generation. For instance, Damm et al. combined silver micro and nanoparticles with the film used for packaging and studied their effectiveness against *E. coli*. The addition of nanoparticles to the film effectively halted the growth of *Escherichia coli*. Nevertheless, around 80% of the bacteria were eliminated by the film that included silver microparticles (Al-Naamani, Dobretsov, and Dutta 2016). For use in food packaging, Mohammed et al. mixed sodium alginate films with silver nanoparticles. Researchers have shown that silver nanoparticles effectively inhibit the growth of *Staphylococcus aureus* and *Escherichia coli*. When it comes to fresh produce, multiple writers have proposed various active packaging films that include silver nanoparticles. An edible antibacterial film made of alginate and silver nanoparticles was suggested by Mohammed et al. for sterilised carrots and pears. Another option for packaging fresh asparagus spears is the nanocomposite film of silver nanoparticles/polyvinyl pyrrolidone (Abo-Elseoud et al. 2018).

17.8.2 Smart Packaging Systems

The main component of smart packaging systems is the ability to detect potential food contamination using nanosensors, which are mostly created using nanoparticles. Potentially useful for tracking physical, chemical, and biological changes brought about by food preparation and preservation, nanosensors based on nanoparticles hold much promise in this area. Smart packaging that makes use of these sensors can detect pollutants, poisons, and foodborne diseases. In addition, information about the quality of packed food items during storage and transportation can be conveniently tracked with smart packaging that includes nanosensors and indicators (Robles-García et al. 2016). Because the integrated sensors in smart packaging record all reactions to changes caused by internal or external environmental stimuli, it has been seen that this technology aids in maintaining food quality during distribution. It is usual practice to attach some of these indicators to food packaging to track changes in time, temperature, and package quality (freshness). Indicators allow for the tracking of product changes along the production and supply chain, which aids in quality control and extends the product's shelf life. Further applications for nanobarcodes in smart packaging include identity badges (Bhattacharya et al. 2007).

There is tremendous potential for nanotechnology in food packaging. Utilizing concepts and technology from nanoscience, smart packaging might mend tiny holes and tears, react to changes in the environment (such as changes in temperature and moisture), and notify the buyer if the food is tainted). The "hybrid system" films used for food packaging contain a dizzying array of silicate nanoparticles. Because they drastically cut down on the passage of gases like oxygen and water vapor, they keep food from drying out or going bad (Riaz et al. 2018). Scientists are aiming to create novel "on-demand foods" powered by nanotechnology that can hydrate cells only when they need them. Interactive foods, which incorporate nanoencapsulated chemicals, will enable customers to tailor the

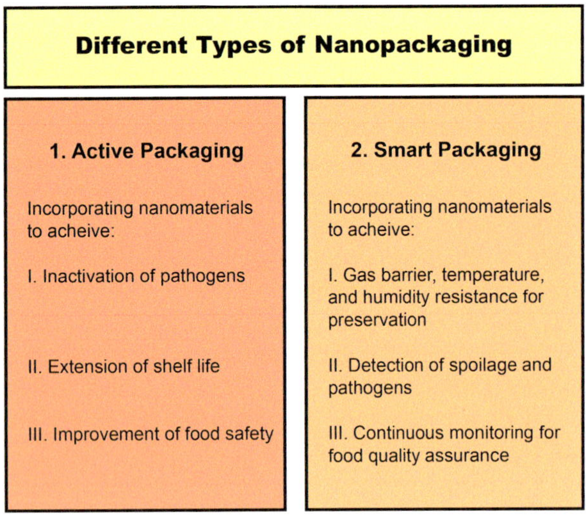

FIGURE 17.3 Different types of nanopackaging in food.

meal to their specific dietary requirements and personal preferences. A potent antibacterial product with possible uses in the food business, Nanocid, was their first commercially available nanotechnology product (Figure 17.3) (Jafarizadeh-Malmiri et al. 2019).

17.9 CHALLENGES AND SAFETY CONSIDERATIONS

17.9.1 Regulatory Frameworks for Accessing the Safety of Nanofood

Food regulators should also pay attention to the health risks associated with processing foods in a way that creates nanoparticles and nanoscale emulsions. To ascertain the necessity for associated new food safety regulations, it is necessary to examine the possibility that these foods may present new health hazards. The food sector is in dire need of regulatory frameworks that can handle the potential dangers of nanofood and nanotechnology (Chau, Wu, and Yen 2007). Additionally, governments need to address the larger ethical, social, economic, and civil rights concerns raised by nanotechnology. Public participation in nanotechnology decision-making is crucial to guarantee democratic management of these emerging technologies in the crucial food and agriculture sector (Biswas et al. 2022).

Chemicals in the form of nanoparticles or nanotubes should be classified as novel substances, according to the U.K.RS/RAE 2004. Before their approval for use in food items, the components of these nanoparticles must pass a comprehensive safety evaluation by the appropriate scientific advisory group. In addition, food items should make it clear on their ingredient labels when nanoparticle materials have been included. Additionally, it is crucial to minimize the discharge of nanoparticles and nanotubes into the environment (Jafarizadeh-Malmiri et al. 2019). There should be defined safety criteria and testing processes in place before new nanotechnology is introduced, according to EU food and food packaging laws. When it comes to nanofoods and the majority of food packaging in the United States, the FDA is in charge, whereas the EPA is in charge of organic chemical regulation. Nevertheless, nanoparticles have not been officially recognized as novel chemicals by either the EPA or the FDA, and no further control of them has been mandated (Dimitrijevic et al. 2015).

17.9.2 Potential Risks Associated with the Use of Nanomaterials in Food

As is widely known, there are many benefits to using different nanomaterials in the food business; yet because of their cytotoxic effects, in addition, they pose serious dangers to ecosystems worldwide, as well as to people's health and the environment. The usage of nanomaterials has recently given rise to some grave concerns, especially for those without any harmful elements in their composition. Because of their small size and subcellular interactions with cells, nanomaterials carry an inherent risk (Bumbudsanpharoke and Ko 2015). For instance, certain nanoparticles can create health issues in both people and animals by penetrating the skin. Even in plants, nanoparticles can alter the genome and proteome and can impact the rate of growth. Research has demonstrated that both single-walled and multiwalled carbon nanotubes can cause oxidative stress and fibrosis in the lungs of model animals including rats and mice. To understand the processes of toxicity of various nanomaterials to human health and the environment, it is necessary to first identify the many entry points and routes of exposure that nanomaterials used in food might take into the human body. Many processed meals contain nanomaterials, which can enter the body through the mouth, skin, and lungs, either intentionally or unintentionally (Vandermoere et al. 2011). Our bodies absorb chemicals, water, and nutrients mostly by oral captivation. There are undoubtedly a lot of nanoparticles in different foods nowadays, and, likely, the digestive system is constantly in close contact with these nanomaterials. Orally consumed nanoparticles go from the mouth to the stomach and intestines, where they pose a major risk to human health (Ravichandran 2010).

Although nanotechnology has many potential uses, there are still many unanswered questions about its safety, appropriate use, and social absorption. Because of their size and quantum characteristics, manufactured nanoparticles may pose particular threats to the environment and human health according to an increasing number of scientific studies and government investigations. Because of their smaller size, nanoparticles appear to be a more hazardous class of material. Particles can become more chemically reactive when they are reduced to the nanoscale because they have a greater surface area (Doyle and Erickson 2006). A material that may be benign at larger scales may take on dangerous properties at the nanoscale as particle size reduces and reactivity rises. Because of their heightened reactivity, nanoparticles may damage living tissue by releasing "free radicals" that could be harmful to people's health. One of the main worries is that the engineered nanomaterials will eventually find their way into the waste stream, landfills, and incinerators, and then the air, soil, and water. Thus, it would be crucial to take into account how different kinds of nanomaterials will be handled and disposed of after their usage, as well as how the regulatory system will handle managing them at different points in their life cycle (Verma et al. 2011).

It is significant to remember that applications of nanotechnology have the potential to create simple-to-use gadgets that lessen reliance on human abilities in numerous ways, hence lowering the risk to humans. Through the use of alternative (renewable) energy sources and filters or catalysts to clean up and reduce pollution, this technology will also indirectly contribute to environmental protection. This technology has a huge chance of improving society through its use in food systems and agriculture. Apart from enhancing the security of tracking systems and detecting contaminants, optimizing agricultural and food systems procedures can also minimize pesticide usage and boost food safety for end users (Bowman and Hodge 2006).

17.9.3 Strategies for Risk Mitigation and Ensuring Consumer Safety

The ceramic sealer Magic Nano, which may be sprayed onto surfaces to prevent dirt buildup, has been the subject of a risk assessment analysis. The medication was withdrawn from sale in March 2006 because more than 110 European consumers experienced respiratory problems following its use (Verma et al. 2011). The risk assessment and management of nanoparticles can be guided by a basic framework that was created. Consequently, research on the hazards of nanoparticles to humans and the environment has yielded several potential reasons. In toxicological science, a forum series of seven papers were published in 2005 and 2006 regarding research methodologies for the safety evaluation of nanomaterials. To evaluate the potential interactions between nanoparticles and biological systems, several methods for characterizing these tiny objects inside the body have been detailed. It is crucial to know how biodegradable nanoparticles are and what consequences they may have on human health (Nielsen 1967). Surface area, concentration, aggregation, dissolving layer characteristics, adsorbing species, surface energy, surface shape, and surface area are among the many parameters that influence dissolution. Humans may be exposed to nanoparticles through the skin, lungs, and digestive system, according to the risk assessment of nanostructures and nanoparticles. Few studies have investigated the possibility of nanoparticle absorption from the intestine into the bloodstream as a result of the incorporation of nanostructured materials into food, water, and medications. As the use of sunscreens and cosmetics has skyrocketed in recent years, the skin has come under intense scrutiny as a possible entry point for nanoparticles. Nanoparticles can cross the epidermis and enter the bloodstream, but little is known about the dangers they pose. Another major way people can be exposed is by breathing in particles in the air (Walters, Pool, and Somerset 2016).

Controlled release of different bioactive agents, such as food ingredients and nutraceuticals, at the correct location and time, is possible with nanoliposomes, which have numerous uses in various scientific and technological domains, such as gene delivery and medicine (Figure 17.4). Consequently, they improve the encapsulated material's efficacy and cellular absorption. Nanoliposomes can stabilize volatile, reactive, or otherwise unstable additives like vitamins, enzymes, antioxidants, slimming agents, etc. Many nations have strict regulatory restrictions in place, so consumers may rest easy knowing that only safe products and uses of nanotechnologies will reach store shelves. To protect consumers from any dangers, however, some developing nations may require practical regulatory control to guarantee a case-by-case pre-market safety assessment of products generated from nanotechnology (Bowman and Hodge 2006). Smaller and more sensitive devices based on nanomaterials have emerged in recent years, thanks to the expansion of nanobiotechnology. When it comes to preparing, preserving, and packing food, these nanomaterial devices are indispensable. Additionally, by shielding them from gases, moisture, and lipids, nanomaterials extend the time that foods can be kept. They provide a more effective means of transporting bioactive substances. Nanosensors and other nanomaterials also play an important role in keeping food safe and high-quality (Sothornvit 2019). The field of nanobiotechnology is advancing at a rapid pace, but it is not without its obstacles. As a means of mitigating the risks connected with nanomaterial and nanodevice use in food production, research into these topics must prioritize environmental impacts and safety concerns. Lastly, cutting-edge nanomaterials and nanodevices have started a

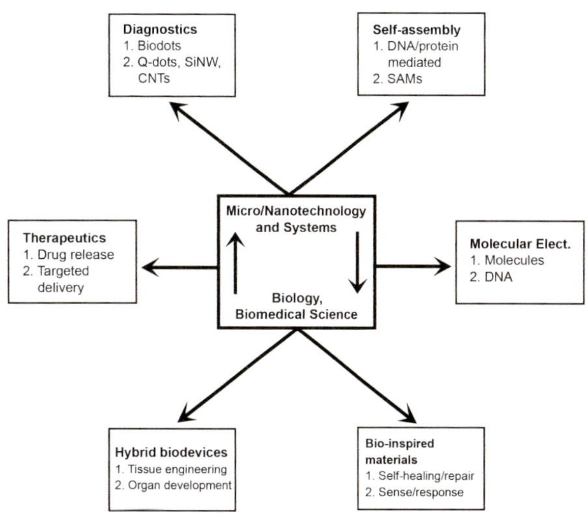

FIGURE 17.4 Nanotechnological systems and their application in biomedical sciences.

revolution in the food business, thanks to nanotechnology, which is still in its early stages. Many sectors of the food industry are investigating novel uses for nanotechnology, which still has a lot of untapped potential. Emerging security and safety concerns must be thoroughly investigated and resolved going forward (Ball and Goats 1996).

17.10 TOXICOLOGICAL ENDPOINTS AND OUTCOMES

For nutritional fortification, increased bioavailability, and preservation, food processing heavily employs inorganic and organic nanomaterials. To determine their toxicity, researchers mostly use in vitro and in vivo studies that have been published in scientific journals. It is possible to determine the absorption dose and hazardous potential of nanoparticles by characterization, particularly with immunological testing. Several recent investigations have shown the toxicological effects of nanoparticle exposures (Roebben et al., n.d.). The possible toxicity and ecotoxicity of nanoparticles have also been studied, in addition to the advantages of nanotechnology. Nanoparticles can translocate across the lymphatic and circulatory systems, bypass biological barriers, and ultimately reach the organs and tissues because of their diminutive size. Nanoparticle exposure to human tissues has the potential to cause major health problems. When nanomaterials enter the body, they can produce oxidative stressors that harm cells and tissues irreversibly due to their size, relatively large surface area, and chemical makeup that makes them more reactive (Walters, Pool, and Somerset 2016). There is concern that nanoparticles, once released into the environment, can linger for a long time and be ingested by living things. This poses toxicological concerns to ecosystems and the crops and foods we eat. Food nano ingredients can enter the bloodstream after partial digestion or for encapsulated nanoparticles to pass past intestinal epithelial and trigger immune reactions (Rezaei, Fathi, and Jafari 2019). Oxidative stress, organ inflammation, immunological, cardiovascular, and pulmonary problems, deposition in bodily tissues, and toxicity can all result from nanoparticles that are absorbed through the skin or the air (Di Sia 2017).

According to Amparo et al. (2018), nanofood products on the market can be safely manufactured, imported, and consumed by following regulatory guidelines. Each step in the life cycle of nanoparticle-integrated food materials—from creation to processing to packaging to consumption—raises unique health and safety concerns. Regulating nanoparticles in the food industry is a bit of a gray area, as no universally applicable laws address this issue. Nanomaterials in food: regulation and safety concerns (V. K. Bajpai et al. 2018). In addition, the evaluation of nanoproduct safety is completely unregulated in many nations. The market potential of nanoproducts produced by this new technology is being constrained by a lack of rules for their correct handling and an inadequate understanding of how to evaluate risks to human and environmental health (Bazana, Codevilla, and de Menezes 2019). Before they may be approved for use, nanotechnology-fabricated food additives and ingredients must undergo safety testing. According to Kar et al. (2021), the Food and Drug Administration has released new safety evaluation guidelines. In their in-depth analysis of the laws governing the proper management of nanomaterials in food. To define the size of the particle, the EC definition uses the range of 1 to 100 nm and is widely applicable through regulatory units. It is the outside dimension of single nanoparticles that are either unbound or aggregated (Roebben et al., n.d.). According to Linsinger et al. (2013), the best way to find out how big or small nanomaterials are is to use a combination of valid methodologies (Di Sia 2017).

The widespread usage of nanoparticles in many different sectors has prompted discussions on the appropriate regulatory classification for nano-based goods and nanomaterials. The broad consensus is that product category is the basis for the regulatory approach to nanotechnology (Bazana et al. 2019). Nanomaterials are finding more and more uses in foodstuffs, as described in the section on nanotechnology's impact on the food industry, and not just to give products a unique appearance. Regulatory bodies' stances determine whether or not the current food legislation requires a precise description of nanomaterials. There are now two main schools of thought: the inclusive and the exclusive. The current laws do, in theory, address nanomaterials and their potential health and environmental hazards, according to advocates of an inclusive approach (Gert Roebben et al.). According to Fender (2008), the majority of laws do not specify particle size when defining a subject. Only if nano-based products pose a distinct threat to human health would the question of targeted regulation emerge. It is thought that the government would rely on current laws to regulate nanotechnology items and processes due to the lack of scientific facts. Interface for managing product security and operations (A. Bajpai and Carlotti 2019). Despite product safety being traditionally seen as a technical issue involving regulators, epidemiologists, design engineers, scientists, quality management, safety engineering, and ergonomics, there is a growing recognition that operations management can provide innovative and efficient approaches to managing product security and safety (Campbell 2014).

Lewis (2003) used the term "operational risk" to characterize the potential for an operation to negatively impact various stakeholders, whether they are internal or external. After delving into a theoretical model that fused operations management theories with risk management through case studies, he discovered that effective risk control is more similar to service quality management than process control. He advocated for a three-pronged approach to risk management: readiness, reaction, and recovery. To address disruptions in the supply chain, multiple individuals subsequently adjusted the notion of risk management (Micheletti et al. 2017). Chopra and Sodhi (2004) used stress testing to isolate potential threats to the supply chain and developed individualized plans to counteract them. This strategy adhered to the fundamentals of risk management, which involve cataloging any dangers, weighing the pros and cons of each, and then taking the necessary steps to lessen the impact of any threats (Chopra, Sodhi, and Lücker 2021). Total quality management (TQM), supply chain agility, optimization, contingency planning, information sharing, collaboration, modularity, and flexibility were some of the operational principles that Kleindorfer and Saad (2005) used to manage disruptive risk in a supply chain. Of course, this is far from a complete list of everyone who has tried to solve the issue of product security and safety by applying concepts from operations management. Some examples include Crosby's cost of quality paradigm, DMAIC, a framework created by Six Sigma, and Lee's and Whang's (2005) total quality management system, all of which aim to ensure supply chain security while reducing costs. Another area where Six Sigma is utilized by Kumar and Schmitz (2011) is recall management. The 3R strategy—readiness, responsiveness, and recovery—is based on continuous improvement and is used by Pyke and Tang (2010) to manage product safety and handle recalls. Using a combined approach to operations management and risk assessment, we will now detail the five industries' unique product security and safety problems. Methods for mitigating risks in these domains are also covered (Chu et al. 2018).

This section provides a synopsis of the food, pharmaceutical, medical device, consumer goods, and automotive industries' product safety and security concerns. The security and safety of food. It should come as no surprise that people have argued about whether or not food is safe for ages. There are a plethora of rules, regulations, standards, procedures, and technology in place today to guarantee that food is safe to eat. But despite this, food safety incidents happen much too often. Several security holes exist in the food supply chain (Whipple et al. 2009). They begin by handling natural items, several of which have a short shelf life and, if not handled properly, could cause harm to customers. Second, many studies have shown that the length, global reach, and interconnectedness of food supply chains increase the likelihood of contamination. Third, there is the possibility of terrorist attacks targeting food and drink goods, as well as the potential of accidental or deliberate adulteration five of the seven broad areas where the United States is susceptible to terrorist attacks are associated with the food supply chain (Aschberger et al. 2014). Ensuring that products safely reach consumers requires meticulous control throughout the whole supply chain. In their study of US food supply chain supplier selection, Voss and Whipple (2009) investigated the relative importance of price, delivery time, product quality, and safety. According to their findings, when choosing a provider, safety is typically not as significant as other factors like pricing, delivery time, and quality (Di Sia 2017). A possible explanation for the prevalence of food safety events, according to their argument, could be this lower priority. Having said that, their findings do show that there are situations in which safety takes precedence, such as when products are imported from other countries. Serious negative outcomes can befall both consumers and companies when food safety is compromised. When accidents cause harm or death, it becomes a catastrophe (Rasmussen et al. 2018). Millions of people are ill each year due to contaminated food, say Thomsen and McKenzie (2001). In the United States alone, 76 million people get sick, and 5,000 people die from food poisoning each year. Consumers will occasionally encounter food safety concerns, according to Thomsen and McKenzie (2001), due to human error and the limitations of food safety technologies (Srivastava 2021). The Peanut Corporation of America's 2008 peanut butter paste Salmonella outbreak and China's 2008 powdered milk melamine adulteration are two high-profile incidents that garnered attention because of their magnitude and severity. Those affected have suffered greatly as a result of these incidents. In February 2009, the Peanut Corporation of America declared bankruptcy. The Chinese case involves Sanlu, the implicated corporation; two of its officers have been sentenced to death, and the general manager and several others are behind bars (Pathakoti, Manubolu, and Hwang 2020).

17.11 FUTURE PERSPECTIVES FOR THE DEVELOPMENT AND IMPLEMENTATION OF NANOFOOD TECHNOLOGIES

A variety of novel GRAS matrices that integrate nutraceutical components and offer regulated release via the oral route can be engineered and developed using food proteins. The nutritional value, availability of renewable sources, and acceptability as naturally occurring, digestible food components are all clear benefits of dietary protein matrices. This study outlines the various ways food proteins can be used to create hydrogel, micro-, or nanoparticle matrices and multicomponent matrices. These matrices can then be customized for specific uses in the creation of new functional food items (Rezaei, Fathi, and Jafari 2019). To manage the rate of release, absorption into the body, and overall effectiveness of bioactive chemicals in food products, it is essential to control the particle size of proteinaceous components. The release rates of the bioactive compounds are just as important as the flavor, aroma, texture, and appearance. Unlike hydrogel and microparticles, which necessitate the release of nutraceutical chemicals from the matrix, nanoparticles may improve nutraceutical absorption via direct uptake by the intestinal epithelium or by adsorption to the GI walls, which prolongs residence duration (Yue, Zhao, and Kuzma 2015). The field of food science and research has made tremendous strides in the use of nanotechnology. To maintain food quality, nanotechnology can help detect pollutants, diseases, and pesticides, as well as with tracking, tracing, and monitoring. Nanotechnology does not face obstacles related to the availability of skilled labor, the expense of analysis, or the acquisition of advanced technological equipment. While some nanosystems have already matured into powerful nanocomponents, others are in the early stages of development. The following domains have the potential for more comprehensive investigation with broad implications. Simultaneous assessment of safety aspects and problems is possible (Chaudhry and Castle 2011).

The innovative concepts of creating biomarkers that are unique to antigens, creating nanocomposite polymeric films by combining nanoparticles, and "smart packaging" are gradually coming to fruition. In the future, there is room for a substantial investigation into potential growth and industrial uses. In addition, sensors that can detect microbes, harmful compounds, and food deterioration are increasingly being integrated into packaging materials using carbon nanotubes (Sahoo et al. 2021). According to Wang and Irudayaraj (2008), smart and intelligent food packaging systems provide a great opportunity for extensive study into potential future packaging materials. The use of nanofibers in food packaging and how different food ingredients interact during nanoencapsulation are also areas that could benefit greatly from more investigation (Esser, Schnorr, and Swager 2012). In the future, the range of functional foods can be greatly expanded with the help of foods derived from nanotechnology. Nanotechnology has the potential to dominate the food processing industry if the laws and regulations designed to address its unique safety concerns are established. Nanotechnology has the potential to solve many societal and industrial problems at both the micro and macro levels, leading experts to predict that it will reach an exponential development rate by 2050 and become the most sophisticated technology in the world.

17.12 CONCLUSIONS

This chapter shows the present and future uses of nanotechnology in the food industries and reveals that they provide numerous advantages to the entire food chain, including novel flavors and textures, better absorption of nutrients and supplements, improved traceability and security of food products, improved preservation of quality and freshness, and a possible decrease in the dietary intake of fat, salt, and other food additives. Many of the advantages have great promise for bettering the nutrition and health of the general population in underdeveloped nations, including enhanced food hygiene and quality, longer shelf life, and water decontamination, among many others. However, most products and applications involving nanotechnology are still in the research and development stage, and the present degree of implementation is quite minimal. Thus, product preservation for consumers and the environment must be considered when using nanomaterials, particularly insoluble and biopersistent nanoparticles, in food applications. Before releasing NPs to the market, it is important to consider safety concerns and environmental impacts. This will help ensure the health of humans. More targeted research into the optimization of variables causing NP toxicity is required, considering the numerous advantages of nanopackaging in food preservation, shelf-life extension, and prevention of foodborne illnesses in humans. Antimicrobial nanopackaging technologies will be advanced through effective industry participation and collaboration with research efforts.

REFERENCES

Abo-Elseoud, Wafaa S., Mohammad L. Hassan, Magdy W. Sabaa, Mona Basha, Enas A. Hassan, and Shaimaa M. Fadel. 2018. "Chitosan Nanoparticles/Cellulose Nanocrystals Nanocomposites as a Carrier System for the Controlled Release of Repaglinide." *International Journal of Biological Macromolecules* 111 (May). Elsevier B.V.: 604–613. https://doi.org/10.1016/j.ijbiomac.2018.01.044.

Acosta, Edgar. 2009. "Bioavailability of Nanoparticles in Nutrient and Nutraceutical Delivery." *Current Opinion in Colloid and Interface Science*. https://doi.org/10.1016/j.cocis.2008.01.002.

Ahamed, Maqusood, Hisham A. Alhadlaq, M. A. Majeed Khan, Ponmurugan Karuppiah, and Naif A. Al-Dhabi. 2014. "Synthesis, Characterization, and Antimicrobial Activity of Copper Oxide Nanoparticles." *Journal of Nanomaterials* 2014. Hindawi Publishing Corporation. https://doi.org/10.1155/2014/637858.

Ahmad, Nabeel, Sharad Bhatnagar, Shyam Dhar Dubey, Ritika Saxena, Shweta Sharma, and Rajiv Dutta. 2017. "Nanopackaging in Food and Electronics." 45–97. https://doi.org/10.1007/978-3-319-53112-0_2.

Alfadul, S. M., and A. A. Elneshwy. 2010. "Use of Nanotechnology in Food Processing, Packaging and Safety–Review." *African Journal of Food, Agriculture, Nutrition and Development* 10 (6).

Al-Naamani, Laila, Sergey Dobretsov, and Joydeep Dutta. 2016. "Chitosan-Zinc Oxide Nanoparticle Composite Coating for Active Food Packaging Applications." *Innovative Food Science and Emerging Technologies* 38 (December). Elsevier Ltd: 231–237. https://doi.org/10.1016/j.ifset.2016.10.010.

Al-Tayyar, Nasser A., Ahmed M. Youssef, and Rashad Al-hindi. 2020. "Antimicrobial Food Packaging Based on Sustainable Bio-Based Materials for Reducing Foodborne Pathogens: A Review." *Food Chemistry*. Elsevier Ltd. https://doi.org/10.1016/j.foodchem.2019.125915.

Amparo, L. R., M. J. F. Rovira, M. M. Sanz, and L. G. Gomez-Mascaraque. 2018. *Nanomaterials for Food Applications*. Elsevier.

Anvar, Amir Ali, Hamed Ahari, and Maryam Ataee. 2021. "Antimicrobial Properties of Food Nanopackaging: A New Focus on Foodborne Pathogens." *Frontiers in Microbiology*. Frontiers Media S.A. https://doi.org/10.3389/fmicb.2021.690706.

Appendini, P. and J. H. Hotchkiss. 1997. "Immobilization of Lysozyme on Food Contact Polymers as Potential Antimicrobial Films." *Packaging Technology and Science: An International Journal* 10 (5): 271–279.

Appendini, Paola, and Joseph H. Hotchkiss. 2002. *Review of Antimicrobial Food Packaging. Innovative Food Science & Emerging Technologies*. Vol. 3. www.bioka.

Arroyo, Betty Jarma, Enayde de Almeida de Melo, Anderson Campos, Lara Lins, Andrelina Pinheiro Santos, and Luis Carlos Boyano-Orozco. 2019. "Bioactive Compounds and Their Potential Use as Ingredients for Food and its Application in Food Packaging." In *Bioactive Compounds: Health Benefits and Potential Applications*, 143–156. Elsevier. https://doi.org/10.1016/B978-0-12-814774-0.00008-6.

Aschberger, Karin, Frans M. Christensen, Hugues Crutzen, Kirsten Rasmussen, Hubert Rauscher, Birgit Sokull-Klüttgen,

Hermann Stamm, and Institute for Health and Consumer Protection. 2014. *Considerations on Information Needs for Nanomaterials in Consumer Products: Discussion of a Labelling and Reporting Scheme for Nanomaterials in Consumer Products in the EU*. Publications Office of the European Union.

Augustin, Mary Ann, and Peerasak Sanguansri. 2009. "Nanostructured Materials in the Food Industry." *Advances in Food and Nutrition Research* 58. Academic Press Inc.: 183–213. https://doi.org/10.1016/S1043-4526(09)58005-9.

Awang, Muhammad Nizam, and Zalina Zakaria. 2019. "Nanotechnology within Halal Legal Framework: Case Study of Nano-Based Food Products." *Journal of Fatwa Management and Research* 17 (July). Universiti Sains Islam Malaysia: 29–43. https://doi.org/10.33102/jfatwa.vol17no1.3.

Bahrami, Akbar, Rana Delshadi, Seid Mahdi Jafari, and Leonard Williams. 2019. "Nanoencapsulated Nisin: An Engineered Natural Antimicrobial System for the Food Industry." *Trends in Food Science and Technology*. Elsevier Ltd. https://doi.org/10.1016/j.tifs.2019.10.002.C:\Users\LENOVO\Downloads\Query Integration (1)\Query Integration\24064-0309-Ref Mismatch Report.docx - LStERROR_18

Bajpai, Ankur, and Stéphane Carlotti. 2019. "The Effect of Hybridized Carbon Nanotubes, Silica Nanoparticles, and Core-Shell Rubber on Tensile, Fracture Mechanics and Electrical Properties of Epoxy Nanocomposites." *Nanomaterials* 9 (7). MDPI AG. https://doi.org/10.3390/nano9071057.

Bajpai, Vivek K., Madhu Kamle, Shruti Shukla, Dipendra Kumar Mahato, Pranjal Chandra, Seung Kyu Hwang, Pradeep Kumar, Yun Suk Huh, and Young Kyu Han. 2018. "Prospects of Using Nanotechnology for Food Preservation, Safety, and Security." *Journal of Food and Drug Analysis*. Elsevier Taiwan LLC. https://doi.org/10.1016/j.jfda.2018.06.011.C:\Users\LENOVO\Downloads\Query Integration (1)\Query Integration\24064-0309-Ref Mismatch Report.docx - LStERROR_19

Ball, David J., and Geoffrey C. Goats. 1996. "Risk Management and Consumer Safety." *International Journal for Consumer and Product Safety* 3 (3). Informa UK Limited: 111–124. https://doi.org/10.1080/09298349608945770.

Bazana, Maiara Taís, Cristiane Franco Codevilla, and Cristiano Ragagnin de Menezes. 2019. "Nanoencapsulation of Bioactive Compounds: Challenges and Perspectives." *Current Opinion in Food Science*. Elsevier Ltd. https://doi.org/10.1016/j.cofs.2019.03.005.C:\Users\LENOVO\Downloads\Query Integration (1)\Query Integration\24064-0309-Ref Mismatch Report.docx - LStERROR_20

Belluco, Simone, Federica Gallocchio, Carmen Losasso, and Antonia Ricci. Accepted Manuscript. *State of Art of Nanotechnology Applications in the Meat Chain: A Qualitative Synthesis*.

Benoit, Danielle S. W., Clyde T. Overby, Kenneth R. Sims Jr, and Marian A. Ackun-farmmer. 2014. *Drug Delivery Systems History of DDS Development*. Saltzman.

Berezin, M. Y. 2015. *Nanotechnology for Biomedical Imaging and Diagnostics: From Nanoparticle Design to Clinical Applications*. John Wiley & Sons.

Bhattacharya, Shantanu, Jaesung Jang, Liju Yang, Demir Akin, Rashid Bashir, and Birck Nanotechnology Center. 2007. *All Rights Reserved. Journal of Rapid Methods & Automation in Microbiology* 15 (1): 1–32.

Biswas, Rahul, Mahabub Alam, Animesh Sarkar, Md Ismail Haque, Md Moinul Hasan, and Mominul Hoque. 2022. "Application of Nanotechnology in Food: Processing, Preservation, Packaging and Safety Assessment." *Heliyon*. Elsevier Ltd. https://doi.org/10.1016/j.heliyon.2022.e11795.

Bowman, Diana M., and Graeme A. Hodge. 2006. "Nanotechnology: Mapping the Wild Regulatory Frontier." *Futures* 38 (9): 1060–1073. https://doi.org/10.1016/j.futures.2006.02.017.C:\Users\LENOVO\Downloads\Query Integration (1)\Query Integration\24064-0309-Ref Mismatch Report.docx - LStERROR_22

Brandelli, Adriano, Cristian M. B. Pinilla, and Nathalie A. Lopes. 2019. "Nanotechnology in Food Preservation." In *Advances in Processing Technologies for Bio-Based Nanosystems in Food*, 277–311. Taylor and Francis. https://doi.org/10.15740/has/fsrj/9.2/441-447.

Bumbudsanpharoke, Nattinee, and Seonghyuk Ko. 2015. "Nano-Food Packaging: An Overview of Market, Migration Research, and Safety Regulations." *Journal of Food Science* 80 (5): R910–R923. https://doi.org/10.1111/1750-3841.12861.

Buzea, Cristina, and Ivan Pacheco. 2017. "Nanomaterial and Nanoparticle: Origin and Activity." 71–112. https://doi.org/10.1007/978-3-319-46835-8_3.

Campbell, Katrina. 2014. *Safefood Report: Nanotechnology in the Agri-Food Industry on the Island of Ireland: Applications, Opportunities and Challenges Maeve Henchion TEAGASC-The Agriculture and Food Development Authority*. https://www.researchgate.net/publication/273575693.

Chau, Chi Fai, Shiuan Huei Wu, and Gow Chin Yen. 2007. "The Development of Regulations for Food Nanotechnology." *Trends in Food Science and Technology*. https://doi.org/10.1016/j.tifs.2007.01.007.

Chaudhry, Qasim, and Laurence Castle. 2011. "Food Applications of Nanotechnologies: An Overview of Opportunities and Challenges for Developing Countries." *Trends in Food Science and Technology* 22 (11). Elsevier Ltd: 595–603. https://doi.org/10.1016/j.tifs.2011.01.001.

Chaudhry, Qasim, Michael Scotter, James Blackburn, Bryony Ross, Alistair Boxall, Laurence Castle, Robert Aitken, and Richard Watkins. 2008. "Applications and Implications of Nanotechnologies for the Food Sector." *Food Additives and Contaminants—Part A*. Taylor and Francis Ltd. https://doi.org/10.1080/02652030701744538.

Chellaram, C., G. Murugaboopathi, A. A. John, R. Sivakumar, S. Ganesan, S. Krithika, and G. Priya. 2014. "Significance of Nanotechnology in Food Industry." *APCBEE Procedia* 8. Elsevier BV: 109–113. https://doi.org/10.1016/j.apcbee.2014.03.010.

Chen, Hongda, James N. Seiber, and Matt Hotze. 2014. "ACS Select on Nanotechnology in Food and Agriculture: A Perspective on Implications and Applications." *Journal of Agricultural and Food Chemistry*. https://doi.org/10.1021/jf5002588.

Chen, Lingyun, Gabriel E. Remondetto, and Muriel Subirade. 2006. "Food Protein-Based Materials as Nutraceutical Delivery Systems." *Trends in Food Science and Technology*. https://doi.org/10.1016/j.tifs.2005.12.011.

Chlumsky, Ondrej, Sabina Purkrtova, Hana Michova, Hana Sykorova, Petr Slepicka, Dominik Fajstavr, Pavel Ulbrich, Jitka Viktorova, and Katerina Demnerova. 2021. "Antimicrobial Properties of Palladium and Platinum Nanoparticles: A New Tool for Combating Food-borne Pathogens." *International Journal of Molecular Sciences* 22 (15). MDPI. https://doi.org/10.3390/ijms22157892.

Cho, Young Hee, and Owen Griffith Jones. 2019. "Assembled Protein Nanoparticles in Food or Nutrition Applications." *Advances in Food and Nutrition Research* 88. Academic Press Inc: 47–84. https://doi.org/10.1016/bs.afnr.2019.01.002.

Choi, J. H., K. W. Park, H. K. Lee, Y. M. Kim, J. S. Lee, and Y. E. Sung. 2003. "Nano-Composite of PtRu Alloy Electrocatalyst and Electronically Conducting Polymer for Use as the Anode in a Direct Methanol Fuel Cell." *Electrochimica Acta* 48 (19): 2781–2789.

Chopra, S., and M. S. Sodhi. 2004. *Managing Risk to Avoid Supply-Chain Breakdown*. MIT Sloan Management Review.

Chopra, Sunil, Man Mohan Sodhi, and Florian Lücker. 2021. "Achieving Supply Chain Efficiency and Resilience by Using Multi-Level Commons." *Decision Sciences* 52 (4). John Wiley and Sons Inc: 817–832. https://doi.org/10.1111/deci.12526.

Chu, Han Wei, Binesh Unnikrishnan, Anisha Anand, Ju Yi Mao, and Chih Ching Huang. 2018. "Nanoparticle-Based Laser Desorption/Ionization Mass Spectrometric Analysis of Drugs and Metabolites." *Journal of Food and Drug Analysis*. Elsevier Taiwan LLC. https://doi.org/10.1016/j.jfda.2018.07.001.

Cleveland, T.E., J. Yu, N. Fedorova, D. Bhatnagar, G.A. Payne, W.C. Nierman, and J.W. Bennett. 2009. "Potential of Aspergillus Flavus Genomics for Applications in Biotechnology." *Trends in Biotechnology* 27 (3): 151–157.

Codina-Torrella, I., B. Guamis, V. Ferragut, and A. J. Trujillo. 2017. "Potential Application of Ultra-High Pressure Homogenization in the Physico-Chemical Stabilization of Tiger Nuts' Milk Beverage." *Innovative Food Science & Emerging Technologies* 40: 42–51.

Coles, R., D. McDowell. and M. J. Kirwan. 2003. *Food Packaging Technology* (Vol. 5). CRC Press.

de Souza, T. S., and H. Y. Kawaguti. 2021. "Cellulases, Hemicellulases, and Pectinases: Applications in the Food and Beverage Industry." *Food and Bioprocess Technology* 14 (8): 1446–1477.

Di Sia, Paolo. 2017. "Nanotechnology Among Innovation, Health and Risks." *Procedia—Social and Behavioral Sciences* 237 (February). Elsevier BV: 1076–1080. https://doi.org/10.1016/j.sbspro.2017.02.158.C:\Users\LENOVO\Downloads\Query Integration (1)\Query Integration\24064-0309-Ref Mismatch Report.docx - LStERROR_24

Dimitrijevic, Mirjana, Nedjeljko Karabasil, Marija Boskovic, Vlado Teodorovic, Dragan Vasilev, Vesna Djordjevic, Natasa Kilibarda, and Nikola Cobanovic. 2015. "Safety Aspects of Nanotechnology Applications in Food Packaging." *Procedia Food Science* 5. Elsevier BV: 57–60. https://doi.org/10.1016/j.profoo.2015.09.015.C:\Users\LENOVO\Downloads\Query Integration (1)\Query Integration\24064-0309-Ref Mismatch Report.docx - LStERROR_25

Dong, Tao G., Shiqi Dong, Christy Catalano, Richard Moore, Xiaoye Liang, and John J. Mekalanos. 2015. "Generation of Reactive Oxygen Species by Lethal Attacks from Competing Microbes." *Proceedings of the National Academy of Sciences of the United States of America* 112 (7). National Academy of Sciences: 2181–2186. https://doi.org/10.1073/pnas.1425007112.

Doyle, M. P., and M. C. Erickson. 2006. "Emerging Microbiological Food Safety Issues Related to Meat." *Meat Science* 74 (1): 98–112. https://doi.org/10.1016/j.meatsci.2006.04.009.

Drusch, S., and S. Mannino. 2009. "Patent-Based Review on Industrial Approaches for the Microencapsulation of Oils Rich in Polyunsaturated Fatty Acids." *Trends in Food Science & Technology* 20 (6–7): 237–244.

Dube, Admire, Ken Ng, Joseph A. Nicolazzo, and Ian Larson. 2010. "Effective Use of Reducing Agents and Nanoparticle Encapsulation in Stabilizing Catechins in Alkaline Solution." *Food Chemistry* 122 (3): 662–667. https://doi.org/10.1016/j.foodchem.2010.03.027.

Dutta, Mintu Maan, and Mridusmita Goswami. 2020. "Coating Materials." 1–30. https://doi.org/10.4018/978-1-7998-4870-7.ch001.

El-Sayed, Samah M., and Ahmed M. Youssef. 2019. "Potential Application of Herbs and Spices and Their Effects in Functional Dairy Products." *Heliyon*. Elsevier Ltd. https://doi.org/10.1016/j.heliyon.2019.e01989.

Espitia, Paula Judith Perez, Nilda de Fátima Ferreira Soares, Jane Sélia dos Reis Coimbra, Nélio José de Andrade, Renato Souza Cruz, and Eber Antonio Alves Medeiros. 2012. "Zinc Oxide Nanoparticles: Synthesis, Antimicrobial Activity and Food Packaging Applications." *Food and Bioprocess Technology*. https://doi.org/10.1007/s11947-012-0797-6.

Esser, Birgit, Jan M. Schnorr, and Timothy M. Swager. 2012. "Selective Detection of Ethylene Gas Using Carbon Nanotube-Based Devices: Utility in Determination of Fruit Ripeness." *Angewandte Chemie—International Edition* 51 (23): 5752–5756. https://doi.org/10.1002/anie.201201042.

Ezhilarasi, P. N., P. Karthik, N. Chhanwal, and C. Anandharamakrishnan. 2013. "Nanoencapsulation Techniques for Food Bioactive Components: A Review." *Food and Bioprocess Technology*. Springer Science and Business Media, LLC. https://doi.org/10.1007/s11947-012-0944-0.

Fender, J. K. 2008. "The FDA and Nano: Big Problems with Tiny Technology." *Chicago-Kent Law Review* 83: 1063.

Fernández-Pan, I., K. Ziani, R. Pedroza-Islas, and J. I. Maté. 2010. "Effect of Drying Conditions on the Mechanical and Barrier Properties of Films Based on Chitosan." *Drying Technology* 28 (12): 1350–1358. https://doi.org/10.1080/07373937.2010.482692.

Fu, Y., H. Lu, G. Nian, P. Wang, N. Lin, X. Hu, H. Zhou, H. Yu, S. Qu, and W. Yang. 2020. "Size-Dependent Inertial Cavitation of Soft Materials." *Journal of the Mechanics and Physics of Solids* 137: 103859.

Gallocchio, Federica, Simone Belluco, and Antonia Ricci. 2015. "Nanotechnology and Food: Brief Overview of the Current Scenario." *Procedia Food Science* 5. Elsevier BV: 85–88. https://doi.org/10.1016/j.profoo.2015.09.022.

Garcia-Gonzalez, L., A. H. Geeraerd, S. Spilimbergo, K. Elst, L. Van Ginneken, J. Debevere, J. F. Van Impe, and F. Devlieghere. 2007. "High Pressure Carbon Dioxide Inactivation of Microorganisms in Foods: The Past, the Present and the Future." *International Journal of Food Microbiology*. https://doi.org/10.1016/j.ijfoodmicro.2007.02.018.

Gokularaman, S., S. A. Cruz, M. M. Pragalyaashree, and A. Nishadh. 2017. "Nanotechnology Approach in Food Packaging-Review." *Journal of Pharmaceutical Sciences and Research* 9 (10): 1743–1749.

Hong, Seyoung, Dong Wook Choi, Hong Nam Kim, Chun Gwon Park, Wonhwa Lee, and Hee Ho Park. 2020. "Protein-Based Nanoparticles as Drug Delivery Systems." *Pharmaceutics*. MDPI AG. https://doi.org/10.3390/pharmaceutics12070604.

Ijaz, Irfan, Ezaz Gilani, Ammara Nazir, and Aysha Bukhari. 2020. "Detail Review on Chemical, Physical and Green Synthesis, Classification, Characterizations and Applications of Nanoparticles." *Green Chemistry Letters and Reviews*. Taylor and Francis Ltd. https://doi.org/10.1080/17518253.2020.1802517.

Jafarizadeh-Malmiri, Hoda, Zahra Sayyar, Navideh Anarjan, and Aydin Berenjian. 2019. "Nanobiotechnology in Food Packaging." In *Nanobiotechnology in Food: Concepts,*

Applications and Perspectives, 69–79. Springer International Publishing. https://doi.org/10.1007/978-3-030-05846-3_5.

Jahangirian, Hossein, Ensieh Ghasemian Lemraski, Thomas J. Webster, Roshanak Rafiee-Moghaddam, and Yadollah Abdollahi. 2017. "A Review of Drug Delivery Systems Based on Nanotechnology and Green Chemistry: Green Nanomedicine." *International Journal of Nanomedicine*. Dove Medical Press Ltd. https://doi.org/10.2147/IJN.S127683.

Jain, Aditi, Shivendu Ranjan, Nandita Dasgupta, and Chidambaram Ramalingam. 2018. "Nanomaterials in Food and Agriculture: An Overview on Their Safety Concerns and Regulatory Issues." *Critical Reviews in Food Science and Nutrition*. Taylor and Francis Inc. https://doi.org/10.1080/10408398.2016.1160363.

Jemil, Ines, Mourad Jridi, Rim Nasri, Naourez Ktari, Rabeb Ben Slama-Ben Salem, Mohamed Mehiri, Mohamed Hajji, and Moncef Nasri. 2014. "Functional, Antioxidant and Antibacterial Properties of Protein Hydrolysates Prepared from Fish Meat Fermented by Bacillus Subtilis A26." *Process Biochemistry* 49 (6). Elsevier BV: 963–972. https://doi.org/10.1016/j.procbio.2014.03.004.

Jha Indira Gandhi Krishi Vishwavidyalaya, Zenu, Shiv Narayan Sharma, Girish Chandel, and A Indira Gandhi. *Nanotechnology: Prospects of Agricultural Advancement*. https://www.researchgate.net/publication/267988722.

Kar, Supratik, Kavitha Pathakoti, Paul B. Tchounwou, Danuta Leszczynska, and Jerzy Leszczynski. 2021. "Evaluating the Cytotoxicity of a Large Pool of Metal Oxide Nanoparticles to Escherichia Coli: Mechanistic Understanding Through In Vitro and In Silico Studies." *Chemosphere* 264 (February). Elsevier Ltd. https://doi.org/10.1016/j.chemosphere.2020.128428.

Khan, A., M. S. Takriff, K. Sanaullah, M. Zwawi, M. Algarni, B. F. Felemban, A. Bahadar, A. Shah, and A. R. H. Rigit. 2020. "Periodic Compression and Cavitation Induced Shear Between Steam-Water Two-Phase Flows for Bio-Materials Degradation." *International Journal of Environmental Science and Technology* 17: 1591–1626.

Khiralla, Ghada M., and Bahig A. El-Deeb. 2015. "Antimicrobial and Antibiofilm Effects of Selenium Nanoparticles on Some Foodborne Pathogens." *LWT* 63 (2). Academic Press: 1001–1007. https://doi.org/10.1016/j.lwt.2015.03.086.

Kleindorfer, P. R., and G. H. Saad. 2005. "Managing Disruption Risks in Supply Chains." *Production and Operations Management* 14 (1): 53–68.

Kumar, Lahir Yogendra. 2015. "Role and Adverse Effects of Nanomaterials in Food Technology." *Journal of Toxicology and Health* 2 (1). Herbert Publications PVT LTD: 2. https://doi.org/10.7243/2056-3779-2-2.

Kumar, S., and S. Schmitz. 2011. "Managing Recalls in a Consumer Product Supply Chain–Root Cause Analysis and Measures to Mitigate Risks." *International Journal of Production Research* 49 (1): 235–253.

Lagnika, Camel, Min Zhang, and Kebitsamang Joseph Mothibe. 2013. "Effects of Ultrasound and High Pressure Argon on Physico-Chemical Properties of White Mushrooms (Agaricus Bisporus) during Postharvest Storage." *Postharvest Biology and Technology* 82 (August): 87–94. https://doi.org/10.1016/j.postharvbio.2013.03.006.

Lee, H. L., and S. Whang. 2005. "Higher Supply Chain Security with Lower Cost: Lessons from Total Quality Management." *International Journal of Production Economics* 96 (3): 289–300.

Lee, Laurence, Joseph Arult, Robert Lenckit, and Francois G. Castaigne. 1996. "A Review on Modified Atmosphere Packaging and Preservation of Fresh Fruits and Vegetables: Physiological Basis and Practical Aspects-Part I." *Packaging Technology and Science* 8.

Lewis, M. A. 2003. "Cause, Consequence and Control: Towards a Theoretical and Practical Model of Operational Risk." *Journal of Operations Management* 21 (2): 205–224.

Leyva-Porras, César, Manuel Román-Aguirre, Pedro Cruz-Alcantar, José T. Pérez-Urizar, and María Zenaida Saavedra-Leos. 2021. "Application of Antioxidants as an Alternative Improving of Shelf Life in Foods." *Polysaccharides* 2 (3). MDPI AG: 594–607. https://doi.org/10.3390/polysaccharides2030036.

Linsinger, T. P. J., Q. Chaudhry, V. Dehalu, P. Delahaut, A. Dudkiewicz, R. Grombe, F. Von der Kammer, E. H. Larsen, S. Legros, K. Loeschner, and R. Peters. 2013. "Validation of Methods for the Detection and Quantification of Engineered Nanoparticles in Food." *Food Chemistry* 138 (2–3): 1959–1966.

Liu, Wenchao, Min Zhang, and Bhesh Bhandari. 2020. "Nanotechnology–a Shelf Life Extension Strategy for Fruits and Vegetables." *Critical Reviews in Food Science and Nutrition*. Taylor and Francis Inc. https://doi.org/10.1080/10408398.2019.1589415.

Malik, Shiza, Khalid Muhammad, and Yasir Waheed. 2023. "Nanotechnology: A Revolution in Modern Industry." *Molecules*. MDPI. https://doi.org/10.3390/molecules28020661.

McClements, David Julian, Glen DeLoid, Georgios Pyrgiotakis, Jo Anne Shatkin, Hang Xiao, and Philip Demokritou. 2016. "The Role of the Food Matrix and Gastrointestinal Tract in the Assessment of Biological Properties of Ingested Engineered Nanomaterials (IENMs): State of the Science and Knowledge Gaps." *NanoImpact*. Elsevier B.V. https://doi.org/10.1016/j.impact.2016.10.002.

Meng, Xianghong, Boqiang Li, Jia Liu, and Shiping Tian. 2008. "Physiological Responses and Quality Attributes of Table Grape Fruit to Chitosan Preharvest Spray and Postharvest Coating during Storage." *Food Chemistry* 106 (2): 501–508. https://doi.org/10.1016/j.foodchem.2007.06.012.

Micheletti, C., M. Roman, E. Tedesco, I. Olivato, and F. Benetti. 2017. "Implementation of the NANoREG Safe-by-Design Approach for Different Nanomaterial Applications." *Journal of Physics: Conference Series* 838. Institute of Physics Publishing. https://doi.org/10.1088/1742-6596/838/1/012019.

Mohammad, Zahra H., Faizan Ahmad, Salam A. Ibrahim, and Sadaf Zaidi. 2022. "Application of Nanotechnology in Different Aspects of the Food Industry." *Discover Food*. Springer Nature. https://doi.org/10.1007/s44187-022-00013-9.

Mohammadi, Ali, Maryam Hashemi, and Seyed Masoud Hosseini. 2016. "Postharvest Treatment of Nanochitosan-Based Coating Loaded with Zataria Multiflora Essential Oil Improves Antioxidant Activity and Extends Shelf-Life of Cucumber." *Innovative Food Science and Emerging Technologies* 33 (February). Elsevier Ltd: 580–588. https://doi.org/10.1016/j.ifset.2015.10.015.

Mousavi Khaneghah, Amin, Seyed Mohammad Bagher Hashemi, and Sara Limbo. 2018. "Antimicrobial Agents and Packaging Systems in Antimicrobial Active Food Packaging: An Overview of Approaches and Interactions." *Food and Bioproducts Processing*. Institution of Chemical Engineers. https://doi.org/10.1016/j.fbp.2018.05.001.

Moustafa, Hesham, Ahmed M. Youssef, Nabila A. Darwish, and Ahmed I. Abou-Kandil. 2019. "Eco-Friendly Polymer Composites for Green Packaging: Future Vision and Challenges." *Composites Part B: Engineering*. Elsevier Ltd. https://doi.org/10.1016/j.compositesb.2019.05.048.

Movassaghian, Sara, Olivia M. Merkel, and Vladimir P. Torchilin. 2015. "Applications of Polymer Micelles for Imaging and Drug Delivery." *Wiley Interdisciplinary Reviews: Nanomedicine and Nanobiotechnology* 7 (5). Wiley-Blackwell: 691–707. https://doi.org/10.1002/wnan.1332.

Mozafari, M. Reza, Chad Johnson, Sophia Hatziantoniou, and Costas Demetzos. 2008. "Nanoliposomes and Their Applications in Food Nanotechnology." *Journal of Liposome Research*. https://doi.org/10.1080/08982100802465941.

Neethirajan, S., and D. S. Jayas. 2011. "Nanotechnology for the Food and Bioprocessing Industries." *Food and Bioprocess Technology* :4 39–47. https://doi.org/10.1007/s11947-010-0328-2.

Nguyen, M. H., N. F. K. Aripin, X. G. Chen, and H. J. Park. 2014. "Methodologies Used for the Characterization of Nano-and Microcapsules." *Nano-and Microencapsulation for Foods* 65–94.

Nielsen, Lawrence E. 1967. "Models for the Permeability of Filled Polymer Systems." *Journal of Macromolecular Science: Part A—Chemistry* 1 (5): 929–942. https://doi.org/10.1080/10601326708053745.

Oehlke, Kathleen, Marta Adamiuk, Diana Behsnilian, Volker Gräf, Esther Mayer-Miebach, Elke Walz, and Ralf Greiner. 2014. "Potential Bioavailability Enhancement of Bioactive Compounds Using Food-Grade Engineered Nanomaterials: A Review of the Existing Evidence." *Food and Function*. Royal Society of Chemistry. https://doi.org/10.1039/c3fo60067j.

Olivas, G. I., and G. V. Barbosa-Cánovas. 2005. "Edible Coatings for Fresh-Cut Fruits." *Critical Reviews in Food Science and Nutrition*. https://doi.org/10.1080/10408690490911837.

Ong, Li Ching, Lei Yin Ang, Sylvie Alonso, and Yong Zhang. 2014. "Bacterial Imaging with Photostable Upconversion Fluorescent Nanoparticles." *Biomaterials* 35 (9): 2987–2998. https://doi.org/10.1016/j.biomaterials.2013.12.060.

Pathakoti, Kavitha, Manjunath Manubolu, and Huey Min Hwang. 2017. "Nanostructures: Current Uses and Future Applications in Food Science." *Journal of Food and Drug Analysis*. Elsevier Taiwan LLC. https://doi.org/10.1016/j.jfda.2017.02.004.

Pathakoti, Kavitha, Manjunath Manubolu, and Huey Min Hwang. 2020. "Mechanistic Insights into TiO2 and ZnO Nanoparticle-Induced Metabolic Changes in Escherichia Coli Under Solar Simulated Light Irradiation." *Water, Air, and Soil Pollution* 231 (1). Springer. https://doi.org/10.1007/s11270-019-4388-2.

Pendergast, Marytheresa M., and Eric M.V. Hoek. 2011. "A Review of Water Treatment Membrane Nanotechnologies." *Energy and Environmental Science*. https://doi.org/10.1039/c0ee00541j.

Pradhan, Neha, Surjit Singh, Nupur Ojha, Anamika Shrivastava, Anil Barla, Vivek Rai, and Sutapa Bose. 2015. "Facets of Nanotechnology as Seen in Food Processing, Packaging, and Preservation Industry." *BioMed Research International*. Hindawi Publishing Corporation. https://doi.org/10.1155/2015/365672.

Pyke, Daniel, and Christopher S. Tang. 2010. "How to Mitigate Product Safety Risks Proactively? Process, Challenges and Opportunities." *International Journal of Logistics Research and Applications* 13 (4): 243–56. https://doi.org/10.1080/13675561003720214.

Rahaiee, Somayeh, Seyed Abbas Shojaosadati, Maryam Hashemi, Sohrab Moini, and Seyed Hadi Razavi. 2015. "Improvement of Crocin Stability by Biodegradeble Nanoparticles of Chitosan-Alginate." *International Journal of Biological Macromolecules* 79 (August). Elsevier B.V.: 423–432. https://doi.org/10.1016/j.ijbiomac.2015.04.041.

Rashidi, Ladan, and Kianoush Khosravi-Darani. 2011. "The Applications of Nanotechnology in Food Industry." *Critical Reviews in Food Science and Nutrition* 51 (8): 723–730. https://doi.org/10.1080/10408391003785417.

Rasmussen, Kirsten, Hubert Rauscher, Agnieszka Mech, Juan Riego Sintes, Douglas Gilliland, Mar González, Peter Kearns, et al. 2018. "Physico-Chemical Properties of Manufactured Nanomaterials—Characterisation and Relevant Methods. An Outlook Based on the OECD Testing Programme." *Regulatory Toxicology and Pharmacology* 92 (February). Academic Press Inc.: 8–28. https://doi.org/10.1016/j.yrtph.2017.10.019.

Rattanapanone, Nithiya, Yuen Lee, Tianxia Wu, and Alley E Watada. 2001. "Quality and Microbial Changes of Fresh-Cut Mango Cubes Held in Controlled Atmosphere." *Hortscience* 36.

Rauscher, H., G. Roebben, A. Mech, N. Gibson, V. Kestens, T. P. J. Linsinger, and J. Riego. 2019. "An Overview of Concepts and Terms Used in the European Commission's Definition of Nanomaterial." https://doi.org/10.2760/459136.

Ravichandran, R. 2010. "Nanotechnology Applications in Food and Food Processing: Innovative Green Approaches, Opportunities and Uncertainties for Global Market." *International Journal of Green Nanotechnology: Physics and Chemistry*. Taylor and Francis Inc. https://doi.org/10.1080/19430871003684440.

Rezaei, Atefe, Milad Fathi, and Seid Mahdi Jafari. 2019. "Nanoencapsulation of Hydrophobic and Low-Soluble Food Bioactive Compounds within Different Nanocarriers." *Food Hydrocolloids*. Elsevier B.V. https://doi.org/10.1016/j.foodhyd.2018.10.003.C:\Users\LENOVO\Downloads\Query Integration (1)\Query Integration\24064-0309-Ref Mismatch Report.docx - LStERROR_29

Riaz, Asad, Shicheng Lei, Hafiz Muhammad Saleem Akhtar, Peng Wan, Dan Chen, Saqib Jabbar, Muhammad Abid, Malik Muhammad Hashim, and Xiaoxiong Zeng. 2018. "Preparation and Characterization of Chitosan-Based Antimicrobial Active Food Packaging Film Incorporated with Apple Peel Polyphenols." *International Journal of Biological Macromolecules* 114 (July). Elsevier B.V.: 547–555. https://doi.org/10.1016/j.ijbiomac.2018.03.126.

Robles-García, Miguel Angel, Francisco Rodríguez-Félix, Enrique Márquez-Ríos, José Antonio Aguilar, Arturo Barrera-Rodríguez, Jacobo Aguilar, Saúl Ruiz-Cruz, and Carmen Lizette Del-Toro-Sánchez. 2016. "Applications of Nanotechnology in the Agriculture, Food, and Pharmaceuticals." *Journal of Nanoscience and Nanotechnology*. American Scientific Publishers. https://doi.org/10.1166/jnn.2016.12925.

Roebben, Gert, Hubert Rauscher, Valeria Amenta, Karin Aschberger, Ana Boix, Sanfeliu Luigi Calzolai, Hendrik Emons, et al. n.d. "Towards a Review of the EC Recommendation for a Definition of the Term 'Nanomaterial' Part 2: Assessment of Collected Information Concerning the Experience with the Definition." https://doi.org/10.2787/97286.

Sahoo, Monalisa, Siddharth Vishwakarma, Chirasmita Panigrahi, and Jayant Kumar. 2021. "Nanotechnology: Current Applications and Future Scope in Food." *Food Frontiers*. John Wiley and Sons Inc. https://doi.org/10.1002/fft2.58.

Salgado, Pablo R., Luciana Di Giorgio, Yanina S. Musso, and Adriana N. Mauri. 2018. "Bioactive Packaging: Combining Nanotechnologies with Packaging for Improved Food Functionality." In *Nanomaterials for Food Applications*, 233–270. Elsevier. https://doi.org/10.1016/B978-0-12-814130-4.00009-9.

Sanchez-Garcia, Maria D., Amparo Lopez-Rubio, and Jose M. Lagaron. 2010. "Natural Micro and Nanobiocomposites with Enhanced Barrier Properties and Novel Functionalities for Food Biopackaging Applications." *Trends in Food Science and Technology*. https://doi.org/10.1016/j.tifs.2010.07.008.

Saulis, Gintautas. 2010. "Electroporation of Cell Membranes: The Fundamental Effects of Pulsed Electric Fields in Food Processing." *Food Engineering Reviews* 2 (2): 52–73. https://doi.org/10.1007/s12393-010-9023-3.

Savolainen, K., L. Pylkkänen, H. Norppa, G. Falck, H. Lindberg, T. Tuomi, M. Vippola, et al. 2010. "Nanotechnologies, Engineered Nanomaterials and Occupational Health and Safety—a Review." *Safety Science* 48 (8): 957–963. https://doi.org/10.1016/j.ssci.2010.03.006.

Sekhon, Bhupinder S. 2010. "Food Nanotechnology-an Overview." *Nanotechnology, Science and Applications*. Bhupinder S. Sekhon, 1–15. https://doi.org/10.2147/nsa.s12187498.

Shafiq, Mehwish, Sumaira Anjum, Christophe Hano, Iram Anjum, and Bilal Haider Abbasi. 2020. "An Overview of the Applications of Nanomaterials and Nanodevices in the Food Industry." *Foods* 9 (2). MDPI Multidisciplinary Digital Publishing Institute. https://doi.org/10.3390/foods9020148.

Shi, Cuiping, Yuning He, Mengzhen Ding, Yifen Wang, and Jian Zhong. 2019. "Nanoimaging of Food Proteins by Atomic Force Microscopy. Part I: Components, Imaging Modes, Observation Ways, and Research Types." *Trends in Food Science and Technology*. Elsevier Ltd. https://doi.org/10.1016/j.tifs.2018.11.028.C:\Users\LENOVO\Downloads\Query Integration (1)\Query Integration\24064-0309-Ref Mismatch Report.docx - LStERROR_32

Singh Sekhon, Bhupinder. 2014. "Nanotechnology in Agri-Food Production: An Overview." *Nanotechnology, Science and Applications*. Dove Medical Press Ltd. https://doi.org/10.2147/NSA.S39406.

Singh, Trepti, Shruti Shukla, Pradeep Kumar, Verinder Wahla, and Vivek K. Bajpai. 2017. "Application of Nanotechnology in Food Science: Perception and Overview." *Frontiers in Microbiology*. Frontiers Media S.A. https://doi.org/10.3389/fmicb.2017.01501.

Sogvar, Ommol Banin, Mahmoud Koushesh Saba, Aryou Emamifar, and Rahman Hallaj. 2016. "Influence of Nano-ZnO on Microbial Growth, Bioactive Content and Postharvest Quality of Strawberries During Storage." *Innovative Food Science and Emerging Technologies* 35 (June). Elsevier Ltd: 168–176. https://doi.org/10.1016/j.ifset.2016.05.005.

Sokamté Tégang, Alphonse, Pierre Désiré Mbougueng, Hippolyte Mouafo Téné, Nikaise Forestine Douanla Nodem, Nakkarike Manjabhat Sachindra, and Léopold Tatsadjieu Ngoune. 2022. "Application of the Spice Afrostyrax Lepidophyllus Mildbr as a Biopreservative and Seasoning Agent for Hot Smoked Fillets of *Pangasius hypophthalmus*." *Journal of Agriculture and Food Research* 10 (December). Elsevier B.V. https://doi.org/10.1016/j.jafr.2022.100449.

Sothornvit, Rungsinee. 2019. "Nanostructured Materials for Food Packaging Systems: New Functional Properties." *Current Opinion in Food Science*. Elsevier Ltd. https://doi.org/10.1016/j.cofs.2019.03.001.

Sridhar, Adithya, Muthamilselvi Ponnuchamy, Ponnusamy Senthil Kumar, and Ashish Kapoor. 2021. "Food Preservation Techniques and Nanotechnology for Increased Shelf Life of Fruits, Vegetables, Beverages and Spices: A Review." *Environmental Chemistry Letters*. Springer Science and Business Media Deutschland GmbH. https://doi.org/10.1007/s10311-020-01126-2.C:\Users\LENOVO\Downloads\Query Integration (1)\Query Integration\24064-0309-Ref Mismatch Report.docx - LStERROR_33

Srivastava, Vivek. 2021. "Ionic Liquid-Mediated Synthesis of Metal Nanoparticles." In *Research Anthology on Synthesis, Characterization, and Applications of Nanomaterials*, 1832–1853. IGI Global. https://doi.org/10.4018/978-1-7998-8591-7.ch076.

Tamayo, Laura, Manuel Azócar, Marcelo Kogan, Ana Riveros, and Maritza Páez. 2016. "Copper-Polymer Nanocomposites: An Excellent and Cost-Effective Biocide for Use on Antibacterial Surfaces." *Materials Science and Engineering C*. Elsevier Ltd. https://doi.org/10.1016/j.msec.2016.08.041.

Tarhan, Özgür. 2020. "Safety and Regulatory Issues of Nanomaterials in Foods." In *Handbook of Food Nanotechnology: Applications and Approaches*, 655–703. Elsevier. https://doi.org/10.1016/B978-0-12-815866-1.00016-9.

Thomsen, M. R., and A. M. McKenzie. 2001. "Market Incentives for Safe Foods: An Examination of Shareholder Losses from Meat and Poultry Recalls." *American Journal of Agricultural Economics* 83 (3): 526–538.

Tongo, Isioma, Ozekeke Ogbeide, and Lawrence Ezemonye. 2017. "Human Health Risk Assessment of Polycyclic Aromatic Hydrocarbons (PAHs) in Smoked Fish Species from Markets in Southern Nigeria." *Toxicology Reports* 4. Elsevier Inc.: 55–61. https://doi.org/10.1016/j.toxrep.2016.12.006.

Vandermoere, Frederic, Sandrine Blanchemanche, Andrea Bieberstein, Stephan Marette, and Jutta Roosen. 2011. "The Public Understanding of Nanotechnology in the Food Domain: The Hidden Role of Views on Science, Technology, and Nature." *Public Understanding of Science* 20 (2): 195–206. https://doi.org/10.1177/0963662509350139.

Verma, Navin K, Pankaj Kumar, Sardar Vallabhbhai Patel, Rakhi Verma, and S S Gaurav. 2011. *Prospects and Promises of Agronanotechnology for Improving Crop Productivity, Food Security, and Economy*. https://www.researchgate.net/publication/362412911.

Voss, D., and J. Whipple. 2009. "Food Supply Chain Security: Issues and Implications." In *Supply Chain Risk: A Handbook of Assessment, Management, and Performance*, 293–305. Boston, MA: Springer US.

Walters, Chavon, Edmund Pool, and Vernon Somerset. 2016. "Nanotoxicology: A Review." In *Toxicology—New Aspects to This Scientific Conundrum*. InTech. https://doi.org/10.5772/64754.

Wang, B., Hu, L. and Siahaan, T.J., 2016. *Drug delivery: principles and applications*. John Wiley & Sons.

Wang, C., and J. Irudayaraj. 2008. *Gold Nanorod Probes for the Detection of Multiple Pathogens*, 397. Birck and NCN Publications.

Weiss, Jochen, Sylvia Gaysinsky, Michael Davidson, and Julian McClements. 2009. "Nanostructured Encapsulation Systems." In *Global Issues in Food Science and Technology*, 425–479. Elsevier. https://doi.org/10.1016/B978-0-12-374124-0.00024-7.

Whipple, J. M., M. D. Voss, and D. J. Closs. 2009. "Supply Chain Security Practices in the Food Industry: Do Firms Operating Globally and Domestically Differ?" *International Journal of Physical Distribution & Logistics Management* 39 (7): 574–594.

Wu, Runrun, Yue Ma, Jianming Pan, Shih Hui Lee, Jinxin Liu, Hengjia Zhu, Runxin Gu, Kenneth J. Shea, and Guoqing Pan. 2018. "Efficient Capture, Rapid Killing and Ultrasensitive Detection of Bacteria by a Nano-Decorated

Multi-Functional Electrode Sensor." *Biosensors and Bioelectronics* 101 (March). Elsevier Ltd: 52–59. https://doi.org/10.1016/j.bios.2017.10.003.

Yildirim, A., A. Mavi, and A. A. Kara. 2001. "Determination of Antioxidant and Antimicrobial Activities of *Rumex crispus* L. Extracts." *Journal of Agricultural and Food Chemistry* 49 (8): 4083–4089. https://doi.org/10.1021/jf0103572.

Youssef, Ahmed M., Fayza M. Assem, Mahmoud E. Abdel-Aziz, Mostafa Elaaser, Osama A. Ibrahim, Mona Mahmoud, and Mohamed H. Abd El-Salam. 2019. "Development of Bionanocomposite Materials and its Use in Coating of Ras Cheese." *Food Chemistry* 270 (January). Elsevier Ltd: 467–475. https://doi.org/10.1016/j.foodchem.2018.07.114.

Yue, Chengyan, Shuoli Zhao, and Jennifer Kuzma. 2015. "Heterogeneous Consumer Preferences for Nanotechnology and Genetic-Modification Technology in Food Products." *Journal of Agricultural Economics* 66 (2): 308–328. https://doi.org/10.1111/1477-9552.12090.

Yun, Xiang, Victor D. Maximov, Jin Yu, Hong Zhu, Alexey A. Vertegel, and Mark S. Kindy. 2013. "Nanoparticles for Targeted Delivery of Antioxidant Enzymes to the Brain after Cerebral Ischemia and Reperfusion Injury." *Journal of Cerebral Blood Flow and Metabolism* 33 (4): 583–592. https://doi.org/10.1038/jcbfm.2012.209.

Zafar, Rabia, Khalid Mahmood Zia, Shazia Tabasum, Farukh Jabeen, Aqdas Noreen, and Mohammad Zuber. 2016. "Polysaccharide Based Bionanocomposites, Properties and Applications: A Review." *International Journal of Biological Macromolecules*. Elsevier B.V. https://doi.org/10.1016/j.ijbiomac.2016.07.102.

Zhang, Hai, Mercedes Hortal, Maria Jordá-Beneyto, Estela Rosa, Marta Lara-Lledo, and Inmaculada Lorente. 2017. "ZnO-PLA Nanocomposite Coated Paper for Antimicrobial Packaging Application." *LWT* 78 (May). Academic Press: 250–257. https://doi.org/10.1016/j.lwt.2016.12.024.

Zhang, Hongcai, and Shunsheng Chen. 2019. "Nanoparticle-Based Methods for Food Safety Evaluation." In *Evaluation Technologies for Food Quality*, 817–835. Elsevier. https://doi.org/10.1016/B978-0-12-814217-2.00032-9.

18 Unveiling the Potential of Essential Oils-Based Food Preservation, Challenges and Role of Nanotechnology in Overcoming Challenges regarding Applications of Essential Oils

Iqbal Ahmed[a,c], *Abdullah Ijaz Hussain*[b], *Ijaz Hussain*[b], *Muhammad Zahid Iqbal*[b], *Yasir Iqbal*[d], *and Sufyan Aslam*[b]

[a]University of Rome Tor Vergata, Via del Politecnico 1, Rome, Italy

[b]MPhil Analytical Chemistry, Government College University Faisalabad, Faisalabad, Pakistan

[c]Government College University Faisalabad, Faisalabad, Pakistan

[d]Department of Chemistry, Baba Guru Nanak University, Nankana Sahib, Nankana Sahib Pakistan

18.1 INTRODUCTION

Foodborne illnesses ordinarily identified as food diseases, which are mainly instigated by ingestion of tainted foodstuffs due to presence of microbes, for instance bacteria, molds, viruses along with ordinary contaminants, biochemical and physical proxies, etc. (Switaj, Winter et al. 2015). Annually, 1 out of 10 persons become sick as a result of foodborne diseases, claiming approximately 33 million lives worldwide. Foodborne diseases are caused by approximately 200 types of pathogens, pharmacological and physical influences. In industrialized states, 20–40% of colonic disorders are caused by foodborne pathogens (Sharif, Javed et al. 2018).

Consideration of the physical and chemical parameters that consequence of the foodstuffs spoilage in anyway, whether they are premeditated or materialized suddenly, has prime value that determine the foodstuff influence on consumer healthiness and insecurity regarding biological aspects (Majumdar, Pradhan et al. 2018). Bio-hazards are usually referred to the microorganism as well as by-products obtained by them, which can cause pernicious health hazards to humans through producing allergies, intoxications, infections and poisoning. Overall with the passage of time incidence rate regarding foodborne illness has been increasing, but it is clearly not tracked in the wake of all foodborne illnesses (Gallo, Ferrara et al. 2020; Theresa, Unni et al. 2021).

Nowadays, fresh agricultural foods as well as supplementary healthy stuffs are in extraordinary ultimatum round the sphere. Customers are identifying and inclining to these foodstuffs' supreme pleasant features, like freshness, nourishing remunerations, essence and pioneering savors with identical appearance later at customary expanse of a period (Donsi, Sessa et al. 2015). In many developed countries food safety has become a global issue. Contagious adulteration is the foremost hazardous factor for the safety of the foodstuff as well as human health. In recent times a considerable rise in pathogenic illness that are produced by microbial contamination has been noticed (Holban and Grumezescu 2018; Nowshad, Mustari et al. 2021). Food insecurity is not only a threatening alarm for human health but also the spoilage of food affects the economy of a country. It is possible to make sure the delivery of food to new and far off market to boost the global economy by prolonging the shelf life of foodstuff (De Corato 2020; Hossain, Skalicky et al. 2021). During this whole process, external factors like temperature, light and humidity can also lead to deprivation of food and to make sure safety of the foodstuff from these factors scientist had designed food packagings (Modi, Timilsina et al. 2021; Ali, Rezende et al. 2023).

All over the world foodborne sicknesses triggered by pathogenic contaminants account for billions of dollars in the manual well-being budget (Hoffman, Maculloch et al. 2015). Every year 25% of foodstuff is deteriorated, which is nearly equal to 1.3 billion tons and costs about one trillion dollar (Liegeard and Manning 2020). So preservation of foodstuff can save billions of dollars annually (Sawaya 2017). According to United Nations report it

DOI: 10.1201/9781032677422-18

is estimated that large portion of food produced is deteriorated having ranges from 30–50%, which by weight accounts approximately two billion tons (Fanzo and Davis 2021). According to a research conducted in 2015 by World Health Organization (WHO), it was estimated that every year six hundred million cases of foodborne illness report and round about 420,000 related deaths occur all around the world (Debnath and Roy 2020). According to a European agency food spoilage by consumers is also a big issue. It has been claimed that in underdeveloped and developed states food chains are affected by wastage of food because consumers purchase too much food. This is the reason that food preservation with modern techniques is a profligate growing area. Originating extraction, preservation and processing methods can increase the food production and will be less harmful to the environment because of less use of fossil energy, harmful materials and solvents, which also leads to prevention of waste (Chemat, Rombaut et al. 2017; Grumezescu and Holban 2018; Rejeb, Keogh et al. 2022). Awareness should be given to the consumers through proper channels how they can minimize food spoilage in their houses. Food-preserving techniques also play a vital role in controlling food contamination and ensuring the safer food utilization (Martindale and Schiebel 2017; Ahmed, Stewart et al. 2021).

People today choose foods that are natural and have undergone little preparation so to maintain freshness of food different preservatives are added to food, but their excess amount can cause serious warnings to human health (Shaw 2018; Flórez-Méndez and López 2022). First it is really imperative to identify the factors that deteriorate foodstuffs. Many factors can be responsible for food spoilage like water having microorganisms, toxins that are present in food stuff, improper storage conditions of food and utilization of pesticides. So it is needed to develop new favorable preservation techniques or methods (Adeyeye 2017). In the modern days, achieving food safety mostly depends on ensuring that food is available, utilized, stable and, most importantly, protected (Roman, Sánchez-Siles et al. 2017; Mbow, Rosenzweig et al. 2020).

18.2 TECHNIQUES FOR FOOD PRESERVATION

There are two approaches for preservation of food that are used from a long period of time. The first and anciently used method to preserve food is physical treatment. Physical method include freeze, heat drying, cold storage, microwave treatment as well as irradiation (Calín-Sánchez, Lipan et al. 2020; Indiarto, Asyifaa et al. 2021). But the chief issue is that food preserved by physical methods is not stable for longer time and microorganisms have developed resistance against them. Due to these reasons other methods for preservation are used (Matthews, Kniel et al. 2017). The thermal method has often been used and is a well-documented method to create the microbial mortality for the fight against diseases causing microorganism, and as a result it avoids danger to human health and increases the food span that is of commercial standards. The food that have a long shelf life and don't need refrigeration can be preserve by using thermal preservation techniques (Teixeira 2014; Augusto, Soares et al. 2018). In the thermal preservation approach the process parameters can be controlled with great ease while treating the food with it. The food may contain some anti-nutritional agents that are also destructed by this method, which is the main advantage of thermal treatment. For instance, in a few legumes trypsin inhibitor are destroyed to finish its anti-nutritional effect. This method also makes easy digestion and absorption of some constituents present in food items, like it makes easier digestion of some proteins as well as the gelatinization of starch (Augusto, Soares et al. 2018; Ibrahim 2020).

In addition to physical methods of preservation, food diligences have been using different types of chemicals to preserve food from different microorganism (Davidson, Taylor et al. 2012; Amit, Uddin et al. 2017). Despite usage of these chemicals to preserve food they also affect the sensory and organoleptic features of certain foodstuffs (Mahmud and Khan 2018; Ben Braïek and Smaoui 2021). However these chemical preservatives have undesirable side effects; they are cancer causing because they produce mutagenic changes, undergo degradation very slowly and are extremely toxic. Due to all these factors, many problems occur including environmental pollution and severe health problems; therefore their use should be minimized (Kumar, Singh et al. 2019). To lessen the uses of these approaches of processing and preservation, bio-based food stabilizers have been established. Bio-based food preservers are natural composites acquired from animals, plants and microbes that increase shelf lives of various foodstuffs (Pisoschi, Pop et al. 2018; Batiha, Hussein et al. 2021). To create new food preservatives against food deterioration, harmful microbes and to experience innovation in food packaging, the foodstuff sector has recently revealed a cumulative desire for natural compounds (Nilsen-Nygaard, Fernández et al. 2021; Novais, Molina et al. 2022). The reason for enhanced utilization of bio-based food preservation approaches is awareness regarding health benefits as well as sustainability (Versino, Ortega et al. 2023). Moreover, naturally acquired materials behave as antibacterial agents and are environmentally friendly, healthful, and secure because they are derived from natural ingredients (Al-Tayyar, Youssef et al. 2020). Nowadays, food safety has grabbed a lot of attention, and the demand for natural antimicrobial agents has enhanced (Mei, Ma et al. 2019; Punia Bangar, Chaudhary et al. 2021). Therefore, this issue has compelled the researchers for utilization of natural additives in foodstuffs to preserve them in accordance with market demand (Vasilaki, Panagiotopoulou et al. 2022).

18.2.1 Essential Oils as Food Preservatives

Essential oils are known as aromatic compounds with an oily appearance. EOs are regarded as volatile hydrophobic secondary metabolites of plant extracts containing a rage of oxygenated and non-oxygenated terpenes hydrocarbon, alcohols, carbonyl compounds, phenols, esters and methyl esters (Jafari, Rashidinejad et al. 2023). Most of the common

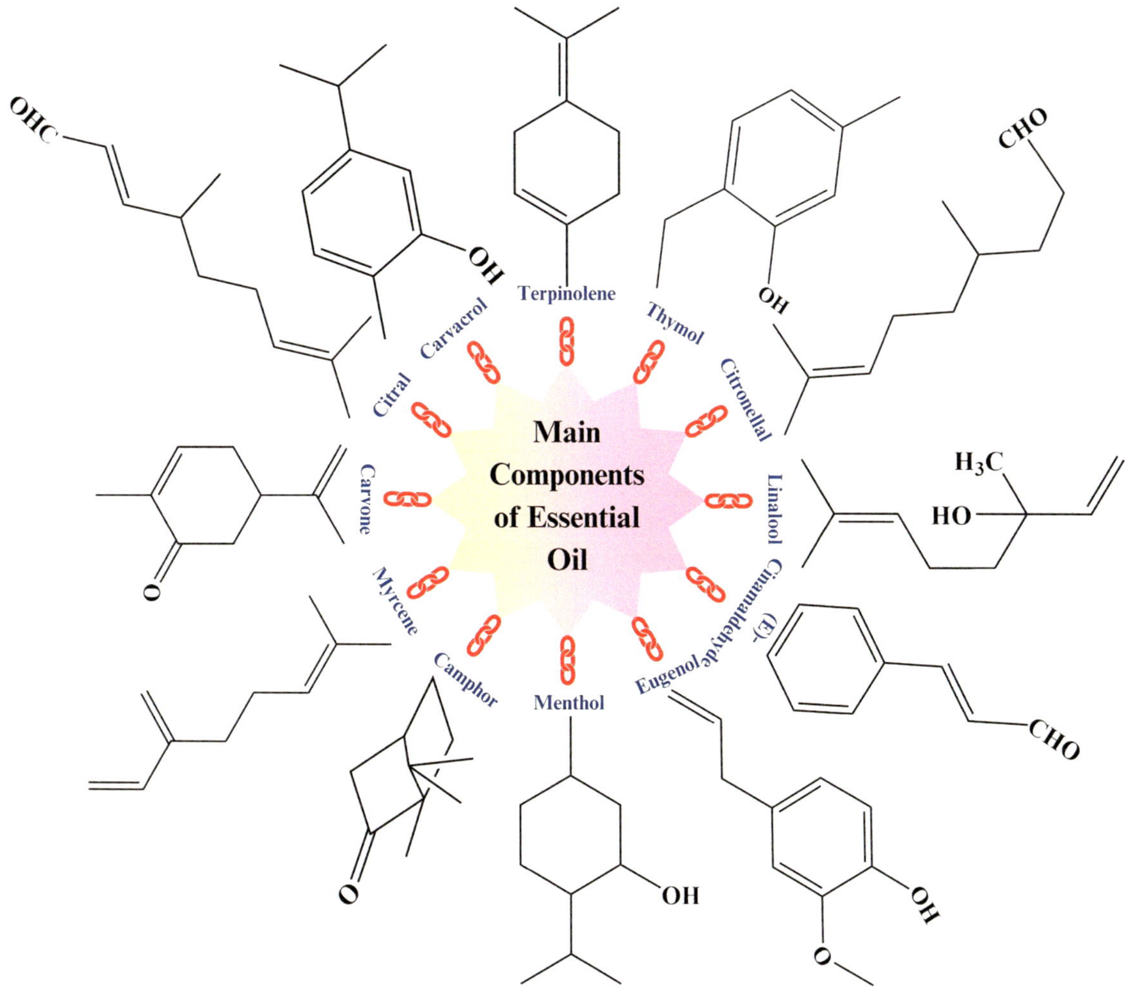

FIGURE 18.1 Components of essential oils.

types of components present in EOs have been presented in Figure 18.1. EOs could be obtained by diverse procedures like steam distillation from several fragments of floras for instance bark, foliage, fruit, wood, spores and stems, some of which have been explained in Table 18.1 (Mahian and Sani 2016). EOs are nontoxic and have been used as flavorings in foodstuff diligence (Ju, Xie et al. 2019). Many EOs provide a wide range of antimicrobial actions because they are effective against many bacteria and fungi (Ahmed, Stewart et al. 2021; Ju, Deng et al. 2022).

EOs are considered necessary to maintain the immune system of plants. So due to the potential of EOs they are used in the medical field to solve many health issues, and due to their antimicrobial activities they are used to gain antimicrobial effects. EOs also possess many medicinal properties. De La Croix first reported that EO as the secondary metabolites. Da La Croix discovered that mainly EOs vapors have the antimicrobial effect (Bhavaniramya, Vishnupriya et al. 2019). From time to time it has been discovered that EOs and phytoconstituents present in them have many biological activities including antibacterial and antifungal activity, which increase their demand to be used as preservative to attain antimicrobial effect (Chouhan, Sharma et al. 2017; Angane, Swift et al. 2022).

Owing to their antimicrobial activities EOs play a vital role in serving humanity in many fields (Napoli and Di Vito 2021). EOs are hydrophobic in nature so due to this nature they can easily disrupt the bacteria. Many mechanisms are proposed for the action of EOs by which they can kill or inhibit the bacterial growth (Angane, Swift et al. 2022). This crucial quality of EOs (hydrophobicity) assists them to fix with triglycerides in mitochondria and cell membranes of microbes, which reduces permeability and organization of the membrane (Chen, Zhang et al. 2021). Due to this less permeable cell membrane ions and cell content flows out, which in turn result in cell death (Jadhav, Pawar et al. 2023). EOs are active in different ways because they can show both single-target and multi-target behaviors (Trifan, Luca et al. 2020). The chief constituents, which are accountable for the antimicrobial action, are carvacrol, eugenol and thymol. All these compounds prevent development of microbes by disturbing bacterial cell membrane, and as a result they produce alternations in proton driving force, electron force, accumulation of the cell material and active transport (El Atki, Aouam et al. 2019; Ju, Xie et al. 2019; Souza, Almeida et al. 2022).

TABLE 18.1
Lists of Various Commonly Applied EOs Along with the Major Constituents and Sources

Plant	Segment of Plant	Major Constituents	References
Oliveria decumbens	Flower buds	Thymol (38.79%), Carvacrol (36.3%), Myristicin (7.75%) and gamma-terpene (4.71%)	(Eftekhari, Ardekani et al. 2019)
Boswellia ovalifoliolata	Leaves	Spathulenol (11.1%), Caryophyllene oxide (9.0%), decyl acetate (6.3%)	(Geetha and Chakravarthula 2018)
Boswellia ovalifoliolata	Bark	Caryophyllene oxide (10.5%) Beta-Farnesene (13.4%) and spathulenol (6.6%)	(Geetha and Chakravarthula 2018)
Salvia officinalis	Leaves	Camphore (33.6%), 1, 8-cineole (22.2%), alpha-thujone (21.4%) and camphene (4.9%)	(El Euch, Hassine et al. 2019)
Lavandula rejdalii	Leaves	Linalool (32.03%), Carvacrol (19.6%), Camphor (11.03%), 1, 8- cineole (9.67%) and lilanyl acetate (9.02%)	(Gharby, Asdadi et al. 2020)
Pelargonium graveolens	Flower stage	Menthol (20.57%), isogeraniol (9.14%), isoborneol (6.96%) and Linalyl acetate (2.97%)	(Al-Mijalli, Mrabti et al. 2022)
Eucalyptus. gunnii	Twigs	1, 8-cineole (65.6%), α-terpinyl acetate (7.6%), ortho-cymene (7.5%), α-pinene (2.9%) and globulol (2.0%)	(Abbaci, Nabti et al. 2023)
Eucalyptus. gunnii	Leaves	1, 8-cineole (83.3%), α-terpinyl acetate (2.5%), α-terpineol (4.6%) and ortho-cymene (3.3%)	(Abbaci, Nabti et al. 2023)
Eucalyptus. gunnii	Fruits	1, 8-cineole (86%), α-terpinyl acetate (4.2%), ortho-cymene (3.3%), α-terpineol (3.5%) and α-pinene (1.1%)	(Abbaci, Nabti et al. 2023)
Cistus ladanifer	Leaves and twig	Viridiflorol (17.74%), trans-pinocarveol (11.02%), bornylacetate (9.38%), p-Cymene (6.11%) and borneol (4.80%)	(El Karkouri, Bouhrim et al. 2021)
Montha arvensis	Leaves	Carvone (23.53%), p-Cymen-2-ol (20.35%), Caryophyllene oxide (18.81%), Borneol (9.56%) and Caryophyllene (1.61%)	(Bibi, Ali et al. 2021)
Citrus limon var. pompia	Leaves	Limonene (29.8%), myrcene (9.7%), beta-Caryophyllene (9.6%), linalool (9.1%) and geraniol (8.9%)	(Fancello, Petretto et al. 2016)
Perilla frutescens	Leaves	Perilla ketone (80.8%), β-caryophyllene (1.6%) and linalool (1.15%)	(Chen, Liu et al. 2020)
Heracleum lehmannianum	Flowers, Leaves, stem	cis-allo-ocimene (17.6%), δ-3-carene (14.2%), phellandrene (10.5%), limonene (7.6%), alpha-cadinol (5.7%), gamma-cadinene (6.2%)	(Mamadalieva, Abdullaeva et al. 2018)
Plectranthus asirensis	Leaves and stem	Alpha-Pinene (46.2%), maaliol (42.8%) and β-caryophyllene (13.3%)	(Mothana, Khaled et al. 2018)
Callistemon viminalis	Leaves and stem	1, 8 Cineole (58.1%), limonene (9.7%), alpha-terineol (9.6%), geranial (6.02%), Gamma-elemene (3.5%) myrcene (2.96%) and alpha- pinene (2.49%)	(Fall, Ngom et al. 2017)

18.2.1.1 Biological Activities of EOs in Food Sector

EOs are natural extracts that possess well-documented antioxidant and antimicrobial characteristics, so they have been a matter of great interest by the food industries for a significant period of time (Basavegowda and Baek 2021). In addition to these activities, other biological potentials of EOs have been elaborated in Figure 18.2. EOs have many phytochemical constituents, and natural extracts are referred to as antimicrobial compounds and may be efficient in various implementations for decreasing the growth and survival of microorganisms. Due to their inherent features EOs are recognize as natural preservatives and are frequently used in food products because they satisfy modern consumer demands for food free of synthetic preservatives (Shaaban 2020; Prakash, Singh et al. 2023). EOs have strong antioxidant and antimicrobial activities, which is why scientists are taking great interest in their solicitations in foodstuff protection (Calo, Crandall et al. 2015).

The utilization of renewable sources as well as biodegradable constituents have recently garnered particular consideration due to the increase in global inhabitants and growing devotion of manufacturing companies and customers to inadequate possessions and ecological safety (Nilsen-Nygaard, Fernández et al. 2021). For example rather than using synthetic packaging materials proteins, lipids, and carbohydrate are employed as polymeric material to make food packaging systems (Mohamed, El-Sakhawy et al. 2020). Antibacterial packaging that includes EOs has

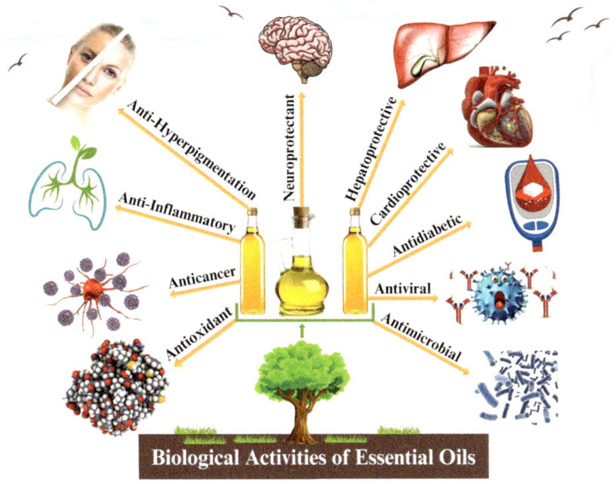

FIGURE 18.2 Activities of essential oils.

gained an increased popularity in food packaging. These active packagings, which can protect foodstuffs from toxicities, are mostly made of various EOs and polymer matrix (Ahari and Soufiani 2021; Siroha and Bangar 2023).

Usually EOs are known as complex combination of various metabolites of secondary importance, i.e., aldehydes, terpenoids, phenolic, ethers and phenylpropenes (Jafari, Rashidinejad et al. 2023). They mainly have the potential of antioxidant and antimicrobial activity. EOs can be effectively used to fight a broad range of the microorganisms that are disease causing such as molds, toxins and bacteria found in the food in the wake of their complex nature (Prakash, Singh, Kedia et al. 2012). Moreover, utmost of these EOs constitute less poisoning and are transitory in the environment, and thanks to these properties they are safe to use with respect to environment and health as compared to the other synthetic chemicals (Prakash and Kiran 2016; Al-Maqtari, Rehman et al. 2021).

In latest ages there have been observed the discovery of EOs and their bioactive constituents, which expressed the high activity against foodborne bacteria, molds and oxidative degradation of food (Kiran, Kujur et al. 2016; Rout, Tambe et al. 2022). According to the recorded documentation EOs are being used effectively as antimicrobial agents to compete with a lot of foodborne pathogens which include *Typhimurium*, *Salmonella*, *Staphylococcus aureus*, *Escherchia coli*, *Listeria monocytogenes* and *Campylobacter* (Kerekes, Vidács et al. 2019; Batiha, Hussein et al. 2021). Jasmine, oregano, celery seeds, carrot seed, ginger roots, clove bud, bay leaf, orange bitter oils, cinnamon and thyme are among the oils that have the strongest antimicrobial effects against strains of *Staphylococcus aureus*. Thyme, oregano, marjoram and cinnamon all spices, bay leaves as well as other ingredients also have some antimicrobial effects against strains *of S. enterica* (Kiran, Kujur et al. 2016; Friedman 2017; Cordery, Rao et al. 2018).

Liang et al. (2011) extracted and examined EOs from 10 herbs normally used against many diseases-causing germs and degradation of microorganisms. They founded higher activity against a wide range of pathogens from the vapors of EOs. Carvacol was the major constituent with the highest activity. All of the investigated EOs were vigorous contrary to Gram-positive bacteria as compare to Gram-negative bacterial cultures, which show that these are the highly resistant bacteria causing the spoilage of the products (Liang et al. 2011).

Cherrat, Espina et al. (2014) extracted EOs from *Myrtus communis* and *L. nobilis* and checked for antimicrobial activity via in vitro analysis against various disease-causing strains. The vapors of EOs inhibited the growth of all the pathogens to some extent, but a few components exhibited possession of higher activity as compared to other ones. Like bacterial strains, the components of EOs obtained from *L. nobilis* and *M. cummunis* exhibited normal to slightly higher activity. In general as compared to Gram-negative bacteria these two EOs were further vigorously contrary to Gram-positive bacteria. *S. aureus* was the only bacteria that was only marginally inhibited by both EOs vapors as compared to *L. monocytogens* and *E. coli* (Cherrat, Espina et al. 2014). Similarly, investigation of black pepper and its main constituent piperin's antimicrobial properties showed that they produce greater mortality to compete with organisms present in the foods having bacterial culture in it (i.e., *Salmonella* and *S. aureus)* (Swathy, Mishra et al. 2018; Abd El-Hack, El-Shall et al. 2022).

EOs are obtained in the form of by-products from citrus compounds, and they are used for antimicrobial action in contradiction of a lot of foodborne mutagens. Among the EOs many have displayed that they possess strong antimicrobial activities (De-Montijo-Prieto, Razola-Díaz et al. 2021; Salanță and Cropotova 2022). In a study, seven EOs were extracted to check their antimicrobial action against eleven strains of *Salmonella*. Orange terpenes, d-limonene and terpenes obtained from its essence all displayed inhibitory action against the selected bacterial culture. So MIC values were calculated for d-limonene as well as orange terpenes, which were recorded to have the MIC value of 1.1% whereas the terpenes obtained from the orange essence were recorded to have the MIC that has the range from .0115% to .06% against the 11 selected strains (O'Bryan, Crandall et al. 2008). In addition to this, Nannapaneni, Chalova et al. (2009) examined seven different fractions obtained from orange oil to check their capability to obstruct development of *Campylobacter* as well as *Acrobacter* (Nannapaneni, Chalova et al. 2009). Similarly, eugenol and carvone have antimicrobial potential against *Fusarium oxysporum* and *Aspergillus flavus* (Stević, Berić et al. 2014). Table 18.2 elaborates on the antimicrobial properties of various EOs components.

Some EOs and the components present in them show their antimicrobial activity in a better way at a low storage temperature (Reyes-Jurado, Franco-Vega et al. 2015). EO acquired as orange juice was used to hinder the bacterial straining in meat products. For this purpose, EO after inoculation was applied to the pieces of meat on the outer surface by spraying at concentration of 3% and 6%. The

TABLE 18.2
Bioactivity of Some Essential Oil Components

Essential Oils	Active Component	Experimental Approaches	Bioactivity	References
Zanthoxylum alatum Roxb	Methyl cinnamate (19.73%), Linalool (56%)	Using PDA medium, antimicrobial activity was evaluated. To achieve the correct concentrations (.25–5.0 micro liter/ml), the necessary volumes of EOs were separately dissolved in petri dishes containing PDA medium.	Exhibited remarkable effectiveness against strains such as *Curvularia lunata*, *Alternaria niger*, *Aspergillus terreus*, *Aspergillus alternate*, *Aspergillus flavus* and Aflatoxin B1 and exceptional antioxidant activity was demonstrated.	(Prakash, Singh, Kedia et al. 2012; Prakash, Singh, Mishra et al. 2012; Prakash, Baskaran et al. 2020)
Cinnamomum zeylanicum	(E)-Cinnamaldehyde (69.15%), Phenol, 2-mrthoxy-3–(2-propenyl-(2.96%),	Antibacterial activities were examined by following nutrient agar well and disk diffusion method. Antifungal was determined by broth microdilution method.	Displayed antibacterial activity against strains like, *S. thyphimurium, E. coli, B. cereus* and antifungal against *C. albicans, C. dubliniensis, C. parapesilosis* and antioxidant potential.	(Valizadeh, Katiraee et al. 2015)
Rosmarinus officinalis L	1, 8- cineole (32.6%), alpha-pinene (24.5%) and beta-pinene (10.2%)	Antifungal activity was conducted by PDA medium by dissolving EO in petri plates. Antioxidant activities were determined by various assays.	Higher antifungal activity against *F. proliferatum.* and *F. moniliforme* and enhanced antioxidant activity.	(Barakat and Ghazal 2016)
Cuminum cyminum L	Cymenen (47%), gamma-terpinene (19.64%), cuminaldehyde (15.03%)	In order to measure antimicrobial activity a poisoned food assay was used. To attain the desired concentration (.1–1µl/ml), the necessary amount of EO was individually dissolved in a petri plate using Potato dextrose broth medium.	Demonstrated excellent antifungal properties against chickpea seed molds like *mycelium*. Also reduces Aflatoxin B1 and fatty acids deterioration.	(Kedia, Prakash, Mishra, Chanotiya et al. 2014; Kedia, Prakash, Mishra and Dubey 2014)
Cananga odorata Hook f. and Thoms	Benzyl benzoate (34%) Linalool (25%) and Benzyl acetate (10%),	Disc and well diffusion methods were used to analyze EOs antimicrobial potential to achieve the necessary concentration, the appropriate amount of EOs was dissolved in appropriate solvents.	Demonstrated significant antibacterial action against different types of bacterial strains other likewise display antifungal potential against various strains also possess potential against other food-rotting microorganisms.	(Kusuma, Arung et al. 2014; Tan, Lee et al. 2015)
Curcuma zedoria Rosc.	Iso-borneol (31.2%), beta-elemene (14.70%), beta-caryophyllene (11.4%) and 1, 8-cineole (9.7%)	Microdilution broth method was performed to assess antimicrobial potential of EO which were diluted in micro-tubes and assess by Elisa plate reader.	Revealed higher antimicrobial potential contrary to *E. coli, S. aureus*, and *candida albicans yeast*.	(Thin, Thinh et al. 2022)
Curcuma longa L.	ar-turmerone (.9–43%), beta-turmerone (5–42.5%), alpha-zingiberene (.25–25%), ar-curcumene (1.2–15.7%)	Various commonly used antioxidant assays were performed to assess antioxidant potential while cytotoxicity as well as anti-inflammatory assay were also conducted to assess biological potential.	Demonstrated significant antioxidant potential and anti-inflammatory action along with cytotoxicity against B16 cells and LNCaP cells.	(Zhang, Yang et al. 2017; de Oliveira Filho, de Almeida et al. 2021)
Mentha spicata L	Carvone DL (60.0%) and Limonene (20%)	Broth microdilution was used to measure antimicrobial activity. To achieve the desired concentration (9.7–2500 micro liter/ml), necessary amount of EO was separately dissolved.	Exhibited strong anti-cholinesterase and antimicrobial efficacy.	(Kedia, Prakash, Mishra, Chanotiya et al. 2014; Kedia, Prakash, Mishra and Dubey 2014)

(*Continued*)

TABLE 18.2 (Continued)

Essential Oils	Active Component	Experimental Approaches	Bioactivity	References
Coriandrum sativum L	Linalool (60–80%), g-terpinene (1–8%), a-pinene (0.2%–8.5%), Camphor (0.9%–5%)	Disc, well and broth diffusion methods were used to analyze antimicrobial efficacy. To achieve the specified concentrations (20%, 25%, 33% and 50%), the necessary amount of essential oil was dissolved in sterile distilled water.	Demonstrated significant antibacterial activity in contrast to numerous bacterial straining comprising *S. aureus*, *E. coli*, *B. cerius* and some pathogenic molds *A. alternate*, *P. oryzae*, *B. oryzae*, *F. oxysprorum*.	(Mandal and Mandal 2015; Al-Khayri, Banadka et al. 2023)
Thymus vulgaris L	Thymol (42.9%) and Alpha-cymene (48%)	Agar disc diffusion mode was utilized to evaluate antimicrobial efficacy.	Show potential against common pathogens including *S. aureus*, *C. jejuni*, *S. enteritidis* and *L. monocytogenes*, all of which have influence in common foodborne diseases.	(Pesavento, Calonico et al. 2015; Galovičová, Borotová et al. 2021)
Cymbopogan citrus (DC.) Stapf.	Geranial (42.23%) and Neral (31.01%)	For the evaluation of antimicrobial activity broth microdilution method was used.	Show strong activity in contrast to *S. aureus*, *E. faecalis*, *E. coli*, *M. catarrhalis* and two yeast species *Candida tropicalis* and *Candida albicans*.	(Singh, Shukla et al. 2010; Ali, Yusuf et al. 2017)
Artemisia dracunculus L	Methyl eugenol (36.01%), Terpinolen (18.91%), and Methyl chavicol (16.6%)	Different assays were performed to assess antioxidant potential of EO. Moreover agar disc diffusion method was employed for analyzing antimicrobial potential.	Demonstrated potent antioxidant potential and antimicrobial potential contrary to yeast, *S. epidermidis*, *E. coli* and *A. flavus* as well as two foodborne molds *C. albicans* and *C. neoformans*.	(Lopes-Lutz, Alviano et al. 2008)

essential oils amazingly decreased progression of *E. coli* in contrast to meat that was not sprayed or the meat that was sprayed and was stored in refrigerator for 90 days. Similarly, it was found that the cold pressed terpenes less Valencia oil exhibit higher efficiency at refrigeration temperature to obstruct the progression of many cultures of *E. coli* separated from the beef (Pittman, Pendleton et al. 2011).

By reviewing the literature, it is clear that on the basis of their flavor and advantageous health characteristics, citrus oils have been included in diets for a long time (EM Mustafa 2015; Bora, Kamle et al. 2020). EOs as the substitutes of synthetic preservers are not only used for food preservation but also have been used in the medical field from ancient times, and well-known investigations have been performed in order to discover the bioactivity of these compounds to fight against bacteria, molds and yeasts (Fisher and Phillips 2008; Settanni, Palazzolo et al. 2012; Jugreet, Suroowan et al. 2020). Roughly 400 compounds derived from these citrus oils are now known, and their relevant capability depends upon the cultivar that is a special citrus in these oils, as well as the extraction and isolation techniques (Tao, Zhang et al. 2009; Mahato, Sinha et al. 2019). Some of the most commonly studied EOs have been reported along with their active components and food-related studies in Table 18.2. It is a matter of fact that EOs are very challenging to disperse in H_2O due to their hydrophobic nature; that's why their large concentrations are required to use in order to be effective in foods. EOs are also degraded when exposed to light, oxygen or high temperature, and consequently their activity is reduced (Turek and Stintzing 2013; El Asbahani, Miladi et al. 2015).

18.2.1.1.1 Mechanism of Action of EOs

Over the years EOs have been employed to combat a variety of bacteria due to their antimicrobial characteristics, but the mechanism of action of these compounds are yet not perfectly determined. A lot of mechanisms have been demonstrated that show chemical components present in the EOs (Burt 2004; Faleiro 2011; Rao, Chen et al. 2019). EOs are combination of many chemical substituents so their effects as antimicrobial agents cannot be determined only on the basis of one component (Bajpai, Baek et al. 2012; Angane, Swift et al. 2022). According to many researchers' points of view, EOs have the ability to enter the cell through bacterial membranes and impart inhibitory activity to function of the cell as well as lipophilic properties. So their antimicrobial activity can be due to this ability (Yap, Yusoff et al. 2021).

The phenolic components present in EOs are also a sign of their antibacterial effectiveness against the bacteria that cause foodborne illnesses (Shapira and Mimran

2007; Ju, Xie et al. 2019). These chemicals have ability to damage cell membranes, which will ultimately impact the cell's ability to operate and thus many internal constituents present in the cell get escaped and the cell becomes dead (Bajpai, Baek et al. 2012; Andrade-Ochoa, Chacón-Vargas et al. 2021). The potential of these chemicals to vary bacteriological cell penetrability, affect cell energy reservoir (ATP), destroy cytoplasmic membranes, affect the development scheme and affect the proton drive energy that may be associated with mechanism of cytoplasmic membrane impairment (Friedly, Crandall et al. 2009; Chouhan, Sharma et al. 2017).

Owing to their hydrophobic character EOs and their constituents have the capacity to segregate internal substances, which makes the cell membrane of microorganisms more permeable (Lopez-Romero, González-Ríos et al. 2015). Few Gram-positive and Gram-negative strains have their growth inhibited by EOs, which are being incorporated into cell membranes (Pateiro, Munekata et al. 2021). Gram-positive bacteria (*S. aureus* or *B. cereus*) in contrast to Gram-negative bacteria (*E. coli or S. enteritidis*) are more vulnerable to the EOs (Chorianopoulos, Kalpoutzakis et al. 2004). Since hydrophobic EOs' components directly interact with cell membranes, it is usually supposed that EOs must be further active instinctively to fight with Gram-positive strains (Angane, Swift et al. 2022). Gram-negative cells, on the other hand, are most resistant to plant EOs because of their hydrophilic cell walls. This top layer helps to prevent hydrophobic chemicals from penetrating (Calsamiglia, Busquet et al. 2007). Both these types of bacteria react with citrus EOs and their constituent parts in the same way. Thymol and carvacrol were said to have different impacts on both types of bacteria. The Gram-negative straining outer membrane may be ruptured by carvacrol as well as thymol, which leads to discharge of lipopolysaccharides (LPS) and increases the membrane permeability to ATP (Burt 2004; Hou, Sana et al. 2022). The mode of action of EO against microbial contaminants has been elaborated in Figure 18.3.

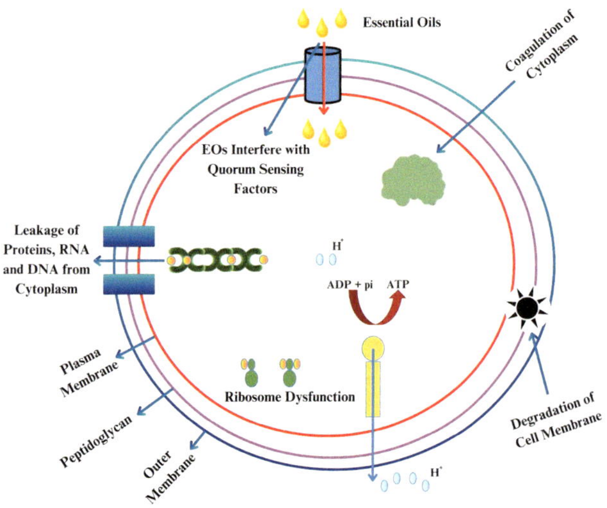

FIGURE 18.3 Mechanism of essential oil.

18.2.2 Reasons for Limited Applications of Pure EOs

Commonly EOs have less solubility in polar solvents (i.e., water) and are highly volatile compounds (El Asbahani, Miladi et al. 2015). Their organoleptic properties (flavor properties) affect the flavor of the matrix material and are affected by higher temperature, oxygen and radiations. It is a big challenge to use pure EO in cognitive food due to their interaction with food matrix (Turek and Stintzing 2013). Encapsulation of EOs comprises of marvelous and authentic approach to cover all these difficulties. In fact, this approach maintains flavor, enhance solubility, and minimize vaporization and physical instability with that of decrease in incorporation with material of food, bioavailability of EOs and improving bioaccessibility as shown in Figure 18.4 (Delshadi, Bahrami et al. 2020; Napiórkowska and Kurek 2022).

Bioactive components present in EO drops are recuperated by coating or cordoned off by homogenous or heterogeneous medium during the encapsulation process to create tiny capsules (Bakry, Abbas et al. 2016; Ferreira and Nunes 2019). These nanoscale capsules have a size of less than

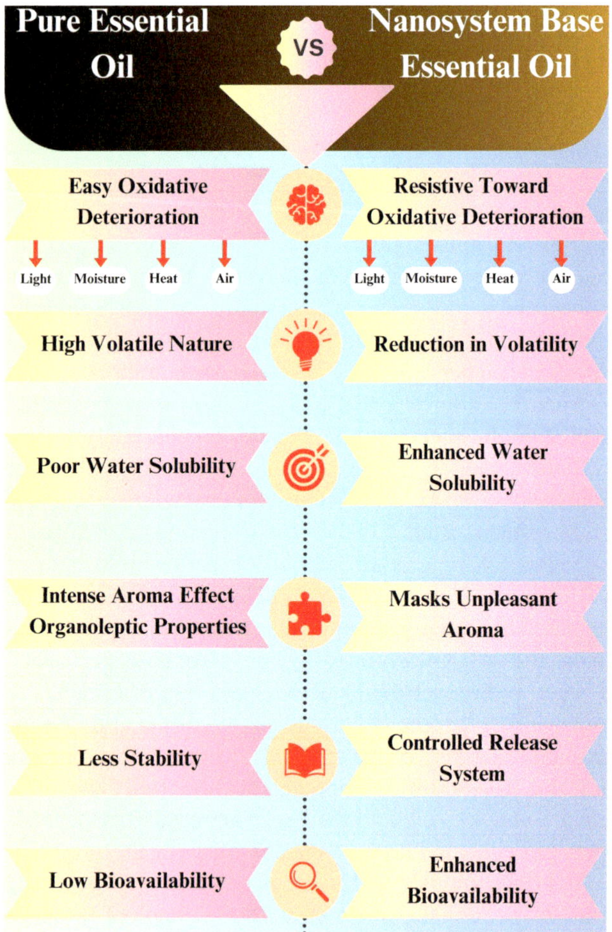

FIGURE 18.4 Comparison of pure and encapsulated essential oils.

1000 nm. Encapsulation is an auspicious approach for sustaining inherent EO potentials throughout time (Saifullah, Shishir et al. 2019). So as to sustain their biological activity, to maximize their operative procedure, to decrease their volatility and to lessen adversative influences of aroma, EOs are compressed in carriage systems, which are acceptable for foodstuffs applications (Al-Maqtari, Rehman et al. 2021). Encapsulation of EOs reduces their volatility while also expanding their surface area in contact with food that are water soluble. Encapsulated EOs are simple to put on food surface where microbes thrive and proliferate (Donsì, Annunziata et al. 2011).

Nanoencapsulation starts with formation of nanoemulsions. Nanoemulsions are materials that are formed by the mixing of oily and aqueous phases. Emulsifiers are primarily used to emulsify nanoemulsions. Additionally nanoemulsions have small drop sizes and large surface areas when they are prepared (Gupta, Eral et al. 2016; Gupta and Variyar 2016; Jadhav, Koli et al. 2020). These properties provide them with a potential advantage over conventional emulsions, like higher bioavailability and good physical stability. For the formation of oil nanoemulsions and oil nanoencapsulation different methods like spray drying, nanoprecipitation, ionic gelation, emulsion-diffusion, interfacial deposition, emulsification, solvent vaporization, use of liposomes, nanostructured lipid carrier and spontaneous emulsification are used (Ferreira and Nunes 2019; Reis, Ambrosi et al. 2022).

There are two major categories of encapsulation like microencapsulation and nanoencapsulation. Both these types have their own properties but nanoencapsulation as compared to microencapsulation has more advantages because in this technique size of obtained material is small, i.e., subcellular size, and thus have the capability to more penetrate and initiate the internal phases (Saifullah, Shishir et al. 2019). As compared to large capsules, the nanosystem has increased potential of EOs in place of the food where more germs are located like the places with high water amount or solid liquid interphases. Nanosystem offers greater surface area per unit volume (Behera, Mittu et al. 2020). Capsule obtained as a result of nanoencapsulation may perform functions like reservoir and a maintained release approach for bioactive components, and this supports their transportation to the bacterial cell (Beyki, Zhaveh et al. 2014).

The European Commission reported that a number of important EOs are intended for their usage in flavoring food matrices. The EOs that are registered as per standards and rules are cinnamaldehyde, limonene, citral, eugenol, vanillin, thymol, carvone and linalool. They all are understood to be safe for human health and have no risk to use by the consumer. All EOs having the GRAS (generally regarded as safe) stated by FDA include thyme, cinnamon, oregano, nutmeg, basil, mustard and clove (Llana-Ruiz-Cabello, Pichardo et al. 2015; Tsitlakidou, Tasopoulos et al. 2023). But there are limitations about how much concentration of EOs can be taken. So before their use a daily intake survey is taken, which is provided by the FDA, it is investigated that EOs and their constituents possess in vitro great potential, but their uses as the food preservers are inadequate since great quantities of EOs are prerequisite to gain desired antimicrobial properties (Calo, Crandall et al. 2015).

In many foodstuffs preservations the EOs components that are hydrophobic in nature are applied by incorporation with food matrix compounds, like fat and starch, as well as proteins (Sharma, Barkauskaite et al. 2021). Moreover the optimum pH value is also important for the EOs components to show efficient activity. It is correspondingly validated that antibacterial potential of EOs also depends on microbial contamination and temperature optimality (Moosavy and Shavisi 2013; Kotba, Bouaichi et al. 2020). In vitro results are significantly explored compared to in vivo, so in vivo is very tough to achieve the best action, and a reduced activity of the applied compound may be observed. Just consider the cilantro oil, which has an explored in vitro activity of inhibition at .019%, but when it is incubated into the food model, even the concentration of 6% cilantro oil is totally failed to hinder the growth of microbial culture present in that food model (Gill, Delaquis et al. 2002). So it is essential to determine the way EOs or their constituents develop interaction when added to the food products.

Moreover, the intense fragrance of the volatile components of the EOs, even at low concentration, damage the organoleptic possessions of foodstuff and consequently make it unacceptable for customers (Mariod 2016; Al-Maqtari, Rehman et al. 2021). In concerns regarding compensation of interaction of EOs for higher activity, if we enhance their concentration, it is also unfortunate because it can be used only in highly spicy foods in which acceptable threshold is relatively larger (Hyldgaard, Mygind et al. 2012). To eliminate this problem many different strategies can be taken into account. One way is that EOs should not be used directly as an ingredient in the food; instead they can be applied in active packaging (Yildirim, Röcker et al. 2018). The second option is to encapsulate the EOs in biodegradable polymers and edible coatings as a result the release rate on the surface of food becomes slow, e.g., meat products, fish and fruit (Stoleru and Brebu 2021; Zhang, Jiang et al. 2022). The benefit of this encapsulation include protection of the volatile components of EOs or active compounds, which will remain on the headspace for a longer period on the food system and its diffusion rate will be decreased (Becerril, Nerín et al. 2020; Yang, Qin et al. 2023). So, encapsulation is a proficient tool to avoid the problems of affecting sensory possessions of foodstuff. These processes of preservations increase physical stability of components by decreasing their volatility and reducing their interaction with the foodstuffs and increasing their antibacterial properties because of enhanced inert cellular endorsement (Donsì, Annunziata et al. 2011).

18.3 NEED FOR ENCAPSULATION OF EOS

Nanoencapsulation and microencapsulation are the two forms of encapsulation. The main difference between microencapsulation and nanoencapsulation is that microencapsulation uses larger particles as compare to nanoencapsulation and has been applied for an extended epoch of time (Saifullah, Shishir et al. 2019). Nanoencapsulation has potential to raise concentration of EOs in the foodstuffs matrix where microbes are favorably situated, like liquid or liquid-solid interfaces, due to their large superficial area and subcellular size (Granata, Stracquadanio et al. 2018).

Due to their chemical and functional characteristics, EOs possess a wide range of uses, including fragrances as well as flavorings in food, personal care products, and cosmetics. Additionally, these EOs' antimicrobial and antioxidant properties are much more significant (Bolouri, Salami et al. 2022). These EOs are primarily composed of mono and sesquiterpene hydrocarbons as well as their oxygenated by-products, though they also contain secondary metabolites from plant sources. Microencapsulation has emerged as a substitute to lessen the volatility, release rate and upsurge stability of these EOs. The material of interest is encapsulated in a polymer matrix during the microencapsulation process, with spray dryer being the most prevalent method (Bakry, Abbas et al. 2016). To prevent chemical as well as somatic responses and retain functional, biological and physicochemical features of the major component, microencapsulation is a method of generating efficient obstruction among central and wall ingredients. Depending on the product's intended use and the processing circumstances, the proper technique and wall material must be chosen (Wang 2015). In contrast, nanoencapsulation, which involves encasing oils in capsules ranging in size from 10–1000 nanometers, not only protects the oils but also significantly boosts their antibacterial, antifungal, antiviral and insecticidal, as well as antiparasitic, activities while maintaining their other medicinal properties (Gupta and Variyar 2016). Because of its subcellular dimensions, the nanometric-size delivery method has prospective to improve the mechanism of passive cellular uptake, thereby lowering the resistance to mass transfer and enhancing a variety of activities (Burt 2004; Weisany, Yousefi et al. 2022). The methods of preserving EOs at nanoscale include different methods, which have been elaborated in Figure 18.5.

18.3.1 Encapsulation Techniques

Techniques for encapsulation encompass a wide variety of subfields within chemistry, including physical chemistry, colloid chemistry, material chemistry, polymer chemistry and also drying and suspension technologies (Figure 18.6).

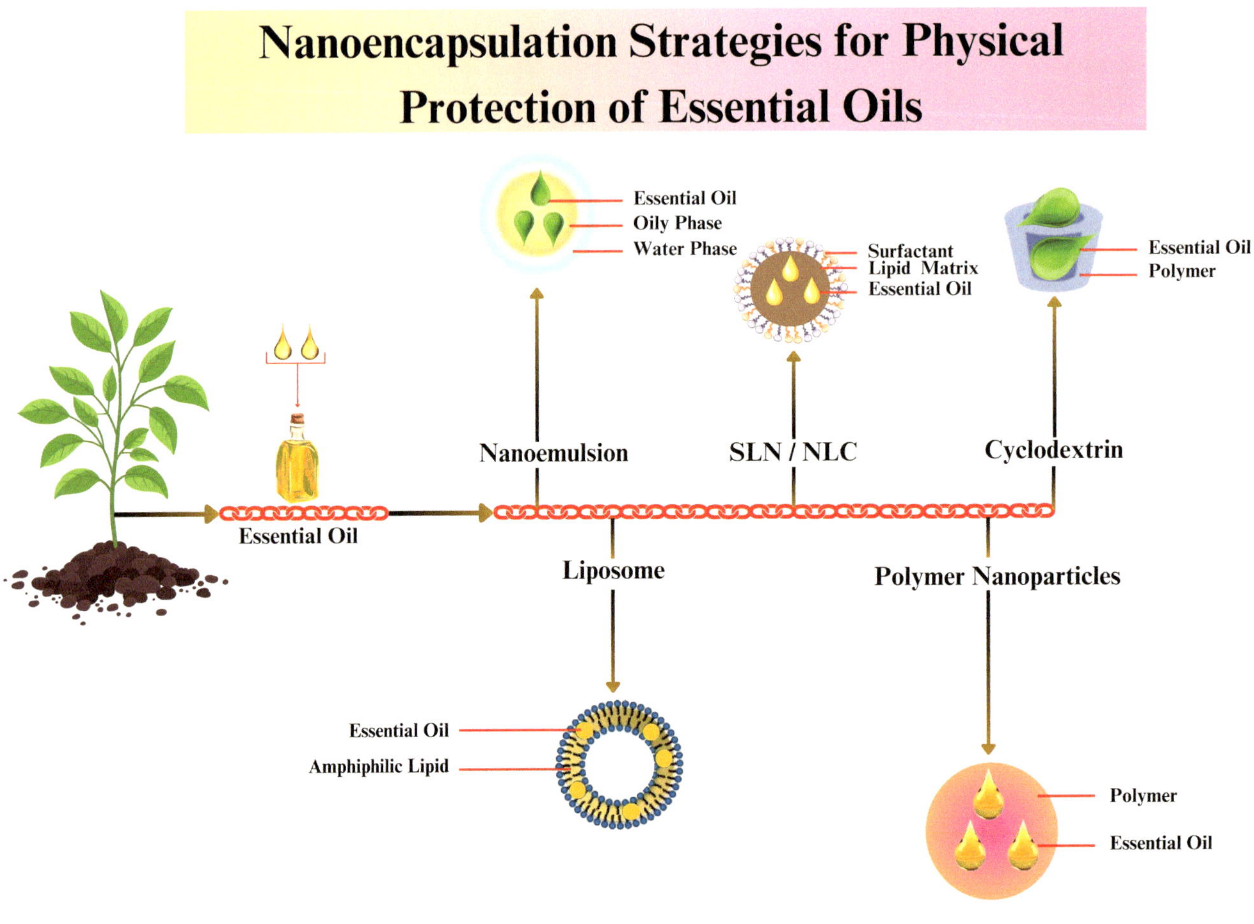

FIGURE 18.5 Nanoencapsulation of essential oils.

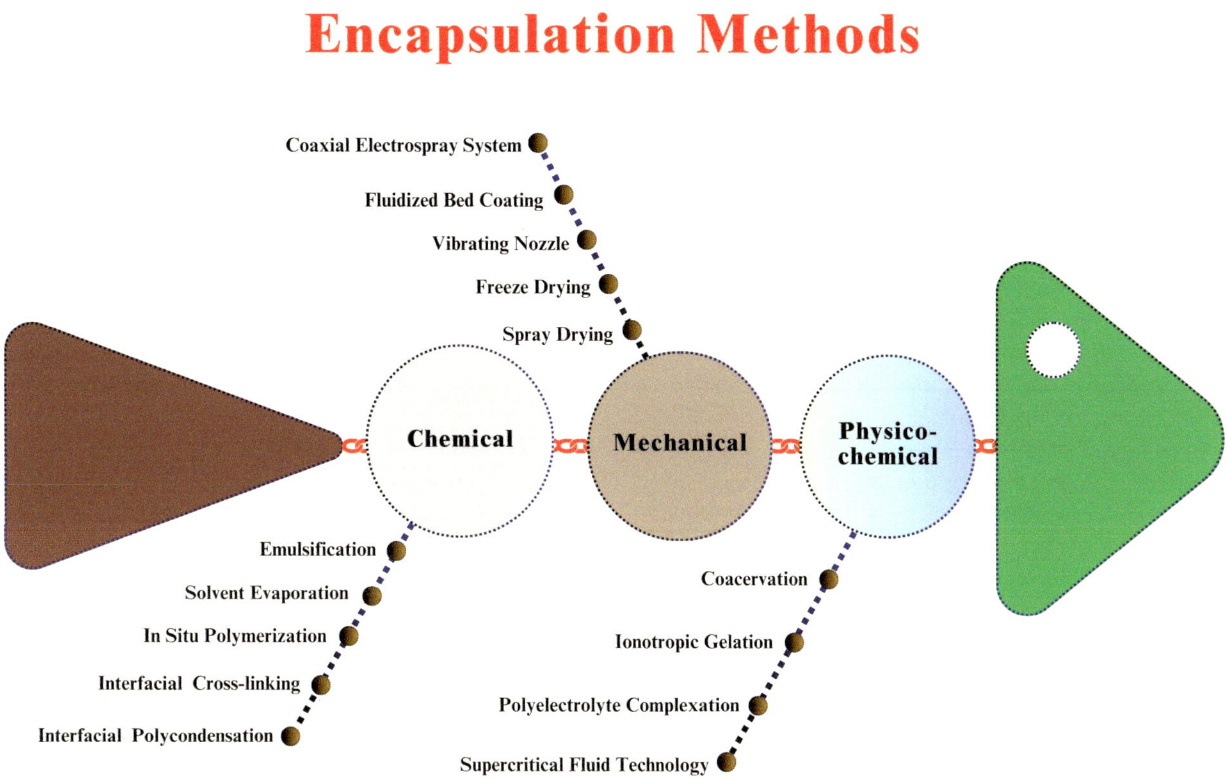

FIGURE 18.6 Nanoencapsulation methods.

The production of capsules can be done by a variety of ways (Martins, Barreiro et al. 2014). As a result of research conducted over the years on a variety of encapsulation techniques, the practice of encapsulation has been divided into two categories: chemical and physical (Kuang, Oliveira et al. 2010; Tyagi, Kaushik et al. 2011).

18.3.1.1 Physical Methods

Since the materials that are used in the beginning of the physical methods are already characterized as a type of polymer, there is no need for polymerization reactions to take place during the physical methods. Therefore, the formation of the shape of the capsules takes place through mechanical means (Tyagi, Kaushik et al. 2011; Martins, Barreiro et al. 2014). The following methods will elaborate on these procedures.

18.3.1.1.1 Extrusion

EOs and other unstable, volatile and organic compounds are ideal candidates for this method of encapsulation. The carbohydrate matrices are ideal candidates for this technique (W Tackenberg and Kleinebudde 2015). There are many uses for this method outside of the food industry, including the cosmetics and pharmaceutical sectors (Uhlemann and Reiß 2010). The extrusion process entails mixing the core material with a liquid. The process then must move over a medium to reach desiccating fluid that will harden the coating (Lee, Ji et al. 2019).

18.3.1.1.2 Lypholization

The lypholization method is commonly used to encapsulate volatile substances like EOs, pharmaceuticals, aroma and flavoring compounds (Cheng, Xie et al. 2017; Yamashita, Chung et al. 2017; Fredes, Becerra et al. 2018). A dry product is the result of a multistage process that includes sublimation, freezing, desorption and storing. Foods' sensory and nutritional properties, as well as their functional characteristics, are maintained throughout the lypholization process because it does not involve heat, so the foods retain their natural flavor and color and are not subject to deterioration from oxidation or chemical changes (Ozkan, Franco et al. 2019).

18.3.1.1.3 Fluidization

During the fluidization process, a fluid moves upward at a high speed over a bed of constituent particles at a speed that is sufficient to keep the particles suspended without causing them to be ejected from the fluid stream. Within a coating chamber, the particles that are going to be encapsulated are fluidized through the use of hot air in this method (KuShaari, Pandey et al. 2006; Pellicer, Fortea et al. 2019). Encapsulation of particles through the use of a fluidized cot or fluidization is regarded as an effective method for aquatic solvable particles, particularly in lower pH, which is also useful for protecting other components of cheered foods (Abbas, Da Wei et al. 2012). In addition to its applications in medicinal diligence, it has also been extensively utilized in food diligence for composing functional additives (Trojanowska, Nogalska et al. 2017).

18.3.1.1.4 Solvent Removal

The technique known as solvent evaporation is primarily predicated on the dissolution of wall material and compound into an appropriate solvent. After being excited to the temperature that is greater than encapsulating material's melting point, the solution is emulsified (Poncelet 2006). In the technique of solvent evaporation, lipids like natural waxes that have low melting point are utilized as coating material. This procedure enables application of hydrophobic as well as hydrophilic units associated with constant phase by which the molecules possess no attraction. It is regularly utilized in encapsulation of peptides and proteins because it enables the application of both types of particles (Lee, Sim et al. 2007). This method of forming capsules is affected by an inclusive range of constraints, comprising volume and nature of solvent, nature and percentage of emulsifier, ratio of volume of phase, rate of solvent removal as well as temperature (Tiwari and Verma 2011).

18.3.1.1.5 Supercritical Fluid Technology

Both solution enhanced dispersion by supercritical fluids (SEDS) as well as supercritical antisolvent (SAS) techniques are relatively new. The supercritical approach, usage of antisolvent liquids as well as supersaturating of compounds are the three primary pillars on which this technique is based (Careno, Boutin et al. 2012). This practice is brought into being by injection of CO_2 into a high-pressure column while it is in the supercritical state. This is trailed through inoculation of emulsion that contains central component, shell material and organic solvent. After that, CO_2 is applied to extract solvent from suspension with only minimal amount of core component being lost (Chattopadhyay, Huff et al. 2006). For the purpose of preparing micro- and nanoparticles, the conventional emulsion techniques have been combined with exclusive characteristics of supercritical fluids through a process known as supercritical fluid extraction of emulsions (SFEE), which is a newer version of this approach (De Aguiar, Silva et al. 2016).

18.3.1.1.6 Spray Dryer

One of the earliest practices for encasing compounds is spray drying. Its basic operation necessitates crafting an aqueous solution with wall and core materials, homogenizing the mixture, atomizing it in a dehydrating compartment, and then dehydrating the particles. This scheme is widely applied in foodstuff diligence and is relatively inexpensive (Jyothi, Prasanna et al. 2010; Jedlińska, Samborska et al. 2019). Due to the decreased packaging, transportation costs, storage space requirements and improved stability of encapsulated compounds, industries are paying more attention in utilization of this approach for production of powdered foods (Chen, Zhong et al. 2013; Liu, Chen et al. 2015; Schuck, Jeantet et al. 2016). Additionally, this approach is versatile and quick, as well as a constant practice that could be used for variety of constituents, such as vitamins, flavors, minerals, oils, etc. (Huang, Sormoli et al. 2018; Nizori, Bui et al. 2020). Polysaccharides like maltodextrin, gums, soy and whey proteins, sodium caseinate, gelatin and chitosan are recurrently used as wall ingredients in this technique (Lee and Wong 2014). This approach has the benefit of operating at low temperature, but one drawback is it requires more processing time (Anwar and Kunz 2011).

18.3.1.2 Chemical Methods

Polymerization reactions using monomers or prepolymers as precursor materials, or chemical interactions, are used to obtain capsules through chemical methods (Poshadri and Aparna 2010). Each of these possesses benefits as well as limitations that have been elaborated in Table 18.3. The following are examples of these techniques.

18.3.1.2.1 Liposomes

Carriers used to transport ingredients and additives for usage in cosmetics and pharmaceuticals as well as foodstuff diligences could get benefits from the liposomes inclusion technique. Liposomes are microscopic spheres composed of an aqueous core enclosed by single or many phospholipid bilayer vesicles (Reza Mozafari, Johnson et al. 2008). EOs, vitamins, dyes, amino acids, microorganisms, enzymes and fatty acids are just some of the components that can be encapsulated in them, and this opens up a wide range of possibilities for usage in foodstuffs diligences (Augustin and Hemar 2009; Singh, Thompson et al. 2012).

18.3.1.2.2 Coacervation

Phase separation is a colloidal phenomenon where the liquefied phase having colloids is isolated from solution as a result of the solubility being decreased through physical or chemical means. The liquid droplets of the new phase eventually coalesce to form a continuous layer, which precipitates to form the capsule's wall. The wall strengthens, and particles are secluded at the end of the process (Timilsena, Akanbi et al. 2019). In foodstuff diligences, this approach is widely applied to encapsulate lipophilic compounds. Despite this fact, this method has great potential for encasing hydrophilic materials. This technique employs a variety of coating materials, including starch, sodium alginate, proteins, gums, etc. (Comunian, Thomazini et al. 2013; Rutz, Borges et al. 2016). This method can be modified to encapsulate vitamins, nutrients and preservatives as well as enzymes. It is highly recommended for encapsulation of EOs and aroma possessing complexes (Gouin 2004).

18.3.1.2.3 Miniemulsion Polymerization

Precipitates of monomers are produced in continuous phase via homogenization in high-powered apparatus.

TABLE 18.3
Principal Benefits and Limitations of Chemical Approaches

Techniques	Essential Oil	Benefits	Limitations	References
Liposomes	Garlic, thyme	Effective control release features, improved stability and a protector of water-soluble active components. Higher bioavailability, biocompatibility and loading effectiveness.	High price, wide range in particle size and low chemical and physical stability. New sophisticated phases must be added for the post-treatment process.	(Trucillo, Campardelli et al. 2018; Ozkan, Franco et al. 2019; Kamkar, Molaee-Aghaee et al. 2021)
Coacervation	Oregano, rose, Sichuan pepper	Lower temperature, regulated discharge of active ingredients, high load capacity (up to 90%) and high repeatability under modest preparation conditions.	Precise adjusting of operative circumstances is necessary. Microcapsule clustering, higher cost and usage of glutaraldehyde for cross-linking with precautions.	(Muneratto, Gallo et al. 2021; Chen, Zhang et al. 2022; Qiu, Zhang et al. 2022)
Miniemulsion polymerization	Oregano, nerol, clove	High transport capacity for active chemicals, excellent control over average drop size and low cost, large-scale application, simple reproducibility and process dependability.	Developing a huge O/W interface is difficult. Monomers and leftover solvents must be removed. Loss of active substances that are water soluble. Toxic effects of the reaction's solvents, by-products and monomers.	(Perignon, Ongmayeb et al. 2015; de Meneses, Marques et al. 2019; Corrado, Di Girolamo et al. 2022)
Ionic gelation	Nettle, flax seed, garlic	Low cost, high encapsulation efficiency and straightforward production control. Microspheres of excellent quality and tiny diameter, mild weather and less temperature are required.	Large size distribution of the particles, dispersion, significant polydispersity (up to 0.5) and heterogeneity. Particle production in an aqueous dispersion demand and regulated agitation.	(Pedroso-Santana and Fleitas-Salazar 2020; Bagheri, Ariaii et al. 2021; Fathi, Samadi et al. 2021; Reis, Zin et al. 2021)

Polymerization reactions or deposits of monomers to be transferred to micelles can take place in such droplets, making them useful as nanoreactors (Landfester 2009). Medications and cosmetic industries (Moreira, de Oliveira et al. 2014) are just two examples where the polymerization encapsulation technique has been put to use. Other examples include fragrance compounds and perfumes, vegetable oil and EOs (Cardoso, Araújo et al. 2013; de Meneses, dos Santos et al. 2017; de Meneses, Marques et al. 2019).

18.3.1.2.4 Ionic Gelation

Cross-linking of polymer in existence of multi-valent cations is the main principle behind the ionic gelation technique. This technique involves continuous stirring of an ionic solution while dripping a hydrophobic-active polymeric or hydrocolloid solution into it (Paques, Sagis et al. 2014). Ionic gelation is used to encapsulate bioactive substances or regulate texture in foods, cosmetics, medications and probiotics (Kurozawa and Hubinger 2017). Chitosan as well as alginate are examples of main coating ingredients used in ionic gelation (Leong, Lam et al. 2016). Ionic gelation is highly advised for encapsulation of foodstuffs components as it uses non-hazardous ingredients, is extremely biocompatible and has virtuous mechanical properties (Patil, Kamalapur et al. 2010). It offers a physically cross-linking process that is reversible through electrostatic interaction, doing away with the need for organic solvents, a potential for chemical components and other negative influences in food streams (Usmiati, Richana et al. 2014).

18.4 EMULSIFICATION AS AN ENCAPSULATING APPROACH

It is generally accepted that encapsulation is a method for stabilizing bioactive components by means of a system that can maintain characteristics and regulate their discharge within predetermined constraints (Silva, Cerqueira et al. 2012; Sonawane, Bhanvase et al. 2020). It is possible to identify an emulsion as a steady dispersal of one fluid in droplets of another immiscible fluid (Yamashita, Chung et al. 2017). The antimicrobial as well as antioxidant properties of EOs can be enhanced through the emulsification process. According to numerous reports in the literature, some researchers consider emulsification to be an encapsulation method. In various studies emulsifying EOs in water by using a stabilizer has been reported as encapsulation approach (Chatterjee and Judeh 2015; Kaade, Ferrando et al. 2019). Though chemical and physical resistance of wall coating material and retaining of essential constituents are hindered as the wall coating material is liquefied (Iqbal, Zafar et al. 2015). Particularly fundamental requirements for edible uses are that encapsulated components must be protected from surrounding environment and not to escape for a specified time duration and also wall material must be permeable, which allows only small molecules to pass through it. Therefore, there must be a tight closure around bioactive compounds for effective functioning; in this scenario emulsification is considered an equally pre-encapsulation method (Alliod, Valour et al. 2018; Doost, Nasrabadi et al. 2020).

On the other hand, in some studies it has been described as a necessary process before performing encapsulation (Sharma, Mann et al. 2017; Zhou, Sun et al. 2018). This practice of emulsification in most cases is followed by the application of different methods that have been proven helpful for enhancing stability or performance of encapsulated constituents. The applications of other approaches in combination with emulsification also result in extended storage stability because EO components are immobilized in compact medium. As a result of these approaches, amalgamation and discharge of the confined components become limited, which further result in enhanced performance and prolonged storage of food products (Bušić, Komes et al. 2018).

A significant number of scientists advocates for the practice of emulsification strategy as an encapsulation approach. The encapsulation approach of lipophilic constituents that are distributed in aqueous medium led to the expansion of a suspension based supply system. However, such an emulsification method has some limitations that are associated with regulation and distribution of shape, structure as well as size of developed particles. These limitations can confine the utilization of this method in some fields such as foodstuffs diligences and medications (Wu, Wang et al. 2015; Kaade, Güell et al. 2020). The efficiency and consistency of encapsulation process is governed by typical particle dimensions of components. Bigger particles are superior compared to small particles in case of protectivity. Despite their superior protectiveness, large particles have a hard spreading ability throughout the food matrix. In addition, the efficiency of encapsulation can be hindered by the presence of particles of very small diameters. Effectiveness, safety, dispersion, and stability are all characteristics that should be taken into account while determining the optimal particle size (Zhou, Sun et al. 2018).

Emulsions are prepared by means of a diverse range of approaches. These are characteristically classified into binary groups: one of these includes membrane emulsification, which has appeared as a conceivable approach for creating innovative blended goods. On the other hand, a conventional emulsification technology is old scheme that includes higher speed liquidizer, ultrasonic homogenization, colloid grinder and a high compression homogenizer (Kobayashi, Wada et al. 2012; Chatterjee and Judeh 2015). The high energy requirement as consequence of higher shear and temperature elevation is the key drawback that restricts the employment of conventional approaches. Only 0.1% of this energy is employed to create the emulsion; the rest is lost as heat (Tadros, Izquierdo et al. 2004). Moreover, the performance of this method is unswervingly influenced by average drop size distribution, which also enhances

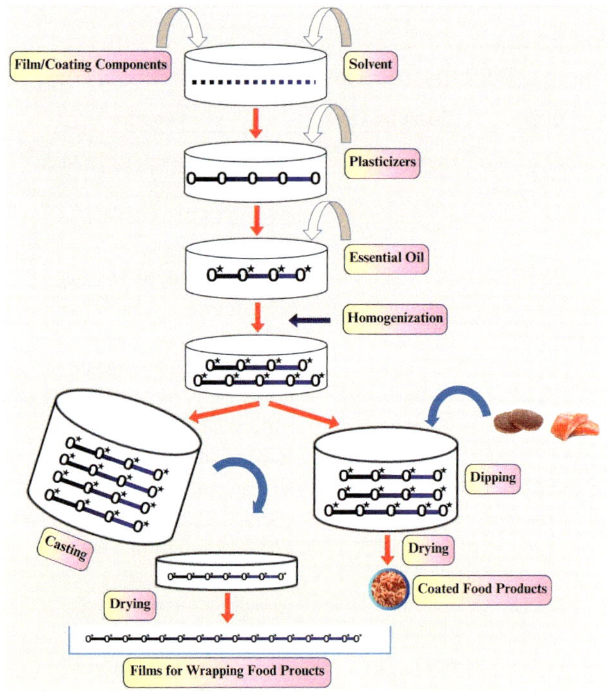

FIGURE 18.7 Preparation method of nanoemulsions.

constancy of emulsions (Chen, Han et al. 2019). The commonly used method of nanoemulsions preparation has been elaborated in Figure 18.7 that has successfully been applied for the preservation of various types of foodstuffs.

18.4.1 Membrane Emulsification

In current ages, this method has gained attention from wide range of diligences (Spyropoulos, Lloyd et al. 2014; Zin, Wu et al. 2019). Contemporary ages have seen an escalation in integer of research devoted to membrane emulsions (Alliod, Valour et al. 2018; Reis, Zin et al. 2021). As compared to conventional technologies like dispersers and mixers, which generate high shear, require a lot of energy, cause a significant rise in temperature and give very little working efficiency over the droplet size distribution; membrane emulsions stand out as a clear choice (De Luca, Di Renzo et al. 2006). The primary benefits of ME are its low-energy requirements, tiny shear rates, low thermal changes, improved droplet size control and scalability (Hancocks, Spyropoulos et al. 2016). Fouling portents on exterior of membrane and pores are main limitations of ME, along with its reliability, short lifetime and higher resistance to mass transfer (Drioli, Ali et al. 2015). The composition of both the continuous and disperse phases as well as operating pressure, membrane pore size distribution and flow of continuous phases are all elements that can limit membrane emulsion. One must also take note of the membrane's substance. Many different materials have been investigated for emulsification, including ceramic, microporous glass, polymeric membrane and sintered metal. Depending on the emulsification process's circumstances, the droplet's average size in a membrane emulsion can be anywhere from 2–10 folds greater as of pore dimension. This is caused by the drop's impartiality from membrane exterior and the pressure given to it (Hao, Gong et al. 2008).

18.4.1.1 Emulsions and Its Types

Oil and water are the most common liquids used to create emulsions. Oil-in-water (O/W) emulsions are categorized by the presence of oil drops disseminated in water or liquid medium (McClements 2011). These emulsion systems are useful for long-distance carriage of hydrophobic bioactive constituents. On the other hand the delivery of hydrophilic active substances is accomplished by means of water drops disseminated in oil, which is recognized as water-in-oil emulsions (W/O). Oil/water/oil (O/W/O) as well as water/oil/water (W/O/W) are further emulsion systems that can be fabricated. The W/O/W systems have enormous oil drops suspended in aqueous media. However, in O/W/O system, the oil phase is disseminated in the water droplets (Huang, Belwal et al. 2019; Callens, Fontaine et al. 2022).

Moreover, depending upon dimension of droplet and permanence, emulsions are classified as microemulsions, nanoemulsions or coarse emulsions (Komaiko and McClements 2016). There is, however, some confusion for interpretation in their sizing description. Coarse emulsions are correspondingly referred to as microemulsions and conventional emulsions. Particles in coarse emulsions are thermodynamically stable and typically have a size distribution greater than 200 nm. Over time, they deteriorate because of exposure to a widespread array of destabilizing impacts. Owing to their increased surface area of these tiny droplets, the encapsulated component's functional properties are enhanced (Huang, Wang et al. 2021). Turbidity in coarse emulsions results from the fact that each droplet's size is roughly equal to wavelength of light, so these emulsions effectively disseminate the incident light and gives the appearance of being opaque. Microemulsions droplets are thermodynamically stable and have sizes of approximately 200 nm or above (Gupta, Eral et al. 2016). However, their reliability is influenced by changes in the environment, such as temperature and composition. These possess weak light-scattering efficiency and optically translucent as their size is lesser than wavelength (Anton and Vandamme 2009).

In contrast to microemulsions, nanoemulsions possess droplet sizes, which range between 100 to 200 nm. Nanoemulsions are both thermodynamically as well as kinetically unstable as phase separation develops with the passage of time, just like conventional emulsions (McClements and Rao 2011; Vladisavljević 2019). Nanoemulsions, in contrast to conventional emulsions, are optically clear, have enhanced bioavailability and are chemically reactive (McClements and Jafari 2018). The amount of surfactants needed to prepare nanoemulsions is less. The stability, optical, rheological as well as discharging ability of bioactive components of nanoemulsions are all affected by the size of the droplets in it. In contrast to conventional emulsions, nanoemulsions preparation is a novel approach in food coatings that falls into class of nanotechnology and is currently creating a lot of attention (Oberoi, Tolun et al. 2019; Sharma, Babaei et al. 2022).

18.5 APPLICATIONS OF NANOSYSTEM-BASED EOS IN FOOD SECTOR

Food product diligences are interested in using EOs as preservers because they are rich in bioactive constituents and possess some extraordinary antibacterial and antioxidant capabilities (Atarés and Chiralt 2016; Tudor, Pițuru et al. 2023). Their use in place of synthetic food additives has recently attracted increasing interest in foodstuff diligences owing to their significant antioxidant as well as antibacterial capabilities. Every EO has its own distinctive physicochemical properties, biological activity and aroma. Thus it's most significant to elect appropriate combination for each application (Angane, Swift et al. 2022). Cinnamon EO is documented as the active valuable plant extract that could be precisely applied both in foodstuff and cosmetic production (Yang, Al-Mahdy et al. 2022). It is elaborated that incorporation of cinnamon EO in cyclodextrin nanosponges results in enhanced antimicrobial packaging system that can preserve various food products. According to various research reports garlic EO possess enhanced antimicrobial potential against different foodborne microbes. Due to this potential this EOs are proving useful in many food diligences that employ them equally in preserving as well as food packagings (Simionato, Domingues et al. 2019; Feltes, Ballen et al. 2023). Moreover these are effectively employed in preservation of different food products in the form of coatings, packaging and thin films. These includes agricultural food products, marine food, dairy products and various others that can be deteriorated due to environment as well as microbial contaminants (Al-Maqtari, Rehman et al. 2021; Masyita, Sari et al. 2022; Rout, Tambe et al. 2022). Benefits of EO-based food preservation systems have been elaborated in Figure 18.8.

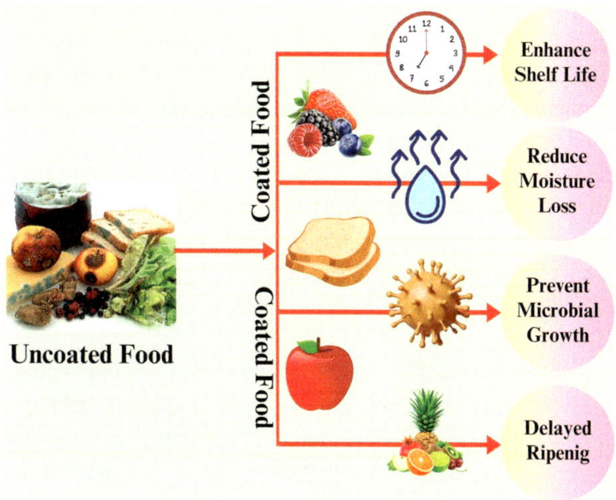

FIGURE 18.8 Difference of coated and uncoated foods.

18.5.1 Encapsulated EOs

Nanoencapsulation is an important approach and advanced technology used to resolve the issues like bioactivity and physical degradation of the actively anticipated components and to minimize discharge proportion of active agents into the external medium that surrounds the nanocapsules by diffusion method moving from higher to lower concentration (Das, Chaudhari et al. 2020; Soni, Maurya et al. 2022). After formation of nanocapsules EO based components gain powdery appearance and their properties become diverse from the large powder particles having similar composition chemically (Plati, Papi et al. 2021). Nanocapsules are a unique system consisting of a polymeric shell that is crowded about a core material. The desired active constituent of core material is intermixed with polymeric wall components. It is fascinating to know that nanocapsules have very small size; their size is thousands folds lesser than bacterium (Vijeth, Heggannavar et al. 2019). The increasing advancement in arena of nanotechnology in previous ages has raised improvement of innovative technologies to release active components, particularly in the living systems (Donsì, Annunziata et al. 2011). Fabrication of the polymeric nanocapsules have been a point of interest for researchers in medicine and foodstuff related sectors and also because the polymeric nanocapsules have functional properties or features as viable release system of the active components (Tahir, Shabir Ahmad et al. 2021).

It has been stated that encapsulation was efficiently applied for the development of active constituents, where extract of tea was encapsulated into the polythene by extrusion method. It was observed that these capsules shielded extracts of green tea for effective release during heating. The releases of a small fraction of the active components at ppb level was affirmed, after which efficiency of the designed packaging was tested with in vivo investigations, which result in prolong life span of renewed minced meat (Wrona, Nerín et al. 2017). However when we use nanocapsules of EOs at large scale or industrial sectors it is essential to deliberate their confines like organoleptic properties, solubility in water and particularly thermal stability. Encapsulation of EOs offers a vital means to shield all these obstacles. Coating these natural components with nanostructured materials that are most resistant, increases their thermal and chemical stability and under some conditions also enhances their solubility in various solvents. Encapsulation of EOs offers the advantages of precise release, enhances cellular bioavailability as well as effectiveness in contradiction of pathogens (Rezaei, Fathi et al. 2019; Jugreet, Suroowan et al. 2020; Yousefi, Weisany et al. 2023).

Weisany, Samadi et al. (2019) described the encapsulation of *Thymus daenensis* and dill EOs in copper NPs and evaluated their antifungal potential contrary to selected strains of microbes. These EOs inhibit the growth *of C. nymphaeae mycelium*. The maximum activity was achieved at a concentration of 720 and 1000 ppm after a period of nine days of the incubation (Weisany, Samadi et al. 2019). Dávila-Rodríguez, Lopez-Malo et al. (2019) tested antibacterial potential of the EOs of cinnamon (CEO), oregano (OEO) and rosemary (REO). A high-frequency ultrasound technique was applied for encapsulation, and they were tested on fresh celery culture. A greater efficiency (EE>80%) was obtained after performing encapsulation (Dávila-Rodríguez, Lopez-Malo et al. 2019). Shetta, Kegere et al. (2019) also performed encapsulation of peppermint and green tea oil with NPs of chitosan by using ionic gelation process, which were later also analyzed by different techniques. The obtained NPs had sphere shape and have an average size of 20–60 nm, which were tested by SEM. After characterization these nanomaterials were employed for antioxidant as well as antimicrobial potential against various strains, which prove helpful (Shetta, Kegere et al. 2019).

Donsì, Annunziata et al. (2011) emphasized EOs encapsulation into nano-sized matrices for their integration into the juices of fruits for the purposes of increasing their antibacterial properties and lessen their influences on the superiority of foodstuff. The development of encapsulated nanoemulsions was performed by homogenization technique at pressure of 300 MPa (Donsì, Annunziata et al. 2011). Also, the influence of delivery system on antibacterial property of the terpenes was examined. Analysis by GC-MS indicated the high concentration dispensation for nanoemulsions fabrication that may disturb chemical stability of various active components. The acquired nanosystem was applied to check its antimicrobial activity in orange and pear juices. Results concluded higher antimicrobial properties of the nanoencapsulated system. Moreover, lesser concentration leads to fewer fluctuations in sensory properties (Reis, Ambrosi et al. 2022). Wu, Luo et al. (2012) encapsulated thymol and carvacrol that have same chemical configurations in NPs of zein by using dispersion process. For observation of the structure, morphology and antimicrobial property, three different pH conditioned were offered. NPs were dispersed efficiently in water under all pH conditions. After lypholization samples of neutral and basic condition form NPs while samples from the acidic condition formed bio film (Wu, Luo et al. 2012).

Nielsen, Kjems et al. (2016) discovered that biofilms that are related to food might cause foodborne diseases and degradation; both of these offer many issues on the global scale. They examined the antimicrobial activities of encapsulated and unencapsulated isoeugenol to compete with the biofilms of different bacterial strains present in the carrot juice. They showed that encapsulated isoeugenol possess increased antimicrobial activity to compete with biofilms in culture however not in the juices of carrots. A number of isoeugenol suspensions were covered with the chitosan and biofilms were treated with emulsions that spoiled the structure of biofilm. Further, they predicted that addition of surfactant Tween 80, that is generally utilized to disperse oil in food, supports antimicrobial activity of isoeugenol. These results indicated that the food additives that are general like surfactants might possess an extensive impact on antimicrobial properties of the preservatives (Nielsen, Kjems et al. 2016). Emulsions encapsulation have the advantages of the increased efficiency of the eugenol; however, effect of the encapsulation depends upon characteristics of the food ingredients in which isoeugenol is intended to be used (Carpena, Nuñez-Estevez et al. 2021).

Kujur, Kumar et al. (2020) described the efficiency of the nanoencapsulated chitosan-cinnamic-based *Gaultheria procumbens L.* EO (GPEO) to compete with *Aspergillus flavus*, the Aflatoxin B1 excretion and its method of exploit. Methyl salicylate (95.26%) was investigated to be the main component of the GPEO. Encapsulated GPEO showed an increased antifungal property in comparison to the unencapsulated GPEO and finally inhibited the growth of bacteria at 1.00 microliter/ml. It elucidated that encapsulated GPEO performs primarily targeting the ergo-sterol constituents in membranes of bacteria, leakage of the cellular ions or structural changes in *A. flavus*. Also, results indicate ability of the chitosan-based encapsulated components to upsurge antibacterial proficiency and also reliability of GPEO (Kujur, Kumar et al. 2020).

Rajaei, Hadian et al. (2017) performed encapsulation of clove by following the chitosan (CS)-Myristic acid (MA) nanoemulsions. It was observed that the antioxidant property of the selected encapsulated system was increased due to their radical-scavenging activity. Moreover, the encapsulation also increased the release time of the selected EO. This encapsulated system proves useful for preservation of beef (Rajaei, Hadian et al. 2017). The two most efficient antimicrobials, mustard and cinnamon oil, were tested to check out their ability to inhibit bacterial growth after 24h incubation at a very low concentration of 0.016% (v/v). Moreover the encapsulated EO systems were applied to upsurge shelf life and enhanced antimicrobial activity. Their antimicrobial activity was evaluated and 10 fold increments in the potential in comparison to free or pure EO was observed. Cinnamaldehyde incorporated onto mesoporous silica nanoparticles (MSNPs) exhibited to be a most efficient antimicrobial, decreasing >95% bacterial growth. This system showed the capability to be utilized to protect against bacterial diseases in the crops and to enable the more controlled and efficient degradation of the volatile compounds as the antimicrobials (Cadena, Preston et al. 2018; Drakopoulos, Meca et al. 2020).

Aflatoxin B1 adulteration in the food products has poisoning effects depending upon the dose. Its higher concentration damages the kidney, liver as well as myocardial tissues, while the low dose usage may also be toxic. In the food matrices, AFB1–8,9 exoepoxide can result in the development of the active oxygen that result in peroxidation of lipids that leads to the major losses of quality of food items (Ajmal, Bedale et al. 2022). Dwivedy, Singh et al. (2018) developed llicium verum EO based nanocapsules, which has strong antioxidant, anti-Aflatoxin and antifungal ability and also efficiency to prevent fumigated samples in the strong vessels and as strengthening its approval as natural source based preserving agents as well as shelf life extender of foodstuffs (Dwivedy, Singh et al. 2018).

18.5.2 EO-Based Nanoemulsions

In addition to other encapsulation techniques, nanoemulsions can be helpful to retain bioactive compounds active for use in food medium for longer periods. These nanosystem-based bioactive composites in oil phase in the presence of emulsifier as well as water phase confirm their bioavailability, stability and precise discharge (McClements 2011). These emulsions established distribution systems that are compatible with the foodstuffs and have a negligible impact on organoleptic features of foodstuffs, for instance texture and flavor, as well as appearance. Encapsulating bioactive constituents can also protect them from damaging storage circumstances. Moreover, the utilization of nanoemulsions based delivery systems to foods necessitates the method's economic viability at the industrial level (Lakshmayya, Mishra et al. 2023).

Beluga sturgeon fillets (*Huso huso*), silvery pomfret (*Pamous argenteus*), rainbow trout (*Oncorhynchus mykiss*) fillets, chicken pate and cheese are some food systems in which the EOs are integrated at laboratory scale either directly or with edible coatings (Sneha and Kumar 2022). While EO nanoemulsions infusion is necessary in porous food matrices and the association of EOs-based nanoemulsions is accomplished by direct mixing (Donsi, Sessa et al. 2015). Freshly cut orange slices with edible pectin coatings made with nanoemulsions have improved sensory characteristics. These EOs nanoemulsions have been applied to minimally processed food systems in two main ways: as a soaking or washing solution on the surfaces of foods and as an ingredient in biopolymer-based films or coatings (Radi, Akhavan-Darabi et al. 2018).

Jemaa, Falleh et al. (2017) performed nanoemulsion preparation and determined influence of solution of the *Thymus capitatus* EO by incorporation with the milk to check its quality, which may be spoiled by bacteria. After a period of 24 hours of the inoculation, the growth of bacterial culture was reached to 202 to 103 CFU/ml with pure EO whereas it was confined among 132 to 103 CFU/ml while treating with nanoemulsions. Also, a decrease in the antioxidant property of the milk preserved with the pure EO was larger than that of those treated with nanoemulsions. Furthermore, nanoemulsions prevented the proteins present in the milk from degradation more effectively than that of the pure EO. As a result, the prepared nanosystems are very efficient to prolonged shelf life of milk and its quality (Jemaa, Falleh et al. 2017).

Jemaa, Falleh et al. (2018) performed the nanoemulsions preparation for the effective function of the *Thymus capitatus* EO (thyme EO) as natural foodstuff stabilizers. The biological potentials of EOs were tested after emulsion fabrication. The outcomes revealed an increase in the effect of antimicrobial property of this EO. It was also noticed that the nanoemulsions showed inhibition zones greater

TABLE 18.4
Efficiency of Nanosystem-Based EOs as Food Preservative

EO	Polymer	Food Sample	Activity	References
Cinnamomum zeylanicum	Chitosan	Different food samples	Effective delivery system of bioactive components	(Matshetshe, Parani et al. 2018)
Tea tree oil	Chitosan	Beef samples	Effective antibacterial potential	(Cui, Bai et al. 2018)
Thyme	Chitosan	Sauces	Result in nitrite reduction	(Vafania, Fathi et al. 2019)
Paulownia Tomentosa	Chitosan	Porks	Effective antibacterial potential	(Zhang, Li et al. 2019)
Clove	Chitosan	Pomegranate	Prolonged shelf life	(Hasheminejad and Khodaiyan 2020)
Garlic	Chitosan	Sauces	Enhanced antioxidant as well as antibacterial potential	(Esmaeili, Cheraghi et al. 2020)
Citral	Sodium alginate	Pineapple	Lower respiration and enhanced antimicrobial potential	(Prakash, Baskaran et al. 2020)
Oregano	Sodium alginate	Tomatoes	Enhanced antimicrobial potential	(Das, Vishakha et al. 2020)
Cinnamon	Pullulan	Strawberries	Prolonged shelf life and antimicrobial potential	(Chu, Gao et al. 2020)
Cymbopogon martini and *Mentha spicata*	Starch/cellulose/wax	Strawberries	Prolonged shelf life, antioxidant and antimicrobial potential	(de Oliveira Filho, Albiero et al. 2022)
Ginger	Carnauba wax	Papaya	Reduce weight loss, enhanced flavor retention and prolonged storage period	(Miranda, Sun et al. 2022)
Cymbopogon martini	Carnauba wax	Papaya	Enhanced duration of ripening and prolonged shelf life	(Oliveira Filho, Silva et al. 2022)
Amla EO	Corn starch	Amla fruit	Enhanced antioxidant, antimicrobial anti-browning potential and prolonged shelf life	(Braich, Kaur et al. 2022)
Thyme EO	Whey protein	Zucchini	Reduced weight loss, higher enzyme activity	(Bleoanca, Lanciu et al. 2022)
Basil	Cellulose/corn starch	Mandarin orange	Maintain color, reduce weight loss and prolonged storage period	(Wigati, Wardana et al. 2023)
Cardamom	Carboxymethyl cellulose	Tomatoes	Strong antioxidant, antimicrobial potential and elongated storage period	(Das, Vishakha et al. 2022)
Lavender	Soluble soybean polysaccharide/ carboxymethyl chitosan	Banana	Enhanced mechanical and barrier properties	(Liu, Liu et al. 2022)
Ruta graveolens	Chitosan	Cape gooseberry	Enhanced antioxidant and antimicrobial potential, reduced weight loss	(González-Locarno, Maza Pautt et al. 2020)
Sweat orange	Chitosan	Tomatoes	Enhanced antifungal activity	(Sheikh, Mehnaz et al. 2021)

than that of the free EO (i.e., 15.8 and 9.8 mm, respectively, for *S. aureus*). Moreover, after nanoemulsions formation there was a momentous reduction in the minimum inhibitory concentration of oils to compete with the investigated bacteria than bulk EO. Though, the results obtained by this study proved to be supportive for the development of effective natural preservation process to minimize the spoilage caused by bacteria (Jemaa, Falleh et al. 2018).

18.5.3 EO-Based Food Coatings

The coatings of EOs must adhere to food surfaces to accomplish their anticipated task. Certain types of EOs, such as clove, oregano, lemon and various others, were developed in the form of edible coatings having variable compositions as well as average droplet size with surface charge to improve their in vitro potential by using biopolymers (i.e., chitosan). These modifications result in enhancing their antifungal as well as antibacterial activities (Ramos, Mellinas et al. 2021). In the foodstuffs industries, applications of EOs are considerably more credible when they are contained in an appropriate delivery system that protects them from interaction with other food matrices. To diminish the influence on sensory characteristics, EOs are typically delivered by means of nanotechnology, which enhances their interaction with cell molecules. By reducing degree of inhalation as well as water evaporation, edible coatings extend storage lives of fruitlets and protect them from postharvest damage (Sheikh, Mehnaz et al. 2021).

The need for sustainable and polymeric materials based food preservation system is quickly developing due to higher shelf life potentials (Salgado, Di Giorgio et al. 2021). Petroplastics have appealing features such as flexibility, mechanical strength as well as economic effectiveness, but they are environmentally hazardous owing to their non-biodegradable nature, which is harmful for human well-being. According to some recent investigation, researchers are paying attention toward biodegradable edible food packaging such as polysaccharides and proteins. According to studies, biodegradable edible packaging provides additional support to food throughout storage, such as antibacterial protection, wall qualities, shelf life extension and pollution-free environment conservation (Pandey, Islam et al. 2022).

EOs research is gaining the attention of scientists these days due their antimicrobial activity. EOs have broad variety of handling and safety from infectious contamination and post fruitage damages due to pathogens sensitivity (Chouhan, Sharma et al. 2017). Packing materials play a major character to evade food from deterioration and to keep food preserve from adversarial surrounding conditions. Food coating materials possess bioactive constituents that provide antioxidant possessions to the foodstuffs. To provide safety characteristics and feature preservation to foodstuffs during storage, antimicrobial and antioxidant agents could be amalgamated into foodstuffs packaging substances (Hossain, Skalicky et al. 2021). Different types of food preservation systems have been explained in Table 18.4, which elaborates the potentials of various EOs in various food products preservation.

18.6 CONCLUSION

Food preservation has become a challenging process as the demand for food has risen due to enormous increments in population. Also the hazardous effects associated with synthetic preservatives make these materials limited to use as food additives, so researchers are moving toward natural sources for the preservation of food. In this study, essential oils were considered in relation to protecting the food from microorganisms as these possess enhanced antioxidant and antimicrobial potential. Essential oils are extremely volatile, capable of reacting with foodstuff medium, changing organoleptic taste of food items and sensitive to light, temperature and oxygen. Consequently, there was a critical need to protect these active components into a wall material that could address all these issues. Moreover nanoencapsulation and nanoemulsions strategies are helpful in enhancing their biological properties by managing active components release system that are further helpful in enhancing the storage period of various foodstuffs. So as a whole these are proving vital for food preservation with the help of nanotechnology. Further efforts can be made to preserve food products by using essential oils-based nanomaterials.

REFERENCES

Abbaci, H., E.-h. Nabti, A. M. Al-Bekairi, S. A. Hagras, M. M. Salem-Bekhit, A. Adjaoud, H. A. Alzahrani, L. Bensidhoum, R. Alenazy and A. Piras (2023). "Comparative bioactivity evaluation of chemically characterized essential oils obtained from different aerial parts of *Eucalyptus gunnii* Hook. f.(Myrtaceae)." *Molecules* **28**(6): 2638.

Abbas, S., C. Da Wei, K. Hayat and Z. Xiaoming (2012). "Ascorbic acid: Microencapsulation techniques and trends—a review." *Food Reviews International* **28**(4): 343–374.

Abd El-Hack, M. E., N. A. El-Shall, N. I. El-Kasrawy, M. T. El-Saadony, M. E. Shafi, N. M. Zabermawi, M. S. Alshilawi, M. Alagawany, A. F. Khafaga and R. M. Bilal (2022). "The use of black pepper (*Piper guineense*) as an ecofriendly antimicrobial agent to fight foodborne microorganisms." *Environmental Science and Pollution Research*: 1–14.

Adeyeye, S. A. O. (2017). "The role of food processing and appropriate storage technologies in ensuring food security and food availability in Africa." *Nutrition & Food Science* **47**(1): 122–139.

Ahari, H. and S. P. Soufiani (2021). "Smart and active food packaging: Insights in novel food packaging." *Frontiers in Microbiology* **12**: 657233.

Ahmed, S., A. Stewart, E. Smith, T. Warne and C. Byker Shanks (2021). "Consumer perceptions, behaviors, and knowledge of food waste in a rural American state." *Frontiers in Sustainable Food Systems* **5**: 734785.

Ajmal, M., W. Bedale, A. Akram and J.-H. Yu (2022). "Comprehensive review of aflatoxin contamination, impact on health and food security, and management strategies in Pakistan." *Toxins* **14**(12): 845.

Al-Khayri, J. M., A. Banadka, M. Nandhini, P. Nagella, M. Q. Al-Mssallem and F. M. Alessa (2023). "Essential oil from *Coriandrum sativum*: A review on its phytochemistry and biological activity." *Molecules* **28**(2): 696.

Al-Maqtari, Q. A., A. Rehman, A. A. Mahdi, W. Al-Ansi, M. Wei, Z. Yanyu, H. M. Phyo, O. Galeboe and W. Yao (2021). "Application of essential oils as preservatives in food systems: Challenges and future prospectives–a review." *Phytochemistry Reviews*: 1–38.

Al-Mijalli, S. H., H. N. Mrabti, H. Assaggaf, A. A. Attar, M. Hamed, A. E. Baaboua, N. E. Omari, N. E. Menyiy, Z. Hazzoumi and R. A. Sheikh (2022). "Chemical profiling and biological activities of pelargonium graveolens essential oils at three different phenological stages." *Plants* **11**(17): 2226.

Al-Tayyar, N. A., A. M. Youssef and R. Al-Hindi (2020). "Antimicrobial food packaging based on sustainable Bio-based materials for reducing foodborne Pathogens: A review." *Food chemistry* **310**: 125915.

Ali, M. M., M. A. Yusuf and M. N. Abdalaziz (2017). "GC-MS analysis and antimicrobial screening of essential oil from lemongrass (*Cymbopogon citratus*)." *International Journal of Pharmacy and Chemistry* **3**(6): 72–76.

Ali, S., V. T. Rezende, S. Ullah, E. L. de Paiva, F. G. Tonin, C. H. Corassin and C. A. F. de Oliveira (2023). "Food processing and challenges in the food production and quality: The foodomics approach." *Food Bioscience*: 103217.

Alliod, O., J.-P. Valour, S. Urbaniak, H. Fessi, D. Dupin and C. Charcosset (2018). "Preparation of oil-in-water nanoemulsions at large-scale using premix membrane emulsification and Shirasu Porous Glass (SPG) membranes." *Colloids and Surfaces A: Physicochemical and Engineering Aspects* **557**: 76–84.

Amit, S. K., M. M. Uddin, R. Rahman, S. R. Islam and M. S. Khan (2017). "A review on mechanisms and commercial aspects of food preservation and processing." *Agriculture & Food Security* **6**: 1–22.

Andrade-Ochoa, S., K. F. Chacón-Vargas, L. E. Sánchez-Torres, B. E. Rivera-Chavira, B. Nogueda-Torres and G. V. Nevárez-Moorillón (2021). "Differential antimicrobial effect of essential oils and their main components: Insights based on the cell membrane and external structure." *Membranes* **11**(6): 405.

Angane, M., S. Swift, K. Huang, C. A. Butts and S. Y. Quek (2022). "Essential oils and their major components: An updated review on antimicrobial activities, mechanism of action and their potential application in the food industry." *Foods* **11**(3): 464.

Anton, N. and T. F. Vandamme (2009). "The universality of low-energy nano-emulsification." *International Journal of Pharmaceutics* **377**(1–2): 142–147.

Anwar, S. H. and B. Kunz (2011). "The influence of drying methods on the stabilization of fish oil microcapsules: Comparison of spray granulation, spray drying, and freeze drying." *Journal of Food Engineering* **105**(2): 367–378.

Atarés, L. and A. Chiralt (2016). "Essential oils as additives in biodegradable films and coatings for active food packaging." *Trends in Food Science & Technology* **48**: 51–62.

Augustin, M. A. and Y. Hemar (2009). "Nano- and micro-structured assemblies for encapsulation of food ingredients." *Chemical Society Reviews* **38**(4): 902–912.

Augusto, P. E., B. M. Soares and N. Castanha (2018). "Conventional technologies of food preservation." *Innovative technologies for food preservation*, Elsevier: 3–23.

Bagheri, R., P. Ariaii and A. Motamedzadegan (2021). "Characterization, antioxidant and antibacterial activities of chitosan nanoparticles loaded with nettle essential oil." *Journal of Food Measurement and Characterization* **15**: 1395–1402.

Bajpai, V. K., K.-H. Baek and S. C. Kang (2012). "Control of Salmonella in foods by using essential oils: A review." *Food Research International* **45**(2): 722–734.

Bakry, A. M., S. Abbas, B. Ali, H. Majeed, M. Y. Abouelwafa, A. Mousa and L. Liang (2016). "Microencapsulation of oils: A comprehensive review of benefits, techniques, and applications." *Comprehensive Reviews in Food Science and Food Safety* **15**(1): 143–182.

Barakat, H. and G. Ghazal (2016). "Antifungal and antioxidant activities of rosemary (*Rosmarinus officinalis* L.) essential oil." *Journal of Food and Dairy Sciences* **7**(5): 273–282.

Basavegowda, N. and K.-H. Baek (2021). "Synergistic antioxidant and antibacterial advantages of essential oils for food packaging applications." *Biomolecules* **11**(9): 1267.

Batiha, G. E.-S., D. E. Hussein, A. M. Algammal, T. T. George, P. Jeandet, A. E. Al-Snafi, A. Tiwari, J. P. Pagnossa, C. M. Lima and N. D. Thorat (2021). "Application of natural antimicrobials in food preservation: Recent views." *Food Control* **126**: 108066.

Becerril, R., C. Nerín and F. Silva (2020). "Encapsulation systems for antimicrobial food packaging components: An update." *Molecules* **25**(5): 1134.

Behera, A., B. Mittu, S. Padhi and A. Singh (2020). "Antimicrobial efficacy of essential oil nanoemulsions." *Nanotechnological approaches in food microbiology*, CRC Press: 293–309.

Ben Braïek, O. and S. Smaoui (2021). "Chemistry, safety, and challenges of the use of organic acids and their derivative salts in meat preservation." *Journal of Food Quality* **2021**: 1–20.

Beyki, M., S. Zhaveh, S. T. Khalili, T. Rahmani-Cherati, A. Abollahi, M. Bayat, M. Tabatabaei and A. Mohsenifar (2014). "Encapsulation of *Mentha piperita* essential oils in chitosan–cinnamic acid nanogel with enhanced antimicrobial activity against *Aspergillus flavus*." *Industrial Crops and Products* **54**: 310–319.

Bhavaniramya, S., S. Vishnupriya, M. S. Al-Aboody, R. Vijayakumar and D. Baskaran (2019). "Role of essential oils in food safety: Antimicrobial and antioxidant applications." *Grain & Oil Science and Technology* **2**(2): 49–55.

Bibi, S., S. Ali, I. U. Shahidin, A. Rauf, M. Arif, M. Umar, A. S. Khan, M. Hussain, A. Ghani, H. Ali, A. Jan and M. Shuaib (2021). "Chemical composition and anti-microbial analysis of *Mentha arvensis* L. and *Thymus linearis* Benth. essential oils of leaves." *Romanian Biotechnological Letters* **26**(5): 2893–2900.

Bleoanca, I., A. Lanciu, L. Patraşcu, A. Ceoromila and D. Borda (2022). "Efficacy of two stabilizers in nanoemulsions with whey proteins and thyme essential oil as edible coatings for zucchini." *Membranes* **12**(3): 326.

Bolouri, P., R. Salami, S. Kouhi, M. Kordi, B. Asgari Lajayer, J. Hadian and T. Astatkie (2022). "Applications of essential oils and plant extracts in different industries." *Molecules* **27**(24): 8999.

Bora, H., M. Kamle, D. K. Mahato, P. Tiwari and P. Kumar (2020). "Citrus essential oils (CEOs) and their applications in food: An overview." *Plants* **9**(3): 357.

Braich, A. K., G. Kaur, A. Singh and B. Dar (2022). "Amla essential oil-based nano-coatings of Amla fruit: Analysis of morphological, physiochemical, enzymatic parameters, and shelf-life extension." *Journal of Food Processing and Preservation* **46**(6): e16498.

Burt, S. (2004). "Essential oils: Their antibacterial properties and potential applications in foods—a review." *International Journal of Food Microbiology* **94**(3): 223–253.

Bušić, A., D. Komes, A. Belščak-Cvitanović, A. Vojvodić Cebin, I. Špoljarić, G. Mršić and S. Miao (2018). "Potencijal kombiniranih tehnika emulgiranja i sušenja raspršivanjem za inkapsulaciju polifenola iz lišća ružmarina (*Rosmarinus officinalis* L.)." *Food Technology and Biotechnology* **56**(4): 494–505.

Cadena, M. B., G. M. Preston, R. A. Van der Hoorn, H. E. Townley and I. P. Thompson (2018). "Species-specific antimicrobial activity of essential oils and enhancement by encapsulation in mesoporous silica nanoparticles." *Industrial Crops and Products* **122**: 582–590.

Calín-Sánchez, Á., L. Lipan, M. Cano-Lamadrid, A. Kharaghani, K. Masztalerz, Á. A. Carbonell-Barrachina and A. Figiel (2020). "Comparison of traditional and novel drying techniques and its effect on quality of fruits, vegetables and aromatic herbs." *Foods* **9**(9): 1261.

Callens, K., F. Fontaine, Y. Sanz, A. Bogdanski, K. D 'Hondt, L. Lange, H. Smidt, L. Van Overbeek, T. Kostic and E. Maguin (2022). "Microbiome-based solutions to address new and existing threats to food security, nutrition, health and agrifood systems' sustainability." *Frontiers in Sustainable Food Systems* **6**.

Calo, J. R., P. G. Crandall, C. A. O'Bryan and S. C. Ricke (2015). "Essential oils as antimicrobials in food systems–a review." *Food control* **54**: 111–119.

Calsamiglia, S., M. Busquet, P. Cardozo, L. Castillejos and A. Ferret (2007). "Invited review: Essential oils as modifiers of rumen microbial fermentation." *Journal of Dairy Science* **90**(6): 2580–2595.

Cardoso, P. B., P. H. Araújo and C. Sayer (2013). *Encapsulation of jojoba and andiroba oils by miniemulsion polymerization.*

Effect on molar mass distribution. Macromolecular Symposia, Wiley Online Library.
Careno, S., O. Boutin and E. Badens (2012). "Drug recrystallization using supercritical anti-solvent (SAS) process with impinging jets: Effect of process parameters." *Journal of Crystal Growth* **342**(1): 34–41.
Carpena, M., B. Nuñez-Estevez, A. Soria-Lopez, P. Garcia-Oliveira and M. A. Prieto (2021). "Essential oils and their application on active packaging systems: A review." *Resources* **10**(1): 7.
Chatterjee, S. and Z. M. Judeh (2015). "Encapsulation of fish oil with N-stearoyl O-butylglyceryl chitosan using membrane and ultrasonic emulsification processes." *Carbohydrate Polymers* **123**: 432–442.
Chattopadhyay, P., R. Huff and B. Y. Shekunov (2006). "Drug encapsulation using supercritical fluid extraction of emulsions." *Journal of Pharmaceutical Sciences* **95**(3): 667–679.
Chemat, F., N. Rombaut, A. Meullemiestre, M. Turk, S. Perino, A.-S. Fabiano-Tixier and M. Abert-Vian (2017). "Review of green food processing techniques. Preservation, transformation, and extraction." *Innovative Food Science & Emerging Technologies* **41**: 357–377.
Chen, F., S. Liu, Z. Zhao, W. Gao, Y. Ma, X. Wang, S. Yan and D. Luo (2020). "Ultrasound pre-treatment combined with microwave-assisted hydrodistillation of essential oils from *Perilla frutescens* (L.) Britt. leaves and its chemical composition and biological activity." *Industrial Crops and Products* **143**: 111908.
Chen, K., M. Zhang, B. Adhikari and M. Wang (2022). "Microencapsulation of Sichuan pepper essential oil in soybean protein isolate-Sichuan pepper seed soluble dietary fiber complex coacervates." *Food Hydrocolloids* **125**: 107421.
Chen, K., M. Zhang, B. Bhandari and A. S. Mujumdar (2021). "Edible flower essential oils: A review of chemical compositions, bioactivities, safety and applications in food preservation." *Food Research International* **139**: 109809.
Chen, Q., F. Zhong, J. Wen, D. McGillivray and S. Y. Quek (2013). "Properties and stability of spray-dried and freeze-dried microcapsules co-encapsulated with fish oil, phytosterol esters, and limonene." *Drying Technology* **31**(6): 707–716.
Chen, X., Y. Han, H. Meng, W. Li, Q. Li, Y. Luo, C. Wang, J. Xie, L. Wu and X. Zhang (2019). "Characteristics of the emulsion stabilized by polysaccharide conjugates alkali-extracted from green tea residue and its protective effect on catechins." *Industrial Crops and Products* **140**: 111611.
Cheng, A.-W., H.-X. Xie, Y. Qi, C. Liu, X. Guo, J.-Y. Sun and L.-N. Liu (2017). "Effects of storage time and temperature on polyphenolic content and qualitative characteristics of freeze-dried and spray-dried bayberry powder." *Lwt* **78**: 235–240.
Cherrat, L., L. Espina, M. Bakkali, D. García-Gonzalo, R. Pagán and A. Laglaoui (2014). "Chemical composition and antioxidant properties of *Laurus nobilis* L. and *Myrtus communis* L. essential oils from Morocco and evaluation of their antimicrobial activity acting alone or in combined processes for food preservation." *Journal of the Science of Food and Agriculture* **94**(6): 1197–1204.
Chorianopoulos, N., E. Kalpoutzakis, N. Aligiannis, S. Mitaku, G.-J. Nychas and S. A. Haroutounian (2004). "Essential oils of Satureja, Origanum, and Thymus species: Chemical composition and antibacterial activities against foodborne pathogens." *Journal of Agricultural and Food Chemistry* **52**(26): 8261–8267.
Chouhan, S., K. Sharma and S. Guleria (2017). "Antimicrobial activity of some essential oils—present status and future perspectives." *Medicines* **4**(3): 58.
Chu, Y., C. Gao, X. Liu, N. Zhang, T. Xu, X. Feng, Y. Yang, X. Shen and X. Tang (2020). "Improvement of storage quality of strawberries by pullulan coatings incorporated with cinnamon essential oil nanoemulsion." *Lwt* **122**: 109054.
Comunian, T. A., M. Thomazini, A. J. G. Alves, F. E. de Matos Junior, J. C. de Carvalho Balieiro and C. S. Favaro-Trindade (2013). "Microencapsulation of ascorbic acid by complex coacervation: Protection and controlled release." *Food Research International* **52**(1): 373–379.
Cordery, A., A. P. Rao and S. Ravishankar (2018). "Antimicrobial activities of essential oils, plant extracts and their applications in foods-A." *Journal of Agriculture and Environmental Sciences* **7**(2): 76–89.
Corrado, I., R. Di Girolamo, C. Regalado-González and C. Pezzella (2022). "Polyhydroxyalkanoates-based nanoparticles as essential oil carriers." *Polymers* **14**(1): 166.
Cui, H., M. Bai and L. Lin (2018). "Plasma-treated poly (ethylene oxide) nanofibers containing tea tree oil/beta-cyclodextrin inclusion complex for antibacterial packaging." *Carbohydrate Polymers* **179**: 360–369.
Das, S., A. K. Chaudhari, A. K. Dwivedy, N. Upadhyay, V. K. Singh, A. Singh and N. K. Dubey (2020). "Nanoencapsulation technology: Boon to food packaging industries." *Nanomaterials and Environmental Biotechnology*: 17–40.
Das, S., K. Vishakha, S. Banerjee, S. Mondal and A. Ganguli (2020). "Sodium alginate-based edible coating containing nanoemulsion of *Citrus sinensis* essential oil eradicates planktonic and sessile cells of food-borne pathogens and increased quality attributes of tomatoes." *International Journal of Biological Macromolecules* **162**: 1770–1779.
Das, S. K., K. Vishakha, S. Das, D. Chakraborty and A. Ganguli (2022). "Carboxymethyl cellulose and cardamom oil in a nanoemulsion edible coating inhibit the growth of food-borne pathogens and extend the shelf life of tomatoes." *Biocatalysis and Agricultural Biotechnology* **42**: 102369.
Davidson, P. M., T. M. Taylor and S. E. Schmidt (2012). "Chemical preservatives and natural antimicrobial compounds." *Food Microbiology: Fundamentals and Frontiers*: 765–801.
Dávila-Rodríguez, M., A. Lopez-Malo, E. Palou, N. Ramírez-Corona and M. T. Jiménez-Munguía (2019). "Antimicrobial activity of nanoemulsions of cinnamon, rosemary, and oregano essential oils on fresh celery." *Lwt* **112**: 108247.
De Aguiar, A. C., L. P. S. Silva, C. A. de Rezende, G. F. Barbero and J. Martínez (2016). "Encapsulation of pepper oleoresin by supercritical fluid extraction of emulsions." *The Journal of Supercritical Fluids* **112**: 37–43.
De Corato, U. (2020). "Improving the shelf-life and quality of fresh and minimally-processed fruits and vegetables for a modern food industry: A comprehensive critical review from the traditional technologies into the most promising advancements." *Critical Reviews in Food Science and Nutrition* **60**(6): 940–975.
De Luca, G., A. Di Renzo, F. P. Di Maio and E. Drioli (2006). "Modelling droplet formation in cross-flow membrane emulsification." *Desalination* **199**(1–3): 177–179.
de Meneses, A. C., P. C. M. dos Santos, T. O. Machado, C. Sayer, D. de Oliveira and P. H. Hermes de Araujo (2017). "Poly (thioether-ester) nanoparticles entrapping clove oil for antioxidant activity improvement." *Journal of Polymer Research* **24**: 1–9.
de Meneses, A. C., E. B. P. Marques, F. V. Leimann, O. H. Gonçalves, R. P. Ineu, P. H. H. de Araújo, D. de Oliveira and C. Sayer (2019). "Encapsulation of clove oil in nanostructured lipid carriers from natural waxes: Preparation,

characterization and in vitro evaluation of the cholinesterase enzymes." *Colloids and Surfaces A: Physicochemical and Engineering Aspects* **583**: 123879.

De-Montijo-Prieto, S., M. d. C. Razola-Díaz, A. M. Gómez-Caravaca, E. J. Guerra-Hernandez, M. Jiménez-Valera, B. Garcia-Villanova, A. Ruiz-Bravo and V. Verardo (2021). "Essential oils from fruit and vegetables, aromatic herbs, and spices: Composition, antioxidant, and antimicrobial activities." *Biology* **10**(11): 1091.

de Oliveira Filho, J. G., B. R. Albiero, Í. H. Calisto, M. R. V. Bertolo, F. C. A. Oldoni, M. B. Egea, S. B. Junior, H. M. C. de Azeredo and M. D. Ferreira (2022). "Bio-nanocomposite edible coatings based on arrowroot starch/cellulose nanocrystals/carnauba wax nanoemulsion containing essential oils to preserve quality and improve shelf life of strawberry." *International Journal of Biological Macromolecules* **219**: 812–823.

de Oliveira Filho, J. G., M. J. de Almeida, T. L. Sousa, D. C. dos Santos and M. B. Egea (2021). "Bioactive compounds of turmeric (*Curcuma longa* L.)." *Bioactive compounds in underutilized vegetables and legumes,* Springer Nature: 297–318.

Debnath, K. and P. Roy (2020). "Safe food for all: Key challenges and process requirements." *Zero hunger. Encyclopedia of the UN sustainable development goals*, Springer, Cham.

Delshadi, R., A. Bahrami, A. G. Tafti, F. J. Barba and L. L. Williams (2020). "Micro and nano-encapsulation of vegetable and essential oils to develop functional food products with improved nutritional profiles." *Trends in Food Science & Technology* **104**: 72–83.

Donsì, F., M. Annunziata, M. Sessa and G. Ferrari (2011). "Nanoencapsulation of essential oils to enhance their antimicrobial activity in foods." *LWT-Food Science and Technology* **44**(9): 1908–1914.

Donsi, F., M. Sessa and G. Ferrari (2015). *Essential oil nanoemulsions to prolong the shelf life of solid food products*. Proceedings of the 29th EFFoST International Conference–Food Science Research and Innovation: Delivering sustainable solutions to the global economy and society, Elsevier.

Doost, A. S., M. N. Nasrabadi, V. Kassozi, H. Nakisozi and P. Van der Meeren (2020). "Recent advances in food colloidal delivery systems for essential oils and their main components." *Trends in Food Science & Technology* **99**: 474–486.

Drakopoulos, D., G. Meca, R. Torrijos, A. Marty, A. Kägi, E. Jenny, H.-R. Forrer, J. Six and S. Vogelgsang (2020). "Control of Fusarium graminearum in wheat with mustard-based botanicals: From in vitro to in planta." *Frontiers in Microbiology* **11**: 1595.

Drioli, E., A. Ali and F. Macedonio (2015). "Membrane distillation: Recent developments and perspectives." *Desalination* **356**: 56–84.

Dwivedy, A. K., V. K. Singh, B. Prakash and N. K. Dubey (2018). "Nanoencapsulated *Illicium verum* Hook. f. essential oil as an effective novel plant-based preservative against aflatoxin B1 production and free radical generation." *Food and Chemical Toxicology* **111**: 102–113.

Eftekhari, M., M. R. S. Ardekani, M. Amin, F. Attar, T. Akbarzadeh, M. Safavi, E. Karimpour-Razkenari, M. Amini, M. Isman and M. Khanavi (2019). "*Oliveria decumbens*, a bioactive essential oil: Chemical composition and biological activities." *Iranian Journal of Pharmaceutical Research: IJPR* **18**(1): 412.

El Asbahani, A., K. Miladi, W. Badri, M. Sala, E. A. Addi, H. Casabianca, A. El Mousadik, D. Hartmann, A. Jilale and F. Renaud (2015). "Essential oils: From extraction to encapsulation." *International Journal of Pharmaceutics* **483**(1–2): 220–243.

El Atki, Y., I. Aouam, F. El Kamari, A. Taroq, A. Gourch, B. Lyoussi and A. Abdellaoui (2019). "Antibacterial efficacy of thymol, carvacrol, eugenol and menthol as alternative agents to control the growth of nosocomial infection-bacteria." *Journal of Pharmaceutical Sciences and Research* **11**(2): 306–309.

El Euch, S. K., D. Hassine, S. Cazaux, N. Bouzouita and J. Bouajila (2019). "*Salvia officinalis* essential oil: Chemical analysis and evaluation of anti-enzymatic and antioxidant bioactivities." *South African Journal of Botany* **120**: 253–260.

El Karkouri, J., M. Bouhrim, O. M. Al Kamaly, H. Mechchate, A. Kchibale, I. Adadi, S. Amine, S. Alaoui Ismaili and T. Zair (2021). "Chemical composition, antibacterial and antifungal activity of the essential oil from *Cistus ladanifer* L." *Plants* **10**(10): 2068.

EM Mustafa, N. (2015). "Citrus essential oils: Current and prospective uses in the food industry." *Recent Patents on Food, Nutrition & Agriculture* **7**(2): 115–127.

Esmaeili, H., N. Cheraghi, A. Khanjari, M. Rezaeigolestani, A. A. Basti, A. Kamkar and E. M. Aghaee (2020). "Incorporation of nanoencapsulated garlic essential oil into edible films: A novel approach for extending shelf life of vacuum-packed sausages." *Meat Science* **166**: 108135.

Faleiro, M. L. (2011). "The mode of antibacterial action of essential oils." *Science against Microbial Pathogens: Communicating Current Research and Technological Advances* **2**: 1143–1156.

Fall, R., S. Ngom, D. Sall, M. Sembène and A. Samb (2017). "Chemical characterization of essential oil from the leaves of *Callistemon viminalis* (DR) and *Melaleuca leucadendron* (Linn.)." *Asian Pacific Journal of Tropical Biomedicine* **7**(4): 347–351.

Fancello, F., G. L. Petretto, S. Zara, M. L. Sanna, R. Addis, M. Maldini, M. Foddai, J. P. Rourke, M. Chessa and G. Pintore (2016). "Chemical characterization, antioxidant capacity and antimicrobial activity against food related microorganisms of *Citrus limon* var. pompia leaf essential oil." *LWT-Food Science and Technology* **69**: 579–585.

Fanzo, J. and C. Davis (2021). *Global food systems, diets, and nutrition*, Springer.

Fathi, M., M. Samadi, H. Rostami and K. Parastouei (2021). "Encapsulation of ginger essential oil in chitosan-based microparticles with improved biological activity and controlled release properties." *Journal of Food Processing and Preservation* **45**(4): e15373.

Feltes, G., S. C. Ballen, J. Steffens, N. Paroul and C. Steffens (2023). "Differentiating true and false cinnamon: Exploring multiple approaches for discrimination." *Micromachines* **14**(10): 1819.

Ferreira, C. D. and I. L. Nunes (2019). "Oil nanoencapsulation: Development, application, and incorporation into the food market." *Nanoscale Research Letters* **14**: 1–13.

Fisher, K. and C. Phillips (2008). "Potential antimicrobial uses of essential oils in food: Is citrus the answer?" *Trends in Food Science & Technology* **19**(3): 156–164.

Flórez-Méndez, J. and J. López (2022). "Food additives: Importance, classification, and adverse reactions in humans." *Natural Additives in Foods*: 1–31.

Fredes, C., C. Becerra, J. Parada and P. Robert (2018). "The microencapsulation of maqui (*Aristotelia chilensis* (Mol.) Stuntz) juice by spray-drying and freeze-drying produces powders with similar anthocyanin stability and bioaccessibility." *Molecules* **23**(5): 1227.

Friedly, E., P. Crandall, S. Ricke, M. Roman, C. O'Bryan and V. Chalova (2009). "In vitro antilisterial effects of citrus oil fractions in combination with organic acids." *Journal of Food Science* **74**(2): M67–M72.

Friedman, M. (2017). "Antimicrobial activities of plant essential oils and their components against antibiotic-susceptible and antibiotic-resistant foodborne pathogens." *Essential oils and nanotechnology for treatment of microbial diseases*, CRC Press: 14–38.

Gallo, M., L. Ferrara, A. Calogero, D. Montesano and D. Naviglio (2020). "Relationships between food and diseases: What to know to ensure food safety." *Food Research International* **137**: 109414.

Galovičová, L., P. Borotová, V. Valková, N. L. Vukovic, M. Vukic, J. Štefániková, H. Ďúranová, P. Ł. Kowalczewski, N. Čmiková and M. Kačániová (2021). "*Thymus vulgaris* essential oil and its biological activity." *Plants* **10**(9): 1959.

Geetha, V. and S. N. Chakravarthula (2018). "Chemical composition and anti-inflammatory activity of Boswellia ovalifoliolata essential oils from leaf and bark." *Journal of Forestry Research* **29**: 373–381.

Gharby, S., A. Asdadi, M. Ibourki, A. Hamdouch, T. Ainane and L. A. I. Hassani (2020). "Chemical characterization of the essential oil from aerial parts of *Lavandula rejdalii* Upson & Jury, a medicinal plant endemic to Morocco." *Journal of Essential Oil Bearing Plants* **23**(6): 1422–1427.

Gill, A. O., P. Delaquis, P. Russo and R. Holley (2002). "Evaluation of antilisterial action of cilantro oil on vacuum packed ham." *International Journal of Food Microbiology* **73**(1): 83–92.

González-Locarno, M., Y. Maza Pautt, A. Albis, E. F. López and C. D. Grande Tovar (2020). "Assessment of chitosan-rue (*Ruta graveolens* L.) essential oil-based coatings on refrigerated cape gooseberry (*Physalis peruviana* L.) quality." *Applied Sciences* **10**(8): 2684.

Gouin, S. (2004). "Microencapsulation: Industrial appraisal of existing technologies and trends." *Trends in Food Science & Technology* **15**(7–8): 330–347.

Granata, G., S. Stracquadanio, M. Leonardi, E. Napoli, G. M. L. Consoli, V. Cafiso, S. Stefani and C. Geraci (2018). "Essential oils encapsulated in polymer-based nanocapsules as potential candidates for application in food preservation." *Food Chemistry* **269**: 286–292.

Grumezescu, A. M. and A. M. Holban (2018). *Food safety and preservation: Modern biological approaches to improving consumer health*, Academic Press.

Gupta, A., H. B. Eral, T. A. Hatton and P. S. Doyle (2016). "Nanoemulsions: Formation, properties and applications." *Soft Matter* **12**(11): 2826–2841.

Gupta, S. and P. S. Variyar (2016). "Nanoencapsulation of essential oils for sustained release: Application as therapeutics and antimicrobials." *Encapsulations*, Elsevier: 641–672.

Hancocks, R., F. Spyropoulos and I. Norton (2016). "The effects of membrane composition and morphology on the rotating membrane emulsification technique for food grade emulsions." *Journal of Membrane Science* **497**: 29–35.

Hao, D.-X., F.-L. Gong, G.-H. Hu, Y.-J. Zhao, G.-P. Lian, G.-H. Ma and Z. Su (2008). "Controlling factors on droplets uniformity in membrane emulsification: Experiment and modeling analysis." *Industrial & Engineering Chemistry Research* **47**(17): 6418–6425.

Hasheminejad, N. and F. Khodaiyan (2020). "The effect of clove essential oil loaded chitosan nanoparticles on the shelf life and quality of pomegranate arils." *Food Chemistry* **309**: 125520.

Hoffmann, Sandra, Bryan Maculloch and Michael Batz (2015). *Economic burden of major foodborne illnesses acquired in the United States, EIB-140*, U.S. Department of Agriculture, Economic Research Service.

Holban, A. M. and A. M. Grumezescu (2018). *Microbial contamination and food degradation*, Academic Press London, United Kingdom.

Hossain, A., M. Skalicky, M. Brestic, S. Mahari, R. G. Kerry, S. Maitra, S. Sarkar, S. Saha, P. Bhadra and M. Popov (2021). "Application of nanomaterials to ensure quality and nutritional safety of food." *Journal of Nanomaterials* **2021**: 1–19.

Hou, T., S. S. Sana, H. Li, Y. Xing, A. Nanda, V. R. Netala and Z. Zhang (2022). "Essential oils and its antibacterial, antifungal and anti-oxidant activity applications: A review." *Food Bioscience* **47**: 101716.

Huang, H., T. Belwal, S. Liu, Z. Duan and Z. Luo (2019). "Novel multi-phase nano-emulsion preparation for co-loading hydrophilic arbutin and hydrophobic coumaric acid using hydrocolloids." *Food Hydrocolloids* **93**: 92–101.

Huang, H., D. Wang, T. Belwal, L. Dong, L. Lu, Y. Zou, L. Li, Y. Xu and Z. Luo (2021). "A novel W/O/W double emulsion co-delivering brassinolide and cinnamon essential oil delayed the senescence of broccoli via regulating chlorophyll degradation and energy metabolism." *Food Chemistry* **356**: 129704.

Huang, X., M. E. Sormoli and T. A. Langrish (2018). "Review of some common commercial and noncommercial lab-scale spray dryers and preliminary tests for a prototype new spray dryer." *Drying Technology* **36**(15): 1900–1912.

Hyldgaard, M., T. Mygind and R. L. Meyer (2012). "Essential oils in food preservation: Mode of action, synergies, and interactions with food matrix components." *Frontiers in Microbiology* **3**: 12.

Ibrahim, O. O. (2020). "Thermal and nonthermal food processing technologies for food preservation and their effects on food chemistry and nutritional values." *EC Nutrition* **15**: 88–105.

Indiarto, R., A. H. Asyifaa, F. C. A. Adiningsih, G. A. Aulia and S. R. Achmad (2021). "Conventional and advanced food-drying technology: A current review." *International Journal of Science & Technology Reserach* **10**(1): 99–107.

Iqbal, M., N. Zafar, H. Fessi and A. Elaissari (2015). "Double emulsion solvent evaporation techniques used for drug encapsulation." *International Journal of Pharmaceutics* **496**(2): 173–190.

Jadhav, R., P. Pawar, V. Choudhari, N. Topare, S. Raut-Jadhav, S. Bokil and A. Khan (2023). "An overview of antimicrobial nanoparticles for food preservation." *Materials Today: Proceedings* **72**: 204–216.

Jadhav, R. P., V. W. Koli, A. B. Kamble and M. A. Bhutkar (2020). "A review on nanoemulsion." *Asian Journal of Research in Pharmaceutical Science* **10**(2): 103–108.

Jafari, S. M., A. Rashidinejad and J. Simal-Gandara (2023). *Handbook of food bioactive ingredients: Properties and applications*, Springer Nature.

Jedlińska, A., K. Samborska, D. Witrowa-Rajchert, A.-M. Seuvre and A. Voilley (2019). "Industry-scale spray-drying microencapsulation of orange aroma." *International Agrophysics* **33**(3): 397–405.

Jemaa, M. B., H. Falleh, M. A. Neves, H. Isoda, M. Nakajima and R. Ksouri (2017). "Quality preservation of deliberately contaminated milk using thyme free and nanoemulsified essential oils." *Food Chemistry* **217**: 726–734.

Jemaa, M. B., H. Falleh, R. Serairi, M. A. Neves, M. Snoussi, H. Isoda, M. Nakajima and R. Ksouri (2018). "Nanoencapsulated *Thymus capitatus* essential oil as natural preservative." *Innovative Food Science & Emerging Technologies* **45**: 92–97.

Ju, J., Y. Deng, C. J. Li and M. Li (2022). "Antibacterial activity of essential oil in food system." *Essential oils: Applications and trends in food science and technology*, Springer: 23–41.

Ju, J., Y. Xie, Y. Guo, Y. Cheng, H. Qian and W. Yao (2019). "The inhibitory effect of plant essential oils on foodborne pathogenic bacteria in food." *Critical Reviews in Food Science and Nutrition* **59**(20): 3281–3292.

Jugreet, B. S., S. Suroowan, R. K. Rengasamy and M. F. Mahomoodally (2020). "Chemistry, bioactivities, mode of action and industrial applications of essential oils." *Trends in Food Science & Technology* **101**: 89–105.

Jyothi, N. V. N., P. M. Prasanna, S. N. Sakarkar, K. S. Prabha, P. S. Ramaiah and G. Srawan (2010). "Microencapsulation techniques, factors influencing encapsulation efficiency." *Journal of Microencapsulation* **27**(3): 187–197.

Kaade, W., M. Ferrando, A. Khanmohammed, C. Torras, S. De Lamo-Castellví and C. Güell (2019). "Low-energy high-throughput emulsification with nickel micro-sieves for essential oils encapsulation." *Journal of Food Engineering* **263**: 326–336.

Kaade, W., C. Güell, A. Ballon, J. Mellado-Carretero, S. De Lamo-Castellví and M. Ferrando (2020). "Dynamic membranes of tunable pore size for lemon oil encapsulation." *LWT* **123**: 109090.

Kamkar, A., E. Molaee-Aghaee, A. Khanjari, A. Akhondzadeh-Basti, B. Noudoost, N. Shariatifar, M. A. Sani and M. Soleimani (2021). "Nanocomposite active packaging based on chitosan biopolymer loaded with nano-liposomal essential oil: Its characterizations and effects on microbial, and chemical properties of refrigerated chicken breast fillet." *International Journal of Food Microbiology* **342**: 109071.

Kedia, A., B. Prakash, P. K. Mishra, C. Chanotiya and N. K. Dubey (2014). "Antifungal, antiaflatoxigenic, and insecticidal efficacy of spearmint (*Mentha spicata* L.) essential oil." *International Biodeterioration & Biodegradation* **89**: 29–36.

Kedia, A., B. Prakash, P. K. Mishra and N. Dubey (2014). "Antifungal and antiaflatoxigenic properties of *Cuminum cyminum* (L.) seed essential oil and its efficacy as a preservative in stored commodities." *International Journal of Food Microbiology* **168**: 1–7.

Kerekes, E. B., A. Vidács, M. Takó, T. Petkovits, C. Vágvölgyi, G. Horváth, V. L. Balázs and J. Krisch (2019). "Anti-biofilm effect of selected essential oils and main components on mono- and polymicrobic bacterial cultures." *Microorganisms* **7**(9): 345.

Kiran, S., A. Kujur and B. Prakash (2016). "Assessment of preservative potential of Cinnamomum zeylanicum Blume essential oil against food borne molds, aflatoxin B1 synthesis, its functional properties and mode of action." *Innovative Food Science & Emerging Technologies* **37**: 184–191.

Kobayashi, I., Y. Wada, Y. Hori, M. A. Neves, K. Uemura and M. Nakajima (2012). "Microchannel emulsification using stainless-steel chips: Oil droplet generation characteristics." *Chemical Engineering & Technology* **35**(10): 1865–1871.

Komaiko, J. S. and D. J. McClements (2016). "Formation of food-grade nanoemulsions using low-energy preparation methods: A review of available methods." *Comprehensive Reviews in Food Science and Food Safety* **15**(2): 331–352.

Kotba, I., A. Bouaichi, H. Lougraimzi, K. Habbadi, A. Errifi, A. O. Touhami and A. Douira (2020). "Effect of temperature, pH and essential oils on the mycelial growth of *Rhizoctonia solani* Kühn (Cantharellales: Ceratobasidiaceae) isolates." *Journal of Microbiology, Biotechnology and Food Sciences* **10**(3): 461–466.

Kuang, S. S., J. C. Oliveira and A. M. Crean (2010). "Microencapsulation as a tool for incorporating bioactive ingredients into food." *Critical Reviews in Food Science and Nutrition* **50**(10): 951–968.

Kujur, A., A. Kumar, A. Yadav and B. Prakash (2020). "Antifungal and aflatoxin B1 inhibitory efficacy of nanoencapsulated *Pelargonium graveolens* L. essential oil and its mode of action." *Lwt* **130**: 109619.

Kumar, N., A. Singh, D. K. Sharma and K. Kishore (2019). "Toxicity of food additives." *Food safety and human health*, Elsevier: 67–98.

Kurozawa, L. E. and M. D. Hubinger (2017). "Hydrophilic food compounds encapsulation by ionic gelation." *Current Opinion in Food Science* **15**: 50–55.

KuShaari, K., P. Pandey, Y. Song and R. Turton (2006). "Monte Carlo simulations to determine coating uniformity in a Wurster fluidized bed coating process." *Powder Technology* **166**(2): 81–90.

Kusuma, I. W., E. T. Arung and Y.-u. Kim (2014). "Antimicrobial and antioxidant properties of medicinal plants used by the Bentian tribe from Indonesia." *Food Science and Human Wellness* **3**(3–4): 191–196.

Lakshmayya, N., A. K. Mishra, Y. K. Mohanta, J. Panda, B. Naik, B. Mishra and R. S. Varma (2023). "Essential oils-based nano-emulsion system for food safety and preservation: Current status and future prospects." *Biocatalysis and Agricultural Biotechnology*: 102897.

Landfester, K. (2009). "Miniemulsion polymerization and the structure of polymer and hybrid nanoparticles." *Angewandte Chemie International Edition* **48**(25): 4488–4507.

Lee, D.-H., G.-S. Sim, J.-H. Kim, G.-S. Lee, H.-B. Pyo and B.-C. Lee (2007). "Preparation and characterization of quercetin-loaded polymethyl methacrylate microcapsules using a polyol-in-oil-in-polyol emulsion solvent evaporation method." *Journal of Pharmacy and Pharmacology* **59**(12): 1611–1620.

Lee, S. J. and M. Wong (2014). "Nano-and microencapsulation of phytochemicals." *Nano-and microencapsulation for foods*, Wiley Online Library: 117–165.

Lee, Y., Y. R. Ji, S. Lee, M.-J. Choi and Y. Cho (2019). "Microencapsulation of probiotic Lactobacillus acidophilus KBL409 by extrusion technology to enhance survival under simulated intestinal and freeze-drying conditions." *Journal of Microbiology and Biotechnology* **29**(5): 721–730.

Leong, J.-Y., W.-H. Lam, K.-W. Ho, W.-P. Voo, M. F.-X. Lee, H.-P. Lim, S.-L. Lim, B.-T. Tey, D. Poncelet and E.-S. Chan (2016). "Advances in fabricating spherical alginate hydrogels with controlled particle designs by ionotropic gelation as encapsulation systems." *Particuology* **24**: 44–60.

Liegeard, J. and L. Manning (2020). "Use of intelligent applications to reduce household food waste." *Critical Reviews in Food Science and Nutrition* **60**(6): 1048–1061.

Liu, J., Y. Liu, S. Shao, X. Zheng and K. Tang (2022). "Soluble soybean polysaccharide/carboxymethyl chitosan coatings incorporated with lavender essential oil: Structure, properties and application in fruit preservation." *Progress in Organic Coatings* **173**: 107178.

Liu, W., X. Chen and C. Selomulya (2015). "On the spray drying of uniform functional microparticles." *Particuology* **22**: 1–12.

Llana-Ruiz-Cabello, M., S. Pichardo, S. Maisanaba, M. Puerto, A. I. Prieto, D. Gutierrez-Praena, A. Jos and A. M. Cameán

(2015). "In vitro toxicological evaluation of essential oils and their main compounds used in active food packaging: A review." *Food and Chemical Toxicology* **81**: 9–27.

Lopes-Lutz, D., D. S. Alviano, C. S. Alviano and P. P. Kolodziejczyk (2008). "Screening of chemical composition, antimicrobial and antioxidant activities of Artemisia essential oils." *Phytochemistry* **69**(8): 1732–1738.

Lopez-Romero, J. C., H. González-Ríos, A. Borges and M. Simões (2015). "Antibacterial effects and mode of action of selected essential oils components against *Escherichia coli* and *Staphylococcus aureus*." *Evidence-Based Complementary and Alternative Medicine* **2015**.

Lv, F., H. Liang, Q. Yuan and C. Li (2011). "In vitro antimicrobial effects and mechanism of action of selected plant essential oil combinations against four food-related microorganisms." *Food Research International* **44**(9): 3057–3064.

Mahato, N., M. Sinha, K. Sharma, R. Koteswararao and M. H. Cho (2019). "Modern extraction and purification techniques for obtaining high purity food-grade bioactive compounds and value-added co-products from citrus wastes." *Foods* **8**(11): 523.

Mahian, R. A. and A. M. Sani (2016). "Essentialoils in food systems: A systemic review." *International Journal of PharmTech Research* **9**(6): 409–416.

Mahmud, J. and R. A. Khan (2018). "Characterization of natural antimicrobials in food system." *Advances in Microbiology* **8**(11): 894.

Majumdar, A., N. Pradhan, J. Sadasivan, A. Acharya, N. Ojha, S. Babu and S. Bose (2018). "Food degradation and foodborne diseases: A microbial approach." *Microbial contamination and food degradation*, Elsevier: 109–148.

Mamadalieva, N. Z., N. S. Abdullaeva, T. Rosenau, M. Fakhrutdinova, S. S. Azimova and S. Böhmdorfer (2018). "Composition of essential oils from four Apiaceae and Asteraceae species growing in Uzbekistan." *Natural Product Research* **32**(9): 1118–1122.

Mandal, S. and M. Mandal (2015). "Coriander (*Coriandrum sativum* L.) essential oil: Chemistry and biological activity." *Asian Pacific Journal of Tropical Biomedicine* **5**(6): 421–428.

Mariod, A. A. (2016). "Effect of essential oils on organoleptic (smell, taste, and texture) properties of food." *Essential oils in food preservation, flavor and safety*, Elsevier: 131–137.

Martindale, W. and W. Schiebel (2017). "The impact of food preservation on food waste." *British Food Journal* **119**(12): 2510–2518.

Martins, I. M., M. F. Barreiro, M. Coelho and A. E. Rodrigues (2014). "Microencapsulation of essential oils with biodegradable polymeric carriers for cosmetic applications." *Chemical Engineering Journal* **245**: 191–200.

Masyita, A., R. M. Sari, A. D. Astuti, B. Yasir, N. R. Rumata, T. B. Emran, F. Nainu and J. Simal-Gandara (2022). "Terpenes and terpenoids as main bioactive compounds of essential oils, their roles in human health and potential application as natural food preservatives." *Food Chemistry: X* **13**: 100217.

Matshetshe, K. I., S. Parani, S. M. Manki and O. S. Oluwafemi (2018). "Preparation, characterization and in vitro release study of β-cyclodextrin/chitosan nanoparticles loaded *Cinnamomum zeylanicum* essential oil." *International Journal of Biological Macromolecules* **118**: 676–682.

Matthews, K. R., K. E. Kniel and T. J. Montville (2017). *Food microbiology: An introduction*, John Wiley & Sons.

Mbow, C., C. Rosenzweig, L. Barioni, T. Benton, M. Herrero, M. Krishnapillai and A. Diouf (2020). *Food security (No. GSFC-EDAA-TN78913)*, IPCC.

McClements, D. J. (2011). "Edible nanoemulsions: Fabrication, properties, and functional performance." *Soft Matter* **7**(6): 2297–2316.

McClements, D. J. and S. M. Jafari (2018). "General aspects of nanoemulsions and their formulation." *Nanoemulsions*, Elsevier: 3–20.

McClements, D. J. and J. Rao (2011). "Food-grade nanoemulsions: Formulation, fabrication, properties, performance, biological fate, and potential toxicity." *Critical Reviews in Food Science and Nutrition* **51**(4): 285–330.

Mei, J., X. Ma and J. Xie (2019). "Review on natural preservatives for extending fish shelf life." *Foods* **8**(10): 490.

Miranda, M., X. Sun, A. Marín, L. C. Dos Santos, A. Plotto, J. Bai, O. B. G. Assis, M. D. Ferreira and E. Baldwin (2022). "Nano- and micro-sized carnauba wax emulsions-based coatings incorporated with ginger essential oil and hydroxypropyl methylcellulose on papaya: Preservation of quality and delay of post-harvest fruit decay." *Food Chemistry: X* **13**: 100249.

Modi, B., H. Timilsina, S. Bhandari, A. Achhami, S. Pakka, P. Shrestha, D. Kandel, D. B. Gc, S. Khatri and P. M. Chhetri (2021). "Current trends of food analysis, safety, and packaging." *International Journal of Food Science* **2021**.

Mohamed, S. A., M. El-Sakhawy and M. A.-M. El-Sakhawy (2020). "Polysaccharides, protein and lipid-based natural edible films in food packaging: A review." *Carbohydrate Polymers* **238**: 116178.

Moosavy, M.-H. and N. Shavisi (2013). "Determination of antimicrobial effects of nisin and Mentha spicata essential oil against *Escherichia coli* O157: H7 under various conditions (pH, temperature and NaCl concentration)." *Pharmaceutical Sciences* **19**(2): 61–67.

Moreira, T. S. A., M. A. M. de Oliveira, L. M. T. da Rocha e Lima, M. N. de Souza and J. C. C. da Silva Pinto (2014). *Synthesis of nanoparticles loaded with tamoxifen by in situ miniemulsion RAFT polymerization*. Macromolecular Symposia, Wiley Online Library.

Mothana, R. A., J. M. Khaled, O. M. Noman, A. Kumar, M. F. Alajmi, A. J. Al-Rehaily and M. Kurkcuoglu (2018). "Phytochemical analysis and evaluation of the cytotoxic, antimicrobial and antioxidant activities of essential oils from three Plectranthus species grown in Saudi Arabia." *BMC Complementary and Alternative Medicine* **18**: 1–10.

Muneratto, V. M., T. C. B. Gallo and V. R. Nicoletti (2021). "Oregano essential oil encapsulation following the complex coacervation method: Influence of temperature, ionic strength, and pH on the release kinetics in aqueous medium." *Ciência e Agrotecnologia* **45**.

Nannapaneni, R., V. I. Chalova, P. G. Crandall, S. C. Ricke, M. G. Johnson and C. A. O'Bryan (2009). "Campylobacter and Arcobacter species sensitivity to commercial orange oil fractions." *International Journal of Food Microbiology* **129**(1): 43–49.

Napiórkowska, A. and M. Kurek (2022). "Coacervation as a novel method of microencapsulation of essential oils—a review." *Molecules* **27**(16): 5142.

Napoli, E. and M. Di Vito (2021). "Toward a new future for essential oils." *MDPI* **10**: 207.

Nielsen, C. K., J. Kjems, T. Mygind, T. Snabe, K. Schwarz, Y. Serfert and R. L. Meyer (2016). "Enhancing the antibacterial efficacy of isoeugenol by emulsion encapsulation." *International Journal of Food Microbiology* **229**: 7–14.

Nilsen-Nygaard, J., E. N. Fernández, T. Radusin, B. T. Rotabakk, J. Sarfraz, N. Sharmin, M. Sivertsvik, I. Sone and M. K. Pettersen (2021). "Current status of biobased and biodegradable food packaging materials: Impact on food quality and effect of innovative processing technologies." *Comprehensive Reviews in Food Science and Food Safety* **20**(2): 1333–1380.

Nizori, A., L. T. T. Bui, F. Jie and D. M. Small (2020). "Spray-drying microencapsulation of ascorbic acid: Impact of varying loading content on physicochemical properties of microencapsulated powders." *Journal of the Science of Food and Agriculture* **100**(11): 4165–4171.

Novais, C., A. K. Molina, R. M. Abreu, C. Santo-Buelga, I. C. Ferreira, C. Pereira and L. Barros (2022). "Natural food colorants and preservatives: A review, a demand, and a challenge." *Journal of Agricultural and Food Chemistry* **70**(9): 2789–2805.

Nowshad, F., N. Mustari and M. S. Khan (2021). "Overview of microbial contamination of foods and associated risk factors." *Techniques to measure food safety and quality*, Springer, Cham.

O'Bryan, C., P. Crandall, V. Chalova and S. Ricke (2008). "Orange essential oils antimicrobial activities against *Salmonella spp*." *Journal of Food Science* **73**(6): M264-M267.

Oberoi, K., A. Tolun, K. Sharma and S. Sharma (2019). "Microencapsulation: An overview for the survival of probiotic bacteria." *The Journal of Microbiology, Biotechnology and Food Sciences* **9**(2): 280.

Oliveira Filho, J. G. d., G. d. C. Silva, F. C. A. Oldoni, M. Miranda, C. Florencio, R. M. D. d. Oliveira, M. d. P. Gomes and M. D. Ferreira (2022). "Edible coating based on carnauba wax nanoemulsion and *Cymbopogon martinii* essential oil on papaya postharvest preservation." *Coatings* **12**(11): 1700.

Ozkan, G., P. Franco, I. De Marco, J. Xiao and E. Capanoglu (2019). "A review of microencapsulation methods for food antioxidants: Principles, advantages, drawbacks and applications." *Food Chemistry* **272**: 494–506.

Pandey, V. K., R. U. Islam, R. Shams and A. H. Dar (2022). "A comprehensive review on the application of essential oils as bioactive compounds in Nano-emulsion based edible coatings of fruits and vegetables." *Applied Food Research* **2**(1): 100042.

Paques, J. P., L. M. Sagis, C. J. van Rijn and E. van der Linden (2014). "Nanospheres of alginate prepared through w/o emulsification and internal gelation with nanoparticles of CaCO3." *Food Hydrocolloids* **40**: 182–188.

Pateiro, M., P. E. Munekata, A. S. Sant'Ana, R. Domínguez, D. Rodríguez-Lázaro and J. M. Lorenzo (2021). "Application of essential oils as antimicrobial agents against spoilage and pathogenic microorganisms in meat products." *International Journal of Food Microbiology* **337**: 108966.

Patil, J., M. Kamalapur, S. Marapur and D. Kadam (2010). "Ionotropic gelation and polyelectrolyte complexation: The novel techniques to design hydrogel particulate sustained, modulated drug delivery system: A review." *Digest Journal of Nanomaterials and Biostructures* **5**(1): 241–248.

Pedroso-Santana, S. and N. Fleitas-Salazar (2020). "Ionotropic gelation method in the synthesis of nanoparticles/microparticles for biomedical purposes." *Polymer International* **69**(5): 443–447.

Pellicer, J. A., M. I. Fortea, J. Trabal, M. I. Rodríguez-López, J. A. Gabaldón and E. Núñez-Delicado (2019). "Stability of microencapsulated strawberry flavour by spray drying, freeze drying and fluid bed." *Powder Technology* **347**: 179–185.

Perignon, C., G. Ongmayeb, R. Neufeld, Y. Frere and D. Poncelet (2015). "Microencapsulation by interfacial polymerisation: Membrane formation and structure." *Journal of Microencapsulation* **32**(1): 1–15.

Pesavento, G., C. Calonico, A. Bilia, M. Barnabei, F. Calesini, R. Addona, L. Mencarelli, L. Carmagnini, M. Di Martino and A. L. Nostro (2015). "Antibacterial activity of Oregano, Rosmarinus and Thymus essential oils against *Staphylococcus aureus* and *Listeria monocytogenes* in beef meatballs." *Food Control* **54**: 188–199.

Pisoschi, A. M., A. Pop, C. Georgescu, V. Turcuş, N. K. Olah and E. Mathe (2018). "An overview of natural antimicrobials role in food." *European Journal of Medicinal Chemistry* **143**: 922–935.

Pittman, C. I., S. Pendleton, B. Bisha, C. A. O'Bryan, K. E. Belk, L. Goodridge, P. G. Crandall and S. C. Ricke (2011). "Activity of citrus essential oils against *Escherichia coli* O157: H7 and *Salmonella spp.* and effects on beef subprimal cuts under refrigeration." *Journal of Food Science* **76**(6): M433–M438.

Plati, F., R. Papi and A. Paraskevopoulou (2021). "Characterization of oregano essential oil (*Origanum vulgare* L. subsp. hirtum) particles produced by the novel nano spray drying technique." *Foods* **10**(12): 2923.

Poncelet, D. (2006). "Microencapsulation: Fundamentals, methods and applications." *Surface chemistry in biomedical and environmental science*, Springer: 23–34.

Poshadri, A. and K. Aparna (2010). "Microencapsulation technology: A review." *Journal of Research Angrau* **38**(1): 86–102.

Prakash, A., R. Baskaran and V. Vadivel (2020). "Citral nanoemulsion incorporated edible coating to extend the shelf life of fresh cut pineapples." *Lwt* **118**: 108851.

Prakash, B. and S. Kiran (2016). "Essential oils: A traditionally realized natural resource for food preservation." *Current Science* **110**(10): 1890–1892.

Prakash, B., P. Singh, A. Kedia and N. Dubey (2012). "Assessment of some essential oils as food preservatives based on antifungal, antiaflatoxin, antioxidant activities and in vivo efficacy in food system." *Food Research International* **49**(1): 201–208.

Prakash, B., P. Singh, P. K. Mishra and N. Dubey (2012). "Safety assessment of *Zanthoxylum alatum* Roxb. essential oil, its antifungal, antiaflatoxin, antioxidant activity and efficacy as antimicrobial in preservation of *Piper nigrum* L. fruits." *International Journal of Food Microbiology* **153**(1–2): 183–191.

Prakash, B., P. P. Singh, V. Gupta and T. S. Raghuvanshi (2023). "Essential oils as green promising alternatives to chemical preservatives for agri-food products: New insight into molecular mechanism, toxicity assessment, and safety profile." *Food and Chemical Toxicology*: 114241.

Punia Bangar, S., V. Chaudhary, N. Thakur, P. Kajla, M. Kumar and M. Trif (2021). "Natural antimicrobials as additives for edible food packaging applications: A review." *Foods* **10**(10): 2282.

Qiu, L., M. Zhang, B. Adhikari and L. Chang (2022). "Microencapsulation of rose essential oil in mung bean protein isolate-apricot peel pectin complex coacervates and characterization of microcapsules." *Food Hydrocolloids* **124**: 107366.

Radi, M., S. Akhavan-Darabi, H. R. Akhavan and S. Amiri (2018). "The use of orange peel essential oil microemulsion and nanoemulsion in pectin-based coating to extend the shelf life of fresh-cut orange." *Journal of Food Processing and Preservation* **42**(2): e13441.

Rajaei, A., M. Hadian, A. Mohsenifar, T. Rahmani-Cherati and M. Tabatabaei (2017). "A coating based on clove essential oils encapsulated by chitosan-myristic acid nanogel efficiently enhanced the shelf-life of beef cutlets." *Food Packaging and Shelf Life* **14**: 137–145.

Ramos, M., C. Mellinas, I. Solaberrieta, M. C. Garrigós and A. Jiménez (2021). "Emulsions incorporated in polysaccharide-based active coatings for fresh and minimally processed vegetables." *Foods* **10**(3): 665.

Rao, J., B. Chen and D. J. McClements (2019). "Improving the efficacy of essential oils as antimicrobials in foods: Mechanisms of action." *Annual Review of Food Science and Technology* **10**: 365–387.

Reis, D. R., A. Ambrosi and M. Di Luccio (2022). "Encapsulated essential oils: A perspective in food preservation." *Future Foods* **5**: 100126.

Reis, D. R., G. Zin, E. Lemos-Senna, A. Ambrosi and M. Di Luccio (2021). "A modified premix method for the emulsification of spearmint essential oil (*Mentha spicata*) by ceramic membranes." *Surfaces and Interfaces* **26**: 101328.

Rejeb, A., J. G. Keogh and K. Rejeb (2022). "Big data in the food supply chain: A literature review." *Journal of Data, Information and Management* **4**(1): 33–47.

Reyes-Jurado, F., A. Franco-Vega, N. Ramírez-Corona, E. Palou and A. López-Malo (2015). "Essential oils: Antimicrobial activities, extraction methods, and their modeling." *Food Engineering Reviews* **7**: 275–297.

Reza Mozafari, M., C. Johnson, S. Hatziantoniou and C. Demetzos (2008). "Nanoliposomes and their applications in food nanotechnology." *Journal of Liposome Research* **18**(4): 309–327.

Rezaei, A., M. Fathi and S. M. Jafari (2019). "Nanoencapsulation of hydrophobic and low-soluble food bioactive compounds within different nanocarriers." *Food Hydrocolloids* **88**: 146–162.

Roman, S., L. M. Sánchez-Siles and M. Siegrist (2017). "The importance of food naturalness for consumers: Results of a systematic review." *Trends in Food Science & Technology* **67**: 44–57.

Rout, S., S. Tambe, R. K. Deshmukh, S. Mali, J. Cruz, P. P. Srivastav, P. D. Amin, K. K. Gaikwad, E. H. de Aguiar Andrade and M. S. de Oliveira (2022). "Recent trends in the application of essential oils: The next generation of food preservation and food packaging." *Trends in Food Science & Technology* **129:** 421–439.

Rutz, J. K., C. D. Borges, R. C. Zambiazi, C. G. da Rosa and M. M. da Silva (2016). "Elaboration of microparticles of carotenoids from natural and synthetic sources for applications in food." *Food Chemistry* **202**: 324–333.

Saifullah, M., M. R. I. Shishir, R. Ferdowsi, M. R. T. Rahman and Q. Van Vuong (2019). "Micro and nano encapsulation, retention and controlled release of flavor and aroma compounds: A critical review." *Trends in Food Science & Technology* **86**: 230–251.

Salanță, L. C. and J. Cropotova (2022). "An update on effectiveness and practicability of plant essential oils in the food industry." *Plants* **11**(19): 2488.

Salgado, P. R., L. Di Giorgio, Y. S. Musso and A. N. Mauri (2021). "Recent developments in smart food packaging focused on biobased and biodegradable polymers." *Frontiers in Sustainable Food Systems* **5**: 630393.

Sawaya, W. N. (2017). "Impact of food losses and waste on food security." *Water, energy & food sustainability in the Middle East: The sustainability triangle,* Springer Cham: 361–388.

Schuck, P., R. Jeantet, B. Bhandari, X. D. Chen, I. T. Perrone, A. F. de Carvalho, M. Fenelon and P. Kelly (2016). "Recent advances in spray drying relevant to the dairy industry: A comprehensive critical review." *Drying Technology* **34**(15): 1773–1790.

Settanni, L., E. Palazzolo, V. Guarrasi, A. Aleo, C. Mammina, G. Moschetti and M. A. Germanà (2012). "Inhibition of foodborne pathogen bacteria by essential oils extracted from citrus fruits cultivated in Sicily." *Food Control* **26**(2): 326–330.

Shaaban, H. A. (2020). "Essential oil as antimicrobial agents: Efficacy, stability, and safety issues for food application." *Essential oils-bioactive compounds, new perspectives and applications,* IntechOpen: 1–33.

Shapira, R. and E. Mimran (2007). "Isolation and characterization of Escherichia coli mutants exhibiting altered response to thymol." *Microbial Drug Resistance* **13**(3): 157–165.

Sharif, M. K., K. Javed and A. Nasir (2018). "Foodborne illness: Threats and control." *Foodborne diseases*, Elsevier: 501–523.

Sharma, K., A. Babaei, K. Oberoi, K. Aayush, R. Sharma and S. Sharma (2022). "Essential oil nanoemulsion edible coating in food industry: A review." *Food and Bioprocess Technology* **15**(11): 2375–2395.

Sharma, M., B. Mann, R. Sharma, R. Bajaj, S. Athira, P. Sarkar and R. Pothuraju (2017). "Sodium caseinate stabilized clove oil nanoemulsion: Physicochemical properties." *Journal of Food Engineering* **212**: 38–46.

Sharma, S., S. Barkauskaite, A. K. Jaiswal and S. Jaiswal (2021). "Essential oils as additives in active food packaging." *Food Chemistry* **343**: 128403.

Shaw, I. C. (2018). *Food safety: The science of keeping food safe*, John Wiley & Sons.

Sheikh, M., S. Mehnaz and M. B. Sadiq (2021). "Prevalence of fungi in fresh tomatoes and their control by chitosan and sweet orange (*Citrus sinensis*) peel essential oil coating." *Journal of the Science of Food and Agriculture* **101**(15): 6248–6257.

Shetta, A., J. Kegere and W. Mamdouh (2019). "Comparative study of encapsulated peppermint and green tea essential oils in chitosan nanoparticles: Encapsulation, thermal stability, in-vitro release, antioxidant and antibacterial activities." *International Journal of Biological Macromolecules* **126**: 731–742.

Silva, H. D., M. Â. Cerqueira and A. A. Vicente (2012). "Nanoemulsions for food applications: Development and characterization." *Food and Bioprocess Technology* **5**: 854–867.

Simionato, I., F. C. Domingues, C. Nerin and F. Silva (2019). "Encapsulation of cinnamon oil in cyclodextrin nanosponges and their potential use for antimicrobial food packaging." *Food and Chemical Toxicology* **132**: 110647.

Singh, H., A. Thompson, W. Liu and M. Corredig (2012). "Liposomes as food ingredients and nutraceutical delivery systems." *Encapsulation technologies and delivery systems for food ingredients and nutraceuticals*, Elsevier: 287–318.

Singh, P., R. Shukla, A. Kumar, B. Prakash, S. Singh and N. K. Dubey (2010). "Effect of *Citrus reticulata* and *Cymbopogon citratus* essential oils on Aspergillus flavus growth and aflatoxin production on *Asparagus racemosus*." *Mycopathologia* **170**: 195–202.

Siroha, A. K. and S. P. Bangar (2023). *Biopolymer-based films and coatings: Trends and challenges*, CRC Press.

Sneha, K. and A. Kumar (2022). "Nanoemulsions: Techniques for the preparation and the recent advances in their food applications." *Innovative Food Science & Emerging Technologies* **76**: 102914.

Sonawane, S. H., B. A. Bhanvase, M. Sivakumar and S. B. Potdar (2020). "Current overview of encapsulation." *Encapsulation of active molecules and their delivery system*, Elsevier: 1–8.

Soni, M., A. Maurya, S. Das, J. Prasad, A. Yadav, V. K. Singh, B. K. Singh, N. K. Dubey and A. K. Dwivedy (2022). "Nanoencapsulation strategies for improving nutritional functionality, safety and delivery of plant-based foods:

Recent updates and future opportunities." *Plant Nano Biology* **1**: 100004.

Souza, V. V. M. A., J. M. Almeida, L. N. Barbosa and N. C. C. Silva (2022). "Citral, carvacrol, eugenol and thymol: Antimicrobial activity and its application in food." *Journal of Essential Oil Research* **34**(3): 181–194.

Spyropoulos, F., D. M. Lloyd, R. D. Hancocks and A. K. Pawlik (2014). "Advances in membrane emulsification. Part A: Recent developments in processing aspects and microstructural design approaches." *Journal of the Science of Food and Agriculture* **94**(4): 613–627.

Stević, T., T. Berić, K. Šavikin, M. Soković, D. Gođevac, I. Dimkić and S. Stanković (2014). "Antifungal activity of selected essential oils against fungi isolated from medicinal plant." *Industrial Crops and Products* **55**: 116–122.

Stoleru, E. and M. Brebu (2021). "Stabilization techniques of essential oils by incorporation into biodegradable polymeric materials for food packaging." *Molecules* **26**(20): 6307.

Swathy, J., P. Mishra, J. Thomas, A. Mukherjee and N. Chandrasekaran (2018). "Antimicrobial potency of high-energy emulsified black pepper oil nanoemulsion against aquaculture pathogen." *Aquaculture* **491**: 210–220.

Switaj, T. L., K. J. Winter and S. R. Christensen (2015). "Diagnosis and management of foodborne illness." *American Family Physician* **92**(5): 358–365.

Tackenberg W, M. and P. Kleinebudde (2015). "Encapsulation of liquids via extrusion-a review." *Current Pharmaceutical Design* **21**(40): 5815–5828.

Tadros, T., P. Izquierdo, J. Esquena and C. Solans (2004). "Formation and stability of nano-emulsions." *Advances in Colloid and Interface Science* **108**: 303–318.

Tahir, A., R. Shabir Ahmad, M. Imran, M. H. Ahmad, M. Kamran Khan, N. Muhammad, M. U. Nisa, M. Tahir Nadeem, A. Yasmin and H. S. Tahir (2021). "Recent approaches for utilization of food components as nano-encapsulation: A review." *International Journal of Food Properties* **24**(1): 1074–1096.

Tan, L. T. H., L. H. Lee, W. F. Yin, C. K. Chan, H. Abdul Kadir, K. G. Chan and B. H. Goh (2015). "Traditional uses, phytochemistry, and bioactivities of *Cananga odorata* (Ylang-Ylang)." *Evidence-Based Complementary and Alternative Medicine* **2015**.

Tao, F., Z. Zhang, J. Liu and M. Yokozawa (2009). "Modelling the impacts of weather and climate variability on crop productivity over a large area: A new super-ensemble-based probabilistic projection." *Agricultural and Forest Meteorology* **149**(8): 1266–1278.

Teixeira, A. A. (2014). "Thermal food preservation techniques (pasteurization, sterilization, canning and blanching)." *Conventional and advanced food processing technologies*, Wiley Online Library: 115–128.

Theresa, M., A. S. Unni, A. Geevarghese, S. K. Sebastian, S. Pareek and R. E. Krishnankutty (2021). "Foodborne pathogens and food-related microorganisms." *Sequencing Technologies in Microbial Food Safety and Quality*, CRC Press: 25–108.

Thin, D. B., B. B. Thinh and D. H. Hanh (2022). "Chemical composition and antimicrobial activity of essential oils from leaves and rhizomes of *Curcuma zedoaria* obtained via supercritical fluid extraction." *Nexo Revista Científica* **35**(04): 1091–1098.

Timilsena, Y. P., T. O. Akanbi, N. Khalid, B. Adhikari and C. J. Barrow (2019). "Complex coacervation: Principles, mechanisms and applications in microencapsulation." *International Journal of Biological Macromolecules* **121**: 1276–1286.

Tiwari, S. and P. Verma (2011). "Microencapsulation technique by solvent evaporation method (Study of effect of process variables)." *International Journal of Pharmacy & Life Sciences* **2**(8).

Trifan, A., S. V. Luca, H. Greige-Gerges, A. Miron, E. Gille and A. C. Aprotosoaie (2020). "Recent advances in tackling microbial multidrug resistance with essential oils: Combinatorial and nano-based strategies." *Critical Reviews in Microbiology* **46**(3): 338–357.

Trojanowska, A., A. Nogalska, R. G. Valls, M. Giamberini and B. Tylkowski (2017). "Technological solutions for encapsulation." *Physical Sciences Reviews* **2**(9): 20170020.

Trucillo, P., R. Campardelli and E. Reverchon (2018). "Production of liposomes loaded with antioxidants using a supercritical CO2 assisted process." *Powder Technology* **323**: 155–162.

Tsitlakidou, P., N. Tasopoulos, P. Chatzopoulou and I. Mourtzinos (2023). "Current status, technology, regulation and future perspectives of essential oils' usage in food and drink industry." *Journal of the Science of Food and Agriculture* **103**(14): 6727–6751.

Tudor, L., M.-T. Pițuru, R.-A. Gheorghe-Irimia, C. Șonea and D. Tăpăloagă (2023). "Optimizing milk production, quality and safety through essential oil applications." *Farmacia* **71**(5).

Turek, C. and F. C. Stintzing (2013). "Stability of essential oils: A review." *Comprehensive Reviews in Food Science and Food Safety* **12**(1): 40–53.

Tyagi, V., S. Kaushik, S. Tyagi and T. Akiyama (2011). "Development of phase change materials based microencapsulated technology for buildings: A review." *Renewable and Sustainable Energy Reviews* **15**(2): 1373–1391.

Uhlemann, J. and I. Reiß (2010). "Product design and process engineering using the example of flavors." *Chemical Engineering & Technology: Industrial Chemistry-Plant Equipment-Process Engineering-Biotechnology* **33**(2): 199–212.

Usmiati, S., N. Richana, D. Mangunwidjaja, E. Noor and E. Prangdimurti (2014). *The using of ionic gelation method based on polysaccharides for encapsulating the macromolecules–a review*. International conference on food security and nutrition.

Vafania, B., M. Fathi and S. Soleimanian-Zad (2019). "Nanoencapsulation of thyme essential oil in chitosan-gelatin nanofibers by nozzle-less electrospinning and their application to reduce nitrite in sausages." *Food and Bioproducts Processing* **116**: 240–248.

Valizadeh, S., F. Katiraee, R. Mahmoudi, T. Fakheri and K. Mardani (2015). "Biological properties of *Cinnamomum zeylanicum* essential oil: Phytochemical component, antioxidant and antimicrobial activities." *International Journal of Food Safety Nutrition and Public Health* **6**(3): 174–184.

Vasilaki, A., E. Panagiotopoulou, T. Koupantsis, E. Katsanidis and I. Mourtzinos (2022). "Recent insights in flavor-enhancers: Definition, mechanism of action, taste-enhancing ingredients, analytical techniques and the potential of utilization." *Critical Reviews in Food Science and Nutrition* **62**(32): 9036–9052.

Versino, F., F. Ortega, Y. Monroy, S. Rivero, O. V. López and M. A. García (2023). "Sustainable and bio-based food packaging: A review on past and current design innovations." *Foods* **12**(5): 1057.

Vijeth, S., G. B. Heggannavar and M. Y. Kariduraganavar (2019). "Encapsulating wall materials for micro-/nanocapsules." *Microencapsulation-processes, technologies and industrial applications*, IntechOpen: 1–19.

Vladisavljević, G. T. (2019). "Preparation of microemulsions and nanoemulsions by membrane emulsification." *Colloids and Surfaces A: Physicochemical and Engineering Aspects* **579**: 123709.

Wang, B. (2015). *Development of spray drying technology for microencapsulation of bioactive materials*, Auburn University.

Weisany, W., S. Samadi, J. Amini, S. Hossaini, S. Yousefi and F. Maggi (2019). "Enhancement of the antifungal activity of thyme and dill essential oils against *Colletotrichum nymphaeae* by nano-encapsulation with copper NPs." *Industrial Crops and Products* **132**: 213–225.

Weisany, W., S. Yousefi, N. A.-r. Tahir, N. Golestanehzadeh, D. J. McClements, B. Adhikari and M. Ghasemlou (2022). "Targeted delivery and controlled released of essential oils using nanoencapsulation: A review." *Advances in Colloid and Interface Science* **303**: 102655.

Wigati, L. P., A. A. Wardana, F. Tanaka and F. Tanaka (2023). "Application of pregelatinized corn starch and basil essential oil edible coating with cellulose nanofiber as Pickering emulsion agent to prevent quality-quantity loss of mandarin orange." *Food Packaging and Shelf Life* **35**: 101010.

Wrona, M., C. Nerín, M. J. Alfonso and M. Á. Caballero (2017). "Antioxidant packaging with encapsulated green tea for fresh minced meat." *Innovative Food Science & Emerging Technologies* **41**: 307–313.

Wu, T., H. Wang, B. Jing, F. Liu, P. C. Burns and C. Na (2015). "Multi-body coalescence in Pickering emulsions." *Nature Communications* **6**(1): 5929.

Wu, Y., Y. Luo and Q. Wang (2012). "Antioxidant and antimicrobial properties of essential oils encapsulated in zein nanoparticles prepared by liquid–liquid dispersion method." *LWT-Food Science and Technology* **48**(2): 283–290.

Yamashita, C., M. M. S. Chung, C. dos Santos, C. R. M. Mayer, I. C. F. Moraes and I. G. Branco (2017). "Microencapsulation of an anthocyanin-rich blackberry (*Rubus spp.*) by-product extract by freeze-drying." *Lwt* **84**: 256–262.

Yang, T., W. Qin, Q. Zhang, J. Luo, D. Lin and H. Chen (2023). "Essential-oil capsule preparation and its application in food preservation: A review." *Food Reviews International* **39**(7): 4124–4158.

Yang, Y.-L., D. A. Al-Mahdy, M.-L. Wu, X.-T. Zheng, X.-H. Piao, A.-L. Chen, S.-M. Wang, Q. Yang and Y.-W. Ge (2022). "LC-MS-based identification and antioxidant evaluation of small molecules from the cinnamon oil extraction waste." *Food Chemistry* **366**: 130576.

Yap, P. S. X., K. Yusoff, S.-H. E. Lim, C.-M. Chong and K.-S. Lai (2021). "Membrane disruption properties of essential oils—A double-edged sword?" *Processes* **9**(4): 595.

Yildirim, S., B. Röcker, M. K. Pettersen, J. Nilsen-Nygaard, Z. Ayhan, R. Rutkaite, T. Radusin, P. Suminska, B. Marcos and V. Coma (2018). "Active packaging applications for food." *Comprehensive Reviews in Food Science and Food Safety* **17**(1): 165–199.

Yousefi, S., W. Weisany, S. E. Hosseini and M. Ghasemlou (2023). "Mechanisms of nanoencapsulation to boost the antimicrobial efficacy of essential oils: A review." *Food Hydrocolloids*: 109655.

Zhang, H., X. Li and H. Kang (2019). "Chitosan coatings incorporated with free or nano-encapsulated Paulownia Tomentosa essential oil to improve shelf-life of ready-to-cook pork chops." *Lwt* **116**: 108580.

Zhang, L., Z. Yang, F. Chen, P. Su, D. Chen, W. Pan, Y. Fang, C. Dong, X. Zheng and Z. Du (2017). "Composition and bioactivity assessment of essential oils of *Curcuma longa* L. collected in China." *Industrial Crops and Products* **109**: 60–73.

Zhang, W., H. Jiang, J.-W. Rhim, J. Cao and W. Jiang (2022). "Effective strategies of sustained release and retention enhancement of essential oils in active food packaging films/coatings." *Food Chemistry* **367**: 130671.

Zhou, Y., S. Sun, W. Bei, M. R. Zahi, Q. Yuan and H. Liang (2018). "Preparation and antimicrobial activity of oregano essential oil Pickering emulsion stabilized by cellulose nanocrystals." *International Journal of Biological Macromolecules* **112**: 7–13.

Zin, G., J. Wu, K. Rezzadori, J. C. C. Petrus, M. Di Luccio and Q. Li (2019). "Modification of hydrophobic commercial PVDF microfiltration membranes into superhydrophilic membranes by the mussel-inspired method with dopamine and polyethyleneimine." *Separation and Purification Technology* **212**: 641–649.

19 Advancements and Ethical Implications of Nanomaterials in Food Preservation and Safety

Farwa Batool[a], *Abdul Wahab*[b,c], *Hajira Bibi*[d], *Murad Muhammad*[c,e], *Gholamreza Abdi*[f], *Muhammad Asif Raza*[c,g], *Ghayyas Us Din*[c,g], *and Wajid Zaman*[h]

[a]Department of Botany, Lahore College for Women University, Lahore, Pakistan

[b]Shanghai Center for Plant Stress Biology, CAS Center for Excellence in Molecular Plant Sciences, Chinese Academy of Sciences, Shanghai, China

[c]University of Chinese Academy of Sciences, Beijing, China

[d]Department of Biochemistry, Quaid-i-Azam University Islamabad, Islamabad, Pakistan

[e]State Key Laboratory of Desert and Oasis Ecology, Key Laboratory of Ecological Safety and Sustainable Development in Arid Lands, Xinjiang Institute of Ecology and Geography, Chinese Academy of Sciences, Urumqi, China

[f]Department of Biotechnology, Persian Gulf Research Institute, Persian Gulf University, Bushehr, Iran

[g]CAS Key Laboratory of Molecular Virology & Immunology, Institutional Center for Shared Technologies and Facilities, Pathogen Discovery and Big Data Platform, Shanghai Institute of Immunity and Infection, Chinese Academy of Sciences, Shanghai, China

[h]Department of Life Sciences Yeungnam University, Gyeongbuk, Republic of Korea

19.1 INTRODUCTION

The growth of the worldwide population, predicted to reach an estimated 9.7 billion by the year 2050, poses a difficult challenge to the international public [1]. This swift demographic growth imposes dependable pressure on the surviving food systems, requiring the development and implementation of active food protection methods [2]. The difficulties of climate change, food security, and inadequate natural resources multiply the imperative to bear a growing population [3]. In this context, food preservation is not just a matter of spreading the shelf life cycle of food products. It is a problematic factor in the fight to combat food surplus, guarantee a stable source of nutritious food, and lessen the environmental influences of food manufacture [4]. In recent decades, traditional methods of food preservation, such as canning, freezing, and drying, have become widely adopted.

In comparison, the perseverance and scale of the current food demands require innovative energy-efficient, adaptable, and sustainable approaches to various socioeconomic and climatic conditions [5]. Recreating these food preservation methods has resulted in the exploration of innovative technologies, such as natural plant-based preservatives, high-pressure processing, and vacuum packing [6]. These technologies aim to preserve food with good taste, safety, and nutritional values with decreased energy consumption and minimal carbon footprints of food preservation methods [7]. In addition to technological innovation, there is an increasing demand for systemic alterations in the global food supply chains. These demanded alterations involve educating distribution networks to minimize postharvest damages, reducing food wastage at the consumer level, and increasing local food production to minimize long-distance transportation costs [8]. These practical measures can significantly subsidize the food industry and express demographic alterations when combined with innovative food preservation techniques [9]. In addition, the role of governance and policy in resolving the global population challenges cannot be ignored. Worldwide organizations, governments, and private bodies need to cooperate in developing favorable environments for approving sustainable approaches to food preservation [10]. This involves investment in research and development, giving incentives for sustainable conduct, and developing regulatory outlines to confirm food quality and safety [11]. Global population growth is a complicated challenge that involves approximately all-inclusive techniques. Addressing this challenge efficiently requires technological advancements, sympathetic policies, and systemic advancements in food supply chains [12]. This effort relies on healthy and sustainable food

DOI: 10.1201/9781032677422-19

protection technologies to meet the nutritional needs of rising populations and preserve the planet for future generations. In the research, researchers expressed their thoughts that sustainable and nutritious food preservation technologies are essential to meeting the nutritional needs of a growing population and preserving the planet for future generations [13].

19.1.1 Traditional Food Preservation Methods

Conventional food preservation methods such as canning, dehydration, chemical additives, and fermentation have been vitally significant in the past for societal food consumption [14]. These methods have been effective in enhancing the shelf life of food and have played an essential role in ensuring food accessibility and food security, particularly during food shortages [15]. Each method involves distinct tools to inhibit the growth of food-spoiling pathogens and preserve food products [16]. Drying is one of the ancient food preservation methods that removes moisture from food, which is crucial for the growth of food-spoiling pathogens. By removing the water content from food material, the dehydration or drying method significantly halts the process of food decay and inhibits the growth of bacteria, molds, and yeast [17]. This method usually keeps herbs, fruits, veggies, and meats fresh. The things that come out of this process are light and last longer. These things are best for storing and transporting. People often use this method to keep meat, fruits, veggies, and herbs fresh. The food items from this process are light and last a long time. This is the best way to store and carry these dried foods [18]. In comparison, the canning method involves storing food products in airtight jars and cans and heating them at a specific temperature, killing microorganisms that cause food spoilage and disease. The heating method also denatures enzymes responsible for food deterioration over time [19]. After heating the food at a particular temperature for a specific time, the container is sealed, and a vacuum is created that prevents the entry and growth of new microorganisms in the food. Canning can effectively preserve several food products, such as meats, vegetables, fruits, and dairy [20]. Another common technique for preserving food is fermentation, which involves using bacteria, yeasts, and molds to convert organic substances like sugars into acids, gases, and alcohols. By increasing the food's aromas and textures, this technique lowers the pH and adds alcohol content, two factors that make it harder for infections to survive [21]. Commonly fermented food products include yogurt, sauerkraut, curd, soy sauce, and cheese. Fermentation also adds the benefit of increased nutritional value to food by increasing the availability of vitamins and minerals and improving digestibility [22]. Using chemicals for food preservation is the latest approach in which chemical compounds such as sodium benzoates, potassium sorbate, and other additives are added to food products to prevent the growth of bacteria, yeast, and molds [23]. These chemical preservatives are predominantly used in condiments, beverages, and processed food products. While chemical preservatives preserve food quality and shelf life, there is a controversy over their health and safety, driving consumer demand for natural alternatives [24]. Each of these food preservation techniques has unique advantages and disadvantages. They have significantly influenced and contributed to human culture and diets throughout history and are continuously employed in the modern food industry and schemes [25]. The fast-growing global population, changing global land spacing, health awareness, and environmental concerns drive demand to upgrade conventional food preservation methods by integrating them with the latest technologies to ensure food conservation, safety, and sustainability [26].

19.1.2 Limitations of Traditional Methods

Conventional methods for food storage have some intrinsic limitations despite their widespread use and horticultural significance, which has led to distinct customer demands from the modern food supply chains and the food industry [9]. These limitations include nutritional loss, environmental impact, changing customer perceptions, and inadequate preserving methods for certain perishable and sensitive food products [27]. Furthermore, disposing of food products containing these chemical additives potentially contributes to environmental pollution. These chemicals might leach into the underground water reservoirs and soil, potentially affecting the ecosystem and wildlife [28]. Procedures such as canning can cause the thermal degradation of crucial nutrients in food. Vitamins such as C and B are usually heat-sensitive and susceptible to thermal degradation during food canning procedures [29]. This can result in a product having a longer shelf life but fewer nutrients than its fresh forms. Similarly, various heating techniques may change or impair the food's flavor and quality, giving it a lower sensory value [30]. Traditional preservation approaches frequently fall short in efficiently conserving bioactive compounds and micronutrients that are subtle to chemical treatments and sensitive to heat [31]. For illustration, canning or chemical preservers may unfavorably affect specific phytonutrients found in vegetables and fruits [32]. This constraint is chiefly important in the setting of functional foods and nutraceuticals, where the protection of these bioactive agents is essential to the health-promoting properties of the product [33]. Recently, consumers have shown an increasing preference for foods free from additives and synthetic chemicals [34]. This change in customer behavior is obsessed with alarms over the latent health effects of these materials and broader attention to organic and natural products [35]. As a result, food manufacturers are progressively pressurized to discover substitutive conservation approaches that support customer expectations for 'clean label' food products [36]. In answer to these restrictions, there is increasing attention to emerging alternative safeguarding techniques that can overcome these tasks. Approaches such as high-pressure processing, cold pasteurization, and natural additives are being discovered and approved [33]. These fresher technologies aim

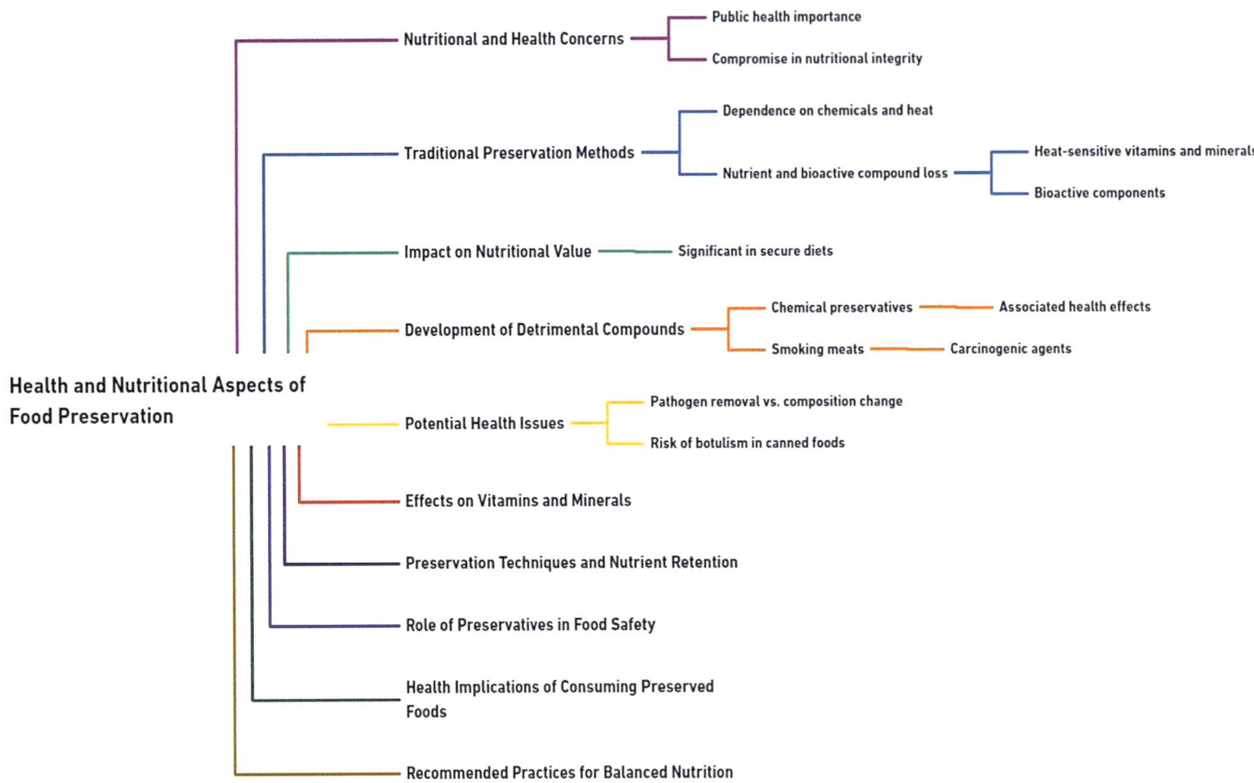

FIGURE 19.1 Health and nutritional aspects of food preservation. This figure was created with Whimsical.com (2024).

to preserve food's nutritious and sensory textures, reduce ecological influences, and address consumer expectations for natural and beneficial products [37]. As the worldwide population rises and the mandate for supportable food systems upsurges, the development of food preservation systems remains a vital part of attention [38].

19.1.3 Health and Nutritional Aspects

Food preservation's nutritional and health aspects are of an excellent position in public health concerns [39]. The customary food preservation procedures, though operative in spreading shelf life and safeguarding safety, often cooperate with the nutritional reliability of food due to their dependence on chemicals and heat [40]. The ruin of essential nutrients and bioactive compounds is a noteworthy concern in upholding the overall excellence of the food source [41]. High-heat procedures like canning or chemical preservatives can lead to losing heat-sensitive minerals, vitamins, and bioactive components. This decrease in nutritional value can be significant, particularly in addressing a secure diet for the populace [42]. Conventional food preservation methods may occasionally adversely affect the development of potentially harmful substances. For instance, smoking meat can release carcinogenic chemicals, and certain chemical preservers have been linked to harmful health consequences (Figure 19.1). Although traditional approaches aim to remove pathogens and decomposed organisms, sometimes they can change the composition of the food in a manner that might cause health issues, such as the potential for botulism in improperly canned foods [43].

19.1.4 Regulatory and Ethical Frameworks

Developing new equipment in foodstuff preservation, like the application of nanomaterials, underlines the necessity for progress in prevailing regulatory structures [44]. These innovative approaches pose exclusive challenges to ensuring security and addressing ethical consequences [45]. Existing regulations may not control the security of evolving preservation technologies. For instance, using nanoparticles in food preservation raises concerns about the long-term effects of these materials on human health and the environment [9]. There is a need for complete analysis and assessment procedures to evaluate the safety of these technologies before extensive adoption [46]. Progressive food conservation skills raise ethical issues. Issues including socioeconomic implications (on traditional farming and food preparation), consumer right to information, and labelling foods processed with novel technology must be examined [44].

19.1.4.1 Updating Regulatory Frameworks

Regulatory organizations must adjust to these technical progressions by updating guidelines and standards [47]. This includes safeguarding that novel preservation systems are secure for consumption or do not harmfully influence the environment and socioeconomic circumstances [48]. Association between policymakers, scientists, and industry leaders is essential in making policies that are both active, effective, and ethically sound [49]. In summary, traditional food maintenance methods have been foundational in confirming food safety; their influence on food's health and nutritional features cannot be ignored [50]. New conservation technologies can alleviate these limits but require careful monitoring and ethical considerations. For global food system sustainability, this machinery must be protected, nutritionally honest, and used morally [51].

The constraints of traditional food preservation techniques become more problematic when considered in the context of a growing global population, complex supply chains, and environmental and health concerns. These challenges underline the need for advanced approaches to food preservation [52]. Nanotechnology is developing as a predominantly talented field in this situation, proposing potentially transformative and novel solutions. The distinctive characteristics of nanomaterials, including enhanced reactivity and increased surface area—can be leveraged to develop more effective and efficient preservation procedures [53]. These could sort from refining the hurdles of food packing to developing nanoencapsulated food additives that deal with measured discharge of preservers and nutrients. Equally, the approval of nanotechnology in foodstuff preservation is not without its intricacies [54]. Strictly assessing these advancements' technical worth, ecological impact, and moral ramifications calls for a multilayered approach. Issues like the long-term impacts of nanomaterials on societal well-being, their potential environmental impact, and moral concerns about consumer rights and cataloguing must be thoroughly considered [55].

19.1.5 Objective and Scope

This chapter offers a comprehensive analysis of these interconnected aspects, proposing awareness of both the potential profits and challenges of integrating nanotechnology in food preservation and conservation. It circles the stage for in-depth negotiations on the progressions in using nanomaterials, the applications of these technologies in food maintenance, and the moral concerns that must be noted regarding their utilization. By investigating these elements in aggregation, this dialogue intends to contribute to an accomplished understanding of the forthcoming food preservation in a developing worldwide landscape. Numerous advancements in nanotechnology have shaped its application in keeping hygienic food conditions and its effectiveness and safety in retaining food's natural qualities. This chapter demonstrates the different applications of nanotechnology to food preservation.

19.2 ADVANCEMENTS IN NANOMATERIALS FOR FOOD PRESERVATION

19.2.1 Historical Context

Nanotechnology has recently gathered massive attention in a recent scientific dissertation, tracing its lineage to a groundbreaking lecture by Richard Feynman in 1959 titled, "There Is Plenty of Room at the Bottom" [56]. Feynman proposed the prospect of exploiting atoms and molecules, giving the theoretical basis for the development of nanotechnology. Over the subsequent years, the field flourished into multidisciplinary domains encompassing biology, chemistry, physics, and engineering. Food preservation is a desperate global concern, dealing with ecological challenges and population growth that these developments have significantly helped [57]. The introduction of nanotechnology in food preservation resulted in a paradigm shift. The preliminary utilizations were experimental, but the investigation became a significant technological revolution over time [58]. Food preservation with nanotechnology is not just a curious scientific endeavor but a distinct force focused on creating safe and effective food preservation programs. The remarkable path that nanotechnology took from Feynman's theoretical ideas to the actual application of nanotechnology in food preservation proves that this field is seeing rapid technical advancement [59].

19.2.2 Types of Nanomaterials

Carbon nanotubes are a distinct class of nanomaterials recognized for their outstanding mechanical strength and electrical conductivity [60]. Their primary use in the food industry is to strengthen packaging resources. By incorporating carbon nanotubes into plastics and other polymers, the resulting compound material exhibits high mechanical strength and improved obstacle properties [61]. This improvement in packaging materials enhances the shelf life of packed food products by offering better defense against environmental factors [62]. Lipid-based nanomaterials encapsulate certain bioactive compounds more sensitive to environmental factors such as light, heat, oxygen, antioxidants, and vitamins [63]. These nanocarriers effectively shield these bioactive compounds from degradation, facilitating their controlled release, maintaining the nutritional value of food products, and contributing to enhanced shelf life. Polymeric nanoparticles have various applications as an alternative delivery method to lipid-based nanocarriers. Polymers that may contain various nutrients and bioactive chemicals include polylactic acid (PLA) and polylactic-co-glycolic acid (PLGA). Their polymeric structure ensures that food's nutritional content is preserved for an extended time by allowing encapsulated bioactive chemicals and components to be measured and gradually released. These nanomaterials significantly contribute to food preservation, presenting promising and innovative results to enhance the shelf life of food products, ensuring food security with high nutritional value. Their advancement, application, and continuous growth have marked a noteworthy footstep in addressing the challenges posed by traditional approaches to food preservation.

19.2.3 Implications and Innovations

The integration of nanomaterials into the realm of food conservation resulted in a wave of innovations, revolutionizing conventional approaches and offering novel competencies: The use of nanomaterials in "active packaging" is a noteworthy advancement that keenly interlinks with food products, promising the preservation of food materials and improving their worth [64]. For instance, incorporating silver nanoparticles into food packaging materials offers antimicrobial surroundings, significantly reducing the risk of fungal and bacterial contamination inside the packaging. This prolongs the shelf life of food materials and guarantees their safety and quality over time [65]. Novelties to develop "smart packaging" incorporate nanotubes to improve food quality. This technology can detect the prevalence of contamination in the food due to pathogenic decomposition and alert customers via visual indicators or other signals [66]. Such innovative packaging ensures that the customer is aware of the safety and freshness of food, which is vital to improve decision-making regarding food consumption and reduce the threat of foodborne diseases [67]. The use of lipid-based and polymeric nanomaterials in the nutrient enhancement of packaged food is a distinguished advancement [68]. Staple foods with higher nutritional content can be preserved by encasing essential components like iron or vitamin A in nanomaterials. This method is beneficial since it preserves the food's flavor and beauty while satisfying consumers, especially in places with dietary deficiencies [69]. The ability of nanomaterials to replace chemical additives or artificial preservatives is a hot area of research [70]. Food preservation can also be accomplished with nano-sized substances made from natural resources, such as oils and plant extracts [71]. These natural nanomaterials reduce the health hazards associated with synthetic food preservatives, realizing healthier and harmless options for food preservation [72]. These practices of nanomaterials in food preservation showcase nanotechnology's potential to modify the approaches followed in food preservation to guarantee food quality, nutritional value, and safety [73]. Due to the growing worldwide population and environmental concerns, these nanomaterials could contribute to food preservation methods that are more efficient, nontoxic, and ecologically stable [74]. The developments in nanomaterials for foodstuff preservation signify a convergence of modernization, utility, and the growing necessities of global safekeeping. Nanotechnology transforms food protection by extending shelf life, boosting nutrition, and offering safe, sustainable alternatives [75]. However, it is dominant that these developments are complemented by demanding safety assessments and ethical deliberations, features that this chapter will discover in subsequent sections [76]. Current developments in nanomaterials for nutritional preservation propose a compelling blend of technical revolution and practical applicability [77]. These materials exemplify the perspective to address some of the most persistent challenges in food preservation nowadays, fluctuating from microbial contamination to nutritional degradation [78]. As the field progresses, it opens new panoramas for research, supervisory appointment, and market acceptance, making it an obligatory constituent of the forthcoming food conservation by technological innovations [79].

19.2.4 Nano-Fabrication Techniques

The technological revolutions in nano-fabrication systems have suggestively transformed the landscape of synthesizing nanomaterials, presenting more efficient and precise systems [80]. Two novel and distinct techniques, laser ablation and atomic layer deposition (ALD), present this development. ALD is a sophisticated method of synthesis of nanomaterials. It works through a sequential and self-limiting method that ensures one atomic layer at a time [81]. This method produces very conformal and thin films, enabling remarkable composition and thickness control. Due to its accuracy, the ALD approach is beneficial for producing nanomaterials for food preservation, including metal oxide nanoparticles widely recognized for their antibacterial qualities [82]. The ability of the ALT technique to alter these materials at the atomic level further refines the compatibility and efficacy of the nanomaterials with several food preservation practices [83]. This procedure employs lasers to effectively remove material from solid or liquid surfaces to develop nanoparticles [84]. One of the vital advantages of laser ablation is its ability to produce nanoparticles without using chemical precursors [55]. This feature is crucial in food preservation approaches as it minimizes the possibility of chemical contagion of food items. Laser ablation is an eco-friendly, clean, and effective technique of nanoparticle synthesis, allowing the formation of pure and uncontaminated nanomaterials [85]. Both procedures demonstrate crucial advancement in nanotechnology, realizing the synthesis of nanoparticles with refined characteristics and functions [86]. The precise deposition of atomic layer techniques in layering and the clean manufacturing procedure of laser ablation are instrumental in utilizing nanomaterials in the food industry for food preservation [87]. These technologies enhance nanomaterials' efficacy from the perspective of food preservation and tackle significant concerns such as food safety and environmental sustainability [88]. As the field of nanotechnology ensures progress, these manufacturing approaches are likely to play an essential part in modelling future novelties in the preservation of food.

19.2.5 Nano-Characterization Techniques

Along with the progress in nano-fabrication, noteworthy progress has also been made in the field of nano-characterization, with methods like atomic force microscopy (AFM) and scanning electron microscopy (SEM) playing essential roles [89]. AFM is an acute instrument for high-resolution imaging at the atomic balance. This method is particularly appreciated when examining nanomaterials' surface morphology and mechanical properties [90]. Considering these qualities is vital for understanding how nanomaterials interrelate with microbial cells, which is an essential factor in determining their efficiency in food preservation [91]. By elaborating the surface details of nanomaterials, AFM aids in modifying their characteristics for specific uses in food preservation [92]. SEM complements AFM by providing high-resolution images focused on the composition and surface of materials [93]. It is beneficial in determining the size distribution of nanoparticles since this is a crucial property affecting their reactivity and, in turn, their success as preservatives. Because SEM assesses the size and shape of nanoparticles, optimizing their usage in food preservation techniques is crucial [94]. Advancements in nano-characterization and nano-fabrication have far-reaching effects on how we store food. Nanomaterials with specific properties that preserve various foods can be created using modern fabrication techniques like ALD [95]. For example, nanoparticles can be engineered to target exact foodborne pathogens. Characterization practices such as AFM and SEM deal with dynamic understandings of nanomaterials' structural and functional characteristics [96]. These understandings are instrumental during the design phase and essential to the growth of more operative and directed nanomaterials for the preservation of food [97].

19.3 APPLICATIONS IN THE FOOD INDUSTRY

19.3.1 Active Packaging

The dawn of active packaging technologies, predominantly those combining nanomaterials, resulted in a significant change in the method of food preservation [98]. Traditional packaging is mainly a passive barrier, separating the food from external ecological influences [99]. However, active packaging plays an additional dynamic role, interrelating directly with the food content and maintaining the interior atmosphere of the package [100]. This contact offers several advantages and innovations. The most significant feature of active packaging involves incorporating materials such as nanoparticles in the food packaging [101]. Silver nanoparticles are appraised for their vigorous antimicrobial properties. Combined with the packaging materials, it can effectively halt microbial growth [102]. The preventive method for microbial growth significantly enhances the shelf life of food products by decreasing the decomposition rate [103]. Nanosensors are incorporated into the food packaging material to ensure food maintenance [104]. These sensors are developed to detect several changes in the food, such as fluctuations in temperature, the presence of pathogens, and variations in gas composition inside the food package [105]. This real-time ability to monitor the internal conditions of the packaging is crucial for food preservation, quality, and safety [106]. An exceptional characteristic of active packaging with nanosensors is their ability to give real-time feedback regarding the packaged food material [107]. These nanosensors can make the customers aware of the freshness and safety of food via visual indicators. For instance, the changed color of the food packaging could reveal that the food has been exposed to unfavorable temperatures that could cause food decomposition and bacterial contamination [108]. The employment of active packaging technologies in the food industry, specifically those with nanomaterials, showcases revolutionary developments in food preservation. Nanotechnological innovations have surpassed the practice of the submissive role of food packaging, now dealing with refined and highly effective measures against food spoilage, contamination, and degradation [109]. In addition, the real-time monitoring abilities of nanosensors in packaging material provide customers with timely awareness about the safety and quality of packaged food, addressing the rising demand for transparency and assurance in food products. These developments help decrease food wastage and align with consumers' demand for safer and long-lasting food products, as shown in Figure 19.2.

19.3.2 Controlled Release Mechanisms

Nanotechnology ability to safely use preservatives and antimicrobials is crucial to food preservation. Therefore, controlled release mechanisms came into operation, which is a robust innovation [110]. The methods that encapsulate preservatives and antibacterial chemicals and release them gradually and under control are known as nanocarriers and nanoparticles. These technologies provide an alternative to directly adding chemicals to food, frequently raising questions about needless overindulgence and potential health risks [111]. The regulated release mechanism is utilized to gradually release the encapsulated bioactive compounds, halting microbial growth and preserving the food for a prolonged period [37]. This controlled release is well-suited to specific triggers such as changes in the food atmosphere, pH levels, presence of microorganisms, and temperature. Consequently, it ensures that preservatives and antimicrobial agents are utilized professionally, tackling specific challenges without overconsumption of preservatives [112]. Reducing the number of additives needed is another significant advantage of using nanotechnology; controlling the release rate and regulating certain conditions may significantly decrease the number of preservatives and antibacterial agents [113]. These support preserving food quality, extending its shelf life, and addressing health distress related to high additive ranges [114]. Additionally, these controlled release systems contribute to the health aspects and safety of food preservation. In a period where consumer awareness concerning their food is at the top, the practice of nanotechnology for minimizing artificial additives supports buyer preferences for safe and more reliable natural food products [115]. In deduction, the controlled release mechanisms supplied by nanotechnology in food conservation exemplify an essential development in addressing the double purposes of safety and effectiveness [116]. This technology confirms that food remains stable and safe over stretched phases while diminishing health issues related to the habit of antimicrobial agents and preservative chemicals. It stances as a tribute to the impending development of nanotechnology to transform food preservation methods, providing the requirements of a health-conscious

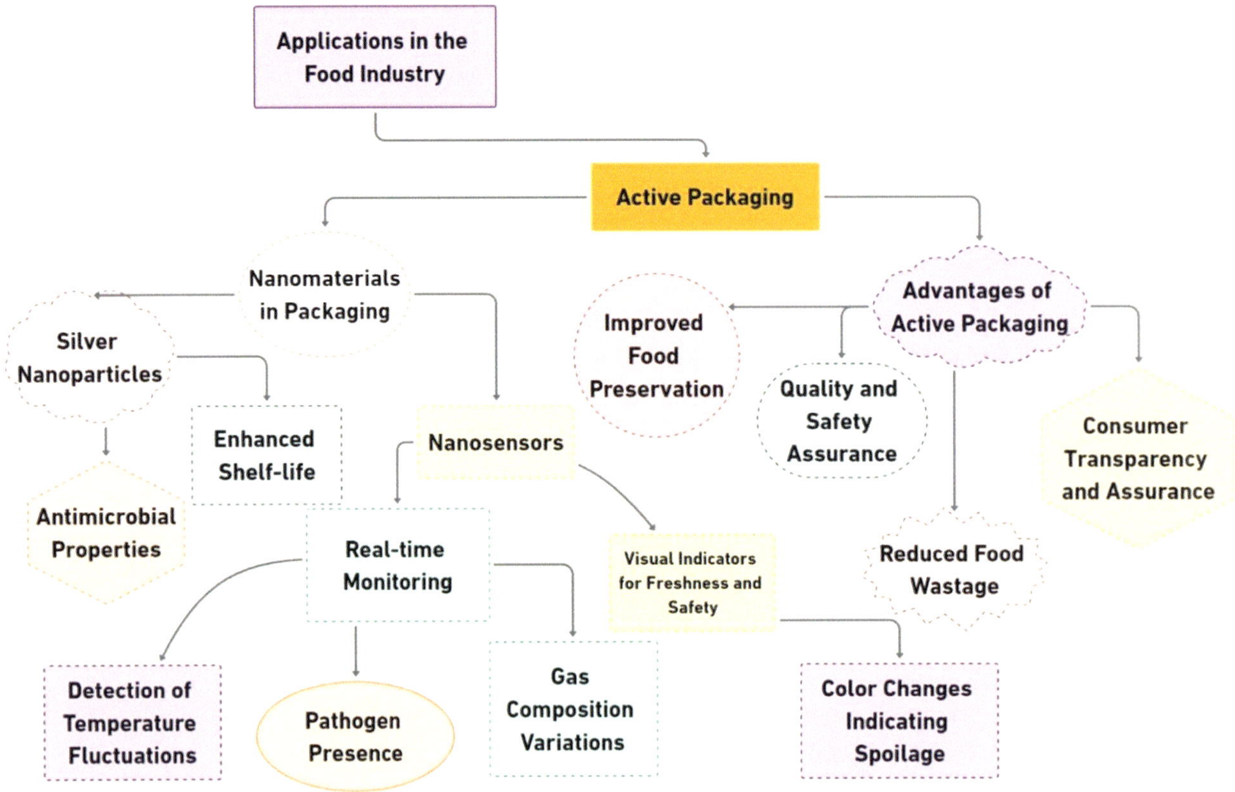

FIGURE 19.2 Role of nanotechnology in the food industry; active packaging plays an additional dynamic role in reducing the waste of food. This figure was created with Whimsical.com (2024).

customer base and paving the way for supplementary sustainable food preservation [117].

19.3.3 Food Processing and Manufacturing

The versatile applications of nanomaterials range well outside the realm of food preservation, permeating numerous features of food processing and manufacturing [118]. For example, in filtration processes, materials like ceramic nanoparticles and carbon nanotubes are engaged for their greater competence in eradicating impurities and extracting components, exceeding the abilities of customary filters [119]. This improved filtration technology is vital for attaining the high purity levels required in food processing. Moreover, nanoparticles are essential in stabilizing emulsions, a necessary procedure in manufacturing an inclusive range of food products, including creams, certain beverages, and mayonnaise [120]. This stabilization is essential to confirming the reliability and texture of these foodstuffs. Also, the advanced use of nanomaterials spreads to flavor improvement [121]. Concocted to carry flavor-enhancing compounds, these nanomaterials can suggestively recover the organoleptic characteristics of food, such as aroma and taste, without disturbing its nutritional profile [122]. This feature of nanotechnology proposes new potentials in food industries, allowing the formation of high-quality products that are safe and attractive to the consumer's palate [123]. Through these various applications, nanomaterials are reforming the land of food treatment and engineering, presenting efficacies and developments that were formerly unachievable [124]. Incorporating nanomaterials in food production has been recognized to be a paradigm-shifting development [125]. The multilayered uses include food preservation through innovative packaging and controlled release mechanisms, successful manufacturing procedures, and product quality. Notably, active packaging technologies propose the twofold advantage of lengthening shelf life while instantaneously providing means for quality reassurance with the help of spoilage indicators [126]. Controlled release mechanisms give a more directed and well-organized application of preservatives, alleviating health risks. Finally, the role of nanomaterials in food handling and manufacturing is increasingly charming and significant, whether in successful filtration efficiency, stabilizing emulsions, or improving flavors [127]. Thus, the practice of nanomaterials in food manufacturing indicates a deep progression that holds the potential to address some of the most severe challenges in global food supply chains nowadays, as shown in Figure 19.3.

19.4 ETHICAL AND HEALTH CONSIDERATIONS

19.4.1 Regulatory Oversight, Ethical Concerns, and Environmental Impact

Utilizing nanomaterials in food preservation raises numerous moral and health concerns that require stringent

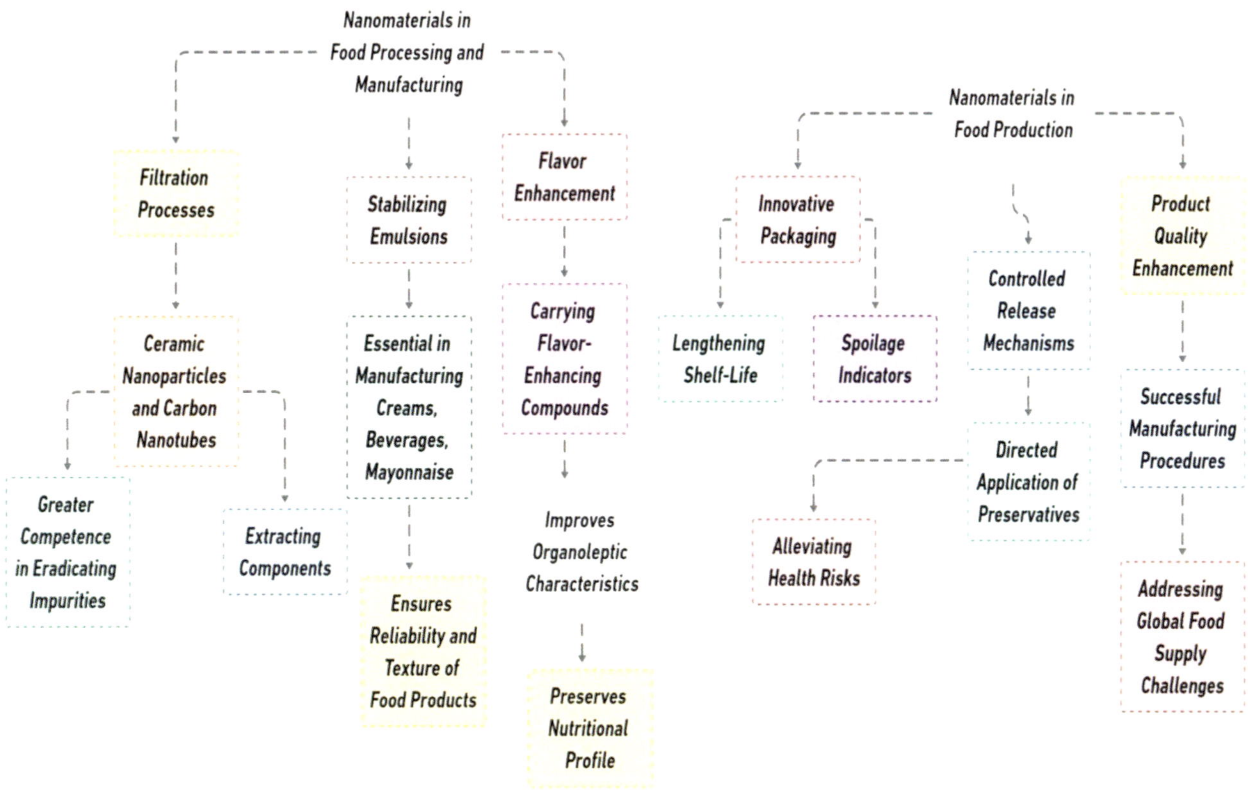

FIGURE 19.3 Versatile applications of nanomaterials range well outside food processing and preservation. This figure was created with Whimsical.com (2024).

regulatory oversight. Regulatory organizations such as the European Food Safety Authority (EFSA) in the European Union and the Food and Drug Administration (FDA) in the United States serve as vanguards in this dome. These officialdoms confirm whether the nanomaterials used in food products meet harsh safety standards [128]. However, the difficulties related to nanoscale relations pose tests for traditional risk assessment approaches. So there is a sharp need for nuanced, inclusive safety measures that ruminate nanomaterials' unique characterization and behaviors, counting their bioavailability, reactivity, and potential for bioaccumulation [129]. While the technological progressions related to nanomaterials propose significant benefits, they also appeal to several ethical contemplations [130]. One of the supreme concerns is conversant agreement. As end users, the common public must be sufficiently alert to incorporate nanomaterials in their food products and the related benefits and risks [131]. Food regulatory authorities and manufacturers need clear labeling and transparent communication policies. Moreover, the long-term effects of nanomaterial intake on human health remain uncertain. As such, continuous investigation must be conducted to determine the long-term security of these materials [132]. The possible environmental effect is an often-overlooked feature of integrating nanomaterials into foodstuffs [133]. Given their nanoscopic magnitudes, these constituents can freely penetrate natural ecosystems. These stances jeopardize the pollution of water bodies and topsoil and possibly cause harm to marine and land-dwelling life [134]. In light of this, any inclusive assessment of nanomaterials for food preservation must also reflect their environmental footprint. This ranges from ethical concerns to human health and ecological comfort [135]. Initiating nanomaterials in food salvation carries a multifaceted network of environmental, health, and ethical considerations requiring rigorous inspection [136]. As nanotechnology advances quickly, regulatory organizations must adapt to ensure comprehensive safety protocols are implemented. Concerns about ethics, primarily related to the recognized agreement and its long-term impact on human health, require further investigation and an open letter [44]. Additionally, the possible ecological effects of nanomaterial usage present an extra layer of complexity that must be meticulously addressed. As society is on the cusp of the extensive acceptance of nanomaterials in food preservation, an optimized, multidimensional methodology for their assessment is authoritative for accountable placement [137].

19.5 CONCLUSION AND FUTURE DIRECTIONS

Incorporating nanomaterials in food preservation highlights a paradigm shift in tackling the challenges faced by global food supply chains. Modern methods for extending food's shelf life, making it safer, and increasing its nutritious content have emerged with this perspective revolution. Authenticated use of nanomaterials in dynamic and active packaging, controlled and sustained release mechanisms, and innovations in food manufacturing and processing showcase the potential of nanotechnology in revolutionizing food systems and the food industry. However, integrating nanotechnology and nanomaterials into dietary items leads to debates and complexities such as regulatory, environmental, and ethical concerns. Looking into the future, the nanotechnology paradigm is open for further research and development. Regulatory bodies demand proactive approaches to polish and enhance the safety assessment methods related to nanomaterials to guarantee that these innovations align with consumers' health standards. From an ethical perspective, increasing transparency and public engagement is crucial, including utilizing explicit tagging that notifies the customer about the nature and quantity of nanomaterials in the food they consume but also meets the general standards. This step would be a cooperative strength, connecting industrial companies and governmental bodies to address customers' concerns. In light of environmental effects, future research must carefully emphasize the life cycle of nanomaterials. Their biological footmark is crucial for sustainable development and growth in this field. The possibility of generating nature-friendly and biodegradable nanomaterials could be a suggested approach to improving ecological implications linked to their use. Nanotechnology offers thrilling potential in food preservation for addressing the current challenges in food schemes; it also seems cautious regarding health, environmental consequences, and ethics. Continuous research, increased public awareness, adapting regulatory agendas, and considering ecological sustainability will harness the true potential of nanomaterials in food manufacturing, processing, and preservation.

Author contribution: All authors have contributed equally.

Conflict of Interest

The authors declare that they have no conflict of interest.

REFERENCES

1. Skendžić, S., et al., The impact of climate change on agricultural insect pests. *Insects*, 2021. **12**(5): p. 440.
2. Mbow, C., et al., *Food security.* 2020, IPCC.
3. Prosekov, A.Y. and S.A. Ivanova, Food security: The challenge of the present. *Geoforum*, 2018. **91**: p. 73–77.
4. De Corato, U., Improving the shelf-life and quality of fresh and minimally-processed fruits and vegetables for a modern food industry: A comprehensive critical review from the traditional technologies into the most promising advancements. *Critical Reviews in Food Science and Nutrition*, 2020. **60**(6): p. 940–975.
5. Al-Juhaimi, F., et al., Effect of various food processing and handling methods on preservation of natural antioxidants in fruits and vegetables. *Journal of Food Science and Technology*, 2018. **55**: p. 3872–3880.
6. Abera, G., Review on high-pressure processing of foods. *Cogent Food & Agriculture*, 2019. **5**(1): p. 1568725.
7. Marrucci, L., M. Marchi, and T. Daddi, Improving the carbon footprint of food and packaging waste management in a supermarket of the Italian retail sector. *Waste Management*, 2020. **105**: p. 594–603.
8. Otto, S., et al., Food packaging and sustainability–consumer perception vs. correlated scientific facts: A review. *Journal of Cleaner Production*, 2021. **298**: p. 126733.
9. Majid, I., et al., Novel food packaging technologies: Innovations and future prospective. *Journal of the Saudi Society of Agricultural Sciences*, 2018. **17**(4): p. 454–462.
10. Sayem, M.A., Religions and environmental sustainability: Focusing on some practical approaches by John B. Cobb Jr. and Seyyed Hossein Nasr. *Australian Journal of Islamic Studies*, 2021. **6**(1): p. 65–80.
11. Sonnino, R., C.L. Tegoni, and A. De Cunto, The challenge of systemic food change: Insights from cities. *Cities*, 2019. **85**: p. 110–116.
12. Sampson, D., et al., Food sovereignty and rights-based approaches strengthen food security and nutrition across the globe: A systematic review. *Frontiers in Sustainable Food Systems*, 2021. **5**: p. 686492.
13. Nicolétis, É., et al., *Agroecological and other innovative approaches for sustainable agriculture and food systems that enhance food security and nutrition.* A report by the High Level Panel of Experts on Food Security and Nutrition of the Committee on World Food Security. 2019.
14. Joardder, M.U., et al., Food preservation techniques in developing countries. In *Food preservation in developing countries. Challenges and solutions*. 2019. p. 67–125.15.
15. Shajil, S., A. Mary, and C.E. Rani Juneius, Recent food preservation techniques employed in the food industry. *Microbial Biotechnology: Volume 2. Application in Food and Pharmacology*, 2018: p. 3–21.
16. Tavman, S., et al., Food preservation technologies. In *Saving food*. 2019, Elsevier. p. 117–140.
17. Abid, J., et al., Food preservation methods and use of nanomaterials as a preservative. *Journal of Xi'an Shiyou University, Natural Science Edition*, 2022. **18**.
18. Pandiselvam, R., et al., Advanced osmotic dehydration techniques combined with emerging drying methods for sustainable food production: Impact on bioactive components, texture, color, and sensory properties of food. *Journal of Texture Studies*, 2022. **53**(6): p. 737–762.
19. Vannini, L. and L. Siroli, Canned and heat preserved foods. In *The microbiological quality of food.* 2025, Elsevier Science Ltd. pp. 261–271.
20. Ariyamuthu, R., V.R. Albert, and S. Je, An overview of food preservation using conventional and modern methods. *Journal of Food and Nutrition Sciences*, 2022. **10**(3): p. 70–79.
21. Mani, A., Food preservation by fermentation and fermented food products. International *Journal of Academic Research and Development*, 2018. **1**: p. 51–57.
22. Teng, T.S., et al., Fermentation for future food systems: Precision fermentation can complement the scope and applications of traditional fermentation. *EMBO Reports*, 2021. **22**(5): p. e52680.
23. Zanetti, M., et al., Use of encapsulated natural compounds as antimicrobial additives in food packaging: A brief review. *Trends in Food Science & Technology*, 2018. **81**: p. 51–60.
24. Silva, C.C., S.P. Silva, and S.C. Ribeiro, Application of bacteriocins and protective cultures in dairy food preservation. *Frontiers in Microbiology*, 2018. **9**: p. 594.
25. Zang, E., et al., Only plant-based food additives: An overview on application, safety, and key challenges in the food industry. *Food Reviews International*, 2023. **39**(8): p. 5132–5163.

26. Sridhar, A., et al., Food preservation techniques and nanotechnology for increased shelf life of fruits, vegetables, beverages and spices: A review. *Environmental Chemistry Letters*, 2021. **19**: p. 1715–1735.
27. Han, J.W., et al., Food packaging: A comprehensive review and future trends. *Comprehensive Reviews in Food Science and Food Safety*, 2018. **17**(4): p. 860–877.
28. Guillard, V., et al., The next generation of sustainable food packaging to preserve our environment in a circular economy context. *Frontiers in Nutrition*, 2018. **5**: p. 121.
29. Krochta, J.M., Food packaging. In *Handbook of food engineering*. 2018, CRC Press. p. 1031–1124.
30. Pauer, E., et al., Assessing the environmental sustainability of food packaging: An extended life cycle assessment including packaging-related food losses and waste and circularity assessment. *Sustainability*, 2019. **11**(3): p. 925.
31. Modi, B., et al., Current trends of food analysis, safety, and packaging. *International Journal of Food Science*, 2021. **2021**.
32. Giannakourou, M.C. and T.N. Tsironi, Application of processing and packaging hurdles for fresh-cut fruits and vegetables preservation. *Foods*, 2021. **10**(4): p. 830.
33. Wrona, M. and C. Nerín, Analytical approaches for analysis of safety of modern food packaging: A review. *Molecules*, 2020. **25**(3): p. 752.
34. Yan, M.R., S. Hsieh, and N. Ricacho, Innovative food packaging, food quality and safety, and consumer perspectives. *Processes*, 2022. **10**(4): p. 747.
35. Caputo, V., G. Sacchi, and A. Lagoudakis, Traditional food products and consumer choices: A review. In *Case studies in the traditional food sector*. 2018. p. 47–87.36.
36. Siddiqui, S.A., et al., Consumer behavior towards nano packaging-A new trend in the food industry. *Future Foods*, 2022. **6**: p. 100191.
37. Fortunati, E., A. Mazzaglia, and G.M. Balestra, Sustainable control strategies for plant protection and food packaging sectors by natural substances and novel nanotechnological approaches. *Journal of the Science of Food and Agriculture*, 2019. **99**(3): p. 986–1000.
38. Motelica, L., et al., Smart food packaging designed by nanotechnological and drug delivery approaches. *Coatings*, 2020. **10**(9): p. 806.
39. Adeyeye, S.A.O., Food packaging and nanotechnology: Safeguarding consumer health and safety. *Nutrition & Food Science*, 2019. **49**(6): p. 1164–1179.
40. Aworh, O.C., African traditional foods and sustainable food security. *Food Control*, 2023. **145**: p. 109393.
41. Coles, R., D. McDowell, and M.J. Kirwan, *Food packaging technology*. Vol. 5. 2003, CRC Press.
42. Kumar, A., et al., Metal-based nanoparticles, sensors, and their multifaceted application in food packaging. *Journal of Nanobiotechnology*, 2021. **19**(1): p. 256.
43. Fung, F., H.-S. Wang, and S. Menon, Food safety in the 21st century. *Biomedical Journal*, 2018. **41**(2): p. 88–95.
44. Baiano, A., Edible insects: An overview on nutritional characteristics, safety, farming, production technologies, regulatory framework, and socio-economic and ethical implications. *Trends in Food Science & Technology*, 2020. **100**: p. 35–50.
45. Bhat, Z.F., et al., Technological, regulatory, and ethical aspects of in vitro meat: A future slaughter-free harvest. *Comprehensive Reviews in Food Science and Food Safety*, 2019. **18**(4): p. 1192–1208.
46. Sohail, M., D.-W. Sun, and Z. Zhu, Recent developments in intelligent packaging for enhancing food quality and safety. *Critical Reviews in Food Science and Nutrition*, 2018. **58**(15): p. 2650–2662.
47. Fortin, N.D., *Food regulation: Law, science, policy, and practice*. 2022, John Wiley & Sons.
48. Pinotti, L., et al., Insects and former foodstuffs for upgrading food waste biomasses/streams to feed ingredients for farm animals. *Animal*, 2019. **13**(7): p. 1365–1375.
49. Santini, A., et al., Nutraceuticals: Opening the debate for a regulatory framework. *British Journal of Clinical Pharmacology*, 2018. **84**(4): p. 659–672.
50. Garcia, C.V., G.H. Shin, and J.T. Kim, Metal oxide-based nanocomposites in food packaging: Applications, migration, and regulations. *Trends in Food Science & Technology*, 2018. **82**: p. 21–31.
51. Marshall, S., et al., Model-informed drug discovery and development: Current industry good practice and regulatory expectations and future perspectives. *CPT: Pharmacometrics & Systems Pharmacology*, 2019. **8**(2): p. 87–96.
52. Primožič, M., Ž. Knez, and M. Leitgeb, (Bio) Nanotechnology in food science—food packaging. *Nanomaterials*, 2021. **11**(2): p. 292.
53. Cerqueira, M.A., A.A. Vicente, and L.M. Pastrana, Nanotechnology in food packaging: Opportunities and challenges. *Nanomaterials for Food Packaging*, 2018: p. 1–11.
54. Hernández-Muñoz, P., et al., Nanotechnology in food packaging. In *Nanomaterials for food applications*. 2019, Elsevier. p. 205–232.
55. Ashfaq, A., et al., Application of nanotechnology in food packaging: Pros and Cons. *Journal of Agriculture and Food Research*, 2022. **7**: p. 100270.
56. Mukherjee, A. and M. Tiwari, Utilisation of quantum dots in food industry. *Agri Mirror: Future India*, 2020. **1**(5): p. 4–7.
57. Alfei, S., B. Marengo, and G. Zuccari, Nanotechnology application in food packaging: A plethora of opportunities versus pending risks assessment and public concerns. *Food Research International*, 2020. **137**: p. 109664.
58. Loretan, C. and A. Muller, Nano, bits, and Feynman's dream: There's plenty of room at the (molecular) bottom. *Journal of Chemical Education*, 2023. **100**(3): p. 1366–1370.
59. Love, L., et al., *There's plenty of room at the top*. 2021, Elsevier. p. 101727.
60. Huang, Y., et al., Recent developments in food packaging based on nanomaterials. *Nanomaterials*, 2018. **8**(10): p. 830.
61. Chaudhary, P., F. Fatima, and A. Kumar, Relevance of nanomaterials in food packaging and its advanced future prospects. *Journal of Inorganic and Organometallic Polymers and Materials*, 2020. **30**: p. 5180–5192.
62. Emamhadi, M.A., et al., Nanomaterials for food packaging applications: A systematic review. *Food and Chemical Toxicology*, 2020. **146**: p. 111825.
63. Lu, H., et al., A review on polymer and lipid-based nanocarriers and its application to nano-pharmaceutical and food-based systems. *Frontiers in Nutrition*, 2021. **8**: p. 783831.
64. Vermeiren, L., et al., Developments in the active packaging of foods. *Trends in Food Science & Technology*, 1999. **10**(3): p. 77–86.
65. Wyrwa, J. and A. Barska, Innovations in the food packaging market: Active packaging. *European Food Research and Technology*, 2017. **243**: p. 1681–1692.
66. Madhusudan, P., N. Chellukuri, and N. Shivakumar, Smart packaging of food for the 21st century–A review with futuristic trends, their feasibility and economics. *Materials Today: Proceedings*, 2018. **5**(10): p. 21018–21022.
67. Schaefer, D. and W.M. Cheung, Smart packaging: Opportunities and challenges. *Procedia Cirp*, 2018. **72**: p. 1022–1027.

68. Subhashree, S. and P.S. Kumar, New analytical strategies amplified with carbon-based nanomaterial for sensing food pollutants. *Chemosphere*, 2022. **295**: p. 133847.
69. Selvasekaran, P. and R. Chidambaram, Bioaerogels as food materials: A state-of-the-art on production and application in micronutrient fortification and active packaging of foods. *Food Hydrocolloids*, 2022. **131**: p. 107760.
70. Arroyo, B.J., et al., Bioactive compounds and their potential use as ingredients for food and its application in food packaging. In *Bioactive compounds*. 2019, Elsevier. p. 143–156.
71. Amiri, S., et al., Natural protective agents and their applications as bio-preservatives in the food industry: An overview of current and future applications. *Italian Journal of Food Science*, 2021. **33**(SP1): p. 55–68.
72. Petkoska, A.T., et al., Edible packaging: Sustainable solutions and novel trends in food packaging. *Food Research International*, 2021. **140**: p. 109981.
73. Huang, T., et al., Polymeric antimicrobial food packaging and its applications. *Polymers*, 2019. **11**(3): p. 560.
74. Carocho, M., P. Morales, and I.C. Ferreira, Antioxidants: Reviewing the chemistry, food applications, legislation and role as preservatives. *Trends in Food Science & Technology*, 2018. **71**: p. 107–120.
75. Mei, J., X. Ma, and J. Xie, Review on natural preservatives for extending fish shelf life. *Foods*, 2019. **8**(10): p. 490.
76. Papadochristopoulos, A., et al., Natural anti-microbials for enhanced microbial safety and shelf-life of processed packaged meat. *Foods*, 2021. **10**(7): p. 1598.
77. Shahid, S., et al., Role of active food packaging developed from microencapsulated bioactive ingredients in quality and shelf life enhancement: A review. *Journal of American Science*, 2021. **17**(2): p. 12–28.
78. Kumar, K.V.P., J. Suneetha, and B.A. Kumari, Active packaging systems in food packaging for enhanced shelf life. *Journal of Pharmacognosy and Phytochemistry*, 2018. **7**(6): p. 2044–2046.
79. Pérez-Santaescolástica, C., et al., Active edible coatings and films with Mediterranean herbs to improve food shelf-life. *Critical Reviews in Food Science and Nutrition*, 2022. **62**(9): p. 2391–2403.
80. Abdulagatov, I.M., et al., Antibacterial food packaging nanomaterial based on atomic layer deposition for long-term food storage. *Journal of Food Science and Technology*, 2023: p. 1–11.
81. Dias, V., et al., Atomic layer deposited TiO_2 and Al_2O_3 thin films as coatings for aluminum food packaging application. *Materials*, 2019. **12**(4): p. 682.
82. Munoz-Rojas, D., et al., Speeding up the unique assets of atomic layer deposition. *Materials Today Chemistry*, 2019. **12**: p. 96–120.
83. Pereira, A., et al., Magnetic nanotubes obtained from atomic layer deposition coated electrospun nanofibers. *Journal of Vacuum Science & Technology B*, 2018. **36**(6).
84. Sportelli, M.C., et al., The pros and cons of the use of laser ablation synthesis for the production of silver nano-antimicrobials. *Antibiotics*, 2018. **7**(3): p. 67.
85. Shabir, I., et al., Laser beam technology interventions in processing, packaging, and quality evaluation of foods. *Measurement: Food*, 2022. **8**: p. 100062.
86. Ionin, A., et al. Combatting bacterial biofilms and bacterial plankton for medicine and food industry via laser nanotechnology. In Сборник трудов конференции *"International Conference on Advanced Laser Technologies (ALT)"*. 2021, Федеральное государственное бюджетное учреждение науки Федеральный. . . .
87. Ghosh, T., G.B. Raj, and K.K. Dash, A comprehensive review on nanotechnology based sensors for monitoring quality and shelf life of food products. *Measurement: Food*, 2022: p. 100049.
88. Khan, Y., et al., Classification, synthetic, and characterization approaches to nanoparticles, and their applications in various fields of nanotechnology: A review. *Catalysts*, 2022. **12**(11): p. 1386.
89. Balea, A., et al., Nanocellulose characterization challenges. *BioResources*, 2021. **16**(2): p. 4382.
90. Wang, H., et al., Vacuum-steam pulsed blanching (VSPB) softens texture and enhances drying rate of carrot by altering cellular structure, pectin polysaccharides and water state. *Innovative Food Science & Emerging Technologies*, 2021. **74**: p. 102801.
91. Das, D., et al., Green chemistry inspired formation of bioactive stable colloidal nanosilver and its wide-spectrum functionalised properties for sustainable industrial escalation. *Results in Chemistry*, 2022: p. 100533.
92. Sharma, S., et al., Nanostructured materials for food applications: Spectroscopy, microscopy and physical properties. *Bioengineering*, 2019. **6**(1): p. 26.
93. Kaminskyj, S.G. and T.E. Dahms, High spatial resolution surface imaging and analysis of fungal cells using SEM and AFM. *Micron*, 2008. **39**(4): p. 349–361.
94. Nazari, M., et al., Garlic essential oil nanophytosomes as a natural food preservative: Its application in yogurt as food model. *Colloid and Interface Science Communications*, 2019. **30**: p. 100176.
95. Nunes, S.P., et al., Thinking the future of membranes: Perspectives for advanced and new membrane materials and manufacturing processes. *Journal of Membrane Science*, 2020. **598**: p. 117761.
96. Mayeen, A., et al., Morphological characterization of nanomaterials. In *Characterization of nanomaterials*. 2018, Elsevier. p. 335–364.
97. Barhoum, A. and M.L. García-Betancourt, Physicochemical characterization of nanomaterials: Size, morphology, optical, magnetic, and electrical properties. In *Emerging applications of nanoparticles and architecture nanostructures*. 2018, Elsevier. p. 279–304.
98. Yildirim, S., et al., Active packaging applications for food. *Comprehensive Reviews in Food Science and Food Safety*, 2018. **17**(1): p. 165–199.
99. Poyatos-Racionero, E., et al., Recent advances on intelligent packaging as tools to reduce food waste. *Journal of Cleaner Production*, 2018. **172**: p. 3398–3409.
100. Janjarasskul, T. and P. Suppakul, Active and intelligent packaging: The indication of quality and safety. *Critical Reviews in Food Science and Nutrition*, 2018. **58**(5): p. 808–831.
101. Deshmukh, S.P., et al., Silver nanoparticles as an effective disinfectant: A review. *Materials Science and Engineering: C*, 2019. **97**: p. 954–965.
102. Hoseinnejad, M., S.M. Jafari, and I. Katouzian, Inorganic and metal nanoparticles and their antimicrobial activity in food packaging applications. *Critical Reviews in Microbiology*, 2018. **44**(2): p. 161–181.
103. Almasi, H., M. Jahanbakhsh Oskouie, and A. Saleh, A review on techniques utilized for design of controlled release food active packaging. *Critical Reviews in Food Science and Nutrition*, 2021. **61**(15): p. 2601–2621.
104. Shawon, Z.B.Z., M.E. Hoque, and S.R. Chowdhury, Nanosensors and nanobiosensors: Agricultural and food technology aspects. In *Nanofabrication for smart nanosensor applications*. 2020, Elsevier. p. 135–161.

105. Singh, R. and S. Kumar, Nanomaterials: Plethora of opportunities as smart packaging, preserving, and processing agent in food industry. In *Nanotechnology advancement in agro-food industry*. 2023, Springer. p. 43–79.
106. Adetunji, C., et al., Prospect of utilizing novel nanomaterials for effective food security, safety, packaging, and preservation. *The Science of Nanomaterials: Basics and Applications*, 2022: p. 273.
107. Sharma, A., et al., Nanoparticles based nanosensors: Principles and their applications in active packaging for food quality and safety detection. *Biochemical Engineering Journal*, 2023. **193**: p. 108861.
108. Chinchkar, A.V., et al., Nanotechnology applications for quality determination of RTE and packaged food. In *Nanotechnology applications for food safety and quality monitoring*. 2023, Elsevier. p. 265–288.
109. Mohammadpour, Z. and S.M. Naghib, Smart nanosensors for intelligent packaging. In *Nanosensors for smart manufacturing*. 2021, Elsevier. p. 323–346.
110. Rahmati, F., et al., New insights into the role of nanotechnology in microbial food safety. *3 Biotech*, 2020. **10**: p. 1–15.
111. Maurya, A., et al., Essential oil nanoemulsion as eco-friendly and safe preservative: Bioefficacy against microbial food deterioration and toxin secretion, mode of action, and future opportunities. *Frontiers in Microbiology*, 2021. **12**: p. 751062.
112. Zhang, C., et al., Typical application of electrostatic layer-by-layer self-assembly technology in food safety assurance. *Trends in Food Science & Technology*, 2022. **129**: p. 88–97.
113. Zhang, X., et al., Emerging chitosan-essential oil films and coatings for food preservation-A review of advances and applications. *Carbohydrate Polymers*, 2021. **273**: p. 118616.
114. Zhao, L., et al., Application of carbon dots in food preservation: A critical review for packaging enhancers and food preservatives. *Critical Reviews in Food Science and Nutrition*, 2023. **63**(24): p. 6738–6756.
115. Seidi Damyeh, M., et al., An insight into curcumin-based photosensitization as a promising and green food preservation technology. *Comprehensive Reviews in Food Science and Food Safety*, 2020. **19**(4): p. 1727–1759.
116. Lv, M., et al., Precision release systems of food bioactive compounds based on metal-organic frameworks: Synthesis, mechanisms and recent applications. *Critical Reviews in Food Science and Nutrition*, 2022. **62**(15): p. 3991–4009.
117. Kumar, N., et al., Recent trends in edible packaging for food applications—perspective for the future. *Food Engineering Reviews*, 2023: p. 1–30.
118. Peerzada, J.G., et al., Advancements in eco-friendly food packaging through nanocomposites: A review. *Polymer Bulletin*, 2023: p. 1–40.
119. Chadha, U., et al., A review of the function of using carbon nanomaterials in membrane filtration for contaminant removal from wastewater. *Materials Research Express*, 2022. **9**(1): p. 012003.
120. Sahoo, M., et al., Nanotechnology: Current applications and future scope in food. *Food Frontiers*, 2021. **2**(1): p. 3–22.
121. Zhu, M., et al., Recent development in food emulsion stabilized by plant-based cellulose nanoparticles. *Current Opinion in Colloid & Interface Science*, 2021. **56**: p. 101512.
122. Aslan Gönül, B. and N. Özkan, *Determining the relationship between sustainable nutrition and tendencies of Orthorexia Nervosa in university students*.
123. Yadav, S.K., et al., Use of nanotechnology in agri-food sectors and apprehensions: An overview. *Seed Research*, 2019. **47**(2): p. 99–149.
124. Srivastava, S., Ethical issues regarding the use of nanobiotechnology-based products. In *Nanobiotechnology for the livestock industry*. 2023, Elsevier. p. 435–473.
125. Solangi, N.H., et al., Mechanism of polymer composite-based nanomaterial for biomedical applications. *Advanced Industrial and Engineering Polymer Research*, 2023. **7**(1): p. 1–19.
126. Wang, T., et al., Nanozyme: An emerging tool for food packaging. *Food Control*, 2023: p. 110104.
127. Jagtiani, E., Advancements in nanotechnology for food science and industry. *Food Frontiers*, 2022. **3**(1): p. 56–82.
128. Kaur, K., et al., Nanotechnology at workplace: Risks, ethics, precautions and regulatory considerations, in *Modern nanotechnology: Volume 2: Green synthesis, sustainable energy and impacts*. 2023, Springer. p. 429–453.
129. Yang, T. and T.V. Duncan, Challenges and potential solutions for nanosensors intended for use with foods. *Nature Nanotechnology*, 2021. **16**(3): p. 251–265.
130. Singh, A., et al., *Nanopriming approach to sustainable agriculture*. 2023, IGI Global.
131. Coles, D. and L.J. Frewer, Nanotechnology applied to European food production–A review of ethical and regulatory issues. *Trends in Food Science & Technology*, 2013. **34**(1): p. 32–43.
132. Engelmann, W., R. von Hohendorff, and D.W.S. Leal, Regulations for using nanotechnology in food and medical products. In *Biopolymer-based nano films*. 2021, Elsevier. p. 387–411.
133. Westlake, J.R., et al., Biodegradable active packaging with controlled release: Principles, progress, and prospects. *ACS Food Science & Technology*, 2022. **2**(8): p. 1166–1183.
134. Hu, C., et al., Phase change materials in food: Phase change temperature, environmental friendliness, and systematization. *Trends in Food Science & Technology*, 2023: p. 104167.
135. Norfaryanti, K., Z. Ainun, and S. Zaiton, The assessment of supply chains, business strategies, and markets in biodegradable food packaging. *Bio-based Packaging: Material, Environmental and Economic Aspects*, 2021: p. 437–451.
136. Gizaw, Z., Public health risks related to food safety issues in the food market: A systematic literature review. *Environmental Health and Preventive Medicine*, 2019. **24**: p. 1–21.
137. Thiruvengadam, M., G. Rajakumar, and I.-M. Chung, Nanotechnology: Current uses and future applications in the food industry. *3 Biotech*, 2018. **8**: p. 1–13.

20 Economic and Sustainability Considerations in Nanotechnology-Based Food Preservation

Hira Zulfiqar[a], *Iqbal Ahmed*[b,c], *Mehak Gull*[d], *Iqra Rani*[c], *Muhammad Furqan Farooq*[c,e], *and Ijaz Hussain*[c]

[a]PhD Chemistry Student, University of Bologna Italy, Italy

[b]University of Rome Tor Vergata, Via del Politecnico 1, Rome, Italy

[c]Government College University Faisalabad, Faisalabad, Pakistan

[d]University of Central Punjab, Department of Chemistry, Venue 1 Khayaban-e-Jinnah, Pir Mansur Johar Town, Lahore, Pakistan

[e]University of Salerno, Department of Chemistry and Biology, Via Giovanni Paolo II, Fisciano SA, Italy

20.1 INTRODUCTION

The demand of food has been escalating dramatically, which occurs due to uncontrolled grow of population all over the world (Davis, Gephart et al. 2016). Because of this rising demand, overall foods diligences are obscuring extensive intimidation from customers to accomplish their apprehensions regarding amount, vitality, primacy as well as healthiness. Researcher and food diligences are making continuous efforts with the passage of time to meet demand of food consumption either by making innovation in food processing or developing more sustainable methods for preservation of huge amount of food products for extended period of time from spoiling which may occur due to different reasons some of which have also been elaborated in Figure 20.1 (Sridhar, Ponnuchamy et al. 2021). Different sort of methodologies have been established and utilized for the preservation of food products. During ancient times mankind used to preserve freshly obtained meat in caves where it was kept until its consumption. Then in recent past human started to preserve food products by drying in sun light, roasting, steaming and fermentation approaches, which nowadays are applied in more improved ways (Chellaram, Murugaboopathi et al. 2014).

In addition to these methods, salting and synthetic additives are also the most commonly used approaches to preserve food products. As a result of these food preservation methods, any type of microbial contaminants that may pose health risks can be killed. Each of these approaches has its own mechanism of preservation, which in turn can cause food storage up to varied periods of time. To preserve food for an extended duration without any nutritional loss is the target of every food industry so that they can cope with the demands of food consumption. Various forms of materials have been developed that are low cost, operative, prompt, and stable (Janjarasskul and Suppakul 2018). These materials can be helpful to preserve the vitality of food products according to the demand of food diligences, consumers, and food legalization authorities. Mostly these materials are applied in the form of packaging that protects food from environmental exposure, contaminants, and microbial attack. Many types of packaging materials have been fabricated in recent times that not only preserve food but also guarantee the accessibility and superiority of foodstuffs for extended time (Pradhan, Singh et al. 2015; Saha 2022). These materials include clay, polymers/biopolymers as well as nanomaterials, which are emerging groups of materials for food packaging and preservation in more precise and sustainable manners (Dainelli, Gontard et al. 2008; Vanderroost, Ragaert et al. 2014; He, Deng et al. 2019).

20.2 ROLE OF NANOTECHNOLOGY

Nanotechnology appears as a prominent approach that links numerous frontiers to fabricate revolutionary products. It is also recognized as an evolving field of science and engineering for its diverse choice of utilization. Nanotechnology elaborates the production, design, and application of configurations that are created by manipulating the size and structure of materials at scale of 10^{-9} m (Chandra 2016; Agriopoulou, Stamatelopoulou et al. 2020). The National Nanotechnology Initiative (NNI) elaborated this term as "the development and analyzing matter at wavelengths ranging generally from 1 to 100 nm, where distinctive spectacles provide innovative applications". The study of nanotechnology encompasses several scientific disciplines, such as chemistry, biology, physics, and engineering disciplines (Neethirajan and Jayas 2011). It also deals with the depiction and operation of nanoscale molecules as well as their exceptional potentials that were absent in their

DOI: 10.1201/9781032677422-20

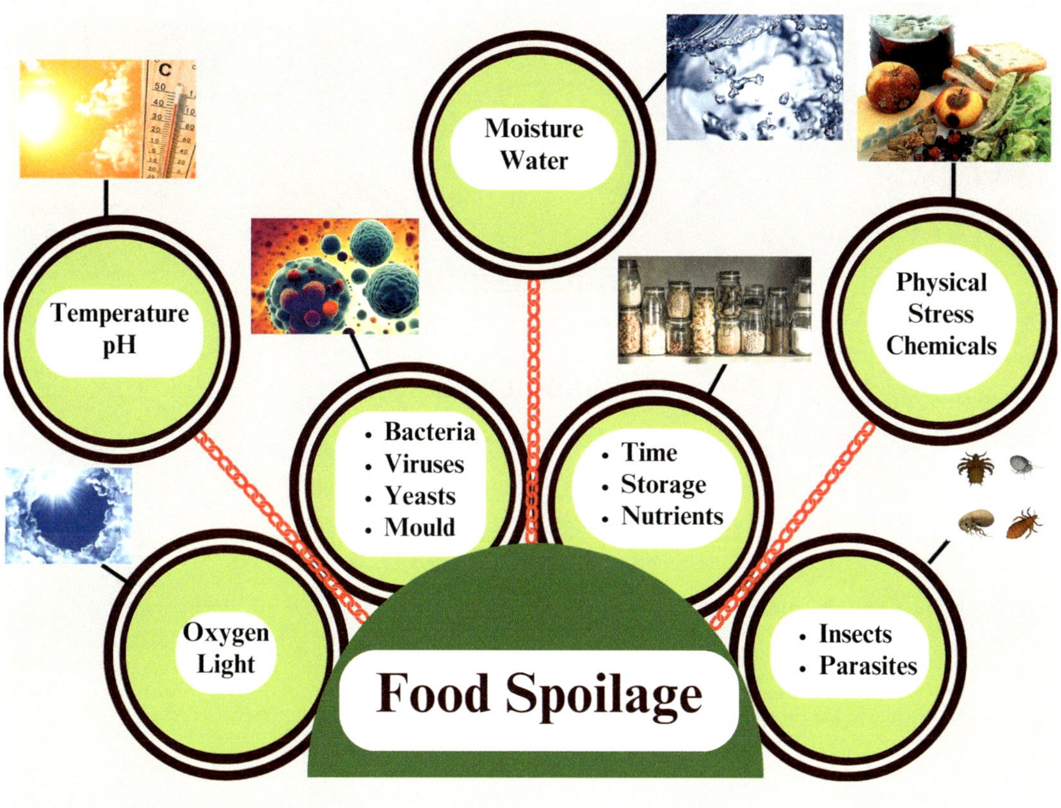

FIGURE 20.1 Food-spoiling factors.

primary form of material. Because of these special qualities of nanomaterials, nanotechnology researchers globally are concentrating predominantly on fabricating novel, cutting-edge, and innovative produces by utilizing green and sustainable approaches (Shende, Rajput et al. 2021).

Food nanotechnology originated with the process of pasteurization, which was developed by Pasteur to destroy germs that cause spoiling (i.e., these germs possess the size range of about 1000 nm). This technique marked the beginning of an upheaval in the manufacturing of food and an enhancement of food quality. Eventually, Watson's and Crick's model of DNA structure, which also ranges in nanoscale, opened the door to applications in agriculture, biotechnology, medicine, and engineering. Furthermore, the discovery of 1 nm-sized carbon nanotubes marked a breakthrough in development and ushered in the era of nanoscience. Carbon has a ton of promise in many industries, including the food industry (Chellaram, Murugaboopathi et al. 2014).

As compared to other countries, European states have considerably augmented utilization of nanostructured materials in food diligences to enhance safety, vitality, and shelf lives of various foodstuffs. Nanotechnology-based marketing in Europe earned total revenue of approximately 2.5 million dollars in 2015, which has been reached to a target of 9 million dollars in 2022 (Primožič, Knez et al. 2021). Various kinds of nanoparticles (NPs) possess antibacterial potential that proves useful in diverse areas of everyday life, some of which include textiles, water purification, pharmaceuticals, and food packing (Seil and Webster 2012; Mihindukulasuriya and Lim 2014).

Food diligences take numerous advantages from nanotechnology in a number of ways, including pathogen identification and treatment, targeted delivery, foodstuff packaging, and the distribution of bioactive components in food supplies to targeted areas. Several key connections among nanotechnology and foods include food security, coloring, nutritional absorption, preservation, nutrition, processing, delivery, flavoring, deterioration and contaminants detection, food functionality, environmental protection as well as economical storage. The effectiveness, bioavailability, nutritional content, and molecular composition of innovative components and products may differ at the nanoscale (Jagtiani 2022). Numerous unique characteristics such as color, strength, diffusivity, toxicity, thermodynamics, solubility, magnetism, and optics are displayed by nanostructured materials, which have also been shown in Figure 20.2. Moreover, many industries like medical, agricultural, cosmetics, clothing, public health, and food industries are using nanomaterials because they have the ability to increase bioavailability and solubility (Fu 2014; Jagtiani 2022).

In food diligences, nanostructured materials (NSMs) are used for nanosensing, novel packaging, and food components encapsulation. The usage of NMs in food diligences offer many advantages, particularly to enhance the possibility for developed and developing nations to elevate food products' vitality, protection, and preservation for longer

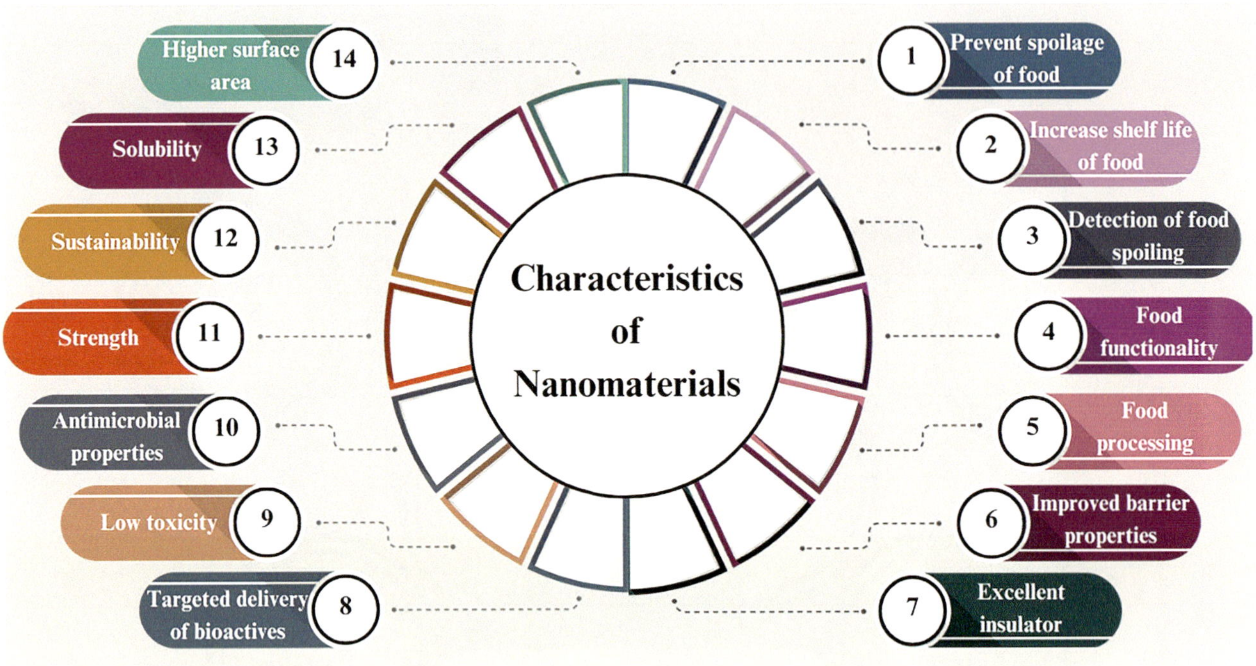

FIGURE 20.2 Characteristics of nanomaterials.

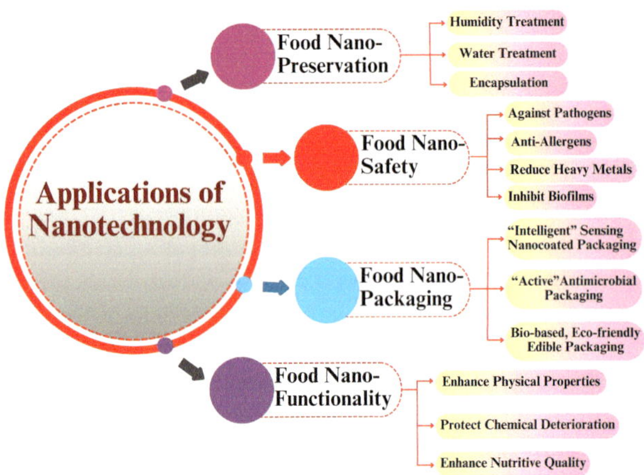

FIGURE 20.3 Applications of nanotechnology.

periods of time. Furthermore, reducing waste of foodstuffs through utilization of NMs in the packaging/preservation of foods is another goal of sustainability. It has also been shown that by extending the nutritional value of food products these nanostructures successfully improve food safety. Common applications of nanotechnology have also been explained in Figure 20.3. Among the most significant challenges of nanotechnology is making improvement of affordable processing techniques and formulations that are harmless for human or society, which is most imperative (Jagtiani 2022). The use of nanotechnology in foodstuffs production has produced a wider variety of new items with improved sensory qualities, flavor, texture, stability, and other aspects of food quality (Anandharamakrishnan and Parthasarathi 2019). Scientists and business associations have already emphasized the prospective utilization of nanostructured materials in almost every facet of the foodstuff industry, especially in agriculture to food preparation, safety, nutrient delivery, and packaging (Sahoo, Vishwakarma et al. 2021). In recent past, the impending remunerations of employing NPs in food superiority assurance, food safety implementation, packaging advancement, and fabrication of foodstuffs with altered utility and nourishment have been the subject of numerous studies (Biswas, Alam et al. 2022).

20.3 BENEFITS OF NANOTECHNOLOGY IN FOOD SECTOR

A distinctive and unambiguous nanomaterial is mostly recommended to employ as a preserving material for specific type of food imperative to adapt convincing and plentiful features since there are several advantageous uses of foods associated with nanotechnology (Sandoval 2009). Nanotechnologies outperform traditional food processing methods due to longer food product shelf lives, lower risk of contamination, and higher-quality food production. Also these materials are quite helpful in maintaining nutritional potential deprived of fluctuating experimental as well as generous possessions (He and Hwang 2016; Das, Livingstone et al. 2018). Common features and type of preservation materials as well as methods have been elaborated in Figure 20.4. Bacteriological adulterations as well as food putrefaction are core bases that shorten foodstuffs' shelf lives, lower foodstuff superiority and impacts food safety, and raise risk of foodborne illnesses. As a consequence, nanostructured materials are employed to develop barrier properties against microorganisms and mechanical properties of packaging systems that are employed for food

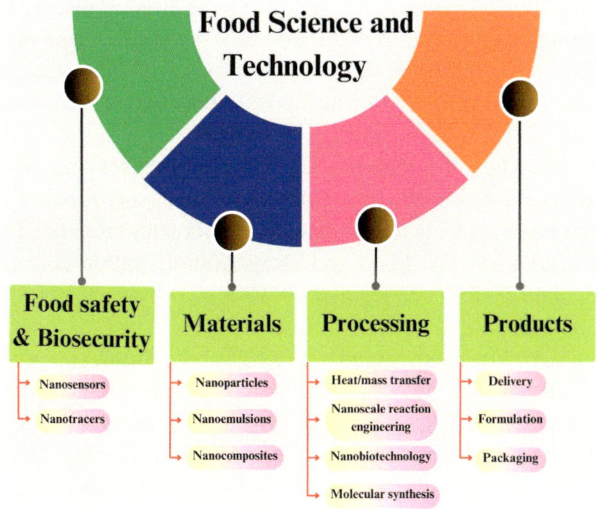

FIGURE 20.4 Types of materials for food applications.

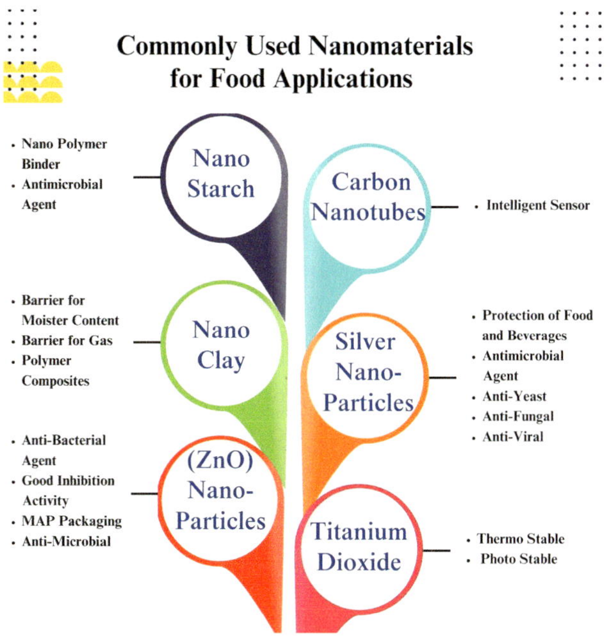

FIGURE 20.5 Nanomaterials for food applications.

preservation, which may be in the form of active or intelligent packagings (Jagtiani 2022).

Moreover, the presence of various NPs in matrices of packaging system can prevent decomposition of foodstuffs as they possess potential to kill or minimize the growth of numerous pathogens. These nanomaterials (NMs) may be plagiaristic from naturally occurring materials as well as are synthesized artificially by using inorganic and organic substances. Most naturally derived NMs that are employed as antimicrobial agents include essential oils (EOs), enzymes, and antibodies. On the other hand, artificially synthesized NMs include silver nanoparticles (AgNPs), graphene, ZnO, as well as TiO_2, which are reported as antimicrobial agents and are commonly employed during processing of foodstuffs that results in enhanced shelf life during storage (Khaneghah, Hashemi et al. 2018). Most of the commonly used NMs for food applications have also been elaborated in Figure 20.5. These NMs possess different mechanism of action against different microorganisms, which may be in the form of disruption of cell membranes (Bondarenko, Sihtmäe et al. 2018); destruction of specific biomolecules, which are beneficial for survival of microbes (Biswas, Tiimob et al. 2019); variations in metabolic activities; and employing oxidative strain (Hoseinnejad, Jafari et al. 2018).

In various food diligences these diverse forms of NMs are extensively applied as packaging material to preserve foods from microbial attacks as they possess antimicrobial activity against pathogens. The antimicrobial potential of NMs is mainly inclined by their configuration, form, dimension, concentration, reliability, distribution, and sprinkling capability, which may be applied lonely as well as in association with some matrices based upon their function, mechanism of action, particular pathogen detection, and various types of food preservative systems. Moreover these may include food safety assessment, encapsulation, food quality maintenance or enhancement, etc. (Weiss, Takhistov et al. 2006). So, as a whole, development of this field of technology has paved the ways for preservation of food products in different diligences (Ravichandran 2010).

Furthermore, NMs comprise numerous benefits as compared to conventional packaging systems for food preservation; these include solubility, infusibility, their antimicrobial potential, stability against higher temperature, enhanced durability, barrier and mechanical properties, recovery, and optical features (Bumbudsanpharoke and Ko 2015; Mustafa and Andreescu 2020). All these features result in a supreme material development for food preservation as well as dispensation of numerous foodstuffs such as beef, cheese, cornflakes, fast foods, dairy products, carbonated beverages, and agricultural food products (Trujillo, Ávalos et al. 2016).

Various studies reported the applications of synthetic as well as natural source based NMs in food preservation systems. For example, bio-nanoencapsulated quercetin-based NMs were developed to improve the storage period of tomatoes, which implies that this approach can also be applied to lengthen the shelf lives of further agricultural products (Nile, Baskar et al. 2020). In addition to extending storage periods even after released packaging, these are utilized as vehicles for carrying flavors, anti-browning proxies, enzymes, antioxidants, and additional bioactive compounds (Weiss, Takhistov et al. 2006; Ongaratto, Vital et al. 2022; de Sousa, Schlogl et al. 2023). It has also been revealed that the addition of clay NPs to polylactic acid (PLA) and ethylene-vinyl alcohol copolymer improve the barrier against oxygen as well as are helpful in enhancing the storage period of foodstuffs (Lagaron, Cabedo et al. 2005). Moreover, application of phyto-glycogenoctenyl NPs combined with the polylysine has effectively extended the food product's shelf life (Scheffler, Wang et al. 2010).

In addition to these types of NMs, AgNPs possess distinctive possessions, a substantially prevailing frontage plasma feature, a higher surface area, a systematized catalytic activity, and an astonishing and comprehensive antimicrobial potential, which are always helpful against numerous bacterial pathogens (Amirsoleimani, Khalilzadeh et al. 2018). Similarly, TiO_2 NPs also possess exceptional antimicrobial potential, lower toxicity, hindrance against UV radiations, and adaptation of color in different food-preserving systems (Garcia, Shin et al. 2018). The use of nanosensors extends the storage period of foodstuffs by keeping out gases and enzymes as well as by reducing the need for synthetic additives. It similarly aids in the removal of ethylene, which ripens food and results in extending the shelf lives of food products. Similarly, chicken breast fillets' shelf life was extended by nanocomposites, which also inhibited bacterial growth and lipid oxidation (Nile, Baskar et al. 2020).

Currently, inefficient global agricultural production of food is partly caused by the inappropriate use of fertilizers and food waste. So pathogens are being controlled by nanocomposites to diminish food wastage and advance foodstuff protection. As a whole, applications of nanotechnology and its characteristics at the nanoscale for controlling pathogens along the food supply chain are prevailing. These applications cover the management of human infectious agents, waterborne pathogens, the eradication of bacteria that cause spoiling on food crops, fresh vegetables, and the cleaning of contaminated areas used in the processing and preparation of food (Rodrigues, Demokritou et al. 2017). Enormous aggregates of shells, pseudostem, and other foliar remnants are produced by banana waste, and these materials damage the environment and are difficult to decompose. Cellulose is one of the most propitious substances derived from agricultural waste; it is found in large quantities in nature and may be derived practically from any type of plant. This substance can be obtained by the large number of agricultural waste banana, specifically explored a potential recovery of the nanocellulose, which could be helpful in numerous applications (Alvarado, Guzmán et al. 2017).

Human health is being deleteriously obstructed by nutrients deficiencies in food products, particularly for those living in rural regions. Nanotechnology may offer the most feasible way to address this issue (Elemike, Uzoh et al. 2019). Although most traditional fertilizers have low nutrient absorption and uptake efficiency, plants nonetheless receive nutrients from them. As a result, nanofertilizers are deliberated for targeted delivery as well as challenging to lose. Hence, agricultural product's nutrient percentages could be heightened by utilizing NPs based nutrients that could be straightforwardly utilized by plants (Dimkpa and Bindraban 2016; Thimiri Govinda Raj and Khan 2016).

Besides these direct remunerations regarding foodstuffs safety, NMs likewise convey ecological reimbursements for food-preserving materials. On the other hand, NMs-based packaging systems are lighter, certainly recyclable, decreasing the depletion of natural sources, have lesser wastage production, are eco-friendly, and are sustainable for food packagings (Aswathi, Meera et al. 2023). Moreover, the utilization of NMs in various fields like safety and vitality of foodstuffs is an apprehension of an excessive level and must all the time be documented owing to their direct impacts on customer well-being (Alfadul and Elneshwy 2010; Hamad, Han et al. 2018). Though, the well-being of NMs-based preserving system is multifaceted issue, as nanostructured-based materials from packaging can transfer into the food streams. Moreover, some of the NPs based preserving systems can be poisonous for creatures as well as surroundings. Regulation on NMs based packaging fluctuates from one to other country. The Food and Drug Administration (FDA) in United States (US) legalizes NMs-based food packaging system that are directly associated with foodstuff matrices. Similarly, Europeans standardize NMs based food-preserving systems, but European regulations are most inclusive as compared to US legislation. Consequently, the vitality of NMs-based material in food diligences is a growing matter, but there is still further need of exploration of threats as well as remunerations of NMs (Amenta, Aschberger et al. 2015; Bumbudsanpharoke and Ko 2015; Mitrano and Wohlleben 2020).

At present, in foodstuffs diligences approximately 417 type of NMs have been developed by 190 enterprises in different countries for public applications so that food can be preserve for longer period (Gómez-Llorente, Hervás et al. 2022). Among these different NMs, 125 types of NMs are considered as innovative food packaging materials, which is analogous to nearly 30% of the overall novelty in numerous food diligences. Moreover, many of these food-preserving NMs encompass clays, Ag, and ZnO NPs, whose chief features include barrier properties against gases, antimicrobial potential, and strong mechanical properties (Pires, Rodrigues et al. 2023).

20.4 THE ADVENT AND POTENTIAL OF BIOPOLYMER-BASED NANOMATERIALS FOR FOOD APPLICATIONS

The ideologies, processes, and methods of nanotechnology make sure the applications in food diligences, which results in the formation of novel packaging materials, flavorings, and encapsulating ability of nutrients (Prasad, Sahoo et al. 2022). All these developments result in addressing the particular public apprehensions regarding multifarious concerns of foodstuff preservation. Consequently, numerous research in nanotechnology have been performed (i.e., active and intelligent packaging), which make sure an enhanced vitality of foodstuffs (King, Cole et al. 2017) as well as maintenance of marketplace in accomplishing functional and resilient food-preserving systems. Moreover, bio-based NMs-based packaging system can similarly be fabricated, which reveal eco-friendly and sustainable features and are popular regarding overcoming existing ecological apprehensions. These include biopolymers that can be plagiaristic from renewable resources, for instance plants, animals, and microbes (Ortega, Versino et al. 2022). These polymeric materials possess distinctive features like biodegradability and lower ecological influences as a result of their degradation, which make them sustainable alternate of petroplastics-based polymeric materials (Wang, Yuan et al. 2021). Consequently, applications of biopolymers-based NMs are fascinating to consider owing to their usage as sustenance or platform for production of nanosystems, which subsidize in the improvement of packagings and play crucial role in fabrication of novel solutions of food safety some of which have been illustrated in Table 20.1 (Webber, Appel et al. 2016).

The integration of various forms of NPs, for instance metal-based NPs; metal oxide-based NPs; hybrid, silicon, or carbon-based nanofillers keen on polymeric backbone, is impressive methodology to advance activity of preserving system that is due to the presence of these NMs. For instance, the presence of these NMs in a polymeric stream result in enhancement of resilience against creep, rigidity/abrasion resistance, oxidation, and barrier features as compared to the pure or native polymeric materials (Kamigaito 1991). Moreover, the presence of various kinds of NMs results in the development of films that possess better properties as compared to only polymeric-based packagings. Therefore, polymer-based nanocomposites materials are proving as state-of-the-art materials, which can actively counter the challenges regarding food preservation (Reig, Lopez et al. 2014).

In previous studies, various kinds of polymer-based nanocomposites have been fabricated by scattering NPs

TABLE 20.1
Applications of Biopolymer-Based Nanomaterials for Food Preservation

Nanomaterials	Polymer	Food Sample	Major Activity	References
TiO_2 NPs	Polylactic acid	Pork	Exhibit stronger antibacterial potential	(Li, Chen et al. 2020)
Sulfur NPs	Chitosan	Plasticizers	Enhanced antimicrobial potential against pathogens	(Shankar and Rhim 2018)
ZnO NPs	Gelatin	Fish	Reduce lipid peroxidation	(Kim, Viswanathan et al. 2022)
AgNPs	Polyamide 6	Water	Inhibit *E. coli* growth	(Damm, Münstedt et al. 2008)
AgNPs	Sodium Alginate	Carrots and pears	Enhanced shelf life	(Mohammed Fayaz, Balaji et al. 2009)
Soybean polysaccharides	Nisin	Tomatoes	Enhanced antimicrobial activity	(Luo, Wu et al. 2019)
AgNPs	Wild mushroom species *Ganoderma sessiliforme*		Antibacterial and anticancer activity	(Mohanta, Nayak et al. 2018)
Titanium silica oxide	Polylactic acid	Petri plates having *E. coli* and *S. aureus* bacterial colonies	Enhanced activity as compared to pure polylactic acid films	(Teamsinsungvon, Ruksakulpiwat et al. 2022)
β-cyclodextrin-TiO_2	polylactic acid and cellulose acetate	Petri plates having *E. coli* and *S. aureus* bacteria	Antibacterial activity	(Goñi-Ciaurriz and Vélaz 2022)
Sulfur NPs	Gelatin/cellulose		UV- protection and antibacterial activity	(Roy and Rhim 2022)
Sulfur quantum dots	Alginate	Bread packaging	Moisture barrier properties, enhanced antioxidants, and antimicrobial activity	(Riahi, Priyadarshi et al. 2022)
AgNPs	Chitosan/Polyvinyl alcohol	Strawberries	Moisture barrier properties and antibacterial potential	(Yang, Liu et al. 2023)
Lignin NPs and acylated soy protein nanogel	Chitosan	Different food samples	Improved barrier, mechanical as well as antibacterial properties	(Zou, Ismail et al. 2023)
ZnO NPs	Chitosan	Food additives	Enhanced antibacterial potential against *E. coli* and *S. aureus*	(Zhang, Zhang et al. 2021)
Copper sulfide NPs	Limonene Essential oil Pullulan/ Carrageenan	Food samples	Enhanced UV—barrier and antibacterial potential	(Roy and Rhim 2020)
ZnO NPs	Carboxymethyl cellulose/ chitosan	Bread	Strong activity against molds and yields	(Noshirvani, Ghanbarzadeh et al. 2017)
ZnO-Ag nanocomposites	PHBV/Chitosan	Meat	Antibacterial activity against *E. coli*	(Zare, Namratha et al. 2019)
ZnO-SiO_2 nanocomposites	Polyvinyl alcohol/Chitosan/ low-density polyethylene	Bread	Enhanced antibacterial potential against *E. coli* and *S. aureus*	(Al-Tayyar, Youssef et al. 2020)

(Continued)

TABLE 20.1 (*Continued*)
Applications of Biopolymer-Based Nanomaterials for Food Preservation

Nanomaterials	Polymer	Food Sample	Major Activity	References
MoS_2/Ag nanocomposites	Chitosan	Apple skin	Antibacterial activity against *S. uvarum* and *niger*	(Zhang, Mou et al. 2019)
Ag/Graphene/ TiO_2	Polylactic acid	Cheese	Enhanced moisture barrier properties	(Peter, Cozmuta et al. 2021)
Graphene	Polylactic acid	Potato chips	Enhanced moisture barrier properties	(Sharma, Biswal et al. 2018)
LDH nanomaterials	Polylactic acid	Films	Enhanced oxygen barrier properties	(Kim, Charoensri et al. 2022)

into polymeric media. The most commonly used materials include clays, silicon NPs, graphene, carbon nanotubes (CNTs), starch and cellulose-based nanofiber, as well as chitosan NPs (de Freitas, da Silva et al. 2022; Jafarzadeh, Forough et al. 2022; Ganapathy, Muthukumaran et al. 2023; Lim, Soon et al. 2023; Trinh, Chang et al. 2023). The integration of the aforementioned NMs with polymeric media influences the barrier features of film-based packaging by two different methods as these materials are resistant against gases. First, it generates a twisting pathway for diffusion of gas molecules that causes navigation of molecules around them as an alternative of succeeding a straight pathway vertical to the surface of the film. As a result existence of these NMs elongates diffusion pathway (i.e., difficulty in passage of gas molecules) over the film surface (Choudalakis and Gotsis 2009). Second, NMs can influence barrier features by bringing fluctuations in polymeric medium. Satisfactory interfaces among NMs and polymer molecules can moderately restrain polymeric chains adjacent to NMs. Therefore, gas molecules transferring from end-to-end interfacial constituencies face hindrance in movement through surfaces that result in enhancing barrier properties of material (Picard, Gauthier et al. 2007).

In addition to barrier properties, presence of NPs into the polymeric media helps in modifying the mechanical durability of polymer-based packaging system for preserving food for longer time (Hoyos-Merlano, Borroni et al. 2022). Also these NMs configuration, size, and dimension influence Young's as well as shear modulus (Aparna, Venu et al. 2023) and constant of thermal expansion (Choi, Yu et al. 2011; Sim, Lee et al. 2022). Moreover, structure, size, and configuration of NMs can influence intermolecular forces contained by polymeric medium that consequence in fluctuations of mechanical and thermal features. Though it is significant to remember that each type of polymer-based nanocomposites system is distinctive, and their features can be anticipated by using common methods. Also combination of NPs with polymeric material demonstrates capacity in accomplishing easy dispensation (de Sousa, Schlogl et al. 2023). Though numerous challenges, predominantly those which are associated with scattering and handling of these NMs, keep on to become overwhelmed (Xiong, Grant et al. 2018).

A NMs-assisted smart packaging system provides identical perseverance as traditional packaging materials and suggests distinctive benefits (Chaudhry and Castle 2011), comprising enhanced barrier properties, superior sensitivity, and prompt receptiveness as well as sustainability (Rodrigues, Demokritou et al. 2017). As a result of these features, NMs-based nanopackaging systems subsidize to additional active food safety, prolonging shelf life as well as retaining its vitality throughout storage, transport, and delivery to consumers (Majid, Nayik et al. 2018). NAFIGATE Company, of Ostrava, Czechia, is the only company that presently deals with biopolymer-based foodstuffs packaging in combination with NMs and is most commonly identified as nanotechnology-assisted bio-packaging. These food packaging systems provide sustainability for food preservation, which is due to the combination of biopolymers and NMs as biopolymers possess biodegradable remunerations and NMs result in improved characteristics of the material (de Sousa, Schlogl et al. 2023).

20.5 THE LEGISLATION, SAFETY, AND TOXICITIES OF NANOTECHNOLOGY

Nanotechnology is an effective implement for cultivating vigorous constituents and formulating novel materials, coatings, and packagings (Adeyeye and Ashaolu 2021). But there's still one unreciprocated query: how safe or nontoxic are NMs? Is the novel NMs packed with risky configurations materials? Do food regulatory administrations have any regulations or criteria that allow the marketing of these foodstuffs? What kind of testing are required for these items to be deemed safe? The flavoring agents and polymeric medium have a direct impact on the toxicity of polymeric coatings or packaging systems. Over the past few decades, the impact of both elements on packaged food has been assessed, and certain nations have put laws for additives and polymers utilization during preservation. Every state has administrations to oversee and control the domestic production, use, and exchange of food items. For instance, the European Union has the REACH (Registration, Evaluation, Authorization and Restriction of Chemicals) gradient, which is created and revised by the European Chemicals Agents (ECHA) in accordance with EU 10/2011, the Commission Regulation EC 1935/2004, and 1907/2006 of the European Council and Parliament. A lot of food items have usually been recognized as non-toxic by Food and Drugs Administration (FDA) in United States under the sections of 409 and 201 of the Federal Drug, Cosmetic, and Food Act. A list by The Inventory of Existing Chemical Substances in China (IECSC) is presented by China. In Brazil, the Agency of National Health Surveillance (ANVISA) is present for these purposes. These organizations control how materials and goods are used and merchandized within their own countries (Mitrano and Wohlleben 2020).

As a result, improvement in nanotechnology has steered the creation of new laws and regulations to manage the applications, production, and handling of NMs. Indeed, a "Guidance on the Hazard Valuation of NMs to be Utilized in Foodstuff and Food Production Cycle: Social as Well as Animal Healthiness" was published by the European Food Safety Authority (EFSA) in 2021 (EFSA Scientific Committee 2011). The addition of NMs to biopolymers as well as polymeric media will alter the packaging's properties. Remember, if we alter the concentration and nanostructure, we will get different properties of packaging. The substance's physicochemical features like electrical conductivity, viscosity, opacity, water solubility, thermal stability, tensile strength, thermal conductivity, and elastic modulus can be modified by these modifications. Furthermore, they may give more advantages by acting as antibacterial agent during food preservation (Ghanbarzadeh, Oleyaei et al. 2015; Sothornvit 2019).

Until now, more research in the discipline of nanostructure toxicology is required. The morphology, size, composition, and geometry of nanostructures can all have an impact on its toxicity (Ortega, Versino et al. 2022). Thus, identifying the nanostructure's toxicity is a challenging task. The toxicity of several nanostructures has been the subject of contradiction. Some scientists described the non-toxicological impact for specific types of nanostructures while others reported toxicity for same nanostructures (Sukhanova, Bozrova et al. 2018; Carreiró, Oliveira et al. 2020; Pal, Sarkar et al. 2021; Haghighat, Alijani et al. 2022). Because of the intricacy of toxicological research, we are facing the deficiency of unified law regarding the proper application of nanostructures. The World Health Organization (WHO) as well as The Food and Agriculture Organization of the United Nations (FAO) organized an international conference in 2009 to introduce possible hazards linked by unintentional application of NMs in agriculture sectors (de Sousa, Schlogl et al. 2023). This was because nanotechnology has facilitated progress in many novel goods, innovative approaches to those challenges that affected the market, and advancements of foods. But, improper use of NMs may result in health issues like kidney problems, vasculitis, dermatitis, and colon cancer (Harish, Tewari et al. 2022).

Moreover, the circumstances as well as types of used NMs always change the results or impacts. Additionally, as the debate over the health risks of NMs continues, every state is sharing a list of nanostructures that are allowed or banned for use as packaging additives, along with the highest concentrations that can be utilized in foodstuffs preservation. The European Union has permitted the use of silicon dioxide, carbon black, and titanium nitride nanostructures with respect to EC 10/2011. The NMs such as zinc oxide, nanoclay, aluminum, silver oxide, and carbon black have been approved sanction for trade in United States. Since December 2019, RDC No. 326 in Brazil has permitted the use of copolymers in nano-forms, ZnO NPs treated with or without 3-(methacryloxy) propyl trimethoxysilane, and titanium nitride NPs (Zenebom and Pascuet 2005; Bumbudsanpharoke and Ko 2015). Society is already utilizing nanotechnology-based coatings and packagings, and federal departments are gradually creating a list of rules and testing for the use of nanomaterials within their own states (Adeyeye and Ashaolu 2021).

20.6 ECONOMIC CONSIDERATIONS

20.6.1 Cost-Benefit Analysis

In terms of cost, the nanotechnology-based food preservation method has many advantages over traditional methods. Many polymeric nanocomposites have been synthesized by adding several organic or inorganic nanofillers. More resilient food packaging materials at a lower cost have been developed owing to the integration of NPs into polymers. The incorporation of inert nanoscale fillers, like chitin or chitosan and silica (SiO_2) NPs into the polymer matrix results in improved thermal characteristics, increased strength, and decreased weight (Singh, Shukla et al. 2017). CNTs-based biosensors have similarly generated a lot of attention because of their quick, easy, and affordable detection. They also make sure active utilization to identify toxic substance, microorganisms, as well as supplementary spoiled constituents, which may influence the quality and vitality of food items (Nachay 2007). Owing to the need for classy, costly apparatus, and skilled personnel, conventional molecular diagnostic techniques are not widely employed in the present era. So fast, sensitive, and affordable analytical tools for analyzing contagious toxicities have been established using the special luminescent, magnetic, electrical, and catalytical properties of NMs (Borah and Chetia 2023). Consequently, the detection system based upon NMs for various microbial pathogens is quick, sensitive, and precise and requires little work owing to NMs properties since NPs tend to be more sensitive than their larger counterparts (Nile, Baskar et al. 2020).

When NMs are added to food media, their huge surface-to-volume ratio helps them in scattering through the media, which improves economy and activity more than macroscale supported products (Neme, Nafady et al. 2021). Conventional methods of controlling pests require the use of significant amounts of chemical pesticides, which raise production costs and cause environmental issues. Dilution of pesticides with water that has been nano-treated could significantly increase their efficacy and decrease the amount of chemicals utilized. When it comes to pest management, nanopesticides are more effective than traditional pesticides. Additionally, using these can cut the cost in half when compared to traditional pesticides (Huang, Wang et al. 2014). Starch has also been widely utilized in various foodstuffs packagings for many years owing to economic benefits. The most prevalent substance generated from biomass is starch. This guarantees the nanocomposites material's biocompatibility as well. After thorough testing, it was shown that starch NPs or nanocrystals are highly ideal for foodstuffs utilizations as possible additives in nanoemulsions and distribution systems for bioactive substances (Aslam, Akhtar et al. 2022). Last, it is shown that widespread application of nanotechnology will greatly enhance growth, raise yields, lower fabrication expenses, and minimize postharvest damages by preserving vitality and lengthening of storing of both processed and agricultural products (Sahani and Sharma 2021; Haris, Hussain et al. 2023).

For certain usage in food and beverage goods, nanoemulsions may be more advantageous than conventional emulsions (Yalçınöz and Erçelebi 2018). Generally speaking, nanoemulsions are further consistent with respect to particle accumulation and gravitational separation. Depending on the inactive principle, there are several ways to assemble nanoemulsions that may be classified as low-energy and high-energy methods (Salvia-Trujillo, Soliva-Fortuny et al. 2017). Different kinds of nanoemulsions have further complicated attributes, for example nanostructured multilayer emulsions can generate different encapsulation abilities from a solo delivery method; this can encourage the active components' potential and make it easier for them to release when an action occurs. For example, a low-fat ice cream was created by Nestle and Unilever with the help of nanoemulsion approach (Sneha and Kumar 2022). Nanoencapsulation has been used to create defensive barriers, flavor and scent screening, precise liberation, and superior processing for H_2O insoluble food flavors and ingredients (Thiruvengadam, Rajakumar et al. 2018).

20.6.2 Economic Viability

Food wastage, as well as leftover of foods in Europe, North Africa, North America, and industrialized Asia, account for significant inefficiencies in the agricultural sector (Sawaya 2017). In Europe as well as North America, it is anticipated that 30–50% supply of food goes to waste at whole stages of the food chain: in food processing, customer's ménages, marketing founding, and on smallholdings. In United States, it is anticipated that 31% of postharvest food resources were discarded at consumer and marketing levels in 2010, leading to an estimated economic loss of roughly USD $160 billion. Food wastage is frequently the consequence of pollutants, toxins, and spoilage from bacteria and fungi. Controlling viruses and allergens on food across the farm-to-fork continuum is essential to ensuring food safety (Kang'ethe, Mutua et al. 2021). The benefits and features of nanotechnology applications also involve control of pathogens in agricultural production. These uses cover the management of human infectious agents, airborne and waterborne pathogens, as well as the eradication of bacteria that cause spoiling on fresh produce, food crops, and animals (Rodrigues, Demokritou et al. 2017; Awuchi 2023).

Nanotechnology has claimed to be next industrial revolution and has the potential to affect several economic sectors, such as defense, energy, food, agriculture, and medicine (Babatunde, Denwigwe et al. 2019; Malik, Muhammad et al. 2023). It is anticipated that the succeeding nano-sciences and nanotechnologies will have an impact on every facet of technology, business, environment, and human existence in general. Worldwide, significant and ongoing investments have been made as a result of the economic and societal potential. Nanotechnology has demonstrated the remarkable capacity to pique the interest of businesses, governments, and non-governmental organizations (Bugusu, Mejia et al. 2009; He, Deng et al. 2019). Research has received billions of dollars to promote the science in all areas. The amount of investment differs between different states. The United States is currently regarded as the world's largest investor, but the other nations and sectors of the economy invest more money globally, and competition for the top spot is heating up. The amount of investment also differs throughout the various economic sectors, with energy and defense now leading the way. In addition to government initiatives, commercial industry is thought to be working on nanotechnology research; however, the information from the private sector is not readily accessible to the public, partly due to the unpredictability of consumer and public acceptance as well as the regulatory agencies (Frewer, Norde et al. 2011; Kim, Kadam et al. 2018).

The Advanced Foods and Materials Network (AFMNet) is the regulator of food nanotechnology exploration (Parr-Vasquez, Paulson et al. 2011). Canada has made significant investments in nanotechnology exploration by the way of Networks of Centers for Excellence (NCE) platforms, which aim to organize Canada's research faculty in communal, educational, and secretive divisions for commercial growth and enhancement of superiority of living style. The network's goal is to advance acquaintance and knowledge that produce socially and financially acceptable food preparation methods (Pigford, Hickey et al. 2022). They collaborate with national as well as international research institutions, the government, business, and nonprofit groups. AFMNet got funding of around $3 million in Canadian dollars from 2003 to 2010 (Bugusu, Lay Ma et al. 2011).

The food sector has made nanoemulsions that are cost-effective, label-friendly, and compliant with the law. The most significant are the toxicological apprehensions. Plums have shown to have an intriguing dietary application of essential oil nanoemulsions. Recently, plums' antibacterial potential as well as their physical and chemical alterations was assessed in combination with lemongrass oil nanoemulsions (Kim, Lee et al. 2013). *Salmonella* and *E. coli* colonies were suppressed by the nanoemulsions without affecting the product's smoothness, brittleness, and flavor. Additionally, it has the ability to postpone changes in the concentration or percentage of phenolic compounds as well as reduce the synthesis of ethylene (Amaral and Bhargava 2015).

Crop diseases endanger production of foods at the base of the food cycle. The protection of food is of top concern in states whose economy are heavily dependent on agriculture, particularly when it comes to the cattle sector (Vurro, Bonciani et al. 2010). Some of most common concerns such as dioxin pollution in Ireland and Belgium, swine influenza in the United States, and bovine spongiform encephalopathy (BSE) in the United Kingdom have recently eroded the public's trust in the protection of foodstuffs they consume and caused an immense decline of worldwide markets (Motarjemi 2023). Two approaches to address these problems could be provided by nanotechnology: first, functional nanostructure-based drugs, vaccines, and insecticides could enable involuntary responses to syndrome by initiated discharge in reaction to particular indications; second, nanostructured constructed tools and functionalized sample pre-treatment, abstraction, and intensified analytical procedures could help in primary identification of pathogenic organisms and damaging entities (Frewer, Norde et al. 2011; Gondal and Tayyiba 2022).

Food management systems of agriculture are made successful by incorporation of functional nanostructure based sensors for initial phase agronomic events analysis, such as crop and livestock crops growing (Mittal, Kaur et al. 2020). Enzyme as well as DNA established sensor systems can be integrated along with communication schemes to provide tailored reaction activities at plantation area, municipal, or national levels. This is made possible by the availability of DNA archives and databases on enzymatic effects of infections, biochemical pollutants, and different pests. Improved water and fuel economy are examples of secondary benefits that can be obtained from independent nanostructured functionalized sensing methods. These benefits are dissimilar to the product itself but thought have an impact on total production (Wilson, Tran et al. 2008; Mittal, Kaur et al. 2020).

The goal of introducing nano-foods to the consumer market is to offer high-quality, safe, and healthier goods for better eating and surviving (Jagtiani 2022). Although the rewards of nano-foods have been well acknowledged, the health hazards related to NPs have not yet been thoroughly investigated. The features of NPs that depend on their size and their subsequent miniaturization present significant opportunities for advancements in nanotechnology (Bouwmeester, Dekkers et al. 2009; Onyeaka, Passaretti et al. 2022). Food-related issues can be resolved and long-term economic gains will result by the introduction of nanotech foods. It is only possible by advances in the understanding of NMs and their potential in the food sector (Yadav, Khan et al. 2013). More food production with economical revenues and intelligent packaging approaches with enhanced storage features for improved foodstuff protection are the outcomes of this technology (Kalpana, Priyadarshini et al. 2019). If we appropriately regulate NMs in foods it will pose a significant positive influence on sustainability and may also result in positive effects on social well-being and atmosphere (Sonkaria, Ahn et al. 2012; Shafiq, Anjum et al. 2020).

20.7 SUSTAINABILITY IMPLICATIONS OF NANOTECHNOLOGY-BASED FOOD PRESERVATION

Over a decade, there has been extraordinary progress in new nanostructures. Furthermore, it is possible that this inclination will persist in upcoming ages. To increase sustainability of agricultural systems, a variety of attractive opportunities have been found for nano-based technology (Rodrigues, Demokritou et al. 2017; Cerqueira and Pastrana 2019). These include advanced methods for identifying and controlling dangerous pathogens and enhancing food safety; water treatment technology for agricultural sectors; nano-based fertilizers; analyzing tools for analyzing substances, quantifying chemical, somatic, or biological characteristics as well as for identifying toxins or microbes in foodstuffs (Prasad, Bhattacharyya et al. 2017; Cerqueira and Pastrana 2019). On the other hand, there are plenty of concerns about their impact on the public health, social life, and environment, which are pillars of sustainable food system as shown in Figure 20.6. In response to these worries, the European Food Safety Authority (EFSA) has established an effective method for actual risk evaluation for utilization of engineered NMs in food's chain. Safety assessment of NMs may include different types of materials that have been shown in Figure 20.7 (EFSA Scientific Committee, More et al. 2021). This study also discusses the absence of reliable techniques for the categorization, identification, and detection of artificial NMs, particularly in complex environments during sampling. The Organization for Economic Co-operation and Development has also addressed this issue (Gomes, Scott-Fordsmand et al. 2021).

Green nano-biotechnology's goal has been sufficiently explained by Maksimović, Omanović-Mikličanin, and Hutchison (Hutchison 2016; Maksimović and Omanović-Mikličanin 2017). These can be summarized into two

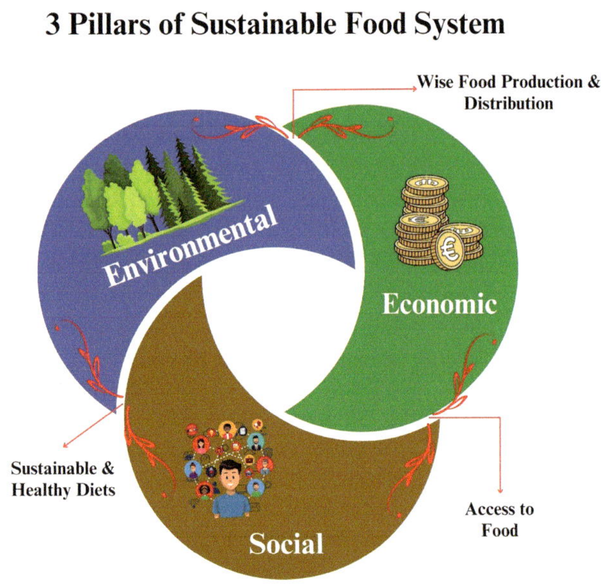

FIGURE 20.6 Sustainability and food preservation.

FIGURE 20.7 Safety assessment.

main points: increasing the beneficial influences on the well-being of people and decreasing their adverse impact on the environment. However, this approach is severely hampered by the previously discussed flawed technique and our insufficient understanding of all the toxic properties that are affecting NMs. Research on organization of different nutrient items at the nanoscale level is a growing sector in nanotechnology. In upcoming years, it will be possible to evaluate food structures and create novel food materials at the nanoscale (Singh, Dutt et al. 2023).

20.7.1 Nano-Based Sensors

The utilization of nanosensors in food systems to detect food spoiling is proving vital help in the fight against harmful microbes and the subsequent decrease in consumer foodborne illnesses (Chinchkar, Singh et al. 2023). Probes and nano-based sensors have demonstrated many advantages for the preservation and protection of food as well as better production in agriculture (Rai, Acharya et al. 2012; Sekhon 2014; Neethirajan, Ragavan et al. 2018). For example, to identify different kinds of infections, poisons, and impurities in food items and packaging products, a lot of nanotechnology-based sensors and instruments are found (Xiang, Li et al. 2016; Neethirajan, Ragavan et al. 2018). Even with a number of outstanding advancements, it is still very difficult to accurately identify certain toxicants and infections in food. It is usually difficult to distinguish between active and dead pathogenic bacteria in the food chain when many distinct pathogens are noticed, and this capacity needs more research (Yazgan, Noah et al. 2014). Moreover, the manufacturing of particular nanosensor types that target a particular function of the food system is extremely difficult (Farahi, Passian et al. 2012).

20.7.2 Nano-Based Control of Pathogens

Systems for monitoring changes in food quality and modifying appropriately are made possible by support of nanotechnology. Commercial applications of nanostructures involve the smart packaging and nano-coating to prolong shelf lives of foodstuffs and protect them from harmful pathogens (Bradley, Castle et al. 2011; Falguera, Quintero et al. 2011; Silvestre, Duraccio et al. 2011). Packing materials with integrated nanosensors can recognize deterioration indicators based on temperature, pH, and gas composition. These analyzing tools enable quick exploit to halt descent and minimize foods wastage by providing instantaneous input on food quality. Antibacterial NMs, including silver NPs or NPs loaded with essential oils, can be found in storage areas or packaging materials. These NPs stop the growth of various pathogens that deteriorate foodstuffs and prolonging the shelf lives of agricultural products (Chaud, Souto et al. 2021; Nair, Mallya et al. 2022; Suvarna, Nair et al. 2022).

Numerous nanoemulsions facilitate the release of various components by a variety of stimulations, including heat, pH, ultrasonic waves, and others. They effectively preserve the flavors and shield them from oxidation and enzymatic processes (Pradhan, Singh et al. 2015). In a study, the copper nanoparticles (CuNPs) were created and utilized to prolong tomatoes' shelf life. The CuNPs application has been shown to enhance accretion of bioactive components, including flavonoids, phenols, lycopene, and vitamin C, leading to the production of high quality tomatoes for human consumption. Nevertheless, there are some hurdles that stop the flawless application of nano-based substances for food safety. The precise procedure of these influences on hazardous microbes is not completely agreed upon. In addition, the consequences of ecological factors like light, pH, temperature, and food waste are also inadequately described for packaging (Hossain, Perales-Perez et al. 2014). To prevent any unintended negative effects on social well-being, the vitality of nano-based packaging substances needs to be demonstrated (Bradley, Castle et al. 2011; Hossain, Perales-Perez et al. 2014).

20.7.3 Smart Nutrient Delivery System

The delayed release of the encapsulated active compounds by the nanocarriers allows for a longer and accurate distribution to the intended site of action (Chariou, Ortega-Rivera et al. 2020). Thanks to this controlled release technology, fewer chemicals are required and applied less frequently. It has reduced the possibility of environmental contamination and applied to enhance the fertility of the soil, productivity, and product quality. Rock phosphate applied in nano-form may enhance the amount of phosphorus available to plants (Suriyaprabha, Karunakaran et al. 2012). Because nanofertilizers enable many crops to withstand stress as well as triple the nutrient usage efficiency (NUE) (Fazelian and Yousefzadi 2022; Nazeer, Ahmad et al. 2023). The mechanism of liberation of bioactives from congenital fertilizers, nanocomposites with organic polymer introduced in films of kaolinite clays can be utilized as cementing materials. In addition to reducing the risk to the environment, this procedure improves nutrient use efficiency. Similar to this, it has been demonstrated that adding mannose and amino acids to a proprietary nanocomposites containing micronutrients like nitrogen, phosphorus, and potassium can improve grain crops' absorption and utilization of nutrients (Hatti, Raghavendra et al. 2020).

20.7.4 Nano-Based Fertilizers

For the transportation of agrochemicals, improvement of nano-empowered ingredients as nanofertilizers is a subject of current research. This makes nanotechnology useful for the proficient transfer of agrochemicals, such as insecticides, herbicides, and fertilizers (Liu and Lal 2014; Kashyap, Xiang et al. 2015; Monreal, DeRosa et al. 2016; Nazeer, Ahmad et al. 2023). In spite of active research, there are typically no methods for guaranteeing targeted transportation of particular organisms using biological substances such as antibiotics and other materials or hormones. The nutritional quality and value of food items, the chemistry of interactions between NMs and green leafy vegetables, and the impact of NMs on the growth of crops are the main challenges that avert the applications of modish nano-empowered fertilizers in agronomy (Bindraban, Dimkpa et al. 2015).

20.7.5 Environmental Impact Assessment

Since the primitive era, man has been attempting to create superior methods for preservation of food. From the caveman who used to store a freshly obtained meal in caves to the invention of refrigeration methods of the 21st century that offered a moist environment to prevent food from spoiling, man has improved greatly. There are a number of methods that have recently been developed to further preserve food from contamination that are also sustainable for the environment. To further lessen pollution in the environment, packing materials are ideally comprised of biodegradable materials (Pradhan, Singh et al. 2015). Nanotechnology aids more operative agricultural production techniques by using less agrochemical, which can reduce chemical utilization as well as environmental pollution (Handford, Dean et al. 2014; Boregowda, Jogigowda et al. 2022).

Certain environmental complications could be resolved through support of nanotechnology. These include methods intended to conserve the environment's quality. Nanoscale carriers, for instance, possess precise discharge mechanisms that permit active constituents to be obtained gradually, thus enhancing their efficiency whereas decreasing quantity provided to be delivered. The public's health will get more advantages from this because of reduced use of agrochemicals due to nanoscale carriers (Chen and Yada 2011; Ditta 2012). Some materials could act as oxygen or ultraviolet scavengers, while silver NPs and coatings could offer antimicrobial qualities. However, biodegradable resources have recently been established via nanotechnology to substitute non-degradable petroplastics-based polymers, which pose a significant worldwide environmental risk (Adeyeye 2019).

NMs have a lot of promises to recover resources. Throughout the whole supply chain, there are streams of agricultural waste that contain precious embodied resources. Energy and other important resources can be saved by extrication of nutrient enriched wastewater streams and reusing them as stimulant for plant growth. By eliminating oxygen from biomass using innovative catalysts, we can recover energy from plant and animal wastes (Rodrigues, Demokritou et al. 2017). NMs can be used where traditional chemical or reactor-based systems become inefficient. They increase the effectiveness of removing pollutants from water and reduce the energy needed to do so. For instance, nano-enabled direct solar membrane distillation uses a lesser amount of energy as compared to traditional thermal distillation membrane and can capture a broader variety of solar radiation (e.g., using visible light instead of UV light). It also enables a confined superficial heating system by means of light captivating black NPs that encourage plasmonic resonance methods and produce localized heating (Westerhoff, Alvarez et al. 2016). Because of the low frictional water flow through, carbonaceous NMs can produce multifunctional water purification membranes with extremely high water fluidity (Crock, Rogensues et al. 2013).

This technology is favored because of its unique characteristics, which include a steady release mechanism of action, target specificity, accurate exploit on active areas as well as higher surface area. Nanotechnology's achievement could be accredited to its auspicious consequences, energy efficiency, and minimal spatial requirements (Sridhar, Ponnuchamy et al. 2021). The pulsed wire discharge technique (vapor phase method) can yield high output rates and are incredibly energy efficient (Sahoo, Vishwakarma et al. 2021). Compared to conventional sensors, a lot of nanoscale sensors consume less energy, need fewer materials, and are frequently reusable. We have many energy-efficient nanofluidic devices that can desalinate water with a similar efficiency as a large-scale reverse osmosis desalination mechanism (Diallo and Brinker 2011).

20.7.6 Resource Efficiency

There has been a lot of interest and potential for using nanotechnology to protect soil and environmental well-being through the prudent use of necessary inputs to achieve system sustainability and increase resource use efficiency (Kremsa 2021). Despite the fact by eliminating effects on human healthiness resulting from discharges of engineered nanomaterials (ENMs) throughout the life cycle, current life cycle assessment (LCA) analysis can offer assessment of resource efficiencies and relative ecological and social well-being influences of NMs-assisted technology contrasted with traditional technologies (Tsang, Kikuchi-Uehara et al. 2017). Nanotechnology technique offers the following benefits of resource efficiencies as well as the environmental impact assessment.

20.7.6.1 Water Management

More than half of the water lost in farms is gone astray through evaporation, percolation, and runoff. In arid and semiarid areas, super water adsorbents derived from nanotechnology play a crucial role in storing and safeguarding water (Hemalatha, Lakshmi et al. 2023). Reactive groups are present on the surface of mica, an aluminum silicate macromolecule layered structure. In order to lower production costs and enhance the superabsorbent materials' overall water-absorbing capabilities, it is important to implant acrylic acid polymer on top of mica and create nanocomposites made of polymer and ultrafine mica particles (Wu, Lin et al. 2000). Plant-based cellulose nanocrystals or nanofibers are a promising class of adsorbents for water cleaning. Nanocellulose may enhance the characteristics of membranes or stabilize reactive NPs. In field settings, mineral zeolites are also utilized to stop water loss (Hatti, Raghavendra et al. 2020; Shahzad, Ullah et al. 2023).

20.7.6.2 Improved Waste Treatment

Nanotechnology improves the efficacy and efficiency of waste treatment processes. NMs such as NPs and nano-catalysts could be utilized to reduce or change organic contaminants, heavy metals, and industrial wastes (Singh 2023). Catalysis, adsorption, and photocatalysis are examples of wastewater treatment techniques that are more successful and have a less negative environmental impact owing to exclusive features of NMs, like their higher surface area and reactivity. Real-time waste parameter detection and monitoring is possible using nanosensors. These sensors can detect and quantify certain waste components or contaminants, which is essential information for wastewater management and recycling processes. They are commonly used in monitoring systems. Nanoscale sensors enable more sensitive and accurate monitoring, facilitating faster responses and improved wastewater treatment (Chaud, Souto et al. 2021; Ramesh, Janani et al. 2022).

20.7.6.3 Soil Remediation

The degradation of chlorinated, halogenated, and heavy metal compounds has been investigated using carbon-based NMs, metal oxide NPs (TiO_2, Fe_3O_4), bimetallic NPs, and zero-valent iron NPs. A soil binder based on nanotechnology, known as "Soil set", was created by Sequoia Pacific Research of Utah (USA) and was covered by the ETC Group in 2003. Soil set mulch is a fast setting covering that fixes soil together by nanoscale biochemical practices (Hatti, Raghavendra et al. 2020). Other products of soil remediation based on nanotechnology include nano-zeolites for preserving soil, nano-magnets for eliminating soil pollutants, specialized nanoparticles for increasing enzyme release through plant and microbe activation, and NPs for improved discharge of polysaccharides by microorganisms for soil accumulation, humidity maintenance, and carbon buildup (Zhou, Chen et al. 2018; Jha, Pathania et al. 2023).

20.7.6.4 Improved Safety and Compatibility

Nanotechnology can make chemicals safer and more suited by reducing their toxicity to creatures other than the ones for which they are meant to influence (Iavicoli, Leso et al. 2017). The physicochemical properties of the active ingredients can be altered during the encapsulation process, decreasing their bioavailability and lowering their detrimental influences on favorable creatures. This controlled selective killing reduces influence on ordinary predators, pollinators, and other favorable creatures, promoting ecological harmony and biodiversity in many contexts (Singh 2023).

20.7.6.5 Enhanced Stability and Persistence

Nanotechnology increases a chemical's stability and persistence by protecting it from environmental factors like UV radiation and microbial activity that cause deterioration. The nanocarriers act as a physical barrier to prevent the encapsulated active compounds from degrading too soon. Because of their increased stability, chemicals work better for longer periods of time, needing fewer applications and offering a lower risk to the environment (Asif, Islam et al. 2021; Chaud, Souto et al. 2021).

20.7.7 Nanoparticle Toxicity

Recently, research has been done on the historical roots as well as the future nature of nano-toxicity. There are several entry pathways, namely uninterrupted utilizations, to ecological compartments (either unintentionally or via intentionally degradation of products) and wastewater treatment plant alluvium and outflow for the deliverance of engineered NMs into the surroundings (Mueller and Nowack

2008; Maurer-Jones, Gunsolus et al. 2013). Nevertheless, it is difficult to predict the relevant NPs absorptions, which are emancipated at any particular time (Jain, Ranjan et al. 2018). The excessive use of NPs has generated food safety concerns and negative effects on the peoples' health. Bundschuh et al. have recently studied the impacts on aquatic and land systems (Bundschuh, Filser et al. 2018). Many parts of the human physique like lungs, abdominal tracts, and skin are in continuous exposure to the external atmosphere (Buzea, Pacheco et al. 2007). In such circumstances, the important role of solubility, size, chemical composition, aggregation, surface features, and shape have been identified quite early (Nel, Xia et al. 2006).

Additionally, multiple studies on the cells of human lungs have demonstrated size-dependent toxicity (Gliga, Skoglund et al. 2014). An additional imperative aspect manipulating the toxicity is the shape of NPs (Hsiao and Huang 2011; Tarantola, Pietuch et al. 2011; Sahu and Hayes 2017). Many kinds of NPs, depending on their shape and size, may be extremely dangerous to cells due to organelle destruction or oxidative stress. With the successful application of a wide variety of coating items such as citric acid, different surfactants, proteins, and polysaccharides, the impact of surface features on the level of toxicity has also enticed prodigious compact attention (Mbeh, Akhavan et al. 2014). Research indicates that exposure to NPs of different organs can result in specific illnesses within the same organs. For instance, inhaling NPs can lead to lung cancer, emphysema, bronchitis, and neuro-degenerative diseases. Additionally, if these NPs affect the intestinal tract, they can cause diseases related to cancer (Sahu and Hayes 2017). Further, several efforts for risk evaluation have been made to guess the quantity of NPs that reached the atmosphere from many sources. An overview of these attempts has been provided, which demonstrated that a widespread series of NPs are brought together to the environment. To determine eco-toxicological problems associated with the NPs, a multidisciplinary approach that combines computational, theoretical, and experimental techniques could be useful for establishing a risk evaluation methodology (Handy and Shaw 2007; Hegde, Brar et al. 2016).

20.7.8 Operational Approaches

Because of concomitant changes as well as reactive nature of NPs, it is exceedingly challenging to identify NPs in food and other ecological matrices. An efficient investigation would incorporate the sample preparation steps that would support the characterization and identification (Mudalige, Qu et al. 2019). Perfectly, it should eliminate any disruptive materials and maintain the nature as well as state of the NMs. Usually, there are multiple steps like sample homogenization, stabilization, and extraction of NMs, which are involved in the preparation of sample. The kind of solvents utilization in the first phase could have significant effect on the second step. Therefore, in addition to the matrix effect, it is essential to proceeds with interpretation of morphological characteristics of the NMs (Abdolahpur Monikh, Guo et al. 2022). Regarding inorganic NPs, matrix interferences are eradicated through an enzymatic or chemical digestion process that can be facilitated by sonication or microwaves, followed by either a liquid or solid-phase extraction technique. Most of the detection was successfully carried out using spectrometric and microscopic methods. In the case of microscopic methods, electron microscopy can reveal details about the morphological composition and properties of the NMs. But many uncertainties were noticed during sample preparation (Dudkiewicz, Boxall et al. 2015). Preferably, this problem was solved in the liquid samples by using atmospheric (or environmental) scanning electron microscopy, which can be done using a simple technique. Early on, the light-scattering technique was thought to be harmonizing. However, in complex samples, the potential is hindered by the different distribution of NPs sizes. NPs tracking analysis (NTA) only overcame this problem in liquid samples (Jarzębski, Bellich et al. 2017).

Regarding spectrometric techniques, they can be used for NPs identification, quantification, and detection. The most common techniques for detecting particles are single-particle ICP-MS (spICP-MS), matrix-assisted laser desorption/ionization-time of flight (MALDI-TOF), and inductively coupled plasma-mass spectroscopy (ICP-MS) (Laborda, Bolea et al. 2016). However, these approaches do not provide information about particle shape and size. Nonetheless, this issue is overcome by combining with a fractionation approach (Mattarozzi, Suman et al. 2017). Comprehensive characterization of organic NPs like carbon CNTs and fullerenes has been well evaluated; graphene and its oxides, on the other hand, have not received the same level of attention. There is currently very rare published material (Aykaç, Gergeroglu et al. 2021; Jamjoum, Umar et al. 2021).

According to Doudrick, Nosaka et al. (2015), the most successful method was the chemical reduction of graphene oxide, which enabled a suitable isolation and extraction through programmed thermal analysis (PTA). Similar for organic NPs, information about the shape, accumulation degree, and size of organic NPs is obtained via light scattering, NTA, and electron microscopy techniques. Astefanei, Núñez et al. (2015) evaluated the techniques for quantification, separation, detection, and extraction for fullerenes. In complicated matrices, the extraction process (liquid–liquid) was carried out either alone or in conjunction with ultrasounds by using toluene, oxidizing agents, and salts after the removal of surfactants and proteins. For detection purposes, liquid chromatography combined with mass spectrometry (LC-MS) is frequently applied. In addition, the detection, extraction and separation techniques have been mostly reviewed to enumerate and detect CNTs (Petersen, Flores-Cervantes et al. 2016). Multiple approaches, namely electrophoretic or chromatographic or asymmetric flow field-flow fractionation (AF4), may be utilized for separation as well as extraction. Further, thermal gravimetric analysis-mass spectrometry (TGA-MS) and SP-ICP-MS, the quantification techniques, may also be employed. Last, identifying man-made pollution from naturally occurring

NPs is a main issue. Regarding inorganic ones, this can be achieved by figuring out particular ratios like Ce to La and Ti to Fe (Montano, Lowry et al. 2014; Praetorius, Gundlach-Graham et al. 2017).

20.7.9 Measurement Issues

One of the most important requirements for NMs' application is the capacity to identify their existence in food's system at specific time duration (Cockburn, Bradford et al. 2012). This quantification of NMs includes food items preparation, storage, food assimilation, and passage across the alimentary track of human or gastrointestinal system, which is a complex process because of the complexity of the human body and the thermodynamic instability of NMs (Szakal, Roberts et al. 2014). Furthermore, it is unclear which specific components need to be measured. It is always important to figure out whether NMs are incorporated naturally or through external sources as food enhancers and additives. Many aspects contribute to improvement of highly applicable technique for NMs' measurement in foodstuffs, which constitute the case of how NMs are added. Furthermore, the complexity of human alimentary track raises supplementary questions about how to categorize and measure NPs. Several analytical techniques, especially combined, are used to identify the NPs in food items (Szakal, Roberts et al. 2014; Gallocchio, Belluco et al. 2015). These techniques comprise spectroscopy (X-ray powder diffraction spectroscopy, energy-dispersive X-ray spectroscopy), centrifugation, chromatography, microscopy (transmission electron microscopy, scanning electron microscopy), and additional relevant procedures (Tiede, Boxall et al. 2008). Various studies have been conducted that elaborate the detection of various NMs in different types of samples, which have been determined in Table 20.2.

20.8 CONCLUSION

Nanotechnology-based food preservation compromises an auspicious opportunity for resolving precarious challenges in worldwide food diligence. This field has not only revealed its prospective to improve shelf life and vitality of foodstuffs but also its prominence to diminish food waste and provide better access to nutritive food products. Though, assumption of nanotechnology in food preservation must be analyzed with vigilant deliberation of its economic consequences as well as sustainability considerations. From an economic point of view, nanotechnology can play a prime role to escalate manufacturing efficacy, lessen energy depletion, and reduce damages owing to food decomposition. These profits can subsidize toward budget relief for food manufacturers as well as customers. On the obverse sustainability, it is vital to evaluate the ecological and communal influences of nanotechnology. The liable usage of NMs requisite highlight ecological safety, resource efficiency, and decent fabrication practices. Resolving apprehensions associated with NM removal and latent health hazards is

TABLE 20.2
Application of Analytical Techniques for Determination of Nanomaterials

Nanomaterials	Sample	Sample Preparation	Techniques	References
AgNPs	Meat	Sonication and enzyme action	SP-ICP-MS	(Peters, Rivera et al. 2014)
AgNPs	Sock fabric	Acid digestion	ICP-OES	(Benn and Westerhoff 2008)
Cu NPs	Soil	Acid digestion followed by microwave treatment	ICP-MS	(Navratilova, Praetorius et al. 2015)
TiO_2	Water matrices	Filtration	SP-ICP-MS	(Gondikas, Kammer et al. 2014)
TiO_2, Ag and Cu NPs	Water matrices	None	SP-ICP-MS	(Donovan, Adams et al. 2016)
Fullerenes	Wastewater sample	Filtration and sonication	LC-QqLIT-MS	(Farré, Pérez et al. 2010)
Fullerenes	Soil and sediment	Sonication followed by extraction and evaporation	UHPLC-HRMS	(Carboni, Helmus et al. 2016)
Different fullerenes	Sediment and water	Ultrasound extraction	UHPLC-MS/MS	(Astefanei, Núñez et al. 2014)
Graphene and graphene oxide	Wastewater matrices	Digestion	PTA	(Doudrick, Nosaka et al. 2015)
SWCN	Sediment	Sonication	NIRF Spectroscopy	(Schierz, Parks et al. 2012)
AgNPs	Fish Tissues	Enzymatic extraction	ICP-MS	(Clark, Clough et al. 2019)
AuNPs	Shellfish sea food	Alkali digestion method	SP-ICP-MS	(Zhou, Liu et al. 2020)
Ag and TiO_2 NPs	Plant animal and body fluids	Alkali and enzymatic extraction methods	SP-ICP-MS	(Laycock, Clark et al. 2022)
Ag and AuNPs	Human samples	None	SP-ICP-MS, AF4-FFF-ICP-MS, MALS and UV-Spectroscopy	(Bocca, Battistini et al. 2020)
Ag, Au, ZnO, and CeO_2 NPs	Human body fluids	Sonication and Filtration	SP-ICP-MS	(He, Zhang et al. 2020)

vital to make sure sustainable and liable methodology in nanotechnology is in agreement in food sector. In conclusion, incorporation of NMs in food preservation systems grasps momentous ability for improving food safety and decreasing wastage. Though a vigilant and obvious evaluation of economic and sustainability aspects is vital to appreciate these remunerations and maintain the globe and the well-being of upcoming generations.

REFERENCES

Abdolahpur Monikh, F., Z. Guo, P. Zhang, M. G. Vijver, I. Lynch, E. Valsami-Jones and W. J. Peijnenburg (2022). "An analytical workflow for dynamic characterization and quantification of metal-bearing nanomaterials in biological matrices." *Nature Protocols* **17**(9): 1926–1952.

Adeyeye, S. A. O. (2019). "Food packaging and nanotechnology: Safeguarding consumer health and safety." *Nutrition & Food Science* **49**(6): 1164–1179.

Adeyeye, S. A. O. and T. J. Ashaolu (2021). "Applications of nanomaterials in food packaging: A review." *Journal of Food Process Engineering* **44**(7): e13708.

Agriopoulou, S., E. Stamatelopoulou, V. Skiada, P. Tsarouhas and T. Varzakas (2020). *Emerging nanomaterial applications for food packaging and preservation: Safety issues and risk assessment*. Proceedings, MDPI.

Alfadul, S. and A. Elneshwy (2010). "Use of nanotechnology in food processing, packaging and safety–review." *African Journal of Food, Agriculture, Nutrition and Development* **10**(6).

Al-Tayyar, N. A., A. M. Youssef and R. R. Al-Hindi (2020). "Antimicrobial packaging efficiency of ZnO-SiO2 nanocomposites infused into PVA/CS film for enhancing the shelf life of food products." *Food Packaging and Shelf Life* **25**: 100523.

Alvarado, M., N. Guzmán, N. Solís and J. Vega Baudrit (2017). "Recycling and elimination of wastes obtained from agriculture by using nanotechnology: Nanosensors." *International Journal of Biosensors & Bioelectronics* **3**(5): 368–375.

Amaral, D. F. and K. Bhargava (2015). "Essential oil nanoemulsions and food applications." *Advances in Food Technology and Nutritional Sciences Open Journal* **1**(4): 84–87.

Amenta, V., K. Aschberger, M. Arena, H. Bouwmeester, F. B. Moniz, P. Brandhoff, S. Gottardo, H. J. Marvin, A. Mech and L. Q. Pesudo (2015). "Regulatory aspects of nanotechnology in the agri/feed/food sector in EU and non-EU countries." *Regulatory Toxicology and Pharmacology* **73**(1): 463–476.

Amirsoleimani, M., M. A. Khalilzadeh, F. Sadeghifar and H. Sadeghifar (2018). "Surface modification of nanosatrch using nano silver: A potential antibacterial for food package coating." *Journal of Food Science and Technology* **55**: 899–904.

Anandharamakrishnan, C. and S. Parthasarathi (2019). *Food nanotechnology: Principles and applications*, CRC Press.

Aparna, A., G. Venu, A. Sethulekshmi and A. Saritha (2023). "Processing methods of polymer nanocomposites: Influence of processing parameters, nanofiller nature, size, and shape on their functional properties." *Biodegradable and biocompatible polymer nanocomposites*, Elsevier: 35–68.

Asif, M., S. Islam, M. A. Malik, Z. M. Dar, A. Masood, S. Shafi, B. Rashid and S. Sidique (2021). "Nano pesticides application in agriculture and their impact on environment." *Agricultural development in Asia-potential use of nanomaterials and nano-technology*, IntechOpen.

Aslam, S., A. Akhtar, N. Nirmal, N. Khalid and S. Maqsood (2022). "Recent developments in starch-based delivery systems of bioactive compounds: Formulations and applications." *Food Engineering Reviews* **14**(2): 271–291.

Astefanei, A., O. Núñez and M. T. Galceran (2014). "Analysis of C60-fullerene derivatives and pristine fullerenes in environmental samples by ultrahigh performance liquid chromatography–atmospheric pressure photoionization-mass spectrometry." *Journal of Chromatography A* **1365**: 61–71.

Astefanei, A., O. Núñez and M. T. Galceran (2015). "Characterisation and determination of fullerenes: A critical review." *Analytica Chimica Acta* **882**: 1–21.

Aswathi, V., S. Meera, C. A. Maria and M. Nidhin (2023). "Green synthesis of nanoparticles from biodegradable waste extracts and their applications: A critical review." *Nanotechnology for Environmental Engineering* **8**(2): 377–397.

Awuchi, C. G. (2023). "HACCP, quality, and food safety management in food and agricultural systems." *Cogent Food & Agriculture* **9**(1): 2176280.

Aykaç, A., H. Gergeroglu, B. Beşli, E. Ö. Akkaş, A. Yavaş, S. Güler, F. Güneş and M. Erol (2021). "An overview on recent progress of metal oxide/graphene/CNTs-based nanobiosensors." *Nanoscale Research Letters* **16**: 1–19.

Babatunde, D. E., I. H. Denwigwe, O. M. Babatunde, S. L. Gbadamosi, I. P. Babalola and O. Agboola (2019). "Environmental and societal impact of nanotechnology." *IEEE Access* **8**: 4640–4667.

Benn, T. M. and P. Westerhoff (2008). "Nanoparticle silver released into water from commercially available sock fabrics." *Environmental Science & Technology* **42**(11): 4133–4139.

Bindraban, P. S., C. Dimkpa, L. Nagarajan, A. Roy and R. Rabbinge (2015). "Revisiting fertilisers and fertilisation strategies for improved nutrient uptake by plants." *Biology and Fertility of Soils* **51**(8): 897–911.

Biswas, M. C., B. J. Tiimob, W. Abdela, S. Jeelani and V. K. Rangari (2019). "Nano silica-carbon-silver ternary hybrid induced antimicrobial composite films for food packaging application." *Food Packaging and Shelf Life* **19**: 104–113.

Biswas, R., M. Alam, A. Sarkar, M. I. Haque, M. M. Hasan and M. Hoque (2022). "Application of nanotechnology in food: Processing, preservation, packaging and safety assessment." *Heliyon* **8**(11): e11795.

Bocca, B., B. Battistini and F. Petrucci (2020). "Silver and gold nanoparticles characterization by SP-ICP-MS and AF4-FFF-MALS-UV-ICP-MS in human samples used for biomonitoring." *Talanta* **220**: 121404.

Bondarenko, O. M., M. Sihtmäe, J. Kuzmičiova, L. Ragelienė, A. Kahru and R. Daugelavičius (2018). "Plasma membrane is the target of rapid antibacterial action of silver nanoparticles in Escherichia coli and Pseudomonas aeruginosa." *International Journal of Nanomedicine*: 6779–6790.

Borah, J. and A. Chetia (2023). "Nanomaterials for biosensing applications: Concepts and recent advancements." *Sustainable nanomaterials for biosystems engineering: Trends in renewable energy, environment, and agriculture*, Apple Academic Press: 165–189.

Boregowda, N., S. C. Jogigowda, G. Bhavya, C. R. Sunilkumar, N. Geetha, S. S. Udikeri, S. Chowdappa, M. Govarthanan and S. Jogaiah (2022). "Recent advances in nanoremediation: Carving sustainable solution to clean-up polluted agriculture soils." *Environmental Pollution* **297**: 118728.

Bouwmeester, H., S. Dekkers, M. Y. Noordam, W. I. Hagens, A. S. Bulder, C. De Heer, S. E. Ten Voorde, S. W. Wijnhoven,

H. J. Marvin and A. J. Sips (2009). "Review of health safety aspects of nanotechnologies in food production." *Regulatory Toxicology and Pharmacology* **53**(1): 52–62.

Bradley, E. L., L. Castle and Q. Chaudhry (2011). "Applications of nanomaterials in food packaging with a consideration of opportunities for developing countries." *Trends in Food Science & Technology* **22**(11): 604–610.

Bugusu, B., C. Mejia, B. Magnuson and S. Tafazoli (2009). "Global regulatory policies on food nanotechnology." *Food Technology (Chicago)* **63**(5): 24–28.

Bugusu, B., U. V. Lay Ma and J. D. Floros (2011). "Products and their commercialization." *Nanotechnology in the agri-food sector: Implications for the future*, Wiley Online Library: 149–170.

Bumbudsanpharoke, N. and S. Ko (2015). "Nano-food packaging: An overview of market, migration research, and safety regulations." *Journal of Food Science* **80**(5): R910–R923.

Bundschuh, M., J. Filser, S. Lüderwald, M. S. McKee, G. Metreveli, G. E. Schaumann, R. Schulz and S. Wagner (2018). "Nanoparticles in the environment: Where do we come from, where do we go to?" *Environmental Sciences Europe* **30**(1): 1–17.

Buzea, C., I. I. Pacheco and K. Robbie (2007). "Nanomaterials and nanoparticles: Sources and toxicity." *Biointerphases* **2**(4): MR17–MR71.

Carboni, A., R. Helmus, J. R. Parsons, K. Kalbitz and P. de Voogt (2016). "A method for the determination of fullerenes in soil and sediment matrices using ultra-high performance liquid chromatography coupled with heated electrospray quadrupole time of flight mass spectrometry." *Journal of Chromatography A* **1433**: 123–130.

Carreiró, F., A. M. Oliveira, A. Neves, B. Pires, D. Nagasamy Venkatesh, A. Durazzo, M. Lucarini, P. Eder, A. M. Silva and A. Santini (2020). "Polymeric nanoparticles: Production, characterization, toxicology and ecotoxicology." *Molecules* **25**(3731).

Cerqueira, M. Â. and L. M. Pastrana (2019). "Does the future of food pass by using nanotechnologies?" *Frontiers in Sustainable Food Systems* **3**: 16.

Chandra, P. (2016). *Nanobiosensors for personalized and onsite biomedical diagnosis*, IET.

Chariou, P. L., O. A. Ortega-Rivera and N. F. Steinmetz (2020). "Nanocarriers for the delivery of medical, veterinary, and agricultural active ingredients." *Acs Nano* **14**(3): 2678–2701.

Chaud, M., E. B. Souto, A. Zielinska, P. Severino, F. Batain, J. Oliveira-Junior and T. Alves (2021). "Nanopesticides in agriculture: Benefits and challenge in agricultural productivity, toxicological risks to human health and environment." *Toxics* **9**(6): 131.

Chaudhry, Q. and L. Castle (2011). "Food applications of nanotechnologies: An overview of opportunities and challenges for developing countries." *Trends in Food Science & Technology* **22**(11): 595–603.

Chellaram, C., G. Murugaboopathi, A. John, R. Sivakumar, S. Ganesan, S. Krithika and G. Priya (2014). "Significance of nanotechnology in food industry." *APCBEE Procedia* **8**: 109–113.

Chen, H. and R. Yada (2011). "Nanotechnologies in agriculture: New tools for sustainable development." *Trends in Food Science & Technology* **22**(11): 585–594.

Chinchkar, A. V., A. Singh, M. G. Kamble, P. K. Prabhakar, M. Meghwal and A. Sharma (2023). "Nanotechnology applications for quality determination of RTE and packaged food." *Nanotechnology applications for food safety and quality monitoring*, Elsevier: 265–288.

Choi, J., S. Yu, S. Yang and M. Cho (2011). "The glass transition and thermoelastic behavior of epoxy-based nanocomposites: A molecular dynamics study." *Polymer* **52**(22): 5197–5203.

Choudalakis, G. and A. Gotsis (2009). "Permeability of polymer/clay nanocomposites: A review." *European Polymer Journal* **45**(4): 967–984.

Clark, N. J., R. Clough, D. Boyle and R. D. Handy (2019). "Development of a suitable detection method for silver nanoparticles in fish tissue using single particle ICP-MS." *Environmental Science: Nano* **6**(11): 3388–3400.

Cockburn, A., R. Bradford, N. Buck, A. Constable, G. Edwards, B. Haber, P. Hepburn, J. Howlett, F. Kampers and C. Klein (2012). "Approaches to the safety assessment of engineered nanomaterials (ENM) in food." *Food and Chemical Toxicology* **50**(6): 2224–2242.

Crock, C. A., A. R. Rogensues, W. Shan and V. V. Tarabara (2013). "Polymer nanocomposites with graphene-based hierarchical fillers as materials for multifunctional water treatment membranes." *Water Research* **47**(12): 3984–3996.

Dainelli, D., N. Gontard, D. Spyropoulos, E. Zondervan-van den Beuken and P. Tobback (2008). "Active and intelligent food packaging: Legal aspects and safety concerns." *Trends in Food Science & Technology* **19**: S103–S112.

Damm, C., H. Münstedt and A. Rösch (2008). "The antimicrobial efficacy of polyamide 6/silver-nano- and microcomposites." *Materials Chemistry and Physics* **108**(1): 61–66.

Das, M. P., J. R. Livingstone, P. Veluswamy and J. Das (2018). "Exploration of Wedelia chinensis leaf-assisted silver nanoparticles for antioxidant, antibacterial and in vitro cytotoxic applications." *Journal of Food and Drug Analysis* **26**(2): 917–925.

Davis, K. F., J. A. Gephart, K. A. Emery, A. M. Leach, J. N. Galloway and P. D'Odorico (2016). "Meeting future food demand with current agricultural resources." *Global Environmental Change* **39**: 125–132.

de Freitas, A. d. S. M., A. P. B. da Silva, L. S. Montagna, I. A. Nogueira, N. K. Carvalho, V. S. de Faria, N. B. Dos Santos and A. P. Lemes (2022). "Thermoplastic starch nanocomposites: Sources, production and applications–a review." *Journal of Biomaterials Science, Polymer Edition* **33**(7): 900–945.

de Sousa, M. S., A. E. Schlogl, F. R. Estanislau, V. G. L. Souza, J. S. dos Reis Coimbra and I. J. B. Santos (2023). "Nanotechnology in packaging for food industry: Past, present, and future." *Coatings* **13**(8): 1411.

Diallo, M. and C. J. Brinker (2011). *Nanotechnology for sustainability: Environment, water, food, minerals, and climate*, Springer Dordrecht.

Dimkpa, C. O. and P. S. Bindraban (2016). "Fortification of micronutrients for efficient agronomic production: A review." *Agronomy for Sustainable Development* **36**(1): 7.

Ditta, A. (2012). "How helpful is nanotechnology in agriculture?" *Advances in Natural Sciences: Nanoscience and Nanotechnology* **3**(3): 033002.

Donovan, A. R., C. D. Adams, Y. Ma, C. Stephan, T. Eichholz and H. Shi (2016). "Single particle ICP-MS characterization of titanium dioxide, silver, and gold nanoparticles during drinking water treatment." *Chemosphere* **144**: 148–153.

Doudrick, K., T. Nosaka, P. Herckes and P. Westerhoff (2015). "Quantification of graphene and graphene oxide in complex organic matrices." *Environmental Science: Nano* **2**(1): 60–67.

Dudkiewicz, A., A. B. Boxall, Q. Chaudhry, K. Mølhave, K. Tiede, P. Hofmann and T. P. Linsinger (2015). "Uncertainties of size measurements in electron microscopy characterization of nanomaterials in foods." *Food Chemistry* **176**: 472–479.

EFSA Scientific Committee. (2011). "Guidance on the risk assessment of the application of nanoscience and nanotechnologies in the food and feed chain." *EFSA Journal* **9**(5): 2140.

EFSA Scientific Committee, S. More, V. Bampidis, D. Benford, C. Bragard, T. Halldorsson, A. Hernández-Jerez, S. Hougaard Bennekou, K. Koutsoumanis and C. Lambré (2021). "Guidance on risk assessment of nanomaterials to be applied in the food and feed chain: Human and animal health." *Efsa Journal* **19**(8): e06768.

Elemike, E. E., I. M. Uzoh, D. C. Onwudiwe and O. O. Babalola (2019). "The role of nanotechnology in the fortification of plant nutrients and improvement of crop production." *Applied Sciences* **9**(3): 499.

Falguera, V., J. P. Quintero, A. Jiménez, J. A. Muñoz and A. Ibarz (2011). "Edible films and coatings: Structures, active functions and trends in their use." *Trends in Food Science & Technology* **22**(6): 292–303.

Farahi, R. H., A. Passian, L. Tetard and T. Thundat (2012). "Critical issues in sensor science to aid food and water safety." *ACS Nano* **6**(6): 4548–4556.

Farré, M., S. Pérez, K. Gajda-Schrantz, V. Osorio, L. Kantiani, A. Ginebreda and D. Barceló (2010). "First determination of C60 and C70 fullerenes and N-methylfulleropyrrolidine C60 on the suspended material of wastewater effluents by liquid chromatography hybrid quadrupole linear ion trap tandem mass spectrometry." *Journal of Hydrology* **383**(1–2): 44–51.

Fazelian, N. and M. Yousefzadi (2022). "Nano-biofertilizers for enhanced nutrient use efficiency." *Nano-enabled agrochemicals in agriculture*, Elsevier: 145–158.

Frewer, L. J., W. Norde, A. Fischer and F. Kampers (2011). *Nanotechnology in the agri-food sector: Implications for the future*, John Wiley & Sons.

Fu, P. P. (2014). "Introduction to the special issue: Nanomaterials—toxicology and medical applications." *Journal of Food and Drug Analysis* **22**(1): 1.

Gallocchio, F., S. Belluco and A. Ricci (2015). "Nanotechnology and food: Brief overview of the current scenario." *Procedia Food Science* **5**: 85–88.

Ganapathy, V., G. Muthukumaran, P. E. Sudhagar, A. Rashedi, M. N. F. Norrrahim, R. A. Ilyas, K. L. Goh, M. Jawaid and J. Naveen (2023). "Mechanical properties of cellulose-based multiscale composites: A review." *Polymer Composites* **44**(2): 734–756.

Garcia, C. V., G. H. Shin and J. T. Kim (2018). "Metal oxide-based nanocomposites in food packaging: Applications, migration, and regulations." *Trends in Food Science & Technology* **82**: 21–31.

Ghanbarzadeh, B., S. A. Oleyaei and H. Almasi (2015). "Nanostructured materials utilized in biopolymer-based plastics for food packaging applications." *Critical Reviews in Food Science and Nutrition* **55**(12): 1699–1723.

Gliga, A. R., S. Skoglund, I. Odnevall Wallinder, B. Fadeel and H. L. Karlsson (2014). "Size-dependent cytotoxicity of silver nanoparticles in human lung cells: The role of cellular uptake, agglomeration and Ag release." *Particle and Fibre Toxicology* **11**: 1–17.

Gomes, S. I., J. J. Scott-Fordsmand and M. J. Amorim (2021). "Alternative test methods for (nano) materials hazards assessment: Challenges and recommendations for regulatory preparedness." *Nano Today* **40**: 101242.

Gómez-Llorente, H., P. Hervás, É. Pérez-Esteve, J. M. Barat and I. Fernández-Segovia (2022). "Nanotechnology in the agri-food sector: Consumer perceptions." *NanoImpact* **26**: 100399.

Gondal, A. H. and L. Tayyiba (2022). "Prospects of using nanotechnology in agricultural growth, environment and industrial food products." *Reviews in Agricultural Science* **10**: 68–81.

Gondikas, A. P., F. v. d. Kammer, R. B. Reed, S. Wagner, J. F. Ranville and T. Hofmann (2014). "Release of TiO2 nanoparticles from sunscreens into surface waters: A one-year survey at the old Danube recreational Lake." *Environmental Science & Technology* **48**(10): 5415–5422.

Goñi-Ciaurriz, L. and I. Vélaz (2022). "Antibacterial and degradable properties of β-cyclodextrin-TiO2 cellulose acetate and polylactic acid bionanocomposites for food packaging." *International Journal of Biological Macromolecules* **216**: 347–360.

Haghighat, M., H. Q. Alijani, M. Ghasemi, S. Khosravi, F. Borhani, F. Sharifi, S. Iravani, K. Najafi and M. Khatami (2022). "Cytotoxicity properties of plant-mediated synthesized K-doped ZnO nanostructures." *Bioprocess and Biosystems Engineering*: 1–9.

Hamad, A. F., J.-H. Han, B.-C. Kim and I. A. Rather (2018). "The intertwine of nanotechnology with the food industry." *Saudi Journal of Biological Sciences* **25**(1): 27–30.

Handford, C. E., M. Dean, M. Henchion, M. Spence, C. T. Elliott and K. Campbell (2014). "Implications of nanotechnology for the agri-food industry: Opportunities, benefits and risks." *Trends in Food Science & Technology* **40**(2): 226–241.

Handy, R. D. and B. J. Shaw (2007). "Toxic effects of nanoparticles and nanomaterials: Implications for public health, risk assessment and the public perception of nanotechnology." *Health, Risk & Society* **9**(2): 125–144.

Haris, M., T. Hussain, H. I. Mohamed, A. Khan, M. S. Ansari, A. Tauseef, A. A. Khan and N. Akhtar (2023). "Nanotechnology–a new frontier of nano-farming in agricultural and food production and its development." *Science of the Total Environment* **857**: 159639.

Harish, V., D. Tewari, M. Gaur, A. B. Yadav, S. Swaroop, M. Bechelany and A. Barhoum (2022). "Review on nanoparticles and nanostructured materials: Bioimaging, biosensing, drug delivery, tissue engineering, antimicrobial, and agro-food applications." *Nanomaterials* **12**(3): 457.

Hatti, V., M. Raghavendra, Y. Singh, S. Sharma, H. M. Halli and B. Goud (2020). "Application of nano technology in crop production to enhance resource use efficiency with special reference to soil and water conservation: A review." *Journal of Soil and Water Conservation* **19**(4): 375–381.

He, X. and H.-M. Hwang (2016). "Nanotechnology in food science: Functionality, applicability, and safety assessment." *Journal of Food and Drug Analysis* **24**(4): 671–681.

He, X., H. Deng and H.-m. Hwang (2019). "The current application of nanotechnology in food and agriculture." *Journal of Food and Drug Analysis* **27**(1): 1–21.

He, X., H. Zhang, H. Shi, W. Liu and E. Sahle-Demessie (2020). "Fates of Au, Ag, ZnO, and CeO2 nanoparticles in simulated gastric fluid studied using single-particle-inductively coupled plasma-mass spectrometry." *Journal of the American Society for Mass Spectrometry* **31**(10): 2180–2190.

Hegde, K., S. K. Brar, M. Verma and R. Y. Surampalli (2016). "Current understandings of toxicity, risks and regulations of engineered nanoparticles with respect to environmental microorganisms." *Nanotechnology for Environmental Engineering* **1**: 1–12.

Hemalatha, M., G. S. Lakshmi, Megha, Vinita and B. C. Kotibagar (2023). "Applications of nanomaterials for water treatment: Current trends and future scope." *Modern nanotechnology: Volume 1: Environmental sustainability and remediation*, Springer: 145–175.

Hoseinnejad, M., S. M. Jafari and I. Katouzian (2018). "Inorganic and metal nanoparticles and their antimicrobial activity in food packaging applications." *Critical Reviews in Microbiology* **44**(2): 161–181.

Hossain, F., O. J. Perales-Perez, S. Hwang and F. Román (2014). "Antimicrobial nanomaterials as water disinfectant: Applications, limitations and future perspectives." *Science of the Total Environment* **466**: 1047–1059.

Hoyos-Merlano, N. T., V. Borroni, M. J. Rodriguez-Batiller, R. J. Candal and M. L. Herrera (2022). "Nanoreinforcement as a strategy to improve physical properties of biodegradable composite films based on biopolymers." *Food Research International* **162**: 112178.

Hsiao, I.-L. and Y.-J. Huang (2011). "Effects of various physicochemical characteristics on the toxicities of ZnO and TiO2 nanoparticles toward human lung epithelial cells." *Science of the Total Environment* **409**(7): 1219–1228.

Huang, S., L. Wang, L. Liu, Q. Fu and D. Zhu (2014). "Nonchemical pest control in China rice: A review." *Agronomy for Sustainable Development* **34**: 275–291.

Hutchison, J. E. (2016). *The road to sustainable nanotechnology: Challenges, progress and opportunities*, ACS Publications.

Iavicoli, I., V. Leso, D. H. Beezhold and A. A. Shvedova (2017). "Nanotechnology in agriculture: Opportunities, toxicological implications, and occupational risks." *Toxicology and Applied Pharmacology* **329**: 96–111.

Jafarzadeh, S., M. Forough, S. Amjadi, V. Javan Kouzegaran, H. Almasi, F. Garavand and M. Zargar (2022). "Plant protein-based nanocomposite films: A review on the used nanomaterials, characteristics, and food packaging applications." *Critical Reviews in Food Science and Nutrition*: 1–27.

Jagtiani, E. (2022). "Advancements in nanotechnology for food science and industry." *Food Frontiers* **3**(1): 56–82.

Jain, A., S. Ranjan, N. Dasgupta and C. Ramalingam (2018). "Nanomaterials in food and agriculture: An overview on their safety concerns and regulatory issues." *Critical Reviews in Food Science and Nutrition* **58**(2): 297–317.

Jamjoum, H. A. A., K. Umar, R. Adnan, M. R. Razali and M. N. Mohamad Ibrahim (2021). "Synthesis, characterization, and photocatalytic activities of graphene oxide/metal oxides nanocomposites: A review." *Frontiers in Chemistry* **9**: 752276.

Janjarasskul, T. and P. Suppakul (2018). "Active and intelligent packaging: The indication of quality and safety." *Critical Reviews in Food Science and Nutrition* **58**(5): 808–831.

Jarzębski, M., B. Bellich, T. Białopiotrowicz, T. Śliwa, J. Kościński and A. Cesàro (2017). "Particle tracking analysis in food and hydrocolloids investigations." *Food Hydrocolloids* **68**: 90–101.

Jha, A., D. Pathania, B. Damathia, P. Raizada, S. Rustagi, P. Singh, G. M. Rani and V. Chaudhary (2023). "Panorama of biogenic nano-fertilizers: A road to sustainable agriculture." *Environmental Research*: 116456.

Kalpana, S., S. Priyadarshini, M. M. Leena, J. Moses and C. Anandharamakrishnan (2019). "Intelligent packaging: Trends and applications in food systems." *Trends in Food Science & Technology* **93**: 145–157.

Kamigaito, O. (1991). "What can be improved by nanometer composites?" *Journal of the Japan Society of Powder and Powder Metallurgy* **38**(3): 315–321.

Kang'ethe, E. K., F. Mutua, K. Roesel, M. Ntawubizi, C. Kankya, S. Niragira, B. Kilima, P. Gathura, K. K. Kuorwel, A. Kiri and D. Grace (2021). *A review of the food safety architecture in the East African Community: Animal-source foods, fruits and vegetables*, ILRI.

Kashyap, P. L., X. Xiang and P. Heiden (2015). "Chitosan nanoparticle based delivery systems for sustainable agriculture." *International Journal of Biological Macromolecules* **77**: 36–51.

Khaneghah, A. M., S. M. B. Hashemi and S. Limbo (2018). "Antimicrobial agents and packaging systems in antimicrobial active food packaging: An overview of approaches and interactions." *Food and Bioproducts Processing* **111**: 1–19.

Kim, D. Y., A. Kadam, S. Shinde, R. G. Saratale, J. Patra and G. Ghodake (2018). "Recent developments in nanotechnology transforming the agricultural sector: A transition replete with opportunities." *Journal of the Science of Food and Agriculture* **98**(3): 849–864.

Kim, H. J., K. Charoensri, J. A. Ko and H. J. Park (2022). "Effects of layered double hydroxides on poly (vinyl alcohol)/poly (acrylic acid) films for green food packaging applications." *Progress in Organic Coatings* **163**: 106634.

Kim, I. H., H. Lee, J. E. Kim, K. B. Song, Y. S. Lee, D. S. Chung and S. C. Min (2013). "Plum coatings of lemongrass oil-incorporating carnauba wax-based nanoemulsion." *Journal of Food Science* **78**(10): E1551-E1559.

Kim, I., K. Viswanathan, G. Kasi, S. Thanakkasaranee, K. Sadeghi and J. Seo (2022). "ZnO nanostructures in active antibacterial food packaging: Preparation methods, antimicrobial mechanisms, safety issues, future prospects, and challenges." *Food Reviews International* **38**(4): 537–565.

King, T., M. Cole, J. M. Farber, G. Eisenbrand, D. Zabaras, E. M. Fox and J. P. Hill (2017). "Food safety for food security: Relationship between global megatrends and developments in food safety." *Trends in Food Science & Technology* **68**: 160–175.

Kremsa, V. Š. (2021). "Sustainable management of agricultural resources (agricultural crops and animals)." *Sustainable resource management*, Elsevier: 99–145.

Laborda, F., E. Bolea, G. Cepriá, M. T. Gómez, M. S. Jiménez, J. Pérez-Arantegui and J. R. Castillo (2016). "Detection, characterization and quantification of inorganic engineered nanomaterials: A review of techniques and methodological approaches for the analysis of complex samples." *Analytica Chimica Acta* **904**: 10–32.

Lagaron, J., L. Cabedo, D. Cava, J. Feijoo, R. Gavara and E. Gimenez (2005). "Improving packaged food quality and safety. Part 2: Nanocomposites." *Food Additives and Contaminants* **22**(10): 994–998.

Laycock, A., N. J. Clark, R. Clough, R. Smith and R. D. Handy (2022). "Determination of metallic nanoparticles in biological samples by single particle ICP-MS: A systematic review from sample collection to analysis." *Environmental Science: Nano* **9**(2): 420–453.

Li, S., G. Chen, S. Qiang, Z. Yin, Z. Zhang and Y. Chen (2020). "Synthesis and evaluation of highly dispersible and efficient photocatalytic TiO2/poly lactic acid nanocomposite films via sol-gel and casting processes." *International Journal of Food Microbiology* **331**: 108763.

Lim, C. S. S., C. Y. Soon, E. W. C. Chan and C. W. Wong (2023). "Nanofillers to enhance biodegradable composites and their niche applications." *Synthetic and natural Nanofillers in polymer composites*, Elsevier: 215–257.

Liu, R. and R. Lal (2014). "Synthetic apatite nanoparticles as a phosphorus fertilizer for soybean (Glycine max)." *Scientific Reports* **4**(1): 5686.

Luo, L., Y. Wu, C. Liu, L. Huang, Y. Zou, Y. Shen and Q. Lin (2019). "Designing soluble soybean polysaccharides-based nanoparticles to improve sustained antimicrobial activity of nisin." *Carbohydrate Polymers* **225**: 115251.

Majid, I., G. A. Nayik, S. M. Dar and V. Nanda (2018). "Novel food packaging technologies: Innovations and future prospective." *Journal of the Saudi Society of Agricultural Sciences* **17**(4): 454–462.

Maksimović, M. and E. Omanović-Mikličanin (2017). *Towards green nanotechnology: Maximizing benefits and minimizing harm*. CMBEBIH 2017: Proceedings of the International Conference on Medical and Biological Engineering 2017, Springer.

Malik, S., K. Muhammad and Y. Waheed (2023). "Nanotechnology: A revolution in modern industry." *Molecules* **28**(2): 661.

Mattarozzi, M., M. Suman, C. Cascio, D. Calestani, S. Weigel, A. Undas and R. Peters (2017). "Analytical approaches for the characterization and quantification of nanoparticles in food and beverages." *Analytical and Bioanalytical Chemistry* **409**: 63–80.

Maurer-Jones, M. A., I. L. Gunsolus, C. J. Murphy and C. L. Haynes (2013). "Toxicity of engineered nanoparticles in the environment." *Analytical Chemistry* **85**(6): 3036–3049.

Mbeh, D. A., O. Akhavan, T. Javanbakht and M. Mahmoudi (2014). "Cytotoxicity of protein corona-graphene oxide nanoribbons on human epithelial cells." *Applied Surface Science* **320**: 596–601.

Mihindukulasuriya, S. and L.-T. Lim (2014). "Nanotechnology development in food packaging: A review." *Trends in Food Science & Technology* **40**(2): 149–167.

Mitrano, D. M. and W. Wohlleben (2020). "Microplastic regulation should be more precise to incentivize both innovation and environmental safety." *Nature Communications* **11**(1): 5324.

Mittal, D., G. Kaur, P. Singh, K. Yadav and S. A. Ali (2020). "Nanoparticle-based sustainable agriculture and food science: Recent advances and future outlook." *Frontiers in Nanotechnology* **2**: 579954.

Mohammed Fayaz, A., K. Balaji, M. Girilal, P. Kalaichelvan and R. Venkatesan (2009). "Mycobased synthesis of silver nanoparticles and their incorporation into sodium alginate films for vegetable and fruit preservation." *Journal of Agricultural and Food Chemistry* **57**(14): 6246–6252.

Mohanta, Y. K., D. Nayak, K. Biswas, S. K. Singdevsachan, E. F. Abd_Allah, A. Hashem, A. A. Alqarawi, D. Yadav and T. K. Mohanta (2018). "Silver nanoparticles synthesized using wild mushroom show potential antimicrobial activities against food borne pathogens." *Molecules* **23**(3): 655.

Monreal, C., M. DeRosa, S. Mallubhotla, P. Bindraban and C. Dimkpa (2016). "Nanotechnologies for increasing the crop use efficiency of fertilizer-micronutrients." *Biology and Fertility of Soils* **52**: 423–437.

Montano, M. D., G. V. Lowry, F. von der Kammer, J. Blue and J. F. Ranville (2014). "Current status and future direction for examining engineered nanoparticles in natural systems." *Environmental Chemistry* **11**(4): 351–366.

Motarjemi, Y. (2023). Crisis management. *Food safety management*, Elsevier: 971–994.

Mudalige, T., H. Qu, D. Van Haute, S. M. Ansar, A. Paredes and T. Ingle (2019). "Characterization of nanomaterials: Tools and challenges." *Nanomaterials for food applications*, Elsevier: 313–353.

Mueller, N. C. and B. Nowack (2008). "Exposure modeling of engineered nanoparticles in the environment." *Environmental Science & Technology* **42**(12): 4447–4453.

Mustafa, F. and S. Andreescu (2020). "Nanotechnology-based approaches for food sensing and packaging applications." *RSC Advances* **10**(33): 19309–19336.

Nachay, K. (2007). "Analyzing nanotechnology." *Food Technology (Chicago)* **61**(1): 34–36.

Nair, A., R. Mallya, V. Suvarna, T. A. Khan, M. Momin and A. Omri (2022). "Nanoparticles—Attractive carriers of antimicrobial essential oils." *Antibiotics* **11**(1): 108.

Navratilova, J., A. Praetorius, A. Gondikas, W. Fabienke, F. Von der Kammer and T. Hofmann (2015). "Detection of engineered copper nanoparticles in soil using single particle ICP-MS." *International Journal of Environmental Research and Public Health* **12**(12): 15756–15768.

Nazeer, A., F. Ahmad, N. Verma and S. Ahmad (2023). "Targeted delivery of nanopesticides." *Targeted delivery of nanopesticides and nanofertilizers in sustainable agricultural Farming*, Springer: 125–139.

Neethirajan, S. and D. S. Jayas (2011). "Nanotechnology for the food and bioprocessing industries." *Food and Bioprocess Technology* **4**: 39–47.

Neethirajan, S., V. Ragavan, X. Weng and R. Chand (2018). "Biosensors for sustainable food engineering: Challenges and perspectives." *Biosensors* **8**(1): 23.

Nel, A., T. Xia, L. Madler and N. Li (2006). "Toxic potential of materials at the nanolevel." *Science* **311**(5761): 622–627.

Neme, K., A. Nafady, S. Uddin and Y. B. Tola (2021). "Application of nanotechnology in agriculture, postharvest loss reduction and food processing: Food security implication and challenges." *Heliyon* **7**(12).

Nile, S. H., V. Baskar, D. Selvaraj, A. Nile, J. Xiao and G. Kai (2020). "Nanotechnologies in food science: Applications, recent trends, and future perspectives." *Nano-Micro Letters* **12**: 1–34.

Noshirvani, N., B. Ghanbarzadeh, R. R. Mokarram and M. Hashemi (2017). "Novel active packaging based on carboxymethyl cellulose-chitosan-ZnO NPs nanocomposite for increasing the shelf life of bread." *Food Packaging and Shelf Life* **11**: 106–114.

Ongaratto, G. C., A. C. P. Vital and I. N. do Prado (2022). "Embalagens ativas e inteligentes para proteção da carne e seus derivados: Revisão." *Pubvet* **16**: 207.

Onyeaka, H., P. Passaretti, T. Miri and Z. T. Al-Sharify (2022). "The safety of nanomaterials in food production and packaging." *Current Research in Food Science* **5**: 763–774.

Ortega, F., F. Versino, O. V. López and M. A. García (2022). "Biobased composites from agro-industrial wastes and by-products." *Emergent Materials* **5**(3): 873–921.

Pal, K., P. Sarkar, A. Anis, K. Wiszumirska and M. Jarzębski (2021). "Polysaccharide-based nanocomposites for food packaging applications." *Materials* **14**(19): 5549.

Parr-Vasquez, C., A. Paulson and R. Yada (2011). "The advanced foods and materials network: A Canadian portal to excellence in innovative food science and technology." *Trends in Food Science & Technology* **22**(9): 476–479.

Peter, A., L. M. Cozmuta, C. Nicula, A. M. Cozmuta, C. M. Talasman, G. Drazic, A. Peñas, A. J. Calahorro, G. Sagratini and S. Silvi (2021). "Chemical and organoleptic changes of curd cheese stored in new and reused active packaging systems made of Ag-graphene-TiO2-PLA." *Food Chemistry* **363**: 130341.

Peters, R. J., Z. H. Rivera, G. van Bemmel, H. J. Marvin, S. Weigel and H. Bouwmeester (2014). "Development and validation of single particle ICP-MS for sizing and quantitative determination of nano-silver in chicken meat." *Analytical and Bioanalytical Chemistry* **406**: 3875–3885.

Petersen, E. J., D. X. Flores-Cervantes, T. D. Bucheli, L. C. Elliott, J. A. Fagan, A. Gogos, S. Hanna, R. Kägi, E. Mansfield and A. R. M. Bustos (2016). "Quantification of carbon nanotubes in environmental matrices: Current capabilities, case studies, and future prospects." *Environmental Science & Technology* **50**(9): 4587–4605.

Picard, E., H. Gauthier, J.-F. Gérard and E. Espuche (2007). "Influence of the intercalated cations on the surface energy of montmorillonites: Consequences for the morphology and gas barrier properties of polyethylene/montmorillonites nanocomposites." *Journal of Colloid and Interface Science* **307**(2): 364–376.

Pigford, A. A. E., G. M. Hickey and L. Klerkx (2022). "Networks for science-informed innovation in the Arctic: Insights on the structure and evolution of a Canadian research network." *Arctic* **75**(2): 161–179.

Pires, J. R. A., C. Rodrigues, I. Coelhoso, A. L. Fernando and V. G. L. Souza (2023). "Current applications of bionanocomposites in food processing and packaging." *Polymers* **15**(10): 2336.

Pradhan, N., S. Singh, N. Ojha, A. Shrivastava, A. Barla, V. Rai and S. Bose (2015). "Facets of nanotechnology as seen in food processing, packaging, and preservation industry." *BioMed Research International* **2015**.

Praetorius, A., A. Gundlach-Graham, E. Goldberg, W. Fabienke, J. Navratilova, A. Gondikas, R. Kaegi, D. Günther, T. Hofmann and F. von der Kammer (2017). "Single-particle multi-element fingerprinting (spMEF) using inductively-coupled plasma time-of-flight mass spectrometry (ICP-TOFMS) to identify engineered nanoparticles against the elevated natural background in soils." *Environmental Science: Nano* **4**(2): 307–314.

Prasad, R., A. Bhattacharyya and Q. D. Nguyen (2017). "Nanotechnology in sustainable agriculture: Recent developments, challenges, and perspectives." *Frontiers in Microbiology* **8**: 1014.

Prasad, R. D., A. Sahoo, O. P. Shrivastav, N. Charmode, R. Kamat, N. Kajave, J. Chauhan, S. Banga, U. Tamboli and P. MS (2022). "A review on aspects of nanotechnology in food science and animal nutrition." *ES Food & Agroforestry* **8**: 12–46.

Primožič, M., Ž. Knez and M. Leitgeb (2021). "(Bio) Nanotechnology in food science—food packaging." *Nanomaterials* **11**(2): 292.

Rai, V., S. Acharya and N. Dey (2012). "Implications of nanobiosensors in agriculture." *Journal of Biomaterials and Nanobiotechnology* **3**: 315–324.

Ramesh, M., R. Janani, C. Deepa and L. Rajeshkumar (2022). "Nanotechnology-enabled biosensors: A review of fundamentals, design principles, materials, and applications." *Biosensors* **13**(1): 40.

Ravichandran, R. (2010). "Nanotechnology applications in food and food processing: Innovative green approaches, opportunities and uncertainties for global market." *International Journal of Green Nanotechnology: Physics and Chemistry* **1**(2): P72-P96.

Reig, C. S., A. D. Lopez, M. H. Ramos and V. A. C. Ballester (2014). "Nanomaterials: A map for their selection in food packaging applications." *Packaging Technology and Science* **27**(11): 839–866.

Riahi, Z., R. Priyadarshi, J.-W. Rhim, E. Lotfali, R. Bagheri and G. Pircheraghi (2022). "Alginate-based multifunctional films incorporated with sulfur quantum dots for active packaging applications." *Colloids and Surfaces B: Biointerfaces* **215**: 112519.

Rodrigues, S. M., P. Demokritou, N. Dokoozlian, C. O. Hendren, B. Karn, M. S. Mauter, O. A. Sadik, M. Safarpour, J. M. Unrine and J. Viers (2017). "Nanotechnology for sustainable food production: Promising opportunities and scientific challenges." *Environmental Science: Nano* **4**(4): 767–781.

Roy, S. and J.-W. Rhim (2020). "Fabrication of copper sulfide nanoparticles and limonene incorporated pullulan/carrageenan-based film with improved mechanical and antibacterial properties." *Polymers* **12**(11): 2665.

Roy, S. and J.-W. Rhim (2022). "Gelatin/cellulose nanofiber-based functional films added with mushroom-mediated sulfur nanoparticles for active packaging applications." *Journal of Nanostructure in Chemistry* **12**(5): 979–990.

Saha, N. (2022). "Food packaging: Concepts and its significance." *Food packaging: Materials, techniques and environmental issues*, Springer: 1–45.

Sahani, S. and Y. C. Sharma (2021). "Advancements in applications of nanotechnology in global food industry." *Food Chemistry* **342**: 128318.

Sahoo, M., S. Vishwakarma, C. Panigrahi and J. Kumar (2021). "Nanotechnology: Current applications and future scope in food." *Food Frontiers* **2**(1): 3–22.

Sahu, S. C. and A. W. Hayes (2017). "Toxicity of nanomaterials found in human environment: A literature review." *Toxicology Research and Application* **1**: 2397847317726352.

Salvia-Trujillo, L., R. Soliva-Fortuny, M. A. Rojas-Graü, D. J. McClements and O. Martín-Belloso (2017). "Edible nanoemulsions as carriers of active ingredients: A review." *Annual Review of Food Science and Technology* **8**: 439–466.

Sandoval, B. (2009). "Perspectives on FDA's regulation of nanotechnology: Emerging challenges and potential solutions." *Comprehensive Reviews in Food Science and Food Safety* **8**(4): 375–393.

Sawaya, W. N. (2017). "Impact of food losses and waste on food security." *Water, energy & food sustainability in the Middle East: The sustainability triangle*, Springer: 361–388.

Scheffler, S. L., X. Wang, L. Huang, F. San-Martin Gonzalez and Y. Yao (2010). "Phytoglycogen octenyl succinate, an amphiphilic carbohydrate nanoparticle, and ε-polylysine to improve lipid oxidative stability of emulsions." *Journal of Agricultural and Food Chemistry* **58**(1): 660–667.

Schierz, A., A. N. Parks, K. M. Washburn, G. T. Chandler and P. L. Ferguson (2012). "Characterization and quantitative analysis of single-walled carbon nanotubes in the aquatic environment using near-infrared fluorescence spectroscopy." *Environmental Science & Technology* **46**(22): 12262–12271.

Seil, J. T. and T. J. Webster (2012). "Antimicrobial applications of nanotechnology: Methods and literature." *International Journal of Nanomedicine*: 2767–2781.

Sekhon, B. S. (2014). "Nanotechnology in agri-food production: An overview." *Nanotechnology, Science and Applications*: 31–53.

Shafiq, M., S. Anjum, C. Hano, I. Anjum and B. H. Abbasi (2020). "An overview of the applications of nanomaterials and nanodevices in the food industry." *Foods* **9**(2): 148.

Shahzad, A., M. W. Ullah, J. Ali, K. Aziz, M. A. Javed, Z. Shi, S. Manan, M. Ul-Islam, M. Nazar and G. Yang (2023). "The versatility of nanocellulose, modification strategies, and its current progress in wastewater treatment and environmental remediation." *Science of the Total Environment* **858**: 159937.

Shankar, S. and J.-W. Rhim (2018). "Preparation of sulfur nanoparticle-incorporated antimicrobial chitosan films." *Food Hydrocolloids* **82**: 116–123.

Sharma, S., B. K. Biswal, D. Kumari, P. Bindra, S. Kumar, T. Stobdan and V. Shanmugam (2018). "Ecofriendly fruit switches: Graphene oxide-based wrapper for programmed fruit preservative delivery to extend shelf life." *ACS Applied Materials & Interfaces* **10**(22): 18478–18488.

Shende, S., V. D. Rajput, A. Gade, T. Minkina, Y. Fedorov, S. Sushkova, S. Mandzhieva, M. Burachevskaya and V. Boldyreva (2021). "Metal-based green synthesized nanoparticles: Boon for sustainable agriculture and food security." *IEEE Transactions on NanoBioscience* **21**(1): 44–54.

Silvestre, C., D. Duraccio and S. Cimmino (2011). "Food packaging based on polymer nanomaterials." *Progress in Polymer Science* **36**(12): 1766–1782.

Sim, J.-H., S. H. Lee, J.-Y. Yang, W.-C. Lee, C. Mun, S. Lee, S.-G. Park and Y.-R. Cho (2022). "Plasmonic hotspot engineering of Ag-coated polymer substrates with high reproducibility and photothermal stability." *Sensors and Actuators B: Chemical* **354**: 131110.

Singh, O. (2023). "Nanotechnology for sustainability and food security in agriculture: A nanopriming long story in short." *Nanopriming approach to sustainable agriculture*, IGI Global: 315–339.

Singh, R., S. Dutt, P. Sharma, A. K. Sundramoorthy, A. Dubey, A. Singh and S. Arya (2023). "Future of nanotechnology in food industry: Challenges in processing, packaging, and food safety." *Global Challenges* **7**(4): 2200209.

Singh, T., S. Shukla, P. Kumar, V. Wahla, V. K. Bajpai and I. A. Rather (2017). "Application of nanotechnology in food science: Perception and overview." *Frontiers in Microbiology* **8**: 1501.

Sneha, K. and A. Kumar (2022). "Nanoemulsions: Techniques for the preparation and the recent advances in their food applications." *Innovative Food Science & Emerging Technologies* **76**: 102914.

Sonkaria, S., S.-H. Ahn and V. Khare (2012). "Nanotechnology and its impact on food and nutrition: A review." *Recent Patents on Food, Nutrition & Agriculture* **4**(1): 8–18.

Sothornvit, R. (2019). "Nanostructured materials for food packaging systems: New functional properties." *Current Opinion in Food Science* **25**: 82–87.

Sridhar, A., M. Ponnuchamy, P. S. Kumar and A. Kapoor (2021). "Food preservation techniques and nanotechnology for increased shelf life of fruits, vegetables, beverages and spices: A review." *Environmental Chemistry Letters* **19**: 1715–1735.

Sukhanova, A., S. Bozrova, P. Sokolov, M. Berestovoy, A. Karaulov and I. Nabiev (2018). "Dependence of nanoparticle toxicity on their physical and chemical properties." *Nanoscale Research Letters* **13**: 1–21.

Suriyaprabha, R., G. Karunakaran, R. Yuvakkumar, V. Rajendran and N. Kannan (2012). "Silica nanoparticles for increased silica availability in maize (*Zea mays*. L) seeds under hydroponic conditions." *Current Nanoscience* **8**(6): 902–908.

Suvarna, V., A. Nair, R. Mallya, T. Khan and A. Omri (2022). "Antimicrobial nanomaterials for food packaging." *Antibiotics* **11**(6): 729.

Szakal, C., S. M. Roberts, P. Westerhoff, A. Bartholomaeus, N. Buck, I. Illuminato, R. Canady and M. Rogers (2014). "Measurement of nanomaterials in foods: Integrative consideration of challenges and future prospects." *ACS nano* **8**(4): 3128–3135.

Tarantola, M., A. Pietuch, D. Schneider, J. Rother, E. Sunnick, C. Rosman, S. Pierrat, C. Sönnichsen, J. Wegener and A. Janshoff (2011). "Toxicity of gold-nanoparticles: Synergistic effects of shape and surface functionalization on micromotility of epithelial cells." *Nanotoxicology* **5**(2): 254–268.

Teamsinsungvon, A., C. Ruksakulpiwat and Y. Ruksakulpiwat (2022). "Effects of titanium–silica oxide on degradation behavior and antimicrobial activity of poly (lactic acid) composites." *Polymers* **14**(16): 3310.

Thimiri Govinda Raj, D. B. and N. A. Khan (2016). "Designer nanoparticle: Nanobiotechnology tool for cell biology." *Nano Convergence* **3**(1): 1–12.

Thiruvengadam, M., G. Rajakumar and I.-M. Chung (2018). "Nanotechnology: Current uses and future applications in the food industry." *3 Biotech* **8**: 1–13.

Tiede, K., A. B. Boxall, S. P. Tear, J. Lewis, H. David and M. Hassellöv (2008). "Detection and characterization of engineered nanoparticles in food and the environment." *Food Additives and Contaminants* **25**(7): 795–821.

Trinh, B. M., B. P. Chang and T. H. Mekonnen (2023). "The barrier properties of sustainable multiphase and multicomponent packaging materials: A review." *Progress in Materials Science*: 101071.

Trujillo, L. E., R. Ávalos, S. Granda, L. S. Guerra and J. M. País-Chanfrau (2016). "Nanotechnology applications for food and bioprocessing industries." *Biology and Medicine* **8**(3): 1.

Tsang, M. P., E. Kikuchi-Uehara, G. W. Sonnemann, C. Aymonier and M. Hirao (2017). "Evaluating nanotechnology opportunities and risks through integration of life-cycle and risk assessment." *Nature Nanotechnology* **12**(8): 734–739.

Vanderroost, M., P. Ragaert, F. Devlieghere and B. De Meulenaer (2014). "Intelligent food packaging: The next generation." *Trends in Food Science & Technology* **39**(1): 47–62.

Vurro, M., B. Bonciani and G. Vannacci (2010). "Emerging infectious diseases of crop plants in developing countries: Impact on agriculture and socio-economic consequences." *Food Security* **2**: 113–132.

Wang, H.-M., T.-Q. Yuan, G.-Y. Song and R.-C. Sun (2021). "Advanced and versatile lignin-derived biodegradable composite film materials toward a sustainable world." *Green Chemistry* **23**(11): 3790–3817.

Webber, M. J., E. A. Appel, E. Meijer and R. Langer (2016). "Supramolecular biomaterials." *Nature Materials* **15**(1): 13–26.

Weiss, J., P. Takhistov and D. J. McClements (2006). "Functional materials in food nanotechnology." *Journal of Food Science* **71**(9): R107-R116.

Westerhoff, P., P. Alvarez, Q. Li, J. Gardea-Torresdey and J. Zimmerman (2016). "Overcoming implementation barriers for nanotechnology in drinking water treatment." *Environmental Science: Nano* **3**(6): 1241–1253.

Wilson, M. A., N. H. Tran, A. S. Milev, G. K. Kannangara, H. Volk and G. M. Lu (2008). "Nanomaterials in soils." *Geoderma* **146**(1–2): 291–302.

Wu, J., J. Lin, M. Zhou and C. Wei (2000). "Synthesis and properties of starch-graft-polyacrylamide/clay superabsorbent composite." *Macromolecular Rapid Communications* **21**(15): 1032–1034.

Xiang, K., Y. Li, W. Ford, W. Land, J. D. Schaffer, R. Congdon, J. Zhang and O. Sadik (2016). "Automated analysis of food-borne pathogens using a novel microbial cell culture, sensing and classification system." *Analyst* **141**(4): 1472–1482.

Xiong, R., A. M. Grant, R. Ma, S. Zhang and V. V. Tsukruk (2018). "Naturally-derived biopolymer nanocomposites: Interfacial design, properties and emerging applications." *Materials Science and Engineering: R: Reports* **125**: 1–41.

Yadav, S. K., Z. A. Khan and B. Mishra (2013). "Impact of nanotechnology on socio-economic aspects: An overview." *Reviews in Nanoscience and Nanotechnology* **2**(2): 127–142.

Yalçınöz, Ş. and E. Erçelebi (2018). "Potential applications of nano-emulsions in the food systems: An update." *Materials Research Express* **5**(6): 062001.

Yang, D., Q. Liu, Y. Gao, S. Wan, F. Meng, W. Weng and Y. Zhang (2023). "Characterization of silver nanoparticles loaded chitosan/polyvinyl alcohol antibacterial films for food packaging." *Food Hydrocolloids* **136**: 108305.

Yazgan, I., N. M. Noah, O. Toure, S. Zhang and O. A. Sadik (2014). "Biosensor for selective detection of E. coli in spinach using the strong affinity of derivatized mannose with fimbrial lectin." *Biosensors and Bioelectronics* **61**: 266–273.

Zare, M., K. Namratha, S. Ilyas, A. Hezam, S. Mathur and K. Byrappa (2019). "Smart fortified PHBV-CS biopolymer with ZnO–Ag nanocomposites for enhanced shelf life of food packaging." *ACS Applied Materials & Interfaces* **11**(51): 48309–48320.

Zenebom, O. and N. Pascuet (2005). "Ministério da Saúde. Agência Nacional de Vigilância Sanitária." *Métodos físico-químicos para análise de alimentos* **4**.

Zhang, W., Z. Mou, Y. Wang, Y. Chen, E. Yang, F. Guo, D. Sun and W. Wang (2019). "Molybdenum disulfide nanosheets loaded with chitosan and silver nanoparticles effective antifungal activities: In vitro and in vivo." *Materials Science and Engineering: C* **97**: 486–497.

Zhang, X., Z. Zhang, W. Wu, J. Yang and Q. Yang (2021). "Preparation and characterization of chitosan/Nano-ZnO composite film with antimicrobial activity." *Bioprocess and Biosystems Engineering* **44**: 1193–1199.

Zhou, B., X. Chen, Q. Wang, W. Wei and T. Zhang (2018). "Effects of nano carbon on soil erosion and nutrient loss in a semi-arid loess region of Northwestern China." *International Journal of Agricultural and Biological Engineering* **11**(1): 138–145.

Zhou, Q., L. Liu, N. Liu, B. He, L. Hu and L. Wang (2020). "Determination and characterization of metal nanoparticles in clams and oysters." *Ecotoxicology and Environmental Safety* **198**: 110670.

Zou, Z., B. B. Ismail, X. Zhang, Z. Yang, D. Liu and M. Guo (2023). "Improving barrier and antibacterial properties of chitosan composite films by incorporating lignin nanoparticles and acylated soy protein isolate nanogel." *Food Hydrocolloids* **134**: 108091.

21 Adopting the Precautionary Approach to Food Packaging in the Absence of Regulation

Nompumelelo Zungu[a] *and Shalini Singh*[a]
[a]Department of Operations and Quality Management, Durban University of Technology, Durban, South Africa

21.1 INTRODUCTION

This chapter addresses the significant role of packaging within the food sector, along with the positive and negative effects that engineered nanomaterials have on the packaging and the food. It highlights the improvement in performance that emerges with the incorporation of the nanomaterials in the development of packaging material, as well as the concerns regarding the migration of the nanomaterials from the parcel into the food. Due to these concerns and the absence of a regulatory framework dedicated to mitigating health and safety risks, the precautionary principle suggested by regulatory bodies will be the main focus of the chapter.

According to Goldstein (2001), the precautionary principle is defined as a basis for preventive action. Fischer and Ghelardi (2016) state that generally, precautionary principles have been used to evaluate processes, their benefits, and cost-efficiency of interventions to mitigate harm that may occur in organisational activities. These definitions discuss how precautionary principles investigate risk prevention measures or controls. Lastly, this chapter will propose precautionary principles for the use of nanomaterials in food packaging that could be beneficially adopted to prevent the migration of the nanomaterials from the food packaging into the food/consumable product.

21.1.1 Purpose of Food Packaging

The main function of packaging is to suppress spoilage and contamination as well as to provide physical protection. Furthermore, packaging acts as a measure to safeguard the food from temperature external shock, vibration, and microbial contamination. This function is accomplished through the packaging acting as an obstruction or barrier to prevent scavenging oxygen and spoilage-causing gases. It was found that the application of nanotechnology in the form of nanoparticles and polymer-based composites in food packaging proved to be the best solution in this regard (Mohammad et al. 2022).

Nanofood is described as food produced by integrating nanotechnology in the production, processing, storage, and packaging (Onyeaka et al. 2022).

21.1.2 Challenges and Positives Attributes That Come with the Use of Nanomaterials

This section will highlight the various challenges experienced when nanomaterials are incorporated in food packaging and also depict a few positive elements from the use of nanomaterials.

The challenge with nanotechnology-based techniques is to create low-cost and economical processing operations to create nano-delivery systems and to establish effective formulations that are safe for human consumption (Bajpai et al. 2018). The costs incurred with managing nano-based techniques are a constant challenge, however the positive element to this technology is the reduction of food waste and improved food safety. This will come at a higher cost and could limit its adoption by smaller food companies and businesses (Newton, 2023).

With the widespread use of nanostructured materials there has been a growing concern regarding the creation/establishment of biocompatible, harmless, and nontoxic nanostructures from food-grade ingredients using lucid, environmentally friendly and economical strategies, including the layer-by-layer technique (Bajpai et al. 2018).

Moreover, it was found that the flow of carbon dioxide, oxygen, and water vapour varies according to the food matrices and the packaging materials that are used. These intricacies can be managed and overcome by the application of various nanocomposite materials, including polymers (Bajpai et al. 2018).

Nanobiocomposites in food packaging have improved the capability of the food packaging to act as a barrier against gases. Biodegradable polymers strengthened with nanofillers are preferred as they are eco-friendly. There is great scrutiny regarding the consumption of these nano-compounds from the food consumed. Hence, the importance to delve into the migration patterns of these nanoparticles within the human body (Bajpai et al. 2018).

According to (Cushen et al. 2014) cited in (Onyeaka et al. 2022) the performance of migration tests can be used to determine the approval and/or acceptance of adopting nanocomposites in packaged foods. Onyeaka et al. (2022) cite Choi et al. (2018) reports that metals migrate from the

DOI: 10.1201/9781032677422-21

packaging when in contact with the food. Thus, migration processes have three stages: diffusion of the migrated nanoparticle, dissolution of the nanoparticle, and dispersion of the nanoparticle in food.

Onyeaka et al. (2022) cite Huang et al. (2011) stating that time and temperature significantly increase the migration of silver in food modelling. The initial release is probably from encapsulated nanosilver particles on the surface layers of the specimen. The nanosilver is typically released via dual-sorption mechanism, diffusion, and embedding.

Onyeaka et al. (2022) cite Farhoodi et al. (2014) and claim that an investigation on the migration of nanoclay from a polyethylene terephthalate stretch blow-moulded bottle was conducted. The findings showed a connection between the concentration of silicon and aluminium in acetic acid solutions and temperature and time. Due to the increase in time and temperature the migration levels increased.

Research on the migration of copper and silver nanoparticles from the surface of an antimicrobial nanocoated packaging material to food simulants using 1CP-MS and acidic digestion procedures was conducted by Hannon et al. (n.d.) cited Onyeaka et al. (2022). The findings showed that 0.82mg/kg and 0.46mg/kg of Cu and Ag, respectively migrated to the food simulants. Cushen et al. (2013) cited in (Onyeaka et al. 2022) used 1CP-MS to analyse the effects of temperature and time on Cu nanoparticle migration into chicken breast from the polyethylene matrix. They discovered that migration ranges were between 0.24mg/dm^2 and 0.042 mg/dm^2.

Onyeaka et al. (2022) cite Cushen et al. (2013) and note that the European Commission (EC) typically controls the use of nanoparticles in foods. They further state that nanoparticle usage in packaged foods may well not endanger public health (Article 3). However, according to Article 23 of Regulation EC 10/2011, nanoparticles must be assessed on a case-by-case basis prior to getting placed on the market. Even when the corresponding bulk substance is already allowed, their use as nanoparticles must be authorised (ECC 89/109).

21.2 APPLICATION OF NANOTECHNOLOGY IN FOOD PACKAGING

Food waste is a major challenge faced by the world. The World Economic Forum states that about one-third of all food produced in the world is lost or wasted. This results in an estimated one trillion in economic losses each year. There are potential risks with the use of nanomaterial in packaging. There is a possibility that the silver nanoparticle can be ingested if they travel onto the food. Research has demonstrated that when silver nanoparticles enter cells, they can cause inflammation and provoke what is known as oxidative stress, which can be a result of an imbalance of antioxidants and free radicals, leading to free radicals causing damage to cells (Newton, 2023).

21.2.1 Regulations and Safety Issues: Risk Assessment

This section discusses regulations established by the United States of America and Europe regarding different chemicals used in the development food packaging.

Although nanotechnology in food packaging holds great promise, the potential toxicity of the certain types of nanomaterials is a concern. There is some scepticism about the long-term effects on human health and the environment (Newton, 2023). Nanostructures in nutrition or related industries must not create any direct or indirect harm to human health (Amini et al. 2014). Thus, food is classified into different categories such as powder, emulsion, aerosol, liquid, or suspension for more effective analysis and safety management (Ramkumar et al. 2019).

Onyeaka et al. (2022) cite Wagner (2013) saying that the U.S. Food and Drug Administration (FDA) regulates the use of nanoparticles in food packaging. The United States Regulation 258/97 states that nanomaterial must be considered as "fresh food" if it is used as a primary ingredient (Agriopoulou et al. 2020). Onyeaka et al. (2022) cites Wagner (2013) that the producers must acquire pre-market clearance for the use of indirect additives in food. The FDA is required to respond to a written notification from the manufacturers within 120 days under the Food Contact Notification (FCN) System; otherwise, the product may be commercialized without additional clearance.

The Regulation No.1333/2008 of the European Commission states that "food additives" do not apply to the nano-form, and the nanosystem requires appraisal as a different additive. Therefore, it must be evaluated as a new harmless material prior to being placed on the market. The European Food Safety Authority (EFSA) must verify that a safety assessment has been conducted using the latest generation of analytical techniques on mechanically engineered food production. This is in accordance with the European Commission Regulation no. 2015/2283. The risk assessment Guidelines for the Application of Nanosciences and Nanotechnologies in the food chain was created by the EFSA. It emphasizes that further studies are needed to assess the safety of nanomaterials, to characterise them, and to measure their reactivity (Agriopoulou et al. 2020). The guideline titled "On the Risk Assessment of the Application of Nanoscience and Nanotechnologies in the Food and Feed Chain" indicates what physiochemical data must be provided by the manufacturer. It requires the provision of in vitro distribution, absorption, genotoxicity, metabolism, and excretion test results and a 90-day oral toxicity assessment with repeated doses. A chemical may be excluded from these standards if data show no migration or full dissolution or degradation The Committee, E.S (2011) cited in (Onyeaka et al. 2022).

Onyeaka et al. (2022) cite Cushen et al. (2013) and state that if an unapproved chemical is employed, a migration limit of 0.01mg/kg using a functional barrier must be met (Article 14, EC 450/2009). The EU regulation recommends that prepacked food containing engineered nanomaterials must be labelled as such (Agriopoulou et al., 2020).

Onyeaka et al. (2022) cite Wagner (2013) and say there are only three nanoparticles that have been found to be acceptable for the use of plastic-packaged foods. These are: silicon dioxide, carbon black, and titanium nitride.

Onyeaka et al. (2022) cite Products (2009) that titanium nitride may not even be significant in food, and the concentration of carbon black must not be greater than 2.5% w/w in packaged foods. The specified limit does not exist for migration for silicon dioxide (Annex I, EC 10/2011). It is worth noting that pre-market approval is not needed for products that are Generally Recognised as Safe substances (GRAS). Where a manufacturer shows a scientific risk evaluation of a product in a scientific peer-reviewed journal, the manufacturer may commercialise the chemical without first consulting the FDA.

Focus should be made by the organisation on the safety assessment of food or foodstuff prior to the implementation of nanotechnology in the food system to reduce the potential toxicity and health risks associated with nanoparticle. The criteria that an organisation is recommended to follow is as follows: standardise identification and measurement of the nanoparticle in food commodities or foodstuff; standardise administration guidelines of nanoparticles in food commodities or foodstuff; determine the consumers maximum exposure to nanoparticles in food commodities or foodstuff; and implement a standardised technique for analysing nanoparticles in food commodities or foodstuff (Biswas et al. 2022).

21.3 PRECAUTIONARY PRINCIPLES

The aforementioned narrative has put the current thinking around regulations and packaging into perspective. The chapter will now turn to a collection of articles that were sourced from various organisations in order to investigate the best precautionary measures that were undertaken to promote nanosafety. It is anticipated that this section will provide valuable insight from the laboratory, the university, environmental health and safety organisations, and researchers regarding their considerations to precautionary measures.

Due to the limited information on the precautionary principles for the use of nanomaterial in food packaging it was necessary to collect information for the use of nanomaterials from other sectors that were available. Arising from this the precautionary principles for the use of nanomaterial in food packaging will be established.

21.3.1 Precautionary Principles for the Laboratory

It is most likely that the earliest exposure to nanomaterial will take place for workers carrying out discovery research in laboratories or developing production processes and pilots plans (Department of Health and Human services 2012).

The foundation to safely handling engineered nanomaterials depends on the principles of good laboratory practices. Their key considerations were adopted from guidelines provided by Dr. Peter Lichty (Lawrence Berkeley National Laboratory) and statements received from members of the American Chemistry Society (ACS) safety community. The following laboratory safety organisation principles recognise hazards, assess risk, minimise risk, and prepare for emergencies (RAMP) were developed (ACS Chemistry, 2021).

21.3.1.1 Recognising Hazards and Assess Risks

The most important part of occupational health and safety is risk management. Flexible and adaptive risk management programs conducted in the research laboratories can control potential exposures to nanomaterials. An effective program can offer a framework to anticipate the emergence of this technology into laboratory settings. Potential hazards can also be recognized and exposure to nanomaterials can be evaluated. An effective programme can develop controls to prevent or minimise exposure, and it can confirm the effectiveness of those controls (Department of Health and Human Services, 2012). The risk assessment needs to begin with a well-defined description of the work. It needs to involve the subject matter expert (Department of Energy Nanoscale Science Research Centre, 2008). The laboratory is required to consider the dangers of precursor materials in evaluating dangers in processing such as acidity, basicity, oxidizing or reducing strength, and flammability. Moreover, it must perform a thorough hazard assessments and risk assessments for the nanomaterials. It must be noted that consideration should be given to the high reactivity of some ultrafine powdery materials, which poses potential fire and explosion hazards. The laboratory must evaluate nanomaterial migration methods, temperature controls, and other process controls (ACS Chemistry, 2021).

21.3.1.2 Engineering Controls

Taking into consideration the differing synthetic methods and experimental goals, aerosol emission controls should be cautiously developed for the process, tested and properly operated to be effective. The National Institute for Occupational Safety and Health (NIOSH) Guidance, bears in mind that engineering control systems, such as adequate insulation or scrubbing of contaminants, are the preferred controls used in handling and reducing potential exposures (ACS Chemistry, 2021).

There must be consideration of additional controls that will guarantee that the engineered nanoparticles are not released out of the work area onto clothing or other surfaces (Department of Energy Nanoscale Science Research Centre, 2008). Laboratories working with nanomaterials need to have general exhaust ventilation (GEV); however these should not be the only means of controlling nanomaterial exposure. Ventilations rates for general laboratory use range from 4 to 12 air changes per hour (Department of Health and Human Services, 2012). Ventilated enclosures are used as a primary method of engineering control in the nanotechnology industry. High efficiency particulate air (HEPA) filters should be used for exhaust air. The evaluation of laboratory equipment and exhaust systems is important for any nanomaterial contamination before removal, remodelling, or repair. Equipment that was previously used to handle or manufacture nanoparticles must be evaluated prior to disposal or reuse. This is done to evaluate for any nanomaterial contamination (ACS Chemistry, 2021).

The laboratory needs to minimise the dispersal and environmental release of nanomaterials. All manipulations of engineered nanoparticles must be conducted in a glove box, glove bag, chemical fume hood, or other airborne contaminant control system. Before the effluent is released into the general environment it is suggested that the contaminant is removed from the effluent. The laboratory needs to attempt to handle dispersible nanoparticles. Should it not be possible, then it is required that the laboratory personnel conduct and document the results of hazards analysis before using the alternative hazard controls (Department of Energy Nanoscale Science Research Centre, 2008).

21.3.1.3 Administrative Controls

Administrative controls are represented by appropriate signs and labels, access control, and a chemical hygiene plan including a standard operating procedure and hazard assessments. These are highly important aspects of GLP supporting the execution of feasible and effective control solutions. To prevent any exposure to nanomaterials, housekeeping is an important aspect. The work area and equipment are required to be cleaned regularly in order to avoid any exposure spills involving nanomaterials and must be cleaned immediately according to the protocols, or the standard operating procedures, and the use of proper personal protective equipment is required. The employee must avoid the use of flammable solvents for cleaning nanomaterials. To clean dry nanomaterials, one must use damp clothes, or one must wet the powder before wiping it. The use of compressed air or other high-energy techniques, such as brushing or shaking to remove nanomaterials from clothing, should be avoided. Vacuum cleaners need to be tested, and the use of HEPA filtered units are required. Personal protective equipment must be used during any spill cleanup and during any equipment maintenance. The hazardous chemical waste guidelines offer guidance as to how an organisation can dispose of and transport waste nanoparticles (ACS Chemistry, 2021). The Department of Health and Human Services (2012) cite Schulte et al. (2008) that maintaining clean room conditions are some of the administrative controls in some research laboratories. Prior to eating, drinking, smoking, or leaving the workplace, laboratory personnel need to use hand washing facilities (Department of Health and Human Service, 2012).

21.3.1.4 Personal Protective Equipment

When engineering or administrative controls are not realisable or successful in mitigating exposures to acceptable levels, then personal protective equipment should be required (Department of Health and Human Services, 2012). A hazard evaluation needs to be conducted in order to determine the selection and use personal protective equipment appropriate for the level of hazard (Department of Energy Nanoscale Science Research Centre, 2008). An individual working with nanomaterials must use GLP safety practices as presented in the organisation's chemical hygiene plan. Personal protective equipment such as nitrile or chemical resistant gloves, laboratory coats, safety glasses, face shields, and closed-toe shoes must be worn. Skin contact with nanoparticles or any nanoparticle containing solutions must be avoided by using appropriate personal protective equipment (ACS Chemistry, 2021). The use of respirators is required should the engineering controls not be feasible or where the engineering controls may not be considered adequate to prevent exposure (Department of Health and Human Service 2012; ACS Chemistry 2021). Respiratory protection should be worn if nanoparticles powders are handled outside of a HEPA-filtered powdered exhaust laminar flow hood. Professional consultation is required for the selection of the appropriate respirator (ACS Chemistry, 2021).

21.3.1.5 Training and Education

The Department of Health and Human Services (2012) cite OSHA (1994) that all laboratory personnel working with engineered nanomaterials are required to receive training on the associated hazards and risks. The requirements of the OSHA hazard communication standard, 29 CFR 1910.1200 is the minimum training required to address the means to detect the chemicals in the workplace, the hazards associated with those chemicals, and procedures to prevent exposure. Safety training and education should be provided for students and researchers. This should be done to assist the students and researchers to understand the nature of possible workplace exposures, health risks, roots of exposure, and instructions for reporting health symptoms, as well as proper use of engineering controls and maintenance of controls. In preparing for emergencies, the laboratory's overall emergency plan should be followed in response to any nanomaterial emergencies. An individual must start by reviewing the response to fires, spills, and releases and consider what additional steps will be appropriate based on this guidance. Significant exposures or symptoms because of possible exposure should be reviewed by a medical professional, preferably an occupational health specialist (ACS Chemistry, 2021).

21.3.2 Precautionary Measure for a University

21.3.2.1 Risk Assessments

Full and concise risk assessment where information on the hazards and risks of engineering nanomaterials must be reviewed. Data pertaining to the hazards and risks of engineering nanomaterials are on safety data sheets (SDS) of parent chemicals and the SDS of the engineered nanomaterial (The University of Queensland, 2022). According to

Kirkegaard et al. (2020), the risk associated with the nanomaterial is determined by the form of the nanomaterial, such as powder form or slurry form, as well as the duration of time to work with the nanomaterial. (Kirkegaard et al. 2020) cite Sayes (2017) and Tate and Hernandez (2017) that risk management approaches are made based on the hierarchy of controls involving elimination, substitution or modification, containment, or isolation, technical or engineering controls, organisational or administrative controls, personal protective equipment (PPE), and simple instruction.

When conducting a risk assessment consideration of nanoparticle generating processes; activities that may produce nanoparticles as a by-product; the safe handling, use, storage, transport, and waste management are required. Establishing and maintaining a register of nanomaterials is recommended. The register will include the type of research, work, and manufacture involved. Safety data sheets must accompany new or manufactured nanomaterials. Workers need to be provided with appropriate information, training, and supervision. Labelling of nanomaterials is required (The University of Queensland, 2022).

21.3.2.2 Engineering Controls

It is recommended that full enclosures with HEPA or P100 filtration need to be used. The organisation must not use horizontal laminar flow hoods that direct a flow of air into the user's face. The use of ventilated enclosures such as a glove box, fume hoods, biological safety cabinets, or local exhaust ventilations is recommended to be used (The University of Washington, 2018; Stanford University, 2009; California Nanosafety Consortium of Higher Education, 2012).

21.3.2.3 Administrative Controls

It has been advised that the use of nanomaterial in solutions, or attached to a substrate, is recommended; this will minimise the airborne release of engineered nanomaterials. Where engineered nanomaterials are being used, access must be restricted, and signs must be posted in such areas. It is expected that an individual must exercise caution when leaving an operation unattended. It is advised that the work area needs to be lined with absorbent pads. Antistatic paper and floor sticky mats need to be used when working with powder. In order to clean contaminated work surfaces, one must use wet wipes or HEPA vacuums. The storage of food and the consumption of food is not permitted in areas where engineered nanomaterials are being used. The application of cosmetics or the use of mouth pieces are not permitted. Nanomaterials must be stored in sealable and labelled containers, and abbreviations must not be used on the labels (California Nanosafety Consortium of Higher Education, 2012).

Laboratory specific training needs to be attended for safety. A designated area for the use of nanomaterials is required, and there must be an emergency shower located in close proximity. Only purchase the material needed and ensure that all containers are labelled (University of Rhode Island, 2015).

21.3.2.4 Personal Protective Equipment (PPE)

It is required that an individual wears close-toe-shoes that are made from low permeability materials. It is necessary that long-sleeved shirts, long pants, and laboratory coats are also worn. Safety glasses with side shields or chemical splash goggles must be worn. The use of chemical resistance or impermeable gloves is necessary. It needs to be noted that different laboratories may specify different and/or additional PPE (The University of Washington, 2018; Stanford University, 2009; California Nanosafety Consortium of Higher Education, 2012). A permeation or penetration table can be provided by the dermal PPE manufacturer. This table allows the user to select the necessary dermal PPE based on the performance criteria of the specific chemical threats (California Nanosafety Consortium of Higher Education, 2012). It is recommended that laboratory coats or protective clothing are worn along with gloves. Wearing two sets of gloves with extended cuff gloves is advised whenever performing tasks involving engineered nanomaterials. When working with dry particulates, use standard powder-free nitrile gloves, and for solutions containing engineered nanomaterials it is recommended to use gloves that offer protection against the solvent. Closed-to-shoes and face shields are considered as an appropriate method to prevent skin and eye contact with the engineered nanomaterials. It is recommended that airtight woven textiles are used for sleeves, gowns, or suits (The University of Washington, 2018).

21.3.2.5 Disposal

Waste streams containing engineered nanomaterials need to be managed according to the hazardous waste program of the institution. This is already being conducted by Massachusetts Institute of Technology, Virginia Tech, the University of California at Berkeley, the University of Pennsylvania, and the California Institute of Technology (California Nanosafety Consortium of Higher Education 2012; Lewis et al. 2016).

21.3.2.6 Transportation

The use of sealed, double-contained container when transporting nanomaterials inside or outside the building is required (Stanford University, 2009).

21.3.2.7 Buddy System

It is required that an employee communicates with others in the building when working alone in the laboratory; the employee must let the organisation know when he/she arrives and when he/she leaves. The employee must avoid working alone in the laboratory when performing high risk operations (Stanford University, 2009 and University of Rhode Island, 2015).

21.3.2.8 Work Practices

It is important that good work practices are incorporated in order to minimise exposure to nanomaterials. It is suggested that nanomaterials in solutions or attached to substrates are used to minimise exposure. The material safety data sheet or other prior references need to be consulted to see if it is available. All projects or applications that have the potential to produce nanomaterial aerosols must be conducted within a chemical fume hood which has been approved. It is necessary that work areas should be cleaned using either an HEPA filtered vacuum cleaner or a wet wiping method (Stanford University, 2009).

Moreover, GLP, which is found in the chemical hygiene plan, needs to be followed. The GLP will recommend the importance of assessing the hazard of the materials through a hazard analysis. It is recommended to consider the higher reactivity of some engineered nanomaterials with regard to fire and explosion hazards (The University of Washington, 2018).

21.3.2.9 Spill Clean-Up

The following procedures must be followed in the event of an accidental spill of nanomaterials: small spills of nanomaterials containing powder can be cleaned up using a wet clean-up method such as a wet wipe cloth or gauze. The use of conventional vacuum cleaners and dry sweeping is unacceptable (The University of Washington, 2018). The use of a wet cleaning-up method or vacuum cleaners equipped with HEPA filters is required, as sweeping is not permitted. The spill up material must be collected in a securely closed container. The spill clean-up debris must be managed as hazardous waste (Stanford University, 2009). A register of the use of the HEPA vacuum should be maintained in order to avoid the collection of incompatible materials (University of Rhode Island, 2015)

21.3.3 General Precautionary Principles

This section will explore general precautionary principles for the handling of nanomaterials. Different principles established to assist a variety of sectors with the application of precautionary measures will be discussed.

The management of nanomaterials within the workplace should be treated the same as handling other hazardous chemicals, including the provision of information and training of workers. This is required to ensure a safe workplace for workers working with nanomaterials. There is very limited knowledge pertaining to the risks related to nanomaterials, and the occupational exposure limits have not been developed as yet. Hence, the reason for the establishment of precautionary principles in order to keep exposure at a level that the risk can be expected to be under control. The precautionary measures will assist in keeping the exposure level low even with hazardous nanomaterials (OSHWIKI, 2019).

21.3.3.1 Risk Assessments

There is difficulty in conducting risk assessments because of insufficient information on the hazardous properties of the nano-materials and the limitations in the methods and devices that can be used for measuring exposure levels and identifying nanomaterials and emissions sources (OSHWIKI, 2019).

Risk management is divided into four categories that indicate various levels of applied precaution. The four levels are as follows: routine-based (simple), risk-based (complex), precaution-based (uncertainty), and discourse-based (ambiguity). Therefore, an organisation needs to determine the appropriate level for their risk management (Tyshenko et al. 2010). The risks are associated with the category of material and exposures to which people are subjected. Detailed information pertaining to a chemical is required for a full risk assessment. The type of information required is as follows: chemical composition of nanoparticles, material safety data sheet (if available), particle morphology, aspect ratio, particle size distribution, zeta potential, solubility, and the full process descriptions on the production source of the nanomaterial and its use (Groso et al. 2010).

The risk assessment of manufactured nanomaterials should include an inventory of nanomaterials stored in the workplace and information about the health risks of the nanomaterials provided in the SDS. It is encouraged to evaluate the exposure through inhalation, the skin, and ingestion. It is imperative that decisions are made on the necessary actions to reduce exposure and an action plan specifying what action is expected by whom and when. There needs to be consideration of risks for vulnerable workers, such as miners and pregnant or breastfeeding woman, and special action is needed to protect them. It is important to regularly revise the risk assessments and to evaluate the actions to be taken and if necessary, improvements to the action plan must be made (OSHWIKI, 2019).

21.3.3.2 Acting and Managing the Risks

In relation to the discussion at hand there is a hierarchy of measures prescribed by European Occupational Safety and Health legislation. These measures are established to prevent or reduce the exposure of workers to dangerous substances (Article 6 of the chemical agent's directive). This is known as the STOP principle (OSHWIKI, 2019; Lewis et al. 2016):

S stands for **substitution** and/or the complete elimination of dangerous substances
T stands for **technological measures**
O stands for **organisational measures**
P stands for **personal protective measures**

Another source refers to a "TOP" precautionary approach for dealing with nanoparticles is:

T stands for **Technical protective measures**
O stands for **Organisational protective measures**
P stands for **Personal protective measures** *Substitution*

Nanomaterials are often utilised due to their unique technical properties. If the use of nanomaterials cannot be eliminated there is a possibility to handle the nanomaterials in a manner which minimises exposure, for example, handling the nanomaterial in liquid form. This will reduce exposure to the nanomaterials through inhalation. The spraying of nanomaterial in liquid form should be avoided, as nanomaterials may be inhaled in the aerosol (OSHWIKI, 2019).

21.3.3.3 Technological Measures or Technical Protective Measures

It has been found that in principle airborne nanomaterials can be likened to aerosols and can be administered using similar measures and encapsulation and ventilation of the process is an effective way of reducing exposure. Risks of leakages must be managed and in connection with maintenance, repair, and cleaning must also be considered and managed (OSHWIKI, 2019).

Where nanomaterials are handled, enclosed systems are often selected due to the need to protect the process from contamination. Closed systems prevent the emission of nanomaterials to the surrounding environment and to the workers. It is recommended that an enclosed system is used for activities such as measuring manufactured nanomaterials, pouring them into or collecting them from producing or processing equipment, cleaning containers, and waste processing, and there is no potential for exposure (OSHWIKI, 2019). The use of closed equipment is also required (Lewis et al. 2016).

It is recommended that the organisation avoid the generation of dust or aerosols. Should substitution or enclosure not be applied, the suction of dusts or aerosols is recommended for engineering controls such as containment, local exhaust ventilation, and general ventilation in order to reduce exposure. There needs to be a partitioning of the working space and adaptation of the room ventilation. Each workplace should have different control measures, and this is dependent on their requirements and materials being used (OSHWIKI, 2019; Lewis et al. 2016). It should consider the emission source, the risk and the needs to reduce emissions and exposure, as well as the quantity and physical form of the nanomaterial and task duration and frequency (OSHWIKI, 2019).

In order to prevent dispersion of airborne nanomaterials the organisation is recommended to look into local ventilation and general ventilation. This should be looked at in the working area and two adjacent spaces. An appropriate filtration system is proposed to be used in order to remove nanoparticles from the exhaust, and this could be a multistage system with high efficiency particulate air filters or ultra-low penetration a filter (OSHWIKI, 2019).

In order to minimise exposure in the workplace it is proposed that an organisation look into optimizing process design and operational practices, which show that the hazardous by-products and waste generation are minimised. There are four specific safety barriers that can be used to reduce the risk of explosions posed by nanoparticles. Prevention barrier is when an organisation must have reinforcing maintenance procedures, and these should help with the prevention of fugitive emissions, accidental generation of an explosive atmosphere, buildup of static electricity, and accidental ignition sources. Mitigation barrier is the lowering process temperature and pressures to reduce process related risk factors. With the mitigation barrier, through substitution or dilution, the organisation can reduce the severity of nano-powder explosion parameters. With the protection barrier the organisation must increase the degree of protection for workers at risk (OSHWIKI, 2019).

21.3.3.4 Organisational Measures or Organisational Protective Measures

The laboratory needs to have a responsible person in charge (Groso et al. 2010). This person must document the safe procedures and working instructions established for the processing of nanomaterials and make them available in the workplace (OSHWIKI, 2019; Lewis et al. 2016). The documentation of safe procedures and working instructions will offer an appropriate working practice, and it will act as a reference point for continual improvement (OSHWIKI, 2019). The organisation can also minimise the number of workers in the workplace exposed to nanomaterials and also reduce the working hours with potential exposure to nanomaterials. There must be restriction of access to areas where exposure may occur; there must be appropriate use of safety and hazard signs (OSHWIKI, 2019; Lewis et al. 2016). Staff working on nanomaterials permanently require medical surveillance (Groso et al. 2010).

21.3.3.5 Personal Protective Measures

The most important exposure routes are mooted to be through inhalation and skin contact (Groso et al., 2010). Personal protective equipment is used as the last resort should the "STOP" principle and the "TOP" precautionary measures offer insufficient protection (OSHWIKI, 2019; Lewis et al. 2016). It is the organisations' responsibility to ensure that information about suggested personal protective equipment be given in the SDS for chemical products containing nanomaterials. It is important that the correct personal protective equipment is selected in order to ensure good protection against enemy materials (OSHWIKI, 2019).

The suitability of the PPE to protect the user from nanomaterial exposure is important (OSHWIKI, 2022). The hazard levels determine how the body parts should be protected (Groso et al. 2010). A personal protective equipment program should be designed. A good personal protective equipment program will consist of a selection of appropriate personal protective equipment, fitting, training, and maintenance of personal protective equipment (OSHWIKI, 2022).

It is required that the work rate and medical fitness of personal protective equipment be inspected to ensure that it still offers an adequate level of protection and can be used appropriately. Regular trials should be carried out on the personal protective equipment to ensure the users can carry out their work safely with the personal protective equipment and that it does not obstruct them from using other equipment and tools (OSHWIKI, 2022).

Personal protective equipment should include respiratory protection devices and dermal and eye protection (European Commission, 2014). Care should be taken to prevent explosion when nano-powders are generated. The requirements in Directive 99/92/EC should be followed. These represent the minimum requirements for improving the safety, health, and protection of workers potentially at risk from explosive atmospheres (OSHWIKI, 2022).

It is recommended that all employees' protective equipment must have a CE marking (OSHWIKI, 2022). Should the work last for more than two hours it is recommended that a mask with powered air respirator be used. For shorter work periods P3 (EN 143) or FFP3 (EN 149)/P-100 (USA NIOSH) filter/filtering masks are acceptable (Groso et al. 2010).

21.3.3.6 Access Control

In a scenario where exposure to nanomaterials occurs through an accident, the organisation is required to manage the spill. It is the responsibility of an organisation to determine the extent of area reasonably expected to have been affected, and it must be demarcated with barrier tape or the use of another reliable means of restricting entry into the area. Only trained personnel are permitted to clean up smaller spills from the laboratory using the procedure described under the section allocated for cleaning procedures and spills. All personnel exposed to nanomaterials must be referred to the medical department (Department of Energy Nanoscale Science Research Centre, 2008).

21.3.3.7 Dry Materials

Walk off mats such as attaching mats should be positioned where clean-up personnel will exit the affected controlled area. It is recommended to use a wet wipe cleaning method to clean up spills. It is necessary to manage, collect, and dispose of spill clean-up materials as nanomaterial-bearing waste (Department of Energy Nanoscale Science Research Centre, 2008). It is important to avoid using a brush, compressed air, or a standard vacuum cleaner for cleaning up. HEPA-filtered vacuum cleaners are recommended to be used to clean up nanoparticles, and it must be dedicated for this purpose only (European Commission, 2014). The functioning of the HEPA filters must be tested frequently as per the manufacturer's recommendations. The HEPA vacuums should be dedicated for nanomaterials and should be labelled accordingly. Used HEPA filters are required to be appropriately characterized, collected, and disposed of as hazardous or potentially hazardous waste based on the material involved (Department of Energy Nanoscale Science Research Centre, 2008). The HEPA vacuum must be classified as hazardous waste at the end of its life cycle (European Commission, 2014).

21.3.3.8 Product Stewardship

According to Yuvaraj et al. (2020) additional information is required in order for an organisation to provide the precautionary measures. Employers need to confirm with manufactures of chemicals and materials used that they do not contain nanomaterials. If the chemical or material used contains nanotechnology, then the employer is required to identify the nanomaterials and identify where it is used. The results from exposure assessments must be carried out at the workplace. Assessments must be carried out to see what will happen due to exposure. The employer is required to acquire engineering and administrative controls and personal protective equipment for the workers to reduce exposure. Emergency measures in the case of a spillage or release need to be identified.

21.3.3.9 Elimination and Substitution

If it is found that the nanomaterial poses risk to the healthcare workers, then it might not be feasible to eliminate or to substitute the material with another less hazardous alternative as the alternative might not demonstrate the same desired properties and positive effects. It might be possible to avoid the presence of nanomaterials that can be airborne and reduce hazardous behaviour by making changes to the surface of the nanomaterial, for example by coating it in order to adjust the dustiness, solubility, and other properties (OSHWIKI, 2022). The manufactured nanomaterials (MNM) that are classified as carcinogenic or mutagens should be the first to be considered for substitution (European Commission, 2014).

21.3.3.10 Engineering Controls

Processes with the potential of creating dust MNMs or aerosols need to be carried out in an area with efficient local exhaust ventilation (European Commission, 2014). It was found that clean benches with high efficiency particulate air filters are another valuable risk prevention measure for activities such as the preparation of nano drugs; samples of patients' tissues, body fluids, or excrements that could contain nanomaterials; or for the preparation or evaluation/examination of samples using nanomaterials containing analytical chemicals. It is recommended that a high-throughput ventilation system in combination with PPE is used. It is suggested that normal multistage filters with high efficiency particulate air filters or ultra-low penetration air filters to be used in the case of nanomaterials. This will be the final filter before reintroduction of the extracted air (OSHWIKI, 2022).

21.3.3.11 Administrative Controls

Risk prevention measures include workplaces which are specifically dedicated to the handling of nanomaterials that have been separated from other workplaces. These workplaces must be clearly indicated with appropriate signs, and this includes workplaces that have been separated from the other workplaces. Furthermore, the number of workers being exposed to new materials need to be minimised along with the duration of time the workers are exposed to the nanomaterial (OSHWIKI, 2022).

To ensure the safe handling of MNMs, working procedures need to be developed. When working with manufactured nanomaterials care should be taken to minimise inhalation. This can be achieved in two ways, as follows: lowering the environmental concentrations or by reducing exposure times. Therefore, the implementation of job rotation programmes to minimise exposure to the manufactured nanomaterials is important. Workers need to be informed about these programmes through training. Lastly, an emergency management plan needs to be established (European Commission, 2014).

Unauthorized personnel must be prohibited from accessing workspaces. These work areas need to be regularly cleaned, and there needs to be regular monitoring of air concentration levels, and it must be compared to background levels where no handling of nanomaterial occurs (OSHWIKI, 2022).

It is recommended to use existing risk and safety phrases from the EU regulation on the classification, labelling, and packaging of substances and mixtures and warning signs to provide sufficient, pertinent, and specific information on any actual or potential health and safety risks from the use and handling of nanomaterials (OSHWIKI, 2022).

Some general principles should also be followed such as the risk assessment. This should be used when planning for how the work is being conducted. Parties in risk management should be provided not only with risks that are known but also with the assessments and management of nanomaterials at workplaces in which hazard and exposure information is either missing, incomplete, or uncertain. Workers need to be given sufficient training in order for them to have skills and knowledge to carry out the work safely and to protect themselves from exposure to any nanomaterial releases. All workers must be provided with instructions and information, especially when workers are contracted for only one task, and they are not familiar with the chemical risks in general and the risks from nanomaterials specifically. It is suggested that a precautionary approach to risk prevention for nanomaterials is undertaken. All the measures available must be put into effect in accordance with the hierarchy of prevention measures. This will reduce the release of nanomaterials. It is also important that organisations have health service programs with detailed documentation of those situations (OSHWIKI, 2022).

It is important that hygiene rules need to be established for the protection of the employees. Employees must be informed not to eat, drink, or store food in a nanotechnology laboratory. Mouth suction for pipetting or siphoning is not permitted. Hands must be washed regularly to reduce nanoparticle exposure during intake or dermal contact. Gloves must be removed when exiting the laboratory to avoid infecting the doorknobs or when handling common objects such as phones and multi-user computers (Yuvaraj et al. 2020).

21.3.4 Precautionary Measures in the Agricultural Sector

In this section, the precautionary principles have been applied for the agricultural sector. The various points surrounding health and safety around agricultural products will be discussed.

There are key regulations and guidelines related to nanotechnology-based agri-products. The regulations and guidelines represent regulatory oversight, and each country has its own regulatory framework for the use of nanotechnology within the agricultural sector (Kumar et al. 2023).

Risk assessments are required by the regulatory bodies before any approval of nanotechnology-based agri-products. The toxicity levels of the nanomaterial used is evaluated, the potential for environmental release and the health impact are also evaluated (Kumar et al. 2023). Amenta et al. (2015) cite SCENIHR (2009) in stating that the a case-by-case approach for the risk assessment of nanomaterials has been recommended by the EU Scientific Committee on Emerging and Newly Identified Health Risks (SCENIHR).

Amenta et al. (2015) cite Aschberger et al. (2014) that numerous stakeholders including the likes of the European Parliament, EU member states, and NGOs have requested more transparency, traceability, and information regarding the use and possible exposure to nanomaterials.

Every product which contains nanomaterials must indicate this on the label. This labelling helps the consumer to make an informed decision about the product they purchase and use. The developments of international standards were to assure the safety and quality of nanotechnology-based agri-products. Several standards have been developed by the International Organisations for Standardization (ISO) related to nanotechnology including ISO/TS 80004–1, which offers terminology and definitions for nanomaterials. Research and development of nanotechnology-based agri-products are encouraged by regulatory bodies. This will ensure that the products are safe for human health and the environment. Collaborative efforts are required between researchers, manufacturers, regulatory bodies, and consumers in order to ensure overall safe use of nanotechnology in agri-products. There are several safety measures that organisations can take to ensure the safety of nanotechnology-based agri-products. It is the organisation's responsibility to conduct rigorous safety testing. Thorough safety testing to determine any potential risks associated with nanomaterials used in their product is required. It is the organisation's responsibility to correctly label their products to provide consumers with information about the ingredients within the product. The label is required to have ingredients, information on dosage, potential exposure routes, and any associated health risks should be provided to ensure consumer safety. The organisation has the responsibility to conduct assessments to determine the potential environmental impact of their products. It is required that the employees are trained on the proper handling and disposal of nanomaterials to reduce the risk of exposure (Kumar et al. 2023).

It is also encouraged to observe the features of the nanoparticles as they are more inclined to unintentional side effects. The features to be observed are the size, chemical composition and solubility (Amini et al. 2014).

21.3.4.1 Size

Size is an important characteristic that determines the surface area of nanoparticles. The size of the particles is an important factor in observed dermal-cell cytotoxicity in vitro (Amini et al., 2014). Zhang et al. (2022) cite (Sayes et al. 2007) and (Sukhanova et al. 2018) that in order to determine the approach of internalisation of nanoparticles into the cell, the particle size is a crucial determining factor. This influences the distribution of the nanoparticle in live tissues, absorbed nanoparticles in various absorption route could set off an immune system response. The size of the particle allows them to pass through like the central nervous system. Consideration of the size of the nanoparticles in different route of exposure is required in assessing the safety of nanomaterials that are to be in food and food-related industries (Amini et al. 2014).

21.3.4.2 Chemical Composition

Zhang et al. (2022) cite Yang et al. (2009) that the nanoparticles chemical composition and crystal structure are vital in determining their toxicity. During the production of the nanoparticles, many reagents are used that could be toxic. It has been stated that some may remain in the final product and lead to exposure to toxins that are unrelated to the nanomaterials themselves. For example, the remaining reagents and impurities may hinder our understanding of possible side effects of carbon nanotubes. It has been discovered that titanium oxide has three different levels of crystallinity, and each have different cytotoxic effects. Iron ions and impurities can accelerate the oxidative stress in cells (Amini et al. 2014).

21.3.4.3 Surface Structure

The hydrophobicity, charge, roughness, and surface chemistry are characteristics that could change the toxicological effects of absorbed nanoparticles with hydrophilic polymer-like polyethylene glycol (PEG) which decreases the toxic effects of bare particles. Different types of coating or functionalisation groups on the surface of nanoparticles are referred to as surface chemistry. Surface chemistry is one of the most important factors affecting the interaction of nanoparticles and biological systems (Amini et al. 2014). Zhang et al. (2022) cite Yang et al. (2009); Arami et al. (2015); Das et al. (2017); Yu et al. (2018a) that surface modifications of the nanoparticles change the clearance, elimination, and biodistribution. They suggested that it furthermore affects their pharmacokinetics, which is the reason for the nanoparticle's toxicity in the human body.

21.3.4.4 Solubility

Zhang et al. (2022) cite Borm et al. (2006; Zhang et al., 2011; Misra et al., 2012a) that the solubility and agglomeration are factors which affect the nanoparticle toxicity. Zhang et al. (2022) cite Misra et al. (2012b) who showed that solubility of the nanoparticles in surrounding media significantly influences their bioavailability, internalisation process, and toxicity mechanisms. Understanding the toxicity and biological activity of nanoparticles needs an understanding of these factors and many others that must be considered in applying nanotechnology in food and related industries. Therefore, all factors regarding the toxicity and environmental activity of nanoparticles should be investigated (Amini et al. 2014).

Special consideration should be paid to the gastrointestinal absorption and possible side effects of nanoparticles. Methods must be developed because of the unique properties of nanomaterials. Conventional methods like 3–(4,5-dimethylthiazol-2-yl)-2,5-diphenyltetrazolium bromide (MTT) or calcein AM(CAM) have not been useful for evaluation of toxicity in some nanostructures like single-wall carbon nanotubes (SWCNT). It has been suggested that with regard to nanomaterials, several methods for cytotoxic evaluation may be needed. Precise characterisation techniques are required for understanding nanoparticles in humans or laboratory animals. There must be regular and systematic classification of nanomaterials used. The preparation method and synthesis of nanoparticles in food products must also be classified and published. The development of proper in vitro and in vivo models for toxicity assessment of used nanoparticles in food and related industries is required (Amini et al. 2014).

It is recommended that the life cycle of the nanomaterial is understood. There is an expectation for the chemical organisations to provide the MSDS to give a forewarning of possible undesirable side effects of the chemical to human health and environment. The SDS, which extracts reference from the UN (2011) "Globally Harmonised System of Classification and Labelling of Chemicals", is a method of conveying information about a substance or mixture for use in the workplace for the purpose of establishing a chemical control regulatory framework. Some countries start with the registration of nanomaterials prior to regulating them (Karim et al. 2014).

Due to the significant resemblance between asbestos and nanotechnology, the way the global governments have handled asbestos can also be applied in regulating nanotechnology (Karim et al. 2014).

European Commissions Joint Research Committee deduced that the principle scientific obstacles for the application of legislation is the definition of nanomaterials, safety assessment (methodology for risk assessment), control of products containing nanomaterials (market regarding labelling and traceability, analytical methods, sampling, and quality of assurance tools (reference methods and market) (Karim et al. 2014).

Furthermore, the risk and safety assessment of nanomaterials are also important. The court in the USA favoured a case-by-case assessment, due to the various sizes of nanoscale material and how the size alters the behaviour (Karim et al. 2014).

21.4 CONCLUDING REMARKS OF PRECAUTIONARY PRINCIPLES ARISING FROM LITERATURE

Nanotechnology has huge benefits within the food sector, one being the reduction of food waste which will reflect as a growth in the economy. However, there is a paucity of information in the health and safety risks that come with the consumption and handling of nanomaterial. There is also limited information on the precautionary principles for its use in food packaging, which causes researchers to rely on precautionary principles established within various sectors.

Moreover, it is also recommended that due to this limited information that collaborative efforts are made between the manufacturer, supplier and consumer along with the guided help of legislation to ensure that the required health and safety risks are reduced or eliminated.

21.5 SUGGESTED PRECAUTIONARY PRINCIPLES FOR THE USE OF NANOMATERIALS IN FOOD PACKAGING

This chapter presented numerous articles sourced regarding the precautionary measures for the safe use of nanomaterial. Each article elaborated on the safety concerns regarding the use of nanoparticles within the various disciplines. The articles also emphasized that nanomaterials assist significantly in the development of products which will have the desirable effects. However, the challenge arose from safeguarding the consumer and the employee from nanoparticle exposure.

This section of the chapter will establish precautionary principles which can be applied within the food packaging sector. These precautionary principles will assist the research and development (R&D) team of various food packaging organisations to reduce nanomaterial exposure and accidental spills. Lastly, methods on how to handle accidental spills and exposure will also be discussed in the section.

21.5.1 Risk Assessment

The organisation is required to establish a research and development (R&D), multidisciplinary team, which will be responsible for establishing methods for the safe use of nanomaterials. This will assist in ensuring the promotion of employee safety. The R&D team will be responsible for conducting a thorough risk assessment to identify and understand the potential hazards associated with the use of nanomaterials.

The R&D team will be the subject matter experts on the handling of nanomaterials. They will also be responsible for keeping up to date with the latest regulations and safety methods for the use of nanomaterials; and lastly identifying the form of nanomaterial which can be used safely by the organisation.

The R&D team are responsible for the stakeholder engagement. Stakeholder engagement also includes consumers, regulatory authorities or bodies, industry partners and scientific experts to ensure that the nanomaterial is incorporated in a responsible and informed manner.

The R&D team will have to understand the regulations and standards related to the use of nanomaterials. The team will be responsible for the application of the regulations and standards within the work processes. Moreover, it is imperative for the team to stay abreast of any updates and amendments made to the regulations and standards.

The R&D team are required to establish a partnership with the suppliers of the chemicals. This communication needs to be clear, and the safety objectives need to be explained. The suppliers' expectations need to be communicated and they need to be aware that full information pertaining to the chemical will be required prior to the organisation purchasing it. If the chemical contains nanomaterials, it is the supplier's responsibility to inform the R&D team on the presence of nanoparticles within the chemical.

Documentation such as SDS need to be provided with each chemical supplied. The SDS needs to advise on the presence of nanoparticles, the handling, storage, transportation, toxicological information, and safety information. The R&D team need to identify the health and safety risks associated with the use of the chemical.

The nanomaterial needs to be characterized by size, solubility, surface structure, and chemical composition. The R&D team will be responsible for understanding these characteristics, as they are essential for gauging their potential impact on food safety and human health.

Migration studies need to be conducted by the R&D team to determine whether nanomaterials from the food packaging can migrate into the food. This is done by simulating real-life conditions to understand the potential for nanoparticle migrations over time. A comprehensive toxicity testing needs to be conducted to investigate the potential health risks associated with the ingestion of the nanomaterials which may have migrated from the packaging into the food. It is necessary to make efforts to minimise human and environmental exposure to nanomaterials in the food packaging. This can be achieved through the designing of packaging that prevents the release of nanoparticles into the food or environment.

It is the responsibility of the R&D team to identify alternative non-nanotechnology solutions which will still have the same desirable effects to the product. This will be needed should it be difficult to reduce hazardous exposure risk.

The R&D team needs to provide transparency, traceability and information regarding the use and possible exposure to nanomaterials. This can be done through the provision of clear and transparent labelling of food packaging that contains nanomaterials. The consumers have the right to know if nanomaterials are used in the packaging of the food they have purchased.

Ongoing research and monitoring efforts are required. This needs to be conducted adequately to continuously assess the safety and responsible innovative solutions in addressing the potential risks while capitalizing on the benefits of nanotechnology. This will better understand the potential risks and benefits of using nanomaterials in food packaging. It is the R&D team's responsibility to be informed about the latest scientific findings and technological developments in the field.

21.5.2 Engineering Control

The implementation of engineering controls is imperative to help mitigate potential risks associated with the use of nanoparticles in food packaging.

The R&D team is required to work closely with the team responsible for the structure of the building to ensure that the building is appropriately set up to handle the processing of nanomaterials. The design of the building can assist in reducing exposure to the nanomaterials. A designated area to work with nanomaterials needs to be created to assist with the containment of the nanomaterials.

Appropriate containment measures need to be investigated by the R&D team. They are meant to prevent the release of nanoparticles during the manufacturing, handling, and disposal process. This may include enclosed systems, protective barriers, and specialised ventilation systems.

The use of enclosed systems and equipment is encouraged as it aids in preventing the release of nanomaterials into the surrounding environment. These can also be used during production and handling of nanomaterials. This will include using close-loop systems for handling, processing, and packaging nanomaterials. Enclosed systems help minimise exposure and potential contamination of workers and the surrounding environment.

Should the funds be available within the organisation, the R&D team should look into establishing an automated handling and processing system. It is believed that this should reduce direct exposure to the nanomaterial. The processes that could be automated are blending, mixing, and packaging.

The ventilation system is crucial in an organisation working with nanomaterials. A well-designed ventilation system is needed as it plays a crucial role in the containment and capturing of airborne nanoparticles. It has been suggested that normal multistage filters with HEPA filters or ultra-low penetration air filters be used in the case of nanomaterials. Aerosol emission controls should be carefully designed for the processing and testing nanomaterials, and they need to be properly operated in order for them to be effective. The HEPA filters need to be inspected regularly according to manufacturing requirements. The HEPA vacuum needs to be labelled as equipment solely for cleaning nanomaterials. It is required that the vacuum is inspected regularly for contamination.

NIOSH Guidelines and OSHA respiratory protection standard need to be understood and incorporated in the standard operating procedure. The evaluation of laboratory equipment and exhaust systems is important. It must be examined for any nanomaterial contamination before removal, remodelling, or repair. Equipment that was previously used to handle or manufacture nanoparticles must be evaluated prior to disposal or reuse.

Material handling plan needs, for the different chemicals, need to be developed for effective containment measures. Containment measures such as sealed double walled containers or designated storage areas can prevent the release of the nanomaterials during storage, transportation, and processing.

It is recommended that the R&D team optimises the manufacturing process. This will minimise the use of nanomaterials and reduce waste generation. This can be accomplished by adopting alternative techniques and technologies that inhibit the distribution of nanoparticles.

A proper waste management plan needs must be established. The plan will describe suitable methods on how to handle and dispose of nanomaterial-containing waste. This will encompass using secure containment, labelling, and appropriate disposal methods that adhere with relevant regulations.

Is it recommended that the systems implemented are frequently monitored to appraise the concentration of nanoparticles of nanoparticles in the workplace and the surrounding environment. This helps to ensure that the engineering controls are effective, furthermore allowing for early detection of contamination or other potential issues.

21.5.3 Administrative Controls

The R&D team needs to develop and implement a clear standard operating procedure. The standard operating procedure will advise on the handling, processing, and packaging of the nanomaterials. The standard operating procedure should highlight safety protocols, proper use of equipment, and emergency response plans. The standard operating procedure should be implemented correctly.

It is important to implement access control measures to restrict entry to designated areas where nanomaterials are handled and stored. This effort will control and limit exposure to nanomaterials. It is imperative that the facility has the appropriate signage and labelling of demarcated areas. Only authorised personnel will be allowed access into restricted areas. They should be identified, and a proper authorisation procedure should be followed to ensure a controlled work environment.

The R&D team needs to establish an inventory and tracking system to monitor the procurement, use, and disposal of nanomaterials in food packaging. This will help control the quantities being used and facilitate the identification of potential exposures or issues.

An incident reporting and investigation procedure needs to be developed. All spills, accidents, or exposures involving nanomaterials in food packaging will be documented according to the procedure. All incidents need to be investigated correctly and promptly to identify the root cause, and the concomitant corrective action can be taken to prevent future occurrences. Improvements to the current work process will be required to prevent the incident from occurring.

The R&D team is responsible for establishing a hygiene plan. The hygiene plan will highlight acceptable hygiene practices. It can advise the employees on the importance of washing hands after working with nanomaterials. It is also encouraged to refrain from eating or drinking in an area that is dedicated to working with nanomaterials. Furthermore, the washing of hands can also be encouraged prior to eating, smoking, or drinking after handling nanomaterials. It is important to clean the work area thoroughly to prevent contamination and that the surface is cleaned three times to ensure a satisfactory clean area.

It is recommended that the R&D team implement a health monitoring program for employees handling nanomaterials. This program can include medical examination, occupational health assessment and the exposure monitoring for the detection of any potential health effects. This will allow for early interventions. It is recommended that this program be compulsory for every worker in the organisation.

There is a need for clear and concise housekeeping rules to prevent exposure to nanomaterials. There needs to be frequent cleaning of equipment and the work area. This process needs to follow a cleaning procedure within the standard operating procedure. Flammable solvents must not be used, and dry-cleaning methods are also not permissible. Lastly, there needs to be clearly marked chemical waste disposal area and protocol.

Compliance audits need to be conducted regularly to ensure that the employees are adhering to administrative controls established. All issues of non-compliance need to address immediately.

Lastly, it is recommended that the organisation maintain transparent communication channels regarding the use of nanomaterials on food packaging. Documentation pertaining to the risk assessments, standard operating procedures, and safety measures need to be saved on the common work system for the employees to reference at any convenient time.

21.5.4 Personal Protective Equipment

It must be reiterated that personal protective equipment (PPE) is the last resort. The selection of the appropriate gloves is required. Gloves like nitrile or rubber gloves are commonly used, and the R&D team must assess the nature of the nanomaterial to determine the quality of the glove.

It is encouraged that the laboratory personnel wear laboratory coats to prevent nanomaterial exposure to the skin and to protect their clothing from nanomaterial contamination. It is encouraged that the laboratory personnel also wear long-sleeved shirts, and gloves must cover the cuff of the shirt. In order to protect the legs and feet, long pants are required to be worn and close-toe-shoes need to be worn. The laboratory coats and coveralls must not be made from cotton fabric. The organisation should rather procure synthetic fabric that is impermeable.

Safety glasses or face shields must be used when working with nanoparticles. These should be worn to protect the eyes of the employee working with the nanomaterial from splashed or airborne nanoparticles. The chosen protective equipment provides adequate coverage and optical clarity.

Respiratory protection is needed should the ventilation not be adequate. Should the work last for more than two hours it is recommended that a mask with powered air respirator must be used. For shorter work periods P3 (EN 143), or FFP3 (EN 149)/P-100 (USA NIOSH) filter/filtering masks are acceptable. It should be noted that the PPE must not obstruct the employee from using any other equipment within the workplace. There needs to be clearly labelled showers and wash basins in case of exposure.

21.5.5 Disposal

The R&D team needs to adhere to local regulations and standards concerning the disposal of nanomaterials.

It is important to develop an appropriate waste management procedure. A waste management procedure ensures the safe disposal of nanomaterials and minimises the potential for environmental containment. The correct method of disposing of nanomaterial waste needs to be communicated with the team.

The R&D team can research the possibilities of recycling and reusing the nanomaterial, even if options such compostable and biodegradable nanomaterials must be explored. If it is not possible to develop recycling and reusing programs, the nano-waste must be treated as hazardous if it cannot be recycled. Designated containers for nanomaterial waste are to be used to avoid contamination.

It is encouraged that the personnel know that nanomaterial cannot be disposed of down the drain or as normal rubbish. Research is required to determine whether any treatment processes are needed before disposal to neutralize any potential hazards of the nanomaterials. The organisation needs to have sealable plastic bags or containers. The area for the disposal of nanomaterials needs to be demarcated and access must also be controlled. Only trained personnel can handle the disposal of nanomaterials.

Lastly, the HEPA filters, wet wipes, and clothes used to clean spills need to be discarded into the hazardous chemical waste bin.

21.5.6 Training and Education

A training and education programme needs to be established to educate all of the employees about the properties of nanomaterials and associated risks. Information on the life-cycle of the nanomaterials in the packaging from production to disposal must be provided.

Handling procedure needs to be included in the training and education programme. The employees need to be trained in the proper handling techniques to minimise release and exposure. The handling procedure must emphasise that the employee must be wearing the appropriate personal protective equipment whilst handling the nanomaterial.

All employees need to be trained on the importance of the personal protective equipment and how to use it appropriately. The employees need to know how to inspect the personal protective equipment prior to wearing it. This will ensure that it will still offer sufficient protection from nanomaterial exposure.

All employees need to be trained on the acceptable hygiene practices found in the hygiene plan. The purpose of the hygiene plan and its benefits need to be emphasized to the employees. Methods on how to effectively clean the working area, and its benefits need to be emphasized. The employees need to be shown the appropriate cleaning tools needed. Training and education on the housekeeping rules is required to help prevent exposure of nanomaterials.

The employees are required to be developed and trained on emergency procedures for spills or accidental release. The staff need to be trained on how to handle spillage of nanomaterial. It must be emphasized that spillage of nanomaterial needs to be cleaned with a wet wipe or an HEPA vacuum. A protocol for potential exposure incidents needs to be established.

The employees need to be trained on how to regularly monitor the work environment for unintended release of nanomaterials. The R&D team need to continually evaluate practices and to improve safety protocols based on the latest research and guidelines.

All the employees need to be trained on document control. The employees need to know how to maintain and file documents pertaining to the various task from receiving the nanomaterial to the processing of the nanomaterial. They need to be trained on how to maintain records of training, incidents, and waste disposal. If any employee nearly causes and accident or if the employee does cause an accident, then this needs to be documented and the risk management strategy needs to be updated.

21.6 CONCLUSION

In summary, this chapter focussed on highlighting the procedure for the establishment of the precautionary principles for the use of nanomaterials on food packaging in the absence of regulation. The chapter depicted that there is a lack of regulation for this technology and due to its limitless applications, the need for regulation is urgent to prevent risk of exposure to nanomaterials. A great number of collaborative efforts from within the supply chains will be needed in order to limit risk and hazards. Once programmes have been established, a great amount of ongoing research and monitoring will be conducted to accommodate this technology.

REFERENCES

ACS Chemistry (2021) *Nanotechnology Safety Resources*. Available via DIALOG. https://www.acs.org/about/governance/committees/chemical-safety/publications-resources/nanotechnologysafetyresources.html#:~:text=Administrative%20Controls&text=Housekeeping%20is%20an%20important%20aspect,using%20proper% 20personal% 20protective%20equipment. Accessed 07 December 2023.

Agriopoulou S, Stamatelopoulou E, Skiada V, Tsarouhas, P and Varzakas T (2020) Emerging Nanomaterial Applications for Food Packaging and Preservation: Safety Issues and Risk Assessment. *MDPI* 70(1): 1–7. https://doi.org/10.3390/foods_2020-07747.

Amenta V, Aschberger K, Arena M, Bouwmeester H, Moniz F, Brandhoff P, Gottardo S, Marvin H, Mech A, Pesudo L, Rauscher H, Schoonjans R, Vettori M, Weige S and Peters R (2015) Regulatory Aspects of Nanotechnology in the Agri/Feed/Food Sector in EU and Non-EU Countries. *Elsevier* 73(1): 463–476. https://doi.org/10.1016/j.yrtph.2015.06.016.

Amini S, Gilaki M and Karchani, M (2014) Safety of Nanotechnology in Food Industries. *PubMed Central* 6(4): 962–968. https://doi.org/10.14661/2014.962-968.

Bajpai V, Kamle M, Shukla S, Mahato D, Chandra P, Hwang S, Kumar P, Huh Y and Han Y (2018) Prospects of Using Nanotechnology for Food Preservation, Safety, and Security. *FDA* 26(4): 1201–1214. https://doi.org/10.1016/j.jfda.2018.06.011.

Biswas R, Alam M, Sarkar A, Haque I, Hasan M and Hoque M (2022) Application of nanotechnology in food: Processing, preservation, packaging and safety assessment. *PubMed Central* 8(11): 1–16. https://doi.org/10.1016/j.heliyon.2022.e11795.

California Nanosafety Consortium of Higher Education (2012) *Working Safely with Engineered Nanomaterials in Academic Research Settings*. Available via DIALOG. https://ehs.ucr.edu/laboratory/nanomaterials. Accessed 17 December 2023.

Department of Energy Nanoscale Science Research Centre (2008) *Approach to Nanomaterial ES&H*. Available via DIALOG. https://science.osti.gov/-/media/bes/pdf/doe_nsrc_approach_to_nanomaterial_esh.pdf. Accessed 17 December 2023.

Department of Health and Human Services (2012) *General Safe Practices for Working with Engineered Nanomaterials in Research Laboratories*. Available via DIALOG. https://www.cdc.gov/niosh/docs/2012–147/pdfs/2012–147.pdf. Accessed 17 December 2023.

European Commission. 2014. *Guidance on the Protection of the Health and Safety of Workers from the Potential Risks Related to Nanomaterials at Work*. Available via DIALOG. https://ec.europa.eu/social/BlobServlet?docId=13087&landId=en. Accessed 19 December 2023.

Fischer A and Ghelardi G (2016) The Precautionary Principle, Evidence-Based Medicine, and Decision Theory in Public Health Evaluation. *Frontiers* :4 1–7. https://doi.org/10.3389/fpubh.2016.00107.

Goldstein D (2001) The Precautionary Principle Also Applies to Public Health Actions. *PubMed Central* 91(9):1358–1361. https://doi.org/10.2105/ajph.91.9.1358.

Groso A, Petri-Fink A, Magrez A, Riediker M and Meyer T (2010) Management of Nanomaterials Safety in Research Environment. *ResearchGate* 7(1): 40. https://doi.org/10.1186/1743-8977-7-40.

Karim E, Munir A, Yasin H and Muhammad-Sukki F (2014) Nanotechnology within the legal and regulatory framework: An introductory overview. *Malayan Law Journal Articles* :3 1–12.

Kirkegaard M, Kines P, Jeschke K and Jensen K (2020) Risk Perceptions and Safety Cultures in the Handling of Nanomaterials in Academia and Industry. *PubMed Central* 64(5):479–489. https://doi.org/10.1093/annweh/wxaa022.

Kumar R, Suman K, Karmaker S, Mishra V, Lakra S, Saurav G and Mahto B (2023) Regulation and Safety Measures for Nanotechnology Based Agri-Products. *Frontier* :5 1–12. https://doi.org/10.3389/fgeed.2023.1200987.

Lewis A, O'Brien F, Stuart I and Moore K (2016) *Innovating Nanoparticle Safety: Storage, Handling, and Disposal Processes*. Dissertation, Worcester Polytechnic Institute.

Mohammad Z, Ahmad F, Ibrahim S, and Zaidi S (2022) Application of Nanotechnology in Different Aspects of the Food Industry. *Springer* 2(12): 1–21. https://doi.org/10.1007/s44187-022-00013-9.

Newton E (2023) *Application of Nanotechnology in Food Packaging: A Bright Future Ahead*. Available via DIALOG. https://foodsafetytech.com/column/applications-of-nanotechnology-in-food-packaging-a-bright-future-ahead/. Accessed 04 December 2023.

Onyeaka H, Passaretti P, Miri T and Al-Sharify Z (2022) The Safety of Nanomaterials in Food Production and Packaging. *Elsevier* :5 763–774. https://doi.org/10.1016/j.crfs.2022.04.005

OSHWIKI (2019) *Manufactured Nanomaterials in the Workplace: Risks and How to Manage Them*. Available via DIALOG. https://oshwiki.osha.europa.eu/en/themes/manufactured-nanomaterials-workplace-risks-and-how-manage-them. Accessed 08 December 2023.

OSHWIKI (2022) *Nanomaterials in the Healthcare Sector: Occupational Risks and Prevention*. Available via DIALOG. https://oshwiki.osha.europa.eu/en/themes/nanomaterials-healthcare-sector-occupational-risks-and-prevention. Accessed 10 December 2023

Ramkumar C, Vishwanatha A and Saini R (2019) Regulatory Aspects of Nanotechnology for food Industry. https://doi.org/10.1201/9780429425370-7.

Stanford University (2009) *Working Safely with Engineered Nanomaterials*. https://ehs.stanford.edu/wp-content/uploads/Working-Safely-with-Engineered-Nanomaterials.pdf. Accessed 17 December 2023.

The University of Queensland (2022) *Working Safely with Engineered Nanomaterials*. https://ppl.app.uq.edu.au/content/2.70.22-working-safely-engineered-nanomaterials. Accessed 17 December 2023.

The University of Washington (2018) *Guidelines for Safety during Nanomaterials Research*. https://www.ehs.washington.edu/system/files/resources/nanosafeguide.pdf. Accessed 17 December 2023.

Tyshenko M, Farhat N, Lewis R, Shilnikova N, and Krewski D (2010) Applying a Precautionary Risk Management Strategy for Regulation of Nanotechnology. *Research Gate* 7(2/3): 243–264. https://doi.org/10.1504/IJNT.2010.031313.

University of Rhode Island. 2015. *Working Safely with Engineered Nanomaterials*. https://web.uri.edu/wp-content/uploads/sites/190/URI-Working-Safely-With-Engineered-Nanomaterials.pdf. Accessed 17 December 2023.

Yuvaraj M, Yuvaraj V, Arunkumar V, Pandiyan M and Subramanian, K (2020) Nanosafety. *IntechOpen*. https://doi.org/10.5772/intechopen.92290.

Zhang N, Xiong G and Liu Z (2022) Toxicity of metal-based nanoparticles: Challenges in the nano era. *PubMed Central* :10 1–16. https://doi.org/10.3389/fbioe.2022.1001572.

22 Challenges and Limitations of Nanotechnology in Food Preservation

Otobong Donald Akan[a,b], *Chiamaka Linda Mgbechidinma*[c,d], *Egong John Egong*[e], *Chinasa Valerie Olovo*[f,g], *Udoka Onyeka Okonta*[g], *Chibueze Anslem Mgbechidimma*[h], *He Zhu*[i], *and Chuka Charles Emene*[j]

[a]Faculty of Biological Science, Akwa-Ibom State University, Akwa-Ibom State, Uyo, Nigeria

[b]College Food Science and Engineering, Central South University of Forestry and Technology, Changsha, Hunan, China

[c]School of Life Sciences, Centre for Cell and Development Biology and State Key Laboratory of Agrobiotechnology, The Chinese University of Hong Kong, Shatin, New Territories, Hong Kong, China

[d]Department of Microbiology, University of Ibadan, Ibadan, Nigeria

[e]Faculty of Biological Sciences, University of Nigeria, Nsukka, Nigeria

[f]School of Medicine, Jiangsu University, Zhenjiang, China
Faculty of Biological Sciences, University of Nigeria, Nsukka, Nigeria

[g]Faculty of Law and Business University of Northampton United Kingdom

[h]Department of Computer Science, University of Ibadan, Ibadan, Nigeria

[i]College of Beef Cattle, Shandong Agriculture and Engineering University, Jinan, China

[j]Laboratory of Intercellular Communication, Institute of Fundamental Medicine and Biology, Kazan Federal University, Kazan, Republic of Tatarstan, Russia

22.1 INTRODUCTION

Food is an essential life commodity: considering its population-support function and inputs needed for its cultivation, preserving them till postharvest use should be of utmost importance to cut economic loss and ensure that food items retain nutrients, are functional, and are safe (Bhat et al., 2012; Olovo et al., 2021). Down the history line, cultures that came after our hunter-gatherer forebearers must have inhabited climes with fluctuating environmental conditions that must have conditioned them to collect food during clement weather and abundance and store it till times of need (or scarcity) (Hammond et al., 2015). With time, humans began to apply simple, non-complex traditional methods to preserve food items. However, with technological advancements and an increased need to meet global food demand and safety, food-preserving equipment and technologies have been employed to increase food supply and ensure food security and integrity (Barrett & Lloyd, 2012; Hammond et al., 2015; Mgbechidinma, Zheng et al., 2022).

The modernization of the food and agriculture sector roughly mirrors the developments seen in medicine (drug delivery and pharmaceuticals), one of which is the application of nanotechnology (Lopez-Rubio et al., 2019). Recently, nanotechnology has been rapidly transforming and boosting food processing, packaging, storage, transportation, and functionality (Bajpai et al., 2018). By using these nanoparticles, which can successfully penetrate the cell walls of pathogenic and spoilage microbes, microbes can be specifically and precisely eliminated, extending the shelf life of food items. Also, nanotechnology utilized in active and intelligent food packaging releases antimicrobial agents and senses/detects the presence of harmful chemicals or toxins in food products (De-Kruijf et al., 2010; Z. Zhang et al., 2013). Most nanotechnology applications are non-invasive, such as in food packaging, where using polymeric and biopolymeric materials enhanced by nanoparticles results in better thermal, mechanical, and barrier capabilities of packaging materials without changing their optical properties (Sanchez-Garcia et al., 2010). Although it may sound like nanotechnology in food preservation should have wide acclaim, it is not without its challenges and limitations.

The food industry's most promising nanofood venture uses nanoparticles in innovative food packaging; in just nine years (2010 to 2019), the demand for nanopackaging increased fivefold (Ravichandran, 2010; Sahoo et al., 2021; Sarfraz et al., 2021). Notwithstanding the challenges, the promising reports for nanotechnology interventions in food systems establish it as an independent priority research area (Sharma et al., 2017). This chapter elucidates the importance of food preservation, traditional and advanced methods adopted for food preservation and the benefits, challenges, and limitations of nanofood preservation.

DOI: 10.1201/9781032677422-22

22.2 THE IMPORTANCE OF FOOD PRESERVATION

Foods are organic substances derived from plant or animal parts, consumed to nourish and energize the body. They can be classified according to perishability, functions, nutrient contents, and processing mechanisms (Akan et al., 2022; Amit et al., 2017; Bhat et al., 2012). Food classification ensures that consumers take in the right balance of nutrients in their diet, helps producers develop new food products, and improves existing food products. Table 22.1 shows food classifications. All foods will eventually deteriorate in quality (after harvest, slaughter, or manufacture); however, how this happens depends on the food type and composition, formulation, and storage conditions (Gould, 2000). Food items lose quality mainly as a result of inappropriate storage conditions, which can lead to a variety of processes, including chemical (rancidity or Maillard reaction), enzymatic (lipase or amylase), physical (bruises or cuts), and microbiological (Gould, 2000). One or all of these could lead to a food item losing its organoleptic appeal, aesthetic appeal, nutritional value, and safety.

Food preservation methods are designed to guard against food spoilage. Conditions that bring about food spoilage could range from the spontaneous breakdown of food chemicals, insect bites, or an infestation of self-perpetuating microorganisms (Hammond et al., 2015). Of these, microbial infestation poses a greater health risk due to its ability to cause sickness and, in extreme situations, even death from foodborne contamination (Das et al., 2020; Fung et al., 2018). Traditional and advanced preservation methods mainly modify conditions around the food item, inhibiting or slowing down food deterioration processes. When food preservation processes are successful, the food retains its nutrients, safety issues are curbed by controlling spoilage and inhibiting oxidation, and the desired characteristics are maintained (Amit et al., 2017; A. Kumar, 2019). Finally, postharvest preservation boosts the food supply, minimizes food waste, increases shelf life, conserves food characteristics, and reduces contamination and disease spread (Amit et al., 2017; Mgbechidinma, Akan, et al., 2022).

Safe food is essential to protect human health and well-being, the environment, and social and cultural aspects of life. Therefore, preservation methods should start with a complete analysis and understanding of the whole food chain and should be embedded in the growing, harvesting, processing, packaging, and distribution of foods (Amit et al., 2017; Rahman, 2007). A retrospective FAO report on improving safety along the food chain opined that it is more achievable if all stakeholders involved in food production and processing are doing their bit (FAO, 2003). The FAO and similar agencies work with stakeholders to promote the application of good hygienic practices (GHP) and ensure compliance with food safety requirements. Table 22.2 shows suggested food safety tips that could be adopted along the food chain. Due to various factors, including cultural and climatic variations, humans have evolved techniques that increase the shelf life of different food products. Additionally, due to industrialization, innovative and sustainable solutions applicable to food preservation have been developed (Hammond et al., 2015; A. Kumar, 2019). The following section explains different traditional and advanced food preservation methods.

TABLE 22.1
Food Classifications

	Classification	Description	References
1	**Perishability**	There are three categories: ***Perishable*** (foods like milk, meat, and seafood that have within seven days shelf life without proper preservation); ***semi-perishable*** (food items like vegetables, fruits, and potatoes whose shelf life extends to four to six months with minimal preservation techniques); and ***non-perishable*** (natural and processed foods whose shelf life extends past 12 months like dry beans, nuts, four, sugar, canned fruits) foods. These food types have low water contents and high concentrations of osmotically active compounds.	(Amit et al., 2017; Hammond et al., 2015)
2	**Functions**	The function any food item renders to our bodies after consumption is an essential service. Food can **build** and **heal** the body, provide **energy**, **regulate** the body, and provide **protection**. These categories are self-explanatory.	(Akan et al., 2022; Amit et al., 2017)
3	**Nutrient contents**	Although most food items contain more than one nutrient component, they are mostly grouped according to their most-containing nutrient. There are four types of foods based on their contents: (**a**) foods high in carbohydrates, (**b**) foods high in protein, (**c**) foods high in fat, and (**d**) foods high in vitamins and minerals.	(Fern et al., 2015)
4	**Processing mechanisms**	Food can also be categorized based on the techniques used and the extent to which they are processed. (a) *Foods that are* ***unprocessed or minimally processed***; (b) ***food components or ingredients used in processing culinary or industrial food products***; and (c) ***ultra-processed*** food products.	(Amit et al., 2017; Xing et al., 2023)

TABLE 22.2
Preservation Methods throughout the Food Chain

Food Chain	Preservation Tips
Growing	Application of mulch to prevent soil erosion and control the growth of weeds.
	Integrated pest management reduces or eliminates crop damage by pests.
	Crop rotation preserves soil health and reduces the incidence of pests and diseases.
	Application of organic and natural fertilizers to improve soil health and curb harmful chemical buildup in the soil.
Harvesting	Harvesting crops at their peak maturity (optimal time helps crops retain its quality and nutrition).
	Wash and dry crops as soon as they are harvested to remove dirt or debris that could cause bruises and spoilage.
	Store crops in a cool, dry place to prevent mold and bacteria from growing on the food.
Processing	Heat-treat food items, either by cooking, pasteurizing, or sterilizing to kill any bacteria that may be present.
	Add preservatives to the food, such as salt, sugar, or acid, to help prevent the growth of bacteria.
	Pack food in airtight containers to prevent oxygen and moisture from entering the food.
Packaging	Vacuum packaging removes pockets of air from the package and seals it tightly; preventing the growth of microorganisms and the food from losing its flavor or color.
	Modified atmosphere packaging (MAP) involves replacing the air inside the package with a gas mixture less conducive to bacterial growth.
	Use of oxygen- and moisture-resistant packing materials like glass or plastic.
Distribution	Refrigerate or freeze food to slow down the growth and activities of microorganisms.
	Utilize controlled atmosphere transport (CAT) to control the transportation container's nitrogen, carbon dioxide, and oxygen concentrations.
	Utilize methods that reduce vibrations and temperature fluctuations that could damage the food item.

Sources: (FAO, n.d., 2003, 2021; Kiaya, 2014)

TABLE 22.3
Principles of Traditional and Modern Preservation Methods

Traditional Methods	Preservation Principles
Blanching	Food, especially fruits and vegetables, is briefly heated to a high temperature to destroy bacteria. After that, the food is dipped in cold water to halt the cooking process, and it is then refrigerated or canned. The color, texture, and flavor of food items are preserved in this way.
Boiling	Food is heated to a certain temperature and then submerged in boiling water for a predetermined time before being allowed to cool.
Canning	Food items (meat, fruits, and vegetables) are heated and sealed in airtight containers. The heat kills any microbes present in the food; the airtight container prevents the entry of new microbes.
Curing	The addition of substances (like salt or sugar) to food (especially meat) reduces water activity (water available for microbial metabolism -a_w) and makes the food less habitable for microorganisms.
Drying and dehydration	Removal of moisture present in food by exposing it to air and heat (sun drying, oven drying, or freeze drying) inhibits bacterial growth.
Fermentation	Fermentation employs the action of microorganisms to change/alter the chemical composition, taste, and texture of the food. It also improves its nutritional value and shelf life.
Freezing	Foods stored at temperatures below 0 °C (32 °F) halt biological and chemical processes that cause spoilage in food. There will be a need for thawing or defrosting before the food item can be used.
Refrigeration	Refrigeration preserves food at low temperatures (between 0 to 4 °C) to retard chemical and biological processes that induce food spoilage.
Pickling	Foods (such as meat, fruits, and vegetables) are submerged in an acidic or brine solution. The solution destroys microorganisms and gives the food products a sour or tangy flavor. Also, lactic acid bacteria (LABs) are incorporated in food containers to produce lactic acid (lacto-fermentation), which kills spoilage microorganisms and preserves the food.
Smoking	Foods exposed to smoke/heat/combustion reduce water content, avert/inhibit microbial growth, and add flavor to the food.

(Continued)

TABLE 22.3 (*Continued*)
Principles of Traditional and Modern Preservation Methods

Advanced Methods	Preservation Principles
Antimicrobial-packaging technology	Food packaging materials are coated with antimicrobial agents such as essential oils or silver nanoparticles, which prevent the growth of microorganisms on food surfaces. The antimicrobial agents in the packaging materials are slowly released over time as the food ages to kill any microbe in contact with the food surface. The antimicrobial agents' concentration must be kept low enough to protect customers and the food's flavor, texture, and appearance.
Biopreservation	Microorganisms or their products are used to prevent food spoilage. A clear example is the use of lactic acid bacteria (LABs) to preserve cheese and yogurt. LABs convert sugars in food into lactic acid, which lowers the food pH and prevents microbial action on the food. The addition of yeasts species averts spoilage in beer, fungi prevent the growth of molds in bread, and nisin (a peptide that is naturally produced by *Lactococcus lactis*) prevents the growth of *Listeria monocytogenes*—microorganism notorious for causing listeriosis through food contamination.
Chemical food preservatives	Chemical additives (salt, sugar, acids, and preservatives) prevent spoilage. For example, meat can be preserved by salting, fruits can be preserved with sugar, and vegetables and pickles can be maintained with acids (vinegar). Additionally, microbiological growth can be inhibited with the use of preservatives such as sodium benzoate and sorbic acid.
Film packaging	Foods are sealed in airtight containers by sucking out air, which creates a vacuum inside the container and prevents bacteria from growing. Also, a special film that allows the passage of O_2 and CO_2 while preventing other gases can be used. Food items in airtight containers, like fruits, fish, vegetables, meat, etc., have extended shelf life.
Freeze drying (lyophilization)	Food items are frozen at a very low temperature (−50 °C) to avoid ice crystals forming and damaging the food's structural integrity. After that, the meal is placed in a vacuum chamber where water—which is ice crystals—is removed through sublimation (depriving bacteria water required for growth). The lyophilized (dried) food is light in weight and can be preserved for years or decades without spoilage.
Hurdle technology	To extend a food's shelf life, various food preservation techniques—such as drying, salting, and refrigeration—are combined. Microbial agents able to beat one method are inhibited by other methods.
Irradiation	Foods are exposed to ionizing radiation like γ-ray and α-ray to kill microbes and retard the ripening process in fruits/vegetables.
Modified atmosphere	The atmosphere in the food packaging is modified by either flushing the package with N_2 or sealing it with a modified gas mixture (O_2 and CO_2). This modified atmosphere packaging (MAP) environment inhibits the growth of bacteria, preserves food taste and nutritional value, extends its shelf life, and ensures high food quality.
Nonthermal plasma	Helium or argon gas discharges are used to preserve food. The gases pass through an electric discharge and form an excited state of the gas molecules (known as 'plasma'). These preserve food by rupturing the cell walls of microbes or by damaging the cell proteins. Nonthermal plasma technique functions in either an anoxic environment or a vacuum. It does not use heat and maintains the nutritional integrity, texture, and flavor of the food.
Pascalization (high-pressure processing [HPP])	Foods (that are sensitive to heat: juices, meat, fruits, etc.) are preserved by exposing them to very high pressures (87,000 to 100,000 psi). It is either called high-pressure processing or "cold pasteurization" because it does not involve the use of heat. High pressure destroy microorganisms in the food by rupturing the microbial cell wall. HPP maintains the food's nutritional quality and flavor.
Pasteurization	Canned foods are heated at a specific temperature for a particular time (145 °F for 30 seconds, 161 °F for 15 seconds, or 280 °F for 2 to 5 seconds) to kill spoilage microbes without affecting the taste and nutritional value of food.
Vacuum packing	Air is removed from food packaging, creating a vacuum, which inhibits the growth of microbes, especially molds.
Microwave	The microwave uses electromagnetic radiation to generate heat from water molecules in food. This heat is fast, uniform, and effective for pasteurized food. The microwave method can be applied to any type of food.
Ohmic	Electrical current is used to rapidly and evenly heat food by placing the food in an electric field. It uses direct electrical current to heat food and control its temperature. The electric current causes water molecules to vibrate, creating heat that destroys spoilage microbes. Ohmic heating is used explicitly for liquid foods like juice, sauce, etc.
Nanotechnology	Materials with nanoscale dimensions (1 to 100 nm) are used to preserve food and enhance food quality. Nanomaterials are used to create packaging materials with antimicrobial properties capable of preventing microbial growth on food surfaces. This extends the shelf life of food. Nanocapsules also encapsulate and deliver antioxidants that preserve food and protect food from spoilage organisms.
Ultrasonic processing technique	Ultrasonic processing applies high-frequency sound waves above the range of human hearing to disrupt microbial activities. The high sound waves cause the microbial cells to vibrate and rupture. Microbes are killed without affecting the quality and integrity of the food.
Ultraviolet radiation technique	UV radiation is used to preserve food. The UV rays damage the DNA of microbes, thereby killing them. Food items are only exposed to UV radiation for a short period ($\leq$ 1 min). This process affects neither the taste nor the texture of the food.

(*Continued*)

TABLE 22.3 (*Continued*)
Principles of Traditional and Modern Preservation Methods

Advanced Methods	Preservation Principles
Pulse electric field (PEF) technique	To destroy bacteria, high-voltage electric pulses are through food materials placed between two electrodes for a few seconds. PEF uses short, intense pulses of electricity to kill microorganisms in food by bursting their cell wall without affecting the nutritional components of the food.
Ozone treatment	The ozone molecule (O_3) is used as an oxidizing agent to break the cell walls of microbes and kill them. Ozone treatment is effective against viruses, fungi, and bacteria, yet this technique is safe and eco-friendly. It is used to clean/sanitize surfaces and wash fruits and vegetables.

Sources: (Amit et al., 2017; Awuchi, 2023; Bhat et al., 2012; Erkmen & Bozoglu, 2016; Gould, 2000; A. Kumar, 2019; Rahman, 2007; Sridhar et al., 2021)

22.3 TRADITIONAL AND ADVANCED FOOD PRESERVATION METHODS

Cultural changes and travels might have informed the initial non-complex traditional methods to preserve food items. Table 22.3 enlists some principles behind some traditional preservation methods. These methods are typically simple and low-tech agricultural technology based, used to modify the food and its environment—arresting food spoilage due to enzymatic, chemical, physical, and microbiological factors (Hammond et al., 2015; Mgbechidinma et al., 2020). Merely washing food surfaces with clean water can reduce or remove adhering microbes, and techniques such as pasteurization and irradiation could kill off microbes from food substances. Using well-fitted, airtight, and impermeable containers can prevent microbial colonization and infestation (Hammond et al., 2015). Additionally, altering the environment surrounding a food item can effectively prevent or delay spoiling. Placing a perishable food item in an altered temperature environment can reduce or inhibit the growth of microbes. For instance, the microbial growth Q10 value in perishable food ranges from 2.3 to 4.1. Therefore, decreasing the food sample's temperature to 4 °C will delay microbial growth (doubling its starting number only after two days). This growth rate is slow compared to 2.8×10^{14} cfu/g per hour at 34 °C (Hammond et al., 2015).

Another essential food spoilage factor that can be mitigated using traditional preservation methods is inherent water activity (a_w) in foods (Hammond et al., 2015). In food, water is present as either free or bound—it is the free water that is important for the growth and activities of vegetative microbial cells, germination of microbial spores, production of microbial toxins, and activity of certain enzymes (Erkmen & Bozoglu, 2016; Hammond et al., 2015). Processes such as heating, smoking, freeze drying, freeze concentration, and osmotic concentration all reduce the water activity (a_w) of foods—causing osmotic shock to vegetative microbial cells, increasing their lag phase, decreasing their growth rate, and inhibiting enzymatic activities (Erkmen & Bozoglu, 2016). Food items become shelf-stable when their moisture content is reduced to 20–50% or their water activity (a_w) is between 0.60 and 0.85 (Erkmen & Bozoglu, 2016). However, halotolerant and xerotolerant fungi like the *Aspergillus sydowii* and *Aspergillus versicolor* concentrate polyols such as glycerol, sorbitol, mannitol, xylitol, erythritol, and arabitol used for preservation in their cells against low-a_w environments (Erkmen & Bozoglu, 2016).

Food products can now be adequately preserved—retaining their flavor, nutritional content, and overall health functions—through using bioagents. For example, the common herb thyme (*Thymus vulgaris*) contains thymol, a monoterpene that inhibits microbial growth (Hammond et al., 2015). Adding salt and sugar creates unconducive conditions for the growth of spoilage microorganisms in foods (Amit et al., 2017). Certain bacteria like *Lactobacillus* and *Bifidobacterium* ferment raw food items, improving their nutritional value and taste, creating unconducive conditions for spoilage microbes, and extending the food's shelf life (Kwofie et al., 2020; Xing et al., 2023). Consuming fermented food that contains beneficial microbial populations improves the gut health of consumers. Furthermore, microbial enzymes like pectinases and amylases can be used to break down food components, making the food components unavailable to spoilage microbes, hence inhibiting microbial proliferation. Also, these enzymes make carbohydrates in fruits and vegetables easily digestible and improve the overall nutritional value of the food products (Haile & Ayele, 2022; Raveendran et al., 2018).

Conversely, advanced preservation methods retard food spoilage using complex and specialized instrumentations. Table 22.3 enlists the principles behind some advanced preservation methods. These methods have helped prevent significant financial losses along the food chain and decreased the time and energy needed for food processing (Barrett & Lloyd, 2012; Hammond et al., 2015). Nanotechnology is the method of focus, emerging and highly specific, that uses nanomaterials (of size between the order 1 nM = 10^{-9} m or less) to preserve and process food items. Additionally, nanotechnology employed in food preservation is versatile, has a slow release rate, is less toxic, has high solubility, has no pollutant release, is bioavailable, is energy efficient, is thermodynamic, has a high surface area, takes up less space, magnetizable, and uses a variety of nanostructured materials (NSMs) to achieve these (Avella et al., 2005; Bajpai et al., 2018; Sridhar et al., 2021).

22.4 NANOFOOD PROCESSES AND PRESERVATION

In contrast to other sectors, the acceptance and usage of nanotechnology in food systems is slower despite its rapid advancement in medicine, biochemistry, chemical manufacturing, life sciences, new energy, the environment, and transportation (Biswas et al., 2022; Santeramo et al., 2018). However, it provides a good platform for combining state-of-the-art technologies with various applications and advantages in food systems (Sahoo et al., 2021). Depending on the application, nanotechnology applications in food could either be "nano-inside" or "nano-outside" (Santeramo et al., 2018). These distinguished nanoparticles incorporated directly into foods and those only used in food-contact materials. A wide range of nanostructured materials (NSMs) are applied as nanofood technology materials (examples are inorganic metals, metal oxides, and their nanocomposites to nano-organic materials with bioactive agents) (Bajpai et al., 2018; Sharma et al., 2017). Generally

TABLE 22.4
Some Applications of Nanostructures in the Food Systems

	Nanostructures	Description	Applications	Reference
1	**Nanocomposite**	Nanocomposites are made of two or more different materials, one on the nanoscale.	Used in food packaging with oxygen barrier capabilities. Nanocomposite costings also reduce moisture loss from food- extending food shelf life.	(Al-Naamani et al., 2018; Arora & Padua, 2010; Sarfraz et al., 2021)
2	**Nanotubes**	Nanotubes are thin, tube-like (cylindrical-shaped) structures typically made of carbon or other materials.	Carbon nanotube films have enhanced oxygen barrier properties. Bamboo-like zinc oxide nanotubes have antimicrobial properties. Iron oxide nanotube packaging materials absorb ethylene gas released by fruits and vegetables, thus extending the shelf life of food items.	(L. Kumar et al., 2024; Li et al., 2021)
3	**Nanofibers**	Nanofibers are long, thin fibers that are typically made of polymers.	Nanofibers are used for aseptic separation (like water and milk and removing bacteria from fruit juices). Nanofiber coatings with antimicrobial properties are used for food packaging to inhibit the growth of spoilage and pathogenic microbes. Nanofiber scaffolds are used to create edible films that also protect food items from moisture loss and oxidation. Nanofibers can also encapsulate food ingredients to improve their bioavailability and stability.	(Forghani et al., 2021; M. Zhang et al., 2023)
4	**Nanostructured proteins**	Nanostructured proteins are a type of protein that has been structurally modified at the nanoscale.	Nanostructured soy protein isolate films provide a high-water vapor barrier, antibacterial, and better mechanical qualities for food packaging. Nanostructured proteins have also been used to encapsulate food additives and improve stability. Nanostructured proteins can be coupled with other nanomaterials and used for food packaging or as edible coatings for food.	(Bourbon et al., 2019; Mohammadian et al., 2020)
5	**Nanoemulsions**	Nanoemulsions are a type of emulsion, or a mixture of two immiscible liquids, where at least one of the liquids has droplet sizes in the nanometer range.	Nanoemulsion improves the stability of food products by presenting the separation of the water and oil phases. Offers stability to encapsulated bioactive compounds (antioxidants, vitamins, and essential oils) for targeted delivery.	(Borthakur et al., 2016; S. Kumar et al., 2022)
6	**Liposomes**	Liposomes are nanostructures made of a phospholipid bilayer that forms a hollow sphere using natural lipids, polymers, or proteins.	Liposomes increase the bioavailability of vitamins, minerals, etc. Liposomes encapsulate and protect bioactive food ingredients from oxidation and microbial spoilage. It is also used to improve bioavailability and targeted delivery of food ingredients.	(Akram et al., 2023; Huang et al., 2022)

speaking, nanomaterials can be classified as one of four types: zero-dimensional (such as nanoparticles: NPs, quantum dots, nanoclusters, and fullerenes), one-dimensional (such as nanotubes and nanorods), two-dimensional (such as thin films), and three-dimensional (such as nanocomposites and nanofibers) (Sharma et al., 2017). Their properties depend on their nature and size and are primarily generated through ultrasonication, homogenization, microfluidization, and milling processes (Sahoo et al., 2021).

Table 22.4 lists some nanomaterials and their attributes. Nanomaterials have been described as smart, edible, self-cleaning, having better mechanical and barrier qualities, and having antimicrobial attributes. They have been utilized for food packaging, processing, and preservation (Sahoo et al., 2021). Nanomaterials have been fused with antimicrobial compounds and used as packaging materials to control the intrusion of food spoilage microorganisms (De-Kruijf et al., 2010). In okra *Abelmoschus esculentus* samples, the concentration of bacteria and fungi decreased by 63% and 50%, respectively, using a nanocomposite zinc-chitosan antimicrobial polyethylene packing film. The pH, total soluble solids, moisture content, and weight loss characteristics of the okra samples were not impacted by the nanopackaging, according to the results (Al-Naamani et al., 2018). A masterbatch of laponite compounded with ethylene vinyl acetate (EVA) was extruded into thin, low-density polyethylene (LDPE) films; in the end, laponite was not released into food samples (Bott & Franz, 2018). A retrospective review lists some everyday food items (nanostructured mayonnaise, frozen desserts, ice creams, and dairy spreads) that were structurally modified using certain nanomaterials and offered a healthier option than conventional ones (Singh, 2016). A tea incorporated with green synthesized selenium nanoparticles (SeNPs) (as an additive) showed higher antioxidant activity and redox balance (W. Zhang et al., 2018). In a study, cholera toxin molecules specifically bound to the active membrane protein encased in nanodiscs—act as good invasive monitoring for pathogens and their toxins (Tark et al., 2010).

22.5 LIMITATIONS AND CHALLENGES OF NANOFOOD PRESERVATION

Consumers still exhibit neophobia (the fear of new foods) toward foods with "nanotechnology labels." Consumers are skeptical about nanofoods because these are relatively new on the food shelves, adding to the lack of scientific data on possible environmental and public health risks (Santeramo et al., 2018; The Royal Society, 2004). The United States Food and Drug Administration (FDA) does not have specific requirements concerning nanoparticles in foods (FDA, 2009). Thus, additional demonstrations are required to allay consumers' concerns, and if approved, they should be added to the GRAS list. Foods marked with nanotechnology may soon be accepted, just as it took a while for irradiation technology and genetically modified foods to be received.

For nanofood to be accepted, there is a need to conduct series of *in vivo*, *in vitro*, and physicochemical tests and assessments. However, due to ethical, time-consuming, and high-cost issues around *in vivo* tests, well-designed and cost-effective *in vitro* tests can be done to accurately and predictably assess potential risk(s) associated with nanotechnology in food systems.

Another concern is the potential for nanoparticles used in food packaging to migrate or leach into food, breach intestinal walls and cellular barriers, and then move to cells and organs- posing health problems (Echegoyen & Nerín, 2013). Research utilizing rat models demonstrates that inhaled iron oxide nanoparticles were detected in the brain and lung tissues, where they caused inflammatory and stress reactions (Srinivas et al., 2012). Their increased contact surface area and nanomaterials pose a new form of risk in the body that is not apparent with bulkier materials.

Food experts also fear the difficulty of detecting or identifying nanoparticles in food matrices or environmental containers. This bothers their migration and toxicity potentials (Gupta et al., 2024). However, recent techniques can identify nanoparticles that migrated from packaging materials into foods. On the other hand, when it comes to the migration of antimicrobial agents from food packages made with nanomaterials, it could be a good thing to prevent microbial infestation and growth in foods.

With nascent food companies involved with nanotechnology disposing of their waste indiscriminately, soon, nanomaterials (like lead, aluminum, titanium oxide, and cadmium) will accumulate in the soil and water (and their fauna and flora) (Echegoyen & Nerín, 2013; Morones et al., 2005). Once in the environment, these can contaminate food sources and pose health risks to consumers in the food web. However, researchers and scientists are working toward incorporating organic/biodegradable nanoparticles into sustainable food packaging materials to enhance film properties, extend the shelf life of food products, and offer more food safety (Barage et al., 2022).

The techniques and instrumentations involved in the generation of nanoparticles are expensive and need experts. Furthermore, synthesizing these substances with diverse chemicals may produce unfavorable and environmentally unfriendly by-products, a potential environmental risk (Sahoo et al., 2021). Therefore, it is only pertinent to evaluate the possible biosafety and health issues with nanotechnology applications in food systems (Bajpai et al., 2018). Being a novel material in the food system, they should be assessed on a case-by-case basis before specific regulations are promulgated.

Lastly, there are no particular regulations governing the use of nanotechnology in food packaging. This is due to nanotechnology being in its developing stage (EFSA Scientific Committee et al., 2021). The legal framework to cover nanotechnology applications is currently playing catch-up with its use.

22.6 CONCLUSION AND FUTURE PERSPECTIVES

Techniques for preserving food have evolved, and at the moment, nanotechnology is becoming increasingly popular and is used in food systems for various reasons. Successful preservation methods should curb food spoilage/contamination and help retain the nutritional characteristics of preserved food; conversely, nanotechnology processes improve the nutritional and health attributes of foods. However, the acceptance and usage of nanotechnology in food systems are slow due to gaps in knowledge surrounding potential environmental toxicity and health risks. To move past this "slow stage," there is a need for intensified studies on the properties of nanoparticles, accurate quantitative methods for measuring nanoparticles in complex systems, duration of their lifecycle, distribution patterns, and degradation rate. This will consolidate the relationship between nanotechnology and food safety because consumers are accustomed to well-preserved, safe, and toxic-free foods.

Artificial intelligence and machine learning algorithms will prove to be effective in improving food preservation when coupled with nanotechnology in the near future. In addition to analyzing large datasets, identifying patterns, and making predictions, these algorithms can further be optimized to allow for the discovery and development of novel nanomaterials. A new avenue for forecasting nanoparticle interaction processes with other materials is being opened by the marriage of artificial intelligence and food preservative technology. This advancement will ensure proper food safety implementation, increased production efficiency, reduced food waste generation, and better customer acceptability.

Also, governments should legislate more unambiguous laws and frameworks on using nanomaterials for food packaging, monitor food producers for strict compliance with laid-down regulations, and sanction defaulters. Clear and comprehensive information should be made available to consumers regarding the benefits or possible risks (or both) involved in the use/consumption of nano products. If these and more are implemented, consumers will maximize the benefits of applying nanotechnology in food preservation.

REFERENCES

Akan, O. D., Qin, D., Guo, T., Lin, Q., & Luo, F. (2022). Sirtfoods: New Concept Foods, Functions, and Mechanisms. *Foods*, *11*(19), 1–20. https://doi.org/10.3390/foods11192955

Akram, N., Afzaal, M., Saeed, F., Shah, Y. A., Faisal, Z., Asghar, A., Ateeq, H., Nayik, G. A., Wani, S. H., Hussain, M., Asif Shah, M., & Khaneghah, A. M. (2023). Liposomes: A Promising Delivery System for Active Ingredients in Food and Nutrition. *International Journal of Food Properties*, *26*(1), 2476–2492. https://doi.org/10.1080/10942912.2023.2247578

Al-Naamani, L., Dutta, J., & Dobretsov, S. (2018). Nanocomposite Zinc Oxide-Chitosan Coatings on Polyethylene Films for Extending Storage Life of Okra (Abelmoschus esculentus). *Nanomaterials*, *8*(479), 1–14. https://doi.org/10.3390/nano8070479

Amit, S. K., Uddin, M. M., Rahman, R., Islam, S. M. R., & Khan, M. S. (2017). A Review on Mechanisms and Commercial Aspects of Food Preservation and Processing. *Agriculture & Food Security*, *6*(51), 1–22. https://doi.org/10.1186/s40066-017-0130-8

Arora, A., & Padua, G. W. (2010). Review: Nanocomposites in Food Packaging. *Journal of Food Science*, *75*(1), 43–49. https://doi.org/10.1111/j.1750-3841.2009.01456.x

Avella, M., Vlieger, J. J. De, Emanuela, M., Fischer, S., Vacca, P., & Grazia, M. (2005). Food Chemistry Biodegradable Starch/Clay Nanocomposite Films for Food Packaging Applications. *Food Chemistry*, *93*, 467–474. https://doi.org/10.1016/j.foodchem.2004.10.024

Awuchi, C. G. (2023). HACCP, Quality, and Food Safety Management in Food and Agricultural Systems. *Cogent Food & Agriculture*, *9*(1), 1–30. https://doi.org/10.1080/23311932.2023.2176280

Bajpai, V. K., Kamle, M., Shukla, S., Mahato, D. K., Chandra, P., Hwang, S. K., Kumar, P., Huh, Y. S., & Han, Y.-K. (2018). ScienceDirect Prospects of Using Nanotechnology for Food Preservation, Safety, and Security. *Journal of Food and Drug Analysis*, *26*(4), 1201–1214. https://doi.org/10.1016/j.jfda.2018.06.011

Barage, S., Lakkakula, J., Sharma, A., Roy, A., Alghamdi, S., Almehmadi, M., Hossain, M. J., Allahyani, M., & Abdulaziz, O. (2022). Nanomaterial in Food Packaging: A Comprehensive Review. *Journal of Nanomaterials*, *2022*, 1–12. https://doi.org/10.1155/2022/6053922

Barrett, D. M., & Lloyd, B. (2012). Advanced Preservation Methods and Nutrient Retention in Fruits and Vegetables. *Journal of Food Science and Agriculture*, *92*, 7–22. https://doi.org/10.1002/jsfa.4718

Bhat, R., Alias, A. K., & Paliyath, G. (2012). Preservation of Plant and Animal Foods: An Overview. In *Progress in Food Preservation* (First, pp. 603–611). John Wiley and Sons, Ltd. https://doi.org/10.1002/9781119962045.ch28

Biswas, R., Alam, M., Sarkar, A., Haque, I., & Hasan, M. (2022). Application of Nanotechnology in Food: Processing, Preservation, Packaging and Safety Assessment. *Heliyon*, *8*(e11795), 1–16. https://doi.org/10.1016/j.heliyon.2022.e11795

Borthakur, P., Boruah, P. K., Sharma, B., & Das, M. R. (2016). Nanoemulsion: Preparation and its Application in Food Industry. In *Nanotechnology in the Agri-Food Industry*. Elsevier Inc. https://doi.org/10.1016/B978-0-12-804306-6/00005-2

Bott, J., & Franz, R. (2018). Investigation into the Potential Migration of Nanoparticles from Laponite-Polymer Nanocomposites. *Nanomaterials*, *8*(723), 1–13. https://doi.org/10.3390/nano8090723

Bourbon, A. I., Pereira, R. N., Pastrana, L. M., Vicente, A. A., & Cerqueira, M. A. (2019). Protein-based Nanostructures for Food Applications. *Gels*, *5*(1), 1–17. https://doi.org/10.3390/gels5010009

Das, S., Chaudhari, A. K., Singh, A., Deepika, Singh, V. K., Dwivedy, A. K., & Dubey, N. K. (2020). Foodborne Microbial Toxins and Their Inhibition by Plant-Based Chemicals. In *Functional and Preservative Properties of Phytochemicals*. https://doi.org/10.1016/B978-0-12-818593-3.00006-3

De-Kruijf, N., Van-Beest, M., Rijk, R., Sipilainen-Malm, T., Losada, P. P., & De-Meulenaer, B. (2010). Active and Intelligent Packaging: Applications and Regulatory Aspects. *Food Additive and Contaminant*, *19*, 144–162. https://doi.org/10.1080/02652030110072722

Echegoyen, Y., & Nerín, C. (2013). Nanoparticle Release from Nano-Silver Antimicrobial Food Containers. *Food and Chemical Toxicology*, *62*, 16–22. https://doi.org/10.1016/j.fct.2013.08.014

EFSA Scientific Committee, More, S., Bampidis, V., Benford, D., Bragard, C., Halldorsson, T., Bennekou, S. H., Koutsoumanis, K., Machera, K., Naegeli, H., Nielsen, S., Schlatter, J., Schrenk, D., Silano, V., Turck, D., Younes, M., Castenmiller, J., Chaudhry, Q., Cubadda, F., . . . Schoonjans, R. (2021). Guidance on Risk Assessment of Nanomaterials to be Applied in the Food and Feed Chain: Human and Animal Health. *EFSA Journal*, *19*, 1–111. https://doi.org/10.2903/j.efsa.2021.6768

Erkmen, O., & Bozoglu, F. T. (2016). Food Preservation by Reducing Water Activity. In *Food Microbiology: Principles into Practice* (First edit, pp. 44–58). John Wiley and Son Ltd.

FAO. (n.d.). *Improving Food Safety Along the Food Chain*.

FAO. (2003). *FAO's Strategy for a Food Chain Approach to Food Safety and Quality: A Framework Document for the Development of Future Strategic Direction*. FAO.

FAO. (2021). World Food and Agriculture- Statistical Yearbook 2021. In *FAO Statistics* (Vol. 274). Food and Agriculture Organisation of the United Nations. https://doi.org/10.4060/cb4477en

FDA. (2009). *Food Code (2009)*. FDA.

Fern, E. B., Watzke, H., Barclay, D. V, Roulin, A., & Drewnowski, A. (2015). The Nutrient Balance Concept: A New Quality Metric for Composite Meals and Diets. *PLOS One*, *10*(7), 1–19. https://doi.org/10.1371/journal.pone.0130491

Forghani, S., Almasi, H., & Moradi, M. (2021). Electrospun Nanofibers as Food Freshness and Time-temperature Indicators: A New Approach in Food Intelligent Packaging. *Innovative Food Science and Emerging Technologies*, *73*(102804), 1–18. https://doi.org/10.1016/j.ifset.2021.102804

Fung, F., Wang, H. S., & Menon, S. (2018). Food Safety in the 21st Century. *Biomedical Journal*, *41*(2), 88–95. https://doi.org/10.1016/j.bj.2018.03.003

Gould, G. W. (2000). Preservation: Past, Present, and Future. *British Medical Bulletin*, *56*(1), 84–96.

Gupta, R. K., Gawad, F. A. El, Ali, E. A. E., Karunanithi, S., Yugiani, P., & Srivastav, P. P. (2024). Nanotechnology: Current Applications and Future Scope in Food Packaging Systems. *Measurement: Food*, *13*(100131), 1–26. https://doi.org/10.1016/j.meafoo.2023.100131

Haile, S., & Ayele, A. (2022). Pectinase from Microorganisms and Its Industrial Applications. *The Scientific World Journal*, *1881305*, 1–15. https://doi.org/10.1155/2022/1881305

Hammond, S. T., Brown, J. H., Burger, J. R., Flanagan, T. P., Fristoe, T. S., Mercado-Silva, N., Nekola, J. C., & Okie, J. G. (2015). Food Spoilage, Storage, and Transport: Implications for a Sustainable Future. *BioScience*, *65*(8), 758–768. https://doi.org/10.1093/biosci/biv081

Huang, L., Teng, W., Cao, J., & Wang, J. (2022). Liposomes as Delivery System for Applications in Meat Products. *Foods*, *11*(19), 1–17. https://doi.org/10.3390/foods11193017

Kiaya, V. (2014). Post-Harvest Losses and Strategies to Reduce Them. In *Technical Paper on Post-Harvest Losses* (Vol. 149, Issues 3–4).

Kumar, A. (2019). Food Preservation: Traditional and Modern Techniques. *ACTA Scientific Nutrional Health*, *3*(12), 45–49. https://doi.org/10.31080/ASNH.2019.03.0529

Kumar, L., Deshmukh, R. K., Hakim, L., & Gaikwad, K. K. (2024). Halloysite Nanotube as a Functional Material for Active Food Packaging Application: A Review. *Food and Bioprocess Technology*, *17*, 33–46. https://doi.org/10.1007/s11947-023-03092-3

Kumar, S., Singh, N., Devi, L. S., Kumar, S., Kamle, M., Kumar, P., & Mukherjee, A. (2022). Neem oil and its Nanoemulsion in Sustainable Food Preservation and Packaging: Current Status and Future Prospects. *Journal of Agriculture and Food Research*, *7*(100254), 1–9. https://doi.org/10.1016/j.jafr.2021.100254

Kwofie, M. K., Bukari, N., & Adeboye, O. (2020). Probiotics Potential of Yeast and Lactic Acid Bacteria Fermented Foods and the Impact of Processing: A Review of Indigenous and Continental Food Products. *Advances in Microbiology*, *10*, 492–507. https://doi.org/10.4236/aim.2020.109037

Li, Q., Ren, T., Perkins, P., Hu, X., & Wang, X. (2021). Applications of Halloysite Nanotubes in Food Packaging for Improving Film Performance and Food Preservation. *Food Control*, *124*(August 2020), 107876. https://doi.org/10.1016/j.foodcont.2021.107876

Lopez-Rubio, A., Fabra, M. J., & Martinez-Sanz, M. (2019). Food Packaging Based on Nanomaterials. *Nanomaterials*, *9*(1224), 3–5. https://doi.org/10.3390/nano9091224

Mgbechidinma, C. L., Adegoke, C. O., & Ogunbanwo, S. T. (2020). Lactic Acid Bacteria as Bioactive Potential against Selected Resistance Canada Species and Pathogenic Bacteria. *International Journal of Pharmaceutical and Biological Science Archive PubMed*, *8*(2), 19–32. www.ijpba.in

Mgbechidinma, C. L., Akan, O. D., Zhang, C., Huang, M., Linus, N., Zhu, H., & Wakil, S. M. (2022). Integration of Green Economy Concepts for Sustainable Biosurfactant Production—a Review. *Bioresource Technology*, *364*. https://doi.org/10.1016/j.biortech.2022.128021

Mgbechidinma, C. L., Zheng, G., Baguya, E. B., Zhou, H., Okon, S. U., & Zhang, C. (2022). Fatty Acid Composition and Nutritional Analysis of Waste Crude Fish Oil Obtained by Optimized Milder Extraction Methods. *Environmental Engineering Research*, *In Press*. https://doi.org/10.4491/eer.2022.034

Mohammadian, M., Waly, M. I., Moghadam, M., Emam-Djomeh, Z., Salami, M., & Moosavi-Movahedi, A. A. (2020). Nanostructured Food Proteins as Efficient Systems for the Encapsulation of Bioactive Compounds. *Food Science and Human Wellness*, *9*(3), 199–213. https://doi.org/10.1016/j.fshw.2020.04.009

Morones, J. R., Elechiguerra, J. L., Camacho, A., Holt, K., Kouri, J. B., Ramírez, J. T., & Yacaman, M. J. (2005). The Bactericidal Effect of Silver Nanoparticles. *Nanotechnology*, *16*(10), 2346–2353. https://doi.org/10.1088/0957-4484/16/10/059

Olovo, C. V, Udoekong, N. S., & Akan, O. D. (2021). Precision Nutrition, Diet and Gut-Microbiota in Obesity. *Journal of Biotechnology & Bioresearch*, *2*(5), 4–6.

Rahman, M. S. (2007). Principles of Food Preservation. In *Food Processing and Preservation Technology* (Second). CRC Press. https://doi.org/10.1201/9781420017373

Raveendran, S., Parameswaran, B., Ummalyma, S. B., Abraham, A., Mathew, A. K., Madhavan, A., Rebello, S., & Pandey, A. (2018). Applications of Microbial Enzymes in Food Industry. *Food Technology and Biotechnology*, *56*(1), 16–30. https://doi.org/10.17113/ftb.56.01.18.5491

Ravichandran, R. (2010). Nanotechnology Applications in Food and Food Processing: Innovative Green Approaches, Opportunities and Uncertainties for Global Market. *International Journal of Green Nanotechnology: Physics and Chemistry*, *1*(2), P72–P96. https://doi.org/10.1080/19430871003684440

Sahoo, M., Vishwakarma, S., Panigrahi, C., & Kumar, J. (2021). Nanotechnology: Current Applications and Future Scope in Food. *Food Frontiers*, *2*, 3–22. https://doi.org/10.1002/fft2.58

Sanchez-Garcia, M. D., Lopez-Rubio, A., & Lagarón, J. M. (2010). Natural Micro and Nanobiocomposites with Enhanced Barrier Properties and Novel Functionalities for Food Biopackaging Applications. *Trends in Food Science & Technology*, *21*(11), 528–536. https://doi.org/10.1016/j.tifs.2010.07.008

Santeramo, F. G., Carlucci, D., De Devitiis, B., Seccia, A., Stasi, A., Viscecchia, R., & Nardone, G. (2018). Emerging Trends in European Food, Diets and Food Industry. *Food Research International*, *104*, 39–47. https://doi.org/10.1016/j.foodres.2017.10.039

Sarfraz, J., Gulin-Sarfraz, T., Nilsen-Nygaard, J., & Pettersen, M. K. (2021). Nanocomposites for Food Packaging Applications: An Overview. *Nanomaterials*, *11*(1), 1–27. https://doi.org/10.3390/nano11010010

Sharma, C., Dhiman, R., Rokana, N., & Panwar, H. (2017). Nanotechnology: An Untapped Resource for Food Packaging. *Frontiers in Microbiology*, *8*(1735), 1–22. https://doi.org/10.3389/fmicb.2017.01735

Singh, H. (2016). Nanotechnology Applications in Functional Foods; Opportunities and Challenges. *Preventive Nutrition and Food Science*, *21*(1), 1–8. https://doi.org/10.3746/pnf.2016.21.1.1

Sridhar, A., Ponnuchamy, M., Senthil, P., & Kapoor, A. (2021). Food Preservation Techniques and Nanotechnology for Increased Shelf Life of Fruits, Vegetables, Beverages and Spices: A Review. *Environmental Chemistry Letters*, *19*(2), 1715–1735. https://doi.org/10.1007/s10311-020-01126-2

Srinivas, A., Rao, P. J., Selvam, G., Goparaju, A., Murthy, B. P., & Reddy, N. P. (2012). Oxidative Stress and Inflammatory Responses of Rat Following Acute Inhalation Exposure to Iron Oxide Nanoparticles. *Human and Experimental Toxicology*, *31*(11), 1113–1131. https://doi.org/10.1177/0960327112446515

Tark, S.-H., Das, A., Sligar, S., & Dravid, V. P. (2010). Nanomechanical Detection of Cholera Toxin Using Microcantilevers Functionalized with Ganglioside Nanodiscs. *Nanotechnology*, *21*(43), 1–14. https://doi.org/10.1088/0957–4484/21/43/435502

The Royal Society. (2004). Nanoscience and Nanotechnologies: Opportunities and Uncertainties. In *Nanoscience and Nanotechnologies* (Vol. 46). The Royal Society and The Royal Academy of Engineering.

Xing, Y., Huang, M., Olovo, C. V., Mgbechidinma, C. L., Yang, Y., Liu, J., Li, B., Zhu, M., Yu, K., Zhu, H., Yao, X., Bo, L., & Akan, O. D. (2023). Traditional Fermented Foods: Challenges, Sources, and Health Benefits of Fatty Acids. *Fermentation*, *9*(2), 1–23. https://doi.org/10.3390/fermentation9020110

Zhang, M., Ahmed, A., & Xu, L. (2023). Electrospun Nanofibers for Functional Food Packaging Application. *Materials*, *16*(5937), 1–31. https://doi.org/10.3390/ma16175937

Zhang, W., Zhang, J., Ding, D., Zhang, L., Muehlmann, L. A., Deng, S., Wang, X., Li, W., & Zhang, W. (2018). Synthesis and Antioxidant Properties of Lycium barbarum Polysaccharides Capped Selenium Nanoparticles Using Tea Extract. *Artificial Cells, Nanomedicine and Biotechnology*, *46*(7), 1463–1470. https://doi.org/10.1080/21691401.2017.1373657

Zhang, Z., Lin, M., Zhang, S., & Vardhanabhuti, B. (2013). Detection of Aflatoxin M1 in Milk by Dynamic Light Scattering Coupled with Superparamagnetic Beads and Gold Nanoprobes. *Journal of Agricultural and Food Chemistry*, *61*, 4520–4525. https://doi.org/dx.doi.org/10.1021/jf400043zC:\Users\LENOVO\Downloads\Query Integration (1)\Query Integration\24064-0309-Ref Mismatch Report.docx - LStERROR_16C:\Users\LENOVO\Downloads\Query Integration (1)\Query Integration\24064-0309-Ref Mismatch Report.docx - LStERROR_16

Index

For Product Safety Concerns and Information please contact our EU representative GPSR@taylorandfrancis.com Taylor & Francis Verlag GmbH, Kaufingerstraße 24, 80331 München, Germany